Unification in Understanding the Fundamentals of Physical Phenomena

The empirically determined constant, G, of Newton's gravitational mechanics can be transformed from the SI values to a unit value, τ_{0G}, based on a theoretical quantum differential scale solution as a function of isotropic expansion stress, where the symmetric component of the 3D space stress tensor is shown here:

Newton: $$F_G = \frac{M_1 M_2}{d^2} G = \frac{n_{M1} m_0 n_{M2} m_0}{n_{r_0}^2 r_0^2} \left(\frac{r_0^2}{m_0^2} \tau_{0G}\right) = \frac{n_{M01} n_{M02}}{n_{r_0}^2} \tau_{0G} \quad \text{where} \quad \tau_{0G} = d\tau_0 = \frac{A_0}{6\sqrt{3}} dT_0$$

The full 3D space tensor shows the 6 symmetric components responsible for quantum gravity as the trace, and the 12 anti-symmetric components responsible for particle spin, the electromagnetic inductive and capacitive moments, and with bets decay, the coulomb force.

QFT: $$\tau_{ij} = \begin{bmatrix} -d\tau_0 & \tau_0 \cos\omega t & \tau_0 \sin\omega t \\ -\tau_0 \cos\omega t & -d\tau_0 & \tau_0 \\ -\tau_0 \sin\omega t & -\tau_0 & -d\tau_0 \end{bmatrix} - \begin{bmatrix} d\tau_0 & -\tau_0 \cos\omega t & -\tau_0 \sin\omega t \\ \tau_0 \cos\omega t & d\tau_0 & -\tau_0 \\ \tau_0 \sin\omega t & \tau_0 & d\tau_0 \end{bmatrix}$$

At the immediate boundary of the 3D space tensor in flat spacetime of general relativity, the time dimension vanishes and the quantum scale is reduced to a function of the isotropic cosmological constant, Λ, equated to the Hubble expansion rate.

GR: $$G_{\mu\nu} + g_{\mu\nu}\Lambda = \left(\frac{4\pi G}{c^4}\right) 2T_{\mu\nu} \quad \text{in flat spacetime, quantum scale is} > T(\Lambda) = 2T_{ij} = g_{ij}\Lambda$$

Diagram 0 – Femto Scale Torsion diagram shows four phases of a sustained recoil oscillation following an initial ½ π torsional strain.

A spin diagram showing the capacitive and inductive moments in a potential – kinetic energy cycle of rotational oscillation as a function of the Hubble rate.

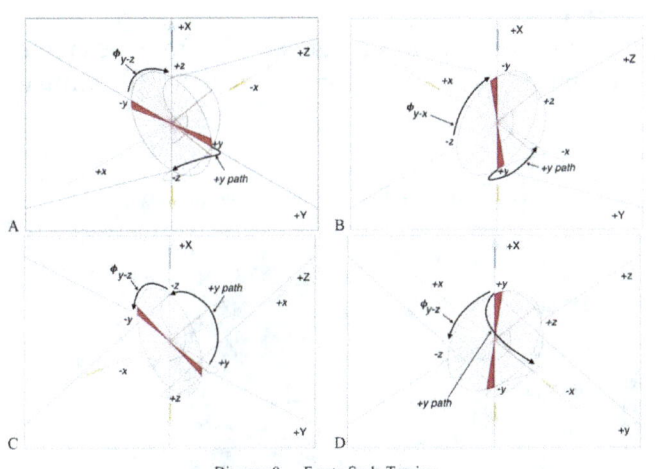

Diagram 0 — Femto Scale Torsion

Unification in Understanding the Fundamentals of Economic Modeling

Economic policy discussions center around how best to provide access to available resources, financial, real, and human, for the ongoing production and consumption needs of the community. Resource supply is limited. Production must be portioned between what is needed for immediate, final consumption and what must be set aside as intermediate products, to replenish real capital for use in the production of final consumable goods and services. If all the production goes into final consumption, an inevitable result will be a draw down in the existing stock of real and perhaps even human capital. An optimum ratio therefore exists in the allocation of expenditures for intermediate and for final production. In equilibrium, that ratio is the golden mean, ϕ or 0.6180…, which is equal to the ratio of final to total production. Growth in productivity is indicated by a ratio above ϕ for intermediate to final, and below ϕ for final to total. While this ratio is apparent globally over the past 50 years, for domestic production, this optimum has diverged significantly.

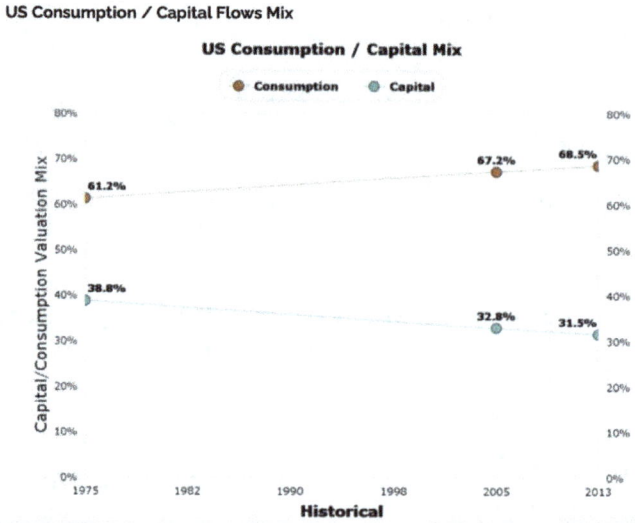

All Capital is Essentially Human.
Created by humans as a mix of God given natural resources, ingenuity, and hard work, capital—especially human—must be continually maintained and replenished if we are to continue to prosper.
Human capital is —
      10 to 20 times the value of real capital
      100 to 200 times the thin veneer of financial capital
      100,000 times the per capita value of gold – of mammon
Hence the wisdom of fostering the development and maintenance of that capital, human life, from the cradle to the grave, not as a source of exploitation for private profit, but as a solid basis for a high quality of spiritual life for the whole community.
National Debt is 1% to 2% of Total Capital.

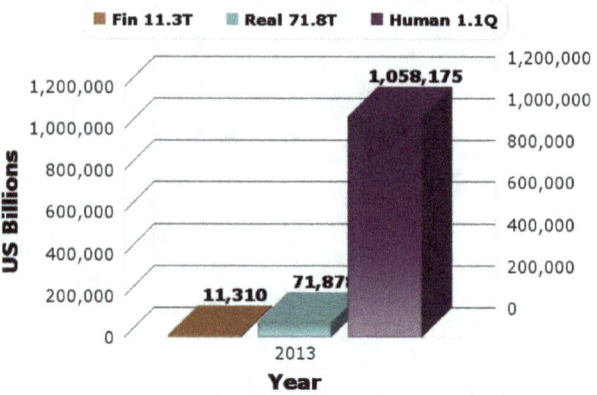

Unification

In

Physics & Political Economy

The Aim & Goal of all Modeling:
Clarity in Navigating Risk & Opportunity

 martin@uniservent.org

 1 (***) ***-****
Text or Voicemail

 PO Box 2358
Southern Pines, NC, 28388
USA

 https://UniServEnt.org

UniServEnt
In Pursuit of Understanding ...

Mission Statement
https://uniservent.org/mission/

Herbert Martin Gibson

Life of Essential Principles
https://uniservent.org/essential-principles/

Appearance of Physical Phenomena
https://uniservent.org/physical-phenomena/

Quality of Political Economy
https://uniservent.org/political-economy/

Change as Risk & Opportunity
https://uniservent.org/disaster-relief/

Copyright © by Martin Gibson 2023 All Rights Reserved
UniServEnt, PO Box 2358, Southern Pines, NC 28388
UniServEnt.org
martin@uniservent.org

UniServEnt
ISBN 978-1-958488-20-1

Mission Statement

Through the development of group affiliation and initiative . . .

To foster understanding that Life itself is the source of all existence from which all physical and metaphysical phenomena proceed; that all such phenomena evolve with purpose, and not from forces of an insentient nature.

This involves an inquiry into Essential Principles of logic of an ecumenical nature which transcends the conventional secular and theological dichotomy.

To further inquiry into the fundamental connections behind the Appearance of complex natural and anthropogenic systems by collaborative means with the intent of understanding the implications of such analysis for humanity.

This involves an inquiry into Physical Phenomena through quantitative logic which unifies the heretofore conventional spacetime-field and quantum-particle duality currently hindering progress in scientific understanding in the physical sciences.

To culture development of the Quality of innovative technology and management systems for the enhancement of the general welfare of humankind and its sustaining environment through the implementation of enlightened private and readily amendable public policy.

This involves an inquiry into Political Economy to distinguish the values truly motivating social change from the traditionally held fiscal and monetary values grounded in a contentious, partisan public versus private narrative.

To respond appropriately to ongoing Change in terms of these three threads of understanding, inquiry, and development.

This involves an immediacy of attention into Risk & Opportunity as it evolves in order to avoid complacency in the face of change going forward.

Dedication

This compilation is dedicated to
my high school chemistry and physics teacher,
Harry Ornstein,
who when asked by me for the most important figure in physics,
responded without hesitation,
"Max Planck,"
and to my college economics professor,
Apostolos Condos,
who introduced me to the fact that a domestic balanced budget was
ergodidiocy,
and to my dad
Herb Gibson
who knew all about impedance.

Contents

Preface   vii

Motivations for this compilation reflecting a study of natural processes necessary to produce a unifying understanding of the risks and opportunities encountered in life.

0.   **Introduction**   1

    The Aim & Goal of all Modeling: Clarity in Navigating Risk & Opportunity, 5/22/23

A statement of the fundamental continuity tying the study of physics with economics; understanding the inertial interactions in natural systems through the intentional capacity of human beings that model those systems for technological advance and well-being. Why kenning this continuity along a spectrum from passive observation to active implementation is important. The potential for pico-scale utilization of lattice catalyzed fusion as a basic energy form of valuation in a modern monetary system suggests itself.

1.   **Monograph 1***   17

    A Condensed Matter Model of Fundamental Particle Genesis as a Function of an Accelerating Cosmic Spacetime Expansion, 9/6/21

A more detailed development of the last monograph is presented with:
    Notes on dimensional analysis and complex wave dynamics
    A classical basis for quantum phenomena as discrete complex waves
    Symmetric components of the wave tensor as quantum gravity
    Antisymmetric components of the wave tensor as spin & charge
    Diagrams of inductive & capacitive dynamics of quantum matter & anti-matter
    Geometric considerations of rotational oscillation and beta decay
    Beta decay as a function of the Hubble rate; the missing mass of beta decay
    Evaluation of elementary charge and the coulomb force
    Model application as lattice inertial confinement & alignment in D+D > He fusion
    The relationship of the tau & muon and the neutron, proton, & electron

2.   **Monograph 2***   109

    Unification – Addressing the Fundamental Problem of Theoretical Physics, 7/15/15

Using dimensional analysis on the Planck and neutron scales,
    A gravitational quantum as a component of Newton's gravitational constant is derived from
       The field equation of general relativity and
       Newton's gravitational law.
    A solution to the general relativistic field equation is defined from this development as
       A fundamental quantum stress–inertial density tensor,
       A quantum metric, and
       A quantum rest mass field equation in flat spacetime

3.   Monograph 3*   129

Simple Harmonic Motion in Classical & Quantum Phase Space, 8/1/13

The same modeling of rest mass as emergent wave-particles from spacetime dilation/contraction, starting from an analysis of simple harmonic and uniform circular motion.
- In two-dimensional phase space, accounting for displacement & momentum
- In three-dimensional phase space, adding potential, energy, & time, from which an invariant Hamiltonian, Lagrangian, & an inertial constant are defined.
- Rotational oscillation as spin space in the generation of spin & charge with beta decay developed in detail
- Verification of this approach in terms of the standard model and general relativity

4.   Monograph 4*   251

Rest Mass Quantization as a Function of Spacetime Exponential Expansion Stress, 9/19/12

Compactification of Time, Geometrization of Quantum Mass and Gravity, and the Fundamental Quantum Metric
- Kinematics and the geometrization of time
- Lorentz covariance
- Geometrization of mass in classical and quantum theories
- Inertial and gravitational mass
- Derivation of Newton's law of gravity from quantum & general relativistic principles
- Analysis of the relationship between the neutron and the Planck scale
- Cosmological implications
- Black hole metrics and the quantum metric

5.   Monograph 5   337

A Dimensional Analysis of the Dimensionless Fine Structure, 12/14/10

6.   Monograph 6*   345

A Classical Complex 4-Wave Foundation to the Cosmic-Quantum Mechanism, 2/1/07

Fundamental Rest Mass Quanta as Simple Harmonic Oscillations of the Spacetime Continuum at Resonant Frequency and Wave Number Driven by Cosmic Expansion

An early rendition of the modeling developed in the first four monographs. It includes greater detail in some aspects of the emergence of rotational oscillation from local isotropic expansion and of a classical wave treatment of beta decay and the missing mass of the process, including its relation to the Lambert W function, as an amplitwist.

7.   Monograph 7   497

Construction of the Natural Numbers from a Real Exponential Field. 7/19/07

Fun bootstrapping a system of naturals from orthogonal power projections of the reals.

8. **Monograph 8**   **507**

    The Browser Economy – An Analysis for determining the Optimization of Investment and Consumption Allocations according to their Valuation in a Market Economy, 2/10/15

Analysis of annual global & domestic consumption expenditures between 1960 and 2013 reveals an empirically determined global optimum ratio of growth as just under the ratio of consumption to total expenditures and thereby just over the ratio of intermediate (capital) to final (consumption) expenditures. That ratio is the golden mean, ϕ, equal to 0.618... .

For the US, sector and combined accounts for 1975, 2005, and 2013 are analyzed for structural changes, with the addition to the federal sector of human capital breakdown at non-market valued 70% and market valued 30% of total, dispelling political concerns of a national debt ceiling.

9. **Monograph 9**   **607**

    A Critique of Neoliberal Economics, Part I – Quantitative Analysis & Assumptions, Capital as Position in Ergodic Economic Modeling, 12/10/20

Hierarchical Position and Focused Rationality in an Analysis of Stocks & Flows,
US Federal Reserve 4Q 1989 & 4Q 2019 Household Income and Net Worth Data.

Quantitative Conclusion:

Given the confirmation of this approach of weighted ergodic modeling for an understanding of the domestic macroeconomic structure as evidenced by the US Fed data presented here, there are only two ways to address the trending hierarchical imbalance that has precipitated and been exacerbated by the current covid-19 crisis; through the ham-handed control by the consumer/producer citizens of the means of production/distribution in order to provide their life needs or by adroit entitlement to the necessary liquidity for purchase of the immediate and long term needs produced by both private and public capital, based on sustainability of the recourse to that production. When the necessary recourse is profitable and not exploitative, private initiative would be a favored approach. When the equally necessary recourse is not privately profitable, public initiative is favored, including where necessary, redirection of entitlement to public use via eminent domain.

10. **Monograph 10**   **647**

    A Critique of Neoliberal Economics, Part II – Qualitative Analysis & Assumptions, Capital as Money, Focused Rationality, and Hierarchical Position, 4/30/21

The Ideals of Omnipotent Money, Omniscient Decision Making, & Omnipresent Freedom bound by Material Conditions

The Dialectics of Neutral Monism in the Historical Philosophy of Marx, Plato, and Christ

Qualitative Critique:

The motivation, organization, resource utilization and participation of individuals and affiliated groups in economic activity—in the interactive production, storage, exchange, and consumption of goods and services—does not happen without much thought. Political economic thinking is a historical process that has focused on certain ongoing observations concerning societal attempts to understand—first, the essential nature and genesis of human life, and second, the experiential nature of knowledge, motivated by emotional concern and trepidation due to observed changes and differentiation in the size, health, technical abilities, access to resources, risk–opportunity of involvement–exploitation, and conflict among endogenous and exogenous populations.

This thinking has been formalized in various cosmological forms, starting in antiquity, to explain and control the forces of nature and the location of human activity within it. With increasing understanding and direction of those thermodynamic forces to mechanical advantage and profit in the rise of the industrial age, and the theoretical application of natural laws to practical capital-intensive manufacturing and innovative research & development in the physical and biological sciences, the growing rationalization by the scientific method has fostered progressively sophisticated attempts at quantitative analysis and modeling of human social organization in the field of economics. The parallel and synergistic development of physical and fiscal technology has produced ground-breaking, if not earth-shaking, advancement in the service of humanity, but this has not been without peril.

In light of the quantitative analysis of Part I of this critique, three of these perils are analyzed in varying detail; the notion of money as omnipotence, the notion of decision-making as omniscience, and the notion of freedom as omnipresence in neoliberal economic thinking.

*These monographs share portions of the development of this condensed matter model of quantum wave-particle generation and the related graphic representation. They were written in reverse order of presentation in the listed Contents, as indicated by the title dates.

Links are to the related material on the UniServEnt.org website and YouTube videos

1. **Link 1:** https://uniservent.org/basic-income-study-slide-0/

   Basic Income Study, online interactive presentation, 4/14/20

Results of the Browser Economy analysis showing changes in US data for 1975, 2005, & 2013 consumption & capital flows and of financial, real, & human capital. Graphic comparisons in 9 slides. There is also a download link to the Basic Income App, which is a FileMaker mobile database solution for doing 'what if' comparisons of funding options for public policy finance based on three alternatives of taxation, borrowing, or fiat issuance of money. The app has been set for a free 30 day trial period, but the user must have a copy of FileMaker Go to operate the app. Claris FileMaker Go is a free download on the Apple App store for use on the iPhone or iPad. The user must have a licensed copy of FileMaker to use the app on a laptop or desktop computer.

2. **Link 2:** https://uniservent.org/ergodidiocy/e-economics/

   Ergodic Economics, spreadsheets and workbook template

These are derived from A Critique of Neoliberal Economics, Part I and were used in the scenarios for various weighted ergodic conditions for a macrostate comprised of 5 iterations of a coin flip producing 32 microstates in what is essentially a Bayesian approach of the model. A link is provided for download of the excel workbook for modeling. Interactive parameter entries for the flips can be weighted globally either, 1) conservative with bilateral symmetry or 2) non-conservative with bilateral asymmetry, and with further modification, on an iteration and individual microstate basis.

3.   **Link 3:** https://www.youtube.com/watch?v=1Z7FXK_2CAc
     "Low Energy Nuclear Reaction", 2016

A graphic depiction of the palladium cuboctahedral lattice in which the spheres represent palladium, deuterium, and helium, shown according to scale of quantum mechanical statistics as the electron cloud structure of the atomic elements, understood in this modeling as coherent wave forms. Lattice infusion, migration, alignment, and nuclear fusion of the deuterium in the palladium crystal in the production of helium is depicted, avoiding the usual deuterium–tritium pathway with relativistic neutron emission.

4.   **Link 4:** https://www.youtube.com/watch?v=QHUMDqv_qy8&t=66s
     "E = mc2, A Wave Interpretation" (no audio), 2016

The mathematics of this phenomenology of quantum rest mass as waves shows that the energy side of the equation is angular frequency, and the mass side is angular wave number. This means that on a quantum level, mass is not 'burned' in an interaction as a material 'substance' releasing energy; rather it is a statement of fact, that interactions resulting in directly unobservable changes in wave number have the identical effect of producing an observable change in frequency.

5.   **Link 5:** https://www.youtube.com/watch?v=FyKRBBjPZXc&t=311s
     "A Classical Complex 4-Wave…", 2015

A step-by-step narrated walk-through of the developed dimensional-analytic model with the emphasis on the variable inertial bulk mechanical nature of the rotational torsion oscillation in producing spin and electromagnetism.

6.   **Link 6:** https://www.youtube.com/watch?v=W8sNgT85xhs&t=630s
     "Quantum Gravity as a derivative of Cosmic Expansion Stress", 2016

A step-by-step narrated walk-through of the developed dimensional-analytic model with the emphasis on the nature of the same oscillation in producing quantum gravity from more of a general relativistic perspective.

7.   **Link 7:** https://www.youtube.com/watch?v=HF-CZkzplrc&t=777s
     "Ergodidiocy and the Nature of Modeling Human Experience", 2021

From the tongue-in-cheek portmanteau of the title, an expression of my frustration in trying to break through the idiocy of left versus right, instinct versus logic, emotion versus reason, or faith versus science, to get to the central point of having an enlightened conversation of how to better address the material, mental, and spiritual priorities of living together on this planet.

Preface

In my twelfth-grade English class on British Literature, Ms. Vale gave us a single collaborative assignment for the whole of one six-week grading period; outline/diagram the essay "Of Studies" by Francis Bacon—again and again and again. I don't recall how many times we did this nor how much of the work was done as homework in addition to the daily discussions in attempting to get to what we took to be the unifying, commonly recognizable point of the essay. We would each offer our perspectives, work to what we thought was a suitable consensus—or not—then have her tear our efforts apart and be instructed to try again. I don't recall any grading of our efforts nor a final determination on how close we got to an acceptable product before the grading period ran out. To most of us it was pointless and frustrating after a short while, as we so often complained. But that was her point, of course, to get us to think, write, and discuss; think, write, and discuss; think, write, and discuss.

I just read the essay again for perhaps the second or third time since that discussion almost sixty years ago. Bacon's point, along with Ms. Vale's, which starts and frames the essay with "Studies serve for delight, for ornament, and for ability…", is to me now quite clear and quite relevant. We start the study of a thing when it first distracts us from our peripheral field of vision; we think we might recognize some feature that would bring us unanticipated pain or pleasure if left unstudied. We bring that appearance into central focus long enough to determine if it is a likely risk of existential pain, an opportunity of essential pleasure, or just meh. Generally, with meh the study ends, and we go back to our prior object of interest before the distraction.

If the predominant recognition is understood to be a risk, then a study for the sake of personal delight or shared show & tell may be warranted, though the interest can become perverse; acquiring at least an existential understanding of the thing as a risk to be avoided should be a goal for all.

If the predominant recognition represents an opportunity for further study, it will often be seen—at least initially—in a unified experience of all three, modeled in a pleasurable way by learning a personal skill in traditional or novel form while receiving accolades for the effort from the community. For the fortunate, such study and a balance of its three benefits may last for a lifetime. For some, study without delight becomes drudgery, show without relevant substance loses its luster, and ability with insufficient interest on the part of the individual or appreciation of the community degrades over time.

For others, a study of one or two of the three may suffice at the start but never complete the trifecta. Acquiring an ability for the sake of ornamentation alone empties the soul; delighting in the pursuit of inauthentic ornamentation in the absence of the declared skill is disingenuous; only pleasure in the pursuit of legitimate understanding of a skill without undue thought to the accolades of the community, as is the presumed case for most non-professionals, would seem to be redemptive.

If we include a career remuneration as an aspect of ornamentation, recognized in the community or not, I presume many college students think first about this aspect of their studies, followed by the enjoyment of the course interactions and the difficulty of the curriculum. There would appear to be some positive correlation between the remuneration and the difficulty of or at least rarity of access to the curriculum, so once the necessary course of studies is completed and given its sufficient compensation, a consistently equivalent motivation presumably would be sought and found in the pleasure and collegial rewards of continuing that and any ancillary study of specialized interest.

Specialization features prominently in remuneration and vice versa, leading to its own lexicological modeling. This affects its field of interest, as well as the intra-collegial, inter-collegial, and society-wide understanding of its place in the ecosphere; and even the reality of the ecosphere itself. While the specific understanding within a given intra-collegial group may be perceived as unified—whereby the term collegial is understood as applicable to any self-identifying group of individuals—the inter-collegial and greater society groupings become ever less so with increasing diversity of studied skill and thereby increasingly lacking in a lexicon essential for communication in addressing the existential needs of a common, unified understanding of life.

Achieving or maintaining intra-collegial unification in the process of studying a specialized area of interest espousing a common canon is not particularly difficult in principle. Excommunication or schism, based on a litmus test in one form or another, will generally take care of the problem—until the litmus test is no longer understood to be a valid indicator of the naturally unifying assumptive principle for which the test is being administered. When the fundamental nature of _Nature_—of the _Universe_ in which we live and move and have our being—is not well-understood and existential conditions are assumed to be essential principles, confusion is allowed to proliferate and in time chaos ensues.

One might ask, who or what is it that 'allows proliferation'? The answer is of course, 'we' do, we human beings; however inspired by our supernal or conspiring through our infernal natures. It is we, in our collective decision-making capacity, based on the sum total of our predilections and prejudicial perspectives in our various collegial capacities, professional and amateur alike, that lead to clarity or confusion.

As I began this compilation of work of the last few decades with this Preface, I realized that the inspiration of Ms. Vale and her class was still with me. Amid the current conspiratorial chaos of all manner of partisanship, one finds the following, above referenced frames; those tyros with a little study, some perversely delighting in the economic policy battles of left and right; those disciples of more in-depth learning, some ostentatiously bowing for position to the philosophical gods of faith or science; and those with a greater experience of living who are humbled, yet remain studiously winnowing the unifying essence of any phenomena from the existential husk from which it must be allowed to germinate, some of whom perceive that to be their lot ad infinitum.

I have tried as much as possible to stick to study of the third path, the phenomenological path, the path to essential, unifying truth. It is a path I have taken because it has always been one that I understood was necessary, at each twist and turn along the way, if I was to better understand Life going forward. Study resulting in understanding is a unifying experience, initially at times as an aha moment of finding a solution to some problem. It is an individual's present recognition of some invariant essential purpose or function re-presented at work or play in an observed form or process, which without such recognition would pass in the consciousness of the observer as an existential phenomenon of unknown or unintended consequence.

Finding such a solution to a problem on any scale is a culmination of transcendent logic; a term to be used and perceived without pretension, transcendence links a problematic phenomenon existent in time and space with its essential solution through a recognition of the axiom functioning in the conscious or subconscious mind of the individual. Such unifying transcendence is more than just a glimpse of nirvana; it _is nirvana_ for the duration of that aha moment of perception and understanding. The nirvana of a buddha is simply the persisting culmination of the comprehensive problem of Life framed in its totality; the ahaaaaammmm…

My last semester at college included a course in phenomenology, a study and understanding of which required its own obscure legend and lexicon such as offered in the work of Edmund Husserl. At first it all seemed weighty, perhaps more precious than precise. In time I came to understand from living a life that the need for precision was real. Phenomenology is a modeling methodology that is essentially subjective from the start. It is a type of modeling that starts with an understanding that all we know is a function of our living consciousness, which is inherently intentional in its nature and intersubjective in communing with living Nature itself.

We find ourselves growing up in the natural world, with which by necessity we must learn to interact in order to survive. Living in a community and world of existent forms and processes, we learn naturally to form models of our interactions, a Weltanschauung based on the traditions and our own unique experience within the community. We come to understand that there are some existential things, some existential forms and processes, that appear to be reduceable to more essential things than others. In time and place, some of these apparently essential things suggest or require a refinement to even more essential things. Each of the aha moments of realizing the essential nature of a studied existential form or process produces a unified understanding nested within an evolving series of prior unifications.

All types of study and modeling are therefore phenomenological; the conscious manipulation in the mind of mental forms and processes suggested by traditional and innovative experience, used in framing solutions for navigating the apparent risks and opportunities we encounter in the world; to determine through the understanding of each and every soul, the difference between the essential truths and the existential facts and illusions of the Life into which we are born. What I have found to be true is that there are aspects of life that are fundamentally changed only by time and other aspects that only we as human beings, human souls, can change. The observation and recognition of forms and processes we encounter and interact with in Life therefore run along a spectrum from the extremes of inertia to those of a unifying intentional, transcendent purpose.

As discussed here, unification is an inherently subjective initiative of phenomenology, an intentional, therefore self-conscious, study of phenomena by the individual in which the connections to the world of things and other thinkers become apparent in terms of their existential and essential nature over time. Whatever unifying connections emerge over time, while subjective and in some degree intersubjectively shared with and as a community, do so because of an essential foundational unity that exists as a potential before it becomes recognized as actual. Before Maxwell produced his equations, allowing the development of wireless communication, radio waves from unrecognized sources penetrated the atmosphere of our planet. Before pheromones were imagined to exist, animals were responding to the unifying signals of their species. Whatever we encounter and draw into our unifying understanding of Life is possible only because the thread of a unified potential has always been there.

I have focused study on three of these threads in my lifetime; on physics as a study of inertia as well-ordered, dynamic change, where inertial, dynamic systems quite literally revolve around the essential notion of mass; on political economy as a study of intended change and its unintended consequences within the human community through the related notions of work and financial leverage; and on a phenomenological study of Life from a perspective that encompasses the other two.

'Unification in Physics & Political Economy' is the result of these studies concentrated sporadically over the past three decades. Note that the title is not 'Unification *of* Physics & Political Economy'. That was not the intention when I approached for my on satisfaction a study of unification in physics in the 1990's or separately, a better understanding of economics and macroeconomic finance in the 2000's. The drafts from the studies of these two separate disciplines were written at various times over this period. They have been compiled here for enhanced delight, ostentation, or ability as the reader sees fit.

Unification in Physics started with a simple attempt to see what would happen if I applied Newton's gravitational law to the interaction between two neutrons in contact in an atomic nucleus, where the coulomb force operating between protons could be disregarded. The results were surprising and started a quantum mechanical condensed matter analysis of the neutron, modeling that particle with deterministic wave mechanics in an elastic inertial bulk as a charge free localized rotation of torsional oscillation. In time it was realized that if the modeling was done within the context of an isotropic, local expansion of the bulk, the modeled emergent properties of the oscillation were conformal with a spacetime of Einstein's general relativistic field equation and the quantum properties of ½ spin, a magnetic moment, and Newton's gravitational constant. The math used for this modeling, as with complex waves in an inertia-elastic medium and stress and strain tensors in a bulk, is by and large supplied from the 'Physics of Waves' by Elmore & Heald. This analysis gives a better understanding of the coulomb force as a wave force, opening up the possibility of latticed confined fusion as linked in the following material.

Unification in Political Economy started with a question from a correspondent about the nature of commoditization of cost-of-living pricing over time despite the technological innovation and productivity increases of the past hundred years. This started an analysis of structural changes in sector accounts of US Federal Reserve data over the past half a century in a correlation of global policy changes, and of global and domestic figures of annual final expenditures as a percentage of GDP over this same period. Of interest was a recognition in the global averages for final expenditures of the significance of an optimization factor for growth found in the ratio of intermediate to final expenditures when equal to the ratio of final to total expenditures as the golden mean, ϕ, at 0.618…. Coupled with the comment on the commoditization of COL, this states what should be obvious, that valuation of such expenditures on a global market basis closes the global system to the lower echelons of the population worldwide. The study on weighted ergodic economics of US 4Q 1989 & 2019 stocks and flows reiterates this conclusion.

Compiling the physics monographs with the economics monographs suggested a unification *of* physical and economic understanding on certain level, which produced the following introduction through an understanding of natural modeling capacity along the inertial to intentional spectrum of human observation and activity. This phenomenological approach echoes a prior effort found in my publication effort of last summer entitled 'The Paros Commune of 1971 to 2021 & Beyond' and the redundant companion piece 'The Paros Commune – 2021 & Beyond' in which the last four chapters offer the same material consisting of three contemplations on the Essential Trigon as a Platonic form, in the inertial, formal, and intentional capacities of Life that generate the material, ideal, and spiritual natures of the embodied soul. This is followed with a final communion on the nature of the soul and the community. The difference between the first and second of these books is that the first includes 'The Paros Commune of 1971' written in 1973 and left unpublished about my two weeks on a Greek isle while contemplating my way forward in the aftermath of my disaffection with Marxist materialist modeling of Life, while retaining the notion of social justice that I still found in the thinking of Platonic, Christian, and eastern teachings. In the aftermath of these few weeks on Paros I encountered a series of spiritual adventures recounted in both these editions that set me on the road indicated in this Preface that I follow to this day.

I learned the value of study in the logic and precision of math; in its delight through a study of music with the influence of my mother; in its ability to create architectural drawings and building from those drawings with the influence of my father; and finally in its capacity to show to myself the veracity or lack thereof in checking my intuitive approach to the problems of Life. The Essential Trigon unifies my understanding on physics and political economy in a way that suffices for me and may be of benefit to other souls. Everyone's unification is unique, while from my perspective connecting to the same source. These studies have helped to make me happy. Perhaps you can find some things herein worthy of study as well.

Introduction

THE AIM & GOAL OF ALL MODELING: CLARITY IN NAVIGATING RISK & OPPORTUNITY

CLARITY IN NAVIGATION REQUIRES CLARITY OF OBSERVATION IN MODELING …

What we know of life, whether that life is by divine design or material evolution or both, is framed largely by the sense of sight—by natural interaction of sunlight and the molecular material that forms the retina of a human eye. The eye is situated in a head for neural proximity to a brain to make mental sense of immediate visual input for the purpose of navigation. The head is placed on a mechanical body in touch with the ground with a sensitivity equal to that of sight, balancing for autonomous navigation in its environment, foraging for chemical stores of energy needed to sustain and reproduce the body, brain, and of course, the eyes, in a continual process of minding itself. Within this sensory spectrum mediating between the hot plasma of the sun and the cold dark soil of the planet, through an innate mix of instinctive, intuitive, and logical cognitive capacities, the mind naturally curates as it generates a variety of mental models of the individual's interaction with and within its environment to aid in its natural, social, and subjective navigation.

The cues for such navigation are mixed. Emotion—as the word implies, a mechanism for elicit motion or evincing a change in state—is an instinctive capacity, motivated to look for natural and social cues in our environment in response to perceived risk and opportunity. Intuition is largely skewed toward an understanding of such motivation in other social beings like ourselves and therefore seeks cues to reliably gauge the intent behind the motivation of others. Logic is largely skewed toward understanding the natural form of things and the processes of their interaction in the environment and therefore tends to minimize the element of intent involved, even in social navigation, looking instead for reliable material cues for navigation.

When neither motivational intent nor inertial cause & effect is readily understood, the default is to rely on instinctive bias—on intuitive faith in interpreting essentially natural material events as intentionally motivated or on logical reason in interpreting the trajectory of socially motivated events as an effect of blind inertial forces. In the default extreme, faith becomes a religion and reason becomes a science, each as dogmatic models for human navigation. In truth, only reliance on an enlightened combination of faith and reason can maintain balanced continuity in an effective charting of the chasm between inertial and intentional change.

UNDERSTANDING THE SCIENCE AND FAITH OF AXIOMATIC MODELING

Among a scientific mix of these models, the standard model of particles as a quantum field theory and general relativity as a spacetime field theory have emerged to provide the most valid understanding, based on the successful application of modeling for the purpose of navigation. Missing is a clear understanding of the coupling of these models—of quanta & spacetime—of interaction of energetic particles in a theater of inertial form and dynamic process in which human interaction has been able to flourish.

In an instinctive attempt to unify understanding, particle physics has logically turned to a theoretical examination of increasingly finer levels of energetic particle interactions, pursuing strings at the Planck scale, while general relativity has intuitively turned to the heavens with increasingly powerful telescopes, hoping to view the cosmos at its beginnings. Yet from such logic and intuition, instead of clarity we are stymied by an inability to access space at the Planck scale for verification of such particle structure and, with the latest reports of the James Webb Space Telescope, to access time at a redshift of early galactic structure for verification of a Big Bang.

Conceptual emphasis on a high energy environment for observing and understanding the genesis of inertial structure—quanta from particle collisions & galaxies from big bang nucleosynthesis—is echoed in the pursuit of an economically viable approach to fusion. While the Planck scale and the redshift horizon may be beyond observation, the sun has been close and accessible for investigation and understanding its nature. It is therefore logical to apply the high energy of magnetic and inertial confinement to emulate the natural effects of solar gravitational confinement for the intended technological development of fusion in future power generation.

THE LOW ORDER–HIGH HEAT AND HIGH ORDER–LOW HEAT OF ENTROPIC STATES

Looking at this from a thermodynamic perspective, it is natural to correlate high energy, or heat, with order, so that cooling becomes associated with entropy as disorder with an increasing unavailability of useful energy for work. The surface of the earth is at greater entropy than the core of the sun, which will continue to burn hydrogen for several billion more years. In a sea of dense hydrogen gas, measured by the availability of energy for work per mass of particles involved, the potential for continued fusion of hydrogen to helium can be seen as a low entropic state, even though the sea of dense hydrogen gas as fuel for fusion might be seen as highly entropic in terms of its general uniformity, in terms of its apparent lack of intrinsic order.

We might posit that the human brain, with the ability through collectively coordinated understanding and effort to release the hidden energy of the atomic nucleus—currently as fission, potentially as fusion—is of many orders of magnitude less entropic with respect to the dust of the earth's surface, than the temperature at the core of the sun is to that of the earth. In fact, developed human consciousness, represented in the mind and brain, is of less entropy—therefore more ordered—than the elemental and molecular configuration of the solar core. Materially, physically, this is because the human ecosystem, like most of the ecosystems of life in general, is by necessity only operational in a well-ordered range of condensed matter, a range of ostensibly greater entropy than the sun, measured by heat and temperature alone. Given the well-ordered differentials of galactic spacetime–quanta, sun–earth, and brain–navigational environment, the living universe is of much lower entropy today than the quark soup from which all these dichotomies are currently modeled as appearing.

The potential to see and touch and otherwise sense change in one's environment, and thereby intentionally form models of material nature in an intention–developed mind—as indicated by the ability to harness the power of nuclear fission—is only possible in a human being of low–heat entropy, working in an equally low temperature environment of highly evolved order. Instinctively, intuitively, or logically, we would not expect such human nature to be found—as materially evolved or divinely created—in either the vast voids of a cold spacetime with little particle interaction or in a solar interior with high heat of particle interactions. The potential for a manifestation of human nature requires a Goldilocks zone for its unfolding, for its evolution.

The seminal point is that the existential fact of an emerging human cognitive capacity with intentionality, witnessed in the increasingly ordered complexity of the living environment, is proof that the capacity for that emergence is inherently necessary as an essential—a potential—as a determinative factor of sentience, though such potential in itself does not appear to suffice for that emergence. Such emergence requires the activation of other necessary conditions—among them the presence of hospitable ambient conditions for life in form and process—to understand those conditions as both necessary and sufficient. The eye and the brain and the body in a hospitable environment are all necessary, but it takes an understanding of the purposeful intent of one's place in the environment to recognize its sufficiency, where intent of a conscious being is the focused aim and purpose is that being's targeted goal of interaction in the environment. While intentional aim at a purposeful goal indicates sufficient capacity, reaching the goal requires tenacious work in an environment of sufficient objective conditions and, sometimes, agency of unknown providence.

WAVES AS STOCHASTIC INTENT & DETERMINISTIC INERTIA IN CONDENSED MATTER

It is mathematically logical and defensible to form a condensed matter model of physical phenomena, unifying general relativity and quantum field theory, one that is axiomatically deterministic and not based on random free parameters. This is not said as a disputation of stochastic determinism at certain stages in a model of quantum interactions. It *is* a disputation of the notion that any empirical model—mentally developed through the human capacity to recognize well-connected purposeful order at every turn in nature—can make logical sense without acknowledging as axiomatic the equally human capacity to understand a priori purpose and intent.

From this perspective, the principal, essential, and too-oft unrecognized axiom of any logical argument is the principal of continuity. Whenever we encounter a discontinuity or a boundary or an asymptote, it is generally a good idea to cogitate further to determine whether that apparent discontinuity is in fact a point of inflection, a path of least action, or a co-ordinate singularity rather than a terminus beyond which there is a meaningless void. For me, the necessary and most effective way to approach such a thinking process is to assume an inherently timeless, continuous potential in a field of observation and interaction that is initially devoid of any 'thing' yet still has the capacity to assume and morph into any form and process I might want, to satisfy my imagination.

My personal though certainly not unique way of pursuing this process is through the mathematical, geometric, topological, and innate capacity of the graphic imagination in 3D modeling using the logic of wave mechanics in a variety of oscillatory forms that can define the discreteness we observe in rest mass quanta without severing the continuity we observe as the leptonic emission of beta decay and the electromagnetic transmission of photons. This can be done while understanding the limits of Heisenberg uncertainty and the Pauli exclusion principle, defining wave activity on a wave potential modification of the density tenets of general relativity.

From a study of classical complex wave mechanics applied to quantum events; from a simple but effective analysis of Newton's gravitational law; from an inherently continuous but variable inertial density in the spacetime of general relativity, quantizable through localized stress and strain under large scale dilation; in various monographs we derive a quantum basis of gravity as the convergence components and the spin and electromagnetic moments as the curl components of a dual tensor oscillation under isotropic stress. This emergent form is recognized as a baryonic particle, the neutron, for which the moments of maximum inductive and capacitive torque provide the quark phenomenology of the standard model. Being unstable in isolation, it decays under ongoing stress into the stable proton and electron. This model effectively unifies general relativity and quantum field theory. This model of the generation of rest mass quanta as an emergent function of the Hubble rate, as linked at UniServEnt.org, answers most of the principal conundrums raised in current physical models.

THE COULOMB FORCE UNDERSTOOD AS A WAVE FORCE — COLD FUSION? ... PERHAPS

Except for the addendum to monograph (3) in the Introduction and Cover Letter, this modeling was essentially completed by the spring of 2007 with the addition of further development of quantum gravity and the quantum metric for the neutron as a quantum sink or black hole. When understood, this model offers salutary possibilities for energy and nanotechnological development, including the theoretical potential for a scalable development of LENR, now generally seen as an example of a pathological science after the failure to replicate the reported results of cold fusion by electrochemists Martin Fleischmann and Stanley Pons in 1989.

The coulomb force as now understood is generally modeled as the interaction of severable 'virtual' point charges free to move within a field, requiring application of exceedingly high energies of lasers or electromagnetic fields in a manner to effect inertial confinement and alignment of the deuterons for purposes of fusion. When that same coulomb force is modeled as the stress force of discrete waves in a continuum and not as a body force of point-like particle interactions—where the strong force is understood as a function of that same wave—the modulated confinement, alignment, and inelastic collision of deuterons without gamma radiation in bound deuterium atoms which have been infused in a palladium lattice becomes understandable, where the palladium and deuterium atom have the same electronegativity on the Pauling scale at 2.20. This is a well-ordered rather than a stochastic process producing fusion as detailed in the enclosed material and in links to the YouTube video of the process.

I did not come across the continued online interest in research into cold fusion until 2015, when my modeling of condensed matter wave-based baryonic mechanics offered a different theoretical approach for understanding the coulomb force. In the process of researching the documentation of LENR, I centered on the nature of the palladium lattice, where palladium in group 10, period 5 of the periodic table is unique with its $5s^0 4d^{10}$ configuration in contrast to what might be anticipated as $5s^2 4d^8$. The free 5s & filled 4D shell, coupled with the common Pauling factor for both deuterium and palladium as depicted in the video, provides a rationale for further research on the referenced modulation in light of this modeling and the continued reporting by independent researchers of anomalous heat and helium production without gamma radiation.

JWST & THE NUCLEOSYNTHESIS OF HYDROGEN & HELIUM WITH NO BIG BANG

In the later 1990's, I began the above outlined analysis of physical phenomena. This research culminated in the last update with a 2021 addendum to monograph (3) pursuant to a thread by Stacy McGaugh, Department of Astronomy at Case Western Reserve and Robert A. Wilson, Emeritus Professor of Pure Mathematics, Queen Mary University of London. This addendum states the exactness, within the standard uncertainty, of the following equation involving the rest masses of five fundamental particles; the neutron(n), proton(p), electron(e), tau(τ), and muon(μ); the first three recognized as fundamental within the domain of condensed matter physics, the last two as transitional states, principally in the study of plasma physics.

$$(1) \quad 5n = 3p + e + \tau + \mu$$

Equation (1) is stated by Robert Wilson as prior art without attribution. The analytical parsing of (2), developed in the addendum, is seminal as far as I know. (2)A adds the missing mass of beta decay, Δm, on the right, necessary to equal the mass of the neutron, n, on the left. (2)B repeats this addition of $e + \Delta m$ twice on the right to equal the mass of two neutron. At (2)C it follows necessarily that the unstated values added to (2)A and (2)B must be subtracted from the remaining particle masses of (1) in order for the third parsed statement to be valid.

$$\begin{aligned}
&\text{A.} \quad 1n = 1p + e + (\Delta m) \\
(2) \quad &\text{B.} \quad 2n = 2p + (2e + 2\Delta m) \\
&\text{C.} \quad 2n = \tau + \mu - (2e + 3\Delta m)
\end{aligned}$$

The missing mass, Δm, represents a complex value as an amplitwist, a conformal amplification and twist in a manifold as developed in the work of Roger Penrose and Tristan Needham. The short quantum-wave explanation of (2) is this:

A. Conservation of spin angular momentum initially found in the neutron, n, requires a spin flip of either p or e or ½ in opposite directions of both. The 'missing' mass-energy of this flip is bound up in the amplitwist of Δm and does not register as a body-force interaction of any observed particle.

B. The mass-energy of beta decay, (e + Δm), is the combined mass-energy shown here, where e is the residual mass-energy resulting from the isotropic stress force operating on the neutron at the time of that decay. It is readily shown that beta decay results from a differential stress force, df_e, on the neutron wave boundary, which divided by the wave speed squared is equal to the differential inertial density at that boundary as a wave strain, recognized as the Hubble rate, dimensionally considered a strain over time. Electron mass, e, is then a correlated—gauged—measure of the Hubble expansion/dilation rate.

C. The mass-energy of the τ and μ are respectively, much greater than and much less than the mass-energy of a neutron, but together equal the mass-energy of 2 neutron plus 2 Hubble strains and 3 amplitwists. The τ and μ are extremely short-lived leptons, perhaps transitional bosons, that contribute the energy of 3 twists in response to the 2 Hubble differentials to produce 2 neutron, which based on this parsing, come in pairs. (2)C can be modeled as a quantum spallation in pairs from the boundary of pre-quantized concentrations of inertial source density occurring in regions of spacetime dilation. With (2)B, this pair production explains both the observed Hubble rate and the predominance of protium, deuterium, and helium found in the universe without recourse to a big bang singularity for explanation.

The modeling of this addendum foreshadows the observational data from the James Webb Space Telescope reported earlier this year. This data indicates a maturity of spiral galaxies that are not readily understood by the current timeline of big bang modeling. As understood, such maturity requires the existence of a gravitational component to such galaxy formation to be in effect at a time prior to the occurrence of a big bang, a modeling contradiction.

OTHER POTENTIAL BREAKTHROUGHS

While I have yet to investigate it in any detail, the work of Mike McCulloch of the University of Plymouth, in which he references quantized inertia, features the Unruh effect in explaining observed galaxy rotation curves without resorting to dark matter as responsible for those observations. His explanation offers a theoretical possibility of technological development, including use as a means of transport. Some of the apparent axioms of his thinking imply an inherent inertial potential of spacetime and resonate with my own.

FROM PHYSICS TO POLITICAL ECONOMY: STOCHASTIC INERTIA & DETERMINISTIC INTENT

While my early interest was in math and the physical sciences, especially astrophysics, my degree work was in economics in the late 1960's at Duke. I learned much of the skill and approach to technical investigation from my father, who was a professional electrical and structural engineer. Starting in 1948, he worked for the TVA, then in the semi-conductor and aerospace industries, before entering industrial design & construction as a private contractor in the mid 1960's. After graduation from college, I worked with him in the industrial design-build business until his passing in 1980.

In the early 1990's I took this skillset into the risk management and catastrophe property insurance claims business as an independent adjuster. This work can be lucrative and, being episodic in nature, freed up time for various creative projects using an engineering approach to reverse-engineering problem solving. Leading up to and in the aftermath of the 2008 recession, I switched investigative interest to political economics for an analysis of the effect of various policy initiatives on US domestic and global economic growth and structural change over the past 50 years.

THE GOLDEN MEAN IN VALUING CONSUMPTION, PRODUCTION, & HUMAN CAPITAL

The first of two unpublished investigations is a study entitled 'The Browser Economy – An Analysis for determining the Optimization of Investment and Consumption Allocations according to their Valuation in a Market Economy', and is linked on the UniServEnt website as listed in the cover letter. A copy of the 'The Browser Economy - Executive Summary' of this research is included with the UniServEnt.org monographs. The intent of both studies is to gain a better understanding of the dynamics behind sector structural changes to provide a guide for policy recommendations and a gauge for instituted policy effectiveness.

In this study, we find that for a given level of liquidity, as quantified by expenditures on final consumption and on capital goods and services, where capital expenditures include both public and private sectors, an optimum equilibrium ratio of 0.618... as the golden mean, ϕ, exists for

(1) final consumption over total expenditures

      equal to that of

(2) capital over final consumption expenditures.

Conditions favorable to overall economic growth, meaning a rise in the general standard of living, are indicated by ratios somewhat below the optimum for (1) and above that figure for (2).

Examination of World Bank data for the period 1970 to 2013 shows a ratio range for the world economy of a few percentage points below (1) and for the OECD nation average of a similar range, before rising above (1) in 2009. Some notable economies trending several points above the target for this duration are Greece, Mexico, and until 2004, Brazil and India. The U.S. trend rose above (1) in 1982 during the Reagan administration with the implementation of supply side policy and has risen gradually—except for most of the Clinton tenure when it stabilized—to a current level of approximately 7 points above the mark.

More significant in this study is the inclusion of an accounting for human capital, 30% market and 70% non-market valued, in the domestic accounting of the US economy. The current absence of quantitative accounting for both MHC & NHC, and the resulting denigration in the public square resulting from not maintaining and enhancing the value of that human capital out of its own store of human value, results in the absurdity in the US and elsewhere of absolutism in fiscal valuation of money over humans.

The study shows that human capital is:

- 10 to 20 times the value of real capital
- 100 to 200 times the thin veneer of financial capital
- 100,000 times the per capita value of monetary gold.

Similar accounting is presumed to be valid in the rest of the world economies.

WEIGHTED ERGODICITY OF POSITIONED DECISION-MAKING IN ECONOMIC MODELING

In 2020, after the start of the Covid pandemic, I became aware of the work of Ole B. Peters, the founder of the London Mathematical Laboratory. He is also an external professor with the Santa Fe Institute, with a PhD in Physics, who has segued in his understanding of statistical mechanics to a study of ergodicity economics, for which ergodic theory first developed in the study of thermodynamics.

The ergodic condition states that some measurable value, if averaged over an extended timeframe or lifetime of an individual microstate(m_a), studied either as an individual element or a grouping of related elements, will be equal to the average value of the entire macrostate(M) at any single point in time. The ergodic condition is therefore generally held to be conservative. For such to be true there must be an interconnectedness such that an increase in one half of M must be offset by a decrease in the other half; an increase in M of a unit, m_1, must be offset by a decrease in some other unit, m_2, or in the sum of units, m_2 to m_n. On the other hand, if all m_n are increasing over time—perhaps some more than others—the value of M and its average will increase; the ergodic condition can still be true but questions if it is still conservative. Either the values of m_a are inflationary, or M is not a closed system; if not, the increase in the valuation of the units of m_a are due to their qualitative growth, either as an inherently open function of m_a or conditionally as a function of the openness of M.

A link to this study, 'A Critique of Neoliberal Economics Part I – Quantitative Analysis & Assumptions, Capital as ~~Power~~ Position in Ergodic Economic Modeling' is in the cover letter. Using US Federal Reserve 4Q 1989 & 4Q 2019 household income and net worth data, I have applied ergodic modeling to stocks and flows in checking for the effect of weighted decision-making based on the focused rationality of microstate decision-makers and/or on the hierarchical position of those decision-makers. However, rather than numerical value generation by an 'infinite' number of flips as an unconnected extant microstate with extrapolation to the macrostate, this modeling consists of a branching series of flips of a coin, where the result of each flip is a stage for the next iteration of flips, so that 5 iterations result in an evolved macrostate of 32 microstates starting with a single microstate flip. Five iterations are sufficient to establish a proof of concept of this approach for understanding the fundamentals of a system's dynamic.

Subject to this methodology, which is essentially Bayesian, this analysis confirms by inference the weighting as being rationally focused, measured as an overall percentage for the macrostate while being individually weighted according to the evolved hierarchical position of each microstate. It points to the vapidity of economic theory which models each decision-making microstate as if it was poised in the marketplace with omnipresent position, omniscient market knowledge, and omnipotent control of monetary value. This linked study starts with the observable fact that every microstate has hierarchically limited market position, limited but more or less focused knowledge of that market, and limited ability to control the money, including the limited ability of government to control the value of money despite its omnipotence at the printing press.

CLARITY IN MODELING — ONTOLOGY VS EPISTEMOLOGY VS PHENOMENOLOGY

Modeling is a natural aspect of thinking that reduces and condenses an experience to its essentials in a mental map for the purpose of mental, social, or physical navigation. In a two- or three-dimensional representational form or process, using graphics, symbolic language of mathematics & set theory, and toy animations & mockups, conscious direction and documentation of the thought process improves understanding on the part of the thinker and dialectical communication of that understanding with others. Such natural modeling finds expression as formal philosophical thinking, where the purpose of formality is to insure logical clarity to the modeled thought.

Modeling that is ontological starts with stating what IS OBJECTIVELY TRUE in the mind of the modeler. It proceeds from that truth as an axiom to the development of a system of logic that embodies that truth. A hiker in the woods sees what looks like a bear in the distance, thinks, "THAT'S A BEAR", and reroutes accordingly. Models that are instinctive, traditional, or are taught and learned by rote tend to be ontological. Such modeling often acts AS IF it is based on a position of omnipotence in defining the system.

Some modeling is epistemological and starts with stating what APPEARS TO BE TRUE, including the means and methodology of consistent validation of that appearance. It proceeds with what would follow or be derived from knowledge based on that consistent appearance and separates the conclusions from experimental data into what is objectively true, appears to be true, appears not to be true, is objectively not true, or is unknown. The same hiker in the woods sees what appears to be a bear in the distance, thinks, "THAT LOOKS LIKE IT MIGHT BE A BEAR," reviews the possible paths for a closer look, determines if possible whether or not it is a bear, and reroutes accordingly. Models that update with experiential data grounded in a valid methodology tend to be epistemological. Such modeling often acts AS IF based on a position of omniscience in viewing the system.

Then there is phenomenological modeling. It starts from the modeler's perception that all investigation of phenomena is based on the SUBJECTIVE EXPERIENCE of the modeler. The responsibility for recognizing and evaluating a phenomenon falls on the skill and experience of the individual in determining whether an existential phenomenon is objectively true/false or apparently true/false; it also adds the perspective of determining if a phenomenon is essentially true/false with respect to being existentially true/false. Thus, an essential false will always be existentially, objectively, false, but can still be apparently true. An essential truth can be existentially, objectively or apparently, true or false. Our hiker, having spent many years as a naturalist in the woods and never seen a bear, sees movement in the distance, thinks, "FROM MY EXPERIENCE, THIS IS WORTH A CLOSER LOOK," being subconsciously aware that, 'I AM ON THE ISLAND OF MAUI, WITH NO PREDATORY ANIMALS', understands there is potential opportunity and no known essential risk to further investigation, reviews the possible paths for a closer look, and reroutes accordingly. My approach; such modeling acts AS IF based on a position of omnipresence in pursuing ontic and epistemic truth.

This indicates the importance of understanding how an axiom is phenomenologically stated as to its essential versus existential truth and how that axiomatic validation is made. Certain truths of medieval Christianity upholding the essential value of human life, promulgated along with models of natural processes as ontic truth, such as the literal six-day creation of an earth centered universe, were essentially destined to collide with the epistemological framework of the scientific method at the dawn of the age of European exploration.

Now the epistemological scientific project has run its course and into the fundamental question of determining if conscious intent is an epiphenomenal function of natural selection as an apparent truth or is an objective expression of an essential truth about the inherent intentional nature of life. Contending with well-meant fear & ignorance, it must decide whether intention is a form of self-delusion—and if so, why—or a subjective capacity to initiate and direct change in a fundamentally counter-entropic project toward increasing order in a condensed matter ecology as required by life, sustained at a well-ordered distance by the fusion furnace of the sun.

Philosophical thinking, from ontological, to epistemological, to phenomenological, as with the biological forms and processes of the ecosystem, has evolved to provide increasingly objective and subjective clarity for navigating the risks and opportunities of life in which we all find ourselves. The technological success of scientific clarity in political and economic innovation brings us to a new inflection point requiring the same degree of clarity in understanding human intentionality for an implementation of enlightened policy.

CONTINUITY OF INERTIA & INTENT: MODELING PHYSICAL PHENOMENA AS INERTIAL

Physics as physical phenomenology exists as a study to differentiate inertia and intent—objective forms from subjective interactive processes. Notably started by the work of Galileo Galilei, refined by Rene Descartes, and codified by Isaac Newton in his three Laws of Motion, this historical project was motivated by the work of Nicolaus Copernicus, born in 1473, with the heliocentric model of the solar system, an inflection point in the understanding of celestial mechanics. It successfully refuted the common mechanical knowledge asserted by Aristotle 2,000 years before, that an inertial body—a body having the property of mass—will come to rest unless it continues to be forced along its path of motion. It implied that the motion of the planets did not require the application of continuous intentional force by a god to stay in motion. It was the start of the scientific project of using experimental methodology to amend common knowledge of life with quantitatively verifiable and technologically implementable mechanical models of physical understanding by intentionally removing subjective intention from the experimental frame of reference and instilling the objective inertial forms and processes of the inertial frame with a functional purpose as an embodied axiomatic logic.

It is gravitationally induced friction that requires action be maintained continually or intermittently in the form of work on a body in contact with the surface or atmosphere of the earth to keep it in motion. Work is the product of the force exerted by an external source in moving the inertial body times the distance over which that force is applied; such work is a measure of the energy expended in doing that work. Such work since antiquity has been associated with the intentional capacity of some sentient, animated being—a domesticated animal, a human, free or slave, or a god, spiritual or corporal. Galileo, with his quantitative experimental study of bodies rolling down an inclined plane, correctly deduced that if the elements of gravity, friction, or other unrecognized interaction were removed from a model, once set in motion said bodies would continue in a straight line of motion indefinitely without any additional impulse or work required. It followed from Galileo's experimental work that an essential intent of the setup must be to close a model by removing any extraneous or unrecognized element of working intent from the mechanism being studied in order to leave only inertial interactions as understandable, reproducible elements of the model for technological application.

We know that Aristotle was qualitatively correct in his observations within that philosopher's historical frame of reference—on the surface of the earth, work is required to keep heavy things in motion—but his lack of quantitative investigation hampered qualitative conclusions that would have allowed technological development. Descartes' refinement at the start of the scientific revolution was to self-consciously branch this study with 'Cogito, ergo sum' into two disciplines as physics and metaphysics, where physics became an empirical study of the field of space and time through experiential observation of physical forms and their interactive processes and metaphysics became a mental imaging study of recognition and manipulation of thought forms and processes embodying purposeful axiomatic logic and innate ideas, human and divine.

To synopsize Descartes, because 'I think, therefore I am' able to observe and recognize innate mental purpose represented in material forms independent of myself, to navigate and to choose if, when, and in what manner to interact with those forms represented in the world. Building on Galileo's work, Descartes produced a law of inertia stating the natural capacity of an inertial body to remain at rest or to stay in motion at a constant velocity in a straight path unless acted upon by an external force that causes the position or velocity to change. As stated, such external body force could be administered by interaction from another material, inertial body or dynamically by a living, animated being. It could also be mediated by a number of extended fields as a strain inducing stress force.

Newton's three Laws of Motion were a culmination of this study in terrestrial and celestial mechanics in which:

1. The first law restates the law of inertia of Galileo and Descartes.
2. The second law dimensionally quantizes a body force as the product of the mass of one of these bodies and its change in velocity as an acceleration, positive or negative, over the duration of that interaction.
3. The third law states that the interactional force from one body will be equal in magnitude and opposite in direction to that of the second body.

When applied to celestial mechanics, the laws of motion produced Newton's Gravitational Law. Further development used Descartes' notion of vortices of various sizes filling all space in modeling of a gravitational field as an extension of substantial flows or by others as a medium of transmission from a concentrated body source according to Newtonian mechanics with a field potential that exists even in the absence of any receiver of that force. In this latter case the field potential might be continuous as a wave bearing stress or quantized as a property of a particle, or both, gauged by the potential of the field. To my understanding, Newton's concept of particulate matter was not well developed, and the composition of celestial bodies was taken as the work of God, not evolved from particle aggregation due to gravity as now modeled. The field concept carried over naturally to the development of electromagnetism with Faraday, Maxwell, and others in the 1800s, where the notion of an electrical charge as an anion or cation was modeled before that of the electron or proton as the corresponding fundamental electrical charge carrier, with the beginnings of quantum mechanics.

With the start of the 1900s, building on the work of Bernhard Riemann's curved manifold geometry and the notion of a geodesic as equivalent to a straight-line inertial path in flat spacetime, and following a Galilean-Cartesian metaphysic, Albert Einstein removed the notion that a gravitational force is operating on a body moving on a geodesic path. In general relativity, spacetime itself is a continuous field devoid of inherent inertial properties; instead, it is the presence of mass–energy that bends an inherently flexible spacetime to mathematically produce the curvature of observed geodesic paths of bodies through that spacetime. However, there is no mechanical explanation of how mass–energy particles couple with the spacetime field to generate the curvature recognized as a gravitational force, unless we are to reverse the time–causality picture to show mass–energy of particles as wave curvature resulting from a bending of inherently flexible spacetime.

Contemporaneously with Einstein, many empirically minded individuals developed quantum mechanics using probabilistic mathematics in a wave–particle duality, eventually growing the standard model of inertial interactions into a quantum field theory, again without understanding the essential inertial connection to the spacetime of general relativity. Significantly, if matter understood as mass-energy in the fundamental units comprising all evolving living and supporting forms is essentially inertial in nature, how does one explain on a quantum level the composite animated, purposeful presence of living individuals connecting to space and interacting over time in navigating the risks and opportunities of the biosphere. The several billion–year project of terrestrial biological evolution, of increasingly ordered life forms and processes arising amid the high entropy of condensed matter, is ostensibly driven by an inherently counter-entropic survival intent, having expended a tremendous amount of work in overcoming the inertia of stellar dust in the process.

The semi-millennial project of scientific inquiry in differentiating inertia from intent is all but complete. The spacetime field of human observation and interaction is now scientifically modeled as a repository of inertially discrete, interacting, charged particles. Both intergalactic spacetime and collider quanta are modeled as devoid of intent except in the understanding of the axiomatic properties acronymically programmed into the models of JWST and CERN. The only thing left to do in completing this project is to model the gravitational force/curvature that links these quanta together. The linked monographs in this writing clearly show that gravitational force/curvature are expressions of an emergent quantum mechanism as localized rotational oscillations producing rest mass in response to dilation/contraction stress and strain. The spacetime of general relativity is the sole source of those quantum forms as a continuum field of elastic inertial density having a wave bearing gauged lattice potential, and its dilation/contraction is the sole source of that quanta's power and energy.

Also needed is a better understanding of the way quanta are biomechanically bound together as coherent, self-replicating life forms to constitute an individual intent for navigating the phenomenology of risk and opportunity found in the natural world. That understanding of individual conscious intent cannot be found in an understanding of a cosmos whose primary principle is unmotivated quantum or general relativistic inertia. With a nod to Gödel's two incompleteness theorems; physical phenomenology states in the first theorem (1) proof, that is, reification of the completeness of the quantitative analysis of inertial properties is not possible from within such an inertially modeled system regardless of inertial consistency, meaning the reality of the intentional capacity of human beings cannot be based on the inertial properties of the system in which it is studied, and in the second theorem, which essentially says the same thing as the first from a different perspective, (2) an inertial system observed to be quantitatively, inertially consistent cannot be proven, that is, reified as an inertial capacity of the system alone. Such reification can only be performed by the subjective aspect in a phenomenological modeling of a system, AS IF that aspect was the peripatetic focus of a soul on the inherently monistic material, mental, and spiritual capacities of nature, without prejudice to the phenomenal nature of such soul, i.e. whether it is essential and relatively immortal or existential and inherently mortal.

The point is not that the applicability of quantitative logic and set theory is constrained to a study of inertial, material systems. In fact, it is applicable to any field of study, including aggregate structures involving intentional beings in physical and economic modeling. Quantitative modeling at some point comes down to counting things in a qualitative set and so to adding or taking them out of such a set; a little imagination gives us the option of grouping things in a set into subsets by dividing a larger set into smaller groups or by multiplying a small set to form a larger set. Simple arithmetic numerical manipulation becomes so facile it is easy to forget that an adjective numeral without a qualitative noun to define and operate on a set is meaningless. Basic arithmetic logic reminds us you can only add like things together without changing the nature of a set. Apples plus oranges remain neither just apples nor just oranges, but fruit. Multiplication is different; you can only multiply or divide different things. Linear meters one way times linear meters another way are square meters. This is meaningful if we don't equate one linear meter with one square meter. But apples times apples are what, square apples? Hardly meaningful.

From the efforts of quantum investigations of the last century, as fundamental stable rest mass particles of condensed matter, protons and electrons and the release of the dynamic interactional potential of the coulomb force are readily understood to be the result of a decay process from a more fundamental form of rest mass, the unstable neutron. Yet without the inertial instability of the neutron to enable the nuclear aggregation of protons within a related electron orbital configuration, the qualitative diversity of elemental material responsible for molecular structure and required for biological forms and energy transformation and utilization processes would never have occurred. This understanding is not a result of seeing the emergence of order from random inertial interactions. Given the predominance of hydrogen in the universe as the fundamental unit from which all other elements are comprised, it is a result of recognizing the deterministic mechanism of simple harmonic motion in nature as the dilation driven neutron which provides structural stability to all forms of matter and thereby to composite molecules with a variety of qualitative properties as selectable components for a multitude of evolving, living intents.

Again, echoing Gödel with respect to incompleteness and consistency in an understanding of proof, where proof is a conscious recognition of a quantitative conformance of an observed inertial condition with its modeled standard quality, there are two threads of continuity in this evolution. First is a continuity of increasingly ordered inertial forms and processes in an evolution along a spectrum of ordered intent, starting with a sea of protium and deuterium plasma at one end and culminating, so far to date, in the DNA of a condensed matter ecosphere as a macrostate at the other, a process that is inherently incomplete. Second is the individualized continuity of instinctive survival intent of a microstate that is naturally self-modeling and consistently (1) recognized in living beings, consciously or subconsciously, as a separate self, (2) identified with the inertial composite complexity of each individual as 'their' body, and (3) utilized in counter-entropic, leveraged work, specially fitted within a system of symbiotic forms, each representing a risk or opportunity to and from others in the biosphere, a system that is consistent with but not 'provable' to another self that is self-recognizing, identifying, and intentionally utilitarian. It is these two axioms of continuity, as applied independently to the entire biosphere and as an individual focal point of awareness among a multitude of such foci, each a dynamic switching of navigational aim of a living being while moving toward a goal along a variety of spectra between inertia and intent.

REVERSE ENGINEERING MODELED INTENTIONALITY

To use a phenomenological game metaphor for GR and QFT, this is like opening a game board of 8 x 8 squares painted with alternating dark and light squares to play a game of chess and finding it furnished with 2 x 15 playing parts from a backgammon set that has more than enough boardmen for a game of checkers. Sure it will work, but if you grew up on checkers, what do you do with the three extra men for each side? If you grew up on backgammon, where do you place the men on the board to set up the game and how do they move? What if you grew up playing chess with the whole board in play and were expecting to find a hierarchy of players?

Neither the gameboard of general relativity nor the playing parts of quantum field theory of themselves direct the play of the game. That would require playing parts with an ability to intentionally aim at a purposeful goal on a well understood field of play. As it is, play is the prerogative of neither the board nor the boardmen, but of the players, though they are constrained in the play by both. Ontic assumptions of boardman placement on the board in setting up the game and epistemic assumptions concerning movement and areas of play may work well. In the case of checkers, it might reveal that the idea is to get to the other side of the board and take all the opposition players captive in the process while ignoring the reserved half of the area in the lighter squares of the board, as would readily be the case if a phenomenological approach to the question was pursued.

The board is obviously intended for a type of game play and the uniformity and size of the boardmen—neatly fitting into the square borders—are obviously designed to function on the board. The extra boardmen? Spares? In the case of chess, for someone with an understanding of natural, social hierarchy, phenomenological intuition might suggest the potential intent to evolve into a system of pawns and rooks and knights and bishops and queens and kings with full range of the board —except of course in the case of the theologically and scientifically trained bishops who are each constrained by their indulgences to non-interacting domains.

CONTINUITY OF INERTIA & INTENT: MODELING ECONOMIC PHENOMENA AS INTENTIONAL

Political economy as an economic phenomenology exists as a study that reverses the order from that of physics in differentiating human intent and inertia. In this case, intent is the socially self-aware process of acquiring, producing, distributing, and utilizing the resources needed to sustain and reproduce human life through an understanding of the life processes of the biosphere. Inertia is the background of natural and developed resources and traditional systems, for use chiefly as they are understood to be required and intended to provide for the support of those processes.

The theoretical advances in physics and their technological application, starting around the time of Copernicus and Galileo, facilitated, and were facilitated by, the impetus for commerce and trade. Starting with the Portuguese a few years after the birth of Copernicus, John II of Portugal sent his explorers, Bartolomeu Dias and Vasco da Gama, south along the coast of Africa with rapacious purpose in search of navigational information for the quickest route to the valued commodities of Asia, for slaves as reproducible productive value—but with unimagined intelligent capacity to assert their own intentions in time—and for gold as money, at the time the most inertially dense measure of material value.

It is worth noting that prior to the publication of his work on a heliocentric solar system, Copernicus published what is reported to be the first work on the quantity theory of money. This theory states that the nominal price level in an economy is directly related to the quantity of money in circulation. Hence, there is no inherent unit value of money as an equivalence measure for setting the purchase price of a basket of commodity goods or services. Over the short term in a market economy, the cost of producing those commodities will reflect their current 'real' pricing, but over time the selling price will reflect the quantity of the money in supply. As the quantity of money increases, ceteris paribus, the price of a basket of goods and services increases and a unit of value relative to the purchase price of those commodities decreases. The inverse is true over time; as the quantity of money decreases, the prices of commodities decrease and the unit value of money increases.

What should be clear from this logic is that the quantity of money as financial capital is not an inherent measure or determinate of a community's productive value. It is not metaphorically applicable as aggregates of energy for work in physical modeling, as with the number of BTUs in a barrel of oil or a watt-hour of electricity. Productive value or productivity is the human capacity to produce the goods and services required for living, qualitatively so if the energy, health, and happiness needs of the community are customarily or otherwise properly maintained. Assuming a stable money supply, productivity is a combined measure of the human and real capital of a community, both of which require a commensurate level of collectively recognized intent, coordination, skill, and need via beneficial technological investment. Financial capital exists only as a socially recognized monetary methodology for allocating real goods and human services in the sustainable production of intermediate and final consumable goods and services.

Monetary methodology evolved principally for that purpose, starting as a method of agreeably accounting and allocating in an agrarian non-market community for day-to-day differences between individual & family production and unmet consumption needs. Verbal agreement to lend and borrow with the expectation of recompense in the near term, in time added the use of tokens of satisfaction and eventually durable coinage in circulation for work performed and for trade outside the immediate community. In time, fractional reserve banking produced a predominance of bank note circulation over precious metal coins, so the money in circulation came to represent a preponderance of debt over credit in the accounts of banks and on the ledgers of private enterprises involved in such banking. In the aggregate, some monetary accounts represent real capital and stock for things already produced, of indeterminate physical depreciation and worth, and some represent accounts for things that have yet to be produced and may never be produced or be fungible, with an economic impact on the community that has yet to be imagined. Financial assets are only as valuable as the current transactional intent of the asset owner operating within the intended agency of a governing authority. And as we know, banks fail, and when they do the insurer of last resort is the community.

While the intent of most monetary authorities may be to provide stability to the value of a unit of currency, a stability in the flow of that currency is also a concern. In a traditional feudal economy, except for the elites, money was a minor factor in the quotidian production, distribution, and consumption patterns of a community, but with the rise of global market economics, money has become an existential concern for everyone, and traditional patterns have all but vanished, particularly in the cities. For those whose employment skills are in surplus in the labor market, compensation is reduced over time to a commodity level, and savings vanish accordingly. Without a viable safety net, disruptions in the supply of money or its valuation produce liquidity and supply crises as experienced in the current Covid pandemic.

There is nothing new in this knowledge. The work of Adam Smith in 1776, Karl Marx in 1867, and John Maynard Keynes in 1936 in different ways point to the importance of understanding the free exercise of individual intent in the production, distribution, and consumption of goods and services, through acknowledging the control over the producers, distributers, and consumers by parties positioned to direct the flow and value of money. Of significance is that such crises are portrayed as a result of maleficent or indolent intent on the one hand or as a lack of resources or collective inertia in dealing with their causes on the other, as if sufficient money in the right hands—or the left—would solve the problem.

But money is not the problem. The WAY money is controlled in flow and value by the positioned parties is the problem. In the US, taxation on income is used as a cudgel for partisan bickering in the current implementation of monetary and fiscal policy, fighting over a pool of money that the right sees as private and the left sees as public in its origin, when it is a mix of debt and credit in both private and public accounts. Both then complain of inflation when the liquidity surfeit of supply chain disruptions are compounded by meeting the unemployment demands for costs of living from a number of public and private sources, including ARM HELOCs, all of which are further compounded by rising FED rates designed to cool down full employment, when what is needed is investment in addressing supply chain issues and in technology affecting growth needs in real and human capital. This includes pursuing the technological potentials of modern monetary policy and a universal basic income as a citizenship dividend of human capital in balancing liquidity needs while maintaining collective interest and engagement in the overall productivity and quality of life.

CONCLUSION AS TO WHAT TIES PHYSICS & ECONOMIC MODELING TOGETHER

An attempt to form a fundamental theoretical connection between the disciplines of physics and political economics for my own satisfaction has been framed by my understanding of history—from trying to look back in time at the nature of various macrostate conditions. Making history, like spinning yarns and weaving fabric for a tapestry or costumes, is a forward-facing process, partially patterned by the colorful, textured weft of individual microstates, fully constrained by the durable warp of the macrostate through which the individuals weave their lives. The storied fabric may not be executed according to any intended pattern, but each weft is purposeful in selecting the fibers it deems most suitable from whatever resource is available at the time to clothe its role in the body politic with a serviceable covering that will last. Likewise, the warp that provides the ecological, political tension required to support the weft as it unwinds over time from the warp beam is purposely drawn, as judged by the invariant continuity in that tension. As for the microstates, presumably little introspection is spent by most individuals envisioning an engineered 'model' of operation of the loom. As long as the durability and interesting color and texture of the cloth serves the intended function of the fabric, whether the material of the fibers in the yarn is defined primarily by the color and texture of the weft or by the economic good and social durability of the warp to the fabric is generally of interest to only a few.

When the material is not so durable or comfortable, nor the fit so functional, history becomes retrospective. Then more individuals are inclined to reverse-engineer a model of the cloth and the loom to understand separately the material nature of the fibers, the method of production and distribution of the fabric, and the purposeful function for which the material was cut, fitted, and sewn. Physics, particularly quantum physics, concerns itself with a technological understanding of the material nature of the fibers. Economics concerns itself with harvesting the fibers and with the production and distribution of the fabric. Politics, or the body politic itself, which for millennia has been organized and managed by individuals in varying degrees of hierarchies in church and state positions, concerning itself with determining how the suit is cut, fitted, and sewn, has only tangentially addressed a need for understanding the engineered structure of the loom. This began to change, in the 15th century when these positions became increasingly filled by individuals with academic and commercial credentials, though as in the case of church and state, not always with individuals of the stated or reputed qualifications.

Geographically born, navigational modeling of commercial and academic experience has specialized these general disciplines further over the past six centuries, with increasing refinement in mapping the investigative scale and operational breadth of the resulting innovations. This specialization has evidenced a significant decrease in the hierarchical position of church and state with its perceived traditional wisdom in deciding how to clothe the body politic, with a shift of influence to a revolution of scientific decision-making in the academic and commercial vanguard. With such technological specialization comes a related lexicon of each subject development with its own set of assumptions and axiomatic understandings and collegial deference to the expertise of other fields. In time, the university becomes divided into the humanities and the sciences, where each excels, if not in a form that is readily recognizable to the other; or to those outside the academy, to either.

The concept that ties together these studies of political economy and physics for me is a fundamental element commonly misunderstood in most current economic theories and missing from most interpretations of the standard model of particle interactions. It is the essential element of human need and purposeful agency, generally unrecognized but continuously present in social and natural interactions.

It is this unrecognized essential continuity that provides a grounding for the recognized existential ties connecting all economic decision-makers in a political economy, connecting all rest mass particles in an observed universe of differentiated plasma and condensed matter for a variety of higher, well-ordered purposes even if the intent is yet unimagined, and connecting the social and the material disciplines of modeling the phenomenal world in its laws of invariance and conservation. It is the axiom of continuity as an individual and group, moving from a sleepful rest of inertia to the full awareness of a working intent.

Inertia is materially formed as rest mass particles of simple harmonic motion in physics and socially incorporated as the basic necessary routines of life, routines which are in themselves somehow insufficient alone for a full appreciation of the promises of economic and cultural tradition. In turn, intent evolves materially and socially in specialized form and process, spiritually and instinctively aimed in a logically understandable mechanism of increasing systemic order amid apparent environmental entropy, toward an intuitively recognizable goal of flourishing as individuals in a collectively organized and directed, life enhancing ecosystem. In the pursuit of opportunity in this ecosystem, logic must embrace wisdom over ignorance, particularly when the ignorance is intentional. In avoiding the risks perceived in the recognized inertial habits and unknown intimidations of our fellow human beings, intuition counsels love and understanding, without giving leave of caution, over dwelling and mongering in fear.

There is always a need for clarity in navigating the changing risk and opportunity of life. The need is heightened by accelerating change that has been implemented by technological insight and now appears to be calling for further material and social innovation on an increasingly disruptive scale in a search of better, cleaner sources of energy and other resources for a sustainable quality of life. Such is the motivating intent, the aim and goal of modeling.

This piece started as an expression of my independent, quantitatively structured investigation into physical and economic phenomena through my experienced understanding of the fundamental human natures of inertial, formal, and intentional capacity in dealing with our world of risk and opportunity while pursuing the satisfaction of basic human needs. In addition to food and shelter, chief among these needs are the technologically generated requirements of supplying energy and environmental sustainability, the holy grail of which is to access the abundant energy of the sun cheaply, without the deleterious effects of having that process occurring at arm's length; this means fusion at room temperature. My enlightened understanding of physical processes indicates that it is worth pursuing.

I may be wrong about the prospects for palladium catalyzed cold fusion as an economically scalable energy technology, but I also know that no one to date has approached the subject with this integrated understanding of what constitute quantum interactions, and until others with the necessary technical skill are willing to help with the necessary theoretical vetting based on that understanding, I believe we will be hard press to garner the necessary experimental interest.

I trust that you will find appropriateness in this discussion and will know how best to proceed with its vetting and disclosure. I thank the reader for your time and welcome any well-defined, innovative work in addressing our energy needs.

A Condensed Matter Model of Fundamental Particle Genesis

as a Function

of an Accelerating Cosmic Spacetime Expansion

Fundamental Rest Mass Quanta as
Simple Harmonic Rotational Oscillations of
the Spacetime Continuum,
Driven by Cosmic Expansion
with
Application of the Analysis to the
Experimental Field of Cold Fusion

By Martin Gibson

August 6, 2019

Martin Gibson
P.O. Box 2358
Southern Pines, NC 28388
910-585-1234

A Condensed Matter Model of Fundamental Particle Genesis as a Function of an Accelerating Cosmic Spacetime Expansion

Martin Gibson

Abstract

The intention of this monograph is to present a foundational model of quantum physical processes as a function of an accelerating, isotropic cosmic expansion. It represents a modification of general relativity as a solution to the gravitational field equation in an initial condition of flat spacetime in which the Einstein curvature tensor vanishes and the stress-energy tensor consists of the dynamic properties of an individual fundamental baryonic particle, the neutron, first as a potential and then as an emergent adiabatic process due to isotropic stress and strain. This fundamental form is developed as an emergent function of the product of the cosmological constant of GR quantified as the Hubble rate with a bi-directional quantum metric, defined herein on a unit cube centered on one of an indefinite number of such centers of isotropic expansion. The isotropic stress is shown to create a torsion strain at the femtometer scale, from which the restorative force initiates a well-defined characteristic rotational oscillation according to the principles of classical wave mechanics.

While the speed of light is held to be invariant in this model, the principle gauge of time is the expansion rate, the source of which operates orthogonal to three-dimensional space. Quantum spin energy, spin angular momentum, and charge are generated by the antisymmetric components and quantum gravity is generated by the symmetric components of the quantum stress-energy double matrix.

Ongoing expansion causes a differential change in inertial density per time unit which is equal to the differential change in mechanical impedance per length unit according to the Hubble rate. These differential drops over time result in a discontinuity at the nodes of the neutron waveform which results in the transmission of a small fraction of the neutron wave energy as the rest mass of the electron and a transfer of the neutron wave momentum as elementary charge, so that beta decay is shown to be tuned to the Hubble rate. With the emission of the electron in a condensed matter state absent ionization, atomic interactions result from the nodal/wave phase interactions of the emitted electron waveform.

This non-stochastic model is developed using dimensional analysis without the addition of extraneous parameters and validated based on the observable invariant properties of the neutron and electron mass and the related reduced Compton wavelengths, the value of h-bar and the speed of light. Newton's gravitational constant is derived and found to be 6.67319×10^{-11} m³kg⁻¹s⁻², close to the 2018 CODATA value of $6.67430(15) \times 10^{-11}$. The Hubble rate is derived and found to have a lower threshold of 73.08 km/Mpc /s, which is interpreted as a dimensionless, compounding strain of 2.36839×10^{-18}s⁻¹. This is within the uncertainty of a recent referenced study by Riess et al, which reports the most precisely defined figure to date at $H_0 = 74.03 +/- 1.42$ km s⁻¹ Mpc⁻¹, validating this approach. It addresses the concerns in that study of the 4.4σ between their figure and the results of the LCDM Planck study at 67.74 km +/-0.46 km s⁻¹ Mpc⁻¹.

The model provides an intuitive grasp of such quantum phenomena and concepts as electron orbitals, tunneling, and nuclear and molecular bonding in the context of condensed matter physical phenomena such as the continued reported experimental results of positive correlations of anomalous heat and helium production in support of cold fusion. This model provides an understanding of physical phenomena that can, among other things, help explicate and expedite the safe development and utilization of palladium catalyzed deuterium nuclear fusion.

Table of Contents

0 — Notes Concerning Dimensional Analysis and Wave Formalism	1
0a — Classical Wave Dynamics	5
1 — A Heuristic Example of a Classical Basis for Quantum Phenomena	8
2 — The Classical Basis for Quantum Phenomena	12
2a — Symmetric Components of the Wave Tensor and Quantum Gravity	13
2b — Anti-Symmetric Components of the Wave Tensor and Spin & Charge	16
2c — Geometric Considerations of Rotational Oscillation and Beta Decay	26
2d — Derivation of Beta Decay as a Function of the Hubble Rate	29
2e — The Missing Mass of Beta Decay	36
2f — Evaluation of Elementary Charge	40
2g — Special Relativity and Muon & Tau Families	43
3 — Condensed Matter Application of this Model	44
4 — Conclusions	53
Bibliography, Citations, and Other Resources	57

Figures, Diagrams, Tables, and Charts

Figure 0	Wave Kinematic Functions	5
Figure 1	Isotropic Expansion Stress over Time	9
Figure 2	Expansion Stress < Inertia, $r_0 = t / m_0$	9
Figure 3	Expansion Stress => Inertia, r_0	9
Figure 4	Expansion Stress in Maximally Dense Space	10
Figure 5	Expansion Stress w/ Differential Vectors	10
Figure 6	Expansion Stress in 3D Manifold	10
Figure 7	Graphic of Emergent Cuboctahedral Lattice Cell	10
Figure 8	Graphic of one half of an Inversphere	11
Diagram 0	Femto Scale Torsion	16
Diagram 1	Rotational Oscillation	17
Diagram 2	Neutron Oscillation	19
Diagram 3	Proton Oscillation	20

Diagram 4	Electron Oscillation	21
Diagram 5	Anti-Proton Oscillation	22
Diagram 6	Positron Oscillation	23
Table 1	Charge and Spin Table for Ordinary Matter for C & L = 1	24
Table 2	Charge and Spin Table for Anti Matter for C & L = 1	25
Figure 9	Superposition of equal area cube and sphere	26
Figure 10	Cross-section of equal area cube and sphere	27
Table 3	Relationship of Hubble rate and particle mass/energy	33
Figure 11	Graph of exponent bases e_n	37
Table 4	Exponential Functions of e_n for $n = 0$ to 3	37
Table 5	Natural log functions of e_n for $n = 1$ to 6	38
Figure 12	Condensed Matter Graphic for N, P, and E spatial relationship	45
Figure 13	Graphic representing covalent bonding	46
Figure 14	Graphic representing electron-positron annihilation	46
Figure 15	Palladium Valence Electron Orbitals $4d^{10}$	46
Table 6	Lattice and Element Parameters for Palladium and Nickel	47
Chart 1	FCC Lattice Properties	49
Chart 2	Cuboctahedral Lattice Configuration	50
Chart 3	Hydrogen Diffusion	51
Chart 4	Fusion Path 1 in the Tetrahedral chamber	52
Chart 5	Fusion Path 2 in the Tetrahedral aperture	52

0 — Notes Concerning Dimensional Analysis and Wave Formalism

An extended note about dimensional analysis as used herein is in order. In theoretical discussion, such analysis makes use of a natural unit of measurement, herein represented by a subscript nought, for any qualitative property where the existence or applicability of the units is understood and supported by observational data. To be verifiable, the natural units of that analysis must then be convertible to a standard system of measurement such as the SI. If a dimensional property such as time or distance is deemed to be fundamentally continuous, as in most interpretations of the nature of spacetime of general relativity, there may be no "natural" unit other than an observed relationship to some familiar observation about the natural world; 1/60 of 1/60 of 1/24 of one solar day of the earth for one second as a unit of time and the distance light travels in one second in a vacuum divided by 299,792,458 for one meter as a unit of distance.

The presumed scale invariant speed of light, c, thereby couples the empirically derived measures of the two fundamental properties of physical kinematic analysis. Analysis is helped in some instances by the normalization of fundamental quantities. In the case of certain derivatives in which a rate of change in a dependent variable as numerator, dividend or antecedent is expressed as a function of a change in the independent variable as denominator, divisor or consequent and in which the dependent change has been evaluated in some system of accounting or calculation, the independent change is generally expressed as a unit of that property, i.e. as 299,792,458 meters per 1 second for the speed of light. For invariants or characteristic modes of a system, analysis can be facilitated by normalization, that is, by rendering the antecedent as a unit value as well; by changing 299,792,458 meters to 1 nouveau-meter, or 3.33… nano-seconds to 1 nouveau-second. The terms here are illustrative only and not intended for use. Such normalization notationally changes c to c_0, though in computation the transition back to standard units such as SI may be made.

This is in fact what happens in what is perhaps the most famous equation of physics, $E = mc^2$, and related wave equations involving use of the speed of light, as will be shown in this development. It is a source of much unnecessary mystification concerning mass and energy, at least to the public. When writers speak of "converting" a small amount of mass into a tremendous amount of energy, they are tacitly and generally unwittingly acknowledging the fact that the speed of light, c, is not normalized. If it was, a unit of mass would simply equal a unit of energy, though the context might indicate a unit of potential energy (mass) equals a unit of kinetic energy, thereby pointing to the fundamental concept of the conservation of energy.

The observations of dynamic analysis, that matter appears to be inherently comprised of discrete and invariant fundamental units of the properties of mass-energy, spin and charge, gives credence to the notion that these properties support ontologically natural units for their qualitative dimensions; either mass or energy that can be converted one to the other as a function of the square of the speed of light, fundamental particle spin angular momentum that arises apparently (but not necessarily in fact) apart from any geometric, dynamic basis for its dimensional property of action or angular momentum, and charge with the fundamental dimensional transformation of potential energy in its static, capacitive state and kinetic energy as current in its inductive, electromotive state.

Despite the energetic connotation of the word, a study of dynamics is grounded in the concept of inertia. It is against the backdrop of stationary objects that energetic ones are measured. Speed can only be measured with respect to an assumed stationary inertial reference frame, be that frame a configuration of fixed objects moving at a constant velocity through space or an abstraction of space itself, so a quantifiable explanation of the interaction between objects requires the concept of mass as a measure of that inertia. Mass can be thought of as a measure of 1) a resistance to change, 2) a delay or retardation factor in the process of a system, 3) a deflection agent to a moving element from an initial path, 4) a concentration of factors governing any such effects, 5) with respect to quantum particles, as a wave number, an inverse measure of a wave length, and perhaps other inertial phenomena. None of these properties can be quantified from experimental data without some reference to space and time, although they may be gauge invariant with respect to a specific location in space and time, and yet the coupling of quantum phenomena with a spacetime continuum remains elusive for conceptual reasons.

In the case of a wave analysis, the coupling of spacetime and quanta is straight-forward, as theoretical models of mechanical, inertial, wave bearing continua are known to support the emergence of characteristic natural and resonant frequencies and wavelengths in response to a driving energy source, once some threshold input level of that energy is reached. A characteristic fundamental frequency and wavelength can then form the basis of a natural time and distance scale, and we might anticipate that a natural unit of mass as derived from the inertial density of such a continuum is indirectly related to that of the time period and the displacement distance of any wave action or oscillation defined by such a characteristic emergence.

Using the angular representations for frequency, ω_0, and wavenumber, κ_0, or the number of periodic wavelengths times 2π per standard unit of length, and inverting for $\theta = 1$, gives a unit length of time, t_0, for one radian of wave activity at fundamental frequency and the corresponding distance unit length, r_0, or fundamental angular wavelength, λ_0, for one radian. The following quotient of frequency over wave number for wave form q gives the following

$$c_0 = \frac{\omega_{0q}}{\kappa_{0q}} = \frac{\theta/t_{0q}}{\theta/r_{0q}} = \frac{r_{0q}}{t_{0q}} = \left(\frac{dr}{dt_0}\right). \tag{1.1}$$

The bracketed last term is stated with the sub-nought to emphasize that while the wave speed can be stated in terms of differentials, as with most derivatives, the dependent variable is virtually always evaluated as a function of a change in one unit of the independent variable, in this case whether it is expressed as one second or one radian of motion of a theoretical clock. In the first instance, dr_{SI} is 299,792,458 meters in the SI system and in the second instance it is equal to the length of a subtending arc of 1 radian, dr_0, in the natural units used here.

It is well known that the square of the speed of wave motion, transverse and longitudinal, is a function of the stress, f_0, and the inertial density, ρ_0, of a wave bearing medium as follows, where the stress is further defined as a force, τ_0, either shearing, τ_s, or tension, τ_t, per cross-sectional area, A_0, of the medium, and inertial density is defined as mass, m_0, per unit volume, r_0^3. In the final term of this equation, A_0 and two of the volume length components are canceled from the term before to leave a statement for a linear component of stress, such as a tension force on a stretched string, over a linear inertial density, λ_0, since the cross section defining the tension stress and also the volume density of the string are the same, so that

$$c_0^2 = \frac{f_0}{\rho_0} = \frac{\tau_0}{A_0} \bigg/ \frac{m_0}{r_0^3} = \frac{\tau_{t0}}{\lambda_0}. \tag{1.2}$$

Note that the string tension component, despite the cross-sectional canceling in the final term, still has the capacity for shearing stress and therefore transverse wave motion as with a sinusoidal wave. Combining (1.1) into (1.2) gives the following

$$c_0^2 = \frac{\omega_0^2}{\kappa_0^2} = \frac{\tau_{t0}}{\lambda_0}. \tag{1.3}$$

Quantum mechanics uses the results of scattering experiments of photons on rest mass particles initially performed by Arthur Compton in 1922 to relate the energy-equivalent of a particle's rest mass to the change in energy of a photon scattered from the particle based on the change in wavelength of that photon and the angle of scattering, where h is Planck's constant and m is the mass of the quantum or

$$\Delta\lambda = \frac{h}{cm_q}(1-\cos\theta)$$
$$\lambda_C = \frac{h}{cm_q}$$
(1.4)

The Compton wavelength, $\lambda_{C,q}$, is therefore an accepted statistically derived parameter of rest mass particles which we will use in the context of dimensional modeling to gauge the natural unit scale. In doing so in the context of the following discussion of fundamental baryonic particles, neutron and proton, as examples of rotational oscillation, we will use the Compton convention as a statistically based physical measure of those quanta's radius of rotation, r_q, but using the reduced Compton or angular measure of the wavelength, $\bar{\lambda}_{C,q}$ lambda-bar, and the corresponding reduced Planck's constant, $\hbar$, h-bar,

$$\bar{\lambda}_{C,q} = r_q = \frac{\hbar}{cm_q}$$
(1.5)

It is assumed in this modeling that the product of the angular velocity of a particle wave phases, ω_q, times a radian-arc of motion that is equal to the reduced Compton and the radius of maximum shear stress of the waveform, is at the speed of light, so that

$$c = \theta\omega = \bar{\lambda}_{C,q}\omega_q = r_q\omega_q$$
(1.6)

and for a fundamental rotational oscillation as derived for this model,

$$c_0 = r_0\omega_0$$
(1.7)

From quantum analysis, we know the dimensional makeup of Planck's reduced quantum of action for use with angular frequency and wave number, which we can express with (1.1) as

$$\hbar = m_0 r_0 c_0$$
(1.8)

Since the length dimension in the last term of (1.1) and in (1.8) can be the same in natural units, and in light of the fact that both h-bar and c_0 are deemed to be invariant, we can rearrange (1.8) in terms of natural units and arrive at the following constant of inertia, ת (tav), which has the value of being a time-independent fundamental gauge of length to mass, showing their essential inverse relationship.

$$\frac{\hbar}{c_0} = m_0 r_0 = ת$$
(1.9)

From quantum analysis we have the following condensed matter identity of a rest mass particle, either baryon or lepton, as

$$m_{0q}c_0^2 = \hbar\omega_{0q} = תc_0\omega_{0q}$$
(1.10)

which with rearrangement shows that in quantum terms mass is simply a proxy for wave number as

$$m_{0q} = \frac{ת}{c_0}\omega_{0q} = תK_{0q} = \frac{ת}{r_{0q}}.$$
(1.11)

While we could continue to use the familiar quotient of h-bar over the speed of light in this discussion, and will do so where it is deemed appropriate for clarity in relating this material to the established dicta of physics, the inertial constant points directly to a physical wave nature of fundamental quanta by virtue of its conversion of angular wavenumber to mass for a particle constrained over time to a discrete location in space, whether that mass is conceived of as a derived property of an inertial substance or substrate or simply as a computational conversion of characteristic inherent energy as determined by a characteristic angular frequency determined from experiment.

It follows for a quantum particle that

$$r_{0q} = \frac{\hbar}{m_{0q}} . \tag{1.12}$$

Thus, for an inertial wave bearing medium of a given inertial density, λ_0, given the inertial constant, the mass of a fundamental particle oscillation, m_0, actually a form of self-oscillation, at an effective natural angular frequency, ω_0, is a direct function of the angular wave number, κ_0, and therefore indirectly of the angular wavelength, $\lambdabar_0 = r_0$.

Clearly then, just as the linear wave equation contains two tacit cross-sectional area dimensions that cancel, (1.3) contains two tacit terms for the inertial constant in each side of the quotient and is in fact the dimensionally correct

$$c_0^2 = \frac{\hbar \omega_0^2}{\hbar \kappa_0^2} = \frac{\tau_{t0}}{\lambda_0} . \tag{1.13}$$

This can be rearranged in a standard form of a wave equation, where the differentials are made explicit and perhaps more recognizable in the middle two terms, as

$$\lambda_0 \equiv \hbar \kappa_0^2 \equiv \hbar \frac{\partial^2 \theta}{\partial x^2} = \frac{1}{c_0^2} \hbar \frac{\partial^2 \theta}{\partial t^2} \equiv \frac{1}{c_0^2} \hbar \omega_0^2 \equiv \frac{1}{c_0^2} \tau_0 . \tag{1.14}$$

A further rearrangement states the important relationship of the mechanical impedance, Z_0, of the continuum or wave bearing medium which is dimensionally the quotient of the fundamental stress force and the wave speed as

$$\lambda_0 c_0 = \frac{\tau_0}{c_0} \equiv Z_0 . \tag{1.15}$$

In keeping with a premise of this development, that of ongoing spacetime isotropic expansion, we state that such expansion inherently leads to a decrease in inertial density of the continuum over time, and assuming an invariant speed of wave motion, to a decrease in the stress force and impedance, so that implicit differentiation by decoupling the wave speed components and rearranging equates the change in density over time to the change in impedance over a displacement

$$\frac{d\lambda_0}{dt} = \frac{d\tau_0}{c_0 dr} = \frac{dZ_0}{dr} . \tag{1.16}$$

This is of principal importance in understanding the Coulomb force and the generation of the electron as it applies to condensed matter.

0a — Classical Wave Dynamics

Various wave properties can then be stated using this ideal analysis and the Euler formalism for the various derivative properties of a complex fundamental torsion wave, where ϕ in this case is a rotating torsional displacement of the wave phase η and ς in the y-z strain cross-section, where the imaginary sense, i, indicates an initial torsion strain at +/- y about Y of ½ pi from z into $+X$ as developed below, and the amplitude, A, is assumed to be equal to the angular wave length, r_0.

$$\phi = \eta + \varsigma = A(\cos\theta \pm i\sin\theta) = Ae^{i\theta} = Ae^{i(\kappa x \pm \omega t)} \quad (1.17)$$

A physical example of this wave will be developed herein. Obviously the ς component lags the η component by ½ pi. In the following sinusoidal plotting of the wave phases and the kinematic functions of the two points, the instances of maximum power in both the direction of storage of potential energy, E, and of release of that energy as kinetic or mechanical energy, M, for each point, η and ς, of either sense, +/-, are shown. The senses indicate the direction of displacement change for η and ς, toward +1 or -1 on the ordinate at the given point in the cycle. These designations will also stand for electrical potential energy or charge, E, and for magnetic energy of electrical current, M. In this development it will be seen that such wave generated mechanical power moments are sustained over time for the oscillation and serve an essential function in the generation of charge and the electron.

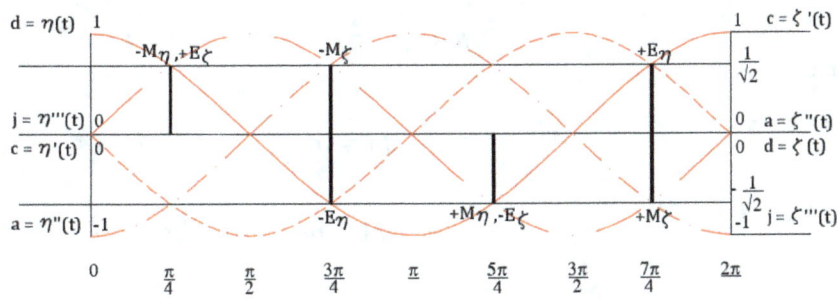

Wave Kinematic Functions

Figure 0

In the Euler formalism, each ½ pi rotation in time and in space is represented by an incremental unit power of ω and κ respectively as a result of each derivation. Orders of integration are represented by negative unit powers, where in cyclic terms one order of integration, ω_0^{-1} or κ_0^{-1}, equates to three orders of derivation, ω_0^3 or κ_0^3. Represented as natural unit values, the following creates a calculus of dynamic invariants for a fundamental rotational oscillation, where we can introduce a compact and convenient notation for each invariant, using the inertial constant as the system gauge or scale factor, where subscripts indicate complex differentiation and super scripts indicate complex integration, with time to the right and space to the left.

Wave Kinematic Time Functions

Displacement, r
$$\phi(t) = Ae^{i\omega t} = r_0$$

Velocity, c
$$\phi'(t) = i\omega_0 A e^{i\omega t} = i r_0 \omega_0$$

Acceleration, a
$$\phi''(t) = -\omega_0^2 A e^{i\omega t} = -r_0 \omega_0^2$$

Jerk, j
$$\phi'''(t) = -i\omega_0^3 A e^{i\omega t} = -i r_0 \omega_0^3$$

Wave Dynamic Time Functions

Inertial constant, ⊓
$$_0\!⊓_0^0 \qquad m_0 \phi(t) = m_0 A e^{i\omega t} = ⊓$$

Transverse Wave Momentum, p_0
(equals raw fundamental charge)
$$⊓_1 \qquad m_0 \phi'(t) = m_0\left(i\omega_0 A e^{i\omega t}\right) = i⊓\omega_0$$

Transverse Wave Force, τ_0
$$⊓_2 \qquad m_0 \phi''(t) = m_0\left(-\omega_0^2 A e^{i\omega t}\right) = -⊓\omega_0^2$$

Transverse Wave Yank, Y_0
$$⊓_3 = ⊓^1 \qquad m_0 \phi'''(t) = m_0\left(-i\omega_0^3 A e^{i\omega t}\right) = -i⊓\omega_0^3$$

Wave Dynamic Space Functions

Inertial constant, ⊓
$$_0\!⊓_0^0 \qquad m_0 \phi(x) = \frac{⊓}{Ae^{i\kappa x}} A e^{i\kappa x} = ⊓$$

Mass, m_0
$$_1⊓ \qquad m_0 \phi'(x) = \frac{⊓}{Ae^{i\kappa x}} i\kappa_0 A e^{i\kappa x} = i⊓\kappa_0$$

Linear Density, λ_0
$$_2⊓ \qquad m_0 \phi''(x) = \frac{⊓}{Ae^{i\kappa x}}\left(-\kappa_0^2 A e^{i\kappa x}\right) = -⊓\kappa_0^2$$

Moment of Inertia, I_0
$$_3⊓ = {}^1⊓ \qquad m_0 \phi'''(x) = \frac{⊓}{Ae^{i\kappa x}}\left(-i\kappa_0^3 A e^{i\kappa x}\right) = -i⊓\kappa_0^3 = -i⊓\kappa_0^{-1}$$

Remaining Wave Dynamic Space-Time Functions

Mechanical Impedance, Z_0 (of the spacetime manifold)	$_1\Pi_1$	$-\hbar\kappa_0\omega_0$
Transverse Momentum Surface Density, p_2	$_2\Pi_1$	$-i\hbar\kappa_0^2\omega_0$
Planck's Quantum of Action, $\hbar$ (Spin Angular Momentum)	$^1\Pi_1$	$\hbar\kappa_0^{-1}\omega_0 = \hbar c_0$
Linear Transverse Force Density, τ_1	$_1\Pi_2$	$-i\hbar\kappa_0\omega_0^2$
Wave Stress, f_0	$_2\Pi_2 = {}^2\Pi^2$	$\hbar\kappa_0^2\omega_0^2$
Spin Energy, E_0	$^1\Pi_2$	$i\hbar\kappa_0^{-1}\omega_0^2$
Mass Frequency Ratio, m_0/ω_0	$_1\Pi^1$	$\hbar\kappa_0\omega_0^{-1} = \dfrac{\hbar}{c_0}$
Yank Surface Density, Y_2	$_2\Pi_3$	$i\hbar\kappa_0^2\omega_0^3$
Wave Power, P_0 (Yank Volume Density, Y_3)	$^1\Pi_3 = {}_3\Pi_3$	$-\hbar\kappa_0^{-1}\omega_0^3 = -\hbar\kappa_0^3\omega_0^3$

1 — A Heuristic Example of a Classical Basis for Quantum Phenomena

We can analyze the local effects of an isotropic expansion of an ideal extended 3-D manifold by considering the heuristic example of the surface of a spherical balloon under expansion. We treat that surface as a two-dimensional manifold without boundary, a 2-sphere. The stress of expansion is directed radially from the center of the balloon from which it is transferred as transverse stress across the surface. Beyond an energy input threshold, the balloon expands with resulting strain, and the stress spreads out radially and equally from each point on the surface. Note that logically, in an elastic medium, strain always implies stress, but stress may not involve strain, just as a force pushing against and moving an object does work, but it does no work without that movement. Strain implies the energy of work.

If we select two points in very close proximity and represent only those stress vectors radiating from the two points, which necessarily incorporate those of the intervening space, we get the situation shown in Figure 1. It is important to emphasize that a point, in fact any n minus 1 reference on an n-manifold, marks a reference location on that manifold but is not itself an element of that manifold, which for a 2-sphere is an area. The two circles shown, two "1-spheres" in topological language, represent reference boundaries of equal stress for the two points chosen, of equal strain potential, and of equal time elapse for a change in stress at the boundaries to be registered at each point and vice versa. Obviously, the further the points are from each other the longer the time required for any such stress, and attendant strain to register.

Figure 1 is a snapshot in time and assumes that the process is ongoing, so that momentum efficiently allows all differential strain of expansion to flow into the continued expansion of the balloon surface. Note that if the vectors represent the stress of expansion, they do not represent the strain, even the accumulated strain, which cannot radiate from a point, a singularity, but can only radiate from an initial condition of an element of the 2-manifold, i.e. a smaller circle, as with the lighter vectors of Figure 3 radiating from the circle out.

Note also in Figure 1 that there is interaction between five stress vectors from each of the two points. The central vector of each radiates to the other point, indicating that with strain the distance between the two points will increase and without strain the pressure, or negative tension stress, between the two will intensify. The vectors to either side of this first common radial vector are tangential to the equipotential circle of the other point, indicating a shear or rotational stress and potential strain at those locations on the reference boundaries, including the areas exterior to the circles from the point of tangency up to their points of crossing. The outer most of the five interacting vectors for each point indicate a region of converging stress and potential strain. This configuration of stress and strain potential represented by the five vectors for each point indicates a potential for simple harmonic oscillation given the necessary and sufficient additional condition for wave mechanics of inertial density.

Not unlike the process of blowing up a real rubber balloon that has an initial physical configuration and does not start from a single point, we might anticipate that initiation of the process requires a threshold level of internal radial stress to be achieved before expansion of the surface as a result of transferred radial-to-surface strain. Figure 2 shows the condition prior to initiation of the process, in which the inertia of a given region of radius r_0 is greater than the transverse stress represented by the radiating vectors inside the circle, and before any expansion strain has occurred. Note that r_0 represents a potential for wave activity based on the properties of the inertial constant as in (1.12).

Figure 3 transposes the inertial vectors and expansion stress vectors from Figure 2 and shows the condition at one of the points on the manifold at the point in time when the inertial force is equal to the expansion force and immediately thereafter. Continued increase in the expansion stress will result in an expansion strain in the value of r_0 or, if the area outside the circle is of less inertial density than that of the inside, in a strain of the area outside the periphery.

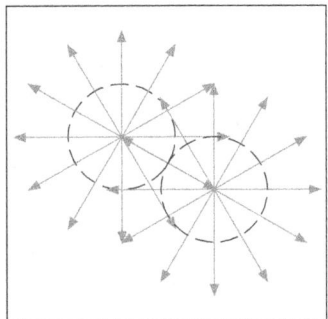 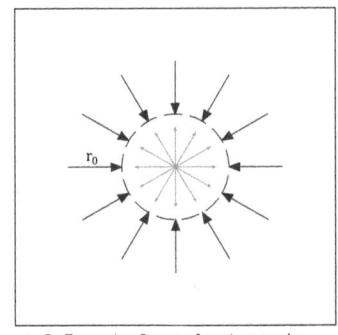 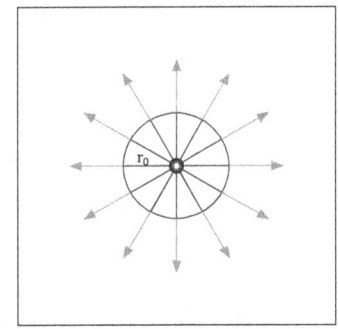

1 - Isotropic Expansion Stress over Time   2 - Expansion Stress < Inertia, $r_0 = \dot{r}/m_0$ 3 - Expansion Stress => Inertia, r_0

Figure 4 shows the general condition of Figure 1 in an area of initial, maximum inertial density of the 2-manifold as indicated by the maximum packing of six identical circles around each other circle, one of which is highlighted here. Under such conditions of density, the triangular interstitial areas become the focus of stress in excess of that required for equilibrium condition of the inflated manifold as in Figure 3. The upper half of the divided figure shows the state when the stress vectors from the adjoining inertial circles converge and concentrate the stress in the center of the interstitial areas. Those vectors that were originally radially in common with an adjacent cell as in Figure 1 are redirected as rotational potential to their common points of tangency. The bottom half of the figure shows the condition a moment latter as the stresses increase in the interstitial regions of the manifold. Such accumulation of energy will lead to rotational and oscillatory stress and potential strain of the circle peripheries and a potential for and eventually effective continued expansion of the 2-manifold counter-centripetally in what will be registered in the interstitial regions as curvature. It will also lead to an expansion of these areas as in Figure 5.

Note there is nothing predetermined about the location of such lattice configuration which is an emergent phenomenon similar to the observed phenomena of Rayleigh-Benard convection which the reader is encouraged to investigate on the internet. In that phenomena, the function of expansion stress is replaced with a heat gradient and the anti-parallel function of inertia is replaced with gravity.

The lower half of Figure 5 shows the eventual expansion strain of the regions between the circular cells in the flat plane of the balloon surface, as well as the emergence of rotational, shear strain in that flat surface about the various cells peripheries, and finally the continued stress and strain in the interstitial areas resulting in curvature which we can see as normal to the flat surface.

To this point, our discussion has focused on the effects of expansion stress on a postulated 2-manifold, the 2-sphere cover of a 3-ball. We now extrapolate that treatment to that of a 3-manifold under expansion, a 3-sphere cover, due to a change in stress over time, which is itself normal to the "surface" of that 3-sphere; this means that the expansion operates as the symmetric components of a 4-stress. This is depicted in Figure 6, which at maximum density takes the configuration of a cuboctahedral lattice (COL). A COL is the same as a face centered cubic (FCC) structure but centered on one of the spherical cells instead of the octahedral central space of the FCC. The COL perspective is achieved by shifting the FCC perspective one half the edge length of a cube to include the four center nodes of the adjacent surfaces, twisting ¼ pi CCW about the top face of the cube, then tilting the top toward the viewer to look along a diagonal axis through the triangular aperture to the central sphere. The circles in this figure represent the diameters of the 2-spheres of each of three representative layers.

In Figure 6, the 3 spheres of the designated top layer center on two edges and one vertex of a defined FCC, so that the view shown is looking along one of four diagonal axes of the configuration at the center sphere. Beyond that are the six 2-sphere cells of the middle layer represent by the six surrounding circular cells of Figures 4 & 5, with the stress dynamics of those figures shown here. The bottom layer with 3 spheres represents the next layer of the 3-manifold, once again represented as four dimensional with respect to the four diagonal axes of the space. The indicators of interstitial curvature and expansion in Figure 5 represent rotational and torsion stress and strain about the center inertial spherical cell in Figure 6.

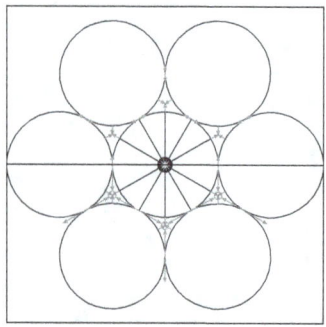 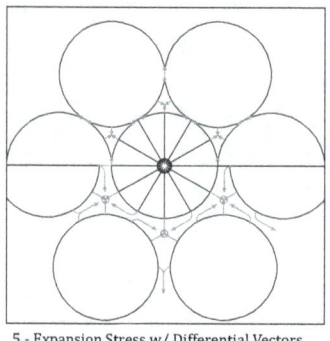 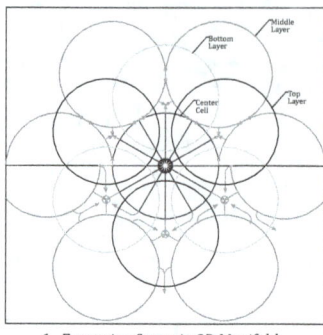

4 - Expansion Stress in Maximally Dense Space   5 - Expansion Stress w/ Differential Vectors   6 - Expansion Stress in 3D Manifold

The configuration is shown as three dimensional and without the stress vectors in the following graphic.

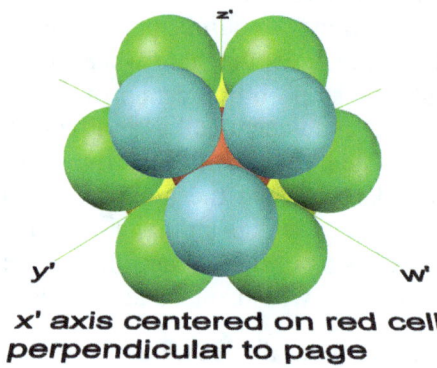

Figure 7 — Graphic of Emergent Cuboctahedral Lattice Cell

The stress of isotropic expansion on a compact 3-manifold (the three dimensional analogy of the surface of a ball) results in an emergent, initial cuboctahedral lattice of rotational stress and strain components from the uniform density in the manifold prior to that expansion. The inertial continuity and local elasticity of the manifold prevents these shear components from rupturing or creating a local internal extrusion strain, resulting instead in oscillation about each inertial center with an expansion and reduced density of the manifold in the interstitial regions. This oscillation can be represented in a couple of ways, which avoid the co-ordinate entanglement problem (not to be confused with quantum entanglement). Instead of such entanglement, rotational strain results in a recoil of the rotational displacement, which is torsional and has less torsional resistance about the diagonal axes, along a path of least action in keeping with (1.17)

A heuristic representation of this rotational oscillation is a sequence of 4, 2/3 pi CCW rotations, facing the cell center, about each of the four diagonal axes extending from the vertices of a face of the represented cube, say the upper face here, a CCW sequence of CCW rotations being equal to a CW sequence of CW 2/3 pi rotations on the opposite ends of the axes on the opposite cubic face. This can be further refined by making the diagonal rotations differential, followed sequentially by a differential rotation of the axis normal to the chosen face, proportionally so as to return all elements of the configuration to their original positions after a full rotation of the surface normal axis.

In this regard, we offer the following 4-dimensional interpretation of a 2-sphere in the context of the above development. Each of the four axes of the cubic diagonals can be defined as the central axis of a pair of pseudo-spheres, one on each side of the central 2-sphere, so that they intersect each other orthogonally at their rims as seen in Figure 8, which shows the top half of such an arrangement.

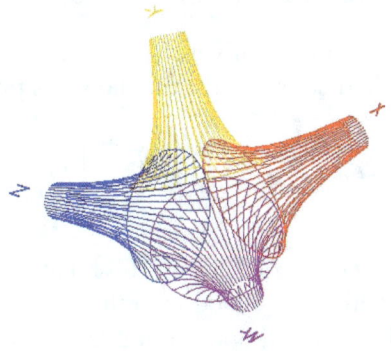

Figure 8 — Graphic of one half of an Inversphere

A pseudo-sphere has constant negative curvature, and the rim intersections will be found to coincide with the intersections of the corresponding three cubic surface axes and the surface of a sphere of curvature related to the positive of the pseudo-spheres; if the pseudo-sphere curvatures are -1, that of the sphere is $+\frac{\sqrt{3}}{2}$. I am calling this contraption of eight pseudo-spheres an inversphere as it inverts the positive curvature of the sphere proportionally to the negative curvatures of the concentric pseudo-spheres. Sequential oscillatory twists of the four axes as described in the development of the cuboctahedral lattice produces the rotation of the center 2-sphere, while straining the adjacent pseudo-spherical surfaces.

From this treatment it becomes apparent that the gravitational components of the resulting wave action are the symmetric components of a matrix defined on the oscillation as in the following section and form the basis of a de Sitter spacetime within the context of general relativity. The electromagnetic, weak and strong forces of quantum modeling are the function of the anti-symmetric components of that matrix and in the manner of an inversphere form the basis of an anti de Sitter quantum spacetime as developed in that section. The de Sitter and anti de Sitter geometries are inherent or curvature potentials of spacetime in this model that emerge as actual quantum effects from the inertia of spacetime in response to expansion tension as defined by the Hubble rate.

2 — The Classical Basis for Quantum Phenomena

The primary physical discipline for an analysis of space and time is general relativity, generally practiced as a classical field theory. The basic field equation for the theory is presented here

$$G_{\mu\nu} + g_{\mu\nu}\Lambda = \left(\frac{4\pi G}{c^4}\right) 2T_{\mu\nu} \qquad (1.18)$$

The first term on the left is the Einstein curvature tensor, $G_{\mu\nu}$, (often stated in further detail as the sum of the Ricci tensor and the Ricci scalar) which is a function of the energy per volume represented by the stress tensor, $T_{\mu\nu}$, on the right. As generally stated the accumulation of mass-energy defined on the right tells spacetime as $G_{\mu\nu}$ how to curve and that curvature in turn responds to tells mass-energy how to move within it. In particular, however, it says that the right side of the equation predominates in curving spacetime if $T_{\mu\nu}$ is great and the left side predominates in controlling the movement of matter/energy if $G_{\mu\nu}$ is great. If both are small at some point in the manifold, we are left with a generally flat spacetime, so that $G_{\mu\nu}$ essentially vanishes and the cosmological constant term, $g_{\mu\nu}\Lambda$, governing spacetime expansion is the predominant feature at that point. The metric, $g_{\mu\nu}$, isotropically distributes the effects of Λ according to the double matrix of (1.20), and their product is effectively a spacetime, oscillating strain. If lambda is small it means that any change is relatively slow, which appears to be the case from our perspective.

The bracketed term on the right represents the isotropic distribution of the effects of the stress-energy tensor according to the constraints of Newton's gravitational constant, G, and at the speed of light. What appears to be generally overlooked or disregarded is that the accumulated effects of Λ can be very great where concentrated locally at a very small scale, even for flat spacetime. At such scale and under such conditions this strain can result in curvature fluctuations as might be associated with oscillation. In any event it would be responsible for the generation of virtual particles and quantum foam of current cosmological thinking.

In this case (1.18) is reduced and transposed to show the space only components of the tensor, T_{ij}, as a function of the expansion strain as

$$T(\Lambda) = 2T_{ij} = g_{ij}\Lambda \qquad (1.19)$$

The oscillating force components, τ_{ij}, of the expression $2T_{ij}$ can be represented by a double matrix as face centered and for the two opposing sides of a unit cube under isotropic stress and strain, where the negative sense of the second equals the addition of the opposite sense or

$$\tau_{ij} = \begin{bmatrix} -d\tau_0 & \tau_0\cos\omega t & \tau_0\sin\omega t \\ -\tau_0\cos\omega t & -d\tau_0 & \tau_0 \\ -\tau_0\sin\omega t & -\tau_0 & -d\tau_0 \end{bmatrix} - \begin{bmatrix} d\tau_0 & -\tau_0\cos\omega t & -\tau_0\sin\omega t \\ \tau_0\cos\omega t & d\tau_0 & -\tau_0 \\ \tau_0\sin\omega t & \tau_0 & d\tau_0 \end{bmatrix} \qquad (1.20)$$

2a — Symmetric Components of the Wave Tensor and Quantum Gravity

The six symmetric differential components represent the quantum gravitational components of the system as demonstrated in the following. The matrix has been configured so that the symmetric differentials are all of the same sense, here centripetal with respect to polar coordinates and the center of a unit cube.

A scalar form of the stress-energy relationship, where T_0 as a 4-dimensional spacetime stress is

$$\frac{1}{6\sqrt{3}} T_0 = f_0 = \frac{\tau_0}{A_0} \tag{1.21}$$

The inverse square root of 3 relates the orthogonality of a fourth dimension of time to the three spatial dimensions and the 6 indicates the six faces of the unit cube. If we think of the time related expansion as operating along one of the four diagonal axes of the cube at any instant of time, its relationship to the cubic faces should be clear. We next want to take the total derivative of the stress on one cubic face as

$$df = \frac{\partial f}{\partial \tau} d\tau + \frac{\partial f}{\partial A} dA$$
$$df = \frac{1}{A} d\tau - \frac{\tau}{A^2} dA \tag{1.22}$$

Assuming for convenience an invariant stress so that df is zero, in natural units gives

$$\frac{\tau_0 + d\tau_0}{A_0 + dA_0} = \frac{\tau_0}{A_0} = f_0 \tag{1.23}$$

confirming the co-equal variance of the units of force and of cross-section in natural units.

Separating the total derivative (1.22) and solving for the differentials in terms of T_0 gives us

$$d\tau_0 = \frac{A_0}{6\sqrt{3}} dT_0$$
$$dA_0 = -\frac{A_0}{T_0} dT_0 \tag{1.24}$$

Note that as stress expressed as a unit or as a derivative mathematically reduces to a force per UNIT of area in any system, natural or otherwise, the derivatives of stress force and of area are expressed with respect to a unit of stress, so that if we can determine the value of A_0 in the first of these differentials, we can quantify the differential stress force.

The dimensional properties of G found in (1.18) are mass, m, length, r, and time, t, of some unknown natural units, though generally stated in terms of the Planck scale, and as required to produce the force of gravity when used in the equation of Newton's gravitational law. While it does not appear to be customarily acknowledged, these units can be understood to be a dimensional reduction as follows, from a fundamental expression of Newton's Law in natural units, where a unit of gravitational stress force is indicated by τ_{0G}, and the two units of mass and square of the separation of massive bodies is converted to natural units by the quotient in the last term,

$$G = \frac{r_0^3}{m_0 t_0^2} = \frac{r_0^2}{m_0^2}\left(\frac{m_0 r_0}{t_0^2}\right) = \frac{r_0^4}{\left(\hbar/c\right)^2}\tau_{0G} = \frac{r_0^4}{\mathsf{n}^2}\tau_{0G} \qquad (1.25)$$

In the final term we have made use of the development of the inertial constant to convert the units of fundamental mass to length, consistent with the conventions of quantum analysis. If we can convert the force term in the final term to a length scale as well, knowing the observed value of G and the inertial constant as h-bar over the speed of light, we can solve the equation and arrive at a value for a fundamental unit of length, and thereby of mass and time as well.

First, Newton's Law can be expressed in terms of these fundamental units as the product of the number of fundamental quanta, n_M, in two gravitational bodies of mass, M_N, divided by the square of their distance of separation, n_r, in terms of fundamental units of length, times a fundamental unit of gravitational stress force

$$F_G = \frac{M_1 M_2}{d^2} G = \frac{n_{M1} m_0 n_{M2} m_0}{n_{r_0}^2 r_0^2}\left(\frac{r_0^2}{m_0^2}\tau_{0G}\right) = \frac{n_{M01} n_{M02}}{n_{r_0}^2}\tau_{0G} \qquad (1.26)$$

so that the value of G unmediated by any displacement in spacetime from the source, which is presumably quantum, and between two fundamental quantum generators of gravity is simply τ_{0G} which equals each of the 6 differentials of the symmetric components of the above matrix or

$$\tau_{0G} = d\tau_0 = \frac{A_0}{6\sqrt{3}} dT_0 \qquad (1.27)$$

and we can substitute the differential here into (1.25) to get the following value for G, in which it is understood that the differential stress equals 1,

$$G = \frac{r_0^4}{\mathsf{n}^2} A_0 df_0 = \frac{r_0^6}{6\sqrt{3}\mathsf{n}^2} dT_0 = \frac{r_0^6}{6\sqrt{3}(\hbar/c)^2} dT = 6.67319\ldots \times 10^{-11}\, m^3 kg^{-1} s^{-2} \qquad (1.28)$$

The current 2018 CODATA value for G is $6.67430(15) \times 10^{-11}$ with standard uncertainty for the last two digits expressed. Solving for r_0, we get a value for a fundamental natural length of $2.1002\ldots \times 10^{-16}$ meters, extremely close to the value of the reduced Compton wavelength of the neutron at $2.1001\ldots \times 10^{-16}$ meters and within the standard uncertainty given the standard uncertainty of Newton's constant. This states that Newton's constant and gravitational law are a function of expansion stress, T_0, of the spacetime manifold.

As indicated in the following development of baryonic wave mechanics, the value of (1.28) would be expected to vary slightly due to the mix of neutrons and protons in any given congregation of baryonic matter and in light of the structural nature of Newton's constant as shown in (1.26).

Thus, from (1.24) we get the following value for the differential stress force responsible for gravity as

$$d\tau_0 = \frac{r_0^2}{6\sqrt{3}} dT_0 = 4.24430\ldots \times 10^{-33}\, Newton \qquad (1.29)$$

where the wave stress force itself, responsible for the nuclear strong force, is

$$\tau_0 = \mathsf{n}\omega_0^2 = \mathsf{n}\frac{c^2}{r_0^2} = \frac{\hbar c}{r_0^2} = 7.16766\ldots \times 10^5\, Newton \qquad (1.30)$$

and the ratio between the two is

$$\frac{\tau_0}{d\tau_0} = 1.68877\ldots x10^{38} \tag{1.31}$$

This ratio should be the same as it is for the change in cross-section given (1.23), maintaining an invariant stress. Thus, we have the following, where it is apparent that the differential area is in fact the Planck area

$$dA_0 = A_0 \frac{d\tau_0}{\tau_0} = r_0^2 \frac{\frac{r_0^2}{6\sqrt{3}} dT_0}{\hbar c/r_0^2} = \frac{r_0^6 dT_0}{6\sqrt{3}\hbar c}$$

$$= \frac{r_0^6 dT_0}{6\sqrt{3}(\hbar/c)^2} \left(\frac{\hbar}{c^3}\right) = \frac{G\hbar}{c^3} \tag{1.32}$$

$$= A_{Planck} = 2.61185\ldots x10^{-70} \, meter^2$$

This development indicates that the Planck scale, which is deemed in current theoretical thinking to be an absolute fundamental quantum scale, is in fact a classical differential related to the quantum stress tensor responsible for gravity and the strong and electromagnetic forces, applicable to the fundamental oscillation of the neutron. Thus, expansion of the cosmic manifold along the rotating diagonal axes of oscillation results in a surface differential, dA_0, and a centripetally directed differential stress force responsible for gravity, $d\tau_0$, which matches the wave stress of the quantum τ_0/A_0. The square root of the inverted differential of the natural log of stress is related to the square root of the linear change of expansion as a dimensionless ratio

$$\frac{r_0}{l_{Pl}} = \sqrt{\frac{A_0}{A_{Pl}}} = \sqrt{\frac{T_0}{dT_0}} = \sqrt{(d\ln T_0)^{-1}} = 1.29952\ldots x10^{19} \, . \tag{1.33}$$

2b — Anti-Symmetric Components of the Wave Tensor and Spin & Charge

We next consider the anti-symmetric components of the matrices in light of the previous description of the emergent rotating torsional oscillation of a quantum system. As indicated by this matrix and developed below, the system generates spin angular momentum and the magnetic dipole field of each of the particles, through the oriented rotation of an inductive moment, L_μ, of each. This explains why there are no magnetic monopoles, as the field source is an axial vector which has inherent polarity.

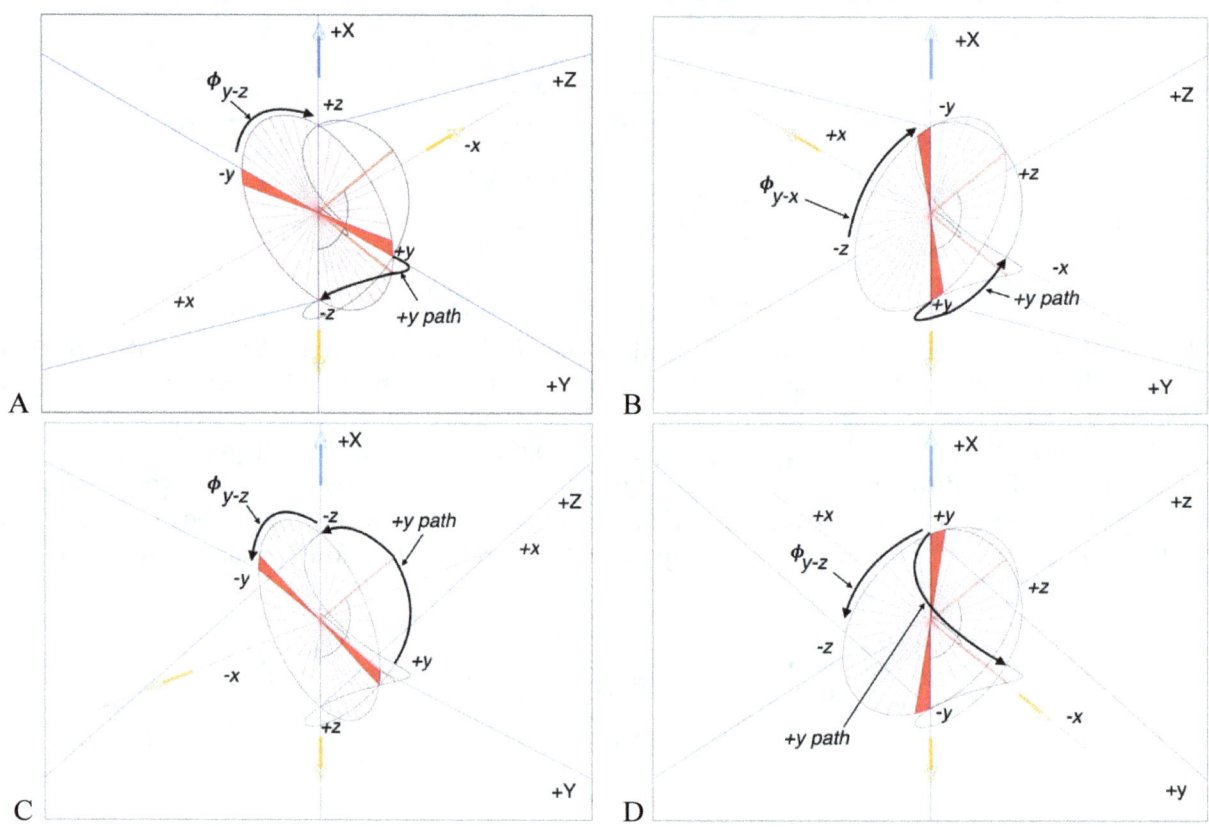

Diagram 0 — Femto Scale Torsion

Initially in X-Z plane (A), recoil under additional torsion stress in X-Y and Y-Z planes results in sustained rotation of ϕ around X axis with y-z oscillation of maximum stress. Path integral of +y point on ϕ_{x-y} is shown for four ½ π phases.

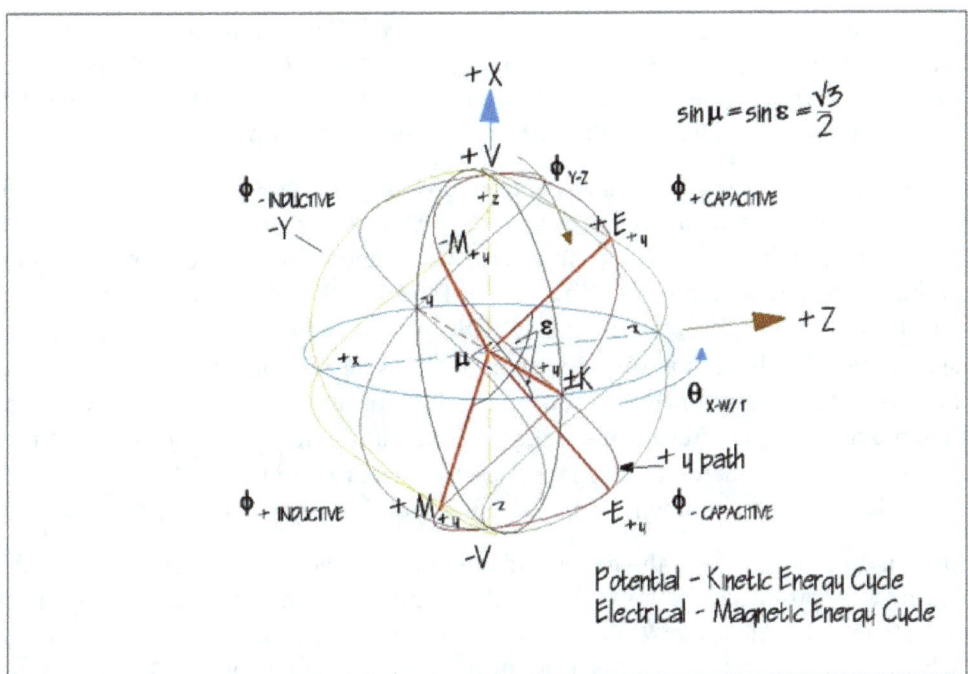

Diagram 1 — Rotational Oscillation

Rotation of on ϕ_{x-y} is shown with path integral of $+y$ point scribing a figure 8 as it oscillates, always crossing the equator to the right to create the spin vector at $+X$. There are a continuous set of such points defined by the ϕ_{x-y} circle making such a path to fill the 2-sphere. They each reach two points of maximum differential potential and kinetic energy/wave momentum, or capacitance (E) and inductance (M), along the path which create a capacitive and inductive moment that circulate as θ, as shown in the following Diagrams 2-6. This results in an invariant inherent Hamiltonian and Lagrangian for the particle and is responsible for all the fundamental quantum properties.

As shown in Diagrams 0 & 1, fixed orthogonal axes are referenced in upper case and the corresponding positions labeled in lower case on the 2-sphere represent the central cell previously discussed after initial torsion strain. It is understood in the following description that any designation of a property being spatially fixed or stationary over time is relative to the dynamics of the system under discussion and is held to move with any translation or rotation of the overall system over multiple cycles of oscillation, which in terms of condensed matter is in the range of 10^{24} hertz. It is not indicative of a fixed spacetime <u>lattice</u> substrate.

As an initial condition, the 2-sphere with central y-z disk, 1D, rotated about an arbitrary Y axis so that the initial $+z$ aligns with the $+X$ axis as shown and that initial $+x$ is rotated anti-parallel to the Z axis. This results in torsion strain in the X-Z plane about the Y axis as shown as a result of the ongoing isotropic stress. Recoil stress acts to reduce the stress and strain by rotation in the Y-Z plane, normal to the twisted disk (disk selection and X-Z axis designation is arbitrary, for illustrative purpose, and could have been any great circle through the 2-sphere with the Y axis as a diameter), where $+z$ and $-z$ represent the points of maximum strain from the Z axis, the locus of torsion equilibrium. Such rotation of 1D, ϕ, has a degree of freedom of rotation about $+Z$, and we have here shown it to be CCW (when facing the interior of the sphere from $+Z$.) This rotation initiates a spin, θ, of 2S on the 1D edge CCW about $+X$ at the same angular frequency as ϕ, so that with each rotation of ϕ and θ, every point, d, on the circumference of 1D passes through the $+/- V$ points on the X axis at maximum displacement and through the points of initial equilibrium, $+/- K_d$, at the points of maximum velocity and recoil wave momentum of the developed rotating torsion wave. Each and every d etches a distinct figure 8 strain path through the theoretical stationary space just above the "surface" of the rotationally oscillating 2S.

The points $+/- V$ are the concentrated, sustained points of maximum potential electro-mechanical energy and the distributed, recurrent collection of $+/- K_d$ points are the sustained loci of maximum kinetic electro-mechanical energy. They are also the loci of maximum and sustained charge at $+/- V$ and of maximum sustained current at Σ+/- K. The points +/- E and +/- M are points on the path of each d of maximum rate of capacitance or charge of potential energy in the direction of +/- V and of maximum rate of inductance or release of kinetic energy in the direction of +/- K respectively. Thus, this configuration constitutes a microscopic LC current, with inertia of the cell as the resistance component of the cycle.

As shown in Diagram 2 for the generated neutron, crossing the two points of equilibrium at +/- K for a given instant of the cycle into their corresponding instant -/+ E produces an instant capacitive torque moment C_ε and crossing the two instant points of +/-M into their corresponding equilibrium -/+ K produces an instant inductive torque moment $L\mu$. While each of the points E and M are fixed over short-range cycles of the system for each point d, C_ε and $L\mu$ rotate with θ at the recoil wave speed, c. This rotational system creates a spin vector for the action of the system at S_L along with an anti-parallel magnetic moment, μ, which is the result of the sustained rotation of θ. This configuration explains the relationship between elemental charge at +/- V as a result of C_ε and the quantum magnetic field as a function of the rotating torque $L\mu$. The reason that there are no magnetic monopoles is immediately clear. In the presence of an external magnetic field B, the moment μ aligns with B and the spin vector S_L precesses.

These torques in turn interact with the nodes and antinodes of the rotating oscillation as shown in the inset of the Neutron C-L Torques at the lower left of the figure. Note that the condition of all the torques is to advance the rotation of both ϕ and θ, making for an inherently unstable condition, given expansion of the manifold which drives the rotation. Over time in response to expansion, the condition in the middle two figures at K results in a transmission of a small portion of the energy of the oscillation as the electron wave and a flip of the S_L spin generally anti-parallel to the $L\mu$ torque occurs, as seen in Diagram 3. The lower left inset of this figure shows the new condition at K which explains the stability of the proton, as the induction torque $L\mu$ is anti-parallel to and retards the rotation of θ and the torque C_ε is anti-parallel to and retards the rotation of ϕ, both effects giving stability to the system. Interestingly, for the anti-proton, Diagram 5, these same influences are both parallel, leading to the instability of the system.

For the electron, shown in Diagram 4, the induction torque $L\mu$ is parallel to both rotations and the capacitive torque C_ε is anti-parallel to both, indicative of the charge and the fact that under expansion, the cosmos is in the process of converting the concentrated elastic potential energy density of the manifold to the distributed kinetic energy of elemental and molecular interaction on both a plasma stellar and condensed matter terrestrial level. Analysis of this modeling produces observables for particle mass, spin, charge, moment, ordinary and anti-matter configurations that are consistent with observed data without the addition of extraneous free parameters. A hint at the nature of the quark phenomenology of baryons can be found, especially in conjunction with the spin and charge tables. Note that the filled and open dots preceding the values of columns 4 through 7 of those tables indicate a centripetal and counter-centripetal sense for the cross-products on each of the listed points as defined at the top of each column.

We will next analyze this stress and strain configuration and how it connects with established thinking about space and time, to show that it naturally and necessarily includes quantum gravity, giving us a complete picture for theoretical physical understanding and its relevance for an understanding of cold fusion.

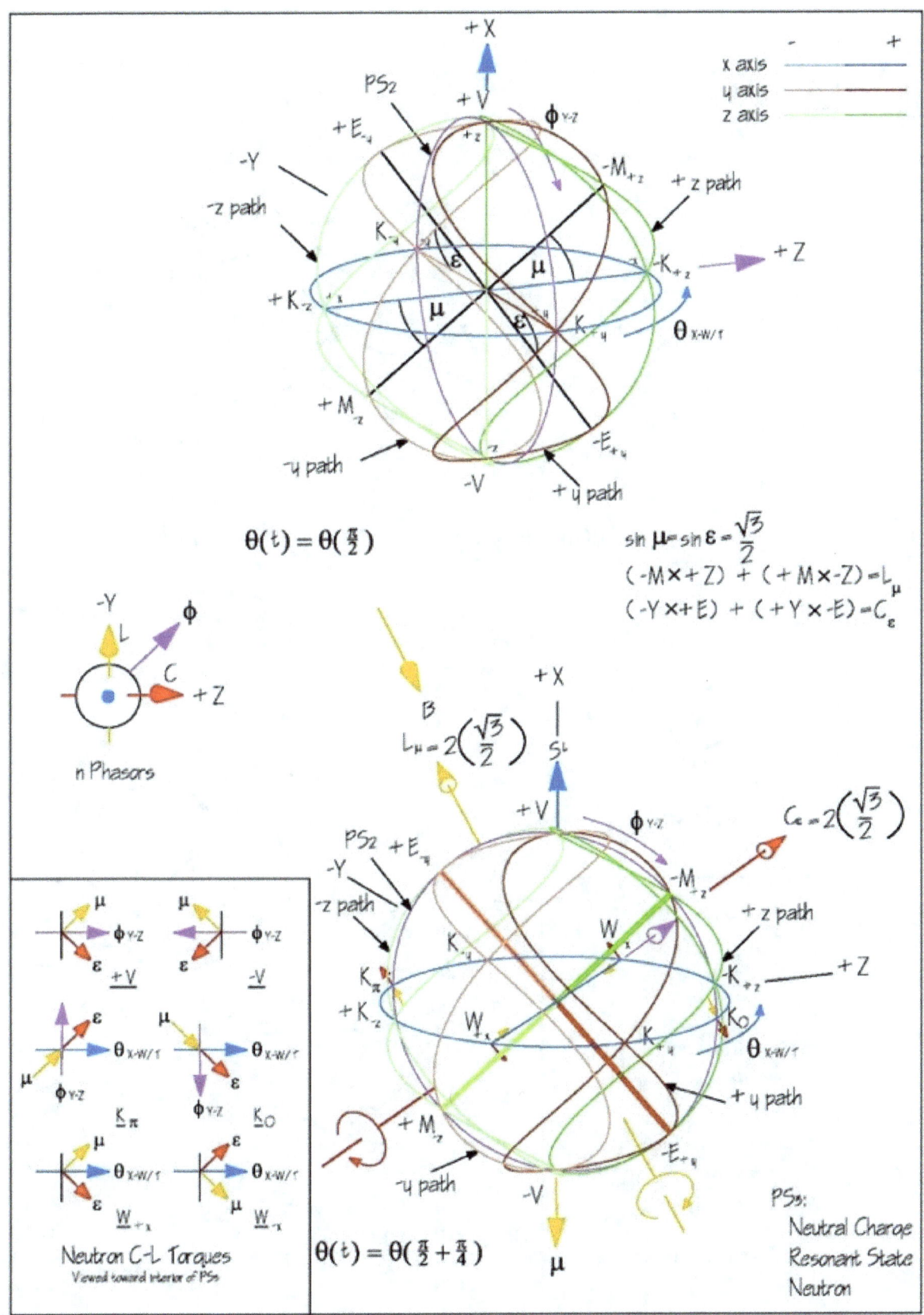

Diagram 2 - Neutron Oscillation

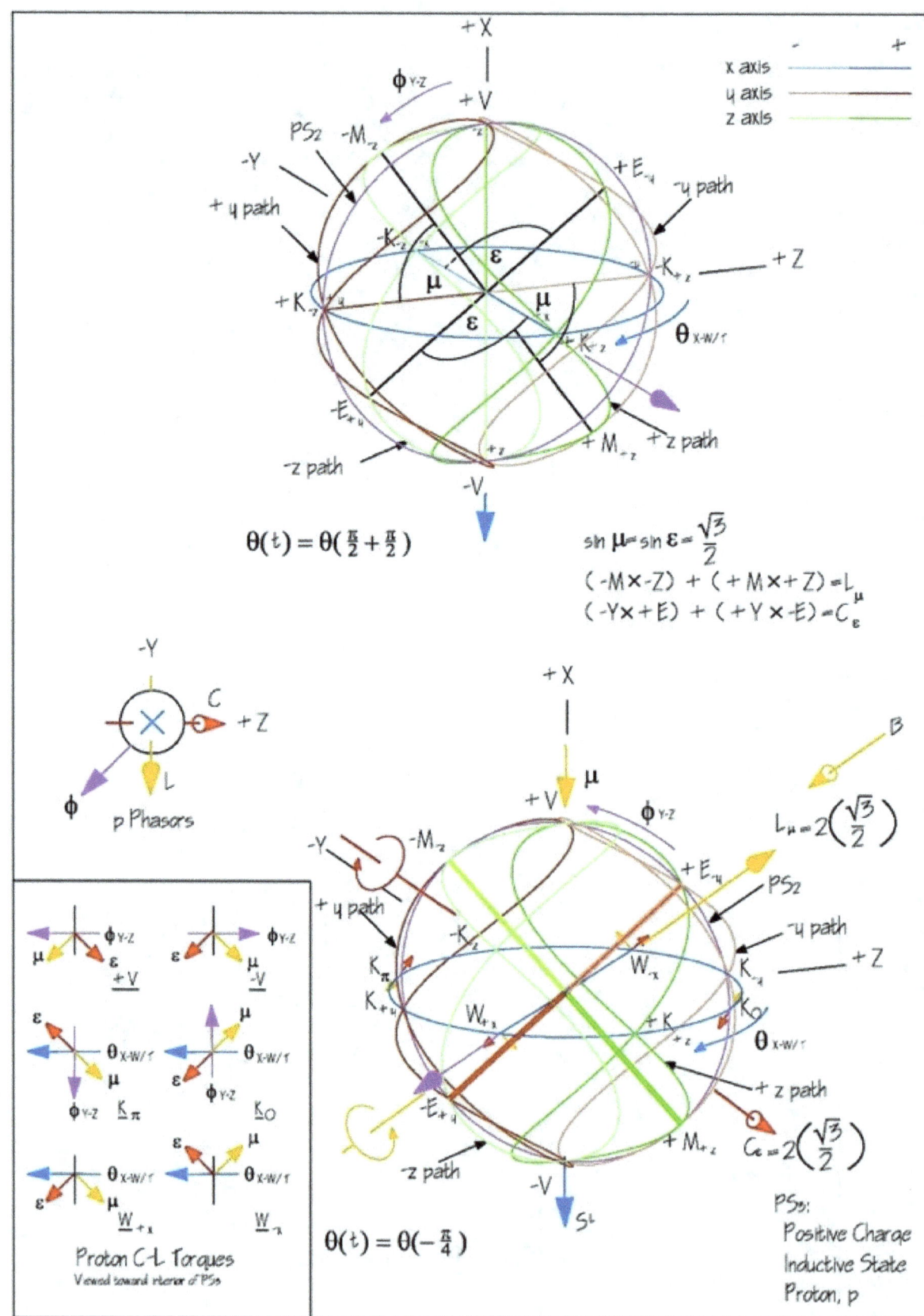

Diagram 3 - Proton Oscillation

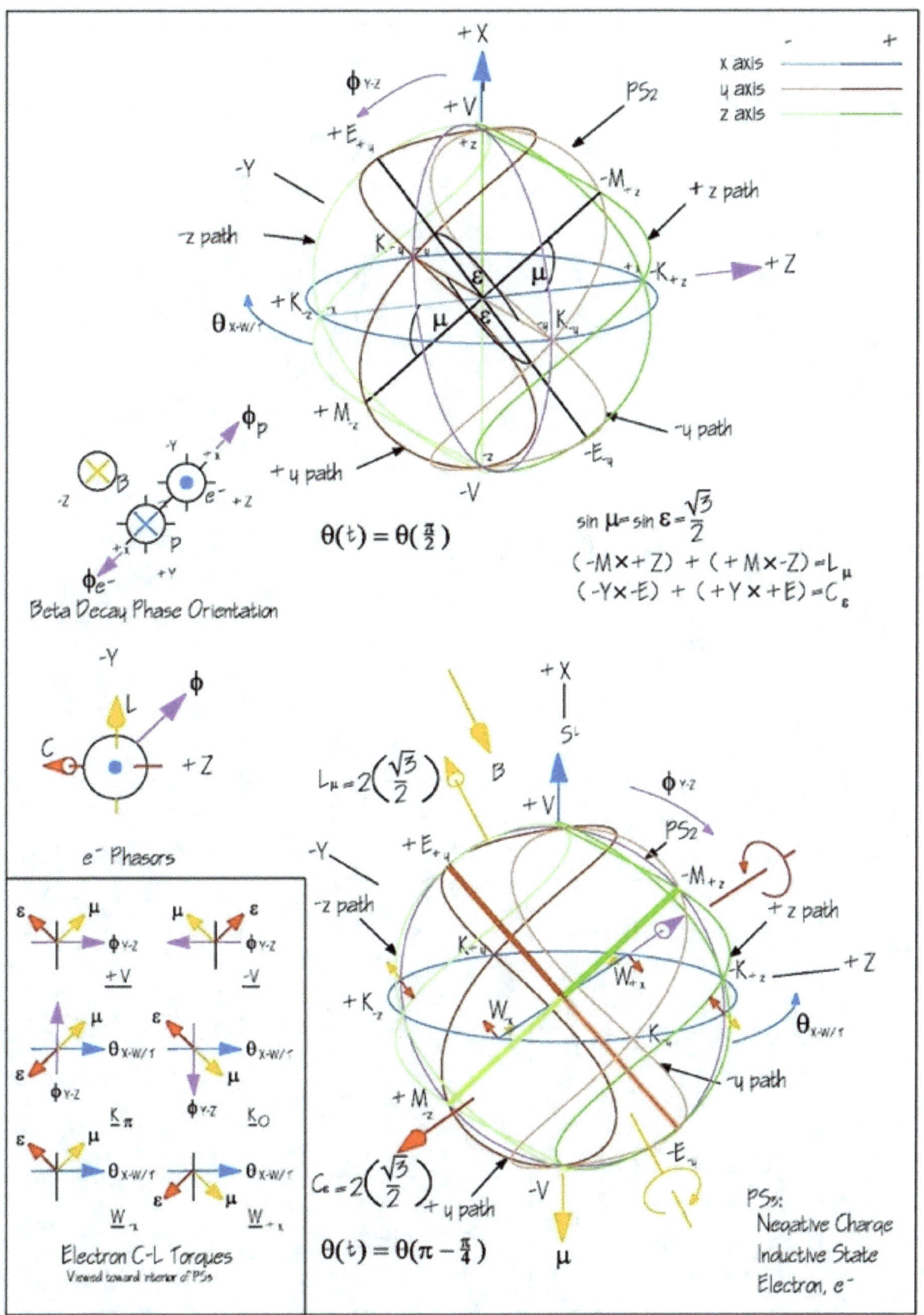

Diagram 4 - Electron Oscillation

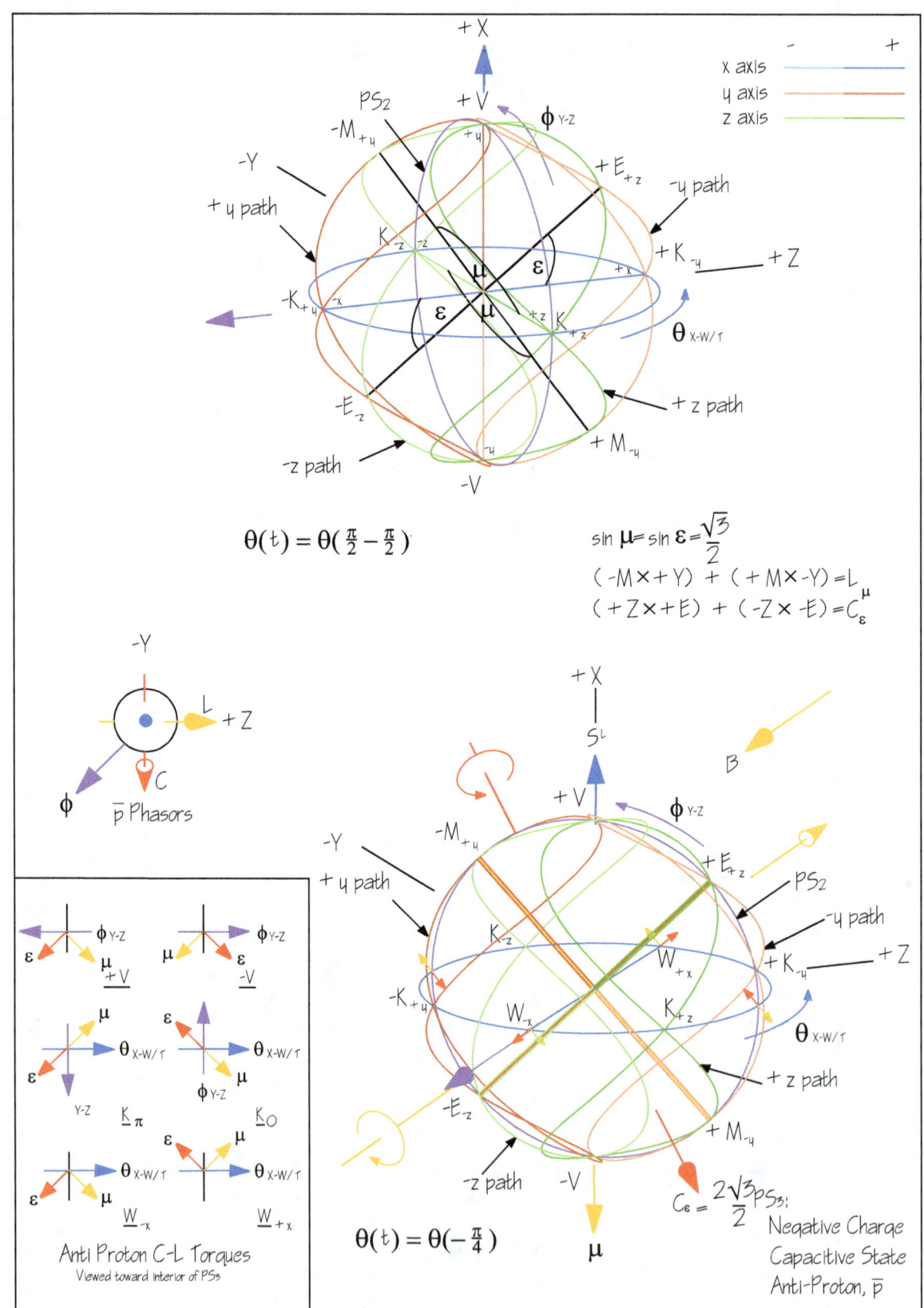

Diagram 5 – Anti-Proton Oscillation

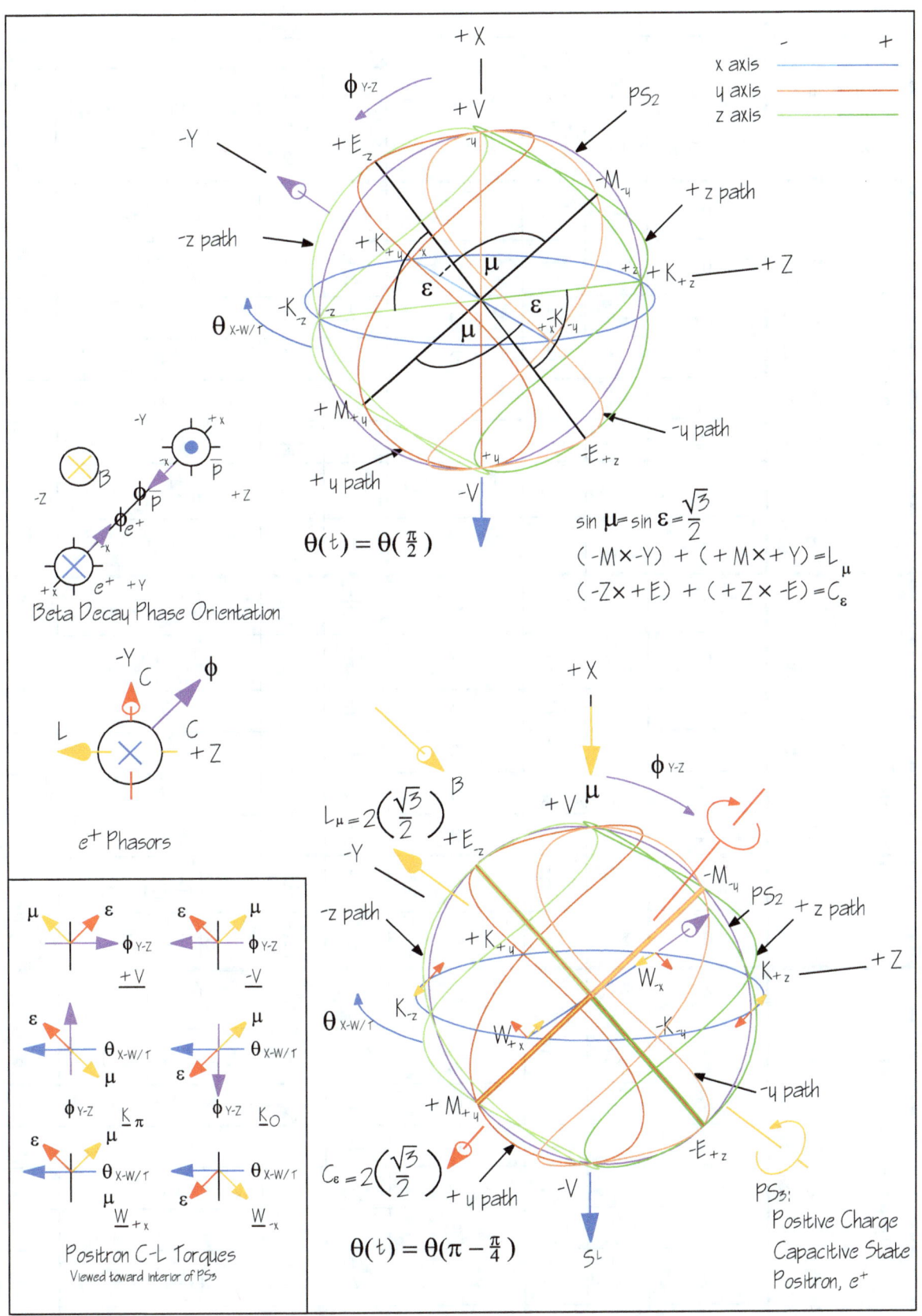

Diagram 6 – Positron Oscillation

		1	2	3	4	5	6	7	8
	Diagram, Node/Antinode, Rotation	$\mu \cdot \varepsilon$ $\mu, \varepsilon = \sqrt{\tfrac{2}{3}}$	$\mu \cdot$ Into θ, ϕ	$\varepsilon \cdot$ Into θ, ϕ	$S = \mu \times \varepsilon$ $\mu, \varepsilon = \sqrt{\tfrac{2}{3}}$	$\mu \times$ Into θ, ϕ	$\varepsilon \times$ Into θ, ϕ	$q = \dfrac{S}{|S|}[\mathrm{T}(\mu \times + \varepsilon \times)]$ $\mathrm{T}=\tfrac{\sqrt{3}}{2}$	Total Charge q_{W-}, q_{W+} $q_{V\pm}, q_{V\mp}$
Diagram 2 – Neutron	W_{+x}-θ	0	$+\tfrac{1}{\sqrt{3}}$	$+\tfrac{1}{\sqrt{3}}$	•$\tfrac{2}{3}$	$\tfrac{1}{\sqrt{3}}$	$\tfrac{1}{\sqrt{3}}$	•$1\left(\bullet\tfrac{1}{2}+\circ\tfrac{1}{2}\right) = 0$	
	W_{-x}-θ	0	$+\tfrac{1}{\sqrt{3}}$	$+\tfrac{1}{\sqrt{3}}$	∘$\tfrac{2}{3}$	$\tfrac{1}{\sqrt{3}}$	$\tfrac{1}{\sqrt{3}}$	∘$1\left(\circ\tfrac{1}{2}+\bullet\tfrac{1}{2}\right) = 0$	0
	K_o-θ	$+\tfrac{2}{3}$	$+\tfrac{1}{\sqrt{3}}$	$+\tfrac{1}{\sqrt{3}}$	0	$\tfrac{1}{\sqrt{3}}$	$\tfrac{1}{\sqrt{3}}$	0	
	K_π-θ	$+\tfrac{2}{3}$	$+\tfrac{1}{\sqrt{3}}$	$+\tfrac{1}{\sqrt{3}}$	0	$\tfrac{1}{\sqrt{3}}$	$\tfrac{1}{\sqrt{3}}$	0	
	K_o-ϕ	$+\tfrac{2}{3}$	$+\tfrac{1}{\sqrt{3}}$	$+\tfrac{1}{\sqrt{3}}$	0	$\tfrac{1}{\sqrt{3}}$	$\tfrac{1}{\sqrt{3}}$	0	
	K_π-ϕ	$+\tfrac{2}{3}$	$+\tfrac{1}{\sqrt{3}}$	$+\tfrac{1}{\sqrt{3}}$	0	$\tfrac{1}{\sqrt{3}}$	$\tfrac{1}{\sqrt{3}}$	0	
	$+V$-ϕ	0	$+\tfrac{1}{\sqrt{3}}$	$+\tfrac{1}{\sqrt{3}}$	•$\tfrac{2}{3}$	$\tfrac{1}{\sqrt{3}}$	$\tfrac{1}{\sqrt{3}}$	•$1\left(\bullet\tfrac{1}{2}+\circ\tfrac{1}{2}\right) = 0$	
	$-V$-ϕ	0	$+\tfrac{1}{\sqrt{3}}$	$+\tfrac{1}{\sqrt{3}}$	∘$\tfrac{2}{3}$	$\tfrac{1}{\sqrt{3}}$	$\tfrac{1}{\sqrt{3}}$	∘$1\left(\circ\tfrac{1}{2}+\bullet\tfrac{1}{2}\right) = 0$	[0]
Diagram 3 – Proton	W_{+x}-θ	0	$-\tfrac{1}{\sqrt{3}}$	$+\tfrac{1}{\sqrt{3}}$	•$\tfrac{2}{3}$	$\tfrac{1}{\sqrt{3}}$	$\tfrac{1}{\sqrt{3}}$	•$1\left(\bullet\tfrac{1}{2}+\bullet\tfrac{1}{2}\right) = +1$	
	W_{-x}-θ	0	$-\tfrac{1}{\sqrt{3}}$	$+\tfrac{1}{\sqrt{3}}$	∘$\tfrac{2}{3}$	$\tfrac{1}{\sqrt{3}}$	$\tfrac{1}{\sqrt{3}}$	∘$1\left(\circ\tfrac{1}{2}+\circ\tfrac{1}{2}\right) = +1$	+1
	K_o-θ	$-\tfrac{2}{3}$	$-\tfrac{1}{\sqrt{3}}$	$+\tfrac{1}{\sqrt{3}}$	0	$\tfrac{1}{\sqrt{3}}$	$\tfrac{1}{\sqrt{3}}$	0	
	K_π-θ	$-\tfrac{2}{3}$	$-\tfrac{1}{\sqrt{3}}$	$+\tfrac{1}{\sqrt{3}}$	0	$\tfrac{1}{\sqrt{3}}$	$\tfrac{1}{\sqrt{3}}$	0	
	K_o-ϕ	$-\tfrac{2}{3}$	$+\tfrac{1}{\sqrt{3}}$	$-\tfrac{1}{\sqrt{3}}$	0	$\tfrac{1}{\sqrt{3}}$	$\tfrac{1}{\sqrt{3}}$	0	
	K_π-ϕ	$-\tfrac{2}{3}$	$+\tfrac{1}{\sqrt{3}}$	$-\tfrac{1}{\sqrt{3}}$	0	$\tfrac{1}{\sqrt{3}}$	$\tfrac{1}{\sqrt{3}}$	0	
	$+V$-ϕ	0	$+\tfrac{1}{\sqrt{3}}$	$-\tfrac{1}{\sqrt{3}}$	∘$\tfrac{2}{3}$	$\tfrac{1}{\sqrt{3}}$	$\tfrac{1}{\sqrt{3}}$	∘$1\left(\circ\tfrac{1}{2}+\bullet\tfrac{1}{2}\right) = -1$	
	$-V$-ϕ	0	$+\tfrac{1}{\sqrt{3}}$	$-\tfrac{1}{\sqrt{3}}$	•$\tfrac{2}{3}$	$\tfrac{1}{\sqrt{3}}$	$\tfrac{1}{\sqrt{3}}$	•$1\left(\bullet\tfrac{1}{2}+\circ\tfrac{1}{2}\right) = -1$	$[-1] = -i1$
Diagram 4 – Electron	W_{+x}-θ	0	$+\tfrac{1}{\sqrt{3}}$	$-\tfrac{1}{\sqrt{3}}$	•$\tfrac{2}{3}$	$\tfrac{1}{\sqrt{3}}$	$\tfrac{1}{\sqrt{3}}$	•$1\left(\circ\tfrac{1}{2}+\bullet\tfrac{1}{2}\right) = -1$	
	W_{-x}-θ	0	$+\tfrac{1}{\sqrt{3}}$	$-\tfrac{1}{\sqrt{3}}$	∘$\tfrac{2}{3}$	$\tfrac{1}{\sqrt{3}}$	$\tfrac{1}{\sqrt{3}}$	∘$1\left(\bullet\tfrac{1}{2}+\circ\tfrac{1}{2}\right) = -1$	-1
	K_o-θ	$-\tfrac{2}{3}$	$+\tfrac{1}{\sqrt{3}}$	$-\tfrac{1}{\sqrt{3}}$	0	$\tfrac{1}{\sqrt{3}}$	$\tfrac{1}{\sqrt{3}}$	0	
	K_π-θ	$-\tfrac{2}{3}$	$+\tfrac{1}{\sqrt{3}}$	$-\tfrac{1}{\sqrt{3}}$	0	$\tfrac{1}{\sqrt{3}}$	$\tfrac{1}{\sqrt{3}}$	0	
	K_o-ϕ	$-\tfrac{2}{3}$	$+\tfrac{1}{\sqrt{3}}$	$-\tfrac{1}{\sqrt{3}}$	0	$\tfrac{1}{\sqrt{3}}$	$\tfrac{1}{\sqrt{3}}$	0	
	K_π-ϕ	$-\tfrac{2}{3}$	$+\tfrac{1}{\sqrt{3}}$	$-\tfrac{1}{\sqrt{3}}$	0	$\tfrac{1}{\sqrt{3}}$	$\tfrac{1}{\sqrt{3}}$	0	
	$+V$-ϕ	0	$+\tfrac{1}{\sqrt{3}}$	$-\tfrac{1}{\sqrt{3}}$	∘$\tfrac{2}{3}$	$\tfrac{1}{\sqrt{3}}$	$\tfrac{1}{\sqrt{3}}$	∘$1\left(\circ\tfrac{1}{2}+\bullet\tfrac{1}{2}\right) = -1$	$[-1] = -i1$
	$-V$-ϕ	0	$+\tfrac{1}{\sqrt{3}}$	$-\tfrac{1}{\sqrt{3}}$	•$\tfrac{2}{3}$	$\tfrac{1}{\sqrt{3}}$	$\tfrac{1}{\sqrt{3}}$	•$1\left(\bullet\tfrac{1}{2}+\circ\tfrac{1}{2}\right) = -1$	

Table 1 – Charge and Spin Table for Ordinary Matter for C & L = 1

		1	2	3	4	5	6	7	8
	Diagram, Node/Antinode, Rotation	$\mu \cdot \varepsilon$ $\mu,\varepsilon = \sqrt{\frac{2}{3}}$	$\mu \cdot$ Into θ,ϕ	$\varepsilon \cdot$ Into θ,ϕ	$S = \mu \times \varepsilon$ $\mu,\varepsilon = \sqrt{\frac{2}{3}}$	$\mu \times$ Into θ,ϕ	$\varepsilon \times$ Into θ,ϕ	$q = \frac{S}{|S|}(\mu \times + \varepsilon \times)$ $T = \frac{\sqrt{3}}{2}$	Total Charge q_{W-}, q_{W+} $q_{V\pm}, q_{V\mp}$
Diagram 2 – Neutron	W_{+x}-θ	0	$+\frac{1}{\sqrt{3}}$	$+\frac{1}{\sqrt{3}}$	$\bullet\frac{2}{3}$	$\bullet\frac{1}{\sqrt{3}}$	$\frac{1}{\sqrt{3}}$	$\bullet 1\left(\bullet\frac{1}{2}+\circ\frac{1}{2}\right)=0$	
	W_{-x}-θ	0	$+\frac{1}{\sqrt{3}}$	$+\frac{1}{\sqrt{3}}$	$\circ\frac{2}{3}$	$\circ\frac{1}{\sqrt{3}}$	$\bullet\frac{1}{\sqrt{3}}$	$\circ 1\left(\frac{1}{2}+\bullet\frac{1}{2}\right)=0$	0
	K_o-θ	$+\frac{2}{3}$	$+\frac{1}{\sqrt{3}}$	$+\frac{1}{\sqrt{3}}$	0	$\bullet\frac{1}{\sqrt{3}}$	$\frac{1}{\sqrt{3}}$	0	
	K_π-θ	$+\frac{2}{3}$	$+\frac{1}{\sqrt{3}}$	$+\frac{1}{\sqrt{3}}$	0	$\bullet\frac{1}{\sqrt{3}}$	$\bullet\frac{1}{\sqrt{3}}$	0	
	K_o-ϕ	$+\frac{2}{3}$	$+\frac{1}{\sqrt{3}}$	$+\frac{1}{\sqrt{3}}$	0	$\bullet\frac{1}{\sqrt{3}}$	$\bullet\frac{1}{\sqrt{3}}$	0	
	K_π-ϕ	$+\frac{2}{3}$	$+\frac{1}{\sqrt{3}}$	$+\frac{1}{\sqrt{3}}$	0	$\frac{1}{\sqrt{3}}$	$\frac{1}{\sqrt{3}}$	0	
	$+V$-ϕ	0	$+\frac{1}{\sqrt{3}}$	$+\frac{1}{\sqrt{3}}$	$\bullet\frac{2}{3}$	$\bullet\frac{1}{\sqrt{3}}$	$\frac{1}{\sqrt{3}}$	$\bullet 1\left(\bullet\frac{1}{2}+\circ\frac{1}{2}\right)=0$	
	$-V$-ϕ	0	$+\frac{1}{\sqrt{3}}$	$+\frac{1}{\sqrt{3}}$	$\circ\frac{2}{3}$	$\frac{1}{\sqrt{3}}$	$\frac{1}{\sqrt{3}}$	$\circ 1\left(\circ\frac{1}{2}+\bullet\frac{1}{2}\right)=0$	[0]
Diagram 5 – Anti Proton	W_{+x}-θ	0	$+\frac{1}{\sqrt{3}}$	$-\frac{1}{\sqrt{3}}$	$\circ\frac{2}{3}$	$\bullet\frac{1}{\sqrt{3}}$	$\bullet\frac{1}{\sqrt{3}}$	$\circ 1\left(\bullet\frac{1}{2}+\bullet\frac{1}{2}\right)=-1$	
	W_{-x}-θ	0	$+\frac{1}{\sqrt{3}}$	$-\frac{1}{\sqrt{3}}$	$\bullet\frac{2}{3}$	$\frac{1}{\sqrt{3}}$	$\frac{1}{\sqrt{3}}$	$\circ 1\left(\bullet\frac{1}{2}+\bullet\frac{1}{2}\right)=-1$	-1
	K_o-θ	$-\frac{2}{3}$	$+\frac{1}{\sqrt{3}}$	$-\frac{1}{\sqrt{3}}$	0	$\frac{1}{\sqrt{3}}$	$\bullet\frac{1}{\sqrt{3}}$	0	
	K_π-θ	$-\frac{2}{3}$	$+\frac{1}{\sqrt{3}}$	$-\frac{1}{\sqrt{3}}$	0	$\bullet\frac{1}{\sqrt{3}}$	$\frac{1}{\sqrt{3}}$	0	
	K_o-ϕ	$-\frac{2}{3}$	$-\frac{1}{\sqrt{3}}$	$+\frac{1}{\sqrt{3}}$	0	$\frac{1}{\sqrt{3}}$	$\frac{1}{\sqrt{3}}$	0	
	K_π-ϕ	$-\frac{2}{3}$	$-\frac{1}{\sqrt{3}}$	$+\frac{1}{\sqrt{3}}$	0	$\bullet\frac{1}{\sqrt{3}}$	$\bullet\frac{1}{\sqrt{3}}$	0	
	$+V$-ϕ	0	$-\frac{1}{\sqrt{3}}$	$+\frac{1}{\sqrt{3}}$	$\bullet\frac{2}{3}$	$\bullet\frac{1}{\sqrt{3}}$	$\bullet\frac{1}{\sqrt{3}}$	$\bullet 1\left(\frac{1}{2}+\frac{1}{2}\right)=+1$	
	$-V$-ϕ	0	$-\frac{1}{\sqrt{3}}$	$+\frac{1}{\sqrt{3}}$	$\circ\frac{2}{3}$	$\frac{1}{\sqrt{3}}$	$\frac{1}{\sqrt{3}}$	$\circ 1\left(\circ\frac{1}{2}+\circ\frac{1}{2}\right)=+1$	$[+1]=+i1$
Diagram 6 – Positron	W_{+x}-θ	0	$-\frac{1}{\sqrt{3}}$	$+\frac{1}{\sqrt{3}}$	$\frac{2}{3}$	$\frac{1}{\sqrt{3}}$	$\frac{1}{\sqrt{3}}$	$\circ 1\left(\circ\frac{1}{2}+\circ\frac{1}{2}\right)=+1$	
	W_{-x}-θ	0	$-\frac{1}{\sqrt{3}}$	$+\frac{1}{\sqrt{3}}$	$\bullet\frac{2}{3}$	$\bullet\frac{1}{\sqrt{3}}$	$\bullet\frac{1}{\sqrt{3}}$	$\bullet 1\left(\bullet\frac{1}{2}+\bullet\frac{1}{2}\right)=+1$	+1
	K_o-θ	$-\frac{2}{3}$	$-\frac{1}{\sqrt{3}}$	$+\frac{1}{\sqrt{3}}$	0	$\frac{1}{\sqrt{3}}$	$\bullet\frac{1}{\sqrt{3}}$	0	
	K_π-θ	$-\frac{2}{3}$	$-\frac{1}{\sqrt{3}}$	$+\frac{1}{\sqrt{3}}$	0	$\bullet\frac{1}{\sqrt{3}}$	$\frac{1}{\sqrt{3}}$	0	
	K_o-ϕ	$-\frac{2}{3}$	$-\frac{1}{\sqrt{3}}$	$+\frac{1}{\sqrt{3}}$	0	$\bullet\frac{1}{\sqrt{3}}$	$\frac{1}{\sqrt{3}}$	0	
	K_π-ϕ	$-\frac{2}{3}$	$-\frac{1}{\sqrt{3}}$	$+\frac{1}{\sqrt{3}}$	0	$\frac{1}{\sqrt{3}}$	$\bullet\frac{1}{\sqrt{3}}$	0	
	$+V$-ϕ	0	$-\frac{1}{\sqrt{3}}$	$+\frac{1}{\sqrt{3}}$	$\bullet\frac{2}{3}$	$\bullet\frac{1}{\sqrt{3}}$	$\bullet\frac{1}{\sqrt{3}}$	$\bullet 1\left(\bullet\frac{1}{2}+\bullet\frac{1}{2}\right)=+1$	$[+1]=+i1$
	$-V$-ϕ	0	$-\frac{1}{\sqrt{3}}$	$+\frac{1}{\sqrt{3}}$	$\circ\frac{2}{3}$	$\circ\frac{1}{\sqrt{3}}$	$\circ\frac{1}{\sqrt{3}}$	$\circ 1\left(\circ\frac{1}{2}+\circ\frac{1}{2}\right)=+1$	

Table 2 – Charge and Spin Table for Anti Matter for C & L =1

2c — Geometric Considerations of Rotational Oscillation and Beta Decay

Electrical charge (conventionally negative) is a transmission of wave momentum as will be described in the following. In the case of the neutron, the rotational oscillation sets up a wave mechanical analogy of an LC current, where the sustained maintenance of a maximum potential energy constitutes charge and the sustained maximum wave kinetic energy constitutes a current sustained within the discrete wave form of the neutron. The potential energy of this charge is therefore concentrated in the +/- V poles of the oscillation and its maximum kinetic energy at +/- K as they move around the θ circle. Ongoing expansion of the manifold leads to a decrease in inertial density in the environment around the wave form and a transmission of the kinetic charge which is localized or focused in the transmitted wave node as a negative charge.

In order for the energy of beta decay, which is quantified as the mass of the electron, to be transmitted from the neutron waveform, the density and impedance at its boundary must decrease sufficient to permit that mass-energy to pass. The electron mass, m_e, is determined according to geometric constraints of the neutron oscillation and is approximately 0.000543867 . . . of the neutron mass, m_0.

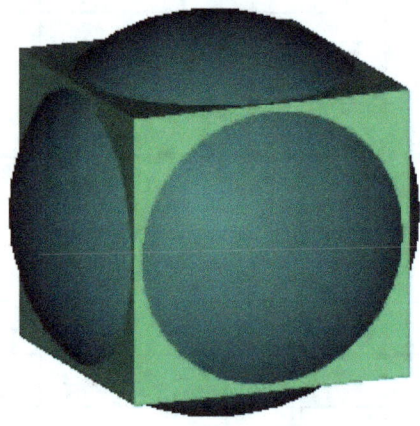

Figure 9

For an intuitive approach, the above figure shows a concentric sphere and cube which have equal individual surface areas. Assuming a sphere of radius r_0 and area A_s and a corresponding cube with area A_c and edge length l_c, we have

$$A_s = 4\pi r_0^2 = 4\pi = 6l_c^2 = A_c$$
$$l_c / r_0 = \sqrt{\tfrac{2}{3}\pi} = 1.4472\ldots$$
(1.34)

Tension stress on the surface of each would be equal, though the sphere represents isotropic stress while the cube represents a breakdown of the orthogonal components of such stress in keeping with the above development. The points midway along each cubic edge are loci of closest stress/strain equivalence between cube and sphere. They are also the points of optimal shear stress and strain in the rotational oscillation, as evidenced by the power moments, E and M. Such stress force operates in an oscillatory manner toward a leading adjacent vertex, directed by the two resultant torques, C and L, aligned with two of the cubic diagonals, toward one or the other of the two vertices beyond the leading adjacent one. These vertices also represent the rotating direction of expansion stress of T_0.

Over time the length of the moments vary as δr_0, in the context of an expanding spacetime, generally in an increasing direction represented by the vertices. The edge of the cube represents a limit for the increase in the moments, which is reflected by an increase in C and L and their orthogonal vector representations ε and μ in the Spin Diagrams and Tables. The result is an increase in the cross-product along the $W_{+x} - W_{-x}$ axis for ϕ, an advance of the moments and a transmission of energy and power at that W_{-x} node as beta decay, where δr_0^2 represents the relative energy and therefore mass of the transmitted oscillation.

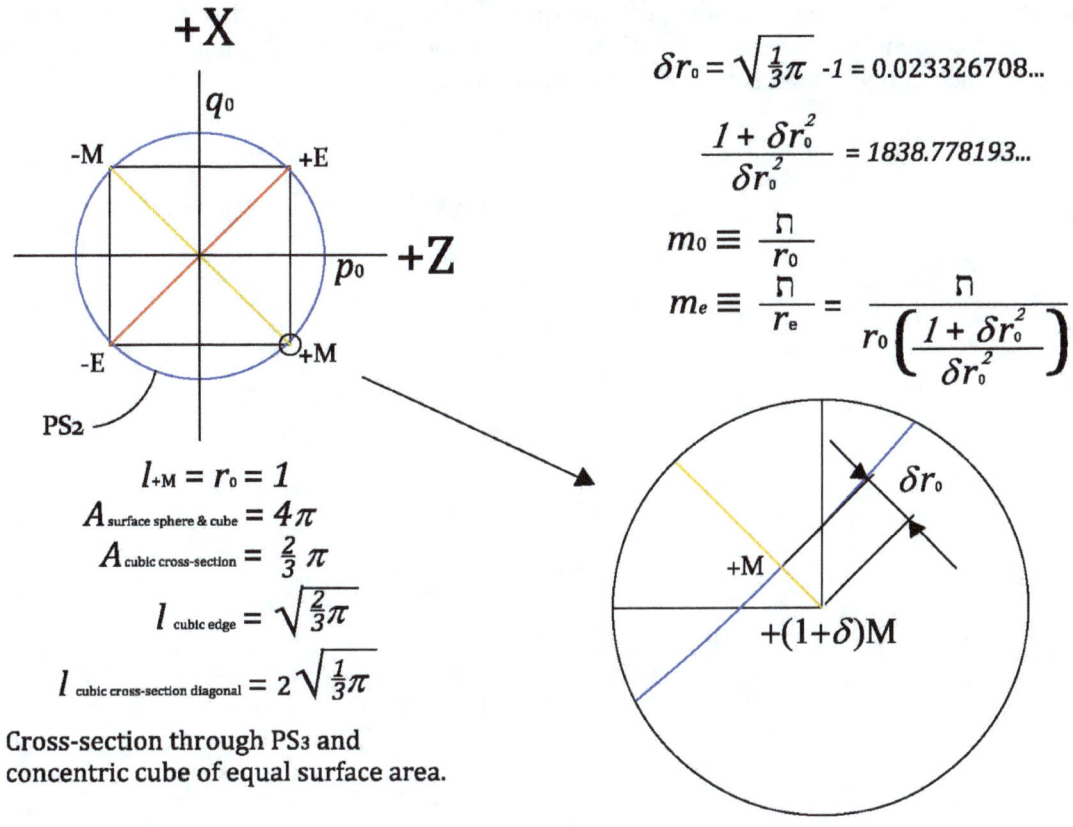

Figure 10

This figure shows a cross-section through the developed 2-sphere structure at the X-Z plane, so that the moments +/-E and +/-M are aligned with the four half diagonals of the cubic cross-section. This indicates the moments rotating in alignment with the mid points of two of the four edges of the upper and lower cubic faces. Each half diagonal length is therefore $\sqrt{\frac{\pi}{3}}$ to the parallel moment length (not strength) of r_0 of 1.

The square of the differential δr_0 is reflected in the cross-product of the differential values of ε and μ as

$$\delta r_0 = \sqrt{\tfrac{\pi}{3}} - 1 = 0.023326708\ldots \qquad (1.35)$$

$$\delta r_0^2 = 0.0005441353061\ldots \qquad (1.36)$$

The ratio of the differential stress to the augmented total according to the resulting strain is

$$\frac{\delta r_0^2}{1+\delta r_0^2} = 0.0005438393841\ldots \qquad (1.37)$$

which when inverted is

$$\frac{1+\delta r_0^2}{\delta r_0^2} = 1838.778193\ldots \qquad (1.38)$$

The 2010 CODATA ratio of the electron to neutron mass is 0.00054386734461(32) or inverted 1838.6836605(11). While outside the stated standard uncertainty shown here, the ratio derived here is within $2.8\ldots \times 10^{-8}$ of the CODATA ratio. Since mass computation presumably uses Newton's gravitational constant somewhere in the standardization of mass and weight, and given that the relative standard uncertainty of that constant at 1.2×10^{-4} is relatively large, it appears that (1.37) and (1.38) are within the relative standard uncertainty of the neutron-electron mass ratio.

2d — Derivation of Beta Decay as a Function of the Hubble Rate

Conventionally, the reduced Compton wavelength of the electron is related to the electron mass and as indicated as a portion of the neutron mass by (1.37) as

$$\lambdabar_{C,e} = r_e = \frac{\hbar}{cm_e} = \frac{\hbar}{m_e} = \frac{m_0 r_0}{m_e} \qquad (1.39)$$

According to this modeling, the mass density within the boundary of the fundamental rotational oscillation is constrained by that oscillation while the region outside the boundary that starts as the interstitial areas between oscillation boundaries decreases with manifold expansion. The rotational oscillation as defined in (1.20) is then responsible for the particle interactions that result in the large aggregations of stellar and galactic matter, while the interstitial areas span the space between particles and define the large voids between the webs of galactic matter. At a minimum the differential change in inertial density of the manifold required for beta-decay represents a loss of potential mass-energy from the region exterior to the neutron oscillation nodes equal to and eventually replaced by that of the transmitted electron by beta-decay from the neutron mass or

$$d\lambda_0 = \frac{\hbar}{cr_e^2} = \frac{\hbar}{r_e^2} = \frac{1}{c^2}\hbar\omega_e^2 = \frac{1}{c^2}\tau_e , \qquad (1.40)$$

where ω_e is the rest mass oscillatory frequency of the electron given by

$$\omega_e = \frac{\theta_e}{t_e} = \frac{c}{r_e} = \frac{c}{\lambdabar_{C,e}} \qquad (1.41)$$

and τ_e is the characteristic wave force of that oscillation. The subscript e in this regard can be thought of as representing the electron, but as a differential value can also stand for the effect of 'expansion' or an 'emergent' condition.

The differential density in the second term of (1.40) represents the decrease in inertia over the distance of a wavelength required to generate a waveform of such mass; the mass of an electron as a differential value distributed over a characteristic differential angular wavelength which we can think of as the initial ground state orbital of an electron in a hydrogen atom; it is stated without further analysis that this value divided by the fine structure constant gives the Bohr radius of the hydrogen atom. For the wave speed to be invariant, the differentials in (1.41) must be equally contravariant, which means that the differential length and time must vary proportionally so that with expansion the wave speed, c_e, is invariant as

$$c_0 = r_e \omega_e = r_e \frac{\theta_0}{t_e} = \frac{r_e}{t_e} = \frac{dr_0}{dt_0} = c_e . \qquad (1.42)$$

For the wave speed to remain invariant with expansion, the unit time standard, with the addition of the differential time standard, must strain as well as

$$c_0 = \frac{r_0}{t_0} = \frac{r_0 + dr_0}{t_0 + dt_0} = \frac{r_e}{t_e} = c_e . \qquad (1.43)$$

This simply says that the wave <u>speed</u> over the distance dr_0 is the same as it is over the distance r_0, and takes a proportional amount of time, whether dr_0 is much less than or much greater than r_0, and whether $r_e = dr_0$ or $r_e = r_0 + dr_0$. Thus, if a unit of distance, natural or arbitrary, increases with expansion of the manifold, say where a new unit of distance $r_{0e} = r_0 + dr_0$, any corresponding <u>unit</u> of time, t_{0e}, must increase proportionally, so that the frequency proportionally decreases, and

$$c = r_0 \omega_0 = r_{0e} \omega_{0e}$$

$$\frac{r_{0e}}{r_0} = \frac{\omega_0}{\omega_{0e}} = \frac{t_{0e}}{t_0} \qquad (1.44)$$

With separation of one of the wave speed components, c, in (1.40), which can be expressed as differentials, a change in inertial density over time is equal to a change in the mechanical impedance over distance as

$$\frac{d\lambda_0}{dt} = \frac{\tau_e}{c} \frac{1}{dr} = \frac{dZ_0}{dr} . \qquad (1.45)$$

We can rearrange this in keeping with (1.15) to show that the wave speed is given by the quotient of the change in impedance over the change in inertial density as

$$c_e = \frac{dr}{dt} = \frac{dZ_0}{d\lambda_0} = \frac{dm/dt_0}{dm/dr_0} = c_0 . \qquad (1.46)$$

Note that we are not using partial derivatives in this case, as dr_0 and dt_0 are dual functions of the expansion of the manifold as with the dual functions of wave number and wave frequency. In the next to last term of (1.46) the dimensional analysis indicates that mechanical impedance is a function of mass over time as inertial density is a function of mass over distance. While dm is a property of the manifold and should be the same for both the impedance and the density terms, and while it may be tempting to think so, it does not follow that $dt = dt_0$ or that $dr = dr_0$, since the first of the differentials in each case represent properties of the manifold to be integrated to unit values t_0 and r_0, while the second in each case are differentials of the unit standard or coefficients of unit strain, and are effectively second order differentials.

Since the values of the inertial constant as Planck's constant over the speed of light and the electron reduced Compton are well determined, we can solve for $d\lambda_0$ in (1.40) and get

$$\frac{\Delta\lambda_0}{t_{0SI}} = 2.358970395...x10^{-18} \frac{kg/m}{s} . \qquad (1.47)$$

The change in density is due to a stretching of the spacetime manifold and not to a loss of energy/inertia within an arbitrary boundary of an expanding region, which is conserved, so this value represents a rate of change of the manifold strain, a dimensionless number, per second. This is essentially what the Hubble rate is, as a measure of spatial expansion, a dimensionless number, per second, H_0, experimentally calculated as roughly 72 to 76 km, ΔR, per megaparsec per second. A megaparsec is $3.0857...x10^{19}$ kilometers, R, so that the spacetime strain per second is given as a Hubble rate, customarily expressed as an expansion velocity per second per megaparsec,

$$H_0 = \frac{\Delta R/t_{0SI}}{R} = \frac{\Delta R/R}{t_{0SI}} = \frac{73 km}{3.0857...x10^{19} km}/s \cong 2.3657...x10^{-18} s^{-1} \qquad (1.48)$$

This indicates that the Hubble rate as a spatial strain is capable of generating the force required for beta-decay. However, we would like something more precise, dimensionally correct, and analytically pleasing.

Returning to (1.45), we can transpose the time and length standard differentials as follows, where integrating the equation for the expansion change in length and time should give the resulting mass change in inertial density-impedance of the manifold

$$d\lambda_0 \int_{unit-length=0}^{1} dr \equiv dZ_0 \int_{unit-time=0}^{1} dt \qquad (1.49)$$

$$\frac{\hbar}{r_e^2} dr_0 \equiv \frac{\hbar \omega_e^2}{r_e \omega_e} dt_0 = \frac{\hbar \omega_e}{r_e} dt_0$$

$$\frac{\hbar}{r_e^2}\left(\frac{t_e}{t_0} r_e\right) \equiv \frac{\hbar \omega_e}{r_e}\left(\frac{r_e}{r_0} t_e\right) \qquad (1.50)$$

where the change in wavelength, r_e, due to expansion is modified by the time strain as

$$dr_0 = \frac{t_e}{t_0} r_e \qquad (1.51)$$

and the change in corresponding time standard, t_e, due to expansion is modified by the length strain as

$$dt_0 = \frac{r_e}{r_0} t_e . \qquad (1.52)$$

Both of these strains represent the Hubble rate, H_0, which we can see in this analysis is an acceleration and not a velocity, since it represents a continuous augmentation of a length or time unit standard. Using the CODATA SI values for the electron and neutron reduced Compton wavelength converted conceptually to rotational wave form radii, r_e and r_0, and related value from the electron angular frequency for time, t_e, we can solve for the change in time standard, which we will call the Hubble time standard, H_t, as

$$dt_0 = \frac{r_e}{r_0 \omega_e} = \frac{r_e}{r_0} \frac{t_0}{\theta_e} = \frac{r_e}{r_0} \frac{t_e}{\theta_0} = 2.36838769...x10^{-18} s = H_t \qquad (1.53)$$

and for the change in the length standard, which we will call the Hubble length standard, H_r, as

$$dr_0 = \frac{\omega_0}{\omega_e} r_e = \frac{t_e}{t_0} r_e = 7.100247672...x10^{-10} m = H_r . \qquad (1.54)$$

The length standard change must be normalized to the time standard by the distance of propagation at wave speed if we want to evaluate in the familiar SI time terms so that in keeping with (1.43) the dimensionless length strain per second is equal to the time strain per second which is equal to the Hubble rate as

$$H_0 = \frac{dr_0}{299,792,458 m}/s = \frac{dt_0}{1s}/s = 2.36838769...x10^{-18}/s \qquad (1.55)$$

Given the constraints of (1.45), (1.46) and (1.49) we find that the product of (1.47) and the Hubble length integral of (1.54) gives the fundamental mass of the system, dm, that of the neutron, which is the product of the Hubble time integral and the change in the mechanical impedance, dZ_0, or

$$d\lambda_0 \int_{H_r=0}^{1} dr_0 \equiv dZ_0 \int_{H_t=0}^{1} dt_0 = dm = m_0 \qquad (1.56)$$

This is the threshold rate of expansion required to drive and sustain the condensed matter quantum system, that of the proton, the electron, and the "missing" mass of beta decay, equal to the mass of the neutron or

$$m_0 = m_n = m_p + m_e + \Delta m_{n-(p+e)} . \tag{1.57}$$

In natural units of mass the three values on the right hand side of the equation are simply the ratios of the particle mass or difference with respect to that of the neutron or

$$1.0 = 0.998623471... + 0.000543867... + 0.000832661... \tag{1.58}$$

The following table gives this model's analytical view of this condition. The top row states the value of the density and impedance differentials due to expansion based on the wave properties of the electron as stated above in (1.49) and whose quotient is the wave speed, c. This is relative to an initial condition at neutron density of

$$\left(c_0 = \frac{Z_0}{\lambda_0} = \frac{2.3908...x10^{-3} \, kg/s}{7.9751...x10^{-12} \, kg/m} \right) = \left(c_e = \frac{dZ_0}{d\lambda_0} = \frac{7.0720...x10^{-10} \, kg/s}{2.3589...x10^{-18} \, kg/m} \right) \tag{1.59}$$

The strain rates for a natural length and time scale in the second and third columns, H_r and H_t, are required to drive the equivalent integral mass values in the first column, given the stated differential density and impedance values. The corresponding Hubble rates, H_0, (1.55), expressed in conventional terms as kilometers per second per megaparsec, are tabulated in the fourth column. The fifth column gives the identity of the ratio between the differential density and H_t and the differential impedance and H_r for the pertinent strain rates. The sixth column gives the ratio of the product of the strains and the resulting mass integrals in accordance with the inverse of the following differential equation

$$\frac{d\lambda_0}{dt_0} = \frac{dZ_0}{dr_0} = \frac{dm_0}{dr_0 dt_0} . \tag{1.60}$$

The relationship of these strains represents an acceleration of H_0 over time, rather it indicates that the Hubble rate is in fact an acceleration and not a velocity. In the context of a model of physical particles as ontologically wave forms with an inherent coupling to cosmic spacetime, as seems required by general relativity, an accelerating spacetime expansion strain implies a differential force on the waveforms, as developed previously. From the perspective of dimensional analysis in column 5, any value greater than 1 indicates a predominance of inertia over changes in space and time and a lack of sufficient force to effect beta decay, much less drive the fundamental, m_0, while values less than 1 indicate a capacity for those spacetime variables to produce changes in the inertial state. With respect to column 6, anything less than 1 is insufficient to drive the whole system as indicated by (1.57) and (1.58). This means that a Hubble rate of 73.0813 km is the threshold required, not just for beta decay, but to drive the cosmos. A recent study by Riess et al determined the most accurate Hubble to date at $H_0 = 74.03$ +/- 1.42 km s^{-1} Mpc^{-1}.

The row marked '*Planck*' is generated based on a Hubble rate of 67.74 +/- 0.46 km s^{-1} Mpc^{-1} from a revised report by the European Space Agency Planck space telescope mission in 2015.

The final two rows show the two theoretical extremes that we might imagine, both evoked by the application of the Planck scale to the analysis. The first of these uses the Planck length and time as the differential values for the Hubble length and time as the smallest conceivable strain based on current theoretical thinking. The result is an inertia/spacetime ratio 25 orders of magnitude above that of unity, indicating an astronomical predominance of inertia over observed spacetime kinematics. The corresponding Hubble rate is the equivalent of 1/10,000th the neutron Compton wavelength per megaparsec per second and would be capable of moving only 3.7e^{-53} kilograms. At the other extreme, to drive the Planck mass would require an expansion rate of approximately 30 megaparsecs per megaparsec per second, which is in the inflationary regime, but is not consistent with current observation.

	$d\lambda_0 = dm/dr$	2.358970395 e^{-18} kg/m		
	$dZ_0 = dm/dt$	7.072015331 e^{-10} kg/s		
Mass integrals with integration of dr_0 & dt_0	Strain values due to cosmic expansion needed to produce wave particle mass integrals shown	Effective Hubble rate (conventional) km/s/mps	Ratio of ΔInertia to ΔSpacetime strain	Ratio of mass integral to m_0 $dr_0 dt_0 / dm_0$
	$H_r = dr_0$ $H_t = dt_0$	$H_0 = H_t$	$d\lambda_0 / H_t = dZ_0 / H_r$	$(d\lambda_0 * H_r)/m_0$
m_0	7.10025 e^{-10} 2.36839 e^{-18}	73.0813	0.996024	1.000000
$m_0 - m_e$	7.09638 e^{-10} 2.36709 e^{-18}	73.0413	0.996570	0.999456
$m_0 - \Delta m$	7.09432 e^{-10} 2.36641 e^{-18}	73.0203	0.996856	0.999167
$m_p = m_0 - (m_e + \Delta m)$	7.09045 e^{-10} 2.36512 e^{-18}	72.9805	0.997400	0.998623
Eq. (1.47)	7.07202 e^{-10} 2.35897 e^{-18}	72.7907	1.000000	0.996018
Planck	6.58131 e^{-10} 2.19529 e^{-18}	67.74	1.074560	0.926912
$\Delta m + m_e$	9.77368 e^{-13} 3.26015 e^{-21}	0.1006	723.5773	0.001376
Δm	5.91209 e^{-13} 1.97206 e^{-21}	0.0609	1196.1961	0.000832
m_e	3.86157 e^{-13} 1.28808 e^{-21}	0.0397	1831.3850	0.000543
	Mega Parsec = 3.08570 e^{19} km			
	Planck scale			
3.7 e^{-53}	1.61612 e^{-35} 5.3908 e^{-44}	1.66 e^{-24}	4.38 e^{25}	2.29 e^{-26}
$m_{PL} = 2.17e^{-8}$	9.22696...e^9 30.77784	9.50 e^{20}	7.66 e^{-20}	1.30 e^{19}

Table 3 — Relationship of Hubble rate and particle mass/energy

This modeling suggests a Hubble exponential strain rate of 73.0813 kilometers per megaparsec per second. That is, a unit of space and co-variant time are currently extended/dilated at this rate. The implication is that space and time are currently expanding exponentially, and such expansion drives the natural frequency as indicated by the conjugate of the frequency, wave number hence mass, as

$$m_0 = \hbar \kappa_0 = dZ_0 H_0 = d\lambda_0 c_0 H_0 \qquad (1.61)$$

Thus, if the Hubble threshold rate of expansion is roughly 73 kilometers per second per mpc, this indicates that every local section of space is moving away from every other at approximately 2.37 x10^{-18} meters per second per meter of separation. However, we would expect this expansion to show up primarily in the large voids between galactic filaments and clusters and not in the galactic environs or filaments of baryonic matter due to the counter effects of gravity and electromagnetism. It follows conventionally that inversion of this number would give us the approximate time since all the matter was at the same locale, though by no means a singularity, and that the universe has been expanding for 4.22 x10^{17} seconds, which is roughly 13.4 billion years.

While this is generally portrayed as an absolute time, it is the position of this model that this figure represents the mean lifetime in the expansion process of the cosmos in a manner analogous to the inverse of a decay function, so that we can look backwards in time since the beginning of such expansion as if it were a decay function. As such, this number represents an expansion via a compounded augmentation of the scale of spacetime itself, as 2.37×10^{-18} meters per second per meter of separation, and not simply an extension of matter within that spacetime, and the following equation for the doubling of spacetime applies, in which: the Hubble rate, H_0, is the eigenvalue of the expansion process analogous in the inverse for past time as a decay rate; H_{mlt} is the Hubble mean lifetime given by the inverse of H_0; giving us the Hubble time or half-life, τ_H as

$$\tau_H = \frac{\ln 2}{H_0} = H_{mlt} \ln 2 = 2.92666...\times 10^{17} s \qquad (1.62)$$

This indicates that space is doubling at a current rate of every 9.274 billion years, measured in terms of today's seconds. If we assume that the wavelength of the cosmic background radiation at approximately 5mm embodies that augmentation, while harkening back to a period of primal beta decay as indicated by the reduced Compton wavelength over 2π of an electron, this represents a doubling of some 31 times, or

$$\frac{\ln\left(\frac{.005/2\pi}{r_e}\right)}{\ln 2} = \frac{\ln 2.060...\times 10^9}{\ln 2} = 30.94...doublings \qquad (1.63)$$

a lifetime in terms of today's measure of time of roughly 287 billion years. If we extrapolate back on the same basis for the expansion over the scale of r_0 to r_e, prior to beta-decay where it may or may not be applicable, we have an additional doubling of 10.84 times or

$$\frac{\ln(1838.6836...)}{\ln 2} = 10.84... \qquad (1.64)$$

or a total number of instances of the Hubble time of 41.78 or 393.47 billion years in current time as

$$(2.927...\times 10^{17})(41.78...) = 1.2227...\times 10^{19} s \qquad (1.65)$$

Finally, if we envision that a current expansion extent or factor, κ_{exp}, can be derived by a comparison of the Planck length and the neutron reduced Compton wavelength, to get the quotient as a dimensionless coefficient of expansion stress as with (1.33) which we can apply to an appropriate linear dimensional property, where we can use it for time in seconds or length in a normalized form as light-seconds

$$\kappa_{exp} = \frac{r_0}{l_p} = \sqrt{\frac{T_0}{dT_0}} = \sqrt{(d\ln T_0)^{-1}} = \frac{2.10019...\times 10^{-16} m}{1.61612...\times 10^{-35} m} = 1.29952...\times 10^{19} s \text{ or } ls \qquad (1.66)$$

which is equivalent to 412 years or light years respectively, we have a close agreement with (1.65). Based on the assumption of κ_{exp} as a coefficient for length, which is doubling for every instance of the Hubble time, we can divide this figure in half, to get a cosmic extension, C_x, of the most recent Hubble time as 206 billion light years or in light seconds

$$C_x = 0.5(\kappa_{exp}) = 6.49763...\times 10^{18} ls \qquad (1.67)$$

Dividing C_x by the Hubble time we get the rate of expansion across an extent of the cosmos over the most recent Hubble time. This indicates that the extremes of the cosmos are "receding" from each other at 22.20 times the speed of light as

$$\frac{C_x}{\tau_H} = \frac{6.49763...x10^{18} ls}{2.92666...x10^{17} s} = 22.20c = \frac{(0.5)\kappa_{exp} H_0}{\ln 2} \ . \tag{1.68}$$

With respect to the Hubble time before the most recent period, the value of (1.67) would be halved with the Hubble time remaining the same and the expansion rate over that period would be $11.10c$, pointing to the observed acceleration of cosmic expansion.

However, if the speed of light is not invariant over time, but rather is continually being renormalized with expansion, so that time as determined by τ_H is also halved, then 22.20c is an invariant of the system. In such a case the Hubble mean lifetime of this interpretation is the comoving or cosmological age of the universe, where the units of time are scaled to the expansion extent of the universe. When the Hubble mean lifetime is interpreted as an inverse of a decay rate, then the cosmological or comoving time is represented as being of the order of (1.65) as the number of instances of doubling of scale times the Hubble time and the universe is best represented as a type of de Sitter universe which emerges from an initial locus of maximum density.

With respect to the period before beta-decay or the last scattering of the standard model cosmology, it is not clear from this extant modeling that rest mass quanta emerged from an initial big bang. Rather it appears likely that such matter emerges in an ongoing manner from a condition of maximum density, including from active galactic inertial centers, i.e. black holes which can be gravitational field sources as well as sinks, and their connecting filaments, in response to the tension stress of expansion, as evidenced by the observance of episodic gamma ray bursts of unknown origin, but attributable by some to active galactic nuclei. In this model, black holes are entities of maximum inertial density and not singularities and the event horizon as defined by an extreme Kerr metric represents a two dimensional locus at which the trajectory of all infalling particles becomes tangential as it approaches the speed of light, and the centripetal motion of the particles vanishes. At this point the inertial density of a particle can be absorbed into the mass of the black hole operating as a sink with the wave form smeared across the surface and/or recycled in its operation as a source by being spun up and ejected into one of the collimated jets as baryonic, and with beta decay, leptonic matter.

From this perspective, if we return to our original 2-D sphere heuristic, at the beginning of cosmic evolution once a threshold differential stress is reached radially across the manifold, the surface of the 2-sphere begins to move radially and to strain with attendant loss of density between macro-sections. These intervening areas, in a 3-D extrapolation, become the voids between galactic clusters and gas webbing that we find today. Spalling and instances of quantum oscillation occur at the peripheries of these voids. Given the preponderance of hydrogen and helium in the present-day universe it appears that such spalling is responsible for most free baryonic and leptonic matter which then circulates to form stars and galactic spaces around center points of gravitational mass, such centers being both without and with remaining inertial sources as active galactic nuclei. The voids continue to expand according to the Hubble rate, while the areas of congregated rest mass, of baryonic/leptonic matter, are not subject to this expansion due to the greater strength of the various quantum wave forces.

2e — The Missing Mass of Beta Decay

We are not quite through with our investigation. While the ratio of neutron-electron mass as developed here is compelling, there is still a matter of the missing mass of beta decay. According to the CODATA ratios, the difference between the neutron-electron mass ratio and proton-electron mass ratio is

$$m_n/m_e - m_p/m_e = 1838.6836... - 1836.1526... = 2.5310... \tag{1.69}$$

Since the relative mass of the electron in this case is 1, there is a relative mass or equivalent energy of 1.530... that is unaccounted for and sometimes referred to as the missing mass of beta decay. If it is assumed that mass is a property that is somehow bound up in the confines of a discrete particle, this is a puzzlement. However, if it is understood to be a measure of the resistance of stress to a straight-line force, therefore a measure of redirection of energy as wave action and of curvature of spacetime strain, the problem vanishes.

The entire waveform of a neutron and proton is confined within the nucleus of an atom in condensed matter and the particle wavelength, as represented by the reduced Compton wavelength, is bound by rotation of the oscillation as developed previously, so that the wave energy is bound to the transverse or anti symmetric component of the wave tensor, except for the relatively minor symmetric components of quantum gravity. The electron wave form has both rotational and translational components, particularly at the time of beta decay, so that the differential changes of exponential expansion stress which generate the decay and the beta particle or electron have both of these components. We can restate the last equation with a little more detail, where the inertial constant as developed previously is used to define mass,

$$\frac{ח}{\lambda_{C,n}} - \frac{ח}{\lambda_{C,p}} - \frac{ח}{\lambda_{C,e_{ij}}} = 0 \tag{1.70}$$

The reduced Compton wavelength of the electron is written with subscripts, ij, to indicate the symmetric and anti symmetric components of the wave form.

Now consider the function

$$W(n) = \ln_0 e_n^n, \text{ where } \ln_n e_n^n = n \tag{1.71}$$

which is related to the Lambert W function, where n can be any real number, though we will only be considering the integers. The significant feature of this function is that it generates a system of natural logs, $\ln_n$, and corresponding exponential bases, e_n, that can be used as normalizing factors, so that

$$\ln_n e_n = 1, \quad \ln_n e_{-n} = -1 \tag{1.72}$$

At $n(0)$, this is simply the natural log and exponential base, and

$$W(0) = \ln_0 e_0^0 = 0 \tag{1.73}$$

In the following Figure 11 we have graphed the significant portion of the natural log and exponential functions. Note the functions mirror each other along the line $y = x$, as do their derivatives. We can define the exponential base, e_0, on both x and y axes by the point on each function at which the lines (blue) whose slopes represent the derivatives intersect each other and the origin of the system. The only other instances of such intersection would be when the functions reach negative infinity along both axes, which of course they never do in the context of Euclidean space. They do on the Riemannian complex sphere, however.

The whole system of e_n, for $n > 0$, occurs in the range $1 < x$ and $y < e_0$, and as n increases, the slope values converge while their intersection moves toward the negative infinities of both x and y. At the points on the curves corresponding to $x = 1$ for the logarithmic and $y = 1$ for the exponential, the derivatives of both functions equal 1 and their slopes are parallel. (In terms of the Riemannian sphere, the slope lines for these derivatives (not shown here) actually form two ellipses about the spherical great circle $x = y$, tangent to a circle centered on the origin of radius $r = 0.707\ldots$ at (-0.5, +0.5) and (+0.5, -0.5) and approaching the +/-x and +/-y infinity intersection asymptotically.) As n decreases the slope intersections shift from negative to positive infinity at their x and y axis asymptotes. Thus $n < 0$ occurs in the range $0 < x$ and $y < 1$.

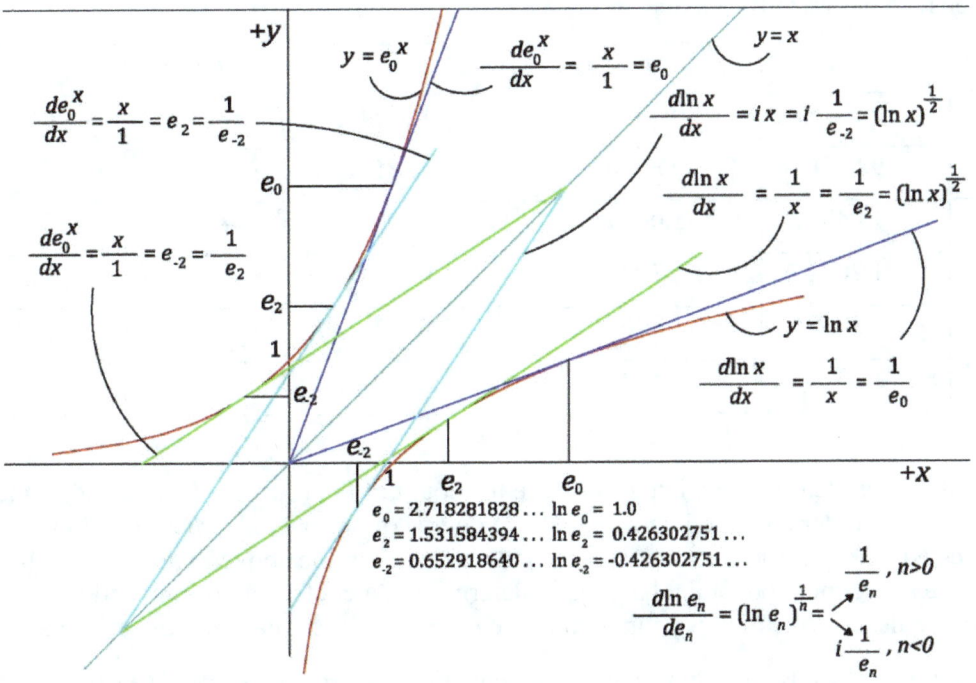

System of exponent bases e_n, shown for e_0 and $e_{\pm 2}$ where $e_{-2} = e_2^{-1}$

Figure 11

$f(n)$ \ n	0	1	2	3	...	∞
e_n	2.718281828..	1.763222834..	1.531584394..	1.419024454..		1
$e_{-n} = e_n^{-1}$	0.367879441..	0.567143291..	0.652918640..	0.704709490..		1
e_n^n	1	1.763222834..	2.345750756..	2.857390779..		∞
$e_{-n}^n = e_n^{-n}$	1	0.567143291..	0.426302751..	0.349969632..		0
$\ln_0 e_n$	1	0.567143291..	0.426302751..	0.349969632..		0
$\ln_0 e_{-n}$	-1	-0.567143291..	-0.426302751..	-0.349969632..		0
$\ln_n e_n$	1	1	1	1		1
$\ln_n e_{-n}$	-1	-1	-1	-1		-1
$\ln_0 e_n^n$ = $W(n)$	0	0.567143291..	0.852605502..	1.049908893..		∞
$\ln_0 e_{-n}^n$ = $W(-n)$	0	-0.567143291..	-0.852605502..	-1.049908893..		-∞
$\ln_n e_n^n$	0	1	2	3		∞
$\ln_n e_{-n}^n$	0	-1	-2	-3		-∞

Table 4

For the range $e_0 < x$ and $y < +\infty$, the slopes of the two derivatives diverge as x increases, and there are no real subscript functions of e_n. Note that for the range n < 0, however, according to the derivative of the natural log with respect to a change in x, the slope has imaginary sense, which generally indicates a rotation of some manner. The table above shows the results of this function for the first three integers, and an assumption of results carried to infinity. The table below shows related results with the introduction of imaginary sense to the various function.

n	$f(n)$	$n \ln_0 e_n$	$in \ln_0 e_n$	$n \ln_0 ie_n$	$in \ln_0 ie_n$
1		0.5671...	i0.5671...	0.5671...+iπ/2(=+1 iπ/2)	-π/2 + i0.5671...
2		0.8526...	i0.8526...	0.8526...+i π(=+2 iπ/2)	-π + i0.8526...
3		1.0499...	i1.0499...	1.0499...+i3 π/2(=+3 iπ/2)	-3π/2 + i1.0499...
4		1.2021...	i1.2021...	1.2021...+i 2π(=+4 iπ/2)	-2π + i1.2021...
5		1.3067...	i1.3067...	1.3067...+i5 π/2(=+5 iπ/2)	-5π/2 + i1.3067...
6		1.4324...	i1.4324...	1.4324...+i 3π(=+6 iπ/2)	-3π + i1.4324...

Table 5

In the final table, it is clear that the integers, n, are the count of the rotations of ½ π and of the powers and hence the number of orders of i, both indications of a degree of orthogonal structure as in Euler, Section 0a. We are interested here specifically in the factor e_2. As a review of the above Figures hopefully makes clear, the value in the subscript exponential bases is in determining a coefficient of proportionality between two related differentials, one of which is a function of the nth-root of the logarithm to the others linear function.

The function in (1.71) finds form in the following equation, where the negative sense in the subscript has the same meaning it does in the superscript exponent, that is it represents inversion.

$$\ln_0 e_n = e_n^{-n} = e_{-n}^n \qquad (1.74)$$

Thus for n = 2, we have the following, where it is understood that e_2 is a normalizing coefficient for any independent variable dimensional property, x,

$$\ln_0 e_2 (e_2)^2 = 1, \quad \ln_0 e_{-2}(e_{-2})^{-2} = -1 \qquad (1.75)$$

and with a unit of property, x_0, and some mathematical operation and rearrangement we can state

$$(\ln_0 e_2)^{\frac{1}{2}} x_0 = e_2^{-1} x_0 = \frac{d \ln_2 e_2 x_0}{de_2 x_0} = \frac{1}{1.53158...} \qquad (1.76)$$

$$(\ln_0 e_{-2})^{\frac{1}{2}} x_0 = ie_{-2} x_0 = \frac{-de_{-2} x_0}{id \ln_2 e_{-2} x_0} = \frac{-0.65291...}{i1.0} = \frac{-1}{i1.53158...} \qquad (1.77)$$

In the above development of the neutron scale for quantum gravity (1.33) and at (1.66) we have an expression of the change in the linear scale of r_0 as the square root of the change in the natural log of the expansion stress scale, T_0. We have modeled quantum mass as a linear function of space, r_0, by the reduced angular wavelength or time by the frequency, ω_0. As developed above in connection with the Hubble rate, using the inertial constant and/or the speed of light we have

$$m_0 = f(H_0) = f(dr_0, dt_0) \qquad (1.78)$$

Stress is modeled as a function of the square of both of these,

$$f_0 = f\left(dr_0^2, dt_0^2\right) = \hbar\omega_0^2 / r_0^2 = \hbar\theta / t_0^2 r_0^2 = \tfrac{\hbar}{c}\theta / t_0^2 r_0^2 \tag{1.79}$$

but as time and distance are dual values as defined by an invariant c, we can treat stress as a functional, where time is a measure, therefore a function of the change in r_0, (we could also use displacement as a measure, therefore a function of a change in t_0), as

$$f_0 = f\left(dr_0^2, c_0^2\left(dr_0^2\right)\right). \tag{1.80}$$

Thus, a change in stress with expansion leads to an increase in r_0, where a preliminary decrease in mass of the fundamental oscillation, the neutron, is equal to the mass of the emitted electron as developed above or

$$m_e = \Delta m_n = m_n\left(\frac{\delta r_0^2}{1+\delta r_0^2}\right) \tag{1.81}$$

The change in stress/energy density of the oscillation is

$$dE_1 = df_0 = f'\left(A_0^{-2}\right) = -\frac{\hbar\omega_0^2}{A_0^2}dA_0 = -\frac{\tau_0}{A_0^2}dA_0 = -\frac{f_0}{A_0}dA_0 \tag{1.82}$$

where it is clear that a change in the log of the stress is inversely equal to a change in the log of the cross-section,

$$d\ln f_0 = \frac{df_0}{f_0} = -\frac{dA_0}{A_0} = -d\ln A_0 \tag{1.83}$$

Obviously, since both stress and cross-section are unit values

$$\ln f_0 = \ln A_0 \tag{1.84}$$

Thus for an exponential expansion of the cosmos, in accordance with (1.83) and using (1.77), we substitute f_0 for A_0 which is the square of $r_0 = x_0$ and get the following where ii indicates the symmetric and ij the anti symmetric or transverse components of the electron wave form

$$\left(\ln f_0\right)^{\tfrac{1}{2}} = ie_{-2}f_0 = \frac{-df_0}{id\ln f_0} = \frac{-de_{-2}f_{ii}}{id\ln_2 e_{-2}f_{ij}} = \frac{-0.65291...f_{ii}}{i1.0f_{ij}} = \frac{-1.0f_{ii}}{i1.53158...f_{ij}} \tag{1.85}$$

The imaginary sense assigned to the natural log differential is an indication of transverse motion and other energy associated with the change in stress, resulting in the change to that of the reduced Compton wavelength of the proton.

The Hamiltonian or total energy of the system resulting from beta decay is therefore the energy of the neutron, which is equal to the energy of the proton plus the energy differential of the electron wave components due to the change in stress of expansion.

$$E_0 - E_p - \left[E_e(df_0) + E_0(id\ln f_0)\right] = 0 \tag{1.86}$$

In terms of mass, as in (1.70) this is accurate to a factor of 2.16×10^{-7}.

$$m_n - m_p - \left[1.0m_e - 1.53158...m_e\right] = 0 \tag{1.87}$$

2f — Evaluation of Elementary Charge

We turn now to the nature of charge. As discussed previously, it as a function of the fundamental quantum oscillation of the neutron, as both the sustained level of maximum potential energy and the simultaneous sustained level of maximum kinetic, transverse wave energy which produces spin and the related quantum magnetic field. We can think of the potential energy as the capacity of positive charge and the induced kinetic energy as negative charge, though the senses are reversed in the case of anti-matter. In terms of phase space, potential energy indicates maximum wave displacement and kinetic energy represents maximum wave momentum.

With beta decay the wave momentum is externalized or transmitted in part from the constrained boundaries of the neutron to form the electron waveform in keeping with the mechanics of the above spin diagrams. While the electron waveform has its own quantization dynamics, for example the maximum wave momentum, along with mass-spin energy, wave force, etc., which are much reduced differential portions of the fundamental system, in total they are the effect of the fundamental wave momentum/charge of the neutron. We will go in to this in detail in a moment.

The SI fundamental charge is the coulomb, C. The coulomb, or ampere per second, is equivalent in mechanical dimensions to one kilogram-meter per second, a measure of momentum. A fundamental unit or elementary charge, e_0, is established as

$$e_0 = 1.60217653(14) \times 10^{-19} \, Coulomb \tag{1.88}$$

As a measure of momentum, in connection with this development and the transmission of momentum with beta decay at W_{-x}, in the spin diagram for the proton, the fundamental unit of conjugate momentum, using angular frequency, is reasonably close to this value at

$$p_0 = \hbar\omega_0 = 5.02130...\times 10^{-19} \, kg \cdot m/s \tag{1.89}$$

Charge is related to each of the two rotational nodes, W_{-x} and W_{+x}, indicating the need to apply semi-periodic frequency, which we can do by dividing (1.89), which is expressed in angular frequency, by π. In addition, the charge generation is conditioned by the product of the momentum and the mechanical impedance of the manifold from (1.59) (not to be confused with the electro-magnetically derived characteristic impedance of the vacuum), which is

$$Z_0 = \hbar\omega_0 r_0^{-1} = 0.002390877...kg/s$$

$$\zeta = \left(\frac{1+Z_0}{\pi}\right) = 0.319070926... \tag{1.90}$$

$$\zeta^2 = \left(\frac{1+Z_0}{\pi}\right)^2 = 0.101806256...$$

where we define the total factor, ζ, and its square for later use. Thus, we would anticipate an elementary charge of

$$e_0 = p_0 \zeta = \hbar\omega_0 \zeta = 1.602152647...\times 10^{-19} \, kg \cdot m/s \tag{1.91}$$

This varies from the established value by a factor of 1.000015…which is in the same order of magnitude as the relative uncertainty for the gravitational constant.

Further development, using the familiar identity for the inverse of the fine structure constant, α, a dimensionless number and therefore the ratio of two like-property magnitudes, as

$$\alpha^{-1} \equiv \hbar c \frac{4\pi\varepsilon_0}{e_0^2} = 137.0359989... \quad (1.92)$$

and the permeability, μ_0, and permittivity, ε_0, relationship, where μ_0 is in units of inductance per meter or henrys per meter which reduces to units of force per current squared or newton's per ampere squared, and ε_0 is in units of capacitance per meter or farads per meter which reduces to ampere squared per newton over the speed of light in vacuo squared, so that

$$\varepsilon_0 = \frac{1}{c^2 \mu_0} \quad (1.93)$$

and with rearrangement in (1.92) gives the following

$$e_0^2 = -\hbar c^2 (\alpha 4\pi\varepsilon_0) = -\hbar \frac{\alpha 4\pi}{\mu_0} = -\hbar \frac{\alpha}{10^{-7}} = -\hbar (\hbar \omega_0^2) \zeta^2 \quad (1.94)$$

It is noted that the value of μ_0 is set by convention in relating charge, q, (of which elementary charge, e_0, is an effective quantum) and current, $i = dq/dt$, resulting in the exactness of the denominator of the next to last term. Since the negative sense of the right terms above can be attributed to the current, therefore charge, squared, it can be incorporated therein, canceling such sense in the charge squared term. This suggests the transparent presence of a current squared argument in (1.94), for which the fine structure constant is a coefficient, since from Ampere's Law for one ampere^2 of current, where the denominator on the right is in newton, we have

$$\mu_0 = 2\pi (2 \times 10^{-7} N) \frac{d}{L} i_0^{-2} = 4\pi \times 10^{-7} \, newton/ampere^2 \quad (1.95)$$

2×10^{-7} newton is the force generated for each meter length, L, of two conductors of infinite length and negligible cross-section and one meter apart, d, in a vacuum with one ampere of constant current flowing in each conductor. The d and L obviously cancel and the i_0^2 component and therefore the force is positive or negative depending on whether the currents are parallel and attractive or antiparallel and repulsive.

Inserting this into (1.94) with some rearrangement gives the ratio of elementary charge squared to current squared as the product of the modified fine structure constant, α' as shown, and the inertial constant. If the fine structure constant is dimensionless and its denominator is a force from the above, then α' is an inverse force, which in terms of this development is the inertial constant times a frequency squared and k is an unknown proportionality factor for the frequency as

$$\frac{e_0^2}{i_0^2} = \frac{\alpha}{10^{-7}} \hbar = \alpha' \hbar = \frac{k^2}{\hbar \omega_e^2} \hbar = \frac{k^2}{\omega_e^2} \quad (1.96)$$

If the force in the last term is the base transverse wave force of the electron as in the above development, then k is an angular measure per unit of elementary charge as,

$$k = \omega_e \frac{e_0}{i_0} = \left(\frac{\theta_e}{s}\right) \frac{e_0}{\frac{ne_0}{s}} = 124.3840198... \frac{\theta_e}{e_0} \quad (1.97)$$

Using this value with (1.96) gives

$$\alpha = \frac{k^2(10^{-7})}{\hbar\omega_e^2} = \frac{k^2(10^{-7}N)}{0.212013671...N} = 0.007297352...\frac{\theta^2}{e_0^2} \qquad (1.98)$$

With another look at (1.94), we get the following relationships between the fundamental wave force and α'

$$\alpha' = \tau_0\zeta^2 = \hbar\omega_0^2\zeta^2 = \zeta\omega_0 e_0 \qquad (1.99)$$

Using our derived value for elementary charge in (1.91) in the first two terms of (1.96) we get the following derived value for the fine structure constant of

$$\alpha_{derived} = .007297134... \qquad (1.100)$$

Comparing with (1.98) once again, in line with our missing mass derivation, this is accurate to a factor of 2.17×10^{-7}.

2g — Special Relativity and Muon & Tau Families

Concerning the compatibility of this model and special relativity, I have written about this extensively elsewhere. Suffice it to say that this model is one of constrained stress/strain in the spacetime manifold, which acts as discrete units of rest mass with derived properties. Each discrete state, remains a wave form and in response to interaction with other states is free to translate and rotate in space according to the ambient energies. It will therefore contract its characteristic strain radius in response to acceleration in keeping with the Fitzgerald-Lorentz length contraction, resulting in an increase in spin energy/mass according to the definition of the inertial constant, ת.

As to the two other families of leptons, the muon and tau, and their theoretical related hadrons, based on their short lifetime and granted my limited knowledge of the experimental background for their theoretical introduction, it is my perception that they are simply the basic states we have discussed, altered by relativistic dynamics and collision interaction. We would expect these states to behave in a generally ordered fashion under constraints of high energy collision and those defined by geometry and mathematics. The evolution of a catalogue of such short-lived phenomenology, while useful, does not indicate the need or wisdom of elevating that phenomenology to ontology. I would grant the status of "fundamental rest mass particle" only to common, stable, relatively long-lived states, of the proton, electron, and including the neutron in nuclear confinement, of course, in keeping with the general conditions of condensed matter physics.

3 — Condensed Matter Application of this Model

The above model development represents rest mass particles as localized torsional oscillations of an expanding spacetime manifold with an inherent elastic potential energy density. From a topological perspective, the spatial 3-sphere or 3-manifold without boundary is itself a boundary within a spacetime 4-manifold in which the time dimension is orthonormal to the 3-dimensional space component. The 3-dimensional analogy is the 2-sphere surface of a rupture-free balloon that expands or contracts radially over time in response to a pressure gradient at its surface and in which the dimension and direction of time is registered by the radial sense of the change. In an expansion phase, the interior of the balloon represents the past and the exterior of the balloon represents the future, while in a contraction phase, the past and future senses are reversed. For an ongoing oscillation there is no absolute time sense indicated, rather only a phase direction of the instant or current state. As the n-sphere boundary of an n_{+1}-manifold can never be reduced to an n_{-1}-dimensional state, any inherent density of the n-sphere must be finite; it can never be infinite or zero. Failure to understand this axiomatic, topical truth is the source of much misunderstanding in current physical theory.

The rest mass particles generated from such change in energy density ultimately produce plasma conditions when congregated by gravitational interaction at a stellar scale, but this is not necessarily the only precursor to condensed matter states. Heat and its measure, temperature, is a convenient analogy for potential energy density, but it bears remembering that heat is a relative condition of particle translational velocity and collision interaction, which itself is an interaction of electromagnetic wave fields of the quantum electron or of the neutron/proton fundamental.

In this model under discussion, quarks are not a separable constituent of such matter states but are rather the signature of oscillatory nodes, antinodes and moments in particle collisions as registered in electronic accelerator recording systems. The fundamental physical structure in this model is the baryon, specifically the neutron, which with ongoing expansion decays into the proton, both of which are ontologically wave forms, but with a particle phenomenology when viewed from the scale of human interaction; mesons, neutrinos, and all leptons and gauge bosons are the transitional result of baryonic decay, of which the electron and photons, and perhaps the neutrino, have longevity.

In the interactions between condensed matter nucleons, i.e. protons and neutrons, it is the nodes and antinodes along with the capacitive and inductive moments that chart their configurations, so in that sense an accounting of nodal/moment "quarks" may be in order, though probably not in the manner of established quantum chromodynamics. As detailed above, sustained quantum rotational strain provides an alignment constraint and mechanism for the congregation of nucleons, and we would anticipate a geometric configuration of nucleons to form an atomic nucleus in the manner of the work of Norman Cook and others, as opposed to a liquid drop or other concepts of the nucleus. We would anticipate these configurations to mimic the constraints of the cuboctahedral lattice of our opening development just as do the crystal structures of many metals such as palladium.

In the referenced graphic representations of this paper it is easy to depict the fundamental oscillations as spheroidal objects due to their inherent rotational dynamics about the θ and ϕ axes, so it is important to emphasize that in this modeling they are not conceptualized as separate objects detached from the space around them; rather they are foci of expansion stress that results in torsion and its eventual axial rotation and the surfaces of the spheroidal graphic images represent the characteristic radii of these rotational stresses and strains.

The spin vectors as represented in the spin diagrams for the neutron and proton are parallel, as shown in Figure 12 as they might be for the deuteron, giving the nuclear system a spin of 1. Their magnetic induction vectors or dipole moments are anti-parallel and therefore have a 0 state. As they have different rotational frequencies, we would anticipate a degree of angular momentum of the system. Finally, we anticipate a torsion connection for transmission of the electron waveform anti-parallel to the ϕ rotation vector of the proton.

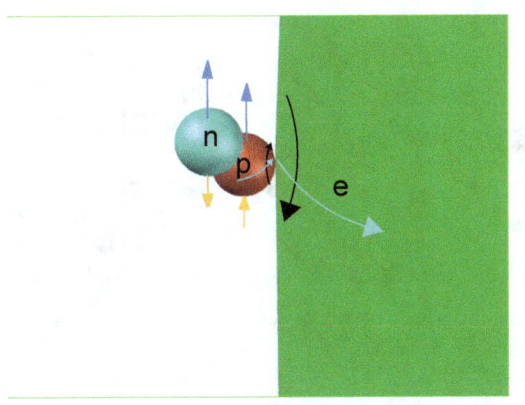 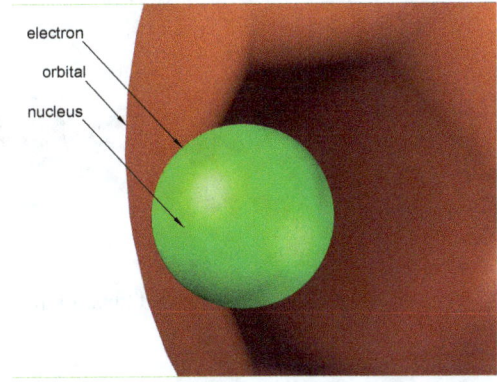

Nucleon, Electron, and Orbital relative spatial relationships. The nucleon and electron figures use the corresponding Compton reduced wavelength as a gauge of torsional wavelength in the drawings. The electron "surface" shown on the left is shown in its entirety on the right in which the nucleons are just a speck at the end of the designated arrow smaller than the period at the end of this sentence. The orbital size is correspondingly larger as seen in the next graphic and is an electromagnetic field effect of the moving electron.

Figure 12

The electron waveform in Figure 12 has been represented here in a general spherical form on the right in green, and in the absence of transient stress, that might be accurate, however, we would anticipate a tendency for elongation as an orbital in the form of a prolate ellipsoid as indicated by the large red graphic here and below, particularly in multi-nucleon atomic structures and within the context of molecular structures. The electron modeled here is not a cloud, though it would resemble the orbital configurations shown in quantum mechanics sources. Such an electron wave mechanism is essentially tethered to the nucleus, short of ionization, free to revolve about the nucleus in the absence of other electrons; in the presence of other electrons in multiple nucleon elements any given electron is constrained to the valence specific orbital arrangement.

The significant thing to focus on within the context of condensed matter is that the coulomb force on a quantum level is not modeled herein as an instant isotropic symmetrical charge field, i.e. over 4π steradians, between a positively charged proton and a negatively charged electron cloud, but rather as a transmission of potential energy and fundamental wave momentum from a "positively charged" capacitive source within the proton waveform as the propagated wave kinetic energy and momentum of the electron in a form of induced "negatively charged" electromagnetic current. The point-like charge of the electron wave form is represented by the twisting node of the positive end of the ϕ rotation vector of the electron which can extend indefinitely and in fact be separated from its source through ionization. In addition to the ongoing transmission of transverse wave momentum that is responsible for the electron waveform, upon initial emission of that electron, at the leading edge of the wave propagation we can anticipate a longitudinal component of the momentum that continues on beyond the boundary of the orbital and is responsible for the phenomena of the neutrino.

In the context of molecular bonding, as in Figure 13, when two ϕ rotational nodes meet, being of the same axial twist, but antiparallel, and therefore of opposite spin state, they reinforce and strengthen the nodal point and form a stabilizing bond. If a positron and electron meet, as in Figure 14, being of opposite twist and antiparallel, the rotational stress opens the nodal structure and destroys the characteristic wave forms with a release of wave energy.

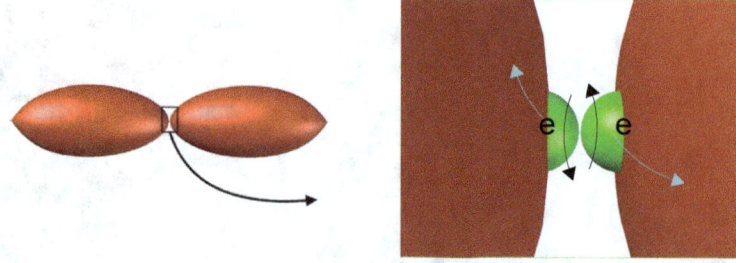

Bonding of two orbitals at point of maximum torsional interaction between electrons as shown in inset enlargement.

Figure 13

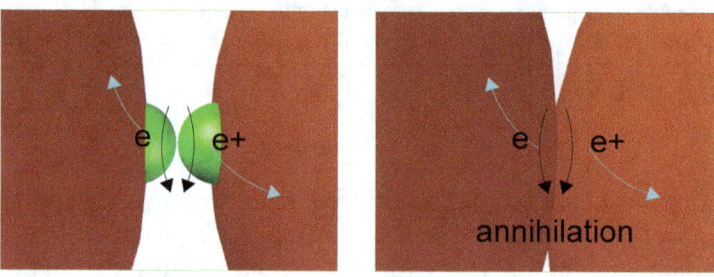

Interaction between electron and anti-electron which is of opposite nodal twist, destroying the nodal structure and releasing the wave energy as antiparallel gamma rays.

Figure 14

The same inherent cuboctahedral stress and strain constraints that produce the neutron, and eventually thereafter the proton, and then nuclear configuration, is found in molecular bonding of metals like palladium that exhibit the face centered cubic crystal structure as seen in the following graphic.

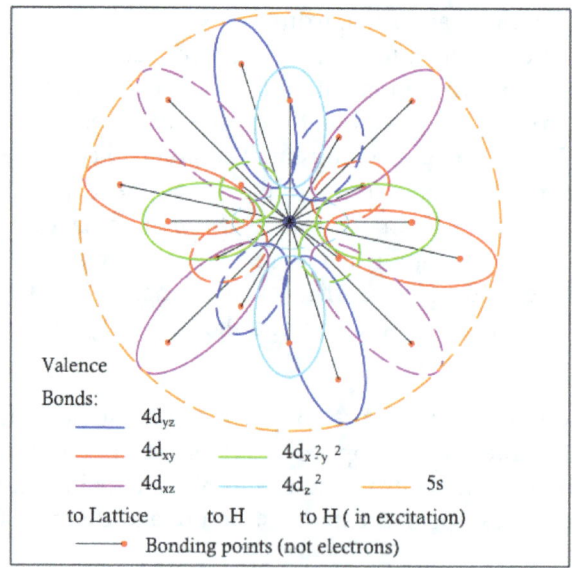

Palladium Valence Electron Orbitals $4d^{10}$
The brown circle represents the 5s excitation that is absent in the atom's ground state.
Figure 15

The $4d_{yz}$, $4d_{xy}$, and $4d_{xz}$ electrons are constrained to the corresponding orbital arrangement of the 4d subshell by cuboctahedral geometry and are occupied in forming the FCC lattice structure. The $4d_{x^2-y^2}$ and $4d_{z^2}$

electrons/orbitals center on the six nodal orientations of the octahedral chambers of the FCC lattice and are available for bonding with any hydrogen absorbed into the lattice. There is no bonding potential for the four nodal orientations within the tetrahedral chambers, however, as there are no orbitals defined in the shell geometry for these locations, EXCEPT in the case of the excitation of one of the deeper (as 4s) electrons to the 5s shell. In that case the subtended solid angle represented by each of the six octahedral chambers around each FCC node, being 1.359347638 steradians(sr) for a total of 8.156085832 sr, leaving a difference from 4π of 4.410284782 sr divided by 8 tetrahedral chambers is 0.551285598 steradians or 4.3869914% of the total 4π sr surface giving a corresponding raw probability of a little less than 5% of such excitation resulting in a bonding potential with any hydrogen in the tetrahedral chamber as a deep electron jumps to the 5s shell. If we assume that the exclusion principle precludes the octahedral spaces from such bonding, the probability is simply 1/8 or 12.5% for any surrounding tetrahedral space.

Table 6 gives some of the parameters of the lattice and the elements involved in this catalytic system.

Lattice and Element Parameters

Lattice statistical data | | | Palladium | Nickel

			Palladium	Nickel
1	Valence shell		*[Kr]$4d^{10}$	[Ar]$3d^8 4s^2$ or [Ar]$3d^9 4s^1$
2	Covalent radius		139(6) pm	124(4) pm
3	d, Covalent radius x 2		278.0 pm	248.0 pm
4	a, Unit cell edge,		393.2 pm	350.7 pm
5	O Chamber width		115.2 pm	102.7 pm
6	T aperture spherical capacity		43.0 pm	38.4 pm
7	T chamber spherical capacity		62.5 pm	55.7 pm

Hydrogen isotopes and Helium 4

8	H covalent radius x 2		62 pm	62 pm
9	D covalent radius x 2		<62 pm	<62 pm
10	He-4 covalent radius x 2		56 pm	56 pm
11	H length as ellipsoid in T aperture		128.9 pm	162 pm

* Unique in [Kr] for lack of 5s subshell, Electronegativity is 2.20 for both Pd and H

Table 6

We next consider the cuboctahedral perspective of the FCC lattice of palladium, in other words with twelve edge centered atoms around a central atom, instead of the 6 face and 8 vertex atoms around a central octahedral void of the FCC. We will consider the valence 4d subshell structure to present a generally spherical atomic component of the lattice. The six octahedral chambers around each center atom offer a potential for bonding with hydrogen infused or absorbed in the bulk or adsorbed in any surface octahedral chambers that are open on one face. If the crystal is faceted along the octahedral axes, as shown in the last cell of the Cuboctahedral Lattice Configuration table, there will be no octahedral opening and any diffusion will require penetration of a tetrahedral aperture and transit of the enclosed T chamber before entry into an interior O chamber, making that diffusion difficult. If the crystal is faceted rectilinearly, surface octahedral openings will be available and will facilitate both adsorption and thereby absorption.

Palladium and hydrogen are somewhat unique in that both have the same electronegativity on the Pauling scale of 2.20, and upon molecular bonding of the electrons, the nuclei of each will seek the same distance from their corresponding nucleus. As palladium is the much more massive atom and is also in a crystal configuration, this means that upon bonding with the palladium, the hydrogen nucleus will move away from the bonded palladium atom. While the $4d_{yz}$, $4d_{xy}$, and $4d_{xz}$ bonds in the idealized crystal are delocalized, that between the Pd and D would be a sigma bond or localized.

In the confines of an O chamber, if it bonds to one of the $4d_{x^2-y^2}$ and $4d_{z^2}$ electron/orbitals of the four surface atoms and not the first interior layer atom, the H or in this case D nucleus has only one of two places to go other than exiting the crystal, and that is to the center of one of the two opposite T apertures leading to a T chamber and eventually to the interior of the bulk. The geometry is such that the D nucleus will thereby be positioned directly in the center of the aperture and in the plane of the three surrounding Pd nuclei by the fact of a common electronegativity of Pd and D.

The "naked" or unshielded D nucleus will be positioned toward the center of the opening by electrostatic force, which is an extension of the wave stress in this model, where it will remain as long as the Pd-D bond is maintained. If there is an electrolytic charge or gas pressure on the crystal sufficient to break the covalent bond to the O chamber Pd atom, the D will have a potential to enter and occupy the T chamber. If the T chamber is occupied, and a pathway into the bulk by that occupant is blocked, entry of the nucleus will be blocked, but if the chamber is empty, the D atom will enter the T chamber. However, it will not be able to bond to the lattice in the absence of an available orbital.

In the case of a D atom occupying a T chamber, if one of the surrounding four Pd atoms is excited to raise one of its electrons to the 5s shell and in the area of the T chamber, the D will be induced to covalent bonding with that Pd electron, and the D nucleus will be accelerated through the T aperture opposite that Pd atom according to the common electronegativity of the two elements. An unshielded D nucleus positioned in the center of the aperture is essentially an inertially confined target for the accelerating D nucleus from the T chamber, and the two nuclei will be positioned to collide. In keeping with the above development of rest mass particles, we can expect the proton of each D nucleus to be oriented toward its bonding electron and for the neutron of each to be opposite that electron direction. This is not due to electrostatic forces as conventionally modeled, but rather is a result of the fact that the electron waveform is being generated by the proton while the neutron is essentially passive to and moved out of the way by that ongoing interaction. The fact that the neutron is on the leading edge of the accelerating D nucleus and on the proximal side of the other, target D nucleus shields the two protons from repulsive near-field coulomb interaction and increases the potential for nuclear fusion with nuclear coincidence which will occur due to wave-spin/moment alignment of the two nuclei.

$D + D > {}^4He$ is not a generally recognized fusion pathway. Because the current model of particle physics tacitly or expressly views matter as comprised of particles that under certain circumstances have wavelike properties, it tends to assume that conservation of energy and momentum in an interaction demand that the energy that is liberated in one such as nuclear fusion be carried off by the particles that are produced by the reaction. Under this thinking, even if $D + D > {}^4He$ did occur, it would require a tremendous velocity on the part of the Helium 4, single product of reaction to be registered, and that is not observed to happen. It does not occur to such a model that if particles are essentially waves, such an interaction can release the energy to the lattice in the form of wave energy, in this case as phononic energy or vibration of the lattice.

We can also expect a probability of fusion occurring when two D enter from two adjacent apertures into a T chamber that has its two other apertures blocked by occupied adjacent O chambers. In this event, there will be a lower probability of the helium escaping from the lattice.

FCC Structure

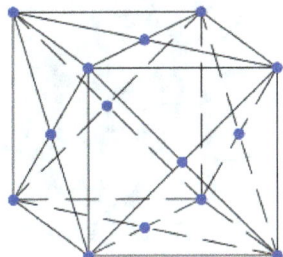

A unit of a face centered cubic lattice depicting a palladium crystal, the blue nodes representing palladium nuclei at the cubic vertices and at the center of each of the cubic faces. In the bulk lattice, 14 nodes are shared with 26 adjacent cubes. In the surface, 13 are shared by 17 cubes.

Octahedral Chamber

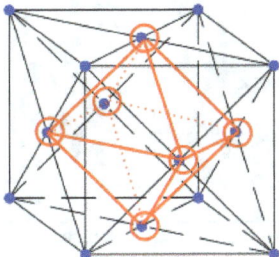

The center of the cube forms an octahedral space with the face centered nodes, but contains no lattice element. It readily absorbs hydrogen, protium and deuterium, at room temperature.

Tetrahedral Chamber

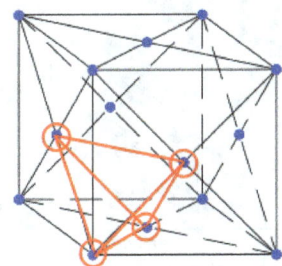

Each corner nuclei forms a smaller tetrahedral chmaber with the three adjacent face centers. Any hydrogen entering the lattice bulk must pass through the T chambers.

O & T Relationship

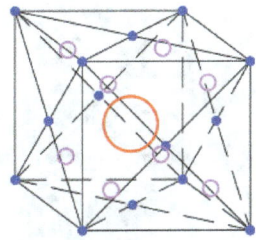

Each octahedral space or O chamber in the bulk is accessed through one of 8 adjacent and smaller tetrahedral spaces or T chambers, the chamber boundaries formed by the orbitals of lattice bonding. In the bulk, there are one O and two T chambers for each palladium atom.

Adjacent ¼ O Chamber

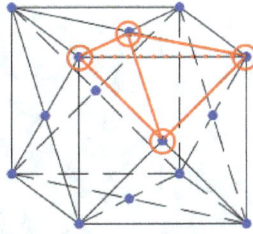

In the bulk, each edge of the cube shares the adjacent two center and two corner nodes to form one fourth of an octahedral space with three adjacent cubes with a total of 18 such cubes. Each bulk cube contains 4 O chambers in its volume. In the surface, the total number of adjacent cubes in the rectilinear configuration is 13, and the O chamber is open at the surface.

Lattice Interlace

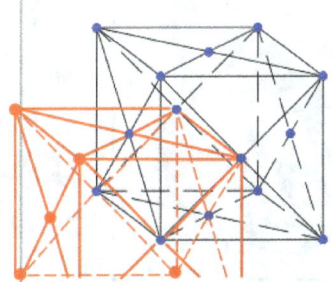

One of 12 interlaced cubes in the bulk sharing and repeating the same nodal configuration as the original cube, but offset in two dimensions by half a cubic length; these are in addition to and interlaced with the other 26.

Coulomb Interaction

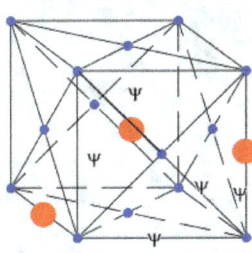

The lattice absorbs 900 times its volume of hydrogen at room temperature equivalent to 70% of the O chambers, assuming one H per chamber. Three random H are shown. The coulomb force, broadly understood to include all electromagnetic interactions and represented by the wave function, psi, is responsible for bonding of the atoms in the lattice. There is one wave function for the system.

Cuboctahedral Form

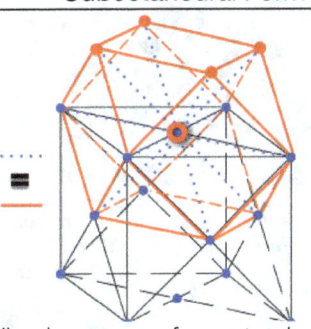

In the bulk, all nodes serve as a face centered node for an arbitrary cube and each such node is also the center of a 14 sided cuboctahedral lattice with 12 vertices equidistant from the center and with 24 equal length edges. Significantly, each such center atom shares 6 O and 8 T chambers with adjacent atoms, the O chambers along the 3 rectilinear axes and the T chambers along the 4 cubic diagonal axes.

Chart 1 — FCC Lattice Properties

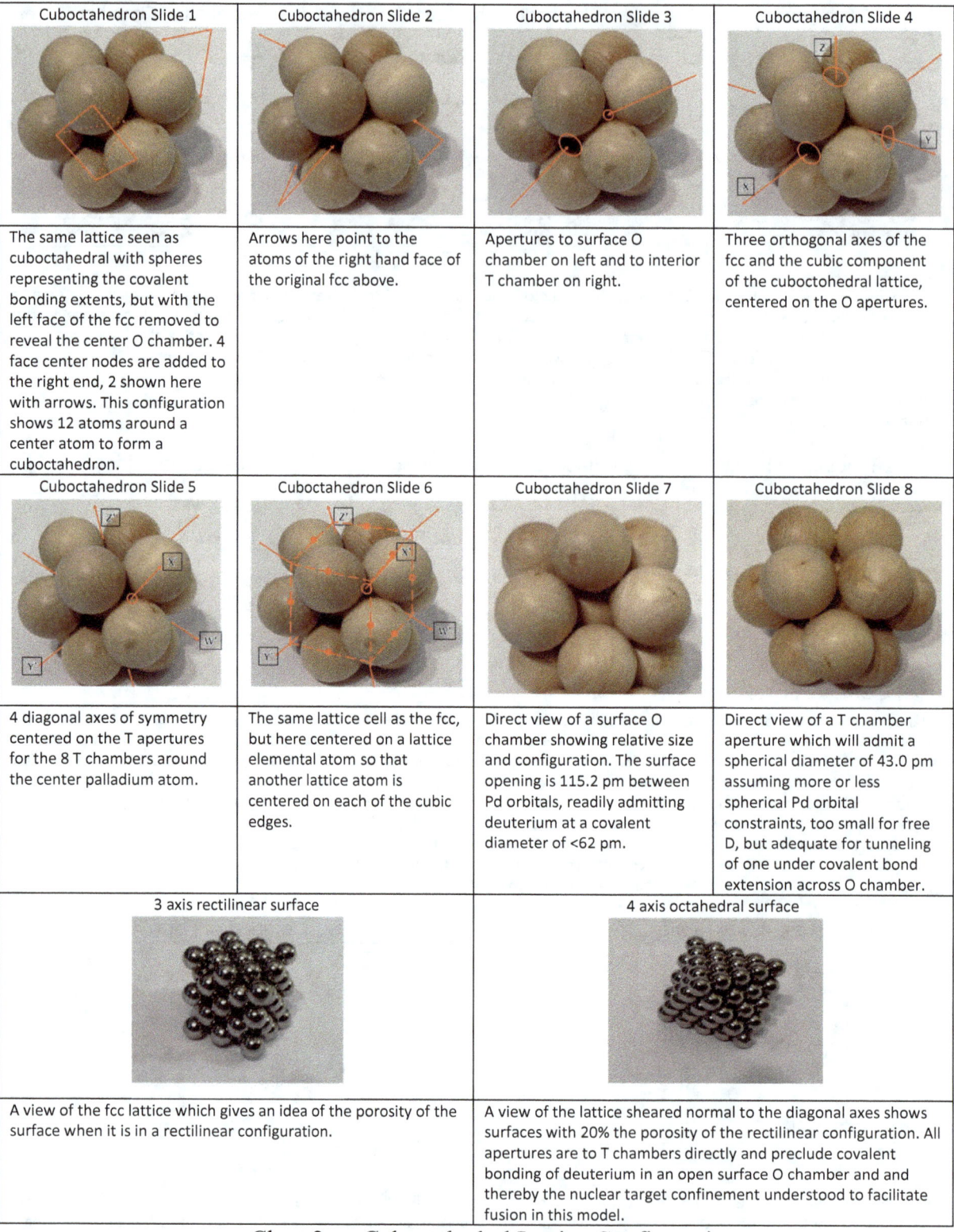

Chart 2 — Cuboctahedral Lattice Configuration

The lattice is physically and geometrically the same as the FCC, but the unit perspective is centered on a nuclear node instead of on an octahedral chamber.

Hydrogen (protium or deuterium) diffusion in the palladium latttice

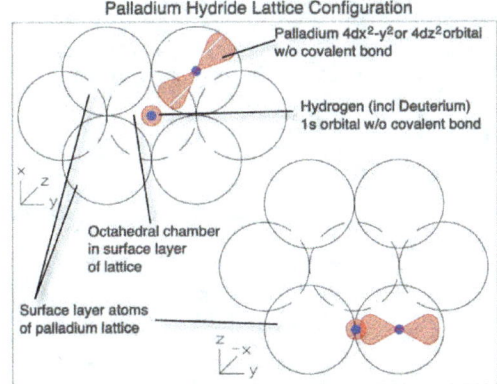

1A

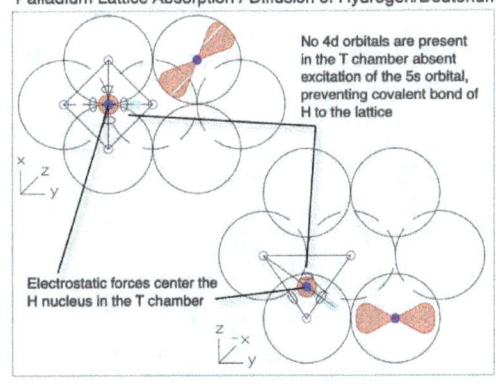

1B

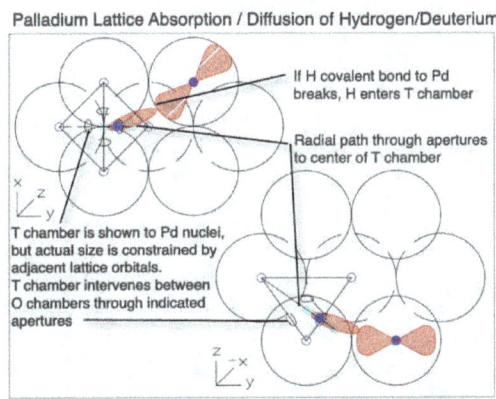

1C

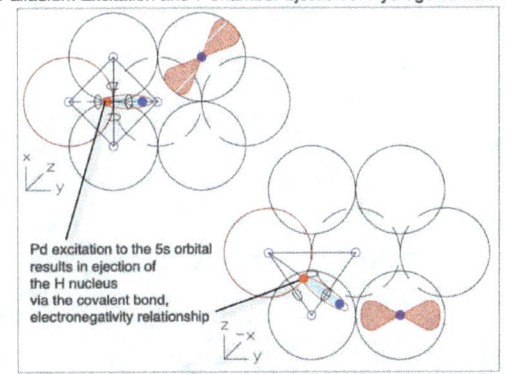

1D

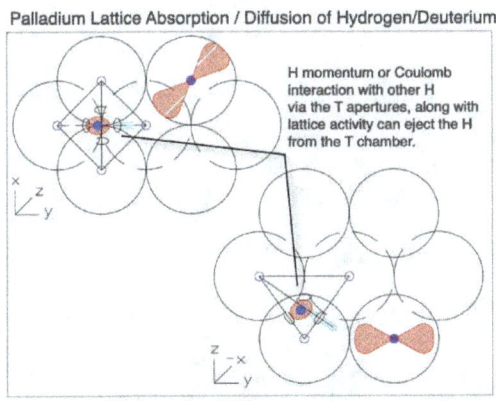

1E

1F

Palladium and hydrogen both have an electronegativity of 2.20 Pauling units which indicates that in this covalent bond the respective nuclei will separate equidistant from the bond or in this case twice the Pd covalent radius. Assuming a uniform charge distribution of the orbital around the three lattice bonds at the tetrahedral aperture, such extension for the H atom projects its nucleus with a high degree of precision to the center of one of the Tetrahedral apertures. If the bond to Pd breaks in the process, the H enters the Tetrahedral chamber.

Chart 3 — Hydrogen Diffusion

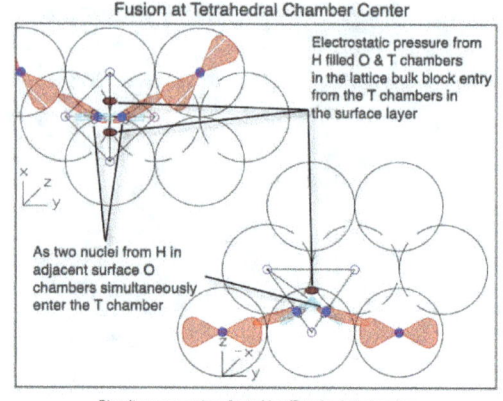

2A   2B

Chart 4 — Fusion path 1 in the Tetrahedral chamber
All hydrogen are assumed to be deuterium.

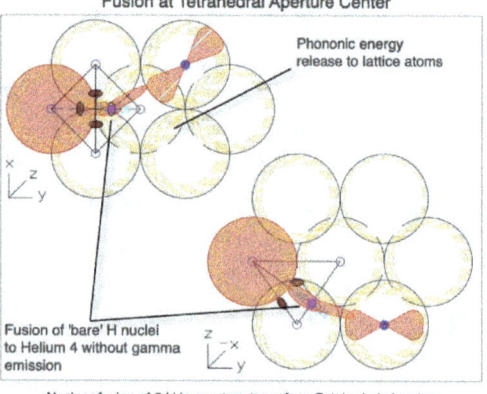

3A   3B

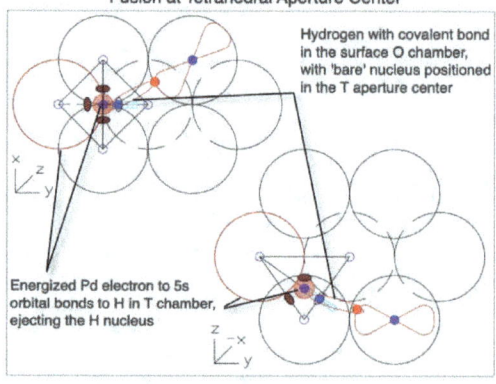

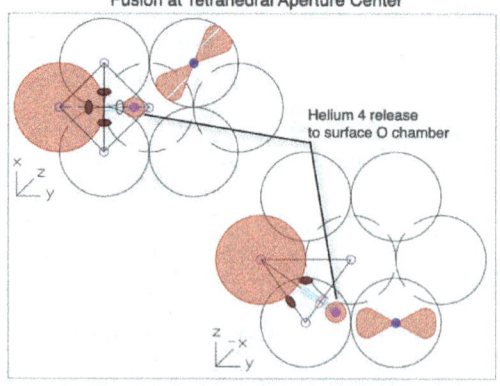

3C   3D

Fusion path 2 appears the most likely for controlled fusion as the common Pauling electronegativity of H and Pd provides precise inertial confinement of the deuterium nucleus in the T aperture of a surface O chamber while in the covalent bond with a surface Pd atom for targeting by another such nucleus ejected by excitation to the 5s orbital of a Pd atom, with each nucleus proton shielded by its neutron. With presumed filling of the bulk chambers, the nuclei merge in the target with a drop in rotational energy transmitted to the bulk and environment as phonic energy and the resulting helium 4 emitted from the surface without gamma emission.

Chart 5 — Fusion path 2 in the Tetrahedral aperture

4 — Conclusions

In 1989, chemists Martin Fleischmann and Stanley Pons reported the results of a table-top experimental electrolysis of heavy water using a palladium electrode in which the recorded heat produced exceeded the energy input plus any known molecular bonding release. They reported some measurement of nuclear byproducts and found the results suggested the possibility of nuclear fusion of hydrogen/deuterium nuclei, catalyzed by the palladium lattice. After initial interest in the announcement from within the physics community, ambiguous and null results from attempts to replicate the process, hampered by a lack of a theoretical understanding of condensed matter nuclear processes that could produce such results, led to a loss of professional interest from within the community.

Interest on the part of a few, from both a technical and business perspective, has led to reports of repeated production of excess heat and in some cases helium production from this and similar experiments, but the ongoing theoretical shortfall and denigration of the effort by various parties has continued to impede the investigation of this possible beneficial and cheap source of energy for human utilization. It is not my intention to go into the historical or technical details of this investigation here; a cursory read of the current article in Wikipedia under "Cold Fusion" appears to give a fair and invective-free overview of the state of the art.

If there is a technically and economically viable avenue to the utilization of a cold fusion process, it is worth pursuing the investigation by correcting the perceived theoretical short-comings and thereby removing one of the major obstacles to increased professional interest in the matter; this work is offered in that spirit.

As I have attempted to convey in this paper, it is my belief that the lack of theoretical understanding of this subject is grounded in the inability of current models to couple the action of the quantum level with the action of spacetime on a cosmic scale. Despite its century of success, general relativity still has no widely accepted explication of the coupling of spacetime and matter at a quantum level; that is, no explication of quantum gravity. And despite its refinement in observation and experimental control down to the nano scale and beyond, the standard model of particle interactions has no explication of this coupling and is at an impasse.

This may be because theoreticians take the results of experiment and try to fit them into the successfully established modeling; when the results fit the model in a straightforward manner, the model is bolstered, but when there is a lack of fit, the response is often to engineer a system of scaffolding and buttresses to keep the theoretical superstructure upright. Sometimes the impetus to enhance the superstructure is valid both for the model and to the satisfaction of the theoretician, but often it can become a set of blinders to a fuller understanding. The history of innovation in general consists of doing more with less, by sometimes revising the foundations and in some cases bulldozing the foundations and rebuilding from the ground up.

Development of this toy model grew out of an initial desire to understand gravity on a quantum level as a function of quantum action and therefore of h-bar. I began with a couple of assumptions, one being that gravity, despite being presented as a result of spacetime curvature by general relativity and not due to a force, still involves an interaction of basic physical particles, where interaction necessarily implies a force, but in this case a stress force and not a body force.. The other was that fundamental particles are some form of more or less stationary waves consisting of oscillating strains of the spacetime fabric that are able to congregate at nuclear density without annihilating each other, though initially I had no idea what that form might be. With respect to such wave forms, h-bar represents the spin angular momentum of the waveform. It quickly became apparent that if the speed of light and h-bar are invariant, the quotient h-bar over the speed of light must be as well, and the resulting time independent inertial constant, п(tav), emerged as a helpful central analytical component of a rest mass quantum model in which particle phenomenology is seen to be based on an underlying wave ontology.

I decided to see according to Newton's gravitational law what the strength of that force might be between two neutrons, the more massive of the two condensed matter nucleons, in "contact" with each other, which to me meant using the reduced Compton wavelength of the neutron as the distance of separation. The result was something like the development in section 2a as quantified in (1.29). Though it initially lacked the differentials and general context as a function of a change in stress, it clearly pointed to the square of the neutron Compton wavelength as the value of what had to be a fundamental interaction. I was surprised when I could find no reference to this phenomenology online, and I went in search of an explanation. I was pointed to a string theorist, the first of many, who seemed mildly perplexed and told me to investigate the relationship of this finding to Planck's constant, h-bar, which I have done.

Along the way, I have made use of certain sources, some of which are listed below. As this is a model of quantum wave phenomena emerging from a classical continuum, chief among these has been <u>Physics of Waves</u>, by William Elmore and Mark Heald, both for the development of the Euler formalism and the complex wave and for their treatment of the stress and strain tensors. Eventually the physical nature of the fundamental rotational oscillation as shown in Diagrams 0 to 6 became apparent, with its relationship to spin, the magnetic moment, the capacitive-inductive nature of the inherent spin energy and eventually it became apparent that something must be energizing the fundamental oscillation. In other words, the inherent spin energy and action, charge, and magnetic moment of the neutron, proton, and electron could be modeled as emergent properties of a constrained wave and not input as free parameters or inferred to some deeper level of ontology of quarks and leptons.

As the accelerating expansion of the cosmos was well publicized, it seemed probable that this acceleration serves as a source; it also explained the predominance of matter over anti-matter, the latter being a feature of a contracting region of spacetime in this model. I have not devoted a lot of study on the nature of the electron, which in this model is derived from the fundamental oscillation of the neutron. In this regard it has never made sense to me how the necessary number of electrons and protons as required to establish charge conservation were generated in a big bang, when if they are both the product of beta decay, charge conservation is straightforward. At some point, from the continuity conditions of the model it became clear that beta decay must be related to the force of expansion operating on the periphery of a free neutron, and analysis showed that it was in fact tuned to a Hubble strain rate of around 73.08 km per Mpc per second, as I prefer to view it dimensionally.

I have used a form of dimensional analysis from the first in this model, which has no extraneous parameters and is not stochastic, and as a result I have been able to avoid some of the pitfalls that I believe ensnare some cosmological investigations. For one, from this perspective neither big bang high energy or quantum mass at the Planck scale has ever held much sway as a principle factor in understanding either the quantum or cosmic system. I don't think of quantum energy in conventional terms of Mev or Joules, but rather dimensionally as angular frequency, as angular wave number times the speed of light. Using the notion of an inertial constant within the context of a quantum wave mechanism, mass is simply wavenumber, ostensibly a time free parameter; by evoking time using the invariant speed of light, energy is simply frequency. In a natural system, for the fundamental oscillation they are both equal to 1. With respect to mass, aggregations of matter are then the summation of the number of particles in a group less any nuclear, atomic, or molecular binding energies that the aggregations liberate.

Applying the inertial constant with the basic breakdown of the Euler derivatives and integrals as stated in section 0a allows a compact way of viewing the dynamic properties of a fundamental quantum wave form. The current big bang notion of having a tremendous amount of energy emerge from a singularity to cool and condense into the form in which we now live requires the addition of a fairly large number of free parameters with no understanding of their ontology to my way of thinking; the ratios of neutron, proton, and electron mass, the underlying quark and leptonic structure, the basis of fermionic half spin and bosonic integer spin, the missing mass of beta decay, the value of the fine structure constant, Newton's gravitational constant. That is not to say that it cannot happen that way; the phenomenology of particles appears to be well known and their interactions statistically well determined, but the lack of understanding of a fundamental mechanism does not mean that the phenomenology does not accurately reflect a yet to be understood ontology.

To my mind this state of affairs offers a less than compelling logic for what we currently observe in the cosmos, which can be addressed with the notion of a non-singularity, i.e. three-dimensional manifold, finite source of initial maximum inertial density that, under an off-manifold stress from a fourth-dimension, expands and is spun up over time like a cosmic flywheel to eventually generate baryonic matter to form stars and condensed matter planets and beings like us who create models. This may be a singular event, or if we are accepting of the notion of conservation of energy (and power), a phase of a current cycle of expansion and contraction that is one of an endless series of such. I believe the possibility that most baryonic matter is generated from active galactic nuclei serving as inertial sources—black holes as sources rather than sinks—and not from a big bang singularity, albeit with a rapid decay of some neutrons to hydrogen in plasma form, with some as deuterium, and then helium and some lithium, deserves study.

While I have not researched the specific computational methodology used in the determinations, dark matter of the ΛCDM model is reported in a 2013 ESA Planck report as 25.8 +/- 0.4% of the total mass/energy content of the cosmos or roughly 5.35 times the baryonic (and leptonic) quantity at 4.82 +/- 0.5%. This content is needed to account for the observed failure of stellar rotational velocities about the galactic centers to decrease with distance from those hubs in accordance with Newtonian dynamics. The alternative is to account for this discrepancy with a modification of Newtonian dynamics such as MOND.

From the perspective of this wave model, the initial inertial density of spacetime is captive to the individual wave action of rotational oscillation, including the symmetric components responsible for quantum gravity as developed here, in the peripheries of generated baryonic matter. In this regard, the individual wave properties of quanta as quantified in the standard model, specifically the reduced Compton wavelengths of the baryons and the electron, are the limiting stress points—nodes, antinodes, capacitive, and inductive moments—of their wave action in this model; the physical waves themselves necessarily extend beyond these local parameters. A phenomenological fact is mentioned in that regard without further analysis. The fundamental rotational oscillation modeled herein for the neutron can be defined, as has been done elsewhere, as an extreme Kerr metric for a quantum black hole. The ratio of the volume within the surface of the ergosphere including the black hole and the volume within the event horizon only of such black hole is 5.52233. It is assumed that the density within the event horizon is 1 and that the density of the region between the ergosurface and the horizon approaches but does not equal 1, therefore the ratio of inertial density between the two will be less than 5.52233, compared to the Planck report above.

In addition, there is nothing in the development of this model that suggests that the spacetime strain on a local scale responsible for baryonic oscillation precludes the occurrence of large-scale torsional strain of the spacetime fabric responsible for galactic rotation. Thus, while the expansion stress as evidenced by the Hubble rate responsible for the oscillation of baryonic matter and registered as exponentially accelerating cosmic expansion in the large voids between galactic webbing, it does not appear to be operating within galactic environments other than in maintaining those oscillations in this model. Within those environments the density of the spatial substrate is maintained by the electromagnetic and gravitational interactions of quanta, while galactic rotational rigidity as evidenced by barred spirals and non-Newtonian dynamics appears to be bolstered by large scale torsional stress and strain. In short, the occurrence of what is dubbed dark matter is indicative of large-scale density differentials between the extra-galactic voids and galactic webbing, without the need to reference another type of particulate matter.

Finally, the customary cosmological constant value normally assigned to the field equation of general relativity and reported in the ESA Planck report of 2013 as corresponding to a dark energy density of 0.693+/-0.013 and revised downward slightly in 2015 to 0.6911+/-0.0062, as a percentage of mass/energy of the total, is at or suggestively close to the natural log of 2 at 0.693147..., indicating that it speaks to the fact that the Hubble rate, instead of being a first order velocity, is an exponential measure of expansion, that is, Hubble is a second order or accelerating rate, as developed above. When it is understood that such expansion is responsible not only for the apparent expansion of the cosmos as evidenced by red shift, but also for driving of all local particle action, as herein developed, the value of the cosmological constant within the context of the field equation of general relativity is seen in a different light. The natural log of 2 as in (1.62) gives a Hubble rate in today's seconds of 9.274 billion years.

The axioms on which this model is based are few and not unreasonable to my thinking grounded in classical wave mechanics, nor do any of their developed implications or application run in the face of observation, as far as I know. This model is offered because after twenty some years of study on this matter it continues to answer many questions concerning physical understanding that I have not been able to find addressed in the extant literature, and I believe if it gets a proper vetting it will be of benefit to the discussion and understanding of cold fusion. I am recently encouraged with regard to the developments of this model of rest mass and derived photonic energy as a function of cosmic accelerating expansion by the recent announcement of the results of the study of Riess et al with respect to a determination of the Hubble rate at 74.03 km +/-1.42 km $^{s-1}$ Mpc^{-1}. It is believed this analysis addresses the concerns in that study of the 4.4σ between their figure and the results of the ΛCDM Planck study at 67.74 km +/-0.46 km $^{s-1}$ Mpc^{-1}, while offering an acceptable alternative to ΛCDM.

In addition to the determination from this model of the Hubble rate, the derivation of the gravitational constant, the explication of the dynamics of rotational oscillation of baryonic matter, the ratios of fundamental condensed matter particle rest mass, the accounting for the missing mass of beta decay, the nature of elementary charge and spin, and the derivation of the fine structure constant offer reason for a thorough review of this model. All photonic energy, which in the standard particle and cosmological models is handled as a free parameter, is not actively addressed in this model as it is held to be predicated on beta decay as a function of electron/neutrino activity in keeping with the structures of quantum mechanics.

Bibliography, Citations, and Other Resources

Astronomy and Astrophysics 338, 856-862 (1998), "Magnetically supported tori in active galactic nuclei", Lovelace, Romanova, and Biermann.

Exploring Black Holes, Taylor and Wheeler, Addison Wesley Longman, Inc, New York, 2000.

The Extravagant Universe, Kirshner, Princeton University Press, Princeton, NJ, 2002.

The Feynman Lectures on Physics :Commemorative Issue, Feynman, Leighton, Sands, Volume I, Addison-Wesley Publishing Company, Inc., Reading, Massachusetts 1963.

Fundamentals of Physics, Fifth Edition, Halliday, Resnick, Walker, John Wiley & Sons, Inc. New York, 1997.

Gravitation, Misner, Thorne, and Wheeler, W.H. Freeman and Company, New York, 1973.

Mathematical Methods for Physicists, Fifth Edition, Arfken and Weber, Harcourt Academic Press, New York, 2001.

Physics of Waves, Elmore and Heald, Dover Publications, Inc., New York, 1985.

This was the primary source for wave, elasticity and tensor equations.

The Six Core Theories of Modern Physics, Stevens, The MIT Press, Cambridge, Massachusetts, 1995.

Three Roads to Quantum Gravity, Smolin, Basic Books, New York, 2001.

The Theoretical Minimum: What You Need to Know to Start Doing Physics, Susskind and Hrabovsky, Basic Books, New York, 2013

National Institute of Standards and Technology, These are the **2002 CODATA recommended values** of the fundamental physical constants, the latest CODATA values available. For additional information, including the bibliographic citation of the source article for the 1998 CODATA values, see P. J. Mohr and

B. N. Taylor, "The 2002 CODATA Recommended Values of the Fundamental Physical Constants, Web Version 4.0," available at physics.nist.gov/constants. This database was developed by J. Baker, M. Douma, and S. Kotochigova. (National Institute of Standards and Technology, Gaithersburg, MD 20899, 9 December 2003).

Table of Nuclides, Nuclear Data Evaluation Lab., Korea Atomic Energy Research Institute (c) 2000-2002, http://yoyo.cc.monash.edu.au/~simcam/ton/index.html

R.R.Kinsey, et al.,*The NUDAT/PCNUDAT Program for Nuclear Data*, paper submitted to the 9 th International Symposium of Capture-Gamma ray Spectroscopy and Related Topics, Budapest, Hungary, Octover 1996.Data extracted from NUDAT database (Jan. 14/1999)

Schwarzschild, Bertram, "Tiny Mirror Asymmetry in Electron Scattering Confirms the Inconstancy of the Weak Coupling Constant", Physics Today, September, 2005

Wapstra, A. H. and Bos, K., "The 1983 atomic-mass evaluation. I. Atomic mass table," Nucl. Phys. A 432, 1-54, 1985, quoted at http://hyperphysics.phy-astr.gsu.edu/hbase/nucene/nucbin2.html

Adam G. Riess, Stefano Casertano, Wenlong Yuan, Lucas M. Macri, and Dan Scolnic , "Large Magellanic Cloud Cepheid Standards Provide a 1% Foundation for the Determination of the Hubble Constant and Stronger Evidence for Physics Beyond ΛCDM", https://arxiv.org/abs/1903.07603

Wikipedia.com for various sources of general information.

Isotropic Expansion Stress on a Unit Space (IESUS)

An Addendum to Section 2g — Muon & Tau Families

of

A Condensed Matter Model of Particle Genesis (CMMPG) as a Function of an Accelerating Cosmic Spacetime Expansion

Fundamental Rest Mass Quanta as
Simple Harmonic Rotational Oscillations of
the Spacetime Continuum,
Driven by Cosmic Expansion
with
Application of the Analysis to the
Experimental Field of Cold Fusion

https://uniservent.org/pp01-condensed-matter-model-of-fundamental-particles/

Martin Gibson

In the latter part of July 2021, through the Twitter presence of Stacy McGaugh, @DudeDarkmatter, of the Department of Astronomy at Case Western Reserve University in Cleveland, Ohio, and his paper "Testing galaxy formation and dark matter with low surface brightness galaxies", I became aware online of the work of Robert A. Wilson from his blog website at https://robwilson1.wordpress.com. Dr. Wilson is Emeritus Professor of Pure Mathematics at the School of Mathematical Sciences, Queen Mary University of London. According to his website, his work has been in group and representation theories, with an interest in its application to the foundations of physics since 2007. That work appears to have intensified with his early retirement in 2016, and has led him to the conclusion that there are fundamental contradictions incorporated in the application of these theories to the physical modeling of general relativity and quantum theory that have prevented a unifying understanding of the physical phenomena on which the modeling is based.

In a monograph by Professor Wilson entitled A GROUP-THEORIST'S PERSPECTIVE ON SYMMETRY GROUPS IN PHYSICS, https://arxiv.org/pdf/2009.14613.pdf, page 18, equation (17) develops the following equation as exact within the standard uncertainty, in which it expresses the following relationship of the observed rest mass for fundamental particles of baryonic and leptonic matter, generally considered as being of invariant rest mass in the standard model of particles. Text communication with Professor Wilson stated that he came across the statement from other sources in his studies; based on the fact that the equation is confirmed by the CODATA 2018 stated values of particle mass as exact within standard uncertainties, we see no need for further academic study of the source of the equation, while expressing gratitude to whoever first noticed it to be the case. If the analysis extant in the downstream development of this equation has priority in a published form, I will remain grateful, though one must wonder why, if this is the case, it has not been given the attention it would logically appear to deserve.

We have inverted and restated the order of the equation from that text to facilitate a better understanding of the process of particle evolution as developed in the work from which this is a continuation, based on the insights this equation offers,

$$5m(n) = 3m(p) + m(e) + m(\tau) + m(\mu). \qquad (1.101)$$

Astrophysical Observation

Based on recent observation and related calculations of scientific sources, there are in the neighborhood of 10^{80} particles of baryonic matter in the known universe; that is, neutrons and protons with correlated electrons, bound or free, along with other rest mass particles. As far as this author is concerned, neutrinos are not rest mass particles for the simple reason that they do not rest. They constantly move at or near the speed of light unless their energy is reabsorbed in the process of interaction with a proton to produce a neutron. In a wave model, neutrinos can be understood as a type of discrete torsional compression wave resulting from the emission of a lepton. Thus, when an electron is emitted from a neutron which transforms into a proton in the process of beta decay, and the electron is stopped in its translational motion by interaction with another particle or field, the torsional compression wave of the medium continues on. The torsional or twisting nature of this compression wave, as with the torsion of rest mass particles from which they originate, are responsible for the property recognized as intrinsic spin in the standard model.

Based on the etymology of the word, a 'particle' is a "minute portion, piece, fragment, or amount" of some greater qualitative *thing*, often invariantly quantized in quantum physics as a discrete value, be it wavelength or its inverse wavenumber or as a proxy for wavenumber, mass, according to a variety of conditions. The standard model of particles as currently presented assumes the thermal environment of a big bang as the condition required for the quantization of various particle fields from their various field sources or perhaps a single source, followed by the secondary composite construction of baryons and eventually as composites of the baryons, the elements. In the logic of the wave model presented here, a thermal environment is a measure of the interactions of particles already created and moving at high velocity; it is not a condition of their quantization, their creation as particles. Instead, the quantization of particles as waves are conditioned by two constraints, the finite inertial density of the wave bearing medium in and of which the waves are quantized and a finite stress, being the function of an expansion stress force operating on the medium and gauged by the cross sectional area on which the force is operating over a range of angles from normal to tangential to that area.

From this perspective, there are only three stable, condensed matter particles in existence in the universe, protons, electrons, and while they remain in nuclear congregation in an atom under certain conditions, the neutron. Absent that nuclear congregation, free neutrons decay into a proton, electron, neutrino, and apparently give up some mass in the process. Understanding the nature of this small amount of 'missing particle mass' is the objective of this analysis.

The universe is very large. According to some cosmological modeling, it would take an observer who did not age or die almost as long as the universe has existed, traveling at the speed of light, to get as far as anyone can see with the strongest telescopes. Yet then again, the universe may be infinite in both extent and age, or better stated, limitless and ageless.

Baryons, on the other hand, are very small. If they were all of neutron size—using the Compton angular wavelength at 2.1×10^{-16} meters as a gauge of particle radius, disregarding any charge, spin or other dynamics—and were laid out in a Euclidean line next to each other, there would be a few more than 10^{15} of them lined up in a meter. This means there would be a bit more than 10^{46} of them in a cubic meter if they were packed to maximum density, after the form of an ideal cuboctahedral lattice. Oddly enough, with a radius of 1.5×10^{11} meters, based on a volume of approximately 10^{34} cubic meters, the 10^{80} known baryons would just fit inside a sphere with a circumference delineated by the earth's orbit around the sun. We might think of such a sphere as a primordial neutron star of earth orbit size.

In such case, the rest of the universe would be a vast empty space, devoid of matter or light. There are said to be 2×10^{12} or two trillion galaxies in the observed universe. If the 10^{80} baryons were evenly distributed among them, this would place a sphere of 10,000-kilometer radius packed with neutrons at the center of each galaxy. Again, if evenly distributed, with an estimate of 10^{22} stars in the universe, this means approximately 10^9 or one billion stars per galaxy, which once again, if all the packed baryons were in the stars instead of the galactic centers, the individual packed stellar cores would have a radius of 10 kilometers.

Obviously, some stars and some galaxies have more mass than these figures and perhaps many more have far less. For our sun's mass, the figure would be about 1.5 kilometers.

This total mass of 10^{80} baryons are anything but evenly distributed, not in the stars, not in the galaxies or galactic groups, and not throughout the extent of the cosmos itself. Virtually all the galactic matter is observed in filaments along what appear to be strands and diaphanous sheets of light transferring matter, connected perhaps by gravity, perhaps by inertia, perhaps by electromagnetic forces. This is often deemed to be supported by a form of dark matter that responds to gravity but does not emit light, separated by large regions apparently devoid of matter or of any energy other than whatever electromagnetic energy transits through these voids, thereby allowing the observation of this distribution.

According to the modeling of general relativity, the threshold size of the event horizon as a radius of an extreme Kerr black hole (KBH) is 2.9 kilometers for a star of 2 solar masses or 3.93×10^{30} kilograms. Conventionally in general relativity, the particles which are inside such black hole are imagined as stretching toward a single point at the center of the KBH.

2.9 kilometers is coincidentally also the radius of the same weight as 3.93×10^{30} kilograms or 2.3×10^{57} neutrons if they were packed as indicated at maximum spherical packing. If this mass of neutrons were either oscillating individually in such a manner that they did not or could not decay and give off electrons or photons or was condensed to a uniform continuous density equal to that of the stated neutron spherical packing, this aggregate sphere would give off no light or other electromagnetic radiation and would therefore be indistinguishable from a conventionally modeled KBH, where the nature of an extreme Kerr Black Hole means that it is spinning at the event horizon and the surface of our sphere, at the speed of light, in this current model either as individual neutrons or as an aggregate mass.

The same simple formula used to calculate the threshold event horizon of a black hole can be used on masses of the same maximum density but lesser quantity such as the mass of the earth, and we will find that the calculated event horizon will be smaller than the actual maximum density.

$$R_{BHH} = \frac{G}{c^2} M \qquad (1.102)$$

Here R_{BHH} is the radius of the black hole horizon, G is Newton's gravitational constant, c is the speed of light and M is the mass of a celestial body. Thus, the calculated black hole radius, R_{BHH}, is linearly related to M, all other parameters being invariant. (Some sources use a coefficient of 2 for the terms on the right in calculating R_{BHH}.)

Using this calculation, the radius of the earth mass at maximum density is 36 meters; that is if all the mass of the earth was collapsed to maximum density, it would fit into a sphere 72 meters in diameter. The calculations for a KBH event horizon based on the earth's mass is 4 millimeters, considerably smaller, and apparently an indicator of the fact that the earth's mass will not collapse into a black hole.

According to the above equation, with an estimated mass for the observed universe of 1.76×10^{53} kilograms, the radius of the black hole horizon is 1.24×10^{26} meters which is 13.1 billion light years, which would indicate that the entire universe is within a black hole event horizon. It is worth noting that perhaps either due to serendipity or some unrecognized causative or computational methodology, even distribution of this quantity of baryons amounts to a density of very closely to 1 baryon per meter. From the above back of the envelope calculations, the average stellar mass is over three times the mass of a threshold black hole.

In the cosmology of general relativity, black holes are universally thought of as being gravitational sinks leading to gravitational collapse, i.e. to a singularity, yet there is nothing in these calculations to suggest that, beyond the scale of two solar masses, everything else is not already within the event horizon of a larger black hole. Clearly something is missing from this thinking.

In the absence in the standard model or general relativity of an understanding of a quantum generation of gravity, the formation of galaxies and of baryons themselves remains an unaddressed mystery. According to current thinking, both the largest and the smallest of material phenomena is modeled as emerging from a hot big bang singularity, apparently for the single simple reason that if the Hubble rate is a velocity measure of isotropic expansion, if we trace this velocity back 13 plus billion years, all 10^{80} baryons must have emerged at the same time from a single point, a point necessarily outside any current understanding of time or space.

But spatial expansion appears to be accelerating, which means among other things that a scenario exists in which the expansion is understood to be logarithmic, which if gauged by the current Hubble rate as an acceleration, indicates that the expanding extent doubles every 9 plus billion years and is much older than generally conceived. Another and different scenario exists in which the Hubble rate is a registration of the force—a dynamic acceleration—of beta decay, with a redshift from energy loss based on the distance from which it is observed, which is in turn interpreted as an expansion of the cosmos.

Perhaps the greatest boost in support of the notion of a big bang is the correlation of nucleosynthesis of the fundamental elements—of hydrogen–protium, deuterium, and tritium, of helium–helium 3 and helium 4, and of lithium 7—all of which is theoretically ascribed to the intense "heat" or high energy of electrons and quarks, the latter of which were created in a theorist's mind to explain the magical condensation of 3 of the right type in the right mix into a neutron or a proton, the latter with just the right energy to bond with an electron. We have shown earlier in the tract for which this one is an addendum, another wholistic explanation for the generation of the three mostly stable particles, explaining the quark phenomena, quantum gravity, and a few other quandaries in the process.

Many unquestioned assumptions go into making the fruit cake that is the standard model as well as those that get whipped into the hard sauce topping that is general relativity—not to mention the couple of dozen free parameters for the ingredients. One of these is Newton's gravitational constant, which is obviously by its function a type of force differential, but a force differential with respect to what? We have shown that it is a quantum force differential with respect to differential stress, which includes in its composition a statement of a fundamental gauge for length and mass. We have shown that this gauge can be found in an inertial constant which is equal to Planck's constant, h-bar, over the speed of light.

From (1.101) we can now get an understanding of the nature of the missing mass of beta decay in the relationship between the gravitational differential force, the strong force as the fundamental baryon wave force, and the electroweak interaction. We will get an understanding that what is generally thought of as a black hole in general relativity is a field of maximum, not infinite, inertial density that can be an inertial source as well as an inertial sink for the extended field of mass–energy potential, in which it is seen that the mass–energy equivalence of Einstein's famous equation is a reduced form of a simple generic wave equation. We will get an understanding that the basic structure of the cosmos proceeds not from a singularity but rather from a quiescent uniform condition of maximum density as indicated above, albeit one with an inherent gauged cuboctahedral lattice potential that separates first into the cosmic filaments and membranes as areas of maximum baryonic wave bearing density and rarefies in the volumes of vast voids that provides tension stress on the filaments and membranes at galactic nodes to produce active galactic and perhaps stellar nuclei and the fundamental light elements just indicated above.

We will look next at the particle structures that result from this interaction between the voids—as they expand and move forward in space—and the galactic centers of maximum neutron density. To those who might point to the introduction of a prime mover in this scenario responsible for the cosmic expansion of the voids as a hand of God, we will only point out that it is no different in quality and far more explicative for the observed cosmic state of affairs than the prime mover that started the modeled big bang.

In free space, over a life of 14 to 15 minutes, a neutron(n) decays through a process, generally known as beta decay, into a proton(p) and an electron(e) both of which are stable in free or condensed matter space as a hydrogen atom, specifically as a protium atom which distinguishes it from the heavier atoms of hydrogen with one or two neutrons in the nucleus as deuterium and tritium respectively. In the high temperature of plasmic space, the stability of proton and electron as discrete particles remains, however in this case as non-binding ionized cations and anions respectively. In contrast, a tau(τ) decays over 300 femtoseconds along an assorted branching, roughly 2/3 of the time into an assortment of mesons, which is a hadron similar to a baryon, 1/6 of the time into a muon, and 1/6 of the time into an electron, with other attendant energetic interactions and particles in the form of neutrinos. A muon(μ) decays over an average of 2 microseconds into an electron, also with the other energetic interactions. The neutrinos are generally considered to be stable as well, but that is harder to qualify or quantify as they are generally considered to travel at very close to if not at the speed of light and in a manner such that they rarely interact with other particles other than in the case of reverse beta decay, generally denoted as inverse beta decay.

It is noted that baryonic matter along with mesons are modeled as being hadronic, that is comprised of an internal structure deemed to be more fundamental than the hadrons themselves. That internal structure is comprised of the more fundamental quarks, necessarily constrained to the particle boundary by a process known as asymptotic freedom, which basically means that the more distant the quarks in a hadron are from each other, the stronger they are attracted to each other as if they were bound together by unbreakable rubber bands which increases in intensity with strain, and the less free they are to move independently with respect to each other. In the case of the generally stable baryons, there are three quarks producing the structure, while in the case of the mesons, which are extremely short lived—much less than a second—there are only two quarks, as a particle and its anti-particle. The leptonic rest mass particles, on the other hand are deemed to be free of any internal structure as comprised by quarks, and are the stable electron, and the much shorter lived, generally transitional, tau and muon. A more concise and understandable treatment of baryonic and leptonic nature is recapitulated in part from the analysis of CMMPG later. It is recommended that the reader who really wants to understand what is going on in regard to the fundamental particles of physics read the analysis both there and in the some of the other treatments of this material on this website. The reader simply cannot understand the subject of quantum rest mass unless they have digested this material. Fortunately, that material is graphically well supported and requires only an intermediate knowledge of classical wave mechanics with the attendant algebra, calculus, and topology.

Development of the Tau–Muon Duplet to Deuterium Interaction Path

The rest mass values used in Wilson are from CODATA 2014 expressed in MeV/c², which can be understood with equal validity as a form of spin energy since $E = mc^2$. We have added an additional column of the CODATA 2018 figures to indicate that there is no overall change other than perhaps a different calibration behind the calculations, once again validating Wilson's calculation:

$$
\begin{array}{lll}
& \text{CODATA 2014} & \text{CODATA 2018} \\
m(e) = & 0.510\,998\,9461(31) = & 0.510\,998\,950\,00(15) \\
m(\mu) = & 105.658\,3745(24) & = 105.658\,3755(23) \\
m(p) = & 938.272\,0813(58) & = 938.272\,088\,16(29) \\
m(n) = & 939.565\,4133(58) & = 939.565\,420\,52(54) \\
\textit{calculated:} & & \\
m(\tau) = & 1776.84145(3) & = 1776.84146(4) \\
\textit{CODATA} & & \\
m(\tau) = & 1776.82(16) & = 1776.86(12) \\
\end{array}
\quad (1.103)
$$

Wilson has calculated the value of τ, above, from the remaining experimentally determined values, since it is known experimentally with less certainty at 1776.86(12), within the standard uncertainty. The virtual identity of the calculated value for τ from 2014 to 2018 indicates a constraint in the overall precision of the value determinations.

This is followed by mass converted to modular ratio values by dividing all CODATA 2014 amounts above by that of each $m(e)$, $m(\mu)$, $m(p)$, $m(n)$, and $m(\tau)$ to arrive at the following modular values.

	$*/e$	$*/\mu$	$*/p$	$*/n$	$*/\tau$
$m(e) =$ 1.0	= 0.004836334	= 0.000544617	= 0.000543867	= 0.000287588
$m(\mu) =$ 206.76828	= 1.0	= 0.112609486	= 0.112454476	= 0.059464131
$m(p) =$ 1836.152674	= 8.88024657	= 1.0	= 0.998623472	= 0.528056144
$m(n) =$ 1838.683662	= 8.89248733	= 1.001378426	= 1.0	= 0.528784030
$m(\tau) =$ 3477.1920	= 16.81686062	= 1.893738026	= 1.891131242	= 1.0

(1.104)

While it is understood that the various particles are discrete and invariant in their various properties as in this case of rest mass-spin energy, a comparison of these modular arrangements offers nothing to suggest that the five particles are comprised of much smaller quantum packages of some discrete invariant size of mass-energy as opposed to being constituted as a characteristic fundamental wavelength or frequency from a wave bearing continuum of variable inertial density. The fact that the five quantities of (1.103) terminate with various statements of standard uncertainty is logically unsupportive of the notion that each particle is comprised of a set quantity of known discrete units; neither does it negate that possibility. They may or they may not be so comprised while subject to the precision of measuring devices, or they may vary continuously within a range of finite extremes as given by the standard uncertainties, all while subject to the same precision of measurement. On the other hand, if discreteness indicates that the particles are essentially the nature of a wave, it gives no indication one way or the other whether the characteristic wave forms are comprised of a much smaller particulate or of a continuous substrate. We will treat them classically as non-particulate if only as an indication that the field of their observed interactions is continuously differentiable. The uncertainty formalism of (1.103) has not been extended into the modular constructions of (1.104) or latter.

This last paragraph indicates that qualitative, non-stochastic constraints exists that are responsible for the various modular distributions, especially as shown with the effective percentages in the $*/n$ and $*/\tau$ columns and the fact that the particles greater than e are not comprised of discrete units of e as shown in the column $*/e$. Such constraints can be understood analytically as discrete in terms of geometry and wave mechanics and the mathematical fundamentals applied to each of these two disciplines.

We parse these parameters of (1.101) over three lines, for reasons that should become clear, where the explicit quality being evaluated above as masses from the first column, $*/e$, of (1.104) is implicit in the following particle designations.

$$5n = \begin{Bmatrix} n \\ 2n \\ 2n \end{Bmatrix} = \begin{Bmatrix} p+e \\ 2p \\ \tau+\mu \end{Bmatrix} = 3p + e + \tau + \mu \qquad (1.105)$$

Assuming conservation of mass–energy in a physical system involving the interactions of the five fundamental particles, the missing mass in the inequalities below of each line is placed in square brackets, which we will think of as different types or instances of transformed or transformational energy or in some cases perhaps as operational catalysts.

$$n > p + e \rightarrow \Delta m = [+1.5310]$$
$$2n > 2p \rightarrow 2\Delta m + 2e = [+5.0620]$$
$$2n < \mu + \tau \rightarrow 3\Delta m + 2e = [-6.5930]$$
(1.106)

From equation (1.57) of CMMPG, the first line of this parsing to include the missing mass of beta decay, Δm, is restated as

$$n = p + e + [\Delta m]$$
(1.107)

Though the term 'decay' and 'decay path' shows up often in this treatment, it should be understood that the notion is in some sense unintentionally pejorative even for the expert in that it implies, based on common usage, a decline in some state or condition from a prior pristine or ideal status, when in fact it is simply an indication of interaction between particles or between particles and their fields. From our dimensional analysis above, using the proportional modular values of $*/e$ we have

$$1838.6836_n = 1836.1526_p + 1.0_e + [1.5310_{\Delta m}]$$
(1.108)

The values for missing mass, Δm, are computed as required to satisfy this equation and is the same for all three of the parsings, based on the comment in (1.112) below. The quantity shown is empirically based and not derived from any deeper analysis of the current standard model.

The second line of (1.106) is a straightforward conclusion of the first line of the parsing from (1.101)

$$2n = 2p + [2(e + \Delta m)]$$
(1.109)

For the two protons of the second line of (1.106) we have

$$2(1838.6836_n) = 2(1836.1526_p) + [2(1.0_e + 1.5310_{\Delta m})]$$
$$2(1838.6836_n) = 2(1836.1526_p) + [2(2.5310_{e+\Delta m})]$$
$$3677.3672_{2n} = 3672.3052_{2p} + [5.0620_{2e+2\Delta m}]$$
(1.110)

Since (1.101) is empirically determined to be exact within the standard uncertainty, the 3 instances of missing mass and 2 electron rest mass in the first two parsings must be equal, but of opposite sense, to the additional mass of the tau and muon as required to balance the equation of the third line with those two neutrons.

Finally, for the third line we have the following, were we transpose the missing masses in the final line

$$2(1838.6836_n) = 3477.1920_\tau + 206.7682_\mu - [2(1.0_e) + 3(1.5310_{\Delta m})]$$
$$2(1838.6836_n) = 3477.1920_\tau + 206.7682_\mu - [2(2.5310_{e+\Delta m}) + (1.5310_{\Delta m})]$$
$$3677.3672_{2n} = 3683.9602_{\tau+\mu} - [6.5930_{2e+3\Delta m}]$$
(1.111)

$$\therefore$$

$$3677.3672_{2n} + [6.5930_{2e+3\Delta m}] = 3683.9602_{\tau+\mu}$$

In classical wave mechanics, for simplicity using the model of an ideal string, the 'mass' of a wave is a measure of the linear inertial density of the string when subjected to a transverse force operating to displace a portion of that string. The more massive the string, given some standard unit of force, the smaller will be the section of the string that can be displaced, represented by a wavelength and inversely by a wavenumber. That same linear density measure of mass, if gauged by the same standard unit of force, will apply whether the string is at rest or has been set in motion by that force, subject to whatever differential conditions are present in that motion to change that density. Therefore, the mass value of 3677.3672_{2n} is present in (1.111) in the initial conditions of neutron density–black hole source material whether it is found in wave form or in the wave substrate.

There is little reason to accept the apparent empirical validity of the value of the Δm without at least some attempt at analysis of the cause for and the structure of the value of 1.5310 or the ratios of the various fundamental particle mass, and we will offer two such analyses which are related to one another. The first of these is found as a mathematical analysis in CMMPG on page 36 at '2e – The Missing Mass of Beta Decay'. The geometric analysis is included in partial form below in this addendum, IESUS, but first the mathematical basis.

The related mathematical analysis establishes a two-dimensional natural exponential component of a three-dimensional differential form of the natural log. Both of these can be found in greater detail in my work on this website, https://uniservent.org/pp08-4-wave-foundation-version-2-2/, with the mathematical analysis in the Appendix D – Exponentiation. The geometric analysis can be found there starting on page 41 and is reproduced in large part later in this addendum. The quantification of 2.531… stated here includes the sum of the linear, 1, and transverse, 1.531…, components, indicated by e and Δm respectively, in both the mathematical analytical results and the empirical observation and establishes a verification and torsional understanding for the nature of the 'missing' mass, so that Δm is in all cases in this wave analysis understood as the twisting portion of the energy embodied in the wave node at a point of wave transmission, customarily thought of as a particle decay.

$$\begin{aligned}
&\text{Mathematical analysis of the natural logs:} &&2.531584394 \\
& &&+0.000596696 \quad \Delta 3.25e^{-7} \\
&\text{Empirical observation:} &&2.5310987698 \\
& &&-0.003449702 \quad \Delta 1.288e^{-6} \\
&\text{Geometric analysis of expanding spacetime:} &&2.527550298
\end{aligned} \quad (1.112)$$

Relative uncertainties are gauged with respect to the neutron rest mass at 1838.683662.

Of interest is the fact that although the totals on each side are exact in (1.101), each parsed line is off by a relevant, and obviously quantized amount, generally referred to as the missing mass attributed to beta decay on the first line and for similar reasons the missing mass and the electron mass for both protons on the second line. In order to balance this missing mass which has been _added_ to the first two lines, we will need to _subtract_ the same amounts from the tau and muon on the third line in the following or transfer it as an addition to the left hand side of (3) to arrive at (3b).

$$\begin{aligned}
&1) \; n = p + e \quad + \left[\Delta m\right] \\
&2) \; 2n = 2p \quad + \left[2e + 2\Delta m\right] \\
&3) \; 2n = \mu + \tau \quad - \left[2e + 3\Delta m\right] \\
&3b) \; 2n + \left[2e + 3\Delta m\right] = \mu + \tau
\end{aligned} \quad (1.113)$$

This third line raises questions about the nature of the entire assortment of quantum particles in the standard model of particles and their interactions. The first line is well recognized as a statement of beta decay, with the second line a statement of two instances of such decay in which the protons are ionized and the electrons along with the differential mass are missing. Both indicate the decay of unstable, perhaps free, neutrons into stable protons and electrons and as parsed appear to indicate a spontaneous process of the release of energy from the nucleon, n, as $e + \Delta m$, with transformation to become p. As indicated in section '2d — Derivation of Beta Decay as a Function of the Hubble Rate', $e + \Delta m$ is anything but spontaneous in the global sense, and represents in tandem a coupling/uncoupling constant in the relationship between the neutron and the protium atom and thereby the rest of composite elemental and molecular matter.

The electron, of either charge, e^- or e^+, are understood as having no internal structure in contrast to the quark structure of the neutron and proton, but what that means from this modeling, CMMPG & IESUS, is that the interaction of the inductive and capacitive torques on the nodal structures of what are essentially the same fundamental, discrete 3 dimensional wave structures of all forms of rest mass quanta behave differently under different continuity conditions associated with the Hubble stress and strain to explain the distinctions we empirically find in baryons and leptons, either as matter and anti-matter, according to the following chart.

Study of these charts and the spin diagrams from which they were taken, indicates the torsional characteristics of the process of particle wave dynamics including the twisting transformation generally referred to as beta decay. This is generally modeled as occurring in the confines of an atomic nucleus, or perhaps for a neutron in free space, but we can also view it as fundamental to the genesis of rest mass particles as part of a process of nucleo-synthesis at the event horizon of a black hole source, a process in reverse to that customarily modeled at the event horizon of an extreme Kerr black hole in general relativity.

In general, black holes are customarily treated as gravitational sinks, but in the absence of an understanding of the nature of quantum gravity, this is an unwarranted assumption of general relativity, which has led to the concept of all matter emerging from an inertial source modeled as a big bang singularity. I will not go into the various reasons that this has never made logical sense to me, but I can say that based on the notion of conservation of energy and other properties, the laws of physics have always been treated as being capable of interpretation with the arrow of time in reverse. As such, the black hole dynamics attributed to gravitational attraction as an inertial sink within the context of an inert background space or spacetime are equally valid if the active source of dynamism is an isotropically expanding space or spacetime against an otherwise inertial source or sources such as active galactic nuclei (AGN) of black hole or neutron density at maximum packing, as a uniform continuum or a gauged lattice potential, which would be my preference.

From this perspective, the third line of (1.113) indicates a more complex process for a series of different reasons. First, the tau and muon are even more unstable and extremely short-lived particles than the neutrons which decay quickly in free space, though they are capable of indefinite stability in suitable congregation with other nucleons. As the rearrangement of the missing mass and e indicates in (1.114), as a reverse decay process, the bracketed mass–energy as an expansion stress must be added to or interact with the density of 2 neutrons on the left in order to produce, apparently in simultaneous manner, the heavier tau and muon, as indicated here, where the e's at each end represent expansion stress sandwiching the $3\Delta m$ on either side of the developing wave forms to produce the tau and muon

$$[e+\Delta m]+n+[\Delta m]+n+[\Delta m+e]=\tau+\mu, \quad (1.114)$$

before quickly decaying into other particles, so that

$$\tau+\mu-[2e+3\Delta m]=? \quad (1.115)$$

If the mass of the various particles is invariant, other than at the time and conditions they are involved in a decay process, these last mathematical statements suggest that the tau and muon are necessarily produced together, as a couplet or duplet, by an energy transformation process, attributed to missing mass, interacting in a manner to redistribute the two mass equivalents of the neutrons on the left to the two transitional states on the right. In addition to (1.114) we can also combine parsing 2 and 3 of (1.113) to get

$$2p + [4e + 5\Delta m] = \tau + \mu, \text{ or}$$
$$n + p + [3e + 4\Delta m] = \tau + \mu$$

(1.116)

→ φ Y-Z	← θ X-W/τ	ε ↙	μ ↘
Oscillating restorative torsional wave force from initial torsional displacement	Resultant rotational spin force from restorative wave oscillation	Capacitive torque from moment of maximum restorative power rotating with spin	Inductive torque from moment of maximum restorative action rotating with spin
Spin Diagrams The diagrams from which these torque charts are taken demonstrate these five conditions in detail as found in the main paper.	**Neutron** Neutral Baryon Resonant Mode Capacitive & Inductive torques operating at maximum reinforce the nodes and antinodes of the φ restorative torsion and θ rotational spin which under increasing Hubble stress results in beta decay		**Beta Decay** Hubble stress drives the electromagnetic force as $e + \Delta m$ decouples the energy of the neutron with a spin flip from an inductive moment advance to emit the neutrino-electron and reduce the spin energy of the neutron to that of a proton
Proton Charged Baryon Positive Inductive Mode With beta decay, μ advances induction over capacitance as positive charge and reinforces restorative force while ε maintains capacitance through spin		**Electron** Charged Lepton Negative Inductive Mode All μ torques reinforce restorative and spin forces and retard capacitive torques ε at all nodes/antinodes leading to the conclusion that leptons lack internal structure	
Anti-Proton Charged Baryon Negative Capacitive Mode With anti-matter, μ retards restorative force while ε retards spin, which explains why anti-matter is not stable		**Positron** Charged Lepton Positive Capacitive Mode All ε & μ torques retard and mitigate the restorative and spin forces, explaining again why anti-matter is not stable under expansion stress.	

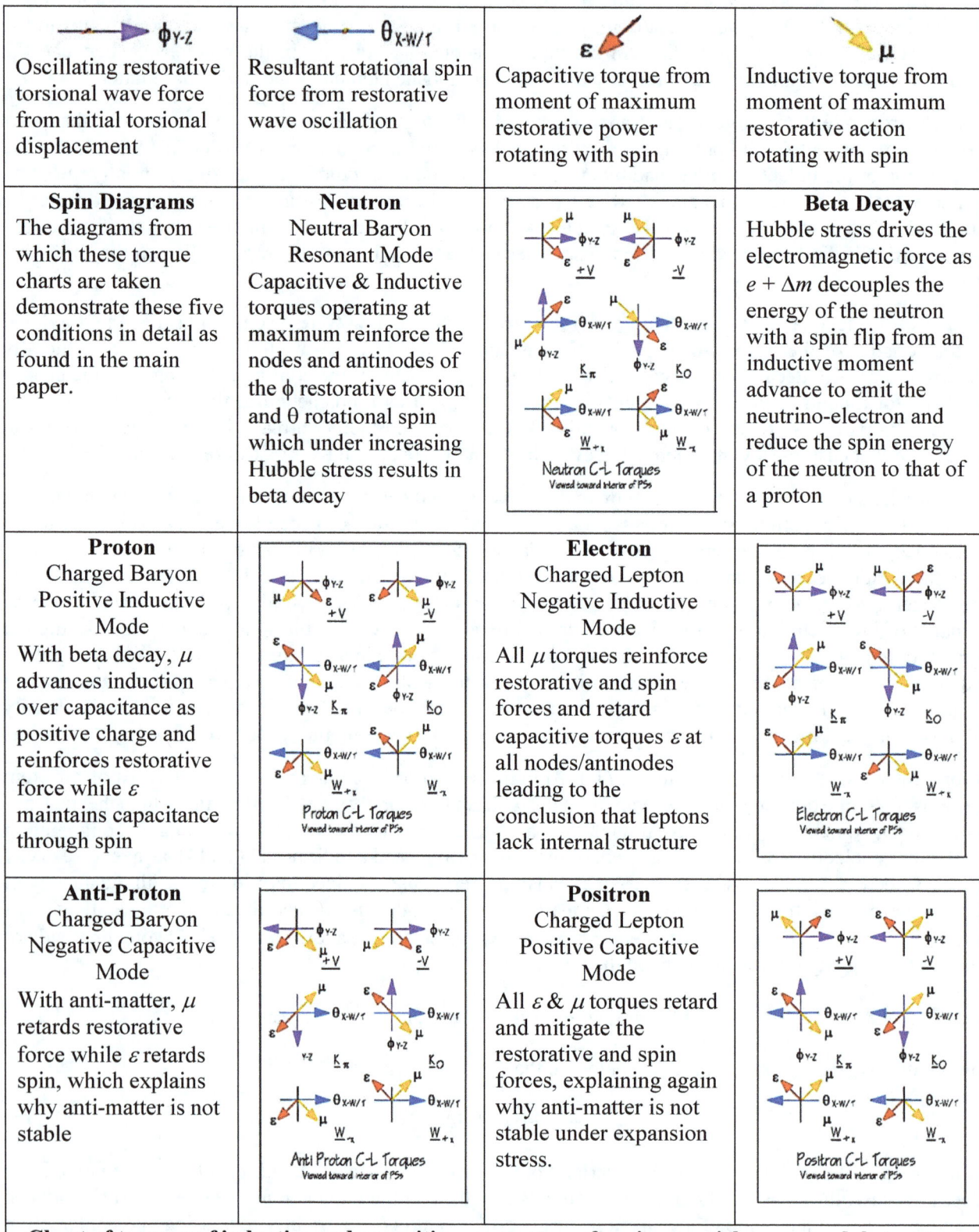

Chart of torques of inductive and capacitive moments on fermion particle-wave nodal structure

These last two equations are representative of particle interactions in the upper atmosphere in the collision of relativistic protons as cosmic rays in collision with protons and neutrons in atomic nuclei. Various other baryonic forms, all heavier than the neutron with spin 1/2, range from the lambda, λ, at 1.17865, to the sigma, Σ, 1.26001, to the xi minus, Ξ^-, at 1.39970 with respect to a neutron modular mass of 1.0. All of these are unstable with a lifetime in the sub-nanosecond range, which indicates that they do not lose their internal structure—stated in terms of their quark structure by the standard model and in terms of the stabilized nodes of rotational oscillation of this model—and revert to either a proton or neutron form at the end of that lifetime by emitting energy as a pion with decay.

The pions then decay as gamma radiation or through the weak interaction, part of the $e + \Delta m$, to produce the muon and then again to produce the electron. Incidentally, in the following geometric analysis, we will see a suggestion of the weak force in action with respect to the mass of the xi minus, Ξ^-, at 1.39566 and the well-known weak mixing angle of weak force decay given as

$$\sin^2\left(\tfrac{1}{2}\sqrt{\tfrac{\pi}{3}}\right) = (0.489628254)^2 = 0.239735827 \tag{1.117}$$

For a sphere of radius = 1, the surface area of the sphere is 4π. If we portion that surface area over each of the twelve cubic edges circumscribed by that sphere, each one centered on the mid-point of each edge, dividing by twelve we have a surface area over each edge of $\pi/3$. The arc length of each of the four sides of each of the twelve surface areas measured along each great circle is $\sqrt{\tfrac{\pi}{3}}$, so that one half of the arc length is $\tfrac{1}{2}\sqrt{\tfrac{\pi}{3}} = 0.511663354$.

We will look at other options in a few minutes. If mass is conceptually thought of as a type of energy producing power as a substance of energy density, we might think that more massive particles come in larger sized packages. In fact, the inverse is the case, and the size of one of these fundamental particles, as designated by their Compton angular wavelength, is inversely related to the rest mass-energy as measured by an angular frequency of the same particle. As a result we can think of the modular mass measure of (1.104) as a measure of frequency, as in the first line of (1.111).

The energy of the tau and muon, therefore, can be understood as the product of an interaction that starts with the energy density of two neutrons. While the empirical literature is deep in the modeling of various decay paths from the baryons to the muon on the way to the generation of an electron, the information is shallow with respect to explanations of the appearance of the tau, conventionally modeled as the interaction of the energy of relativistic electrons and positrons.

Apparently, it is this conventional modeling that has resulted in the designation of the tau as a lepton, despite the fact that roughly 2/3 of the decays are hadronic and decay weakly into an assortment of pions, which are mesons. In the above torque charts and related information, the rotating restorative nodes of the torsion oscillation of ϕ correspond with the mesons, comprised of a quark and an anti-quark pair in the standard model. In our wave modeling of CMMPG, following the pattern of the neutron, the nodes and antinodes of torsional oscillation are orthogonally superimposed to produce rotational oscillation and ½ spin, but with weak decay the superposition is broken, resulting in the realignment of the μ and ε torques associated with either hadronic or leptonic form of the oscillation shown in the above chart.

The question of whether the tau can or should be considered a hadron, perhaps even a baryon prior to weak decay, based on (1.115), is of less interest than whether the muon and tau are created as a pair under certain conditions that give physical meaning to the exactness of (1.114) and the fact, if it represents a transformation process indicative of a deeper symmetry, in which baryonic number is conserved, a tenet of the standard model.

The notion that a proton, as a baryon comprised of 3 quarks and positive charge, and an electron, as a lepton without any internal structure but with equal negative charge, should come together in apparently equal numbers in the wake of their emergence from a big bang singularity, along with the necessary missing mass, in the absence of any deeper understanding, has never been logically compelling. The same logical challenge applies to the concept of a tau and a muon, both leptons of largely disparate energies with no internal structure in the form of the necessary quarks, coming together with a release of identical missing coupling energies to produce the six quarks of the 2 baryons of the standard model, once again in the absence of a compelling mechanism. This does not preclude their independent further decay.

A geometric analysis involves the nature of the mechanical work done by an isotropic stress differential, df_n, in a 3 and 4-dimensional manifold on a unit space as a cube and hypercube from an initial condition of a uniformly continuous inertial field with a property of an emergent lattice potential as gauged by the inertial constant, ת, with initiation of a differential stress. This analysis establishes certain inflection points that are reached as the stress increases or decreases on the unit, or alternatively as that unit expands or contracts so that transverse stress or concomitant strain culminating in rotation is induced about such unit as an emergent phenomena. In the absence of such transverse inflection components, any stress induced strain would be an exclusively divergent and therefore curl free expansion that would amount to a scaling phenomena.

For modeling purposes, the energy embodied in a physical object is proportional to, and in some cases equal to, the work performed in moving that object or portion thereof over a distance a. This includes a force applied in stretching, compressing, expanding, bending, or twisting a portion of some object in such a manner that the portion recoils or redirects its motion as a transverse strain and/or kinetic energy when the direction of the applied force changes. For an object of uniform density, assuming a change in that density due to a change in volume in some portion of the object, the energy change can be understood to be proportionally directed according to the volume change.

As an example, we might consider a unit cube-shaped balloon with an indefinitely flexible surface that is uniformly expanding over time—that is changing in size without changing its geometric definition as a cube—due to an isotropic expansion stress at the three boundary components of the 6 (S) surfaces, 12 (E) edges, and 8 (C) corners of the cube. The expansion stress might be envisioned as inflating the cube by pumping the interior of the cube with fluid or gas, or alternately by evacuating the area uniformly around the cube. At the beginning of the expansion we can consider that it takes more energy to move the 6 faces any differential distance than it takes to move the 12 edges that same differential distance in two directions, and the 8 corners, that same distance in three directions. The work–energy done in each case is a direct function of the volume increase achieved by filling in the displaced components of the cube as they expand.

The details of this analysis are found below at 'The Effect of Isotropic Expansion Stress on a Unit Space'. We are giving an overview of the 3-dimensional unit cube here. The 3D analysis establishes six points of inflection defined by the points at which the ratios of the increases in differential stress and/or volumes of the different components as defined by the stress or displacement, a, are unity. These six ratios are

	Component Predominance	Component Ratio = 1	Differential $dx = a$	Inflection Result
1	E+C over S	$\frac{S}{E+C}$	$0.39564...,\ -1.89564...$	Torsional symmetry break
2	E over S	$\frac{S}{E}$	0.5	Oscillation potential about S
3	C over E	$\frac{E}{C}$	$0.66666...$	Oscillation potential about S
4	C over S	$\frac{S}{C}$	$\pm\frac{\sqrt{3}}{2} = \pm 0.86602...$	Oscillation potential across S
5	S+C over E	$\frac{E}{S+C}$	$0.86602...e^{\pm i\frac{\pi}{6}}$	Oscillation & ½ spin rotation
6	C over S+E	$\frac{S+E}{C}$	$1.89564...,\ -0.39564...$	Weak force decay
		Table 7 – Table of IESUS Inflection Points		

The modeling of these inflection points as an increase in volume is a simple heuristic device for pointing to the energy investment transitions in a unit of space as a function of expansion stress. It need not be interpreted as a local physical expansion strain for the increase in stress to be understood as an increase in energy density or mass equivalence. An increase in such density can be physically understood as an increase in frequency and corresponding decrease in wavelength of a wave particle. As a result, the interpretation of an increase of mass of the τ should not be interpreted as an increase in physical size, but instead as an increase in frequency/decrease in Compton wavelength.

Note that the ratios are between the total stress on each type of component, so that with row 1, the total of stress–strain on 6 cubic surfaces equals the total on the 12 edges and 8 corners or vertices. In all rows, assuming a condition of increasing dx, the antecedent or numerator is decreasing relative to an increasing consequent or denominator in the ratios. The breaking of symmetry and the emergence of oscillation is apparent, at first chaotically before eventually becoming ordered with row 5, all of which have a dx less than 1, as they all indicate an isotropic stress as on a free baryon, which generally precludes the emergence of row 6. With oscillation, the totals for the consequent component stress–strain are not equally distributed across all the relevant components but are maximized and minimized sequentially in a well-ordered manner across those components as with any force oscillation.

By the time the work–energy change at the 12 edges and 8 corners rises to the level at the 6 surfaces, the first inflection point is reached as a will have increased 0.39564 (or for the sake of symmetry decreased by -1.89564) as in (1.140) in the following analysis. The tension stress at the surface of each cubic face is registered as equal to the stress at the perimeter components, the 4 edges and 4 corners, of each face, which are transverse, so that as these perimeter components begin to exceed that of the surface at the inflection point, more energy is invested in the transverse components than in the surface. As a result, the inverse of a, 2.52752, is the ratio of the initial condition to the surface differential stress, where the initial includes the total of tension and transverse stress components. This sets up a condition such that continued increase in a transfers a greater differential of stress to the transverse components and a torsional potential for a breaking of symmetry with the eventual emergence of rotation. This feature is reflected in (1.112) and again in the development of row 6.

By the time the surface has expanded as in (1.141) by a differential length, a, equals 0.5, the work–energy increased at the 6 surfaces will equal the volume increased at the 12 edges, after which the accumulated work in moving the edges will be greater than the work in moving the surfaces; we will assume that the mass difference between the three cube components, face, edge, and corner, in all cases vanishes as the work–energy is invested in the volume and not the three component surfaces.

As with (1.142) by the time the work–energy change at the 12 edges equals that at the 8 corners, differential a will have increased to 2/3 of the edge length.

Most significantly at (1.143), by the time the work–energy change at the 6 surfaces equals that at the 8 corners, a will have increased to +/- 0.86602 = $\pm\frac{\sqrt{3}}{2}$, ½ spin, while at the same time with (1.144) the work–energy at the 12 edges equals the sum of both the 6 surfaces and the 8 corners, and this sets in motion an oscillation of energy flow from surface to edge to corner, with a potential for rotation of the cube.

Under conditions that mitigate free rotational oscillation as at the surface of a black hole source, with (1.145) the work–energy increase at the corners equals the sum of the increase at the faces and edges, a has increased by 1.89564 times the original edge length, and all increasing stress–stain or work–energy thereafter is concentrated in the corners, the cubic vertices. This can happen only when free rotational oscillation of ½ spin is restricted in some manner, as with a particle collision or in the case of emergence of that rotational oscillation from an initial condition of a stellar or galactic black hole inertial source of neutron maximum density as might be found in the galactic filaments of the cosmos. Under such conditions at the event horizon surface of such source, individual particle rotation resulting from the lattice gauge would be prevented until the inflection point of row 6 results in the emission of a tau and muon. These then quickly decay according to the various branching paths which follow.

These work–energy inflection points can be represented as angular frequency potentials in the context of a wave model, and thereby with a mass–angular wave number equivalence. The inverse of such wave number is the angular wavelength, which in the context of a rotating torsional oscillation can be represented as the particle's radius. Thus, an energy differential inflection representing an increase of 1.89564 with respect to a base frequency of 1.0 is equivalent to a reduction in the angular wavelength from 1.0 to 0.527525232. Three relevant parameters of wave phenomena are shown here. Note the difference between the parameters for Row 1 and Row 6 is 1.5 and their inverse is 2 and that these differences are exact. This is related to the predominance of the 3-axis surface stress prior to inflection Row 1 and of its supersession by 4-axis corner stress after Row 6 is reached, reflecting the 3 to 4 axial ratio found in the cuboctahedral lattice potential, that is the three axes through the center of the 6 surfaces and the 4 axes running through the 8 corners.

Row 1: $\qquad (0.395643924...)^{-1} = 2.527525230$

Row 5 (argument): $\qquad \pm i\pi/6 = 0.523598776 = (1.909859317...)^{-1}$ (1.118)

Row 6: $\qquad (1.895643924...)^{-1} = 0.527525232$

This last figure is significant as its representation as an inflection point for energy concentration in the vertices of the cube. Referring to the modular ratios for mass of the tau in (1.104) in the columns for */p at 1.89373 and for */n at 1.89113, if we assume that the figure 1.895643924 as just derived from this cubic analysis (CA) is a measure of the increase in mass/energy at the inflection point, τ_{CA}, we can adjust this figure for the missing mass involved in the following analysis. From the column for */τ using the value of e_τ as a percentage of the value of τ, we calculate the following value for (1.114) of

$$(2+3\Delta m)e_\tau = (6.594753182)(0.000287588) = 0.001896572 \qquad (1.119)$$

which we add to the base value of $\tau = 1.0$ to get 1.001896572. This represents the total percentage of work–energy stress required to produce the tau, plus the 'missing mass' and Hubble stress related to beta decay as developed previously required for the tau and the companion muon. Dividing this into the inflection point value based on a cubic analysis to arrive at a theoretical value for the τ with respect to the initial neutron density, we get a value in the middle of the values based on */p and */n.

$$(1.895643924)_{CA.row6} / 1.001896572 = \tau_{CA}(1.892055505), \text{ where}$$
$$\tau_p(1.893738026) > \tau_{CA}(1.892055505) > \tau_n(1.891131242) \qquad (1.120)$$

From an alternate approach, we can compute the following difference

$$\tau_{CA}(1.895643924) - \tau_n(1.891131242) = 0.004512682 \qquad (1.121)$$

which can be further unpacked as

$$0.004512682 \div ((2.531584394)(0.000543867)) = 3.277552167, \text{ where}$$
$$3.277552167 > (3((2.531584394)(0.000543867))) = \{3(\Delta m+1)e_n\} = \{3\Delta m + 3e\ \} \qquad (1.122)$$

This indicates that the energy represented by τ_{CA} is sufficient to produce τ_n and μ_n, the tau and muon, as in (1.114) based on an excess of missing matter in (1.121) and (1.122).

τ_n, Neutron ratio: $\quad (1.891131242...)^{-1} = 0.528784030$

$$\Delta(\tau_{CA-\tau_n}) = 0.000258309 \quad < m_e(0.000543867)$$

τ_{CA}, Cubic Analysis: $\quad (1.892055505...)^{-1} = 0.528525721 \qquad (1.123)$

$$\Delta(\tau_{CA-\tau_p}) = 0.000469577 \quad < m_e(0.000543867)$$

τ_p, Proton ratio: $\quad (1.893738026...)^{-1} = 0.528056144$

With respect to this modeling, this is true whether the black hole density of the inertial field upon which the Hubble tension stress is operating to produce the tau and muon as a leptonic duplet is comprised of maximumly packed existing neutrons or a pre-emergent, gauged cuboctahedral lattice potential field of the same neutron density, both indicated by the *2n* of (1.114).

To answer the question of (1.115), then we have the following decay paths for a tau–muon duplet

$$2n + \{3\Delta m + 3e\}_{plus} = \tau + \mu - [3\Delta m + 2e] = -[e + \Delta m] + \tau - [\Delta m] + \mu - [\Delta m + e] = \qquad (1.124)$$

A1) Molecular Hydrogen or Protium
$$= 2(p + [\Delta m] + e) \qquad (1.125)$$
$$= H_2 + [2\Delta m]$$

A2) 2 Cations and 2 Anion
$$= 2(p + [\Delta m + e]) \qquad (1.126)$$
$$= 2p^+ + 2e^- + [2\Delta m]$$

A3) Atomic Protium, a Cation and Anion
$$= 1(p + [\Delta m] + e) + 1(p + [\Delta m + e]) \qquad (1.127)$$
$$= H + p^+ + e^- + [2\Delta m]$$

B1) A Neutron and Atomic Protium
$$= n + (p + [\Delta m] + e) \qquad (1.128)$$
$$= n + H + [\Delta m]$$

B2) Atomic Deuterium
$$= ((n + p) + [\Delta m] + e) \qquad (1.129)$$
$$= D = {}^2H + [\Delta m]$$

For composite structures of 2 duplet interactions in the formation of Helium

$$2(\tau + \mu - [3\Delta m + 2e]) =$$

C1) Atomic Tritium = B1 + B2

$$= 2((n+p) + [\Delta m] + e) \quad (1.130)$$
$$= T + H = {}^3H + H + [2\Delta m]$$

D1) Atomic Helium 4 = 2(B2)

$$= 2((n+p) + [\Delta m] + e) \quad (1.131)$$
$$= {}^4He + [2\Delta m]$$

D2) Atomic Helium 3 = B1 + B2

$$= 2((n+p) + [\Delta m] + e) \quad (1.132)$$
$$= {}^3He + n + [2\Delta m]$$

For composite structures of 3 and 4 duplet interactions in the formation of the Lithium we have

$$3(\tau + \mu - [3\Delta m + 2e]) =$$

E2) Atomic Lithium 6 = D1 + B2

$$= 3((n+p) + [\Delta m] + e) \quad (1.133)$$
$$= {}^6Li + [3\Delta m]$$

$$4(\tau + \mu - [3\Delta m + 2e]) =$$

E1) Atomic Lithium 7 = 2(D1)

$$= 4((n+p) + [\Delta m] + e) \quad (1.134)$$
$$= {}^7Li + (p^+ + e^-) + [4\Delta m]$$

These path and interactions are all the result of the high energy emergence of baryonic and leptonic matter from the surface of a neutron density inertial source. In an ideal case of an extreme Kerr black hole source, the tau–muon duplets slough from the surface, generating the electromagnetic field of quasar or other active galactic nuclei along with the collimated relativistic jets comprised of the particles of these various decay paths. In this manner and by this process, the Hubble stress responsible for all quantum properties produces quantum ½ spin, the quantum electromagnetic field and charge, and quantum gravity and all the secondary properties of matter. Gravity plays no appreciable part in the formation of these AGN black holes. The jets then generate the clouds of gas comprised principally of the light elements, which then generate the quantum gravitational and electromagnetic interactions responsible for aggregation into stellar configurations which populate and circulate around the AGN black holes.

Conclusion to the Addendum

Current cosmological modeling quite literally centers around the concept of a big bang singularity as a necessary foundation for the observed abundance of hydrogen and helium in the universe, based on the astrophysical observation of an apparent isotropic expansion at the Hubble rate on a universal scale of baryonic matter located in the stellar systems and interstellar gas comprising the galaxies which are congregated along filaments and across membranes of high inertial density separating vast transparent and apparent voids but for the transiting spectra of electromagnetic energy that belies the illusion of emptiness. Such modeling embodies theoretical contradictions, some of which are unrecognized on both a quantum and cosmological level, that hamper a greater scientific understanding.

This modeling offers an amended solution to this quandary, and makes the following assumptions and development:

1. Any and all observation and activity, astronomical to quantum experimentation, is comprised of and within an extended multi-dimensional manifold of finite, variable stress–potential energy density, appearing as uniformly discrete instances of well-ordered and connected phenomena of invariant inertial properties of matter, quantized as a fundamental rest mass particle, m_0.
2. Designated as the space–time fabric (STF), the operation of such manifold can be defined using the formal structure of classical complex wave mechanics as a three-dimensional wave bearing medium with a cuboctahedral lattice potential, gauged by a time–independent inertial invariant that described by the analysis of Isotropic Expansion Stress on a Unit Space (IESUS) and recognized by the Hubble rate, results in the quantization of m_0 as a discrete rotating torsional oscillation of the STF.
3. While the Hubble phenomena can be and is customarily modeled as an expansion of all ponderable matter from a single locus in such a manner as to leave the observed stochastic distribution of baryonic matter in filaments and membranes of galaxies and nebulae based on inflationary theory, that phenomena can also be modeled as a contraction of large regions of the extragalactic voids within a cosmos of indefinite if not infinite extent away from those voids and into diaphanous webs of stress–potential energy density containing loci of black hole sources sufficient to produce baryonic matter from active galactic and active stellar nuclei. Neither of these distributions is well defined; neither of them can be explained as a function of gravity as they include no quantum understanding of gravity. The second, however, offers a path for an understanding of the predominance of hydrogen and helium in the cosmos in addition to an explanation of quantum gravity that requires no explanation for the stated distribution. Other models are conceivable for the explanation of the concentration of baryonic matter in the galactic filaments and membranes separated by vast voids, that require no gravitational component; the same expansion stress represented by the Hubble rate might be indicated on a soap bubble model, for instance.
4. An alternative to big bang nucleosynthesis in the production of the lighter elements can be found in an understanding of the nature of the filaments and membranes with their entrained baryonic matter. From the above development, these web-like features can be modeled as comprised of neutron star-black hole density source material subject to the Hubble stress at the web surfaces.
5. As IESUS produces a torsion stress as a form of internal friction at the surface of these structures equivalent to that registered at the corners of the unit space equal, to $e + \Delta m$, the tau–muon duplets peel off the surface of the inertial source and quickly decay within a fraction of a second along one of the several paths indicated above to produce the abundance of hydrogen, helium, and lithium gas, along with the assortment of protons, electrons, neutrons, neutrons, and gamma and lower frequency photons.
    a. This fundamental quantization, m_0, is the neutron, in which various properties can be understood as emergent qualities of simple harmonic motion as the rotating torsional oscillation, which produce
        i. Two inductive and two capacitive moments of maximum power and action of the torsional oscillation that with rotation generate a quantum electromagnetic field and a magnetic moment and an internal neutral current, short of beta decay,
        ii. ½ spin angular momentum
        iii. A quantum of gravity as a differential centripetal force as a function of the expansion stress, which forms the basis of Newton's gravitational constant.
    b. The hadrons and leptons created in this process are responsible for the collimated jets and gamma ray bursts of active galactic nuclei, the aggregation of light element gas by gravitational and electromagnetic interaction into stellar formation.

This can be mathematically modeled as follows:

Initial Condition, $M_0 \int_0^\infty df_n$, where $df_n = 2.53158...df_0 = \sum_{antisymmetric}^{symmetric} \begin{matrix} 1.0 df_{ii} \\ i1.5318...df_{ij} \end{matrix}$:

$M_n =$ Inertial-potential energy density of space

$df_n =$ Expansion stress quantized by inertial constant, $\eta = \hbar/c$ where $m(\eta) = \eta/\lambda_{C,n}$

mass-energy of Hubble tension stress $(e) = 1.0 df_{ii}$,

mass-energy of Hubble transverse stress $(\Delta m) = i1.5318...df_{ij}$

$$(\ln f_0)^{\frac{1}{2}} = ie_{-2}f_0 = \frac{-df_0}{id \ln f_0} = \frac{-de_{-2}f_{ii}}{id \ln_2 e_{-2}f_{ij}} = \frac{-0.65291...f_{ii}}{i1.0 f_{ij}} = \frac{-1.0 f_{ii}}{i1.53158...f_{ij}}$$

Evolution of $M_n \int_0^a df_n$:

Single particle beta decay:

$$n - \{df_{ij} + df_{ii}\} = p + [\Delta m] + e$$

Sum of 3 separate particles, with one bound H and 2 ionized H^+ :

$$3n - \sum_{1n}^{3n} \{1 df_{ij} + 1 df_{ii}\} = [3\Delta m + 2e] + (p + e) + 2p$$

2 neutrons at nuclear density congregation under isotropic Hubble stress, where $\{3df_{ij} + 2df_{ii}\} = [3\Delta m + 2e]$ and $(\mu + \tau)$ is a transitional state :

Unstable state w/ expansion | Transitional state | Stable state

$$(2n) + \{3df_{ij} + 2df_{ii}\} = \mu + \tau - [3\Delta m + 2e] = n + p + e + \{\Delta m\} = D$$

↙ ↘ Hyper-stable State

Catalytic stress transfer $\qquad 2n + 2p + 2e = He_4$

↙ ↗

$$(2n) + \{3df_{ij} + 2df_{ii}\} = \mu + \tau - [3\Delta m + 2e] = n + p + e + \{\Delta m\} = D$$

(1.135)

Detail of the effects of Isotropic Expansion Stress on a Unit Space (IESUS)

We can imagine the center of a cube, and later hypercube, as a local center of expansion of physical space according to the Hubble rate. We will integrate the following differentials to compare the contribution made by each boundary order to the change in the corresponding core, in this case a volume. We are interested in the relative contributions of each order over time to the initial unit volume, V, and not to the changing magnitude of the volume itself. We substitute the following boundary placeholder identities for Surface, Edge and vertices (Corner), $1^2 S = x^2$, $1^1 E = x^1$, and $1^0 C = x^0$, so as to maintain proper integration. It will be helpful if we assign a "normal" boundary strain vector to each of these components, which in each case will be in the direction in which the boundary is increasing. Thus

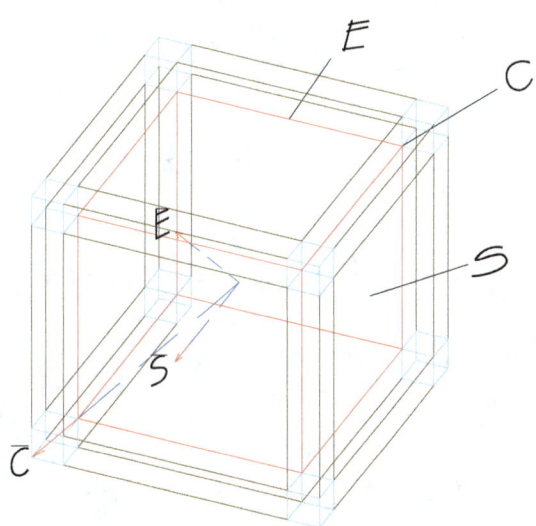

Cubic Expansion

$$|\mathbf{S}| = \left|\sqrt{\tfrac{1}{2}}\,\mathbf{E}\right| = \left|\sqrt{\tfrac{1}{3}}\,\mathbf{C}\right| \tag{1.136}$$

$$|\mathbf{E}| = \left|\sqrt{2}\,\mathbf{S}\right| = \left|\sqrt{\tfrac{2}{3}}\,\mathbf{C}\right| \tag{1.137}$$

$$|\mathbf{C}| = \left|\sqrt{3}\,\mathbf{S}\right| = \left|\sqrt{\tfrac{3}{2}}\,\mathbf{E}\right| \tag{1.138}$$

In the following, no assumption is made about the universal configuration or number of dimensions of the space in which the unit cube is embedded. We are only interested, at least initially, in the local geometry, which is assumed to be flat and therefore Euclidean. Thus, it is background independent.

The integration will be simultaneous on each order, as indicated by the pre-subscript n, in $\int_n dx^n$ so that we have

$$\int_V dV = 6x^2 \int_0^a {}_1dx^1 + 12x^1 \int_0^a {}_2dx^2 + 8x^0 \int_0^a {}_3dx^3$$

$$\int_V dV = 6S\int_0^a dx + 12E\left(\int_0^a dx\right)\left(\int_0^a dx\right) + 8C\left(\int_0^a dx\right)\left(\int_0^a dx\right)\left(\int_0^a dx\right) \quad (1.139)$$

$$\Delta V = 6aS + 12a^2 E + 8a^3 C$$

Solving for the following ratios, all at unity, where the designations S, E, and C are unit names, their dimensional quantities being absorbed in the numerical coefficients of a^n, i.e. 6 square units times a, 12 length units times a^2, 8 point units times a^3, gives the value of a for each equivalence. The ratios have been stated with the highest order in the consequent or denominator so they are decreasing from infinity as dx increases, until unity is reached as stated. We have (showing the negative for the sake of symmetry)

$$\frac{S}{E+C} = \frac{6a}{12a^2+8a^3} = \frac{1}{2a+\frac{4}{3}a^2} = 1 \therefore a = -\tfrac{3}{4} \pm \tfrac{1}{4}\sqrt{21} = 0.39564...,-1.89564... \quad (1.140)$$

$$\frac{S}{E} = \frac{6a}{12a^2} = \frac{\frac{1}{2}}{a} = 1 \therefore a = \tfrac{1}{2} = 0.5 \quad (1.141)$$

$$\frac{E}{C} = \frac{12a^2}{8a^3} = \frac{\frac{2}{3}}{a} = 1 \therefore a = \tfrac{2}{3} = 0.66666... \quad (1.142)$$

$$\frac{S}{C} = \frac{6a}{8a^3} = \frac{\frac{3}{4}}{a^2} = 1 \therefore a = \pm\tfrac{\sqrt{3}}{2} = \pm 0.86602... \quad (1.143)$$

$$\frac{E}{S+C} = \frac{12a^2}{6a+8a^3} = 1 \therefore a = \tfrac{3}{4} \pm i\tfrac{1}{4}\sqrt{3} = \tfrac{\sqrt{3}}{2}e^{\pm i\frac{\pi}{6}} = 0.86602...e^{\pm i\frac{\pi}{6}} \quad (1.144)$$

$$\frac{S+E}{C} = \frac{6a+12a^2}{8a^3} = 1 \therefore a = \tfrac{3}{4} \pm \tfrac{1}{4}\sqrt{21} = 1.89564...,-0.39564... \quad (1.145)$$

If we think of the cube as embedded in an isotropic elastic continuum, which is of some inertial density and under tension, dx represents the work done in displacing or distorting the medium, and by virtue of Gauss' theorem, the integration of that work represents the energy of the distortion. By way of reference, in an ideal elastic medium, the stress operating on the locale is a function of the strain and the elastic modulus as

$$\mathbf{F} = \frac{Y\mathbf{E} - 3\sigma \bar{P}\mathbf{1}}{1+\sigma} \quad (1.146)$$

where $\mathbf{F}$ is the stress tensor, $\mathbf{E}$ is the strain tensor, Y is Young's modulus of elasticity, σ is Poisson's ratio or the negative ratio of lateral to axial or shear to tension strain, $\bar{P}$ is the mean pressure in the medium, and $\mathbf{1}$ is the idemfactor or unit tensor. Assuming a value of σ of -1/3 for an ideal isotropic 3 dimensional medium we have

$$\mathbf{F} = \frac{3}{2}(Y\mathbf{E} + \bar{P}\mathbf{1}) \quad (1.147)$$

The vector fundamental tension stress component is

$$\mathbf{f} = Y\mathbf{e} \qquad (1.148)$$

and is related to the energy distribution by Gauss' theorem for the radial strain

$$E_r = \int_V \nabla \cdot \mathbf{e}_r \, dv = \oint_S \mathbf{e}_r \cdot d\mathbf{S} \qquad (1.149)$$

and Stokes' theorem for the angular or tangential strain

$$E_t = \int_S \nabla \times \mathbf{e}_t \cdot d\mathbf{S} = \oint_{\mathbf{r}} \mathbf{e}_t \cdot d\mathbf{r}_t \qquad (1.150)$$

These boundary order ratios, then, are inflection points indicating the energy contributions and potential energy gradient changes over time among the boundary components. In an ideal static, kinematic case the change in the ratios with an increase in dx would have no functional effect on the components, if dx has the same magnitude for each of them as it increases. This would amount to a simple change of scale. The real solutions above would appear to reflect this static condition. However, in a dynamic condition, we understand that as each ratio decreases below unity and past the inflection point, the magnitude of the consequent exceeds and affects the antecedent or numerator, whose magnitude then becomes a partial function of the consequent. This would appear to be the case for the complex solutions in particular, which correspond with an angular gradient potential of the boundary vectors from that of the antecedent to the direction of that of the consequent.

These evaluations were done with Maple. It is significant that if we convert (1.144) to complex polar notation as in the last term, the modulus is equal to the value for a in (1.143). It is important that we understand that the ratios represent the point at which the change in volume due to the sum totals of all component orders in the antecedent and consequent are equal. It is not the point at which one single component of a given S, E, or C times its appropriate $\int_n dx^n$ is equal to another, since this happens for all at the point where $a = 1$.

In these evaluations, the S component of the strain and hence of the work predominates until (1.140) is reached. At this point, the stress will begin to shift from a predominance of tension to that of shear, meaning there will be a potential for the surface and edge strains to oscillate. As the edges and vertices ring each of the surfaces, the system remains basically stable, however. At the point of (1.141) the edges assume dominance over the surfaces and a gradient is produced for the bulk strain and the tension stress in the direction of the edges and a torsional potential about each of the 3 surface axes. Once again, the 2:1 symmetry of edges to surface maintains stability. At (1.142) the vertices contribute more work than the edges and the strain gradient shifts in their direction. Thus, there is a vector potential from the surfaces to the edges to the vertices. Once more the symmetry between vertices and edges maintains stability.

Jumping to (1.145), at this point the strain contributed by the vertices dominates both of the other components combined and the related stress is greatest at these locations. This would result in a transmission and oscillation of the energy in the form of a weak interaction as a lepton in a particle context, were it not for the unusual and unique condition created by (1.143) and (1.144). The point at which the strains of the vertices come to equal those of the surfaces is also the point at which their combined strain comes to equal that of the edges, as given by the modulus of the latter's ratio. We can assume that the imaginary component of this ratio indicates a rotational component of $\pi/6$ or $30°$, and since the vertices are assuming a predominance over the surfaces at this point, having already exceeded the edge strain, and as there is an imbalance in the number of vertices to surfaces, a necessary break in symmetry ensues.

We can imagine a rotational potential of the surface strain in the direction of the vertices, which by virtue of the asymmetry between S and C, of 3 degrees of rotational freedom and 4 possible rotational axes, results in an eventual rotational strain about one pair of the axes. This is simultaneous with a shift of the Es in the direction of S + C and a dragging of the strains at each of the two axial C poles. This then leads to a rotation of the axial Cs in the direction of one of the three E pairs extending from those two vertices. The equation of (1.144) gives this rotational relationship. The nature of the ambiguous sense in the argument is indicative of the equation of a rotational oscillation and its complex conjugate, when viewed from both senses of its axis, i.e. by rotating it about the real axis, where $\pm$ means plus <u>and</u> minus and not plus <u>or</u> minus, if we adjust the Euler identity to

$$e^{\pm i\theta} = \sin\theta \pm i\cos\theta \tag{1.151}$$

One end of the axis of strain then can be shown as indicated by the "symmetry breaking" in (1.154).

$$12a^2 E = \left(6aS + 8a^3 C\right) \tag{1.152}$$

$$12\left(\tfrac{\sqrt{3}}{2} e^{\pm i\tfrac{\pi}{6}}\right)^2 E = 6\left(\tfrac{\sqrt{3}}{2} e^{\pm i\tfrac{\pi}{6}}\right) S + 8\left(\tfrac{\sqrt{3}}{2} e^{\pm i\tfrac{\pi}{6}}\right)^3 C \tag{1.153}$$

$$e^{-i\tfrac{\pi}{3}} E = \tfrac{1}{\sqrt{3}}\left(e^{+i\tfrac{\pi}{6}} S + e^{+i\tfrac{\pi}{2}} C\right) \tag{1.154}$$

Thus, the strain vector E, rotated in some direction $\tfrac{\pi}{3}$, is equal to $\tfrac{1}{\sqrt{3}}$ of the S and C strains rotated $\tfrac{2\pi}{3}$ in the opposite direction, presumably in the same plane. In fact, this states that C rotates $\tfrac{\pi}{2}$ while S rotates $\tfrac{\pi}{6}$. We can see specifically how these rotations occur in Spin Diagrams 1 and 2 (of CMMPG). We can also see there how a rotation back in time of $\tfrac{\pi}{3}$ equals one forward in time by $\tfrac{2\pi}{3}$ and vice-versa, if their plane of rotation, ϕ, is itself rotating at a constant rate with respect to an orthogonal plane, θ, that is where the two axes intersect at the centers of rotation. However, it is shown there that this corresponds with a rotation of θ, back $\tfrac{\pi}{4}$ and forward $\tfrac{3\pi}{4}$, indicating a variability in the strain velocity.

It should be understood that this cubic structure is simply an expression of the orthogonal tendency for stress equalization and energy conservation. The condition found at (1.143) and (1.144), then becomes a stable dynamic condition of rotational oscillation or spin, within certain parameters of inertial density and mechanical impedance. If the isotropic tension in this situation was sufficient to increase the strain indefinitely, if the medium was to lose its elasticity and become plastic or even rupture, any tendency to oscillate would be overcome by the transfer of energy via strain to the vertices. Local energy would not be conserved, but be drawn away by the strain.

It is essential to extrapolate this scenario to the hypercube, H, to achieve a full understanding. We will skip the integrals but show the results for the corollary of the last line of (1.139) as

$$\begin{aligned}\Delta H &= 8aV + 24a^2 S + 32a^3 E + 16a^4 C \\ &= 1aV + 3a^2 S + 4a^3 E + 2a^4 C\end{aligned} \tag{1.155}$$

There are 25 combinations with corresponding non-ordered permutations or sub-combinations, for the 4-cube; 7 involving all 4 parameters, 12 permutations involving all sub-combinations of 3, and 6 one to one relationships. With the 3-Space, there are 2 single real positive solutions at (1.141) and (1.142), one instance of a complex solution at (1.144), one correspondence between a real and a complex solution at (1.143) and (1.144) where the real value of a in one is equal to the complex modulus in the other, and one instance of a correspondence of solutions with sense inversion, (1.140) and (1.145), that is their solutions have the same magnitude, but of opposite sense. As might be expected, the 4-Space of a hypercube shows significantly more of these symmetries. It should be noted that while an attempt has been made to analyze the ratios qualitatively so that all are represented as decreasing with respect to an increasing dx, they have not all been checked quantitatively, and some may be increasing as shown. In fact, (1.168) and (1.170) are found to be increasing at the point represented by the first positive solution and decreasing at the second. For (1.165) it is worth stating that for every value of the ratio $0.75 < \left(\frac{S}{V+E}\right) < +\infty$, the modulus is ½ and the argument ranges from 0 to ½π.

It is important to remember that a given component in the 3-cube is identical to the same component in the 4-cube, but the relationships between them are different. An edge still is bounded by 2 vertices, but there are 4 edges intersecting at each vertex of the 4-cube. A line segment in an x-y plane is qualitatively no different than one in the z-x or for that matter z-w plane. In fact, a point in 3-space also has a location in n-space, at least in Euclidean n-space. In the following, it is also important to remember that a is not the value of the corresponding ratio, but rather the value found in both antecedent and consequent when the ratio equals 1. The evaluations are based on the following identities in (1.156)

$$V \equiv 1a, S \equiv 3a^2, E \equiv 4a^3, C = 2a^4 \tag{1.156}$$

$$\frac{V}{S}, a = \tfrac{1}{3} \tag{1.157}$$

$$\frac{V}{E}, a = \pm\tfrac{1}{2} \tag{1.158}$$

$$\frac{V}{C}, a = \tfrac{1}{\sqrt[3]{2}}, -\tfrac{1}{2}\left(\tfrac{1}{\sqrt[3]{2}}\right)\ldots \pm i\tfrac{\sqrt{3}}{2}\left(\tfrac{1}{\sqrt[3]{2}}\right) = \tfrac{1}{\sqrt[3]{2}} e^{\pm i \tfrac{2\pi}{3}} = 0.79370\ldots e^{\pm i \tfrac{2\pi}{3}} \tag{1.159}$$

$$\frac{S}{E}, a = 0, \tfrac{3}{4} \tag{1.160}$$

$$\frac{S}{C}, a = 0, \pm\sqrt{\tfrac{3}{2}} \tag{1.161}$$

$$\frac{E}{C}, a = 0, 0, 2 \tag{1.162}$$

$$\frac{V}{S+E}, a = -1, \tfrac{1}{4} \tag{1.163}$$

$$\frac{V+S}{E}, a = -\tfrac{1}{4}, 1 \tag{1.164}$$

$$\frac{S}{V+E}, a = \tfrac{3}{8} \pm i\tfrac{1}{8}\sqrt{7} = \tfrac{1}{2} e^{\pm i 0.722734248\ldots} \tag{1.165}$$

$$\frac{V}{S+C}, a = 0.31290\ldots, -0.15645\ldots + i1.25436\ldots = 1.26408\ldots e^{\pm i 1.694883228\ldots} \tag{1.166}$$

$$\frac{V+S}{C}, a = -1, -0.36602\ldots, 1.36602\ldots \tag{1.167}$$

$$\frac{V+C}{S}, a = -1.36602..., 0.36602..., 1 \tag{1.168}$$

$$\frac{V}{E+C}, -1.85463..., -0.59696..., 0.45160... \tag{1.169}$$

$$\frac{V+C}{E}, a = -0.45160..., 0.59696..., 1.85463... \tag{1.170}$$

$$\frac{V+E}{C}, a = 2.1120..., -0.05604... \pm i0.48331... = 0.48655...e^{\pm i1.686235431...} \tag{1.171}$$

$$\frac{S}{E+C}, a = -2.58113..., 0, 0.58113... \tag{1.172}$$

$$\frac{S+E}{C}, a = -0.58113..., 0, 2.58113... \tag{1.173}$$

$$\frac{E}{S+C}, a = 0, 1 \pm i\tfrac{1}{\sqrt{2}} = \sqrt{\tfrac{3}{2}}e^{\pm i0.615479709...} \tag{1.174}$$

$$\frac{V}{S+E+C}, a = 0.24415..., -1.12207... \pm i0.88817... = 1.43105...e^{\pm i2.472026458...} \tag{1.175}$$

$$\frac{E}{V+S+C}, a = -0.24415..., 1.12207... \pm i0.88817... = 1.43105...e^{\pm i0.669566197...} \tag{1.176}$$

$$\frac{V+E+C}{S}, a = -2.63993..., 0.31996... \pm i0.29498... = 0.43519...e^{\pm i0.744798022...} \tag{1.177}$$

$$\frac{V+S+E}{C}, a = 2.63993..., -0.31996... \pm i0.29498... = 0.43519...e^{\pm i2.396794631...} \tag{1.178}$$

$$\frac{V+S}{E+C}, a = -2.51702..., -0.25673..., 0.77375... \tag{1.179}$$

$$\frac{E+S}{V+C}, a = -0.77375..., 0.25673..., 2.51702... \tag{1.180}$$

$$\frac{V+E}{S+C}, a = 1, \tfrac{1}{2} \pm i\tfrac{1}{2} = \tfrac{1}{\sqrt{2}}e^{\pm i\tfrac{\pi}{4}} \tag{1.181}$$

Once again using Maple, there are a total of 10 couplings involving complex solutions, of which one is exclusively complex and one other has only a zero for the third and real solution. Only one single real positive solution is given. There are, however, 7 corresponding pairs of solutions involving sense inversion, 5 real and 2 complex. Note that all cases of sense inversion involve a combination of one or more components in either the antecedent and/or consequent and the sense change is associated with a transposition of one or two components in each pair. These do not appear to have any special relationship to the conditions of the 3-cube, at first glance, and we have not investigated them further.

There are several, however, that appear to have a direct relationship to some of the ratios of the 3-cube. Two conditions of correspondence are found between a real positive solution and the complex modulus of a complex solution with a positive real component. $(1.161)\left(\frac{S}{C}\right)$ and $(1.174)\left(\frac{E}{S+C}\right)$ are directly related to (1.143) and (1.144) respectively, the real solution and the modulus of the complex of the second two being equal to the product of the first and $\sqrt{2}^{-1}$. The argument of (1.174) is the angle at the center of a cube between a radial normal to an edge of the cube and one extended along a diagonal to a vertex. $(1.158)\left(\frac{V}{E}\right)$ and $(1.165)\left(\frac{S}{V+E}\right)$ are related to $(1.141)\left(\frac{S}{E}\right)_3$ with a common value for their real solutions and the modulus of the complex one. The cosine of the argument of (1.165) is equal to the solution of $(1.160)\left(\frac{S}{E}\right)_4$, which is the same ratio coupling as (1.141). This pairing (1.165) in turn has a modulus equal to the real and imaginary components of an additional complex solution in $(1.181)\left(\frac{V+E}{S+C}\right)$. This latter solution has an argument of $\pi/4$ or 45° which appears to be an extremely stable condition, as found in a sine wave model as the point of maximum power of the wave, where the product of the transverse wave force and transverse wave speed are maximum. It is also the angle of the strain vector E discussed above for the 3-cube, with respect to the plane normal to the spin angular momentum vector as shown in the spin diagrams. In the model developed here, this condition is found to be invariant and rotates about the oscillation's angular momentum vector.

Finally, $(1.174)\left(\frac{E}{S+C}\right)$, $(1.181)\left(\frac{V+E}{S+C}\right)$, and $(1.159)\left(\frac{V}{C}\right)$ are found to be related in a most profound way in the mechanism of the oscillation herein described. The imaginary component of (1.174) equals the modulus of (1.181). Note that (1.159) represents a $\frac{2\pi}{3}$ rotation due to the interplay between the volume and vertex components of strain and a modulus of that strain of $\frac{1}{\sqrt[3]{2}}$. Using the equation for (1.159) or

$$aV = 2a^4C \tag{1.182}$$

$$\tfrac{1}{\sqrt[3]{2}}e^{\pm i\frac{2\pi}{3}}V = 2\left(\tfrac{1}{\sqrt[3]{2}}e^{\pm i\frac{2\pi}{3}}\right)^4 C \tag{1.183}$$

tells us that a rotational oscillation of the 4-volume (boundary) strain V of modulus $\frac{1}{\sqrt[3]{2}}$ by $\frac{2\pi}{3}$ is equal to 4 axial rotations about the vertices of the same modulus and argument, where the 2 in the consequent indicates simultaneous rotations of opposite sense at each end of an axis. The oscillation of V is fourth dimensional, and therefore beyond our direct sensory ken, however, the 4 vertices are not, and we can envision the above consequent, the expression in 3 dimension of this four dimensional rotation, as a sequence of 4, $\frac{2\pi}{3}$ rotations about the 4 diagonals of a 3-cube. This sequence leaves the cube unchanged and avoids the entanglement condition, i.e. the continuity of Euclidean 3-coordinates of the cube are not twisted by the sequence. This condition of limits on the twistability of the continuum strain is a necessary consequence of its inertial/elastic properties. As the rotation of V is continuous, we would imagine that the sequence of 4 rotations is continuous, i.e. the strain rotates from one reference diagonal to another about one of the three surface axes of the 3-cube. We can also envision this as one diagonal axis rotating $\frac{2\pi}{3}$, followed by a 2π rotation of the same sense about one of the adjacent 3-cube surface axes. We can also treat it as a sequence of 4 orthogonal permutations.

106

84

Unification

—:—

Addressing the Fundamental Problem of Theoretical Physics

By Martin Gibson

July 15, 2015

Martin Gibson
P.O. Box 2358
Southern Pines, NC 28388
910-585-1234
martin@uniservent.org

Copyright © Martin Gibson 2021 All Rights Reserved

Unification
Addressing the Fundamental Problem of Theoretical Physics
Martin Gibson

Abstract

The fundamental problem of theoretical physics for some time has been the inability to couple the gravitational interaction and the various quantum interactions in a manner that is consistent with both models. Both general relativity as the best model for the gravitational interaction and quantum field theory as the best model for the strong and electroweak interactions have a long history of proven efficacy. What is missing is the provision of a quantum mechanism for coupling rest mass and light speed quanta with spacetime in a manner that is consistent within itself and with the established models. This monograph examines the subject with an analysis of the Planck scale in the context of the formal structure of the field equation of general relativity.

The Planck Scale - Fundamental, Differential, or Both?

Most attempts at such provision focus on a natural regime or scale at which all basic physical properties are unitary. At such scale length, r_0, time, t_0, and mass as a measure of inertia, m_0, as well as the composite kinematic and dynamic properties such as velocity, acceleration, momentum, force, action, and energy, are treated as fundamental units. We will designate such in this presentation by the subscript naught.

The Planck scale states one such regime in which the fundamental invariants of gravitational theory as Newton's gravitational constant G, relativity as the speed of light c, and quantum theory as Planck's reduced quantum of action $\hbar$, are combined to produce a fundamental length squared value as

$$A_{Planck} = A_0 = r_0^2 = \frac{G\hbar}{c^3} \tag{1.1}$$

Using SI values for the three invariants, this evaluates to $2.61...\times 10^{-70}$ meter squared.

Some additional values in light of the following dimensional analysis are:
- r_0 $1.62...\times 10^{-35}$ meters
- t_0 $5.39...\times 10^{-44}$ seconds
- m_0 $2.18...\times 10^{-8}$ kilograms

General relativity relates mass to length directly according to the following rephrasing of the relationship of (1.1) as

$$\frac{G}{c^2} = \frac{c}{\hbar} A_0 = \frac{r_0}{t_0} \frac{t_0}{m_0 r_0^2} r_0^2 = \frac{r_0}{m_0} \tag{1.2}$$

Therefore, at any scale it is assumed that

$$n m_0 \frac{G}{c^2} = n r_0 \tag{1.3}$$

where n is the same number for mass and length.

On the other hand, quantum theory relates individual particle mass, m_q, and length, r_q, indirectly, where r_q, is equal to the reduced Compton wavelength, λ_{Cq}, of a particle, a statistically derived experimental value as

$$m_q \frac{c}{\hbar} = \frac{1}{r_q} = \frac{1}{\lambda_{Cq}} \tag{1.4}$$

Assuming that at some scale $m_q = m_0$ and $r_q = r_0$, then (1.4) is consistent with (1.3), which recommends the Planck scale as a unifying scale.

In terms of natural units, the observed invariants are expressed as

$$c = \frac{r_0}{t_0} \tag{1.5}$$

$$\hbar = \frac{m_0 r_0^2}{t_0} \tag{1.6}$$

$$G = \frac{r_0^3}{m_0 t_0^2} \tag{1.7}$$

The first two of these are straightforward dimensional properties, the first as a velocity and the second as an action or angular momentum. The gravitational constant requires a bit of analysis, since its dimensional composition does not conform to any well-known physical property. For this we must turn to Newton's gravitational law, which is the context in which G was experimentally derived, though it is utilized in Einstein's gravitational field equation, GFE, stated here as

$$R_{\mu\nu} - \tfrac{1}{2} g_{\mu\nu} R + g_{\mu\nu} \Lambda = \frac{4\pi G}{c^4} 2 T_{\mu\nu} \tag{1.8}$$

Newton's gravitational law expresses the force of gravitational interaction, F_G, between two bodies of celestial size and separation as the product of their individual masses, M_a, and the square of their inverse distance of separation, d, times an experimentally derived constant of proportionality, G as

$$F_G = \frac{M_1 M_2}{d^2} G \tag{1.9}$$

Despite the derivation of the law in application to celestial mechanics, G was first determined to a reasonable degree of precision by a laboratory experiment involving a torsion balance after the manner devised by Henry Cavendish. This of itself indicates a massive particle or quantum basis of the interaction, rather than the operation of a mechanism that is essentially celestial in nature. In the value of G there must be an implied force, τ_0, of some mechanism that effects the gravitational interaction as indicated by the following analysis of (1.7) at the Planck scale

$$G = \frac{r_0^3}{m_0 t_0^2} = \frac{r_0^2}{m_0^2} \left(\frac{m_0 r_0}{t_0^2} \right) = \frac{r_0^2}{m_0^2} \tau_0 \tag{1.10}$$

Based on the values of the Planck scale, this gives the following value for a fundamental unit of force:

$$\tau_0 \quad 1.21\ldots \times 10^{44} \text{ Newton}$$

General relativity assigns such interaction not to a force but to the geometry of spacetime ie. to celestial mechanics, but as it is particle mass and energy that shapes that geometry in general relativity, it leaves open the manner in which such coupling between mass/energy and spacetime is effected. The stress energy momentum tensor of the field equation provides the answer.

From (1.3) and (1.4) we can rearrange (1.9) for any bodies of aggregate mass and aggregate distance in terms of a fundamental scale to get the following

$$G = \frac{n_{r_0}^2 r_0^2}{n_{M_1} n_{M_2} m_0^2} F_{G0} = \frac{r_0^2}{m_0^2} F_{G0} \tag{1.11}$$

In the last term of (1.11) the n values in the middle term have all been set to 1. The question, then is if it makes sense that $\tau_0 = F_{G0}$ and is the same for both expression (1.10) derived from the Planck scale and (1.11) derived from Newton's gravitational law. Keep in mind that in the latter expression, G converts the M_1 M_2 and d^2 terms to natural units that are then multiplied by the imputed fundamental force, F_{G0} to arrive at the interaction strength.

The Fundamental Stress Tensor

To answer this, we turn to Einstein's field equation(1.8), to the right hand term, $2T_{\mu\nu}$, which is the stress energy momentum tensor. While tensors customarily configure the relationship of vector transformations from a common origin, in the case of a stress tensor we prefer to use a unit cube as an origin for the tensor, in the manner of a coordinate singularity. Stress is the product of a force or forces operating on the surface of such cube, or half of such cube as indicated by three, unit surface squares with a common vertex at the point (1,1,1). Table 1 shows this condition. These surfaces may be deformed or strained by an eventual stress, but the initial condition assumes they are square. The coefficient of 2 represents the sum of the original $T_{\mu\nu}$ and its geometrical negative formed by the three, unit surfaces with a common vertex at (0,0,0), thereby completing the unit cube. The indices for the second tensor are oriented antiparallel to those of the first tensor. In particular this configuration represents an isotropic condition.

The μ rows of the 3-dimensional spatial matrix of the complete 4-dimensional spacetime matrix, represent these surfaces by a unit normal vector at each surface. While this component is a vector for mathematical purposes, in a classical or continuum analysis such as general relativity, it is conceptually a cross-sectional area and any force operating on or across that area is understood to be distributed over the whole cross-section. As such, these rows set the scale of the unit cube, including the unit of time which is normalized to the space scales. As with the space unit areas, the time unit is a square of a time length, therefore an acceleration factor as it applies to a force or energy components of the tensor.

In Table 1, the 9 cells in the heavy outline represent the space components of the tensor. Note that there are no time derivatives in these cells, as they represent a snap shot of the unit cube in time. The matrix product for each cell is indicated by the dimensionless strain, ε_{ij}, in the ν direction with respect to the μ unit vector designated by the cross-section or

$$r_j / A_{0|i} = \varepsilon_{ij} \qquad (1.12)$$

The unit stress/energy density, ρ_0, is effectively a stress modulus and the product of the density and the strain gives the tension or shear/torsion stress for each cell, expressed as the ratio of the stress force, τ_j, and the cross-section, $A_{0|i}$. The straight bracket divider is used to separate the unit naught indicator on the left of the subscript from the index values on the right.

The time derivatives are shown in the seven cells with wave borders for $\mu = 0$ and $\nu = 0$. As discussed above, the μ rows are unit scales, but the ν vectors are differentials which are integrated to unit values with respect to the $\mu = 0$ row in this representation. They could be less, but cannot be more than one, as they are normalized to the time unit and if greater than 1 would indicate superluminal speed.

The three μ_{0j} cells are the derivatives of a force operating on the cube with respect to the time differential. The three ν_{i0} cells give the derivative of the cross-section with respect to the time differential of $\nu = 0$, as they would be expected to change with an expansion or contraction of spacetime. Cell T_{00} in the double wave borders gives the time scale derivative as a change in the time scale with the passage of time. Such passing of time is gauged by the change in inertial density of flat spacetime as would be the case for an inertial spacetime as a function of a cosmological constant or the Hubble rate. Without change in density, there is effectively no time. Thus, the acceleration, either expansion or contraction, of the time scale indicated by the product that is T_{00}, is due to a change in unit density over time, $\dot{\rho}_0$.

Thus, an increase in the inertial density results in a positive differential force and a negative differential area, while a decrease in density produces a negative differential force and a positive differential area with respect to tension stress. This does not indicate a violation of energy conservation, since a decrease in density in one area may be offset by an increase in another. In addition, the energy of shear and torsion stress transformed to an increasing frequency of angular motion can add to the increase in density without contraction or expansion.

The ν columns of this 3-D matrix represent the inertial components of a force or forces operating on and across the surfaces. They are vectors comprised of the product of a unit length vector and the unit mass value, which is equal to the inertial invariant, ת, (tav), which is itself equal to h-bar over the speed of light.

If there is a net concentric flux of inertial components, ie. momentum or force, through the surface over time, the cube is a sink. If there is a net radial flux of such components out through the surfaces over time, the cube is a source, and if there is no change over time, the cube is divergence free. Similarly, if there is a net rotation of the surfaces of the cube over time, the cube exhibits curl, whereas if there is no such circulation or rotation, it is curl free.

Displacement >	$\nu = 0$, Time, $\int dt = t_0$	$\nu = 1$, Space, $\int dr_1 = r_{0|1}$	$\nu = 2$, Space, $\int dr_2 = r_{0|2}$	$\nu = 3$, Space, $\int dr_3 = r_{0|3}$
Cross-section, Unit Scale, A_0 V	$\frac{\hbar}{c\, t_{00}} = m_0 c\, t_0 = \mathsf{n}$	$\frac{\hbar}{c\, r_0} r_{0|1} = m_0 r_{0|1} = \mathsf{n}_1$	$\frac{\hbar}{c\, r_0} r_{0|2} = m_0 r_{0|2} = \mathsf{n}_2$	$\frac{\hbar}{c\, r_0} r_{0|3} = m_0 r_{0|3} = \mathsf{n}_3$
$\mu = 0$, Time Scale $A_{0|0} = |t_0^2|$	T_{00}, $\dot{\rho}_0$ Time Acceleration $\partial t_0^2 / t_0 = t_0^2$	T_{01}, $\dot{\tau}_{0|1}$ Differential Force $\frac{-m_0 r_{0|1}^2}{r_0} / t_0^2 = \frac{\mathsf{n}_1}{t_0^2}$	T_{02}, $\dot{\tau}_{0|2}$ Differential Force $\frac{-m_0 r_{0|2}^2}{r_0} / t_0^2 = \frac{\mathsf{n}_2}{t_0^2}$	T_{03}, $\dot{\tau}_{0|3}$ Differential Force $\frac{-m_0 r_{0|3}^2}{r_0} / t_0^2 = \frac{\mathsf{n}_3}{t_0^2}$
$\mu = 1$, Space Scale $A_{0|1} = |r_0^2|$	T_{10}, $\dot{A}_{0|1}$ Differential Unit Area $\partial r_{0|1}^2 / t_0 = \partial A_{0|1} / t_0$	T_{11}, Tension $\rho_0\, r_1 / A_{0|1} = \frac{\tau_1}{A_{0|1}}$	T_{12}, Torsion/Shear $\rho_0\, r_2 / A_{0|1} = \frac{\tau_2}{A_{0|1}}$	T_{13}, Torsion/Shear $\rho_0\, r_3 / A_{0|1} = \frac{\tau_3}{A_{0|1}}$
$\mu = 2$, Space Scale $A_{0|2} = |r_0^2|$	T_{20}, $\dot{A}_{0|2}$ Differential Unit Area $\partial r_{0|2}^2 / t_0 = \partial A_{0|2} / t_0$	T_{21}, Torsion/Shear $\rho_0\, r_1 / A_{0|2} = \frac{\tau_1}{A_{0|2}}$	T_{22}, Tension $\rho_0\, r_2 / A_{0|2} = \frac{\tau_2}{A_{0|2}}$	T_{23}, Torsion/Shear $\rho_0\, r_3 / A_{0|2} = \frac{\tau_3}{A_{0|2}}$
$\mu = 3$, Space Scale $A_{0|3} = |r_0^2|$	T_{30}, $\dot{A}_{0|3}$ Differential Unit Area $\partial r_{0|3}^2 / t_0 = \partial A_{0|3} / t_0$	T_{31}, Torsion/Shear $\rho_0\, r_1 / A_{0|3} = \frac{\tau_1}{A_{0|3}}$	T_{32}, Torsion/Shear $\rho_0\, r_2 / A_{0|3} = \frac{\tau_2}{A_{0|3}}$	T_{33}, Tension $\rho_0\, r_3 / A_{0|3} = \frac{\tau_3}{A_{0|3}}$

Table 1 - $T_{\mu\nu}$, Stress - inertial density tensor

For consistence, in terms of the 4-D spacetime tensor, we can envision an extra-dimensional space unit normal vector representing time which is orthogonal to the three space unit vectors for each of the doublets. For representation in 3-space, this vector would have a resultant length of $\sqrt{3}$ times each space unit vector. We will make use of this in a moment when we look at this form in detail, but first we want to get to where we are going with a scalar short cut.

While stress is essentially a tensor as described above, it can be represented analytically as a scalar force over a cross-sectional area. For a unit volume, or really for some flux through that volume or across its surface in keeping with Gauss theorem, the stress can be equated to the energy density of the volume as follows

$$\rho = \frac{E}{r_0^3} = \frac{W}{r_0^3} = \frac{\tau r_0}{r_0^2 r_0} = \frac{\tau}{r_0^2} = f \qquad (1.13)$$

Here f is the tension stress per face of the cube, τ is the force operating on that surface, W is the work done by that force operating over a length r_0, E is the energy in and at the surface, and ρ is the volumetric density. A change in ρ is represented by the 4-D matrix cell T_{00} which is equal to the sum total of all stress, tension and shear, operating on the 3-D components, f_{ij}, or conventionally T_{ij}, where the i and j indices represent the spatial, but not the time rows and columns.

The time row, $T_{0\nu}$, represents the rate of change in the νth vector with respect to the time scale given by $\mu = 0$ as a time unit vector. In keeping with the cross-sectional aspect of the space unit vectors, the time unit vector can be thought of as a clock face that marks out one unit of time. It essentially provides for a time derivative of the inertial flux vectors. In contrast, the time column, $T_{\mu 0}$, represents the arrow or extension of time, a time integral over which some change to the unit vectors represented by μ may occur. T_{00}, then represents a change in the time scale over time or an inherent acceleration. It therefore represents the kinetic energy of the unit cube, while f_{ij}, as time free, represents the elastic potential energy density of the unit as a snapshot in time both inherently, absent any stress, and given the extant stress/strain state of the system.

T_{0j} represent the force or momentum flux, essentially force differentials at the unit surface and T_{i0} represent the change in density resulting from a change in the cross-sectional areas of the cube.

Taking the total derivative of ρ with respect to the second to the last term of (1.13) gives

$$\frac{\rho}{6\sqrt{3}} = \frac{\tau}{r_0^2} = \frac{\tau}{A_0}$$

$$\frac{d\rho}{6\sqrt{3}} = \frac{\partial \rho}{\partial \tau} d\tau + \frac{\partial \rho}{\partial A_0} dA_0$$

$$\frac{d\rho}{6\sqrt{3}} = \frac{1}{A_0} d\tau - \frac{\tau}{A_0^2} dA_0 \qquad (1.14)$$

$$d\rho = \frac{6\sqrt{3}}{A_0} d\tau - \frac{6\sqrt{3}\tau}{A_0^2} dA_0$$

The $6\sqrt{3}$ figure arises from the above comments as a product of the trace of the doublet tensor and the orthogonal condition of T_{00}.

If energy is to be conserved in the unit, the net change in the density, measured by a variable, yet still unitary volume and surface, will be unchanged, even by an expansion or contraction of the unit volume. This may involve a redirection of radial to rotational strain, from the symmetric to the anti-symmetric components of the matrix. In this context we can state that the two differential components of the last line of (1.14) are necessarily equal. We can separate the two terms as follows

$$d\tau = \frac{A_0}{6\sqrt{3}}d\rho \qquad (1.15)$$

$$dA_0 = \frac{A_0^2}{6\sqrt{3}\tau}d\rho = \frac{A_0}{\rho}d\rho \qquad (1.16)$$

If we are considering this with respect to a fundamental scale, such as the Planck, then the rest of the terms in (1.13) will be unit terms, which we will restate as

$$\rho_0 = \frac{E_0}{r_0^3} = \frac{W_0}{r_0^3} = \frac{\tau_0 r_0}{r_0^2 r_0} = \frac{\tau_0}{r_0^2} = f_0 \qquad (1.17)$$

Doing the same for (1.15) and (1.16) and combining the two gives the invariance of the system as

$$dA_0 = \frac{d\tau_0}{d\rho}\frac{A_0}{\tau_0}d\rho$$

$$\frac{\tau_0}{A_0} = \frac{d\tau_0}{dA_0} = \frac{\rho}{6\sqrt{3}} \qquad (1.18)$$

This of course does not mean that the differentials and the unit values are equal and in fact, $\tau_0 \neq d\tau_0$ and $A_0 \neq dA_0$. What it means is that a unit volume change as a proportion of the total field will result in a proportional change of energy content and an invariant inertial density. Such density can still vary regionally, but the length/time scales will vary with it.

With respect to the value of $d\rho$ in (1.15) and (1.16), as with any derivative in which the value is known, it is equivalent to 1 upon such evaluation, and is required on the right hand side in the following to maintain dimensional consistency of the dependent variable differential. These two equations thus become

$$d\tau_0 = \frac{A_0}{6\sqrt{3}}d\rho_0 \qquad (1.19)$$

$$dA_0 = \frac{A_0^2}{6\sqrt{3}\tau_0}d\rho_0 = \frac{A_0}{\rho_0}d\rho_0 \qquad (1.20)$$

Clearly, the use of the valuation for τ_0 at (1.10) in conjunction with the natural mass and length values of the Planck scale should equal the observed value of G. This is a tremendous amount of force and an even greater amount of stress based on a cross section of the Planck area, perhaps viable at an early cosmic epoch or in an inertial source. Let's examine another option.

Let us assume that the force given above and attributed to a quantum of gravity is a differential force and that the Planck area is a differential cross-sectional area in keeping with the stress tensor analysis above.

Inserting (1.15) into the analysis of Newton's gravitational constant at (1.11) so that $F_{G0} = d\tau_0$ gives

$$G = \frac{r_0^4}{\hbar^2/c^2} \frac{A_0}{6\sqrt{3}} d\rho = \frac{r_0^6}{\hbar^2/c^2} \frac{1}{6\sqrt{3}} d\rho \quad (1.21)$$

where the differential density is a dimensional placeholder equal to 1. Rearrangement gives

$$r_0^6 = 6\sqrt{3}\, G \frac{\hbar^2}{c^2} \frac{1}{d\rho} = 8.58\ldots \times 10^{-95} \text{ meters}^6 \quad (1.22)$$

so that $r_0 = 2.100\ldots \times 10^{-16}$ meters. This is the reduced Compton wavelength of the neutron, $\lambdabar_{Cn}$.

To check ourselves, we can use the Codata value for $\lambdabar_{Cn}$ and other invariant values in (1.21) and we get $6.673198\ldots \times 10^{-11}$. The Codata value is $6.674\,08(31) \times 10^{-11}$.

The gravitational quantum differential is then

$$d\tau_0 = \frac{A_0}{6\sqrt{3}} d\rho = 4.244\ldots \times 10^{-33} \text{ Newton} \quad (1.23)$$

The fundamental force, the strong force, at this scale rather than that given for the Planck scale is determined by the neutron frequency per the following

$$m_n c^2 = \frac{\hbar}{c \lambdabar_{Cn}} c^2 = \hbar \omega_n$$

$$\tau_0 = \tau_n = m_n c \omega_n = 7.167\ldots \times 10^5 \text{ Newton} \quad (1.24)$$

The ratio of the strong force and the gravitational differential is then

$$\frac{\tau_0}{d\tau_0} = 1.688\ldots \times 10^{38} \quad (1.25)$$

This, coupled with the invariance requirements of energy density as stated in (1.18) indicates that the same relationship will hold for the cross-sectional differential, dA_0, given the following, where the fundamental scale area is given by the square of the reduced neutron Compton wavelength as

$$A_0 = r_0^2 = \lambdabar_{Cn}^2 = 4.410\ldots \times 10^{-32} \text{ meter}^2 \quad (1.26)$$

Thus we have

$$dA_0 = \frac{d\tau_0}{\tau_0} A_0 = 2.61\ldots \times 10^{-70} \text{ meter}^2 \quad (1.27)$$

and we see that the Planck scale is a differential scale of the spacetime continuum and not a discrete scale.

The fundamental stress is therefore

$$\frac{\tau_0}{A_0} = 1.625\ldots \times 10^{37} \text{ Newton/meter}^2 \quad (1.28)$$

and the inertial density is

$$\rho = \frac{E}{V} = 1.688...x10^{38} \text{ Joules/meter}^3 \qquad (1.29)$$

This analysis demonstrates that the unification scale for gravity and the other quantum interactions is the neutron scale.

Returning to an analysis of the spacetime stress tensor as developed above, there are essentially three conditions in addition to an initial condition in which stress and strain are absent, therefore absent change and thus time. With respect to the Einstein field equation, all the remaining terms on the right-hand side of the equation are redundant, unitary, or normalized in light of the above analysis. The $2T_{\mu\nu}$ term includes the surface integral represented by the 4π term, G as a gravitational quantum is included as the differential components of the force, and the length and time scales of the speed of light quad are normalized in the four rows and columns of the matrix.

1. Of the three, the first condition is the vacuum state in the absence of transiting messenger or resident rest mass phenomenon. Here, assuming a positive cosmological constant, Λ, as indicated by the Hubble rate, H_0, the unit of spacetime is either expanding or contracting, depending on whether we consider the cosmic manifold to be expanding or fixed over time. In either case, the time row and column values are non-trivial, but the resulting field is flat over the near term, that is it transmits the gravitational force along the normal unit vectors according to the symmetric portion of the tensor, and while we can anticipate fluctuations in the shear components, it is essentially curl free.

2. The second condition is that of a transiting messenger, but no rest mass, phenomenon. In addition to the dilatation of the vacuum state, the tensor will exhibit torsion and directional flux with the transit, but no sustained divergence or curl.

3. The third condition is that of the generation of fundamental rest mass. In this state, the vacuum state has progressed to the point at which transverse or shear strain components of the stress/strain relationship result in orthogonal torsion oscillation which in turn results in rotation of the field stress in the antisymmetric components. Note that it is a rotation of the field stress phases and not a physical rotation, though there is an oscillation and alternation of strain curl normal to the direction of rotation.

   As a phase change between moments of relative maximum displacement and momentum, maximum potential and kinetic energy, this oscillation is responsible for the generation of quantum charge, which is dimensionally a measure of fundamental momentum and which initially remains confined to the unit oscillation. This strain oscillation and stress rotation produces fundamental rest mass particle spin and the magnetic dipole moment. As an instance of confined oscillation, the waveform has a sustained nodal–anti-nodal structure and sustained moments of maximum power and action that rotate with the stress phases. This structure is disturbed over brief time frames by energetic particle collision, producing the perceived quark nature of this structure in accelerator experimental results.

Continued expansion according to $\Lambda = H_0$ produces in a drop in mechanical impedance as an ambient condition of $2T_{\mu\nu}$ which results in a transmission of a small portion of the oscillation energy and power to produce the electron oscillation. The strain state in the fundamental oscillation results in a rotational axis flip and frequency reduction to produce the quantum state of the proton. As a rest mass oscillation, the emitted electron is confined to a sub-light speed trajectory, however the counter-centripetal stress responsible for the emission comprises a signature stress transmission that travels at the speed of light as the neutrino.

While this covers the basic analysis of the right hand side of the GFE (1.8), we need to examine the left hand side. For our purposes here, we are primarily concerned with the case of a single fundamental unit of spacetime, therefore with a flat spacetime, other than the strain produced at the quantum tensor. From a cosmological perspective, (1) is an initial condition of cosmic expansion, from which (3) arises. With continued expansion, and beta decay in (3), (2) arises. We would also state that the condition in the vast voids of space between galactic filaments is primarily a combination of states (1) and (2), perhaps primarily (2), with infrequent instances of (3).

In the case of flat spacetime, the Ricci components which form the Einstein curvature tensor become trivial, leaving only the product, $g_{\mu\nu}\Lambda$, on the left. It is understood, however, that in the galactic environment, and in particular in the regions of galactic black holes, the curvature tensor is hardly trivial. In fact, it is my belief that such black holes, in particular active galactic nuclei, are inertial sources rather than inertial sinks, though some may function in both capacities and that the conditions necessary for the generation of state (3) are found primarily at the surface of such non-singularity black holes, and not in a universal big bang. The expansion of spacetime as indicated by the Hubble rate, which in the full development of this model is an exponential rate, drives this generation of (3).

The Fundamental Quantum Metric

We turn now to the metric, specifically a chargeless extreme Kerr metric in the equatorial plane (the ϕ coordinates are suppressed), in which the angular momentum parameter, a, is equal to the horizon reduced circumference and the geometrized mass, or $a = r_h = M_l$. (Note in the discussion of this section that τ indicates the proper time and not the fundamental force of the prior section, τ_0.) The time-like metric at the horizon is

$$d\tau^2 = \left(1 - \frac{2M_l}{r_h}\right)dt^2 + \frac{4M_l a}{r_h}dt d\theta - \frac{dr^2}{\left(1 - \frac{2M_l}{r_h} + \frac{a^2}{r_h^2}\right)} - \left(1 + \frac{a^2}{r_h^2} + \frac{2M_l a^2}{r_h^3}\right)r_h^2 d\theta^2 \quad (1.30)$$

Substituting for $a = M_l$ gives

$$d\tau^2 = \left(1 - \frac{2M_l}{r_h}\right)dt^2 + \frac{4M_l^2}{r_h}dtd\theta - \frac{dr^2}{\left(1 - \frac{M_l}{r_h}\right)^2} - \left(r_h^2 + M_l^2 + \frac{2M_l^3}{r_h}\right)d\theta^2 \quad (1.31)$$

We make the following observation concerning the dr^2 term. While the conventional interpretation is that the term goes to infinity as the denominator approaches zero, and any infalling test particle transits the horizon, the math can also be interpreted in terms of a limit for radial motion. A mathematical conflation is at work in the formulation, since the differentials are deemed to approach zero in the limit, but are effectively treated as dimensional units, ie. equal to one of some infinitesimal scale. This is necessary since the product of a non-zero co-efficient and a zero differential at the limit would be zero. This is warranted since we find a similar non-zero differential without a coefficient on the left side of the equation.

This is contradicted, however, if the metric component represented by the differential has a natural limit where it is necessarily zero. Thus, if the horizon in an extreme Kerr spacetime represents that limit, dr equals zero at the limit of that horizon coincident with the term in the denominator, the coefficient and the differential cancel. The result is simply -1 as shown below, which when factored gives an imaginary or orthogonal sense, ie. it rotates any differential change into tangency. The horizon, then, is effectively a physical asymptote. Thus, at the event horizon, where $r = r_h = M_l$ this simplifies to

$$d\tau^2 = -dt^2 + 4r_h dtd\theta - (2r)^2 d\theta^2 - dr^2 = (idt - i2r_h d\theta)^2 + (idr)^2 \quad (1.32)$$

This can be factored as a complex number and its conjugate

$$d\tau^2 = \left[(idt - i2r_h d\theta) + i(idr)\right]\left[(idt - i2r_h d\theta) - i(idr)\right] \quad (1.33)$$

or can be simplified as follows,

$$d\tau^2 = \left[(idt - i2r_h d\theta) - dr_h\right]\left[(idt - i2r_h d\theta) + dr_h\right] \quad (1.34)$$

where r_h is the reduced circumference at the horizon and $dr_h = 0$ is a zero vector with respect to the radial, giving a proper time of

$$d\tau = \pm i(dt - 2r_h d\theta) \quad (1.35)$$

If we assume that for bookkeeper time the differential is in the plane of the horizon, and time as developed earlier flows with the rotational motion of the ergo-sphere, so that

$$dt = r_h d\theta \quad (1.36)$$

then the proper time is found to flow orthogonally to that rotational motion, into the negative and positive ϕ coordinates, since

$$d\tau = \mp idt \quad (1.37)$$

This will be significant in our statement of the quantum metric.

From this perspective, at the static limit and the start of the ergo-sphere, where $r = 2M_l$, pure radial motion is no longer possible, and a rotational component or frame dragging element is injected into the equation so that at the event horizon, all motion is rotational as indicated by the "imaginary" or orthogonal senses. Note that if we consider spacetime as an inertio-elastic continuum, frame dragging is simply the wave strain associated with a rotational waveform, be it macrocosmic or quantum. Instead of gravitational collapse, this argues that any incremental matter or light accruing to the inertial sink is smeared out and bound at the horizon.

We now get to the meat of the matter with an expression of the quantum metric, $q_{\mu\nu}$, which we distinguish from the gravitational metric, $g_{\mu\nu}$. The dynamics of the quantum waveform is not extremely complicated, but it does involve some rather lengthy, non-standard analysis using methods of complex classical wave physics extended to 4 dimensions and is beyond the scope of the present discussion. We will simply state that its dynamics prevent the orientation entanglement condition.

With reference to Quantum Inertial Sink Diagram 1, the time-like quantum metric is given as a modified chargeless extreme Kerr metric. The modification is in the ϕ coordinates as shown here, where the quantum mass has been explicitly geometrized as r_{0n},

$$d\tau^2 = \left(1 - \frac{2r_{0n}}{r_{0n}}\right)dt^2 + \frac{4r_{0n}^2}{r_{0n}}dtd\theta - \frac{dr^2}{\left(1 - \frac{r_{0n}}{r_{0n}}\right)^2} - R^2 d\theta^2 \mp \left\{\left(e^{\pm i(\omega_0 t \mp \theta)} L d\phi\right)^2\right\} \quad (1.38)$$

The caveat stated above concerning the limit of radial motion represented by r_{0n} remains. In the last term, the complex exponential is defined as

$$e^{\pm i(\omega_0 t \mp \theta)} = \mathrm{Re}\left(e^{\pm i(\omega_0 t \mp \theta)}\right) = \cos_{ccw}(\omega_0 t + \theta) \text{ or } \cos_{cw}(\omega_0 t + \theta)$$
$$= \cos(\omega_0(+t) - \theta) \text{ or } \cos(\omega_0(-t) + \theta) \quad (1.39)$$

Either the real or the imaginary part could of course be used. The *ccw* term indicates rotation in the upper hemisphere according to the right-hand rule, while the *cw* term indicates clockwise rotation in the bottom hemisphere according to the left hand rule, when viewed from the exterior of the corresponding rotational pole.

The plus and minus curly bracket has the following definition and indicates a flipping of the sign of the $d\phi$ vector, with every π rotation of θ, plus being parallel and minus being anti-parallel with respect to the RHR spin axial vector. It thus performs a function similar to a mathematical spin matrix.

$$\pm\{a\} \equiv \frac{\cos(\omega_0 t - \theta)}{|\cos(\omega_0 t - \theta)|} a, \quad \mp\{a\} \equiv -\frac{\cos(\omega_0 t - \theta)}{|\cos(\omega_0 t - \theta)|} a \quad (1.40)$$

Obviously, θ and ϕ rotate at the same frequency, with the axis of the ϕ rotation rotating in the equatorial plane. This motion avoids the orientation entanglement condition and is necessitated by the assumed continuity condition of a classical spacetime continuum and the density property postulated in this development. When analyzed it is apparent that the motion is that of a transverse wave traveling in tight orbit around the spin axis, its amplitudes inclined toward the poles, analogous to a gravitationally bound, electromagnetic wave, and in fact constitutes the magnetic field of the quantum.

This diagram is a cross-section through the spin axis and shows the relationship of the static limit, the ergo-sphere, and the horizon. The ergo-sphere is the domain of the strong interaction. The transverse or ϕ differential is limited in its motion toward the spin poles to the point on the static limit where $L = 1$.

The metric simplifies at the horizon with no radial motion as
$$d\tau^2 = -dt^2 + 4r_{0n}dtd\theta - R^2d\theta^2 \mp \left\{\cos^2(\omega_0 t - \theta)L^2 d\phi^2\right\} \tag{1.41}$$

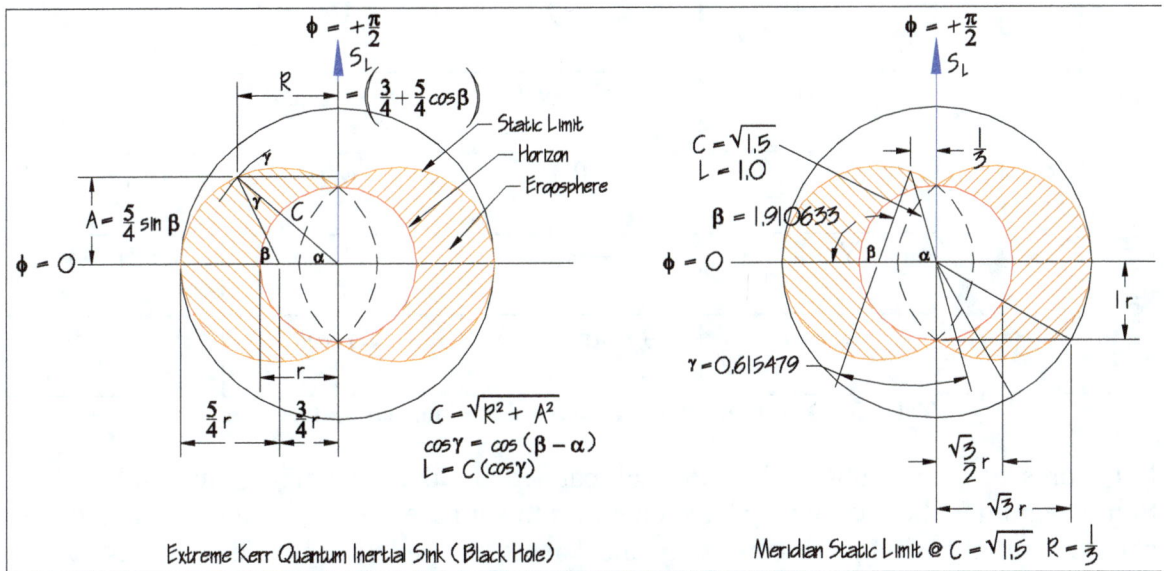

Quantum Inertial Sink 1

From this diagram we have the following coefficient component for ϕ along the meridians at the static limit
$$L = \tfrac{4}{5}r_{0n} + \tfrac{3}{5}R = \tfrac{4}{5}r_{0n} + \tfrac{3}{5}\left(\tfrac{3}{4} + \tfrac{5}{4}\cos\beta\right)r_{0n} = \left(\tfrac{5}{4} + \tfrac{3}{4}\cos\beta\right)r_{0n} \tag{1.42}$$
Substituting this in equation (1.41) simplifies at the horizon along the equatorial plane of a fixed spin axis where $\cos\beta = 1$, as
$$d\tau^2 = \left(idt - i2r_{0n}d\theta\right)^2 \mp \left\{\cos^2(\omega_0 t - \theta)(2r_{0n})^2 d\phi^2\right\} \tag{1.43}$$
The corresponding spacelike metric is
$$d\sigma^2 = -\left(idt - i2r_{0n}d\theta\right)^2 \pm \left\{\cos^2(\omega_0 t - \theta)(2r_{0n})^2 d\phi^2\right\} \tag{1.44}$$
giving the fundamental symmetry
$$d\sigma^2 \equiv -d\tau^2 \tag{1.45}$$
and for the proper time and space, indicating the orthogonal nature of space and time,
$$d\sigma \equiv id\tau \tag{1.46}$$

This can be represented by the following anti-symmetric orthonormal matrix at r_0,

		Direction of ortho normal vector dx_i with respect to		
		X Axis	Y Axis	Z Axis
Vector dx_i originating at	X = +1	0	$+rd\theta$	$+r\sin\omega t d\phi$
	Y = +1	$-rd\theta$	0	$-r\cos\omega t d\phi$
	Z = +1	$-r\sin\omega t d\phi$	$+r\cos\omega t d\phi$	0
	X = -1	0	$-rd\theta$	$-r\sin\omega t d\phi$
	Y = -1	$+rd\theta$	0	$+r\cos\omega t d\phi$
	Z = -1	$+r\sin\omega t d\phi$	$-r\cos\omega t d\phi$	0

Table 3 - Quantum Anti-Symmetric Orthonormal Matrix at r_0

In the presence of an anti-parallel external magnetic field as shown in Quantum Inertial Spin Diagram 2, the quantum spin axis inclines toward the equatorial plane and precesses about its initial position. The resulting coefficients of ½ spin can be seen here. Note also that the Heisenberg "observational" uncertainty is limited by the inverse curvature of the horizon to

$$r_0^2 c = m_{l0} r_0 c = \hbar \qquad (1.47)$$

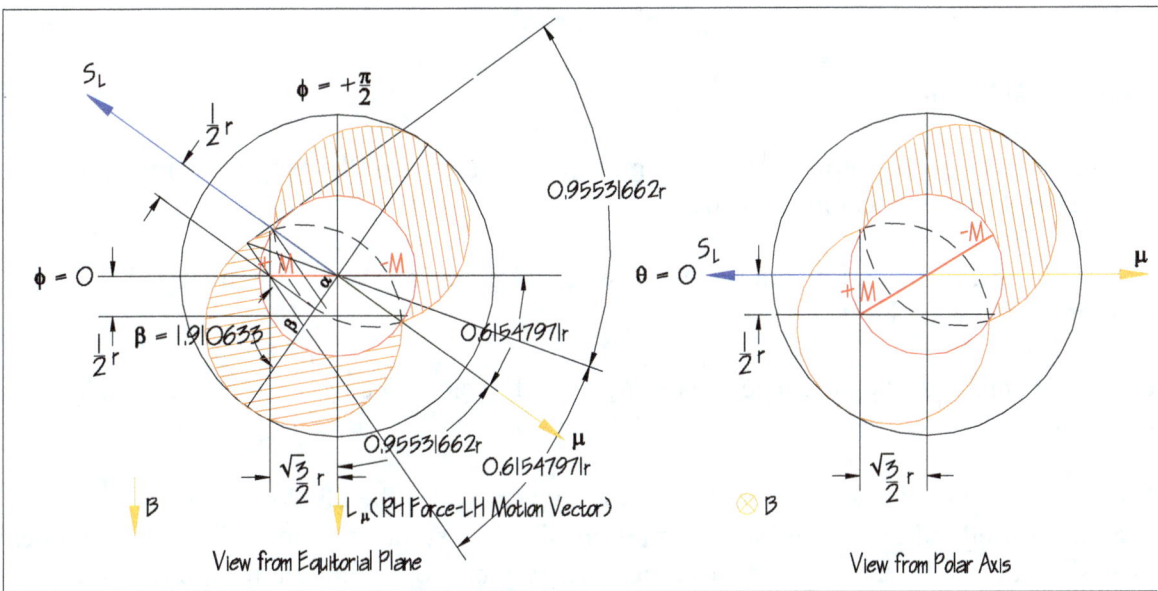

Quantum Inertial Sink 2

This model is elsewhere more fully developed and presented as the 3-D representation of a classical 4-D oscillation. Expansion acts as an EMF that drives the fundamental frequency, both by mechanical analogy and as the actual mechanical or piezoelectric basis for electro-magnetism. The rest-mass quantum is thus a small simple harmonic oscillator, with a potential-kinetic, capacitive-inductive energy cycle, in a general inductive mode during expansion, of which the waveform of ordinary matter is the result. During universal contraction, a capacitive mode ensues, resulting in a predominance of anti-matter.

<u>The Fundamental Rest Mass Field Equation in Flat Spacetime</u>

In flat spacetime in which the Ricci curvature tensor and scalar vanish, the product of the quantum metric, $q_{\mu\nu}$, and the cosmological constant/Hubble rate is equal to the fundamental stress tensor for a rest mass oscillation or

$$q_{\mu\nu} H_0 = 2T_{\mu\nu} \tag{1.48}$$

Here the Hubble rate gauges the expansion and the development of the time row and column of the doublet. In essence, $T_{00} = H_0$, and is a spacetime strain.

Bibliography

General References

M. Jammer, <u>Concepts of Mass in Contemporary Physics and Philosophy</u>, Princeton University Press, Princeton, NJ (2000).

C. W. Misner, K. S. Thorne, and J. A. Wheeler, <u>Gravitation</u>, W.H. Freeman and Company, New York (1973).

C. F. Stevens, <u>The Six Core Theories of Modern Physics</u>, The MIT Press, Cambridge, MA (1995).

National Institute of Standards and Technology, These are the **2002 CODATA recommended values** of the fundamental physical constants, the latest CODATA values available. For additional information, including the bibliographic citation of the source article for the 1998 CODATA values, see P. J. Mohr and B. N. Taylor, "The 2002 CODATA Recommended Values of the Fundamental Physical Constants, Web Version 4.0," available at physics.nist.gov/constants. This database was developed by J. Baker, M. Douma, and S. Kotochigova. (National Institute of Standards and Technology, Gaithersburg, MD 20899, 9 December 2003).

E. F. Taylor and J. A. Wheeler, <u>Exploring Black Holes</u>, Addison Wesley Longman, Inc, New York (2000).

N. J. Cornish, et al, "Constraining the Topology of the Universe", Physical Review Letters, Volume 92, Number 201302 (2004).

A study by R. Eastman, B. Schmidt and R. P. Kirshner in 1994 quoted in <u>The Extravagant Universe</u>, R. P. Kirshner, Princeton University Press, Princeton, NJ (2002).

[8] W. L. Freedman, et al, Astrophysical Journal, 533, 47-72 (2001).

<u>The Feynman Lectures on Physics</u> :Commemorative Issue, Feynman, Leighton, Sands, Volume I, Addison-Wesley Publishing Company, Inc., Reading, Massachusetts 1963.

<u>Fundamentals of Physics</u>, Fifth Edition, Halliday, Resnick, Walker, John Wiley & Sons, Inc. New York, 1997.

<u>Mathematical Methods for Physicists</u>, Fifth Edition, Arfken and Weber, Harcourt Academic Press, New York, 2001.

<u>Physics of Waves</u>, Elmore and Heald, Dover Publications, Inc., New York, 1985. This was the primary source for wave, elasticity and tensor equations.

<u>Visual Complex Analysis</u>, Needham, Oxford University Press, Oxford, 1997.

Simple Harmonic Motion

in

Classical and Quantum Phase Space

-

Fundamental Rest Mass Quanta as
Simple Harmonic Oscillations of
the Spacetime Continuum,
Driven by Cosmic Expansion

By Martin Gibson

August 1, 2013

Martin Gibson
P.O. Box 2358
Southern Pines, NC 28388
910-695-9274
martin@uniservent.com

Copyright © Martin Gibson 2021 All Rights Reserved

Simple Harmonic Motion in Classical and Quantum Phase Space

Abstract

In classical mechanics, natural, including man-made, repetitive motion sustained over extended time frames can be studied using the model of Simple Harmonic Motion (SHM) in which an oscillator and its support framework is deemed to be a closed system in that energy, momentum and related properties of the motion are conserved. In fact no system is completely closed and the energy of such motion is in some measure damped or otherwise lost to the background of the system. Still SHM can be closely approximated by driving the oscillation with controlled energy input. The oscillation of displacement and momentum in SHM can be accounted for succinctly with the use of planar phase space (PS_2) modeling in which the correspondence of the dynamics of linear oscillation and Uniform Circular Motion (UCM) is utilized. Accounting for energy, force, action, and power oscillation is better handled by graphic modeling of sine and cosine wave functions.

In the first section, these models are briefly recapitulated. In the second section they are applied to an analysis of the energy of the oscillation, through the Hamiltonian and Lagrangian approach, with the graphic development of the action, via both Lagrange and Maupertuis, and the power of the oscillation, and their semi-periodic maximum moments. The two models are synthesized in a three-dimensional phase space that we are calling PS_3. In the process of this synthesis it is shown that there is necessarily a component of the dynamics that is present even in the absence of oscillation, an inertial invariant that is both a scalar and vector potential. Finally, the point oscillator of the initial development is replaced by torsion oscillation as a disk, which can be represented by PS_2. The synthesis also suggests a quantum application of the modeling. Section three shows the development of rotation of the action and power moments with attendant torques which sustain the oscillation and result in the property of angular momentum, with an invariant Lagrangian as well as Hamiltonian. A review of the nature of body forces and stress or surface forces models PS_3 as a system of rotating stress force and corresponding strain, and the model is fully developed as an emergent quantum phenomena driven by an expanding spacetime fabric (STF) coupled with necessary geometric constraints. The neutron is shown to be the resonant state of PS_3. Spin and charge as elaborations of the angular momentum is developed, along with beta-decay for both ordinary and anti matter.

The Verification section derives a gravitational quantum, Newton's gravitational constant and law, ties beta-decay to cosmic expansion and thereby predicts the Hubble rate, which is shown to be an exponential rate. In the process the reason for the neutron-electron mass ratio is developed along with the nature of the missing mass of beta decay. Finally, the value of elementary charge is derived, with some interesting observations about the structure of the fine structure constant.

The Conclusion section waxes philosophical, concludes the PS_3 model deserves a proper vetting, and the Asides offer supporting information.

Table of Contents

Introduction	1
Simple Harmonic Motion and Two Dimensional Phase Space (PS$_2$)	3
Simple Harmonic Motion and Three Dimensional Phase Space (PS$_3$)	14
Simple Harmonic Motion and Rotational Oscillation or Spin Space	31
Verification	63
1) Quantum Gravity	63
Gravitational Quantum	64
Newton's Gravitational Constant	65
Quantum Newtonian Law of Gravity	66
2) Quantum PS$_3$ Oscillation States, Cosmic Expansion and Beta Decay	68
Classical Wave Mechanics	68
Beta Decay as a Function of Expansion	71
Neutron/Electron Mass Ratio	72
Derivation of the Hubble Rate, the Expansion Rate of the Cosmos	74
The Missing Mass of Beta Decay	77
Evaluation of Elementary Charge	81
Fine Structure Constant	82
Special Relativity and Muon and Tau Families	83
Conclusions	85
Aside #1, discussion of some implicit assumptions of the Calculus	88
Aside #2, discussion of body force and stress force and stress/strain analysis in three dimensions	92
Aside #3, (taken from an earlier work-in-progress) discussion of the geometric, work/energy constraints of isotropic expansion on a unit cube.	98
Bibliography and Other Resources	113

Figures and Photos

Figure 1	Linear and Circular Motion in Euclidean Space	4
Figure 2	Transition to Phase Space	6
Figure 3	Phase Space with Simple Harmonic Motion	7
Figure 4	Phase Space Viewed from Below (Time reversed)	8
Figure 5	Phase Graph of Momentum and Displacement	12
Figure 6	Phase Space with Hamiltonian	14
Figure 7	Phase Graph of Action	15
Figure 8	Phase Graph of Force and Power	16
Figure 9	Phase Graph of Kinetic and Potential Energy	16
Figure 10	Various Integrals/Areas under the Whole Curves	17
Figure 11	Various Integrals/Areas under the Half Curves	17
Figure 12	Action and Power Integrals with related components	18
Figure 13	Comparison of Lagrange and Maupertuis	19
Figure 14	Action/Power in Phase Space	19
Figure 15	Inertial Invariance Superposition orthogonal to Phase Space	20
Figure 16	Energy Flow in Phase Space	21
Figure 17	Inertial Invariance in Phase Space	22
Figure 18	Phase Time Differentials – Uniform Circular Motion	23
Figure 19	Phase Space & Time Differentials – Uniform Circular Motion	23
Figure 20	Phase Space & Time Functions – Uniform Circular Motion	24
Figure 21	Mapping of Phase graph to 2-D Phase Space	25
Figure 22	Collapse of $\mathcal{K}$ & $\mathcal{V}$ Integrals to Phase Space	26
Figure 23	Phase Space 3 for Reciprocal Path Oscillation	27
Figure 24	Phase Space 3 for Cyclical Path Oscillation	30
Figure 25	Phase Space 3 for Whole Cycle - Cyclical Path Oscillation	31
Figure 26	Spin Diagram 1	32
Photo 1	Cuboctahedral Lattice	37
Photo 2	Tetrahedral Aperture	37
Photo 3	Octahedral Aperture	37
Photo 4-6	Cuboctahedral Lattice extension	39

Figures and Photos (continued)

Figure 27	Spin Generation in PS_3, Initial Strain State	41
Figure 28	Resonant Strain State	42
Figure 29	Spin Diagram 2, Neutron	45
Figure 30	Strain States of Ordinary Matter at Beta-Decay	46
Figure 31	Inductive Strain States, Electron and Proton	48
Figure 32	Spin Diagram 3, Proton	49
Figure 33	Spin Diagram 4, Electron	50
Figure 34	Strain States of Anti Matter at Beta-Decay	51
Figure 35	Capacitive Strain States, Anti Proton and Positron	52
Figure 36	Spin Diagram 5, Anti Proton	53
Figure 37	Spin Diagram 6, Positron	54
Figure 38	Charge and Spin Table for Ordinary Matter	55
Figure 39	Charge and Spin Table for Anti Matter	56
Figure 40	Charge and Spin Table for Ordinary Matter for C & L = 1	57
Figure 41	Charge and Spin Table for Anti Matter for C & L = 1	58
Figure 42	Comparison of Strain States	59
Figure 43	Comparison of Strain States with Parallel Spin	60
Photo 7	Toy Models of Anti Proton, Neutron and Proton	61
Photos 8-11	Close-ups of Toy Models	62
Figure 44	One Half of an Inversphere	67
Figure 45	Superposition of Equal Area Cube and Sphere	72
Figure 46	Electron mass determination in PS_3	73
Figure 47	System of exponent bases e_n	78
Figure 48-49	Table of exponent bases e_n	79
@Conclusion	Matrix of PS_3 Functions and Invariants	87

Introduction

Nature is repetitive. Sure, she always offers something new and is in constant change; yes she changes, but in recurring patterns. Day follows night follows day. Summer follows winter follows summer. Heat follows cold follows heat. Given sufficient breadth of vision most, if not all, linear processes of change prove to be phases of a bigger cycle, where the extents of the cyclic pattern are found to be the extremes of some linear dimension. Birth to youth to maturity to old age to death is linear enough for the individual living creature, but from the perspective of her species, the line resonates in each and every birth.

A realist would say there is small wonder that those stuck upon the terminating line of life should look for comfort in the endless circle. Rather, as one who has had sufficient experience of finding an uncharted route prove to be a roundabout, I would say there is little reason to be satisfied with the boundaries of one dimension if one has yet to reach the vista and added breadth of higher ground.

From the smallest of atoms to the largest of galaxies, nature displays herself in circles and cycles and all manner of recursiveness magnificent to behold. What is more wondrous yet is that all of this can be understood in an ideal, simplified manner, in a fundamental form by the concept of simple harmonic motion (SHM) or simple harmonic oscillation, the changes in position and momentum accounted for by comparing motion about a circle with motion of the same periodic frequency along a simple line segment, its diameter. Some examples of physical processes that clearly embody the concept are pendulums of relatively small arcs of motion, massive objects attached to springs that extend and compress, and stretched strings that vibrate in sinusoidal fashion, each in response to some extraneous impulse that sets them in motion. In reality each of these systems involves inherent components and external connections that drain away the energy of oscillation over time, but in the ideal world of the mind, free from such external and inherent interaction, once set in motion these systems oscillate indefinitely. So we can learn much from an ideal description.

This discourse is intended for a general readership with some level of technical education or experience or the ability to gain the same through self-directed effort. (The Internet is the obvious source of information in this regard. Wikipedia in particular has excellent graphics, including animations, to help explicate the ideas.) It is for those with curiosity and will hopefully have something new for novice and expert alike; therefore, it will take pains to explain some basic concepts in some detail, while not avoiding the use and some assumed knowledge of special language.

In our description of oscillation, we define the above referenced line segment as $2r_0$ long, the diameter of a circle with a radius of r_0. The naught subscript in this discussion indicates a unit value, or in some cases characteristic value of the corresponding property that the letter represents, in this case, a change in position, i.e. a displacement or length along some linear dimension, such as x or y or q. As such, r_0 equals x_0 equals y_0 equals q_0 equals 1 unit of length of some undefined system of measurement, without regard to the direction it is headed.

To the expert, if I depart from a rigorous notation in this piece with respect to vector notation, I apologize. I assume that the reader knows the difference between a length as a *scalar*, i.e. the magnitude of the difference between two positions in space (or time) as measured by an appropriate scale, and length as a *vector*, which adds to this magnitude the direction of the linear difference starting at one end of a standard or the other. Virtually any scalar can be made a vector by taking its gradient, the direction it is likely to change in space or time.

I will mention one refinement in the concept of direction, that of *sense*, which we generally think of as the sign, as +x or -x, of a direction otherwise understood. Thus x_0 and y_0 can be understood as explicit vectors of either sense with respect to the x and y co-ordinates of some rectilinear system, q_0 as a generalized vector that may correspond to any x_0 or y_0 or even z_0 depending upon the defined context, and r_0 as a vector of inherently indeterminate or changeable direction and sense.

Simple Harmonic Motion and Two Dimensional Phase Space (PS$_2$)
Accounting for Displacement and Momentum

Let us consider a simple pendulum, a plumb bob at rest, hanging from a string about a meter long just above a central point, free to move in any angle, θ, of 360 degrees or 2π radians. We place a piece of paper under the bob with a circle of radius $r_0 = x_0$ its diameter clearly marked. For future reference we draw a second diameter 90 degrees or ½ π radians from the first, and we extend both diameters through to the edge of the paper and label one of the diameters +x and the other one +y as in Figure 1. We place the paper so that the center cross hairs of the circle are directly beneath the point of the plumb bob. The one I am using weighs about a pound and if you unscrew the collar that retains the line from its top, it reveals a miniature bob weighing about an ounce, nested away like a Russian doll.

It doesn't matter which way we orient the paper, as indicated by the dotted line axes x'–y' or what we call them. In fact, we can wait until we set the pendulum in motion, swaying back and forth along a gentle arc, before we turn the paper so that the x diameter and axis aligns with the arc of the oscillating plumb bob, designated in Figure 1 as O_l. An ideal pendulum, once set in motion, will swing back and forth forever. "Ideal" means the system which includes the table and the paper and the pendulum and the tripod from which the pendulum is suspended, along with the gravity that makes it all work, is isolated from any other activity which might effect it. The energy in such system is defined as being conserved, i.e. no energy is lost or gained from the system. In reality, the oscillation will cover a smaller and smaller arc over time due to *damping* and other causes, that is to say, it loses energy to the air around it and to the tripod through friction and other forces at the strings attachment. If we are to keep it going, we must *drive* the oscillation by adding a small amount of energy, ideally the amount that is damped away, as we might push a child in a swing to keep her going. In an ideal situation with no damping or driving, or in a controlled setting with driving offsetting the damping, the oscillation operates at its resonant angular frequency, which we designate as ω_0. The angular frequency measured in radians, θ, is related to a cycle or single period of the oscillation, T_0, and the periodic frequency, f_0, by

$$\omega_0 = \frac{2\pi}{T_0} = 2\pi f_0 = \frac{d\theta}{dt}. \tag{1.1}$$

For reasons both practical and arcane, we will use angular frequency in our discussion unless noted otherwise.

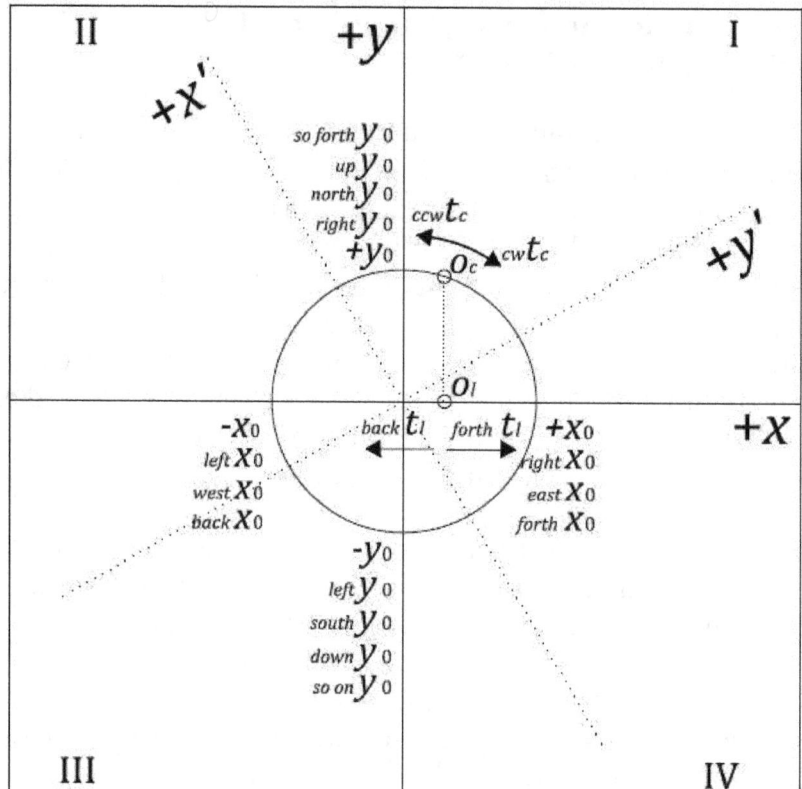

Linear and Circular Oscillation

Figure 1

We watch as the plumb bob swings back and forth along the *x*-axis and across the center point. At some point, since it is not ideal, the extent of its arc will lessen until it extends from the +x_0 to the -x_0 points where the axis crosses the circle. Notice that the terms *plus* x and *minus* x are simply reference designations to help us make *sense* of what is occurring. We just as correctly could have called these points *right* x and *left* x or *east* x and *west* x or *forth* x and *back* x or any of other terms as shown. Nor is it necessary, at this stage, that the *y*-axis have the same sense designations as the *x*-axis, though that will change. If we were surveyors, we might prefer east and west for *x* and north and south for *y*.

An interesting thing we note as the pendulum swings back and forth is that for small arcs, the frequency does not depend on the *amplitude* or extent of the oscillation along the *x*-axis. If we count the number of cycles that occur over a set period of time when we first set the bob in motion and count again after it has decreased to well within the circle, the frequency will be closely the same. We might think that if we change the size of the plumb bob it would affect the frequency, as in my case by removing the larger outer bob to leave the much less massive hidden bobby. In fact we find that it is unchanged.

The frequency, therefore, is not a function of the mass or, generally, the force that set it in motion in the first place. If we shorten the line that is suspending the bobs however, we find that the frequency increases. If we lengthen the line and hold it off of the table, the frequency decreases. We find that the resonant frequency of the oscillation is an inverse function of the length of the line suspending the plumb bob. It is inversely related to the *square root* of the length of the line, l_{pen}, and directly related to the square root of gravitational acceleration at the location of the pendulum, g_{pen}, as

$$\omega_0 = \sqrt{\frac{g_{pen}}{l_{pen}}} \qquad (1.2)$$

Squaring this statement and rearranging gives a statement for that acceleration,

$$g_{pen} = l_{pen}\omega_0^2. \qquad (1.3)$$

It is worthy of note that for the pendulum system, the acceleration of g_{pen} is generally parallel to l_{pen}, or toward the earth, and perpendicular to the angular acceleration represented by ω_0^2, which oscillates about the center of the bob's travel, the point of rest or equilibrium represented by the center of our circle. In the case of O_l, the acceleration vector dips downward near each end of the bob's travel and points upward as it passes its point of equilibrium beneath its pivot. Wikipedia http://en.wikipedia.org/wiki/Pendulum demonstrates this oscillation.

We notice another feature of the system. The plumb bob over time begins to deviate from its arc along the x-axis and begin to follow an elliptical path, clockwise (cw) or counterclockwise (ccw). With a little help, we can nudge it into a circular path in a uniform circular motion (UCM), and we find that for a given length of pendant line, the frequency in the circular path is the same as the frequency along the diametric path. Since the circular path is longer than that of the diameter, this means that the average speed along the circle must be greater than the average along the diameter. We might suppose that the speed is constant along the circle, while obviously the bob stops at each end of its diametric travel. It must accelerate back toward the other end after it stops, and we wonder about its top speed on the way back.

When we study the system carefully, we find that if we had two bobs with equal pendulum length, one traveling the circular path at O_c and one traveling the diameter at O_l, so as magically not to interfere with each other, synchronized so that they both cross the y-axis at the same time, at any point in time the velocity of O_c in the x direction, i.e. projected on to the x-axis, equals the velocity of O_l on that axis. As shown in Figure 2, their speed, sense and direction along x are the same, and the maximum velocity of O_l, when it crosses the y-axis and is instantaneously parallel to the path of O_c, is the same velocity as O_c.

This coincidence means that the *momentum*, p_l, of O_l, which is equal to the *mass* of the bob times its velocity, equals the projected momentum of O_c along the x-axis, provided O_c has the equivalent mass as O_l. Since O_c has a constant angular velocity and tangential speed and the same mass as O_l, it means that the projected momentum of O_c onto any diameter, including the one on the y-axis, is equal to the momentum of O_l along x at some point in time. The momentum of O_c along y is $\pi/2$ out of phase with its momentum along x, so that the magnitude of the momentum of O_c projected onto one axis can be read by its position with respect to the other axis. Thus, when O_l is at the origin or center of the circle and its displacement is 0, O_c is crossing the y-axis and the momentum of O_l at the origin position can be correctly read by the position of O_c on the y-axis.

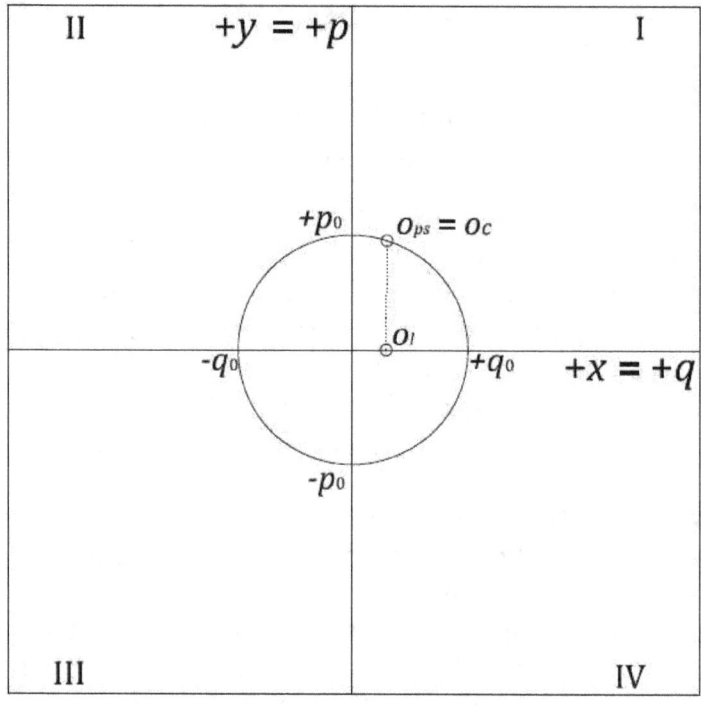

Transition to Phase Space

Figure 2

This provides us with a convenient analytical tool for investigating SHM known as phase space. Since the motion of a body or particle in UCM can be projected onto any arbitrary diameter, we can assign an axis associated with that diameter the designation q as a generalized co-ordinate, and the axis normal or perpendicular to it the designation p for the related or conjugate momentum. Conversely, we can project the linear motion of a one-dimensional oscillator onto an associated circle of UCM with the same effect, substituting O_{ps} for the circulating plumb bob, O_c. In the above diagram, x has become q and y has become p, though we could have done the same thing to any set of orthonormal axes.

There is an important caveat to this statement. I have said the "magnitude of the momentum of O_c ... can be read" and not simply "the momentum of O_c", because momentum is a vector quantity. It has a direction, which is determined by the velocity, the speed and direction of travel of the oscillator. If O_c is traveling either cw or ccw or overhead/underneath for that matter and its projected velocity on x or q is synchronized with O_l, it will make no difference to O_l which way O_c is circling. That is to say, in this case at O_l we cannot tell whether O_c is moving cw or ccw. All we know is that it is moving in the same direction relative to the x-axis as O_l. It does make a difference, however, if we are to use the y-axis to record p_l, the conjugate momentum of O_l. The sense of the momentum p_l indicates the direction O_l is traveling, while the sense of the position of O_l, q_l, indicates whether O_l, is to the left (negative) or right (positive) of the y-axis. As a result, half of the time p and q will be of different sense. They will be $\pi/2$ out of synch or out of phase.

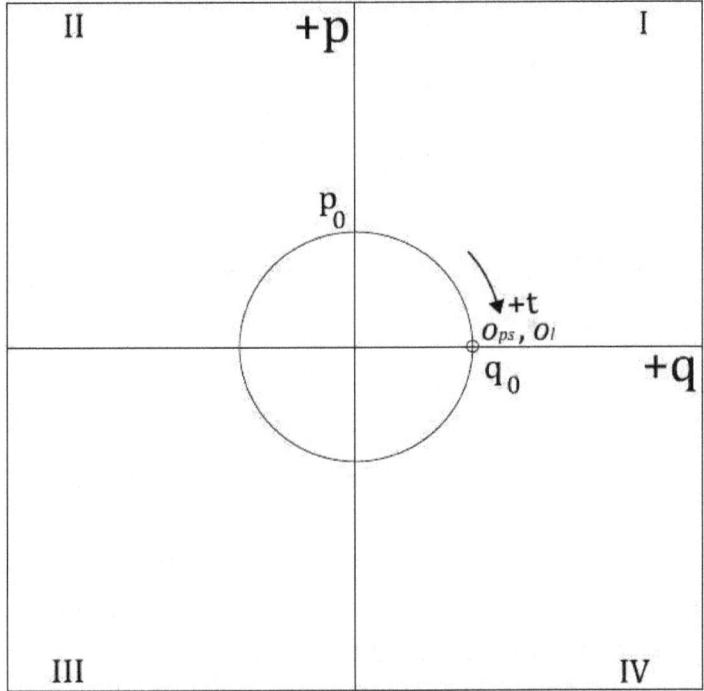

Phase Space with Simple Harmonic Motion

Figure 3

As seen in Figure 3, this dictates that for SHM in phase space using the standard configuration for the positive ordinate (y-axis) and abscissa (x-axis), unlike the paper plane onto which we have projected the traveling point of our plumb bob, O_{ps} can only move in one circular direction, cw, if we are to read the correct sense of the conjugate momentum of O_l from the p-axis component of the position of O_{ps}. Clearly, the rotational sense of O_l, conventionally given as + for ccw and – for cw in the Cartesian system or $+i$ and $-i$ using complex notation, is a subjective value determined by the point of reference of the observer. The same physical oscillation, viewed from the back side of phase space, as in Figure 4 in which the positive and negative sense of the q-axis are transposed, would be seen moving ccw, with the correct relative sense designations for both q and p in all four phases of the oscillation as seen below.

(It is serendipitous that the font images for generalized displacement, q, and its conjugate momentum, p, in non-italicized form of some font styles as here, are the mirror image of each other. If we took this at face value in the following figure, we would be forced to conclude that ccw rotation is not possible in this view either, since q appears to be leading p. In conclusion, the sense, direction and extent of momentum must lead the displacement values by one phase or $\pi/2$ in any representation, since that is what it does physically in SHM.)

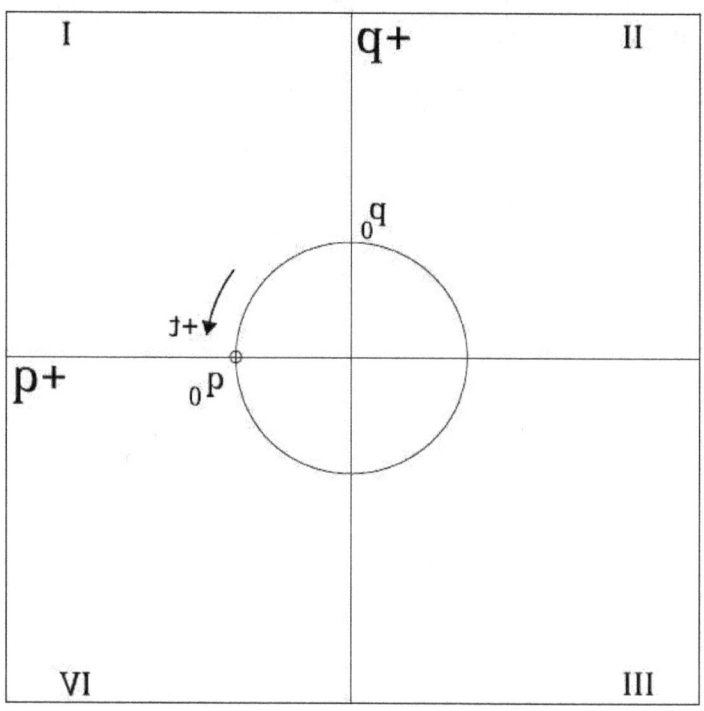

Phase Space Viewed from Below
Time is reversed
Figure 4

While the use of phase space in depicting the position and momentum flow of a simple harmonic oscillator is interesting, a greater interest might be the flow of energy in the system. Such energy is of two types that are transformable one to the other. In general, they are constantly in flux while their total in a closed system remains invariant. With respect to the present discussion of SHM, potential energy, $\mathcal{V}$, is a function of the position of the oscillator, O_l, and kinetic energy, $\mathcal{K}$, is a function of its velocity and its conjugate momentum. Therefore in Figure 3, $\mathcal{V}$ is at a maximum, $\mathcal{V}_0$, when O_{ps} is at $+/-q_0$ where $\mathcal{K}$ momentarily vanishes, while $\mathcal{K}$ is at a maximum, $\mathcal{K}_0$, when O_{ps} is at $+/-p_0$ where $\mathcal{V}$ momentarily vanishes. Clearly, the total energy of the system oscillates as does the oscillator, O_l, and it would be convenient and elucidating if we could map it onto phase space, so we will start with a few equations.

As stated, the potential energy of the system is a function of the displacement of O_l, so in derivative form the function, $\mathcal{V}'(q)$, can be written as the change in $\mathcal{V}$ with respect to a change in q which is equal to the resultant force, τ, which displaces it and which the gravitational field exerts on O_l in returning it toward its equilibrium position, where the force vanishes at $q = 0$, as

$$\mathcal{V}'(q) = \frac{d\mathcal{V}}{dq} = \tau = m_{O_l}\ddot{q} = m_{O_l} q \omega_0^2 \tag{1.4}$$

Here we have used the dotted mechanical notation for time derivatives, where

$$\dot{q} = \frac{dq}{dt} = v, \text{ velocity}$$

$$\ddot{q} = \frac{d^2q}{dt^2} = a, \text{ acceleration} \tag{1.5}$$

$$\dddot{q} = \frac{d^3q}{dt^3} = j, \text{ jerk}$$

The kinetic energy of the system is a function of the momentum of O_l, so in derivative form the function, $\mathcal{K}'(p)$, can be written as the change in $\mathcal{K}$ with respect to a change in p which is equal to the velocity of O_l, where ω_0 momentarily slows to 0 at $\pm q_0$, or

$$\mathcal{K}'(p) = \frac{d\mathcal{K}}{dp} = \dot{q} = q\omega_0 \tag{1.6}$$

The force of (1.4) is related to the gravitational field force responsible for the acceleration found in (1.3), as

$$\tau = m_{O_l}\ddot{q} = \sin\phi\left(m_{O_l} g_{pen}\right) = \sin\phi\left(m_{O_l} l_{pen} \omega_0^2\right) \tag{1.7}$$

where the mass, m, is the mass of O_l and ϕ is the instant angle between the plumb line and its vertical direction at rest. Note that according to (1.2), the two bracketed terms are invariant conditions of the pendulum setup and therefore of the system being examined, but that transverse force, τ, varies sinusoidally with the oscillation, as determined by ϕ, so that

$$q = \sin\phi\, l_{pen}. \tag{1.8}$$

With respect to $\mathcal{V}$, if we assume SHM and no damping or loss of energy, then q_0 is an initial condition determined by the force used to displace the plumb bob and set the oscillation in motion which establishes the angle ϕ, which in the context of phase space is the invariant angle for O_{ps}, or

$$\tau = \tau_0 \sin\phi_{Ops} = m_{O_l} q_0 \omega_0^2 = \left(m_{O_l}\omega_0^2\right)\sin\phi_{Ops} l_{pen} \tag{1.9}$$

A similar statement can be made for $\mathcal{K}$, as

$$c = c_0 \sin\phi_{Ops} = q_0\omega_0 = \sin\phi_{Ops} l_{pen} \omega_0. \tag{1.10}$$

We should note that the conditioning properties and parameters, g_{pen}, l_{pen}, ω_0, as well as the initial condition that establishes ϕ, are not a part of phase space, and in fact are orthogonal to, i.e. outside it, just as the tripod, table, string and earth are outside our Figure 1-4 paper planes. Those familiar with such things will recognize that the term in brackets in (1.9) has the form of the spring constant, k_s, of Hooke's Law for elastic bodies in compression as in a spring-mass mechanism exhibiting SHM. I have not included any graphics of this mechanism, which are easily found on Wikipedia and elsewhere on the Internet.

In the case of such mechanism for which k_s is a measure of the stiffness of the spring material, the restorative force of the spring, τ_0, after compression or extension is equal to k_s times the strain or change in length of the spring, which would be q_0 in phase space. The spring constant then is

$$k_s = m_{O_I}\omega_0^2 \tag{1.11}$$

Unlike the pendulum, for which the restorative force of gravity is operating normal to the plane of phase space and the line of the initiating force, it first appears that the initiating and restorative force are anti-parallel in the case of the spring-mass mechanism; that is, parallel but of opposite sense. Careful analysis shows, however, that in the case of the spring, the equivalent of the gravitational force is actually distributed throughout the spring, normal to its travel, so the mechanisms are analogous. The spring at the points of maximum extension and compression stores the maximum potential energy, V_0, and the oscillating mass, at the point of initial rest, exhibits the maximum kinetic energy, K_0, which are expressed here as characteristic, unit values.

One other mechanism of SHM that exhibits some of the features of both spring-mass and pendulum oscillation is that of a transverse wave on an ideal stretched string. We do not include any harmonics of such wave action in this analysis and assume only a fundamental characteristic frequency in keeping with SHM. Such a mechanism has two primary conditioning properties, tension stress, f_t, or force per unit area, and inertial density, ρ, mass per volume of the oscillating medium. In the case of a string, which is modeled along one dimension, the cross-sectional component of the stress and the planar parallel component of the volume in the density cancel in the following equation and we are left with the tension force parallel to the string, τ_t, and linear inertial density, λ, in their determination of the square of the transverse wave speed as

$$c^2 = \frac{f_t}{\rho} = \frac{\tau_t/A_0}{m/r_0 A_0} = \frac{\tau_t}{\lambda} \tag{1.12}$$

From (1.10) we can see that for a given instance of SHM of frequency, ω_0, we have the following

$$c_0^2 = q_0^2 \omega_0^2 = \frac{\tau_t}{m_{O_I}/q_0} \tag{1.13}$$

$$k_s = m_{O_I}\omega_0^2 = \frac{\tau_t}{q_0} \tag{1.14}$$

showing that the spring constant can be understood as the linear force density of the oscillation medium, be it gravitational field, elastic continuum/body or stretched string, where the linear unit is expressed in terms of the characteristic displacement, q_0. There is a difference, however, between the oscillating string and the other two examples of SHM in that the mass, m, in the first two examples is that of a separate body, the plumb bob and the oscillating weight, both of which generate a body force, F_b, due to their initial acceleration, a, from rest according to Newton's second law of motion

$$F_b = ma. \qquad (1.15)$$

The oscillation media or fields, i.e. the gravitational field/pendulum line and the spring, respond in equal and opposite direction to F_b, according to Newton's third law of motion but as instances of stress force τ_t. In both cases the body force of the massive object moves transverse or normal to the stress force, which is a tension stress force. In the case of the pendulum, this transverse motion is apparent. In the case of the spring, it may be less obvious, but the tension in the coils of the spring or the molecular bonds of an analogous solid elastic body that provides the restorative force is oriented normal to the travel of the massive body.

The relationship between the tension force in the oscillation field, τ_t, and the transverse force of the oscillating body, τ_s, can be described as the operation of a stress field, a type of tensor field. A tensor is a mathematical description of the way forces or velocities or other properties distribute in space and time, as when you swing a mallet down on a tomato and watch it explode out on all sides. Try this. It's fun in a juvenile sort of way.

A tensor is essentially a description of redirection of some conserved property often involving rotations and is therefore an elaboration of trigonometry. A stress field has two essential components with their corresponding forces, tension/compression, directed normal to the surface of a small unit volume of space, and shear, directed along the edges and in the plane of each surface of that unit. Since stresses describe the relationship of a force to some surface area, those stress forces are also called surface forces. A good example of the interaction between a body force and a stress force is a person jumping on a trampoline.

In the case of SHM on an ideal stretched string, the mass is the mass of the string itself, of the oscillation field, and the force it exhibits is a stress force. Whatever type of force may displace the string and initiate the oscillation, its ongoing SHM is an oscillation of stress forces in which a portion of the longitudinal tension stress transforms to transverse tension stress before recoiling in alternating direction to either side of its position of rest. It is like the trampoline bouncing back and forth sans person, like a drumhead. In this case the transverse force of the oscillation is

$$\tau_s = -\cos\theta\tau_0 = \pm i \sin\phi\tau_t \qquad (1.16)$$

There are two generally recognized types of waves on a string, traveling and standing or stationary. Traveling waves are customarily modeled on an indefinitely long string, so that their shape is that of a graph of the sine (or cosine) function of an angle, θ, as it increases from 0 through one cycle of 360 degrees or 2π radian, with the function of q mapped with respect to the ordinate and the angle mapped to the abscissa. If the angle represents an event rotating or cyclically occurring at a uniform rate, i.e. UCM, the abscissa can represent time and time can be measured in radians of either *cyclic or linear* change. A snapshot of a transverse wave *in* time and *over* an interval of space looks like such a graph and a video *over* time exhibits SHM at each point along its length. The result is the familiar sinusoidal wave of Figure 5.

A wave traveling on a string of finite length will eventually reach the string ends, which we assume to be rigidly fastened to immovable objects, so that the oscillation is not damped. The wave is reflected at these ends and begins to travel back down the string toward the other end. Waves traveling in opposite direction reinforce and cancel each other as they interfere, according to the wave half-length, πq_0. The points of cancelation form nodes where the string exhibits no transverse motion, every πq_0, and antinodes or points of maximum displacement at the half points between each node. If we have some very precise grips, we can clamp the nodes in the following graph at $-\pi/2$ and $\pi/2$ along the θ axis and the point at q_0 will continue to oscillate in SHM. If we view this graph as a physical half wave in motion with respect to axis θ, the graph of p between 0 and 2π maps out the conjugate momentum for q, starting from its position at q_0 for $\theta(0)$.

We should not loose sight of the fact that within the context of a one-dimensional oscillator, momentum is always in the same direction as the displacement over time, as seen here, though their magnitudes, p and q, are still out of phase by $\pi/2$. The following phase graph Figure 5 shows both p and q map to the ordinate and the phase propagation, θ, over time maps to the abscissa. The phase space diagram above then rotates the q and θ dimensions from the phase graph clockwise $\pi/2$, or using complex notation, by a factor of $-i$; q mapping to the abscissa and θ mapping to the circle contour at a distance of $|r_0| = |p_0| = |q_0| = +1$ from the Phase Space origin.

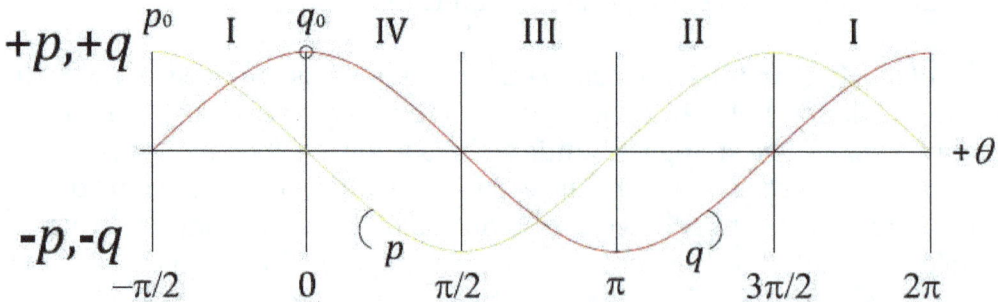

Phase Graph of Momentum, p, and Displacement, q
$p = \sin \theta \, p_0$, $q = \cos \theta \, q_0$
with counter-clockwise phasing and sense of function determined by standard Cartesian co-ordinate system

Figure 5

From this it is straightforward to define a two dimensional phase space, $PS_2 = \theta$, where $\theta = 0$ lies on the positive abscissa and the senses of the trigonometric functions are taken from projection onto the appropriate co-ordinate axis of a complex plane, where $p_0 = +iq_0$

$$r(\theta) = r_0 e^{i\theta} = q_0 \cos\theta + iq_0 \sin\theta \tag{1.17}$$

It is important to the following development to emphasize that θ is essentially a clock and at time $\theta = \pi$, the direction of time in phase space with respect to the conjugate momentum has reversed itself from its initial direction at time $\theta = 0$. In other words, in such a phase space, time oscillates.

Finally, as those musicians who play string instruments will know, a string doesn't simply oscillate in one dimension normal to the string, as in the plane of the phase graph. It is free to move transversely all about the string, and we can imagine an ideal condition under which it moves in UCM about the position of the string at rest. Under these circumstances the point q_0 circulates in the manner of O_c and O_{ps} in the above diagrams, clockwise and counterclockwise depending on the end of the θ axis from which it is viewed.

Simple Harmonic Motion and Three Dimensional Phase Space (PS₃)
Accounting for Potential, Energy and Time

Let's next consider a simple harmonic motion or oscillation (SHM) in PS$_2$, where,

$$p^2 + q^2 = r_0^2 = 1, \qquad (2.1)$$

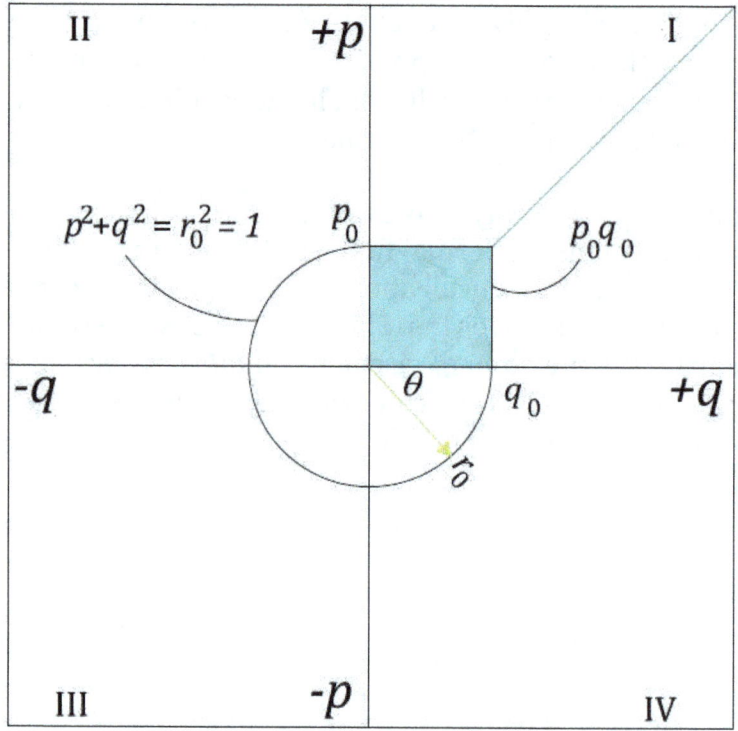

Phase Space with Hamiltonian

Figure 6

In Figure 6, the square designated as the product $p_0 q_0$ represents the total energy or Hamiltonian, $\mathcal{H}$, of the system that cycles about phase space at the cyclical frequency $\omega_0/2\pi$, for an invariant energy of

$$\mathcal{H} = p_0 q_0 \omega_0 = 1. \qquad (2.2)$$

Since the total energy of the system is conserved, this means that the Hamiltonian is equal to the total, at any point in time, of $\mathcal{K}$ and $\mathcal{V}$ which oscillate sinusoidally, as

$$\mathcal{H} = \mathcal{K} + |\mathcal{V}| \qquad (2.3)$$

The absolute value for $\mathcal{V}$ indicates that the potential energy is always negative due to the fact that its components τ and q are always of opposite sense, while $\mathcal{K}$ is always of positive sense, since its components p and c are always of the same sense, as seen in Figure 9. In these graphs p and c are represented by the same line. For both $\mathcal{K}$ and $\mathcal{V}$ the two components are always of the same relative magnitude.

The Hamiltonian is related to another relationship between $\mathcal{K}$ and $\mathcal{V}$, the Lagrangian, $\mathcal{L}$, referring to (1.4) and (1.6), though in this case $\mathcal{K}$ is defined as a function of the rate of change in q over time instead of the momentum, as

$$\mathcal{L} = \mathcal{L}(q,\dot{q}) = \mathcal{K} - |\mathcal{V}| \tag{2.4}$$

and since $\mathcal{K}$ and $\mathcal{V}$ oscillate in magnitude between 0 and 1 out of phase with each other, it is clear that the Lagrangian is not invariant, except in a case of UCM in which $\mathcal{V}$ is the scalar invariant product of $\tau_0 r_0$ and $\mathcal{K}$ is the scalar invariant product of $p_{0\tan} c_{0\tan}$, tangential momentum and velocity. The product $p_0 q_0$ has the units of action, $\mathcal{S}$, which is generally defined as the time integral of the Lagrangian, $\mathcal{L}$,

$$\mathcal{S}_t = \int_{t_i}^{t_f} \mathcal{L}(q,\dot{q}) dt \tag{2.5}$$

but is also defined as by Maupertuis as the displacement integral of an impulse, $\mathcal{J}$,

$$\mathcal{S}_q = \int_{q_i}^{q_f} \mathcal{J}(q) \cdot dq \tag{2.6}$$

the impulse being the time integral of a force which is equal to a change in momentum, Δp,

$$\mathcal{J} = \int_{t_i}^{t_f} \mathcal{F}(t) dt = \Delta p \tag{2.7}$$

Thus, in the context of SHM in phase space, the integral of the change in momentum over the displacement path of the oscillation is equal to the time integral of the Lagrangian over the same path. Let's see how this graphs out using the sine/cosine functions with respect to Maupertuis.

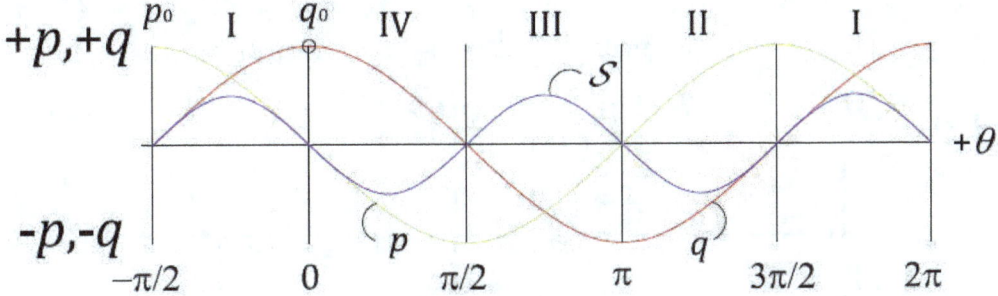

Phase Graph of Action

$\mathcal{S} = pq$

Figure 7

In Figure 7 we see that the action varies sinusoidally at twice the frequency and half the amplitude of the phase cycle, peaking at mid phase in each direction. We can add the force graph to this in Figure 8, which is the inverse of the q graph, and plot the power of the oscillation, $\mathcal{P}$, which is the product of the force and velocity, the latter being the same as the momentum graph or

$$\mathcal{P} = \tau \dot{q} = \tau c \tag{2.8}$$

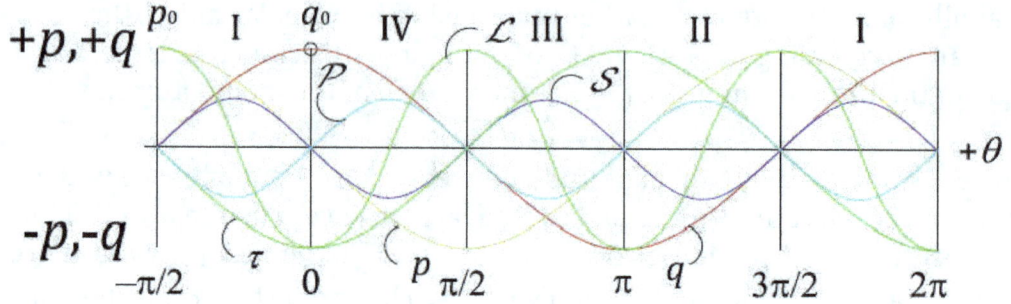

Phase Graph of Force, τ, and Power, $\mathcal{P}$

$\tau = -\cos\theta\, q_0\, m\omega^2 = -m\omega^2 q$

$\mathcal{P} = \tau c$, where $c = p/m = q\omega$

Figure 8

It is clear that the power graph is the inverse of the action graph. We can next plot the kinetic and potential energy curves, which will also give us the Lagrangian, the green curve in the first graph, and the invariant Hamiltonian shown in the second.

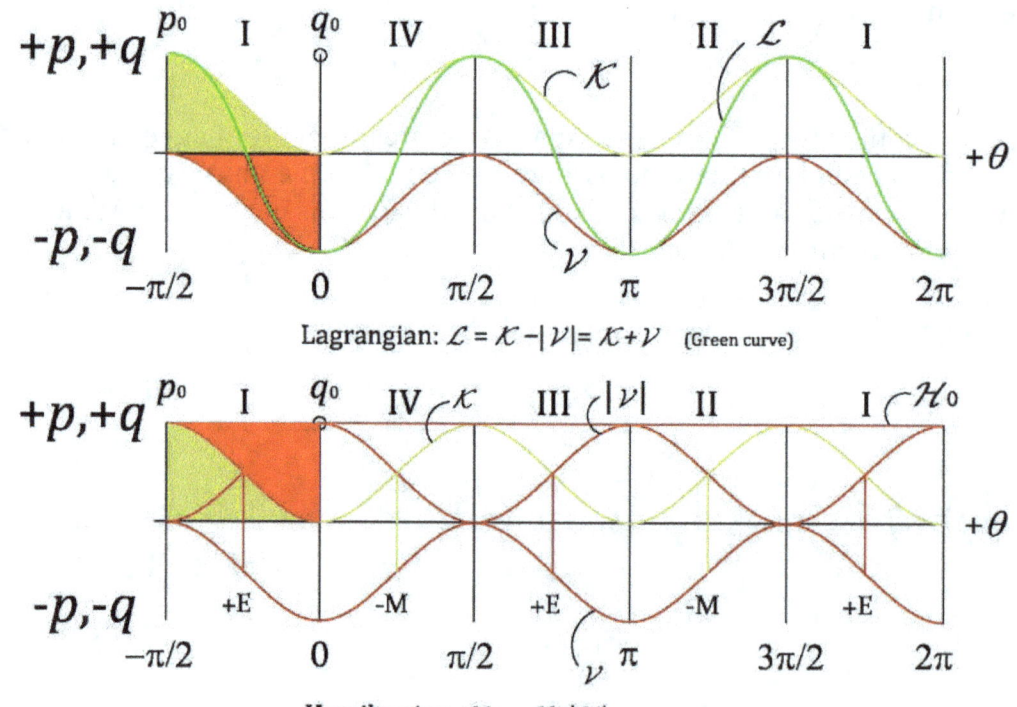

Lagrangian: $\mathcal{L} = \mathcal{K} - |\mathcal{V}| = \mathcal{K} + \mathcal{V}$ (Green curve)

Hamiltonian: $\mathcal{H}_0 = \mathcal{K} + |\mathcal{V}|$ (Invariant red curve)

Phase Graph of Kinetic, $\mathcal{K}$, and Potential, $\mathcal{V}$, Energy

$\mathcal{K} = pc = \sin^2\theta\, p_0 c_0$, $\mathcal{V} = \tau q = -\cos^2\theta\, \tau_0 q_0$

Shaded areas are phase integrals of Kinetic (gold) and Potential (red) Energy.

Figure 9

The next few graphs show some integrals and areas under the curves for reference.

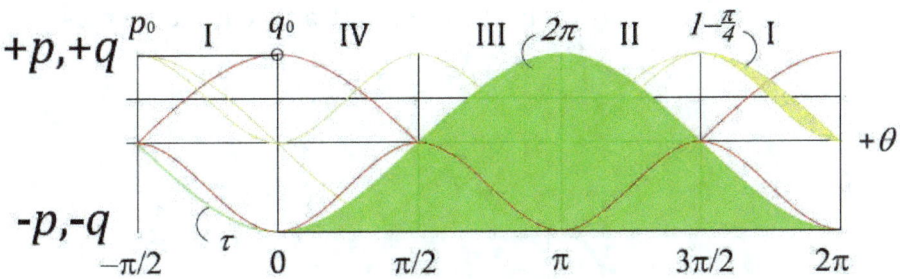

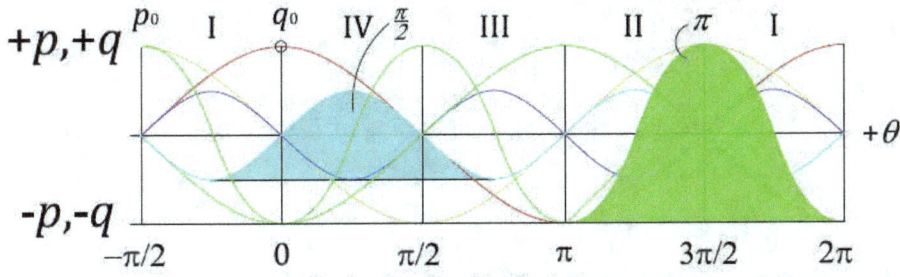

Various Integrals/Areas under the Whole Curves

Figure 10

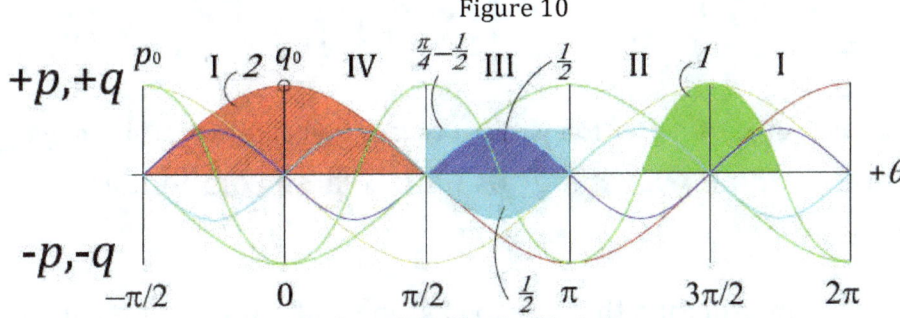

Single and Double Half Cycle Curves
Displacement, Action, Power, & the Lagrangian shown

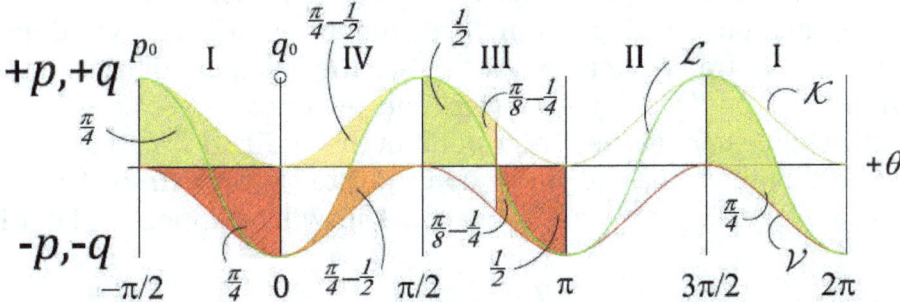

Lagrangian, Kinetic and Potential Energy Half Cycle Curves and Components

Various Integrals/Areas under the Half Curves

Figure 11

Obviously, the action varies over the range of each quadrant, as seen in the following graph, though its total is invariant from quadrant to quadrant, as

$$S = p\int_{-\pi/2}^{0} dq = p\dot{q}\int_{-\pi/2}^{0} dt - \dot{p}q\int_{-\pi/2}^{0} dt = pq = \tfrac{1}{2} p_0 q_0 \qquad (2.9)$$

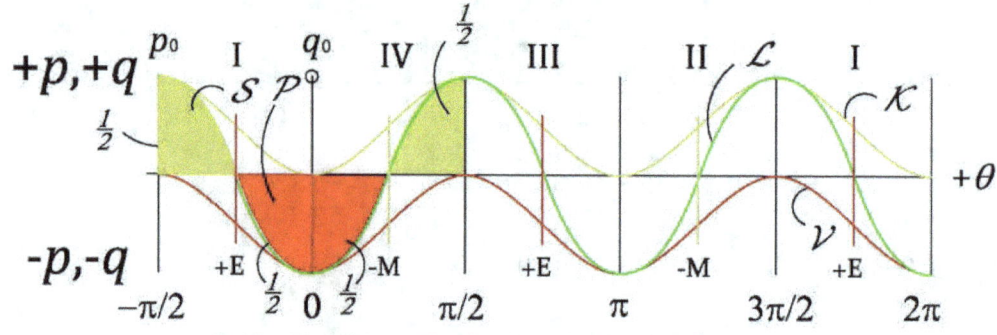

Action (and Power) using the Lagrangian Time Integral

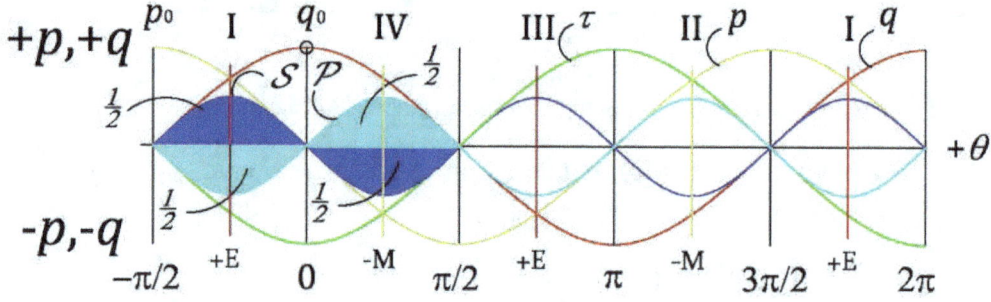

Action (and Power) using the Maupertuis Displacement Integral of Momentum

Action and Power Integrals with related components

Figure 12

The action due to Maupertuis and that due to Lagrange offer different pictures of the energy transformation of the oscillation. While Maupertuis is an integral over the displacement path, the filled areas of the curve show that the action and its inverse, the power of the oscillation are in operation throughout the cycle, i.e. over time, peaking with their rate of transformation greatest, at the half point of each of the four phases. While Lagrange is an integral over time, the filled areas show that the action is concentrated in space at the center point of the oscillator's path as an expression of kinetic energy, with the power concentrated at the extremes of displacement as an expression of potential energy. This will be important later in our discussion.

Next is a direct comparison of the graphs of Maupertuis and Lagrange for one phase of the cycle as

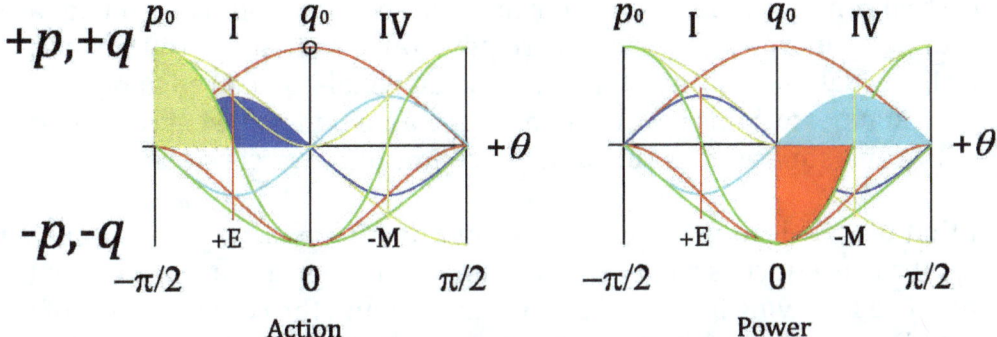

Action   Power

E and M are Moments of maximum action and power and therefore of
maximum transformation between kinetic and potential modes at E and
vice versa at M of energy, mechanical, electromagnetic or other.

Comparison of Lagrange and Maupertuis

Figure 13

We would next like to apply this to phase space to see what it tells us. All images in Figure 14 show the condition in phase I at the moment of maximum action (and power charging) at the point +E, where the designation indicates that it is a potential energy/mechanical analog for electrical charging or capacitance. In Figure 13 above, the –M indicates a moment of maximum power (and negative action) in the kinetic energy/mechanical analog for electrical discharge and magnetic inductance. The moment senses indicate the direction of travel of the oscillator.

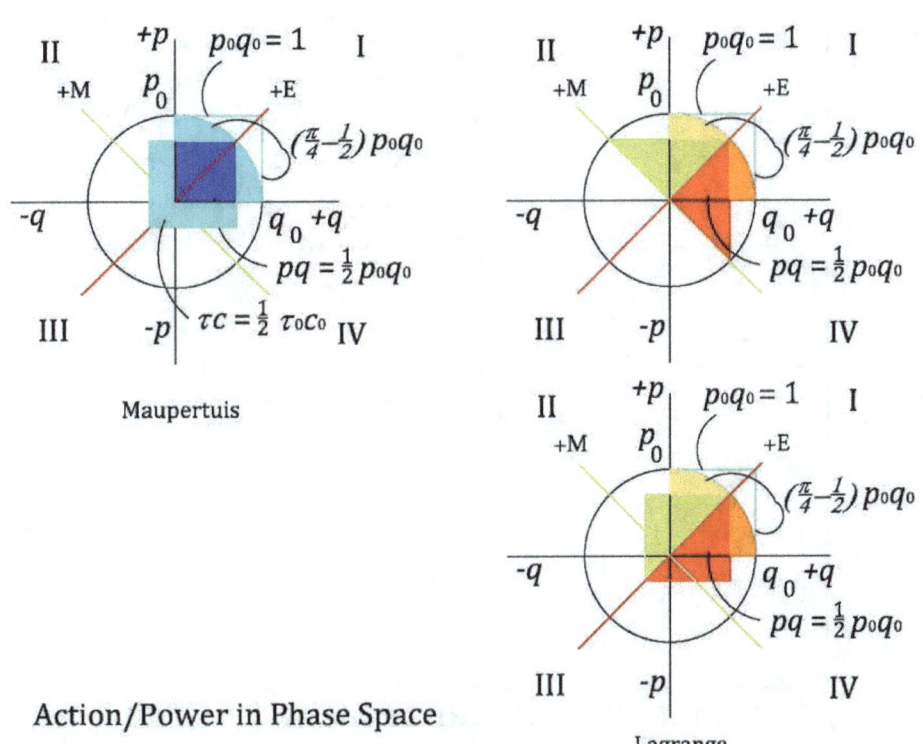

Action/Power in Phase Space

Figure 14

There are two configurations shown for the Lagrange method. The unit squares representing the energy of the system are positioned so that the square area in quadrant I is equal to ½ of the total as are the diagonal-halved red and gold sections of each. For Maupertuis this is equivalent to the area under the curve above. In this scenario, the unit square pivots around the center of the circle clockwise.

The fact that we are dealing with an invariant quantity, p_0q_0, suggests that we might want to plot the product as a hyperbolic curve to phase space in each of the four quadrants. The following Figure 15 shows that beyond the colored range of the kinetic (gold) and potential (red) energies, here along the positive axes, the values for p and q exceed the system constraints as the curves approach each axis asymptotically. The solution is to collapse the hyperbolic curves toward the circle center with a bit of origami, by folding forward along the axes and backward at the lines that cross the curves parallel to the axes. Make a copy and do this, which represents the elements of the oscillatory field that are orthogonal to the phase space. The peach sections are duplications of the adjacent kinetic and potential components, which fold up against them as the four quadrants are brought together.

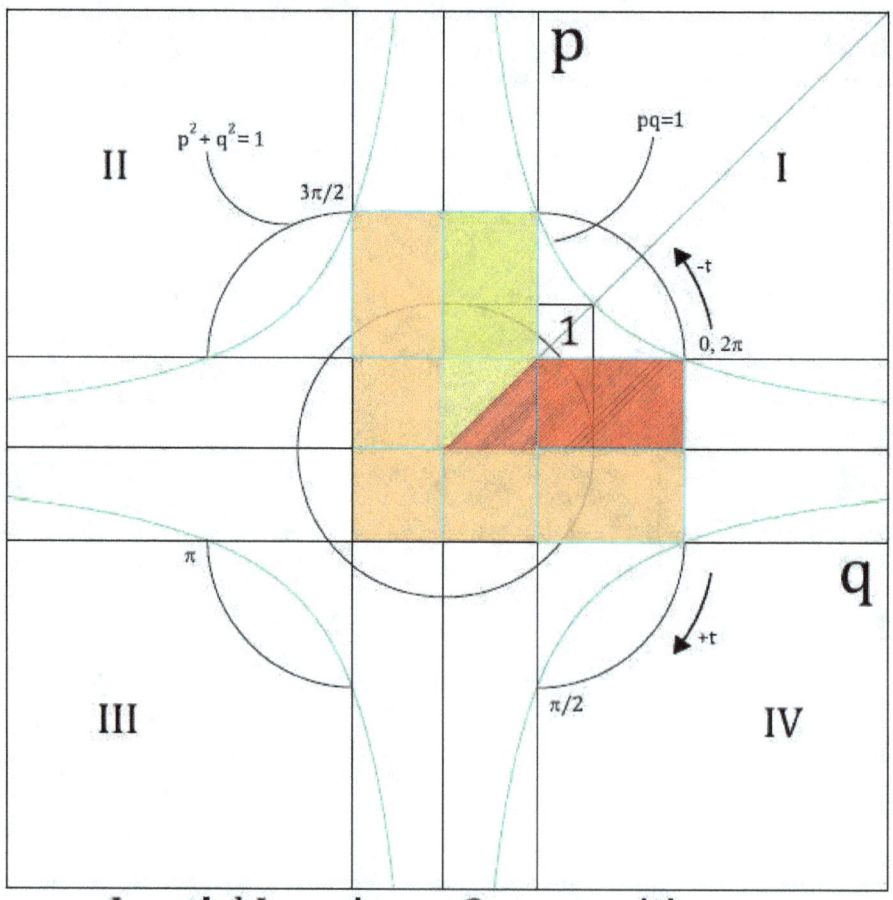

Inertial Invariance Superposition orthogonal to Phase Space

Figure 15

When folded, the center square inside the circle and composed of four smaller squares folds in eight triangles upon itself, and represents the inertial potential of the system. The geometry of this arrangement, which arises naturally in this scenario, is such that when the hyperbolic curves intersect the phase space circle as shown in Figure 16, with the unit square offset as shown, the area of the rectangle when the kinetic is at a maximum is unitary and remains so as the distal vertex moves along the hyperbolic curve between the *p* and *q* axes to the position of maximum potential energy. This energy flow is shown in Figure 16.

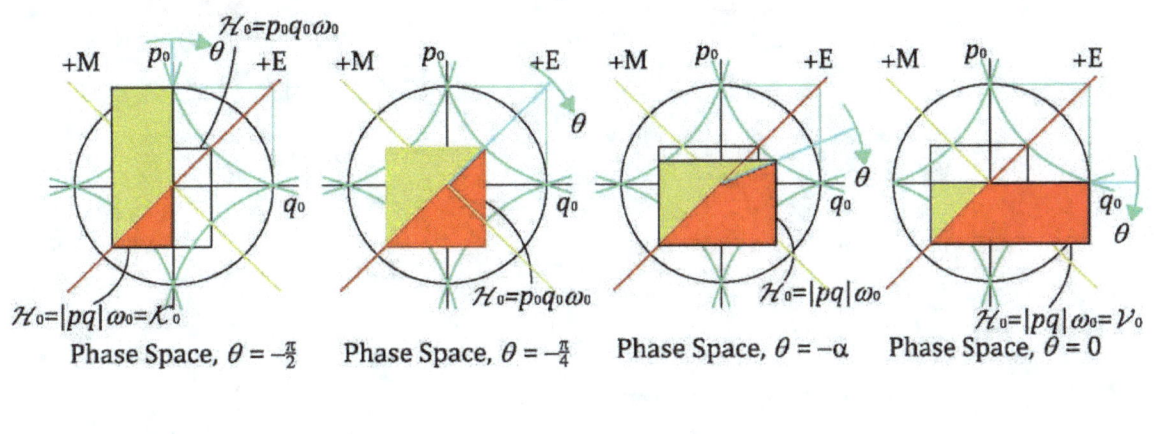

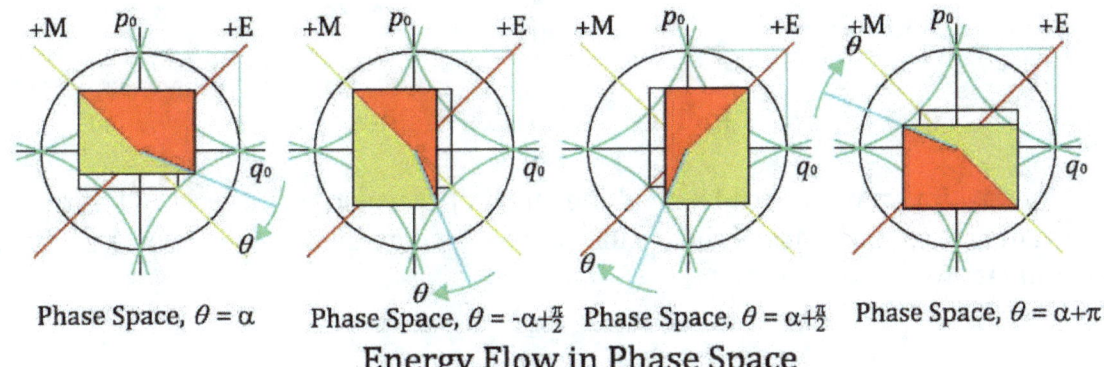

Energy Flow in Phase Space

Figure 16

Here the off-manifold or extra-dimensional components of the energy are shown as θ rotates about the phase space with the motive end of the rectangle moving along the hyperbolic curve in the extant phase. This is indicative of the three dimensional nature of the energy. The importance of this contrivance can be seen in the next chart, Figure 17.

In the first figure at the upper left we have a superposition of the kinetic and potential energies of the oscillation, with the defining physical property dimension components of the energy and phase space shown on the dimension lines. In the upper right we have the condition if the oscillator was stopped at its point of maximum displacement with a retention of potential energy. Manually returning the oscillator or oscillating medium to its point of equilibrium, stops the oscillation, thereby removing all kinetic and potential energy from the system. However we are still left with an invariant inertial <u>potential</u> component of units mass-displacement.

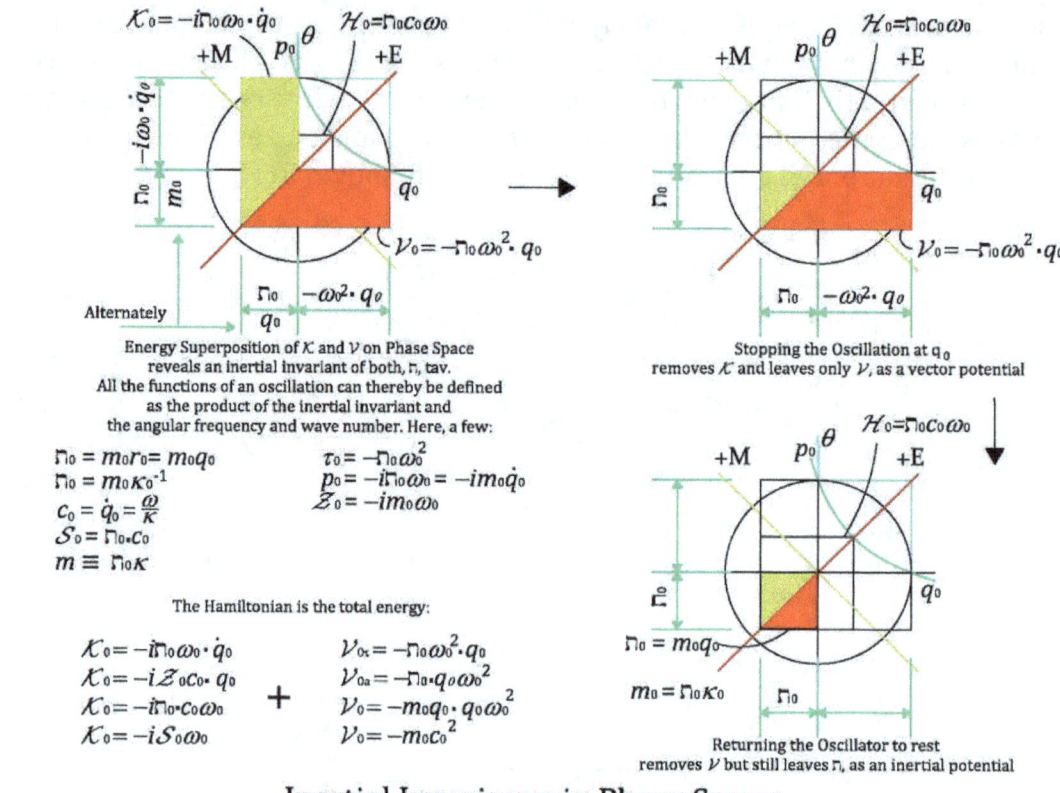

Inertial Invariance in Phase Space

Figure 17

This inertial invariant, which I am calling ת, (tav), can also be understood as the invariant ratio of the mass of the oscillator and its angular wave number, κ_0, which is conjugate to the frequency as

$$\kappa_0 = \frac{\omega_0}{c} \qquad (2.10)$$

$$\therefore m_o = ת\kappa_0 \qquad (2.11)$$

shows that with SHM mass is essentially the wave number of the oscillation. In addition, the invariants S_0 and c_0 are related to the inertial invariant as

$$ת = \frac{S_0}{c_0} \qquad (2.12)$$

In a quantum application, this is

$$ת = \frac{\hbar}{c}. \qquad (2.13)$$

Here $\hbar$ (h-bar) is the invariant quantum of action of quantum theory. The value in this development is that the various functions of the oscillation we have discussed, in addition to several others, can be stated in terms of the inertial constant and the various orders of derivatives and integrals of the frequency and wave number using the Euler identity, without recourse to extraneous properties. Figure 18 gives such frequency derivatives, followed by a second chart combining frequency and wave number derivatives and integrals.

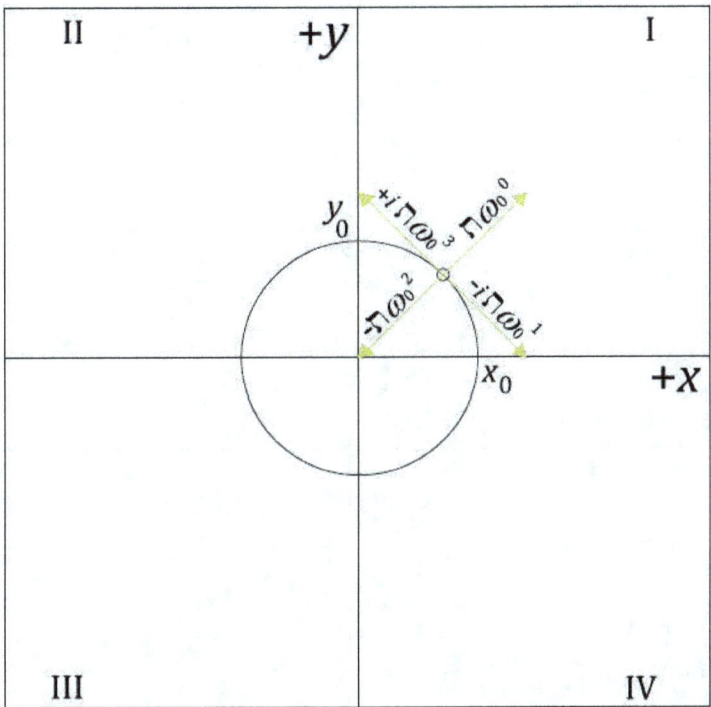

Phase Time Differentials - Uniform Circular Motion

Figure 18

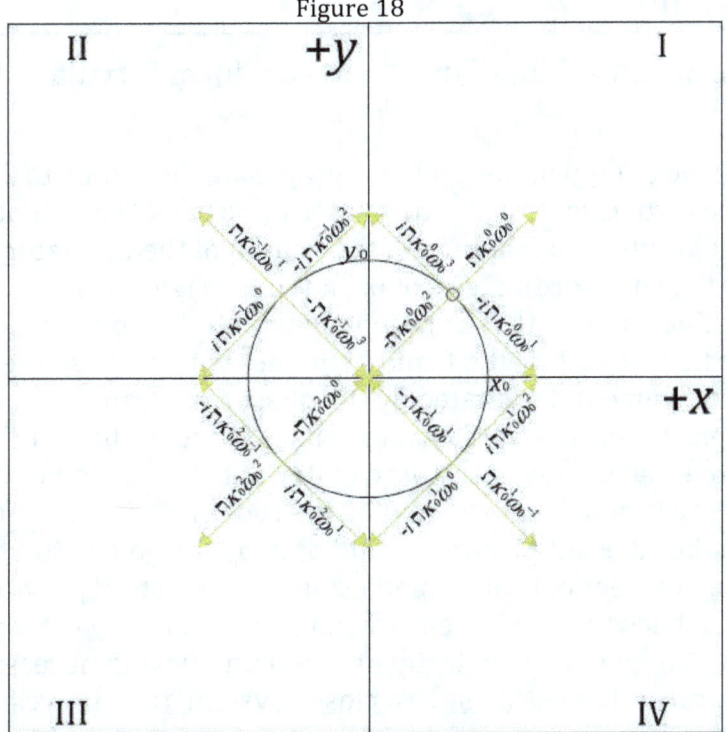

Phase Space & Time Differentials - Uniform Circular Motion

Figure 19

This next chart gives the equivalent properties of the functions. See the Matrix of PS₃ Functions and Invariants chart after the Conclusion section for another representation.

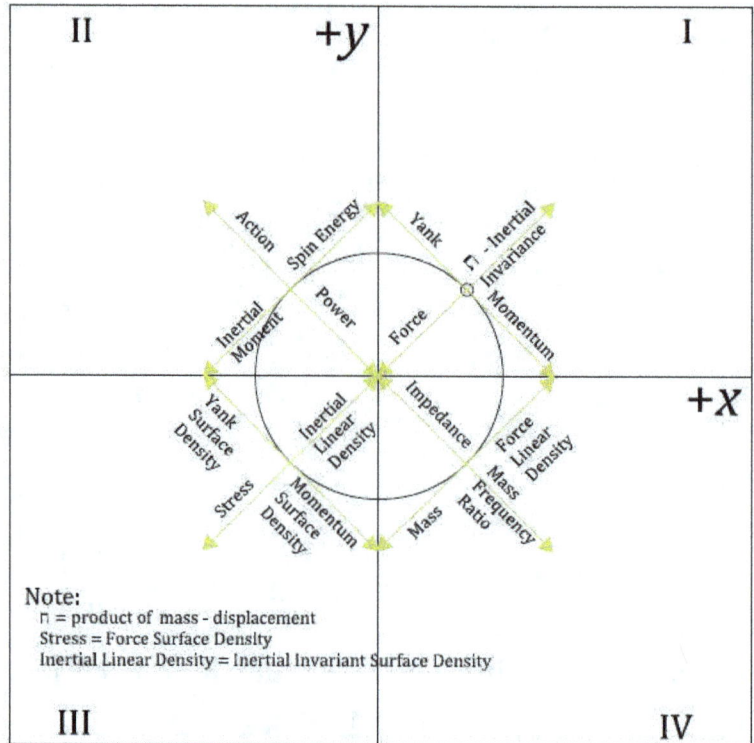

Phase Space & Time Functions - Uniform Circular Motion
Figure 20

Figure 21 maps the energy of the previous graphs in *phase time* to 2-dimensional *phase space*. The arrows for the kinetic phases (gold) and the potential phases (red) in the upper left graph are a mapping of the motion of the oscillator up and down with respect to the space ordinate as at 0, and to the right over linear time, thus completing the sine curve path. The second figure on the upper right reverses the direction of the flow at each of the E and M moments to collapse the first into a cyclical time arrangement designated by the phase/quadrant for superimposing onto a cyclical phase space. The reason for this is found in the fact that the direction of increasing potential energy is always counter to the direction of increasing kinetic energy as in a gravitational field. Following the red arrow from either of the E points in this figure to the adjacent M moment and then by the gold arrows to the next E moment is the equivalent of lifting an object in the terrestrial gravitational field and releasing it to fall back to earth. The work of lifting is done by the action of the wave, twice per cycle. The reversal of polarity here is a unique and necessary feature of SHM, which indicates that <u>within such a closed system, time is cyclical</u>. In the final figure the collapsed graph is superimposed onto phase space, which has transposed q and p axes so that the senses of the two depictions are synchronized with the displacement and the momentum.

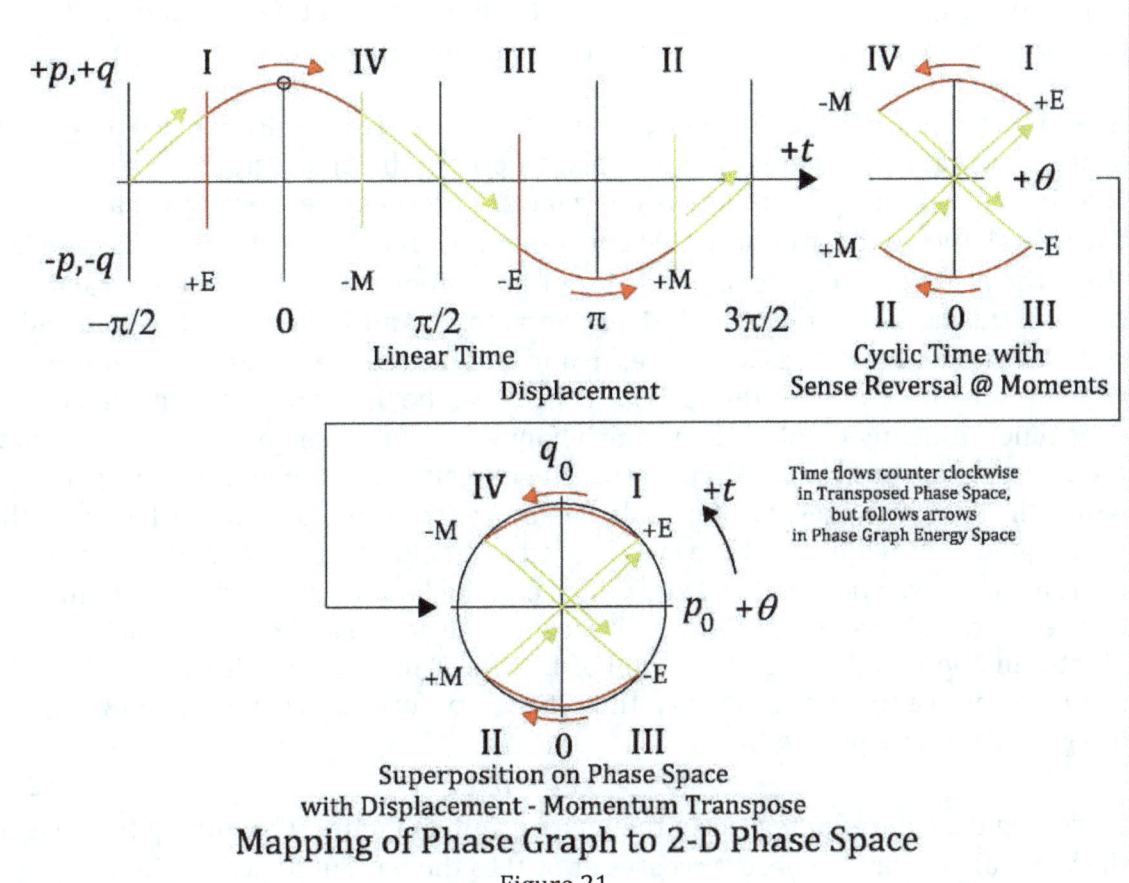

Mapping of Phase Graph to 2-D Phase Space
Figure 21

Thus the potential energy that is a function of displacement is graphically superimposed on the regions of greatest displacement in phase space and the kinetic energy that is a function of velocity is graphically superimposed on the regions of greatest momentum and oriented in the general direction of positive momentum. Note that the rotation over time is now ccw, but this is only for phase I and IV, at which point it reverses from III to II.

We can now begin to clear up some features that may have been apparent to the reader earlier. Figure 6 shows the total energy of the system as equal to the unit square, which overlaps the boundary of the phase space and the single phase or quadrant area of $\pi/4$. In Figure 9 the same total energy is depicted as the straight, therefore invariant line $\mathcal{H}$ equal to one, while its $\mathcal{K}$ and $\mathcal{V}$ components are shown as integrals, as areas under the Lagrangian curve, each equal to $\pi/4$. Obviously some interpretation is in order.

$\mathcal{H}$ is correctly shown as an area in Figure 6 as it is extra-dimensional to the linear dimensions of p and q and in fact is dimensionally the product of these for its two components as

$$\mathcal{H} = \mathcal{K} + \mathcal{V} = \tfrac{1}{2} p\dot{q} + \left|\tfrac{1}{2} \dot{p} q\right| = p_0 q_0 \omega_0 \qquad (2.14)$$

In phase I, $\mathcal{K}$ goes from 1 at p_0 to 0 at q_0 and $\mathcal{V}$ goes from 0 at p_0 to -1 at q_0. At the mid-point between the two, +E, each are equal to ½ $p_0 q_0 \omega_0$.

As shown in Figure 9, the line $\mathcal{H}_0 = p_0 = q_0 = 1$, is the absolute difference between the $\mathcal{K}$ and $\mathcal{V}$ curves at any point in time and any point in the phase graph. The Lagrangian, $\mathcal{L}$, on the other hand is the relative difference between the $\mathcal{K}$ and $\mathcal{V}$ curves, relative to the ground state represented by the θ, time axis. Thus $\mathcal{L}$ is zero when the distance from θ is the same for both components. Though the (straight) curve in this figure for H equals 1, the integral for $\mathcal{K}$ and $\mathcal{V}$ shown by the gold and red areas under the curves are $\pi/4$ each and $\pi/2$ for the area under the Hamiltonian, which is obviously greater than 1. This is because the line represents the value of $\mathcal{H}$ over time, i.e. at any point in time, which happens not to change, and the area under the curve for each of $\mathcal{K}$ and $\mathcal{V}$ represents the magnitude of the phase space over which the energy ranges, $\pi/4$ for each. Looking at the top graph in this Figure of the Lagrangian, it is evident that the areas of red and gold that are not between the Lagrangian curve and the θ axis and which equal $\pi/4$ minus ½ each, as seen in Figure 11, cancel leaving the ½ values each for the action and power, which is net kinetic and potential energy for a total of 1 or the Hamiltonian. We can see the symmetry in the following in which the action is differentiated and the power integrated with respect to time.

$$\mathcal{H} = i\mathcal{S}\omega_0 - i\mathcal{P}\omega_0^{-1} \qquad (2.15)$$

Morphing the colored integrals in the lower graph of Figure 9 by pulling the base of the vertical line for q_0 where it crosses at $\theta(0)$ to the base of p_0 at $\theta(-\pi/2)$, and repeating for the next three phases, then rotating as in Figure 21 gives us the following:

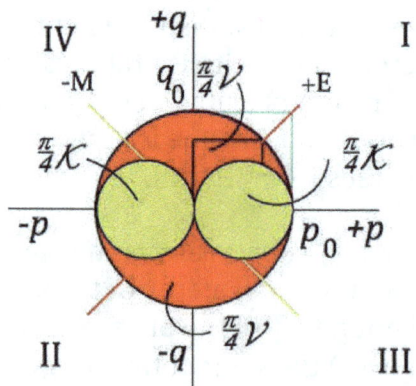

Collapse of $\mathcal{K}$ & $\mathcal{V}$ Integrals to Phase Space
$\pi/8$ of each in each phase

Figure 22

Note that the area of the region in each phase represented by $\mathcal{K}$ and $\mathcal{V}$ are each one half of what they are in the phase graph of Figure 9. We are getting close to a breakthrough. While our phase space representation, PS_2, so far has been two dimensional, the superposition of the energy component of an oscillation suggests we look for a three dimensional representation, PS$_3$, or at least a two dimensional representation in spherical space, i.e. on a topological 2-sphere. In fact, the surface area of a sphere is twice the surface area of a circle if you count both faces of the circle's disk. This means that an octant, which is the equivalent of one quadrant of a one sided phase space, has an area of $\pi/2$, which is the area under the curve of $\mathcal{H}_0$ in Figure 9. In the phase sphere we are creating, the line for the Hamiltonian becomes a geodesic or great circle for the sphere and the area under the Hamiltonian curve for four quadrants or an area of 2π, is one half the spherical area and represents the energy/action/power integrals over one cycle of time and one hemisphere.

To construct PS$_3$ from the Figure 21 bottom figure, we pick up the left and right sides of the energy path from the intersection of the circle and p-axis, folding along the q-axis, and pinch them together at the apex of the hemisphere, like a phase space taco. We also rotate the $+p$-axis and the phase space circle around the q-axis as shown in Figure 23. Note that the point at which the taco is pinched together is r_0 from the

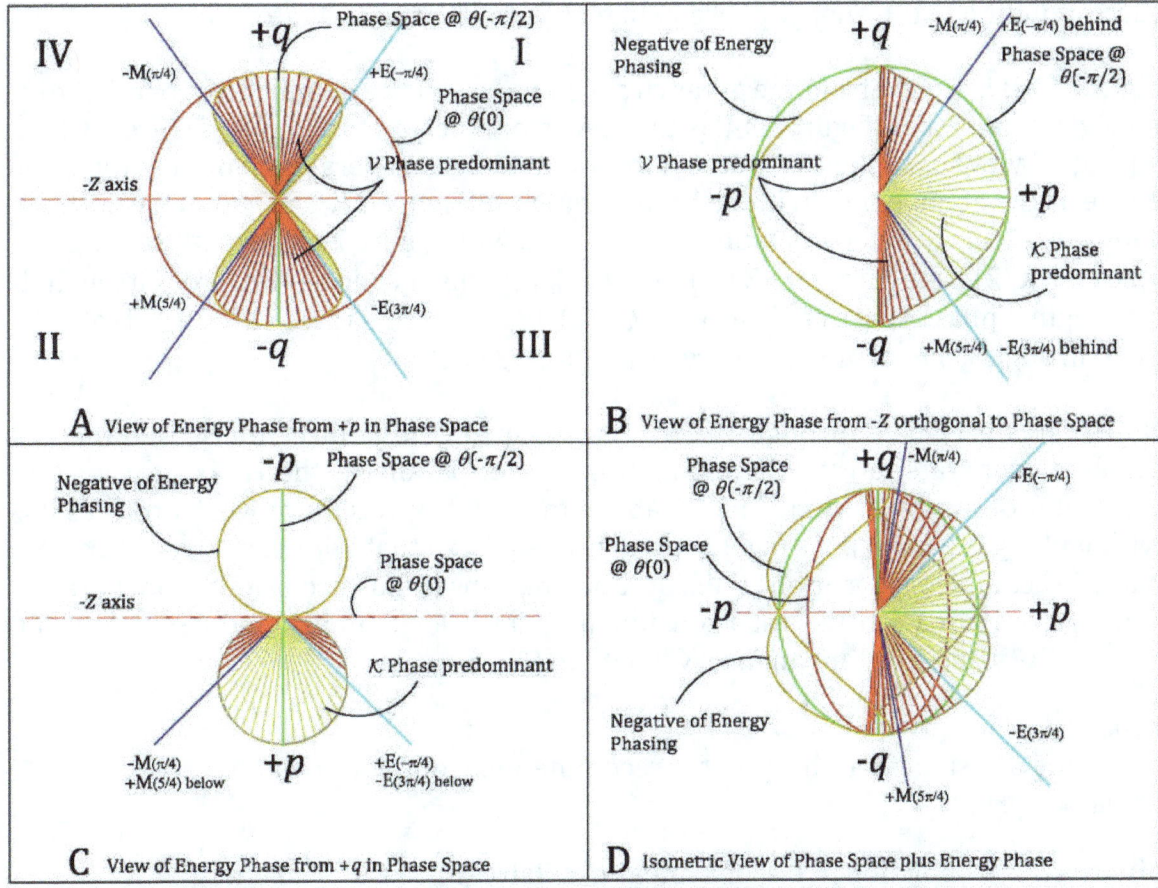

Superposition of Energy Phase Space and conventional Phase Space
for Reciprocal Path Oscillation

Figure 23

center of the rotated phase space and the phase sphere and that it is at the point in phase space where q is zero, that is the oscillator, O_{ps}, is at zero displacement. Note also that the point at +p where the taco is pinched and along the radial behind it, the left and right halves of the phase space are sealed together, creating and separating the upper and lower sections, so that as O_{ps} cycles around the taco edge, it crosses its own path at each half cycle as one might draw a figure 8 with one stroke instead of drawing two circles.

Figure 23-A shows the view from the +p-axis on the surface of the hemisphere. Each of the spokes has length r_0, originating at the center of the space and terminating in the oscillator path for O_{ps}, and is separated from the one on either side by a 5 degree segment of the space. There is no significance to the 5 degree magnitude of separation other than as a convenient way of dividing and depicting the cycle in the creation of the graphic. There is significance, of course, at the transition between the red and gold radii, colored blue and cyan, at $\pi/2$ separation midway along the path in each quadrant, as they represent the E and M moments of energy transitions. The oscillation moves through each of the segments in the same interval of time. The colors are as in the previous graphs, red for potential and gold for kinetic, corresponding to the energy phase being transited by the oscillator O_{ps}.

Figure 23-B is a view from the great circle boundary of the hemisphere, parallel to a newly defined z axis. Figure 23-C is the view from the +q-axis. Finally Figure 23-D is an angled view from the qz plane. O_{ps} follows the same general path on the upper left figure as the arrows shown in the bottom figure of Figure 21. The geometry as shown is accurate and not simply suggestive as with Figure 21. The path starts at $+p_0$ at $\theta(-\pi/2)$, arcs through +E to $+q_0$ at $\theta(0)$, returns through –M to cross itself and the original phase space circle at $+p_0$, $\theta(\pi/2)$, before arcing through –E to $-q_0$ at $\theta(\pi)$, through +M to complete the cycle at $+p_0$, $\theta(3\pi/2)$.

Note that at each pass through $+p_0$, O_{ps} is traveling in the +z direction so that the total path generates an angular momentum vector parallel to the +q-axis through the center of the lower circle in 23-C above the point, +p. As O_{ps} moves, it rotates PS_2 with it along the path, pivoting it around the q-axis, so that with each two octant path, PS_2 completes a π or one half rotation and flips its surface orientation, here from left to right. O_{ps} travels at a constant angular velocity in PS_2, and PS_2 rotates on edge at the same constant angular velocity in PS_3.

<u>Aside #1</u>
A reading of Aside #1 at this point is recommended, but optional.
<u>End of Aside #1</u>

Figure 23 is a depiction in PS_3 of a linear oscillator, O_l, in PS_2. The energy relationships remain the same from our earlier discussion, in which the Hamiltonian is invariant, the Lagrangian oscillates between +1 and -1, and the oscillator passes through four maximum action/power moments, two mechanical capacitive and two mechanical inductive, over each cycle.

It would be interesting to see if there is a corollary in PS_3 to the UCM of O_c in PS_2. Remember that in that instance we had a uniform angular velocity and a constant displacement magnitude of r_0, therefore a constant potential energy with respect to the center of motion of, $\mathcal{V}_0$, and a constant rotational kinetic energy of $\mathcal{K}_0$, therefore an invariant Lagrangian of zero. Every possible point, O_{ci}, on the circle of motion in PS_2 is a $+q_{0i}$, $-q_{0i}$, $+p_{0i}$, and $-p_{0i}$ along with every point in between with respect to some set of axes, q_i-p_i.

We might imagine, therefore, that each of the radii shown in the 4 figures of Figure 23 represents a radius of intersection with respect to different paths along the surface of PS_3, each with its own phase space, PS_{2i}, and its own point of maximum conjugate momentum, p_{0i} at the location $+p_i$, forming a circle, C_{+p} in the zp plane. All share the same radius of maximum potential at $+q_{0i}$ and $-q_{0i}$ and each and every PS_{2i} intersects at and rotates in unison with, i.e. at the same frequency as, the q-axis or rotation θ. We will call this rotation, ϕ, a rotation not of a point in PS_2, but rather of a disk PS_2 normal to θ. At any point in time, four paths of oscillation, whose points of zero displacement on C_{+p} are $\pi/2$ apart, are at one of the four action/power moments, $+E$, $-M$, $-E$, and $+M$, so that these moments rotate in unison with θ, which rotates to the right hand rule as an axial vector parallel to the $+q$-axis. Each of the paths has an anti-path or polar opposite, defined as the path with an equilibrium point π apart on C_{+p}, so that they share the front and back sides of a common PS_2, though the O_{ps} for each are π apart in their cycle. Thus the two back-to-back PS_2's incorporate the instant $+/-$ E's, and $+/-$ M's, which creates an effective armature, a disk formed by the instant PS_2's that rotates on end with θ, and in turn rotates about an axial vector, ϕ, orthogonal to θ, its ends intersecting the circle, C_{+p}, creating two torques of opposite chirality along the circle as it rotates.

It is important to point out that the pair of PS_2's just described, which we will call disk ϕ, are the only "real" PS_2's in the system, as each of the others is simply a 2π circuit with respect to a reference point on the circumference of disk θ designated $+p_i = O_{psi}$, indicating where θ crosses C_{+p} on its path. Each such crossing is unique to the circumference of θ, and there is a 1 to 1 correspondence between their circumferences. Thus the path of each O_{psi} traces a path as seen in Figure 23 on the surface of the sphere PS_3, as disk ϕ rotates on its rim with θ and about its axial vector in PS_3. Thus the θ of our original PS_2 in the zp plane becomes ϕ after rotation into PS_3.

Figure 24 shows the superposition over one cycle of the paths of two such PS_2's separated in ϕ by $\pi/2$. In this representation, the induction moments, $+/-$ M, of the original path of $PS_2(-\pi/2)$ are coterminous in PS_3 with $-/+$ E, the capacitive moments of $PS_2(\pi)$, which leads it in space and lags it in time by $\pi/2$. The co-incidence of these moments is therefore separated in time by one half cycle. The inverse or negative paths of these two PS_2's can be seen in 24-C, completing the cloverleaf pattern. Each PS_2 and its inverse are co-extensive in phase time, but of alternate sense, i.e. they have an angular separation of π in phase space.

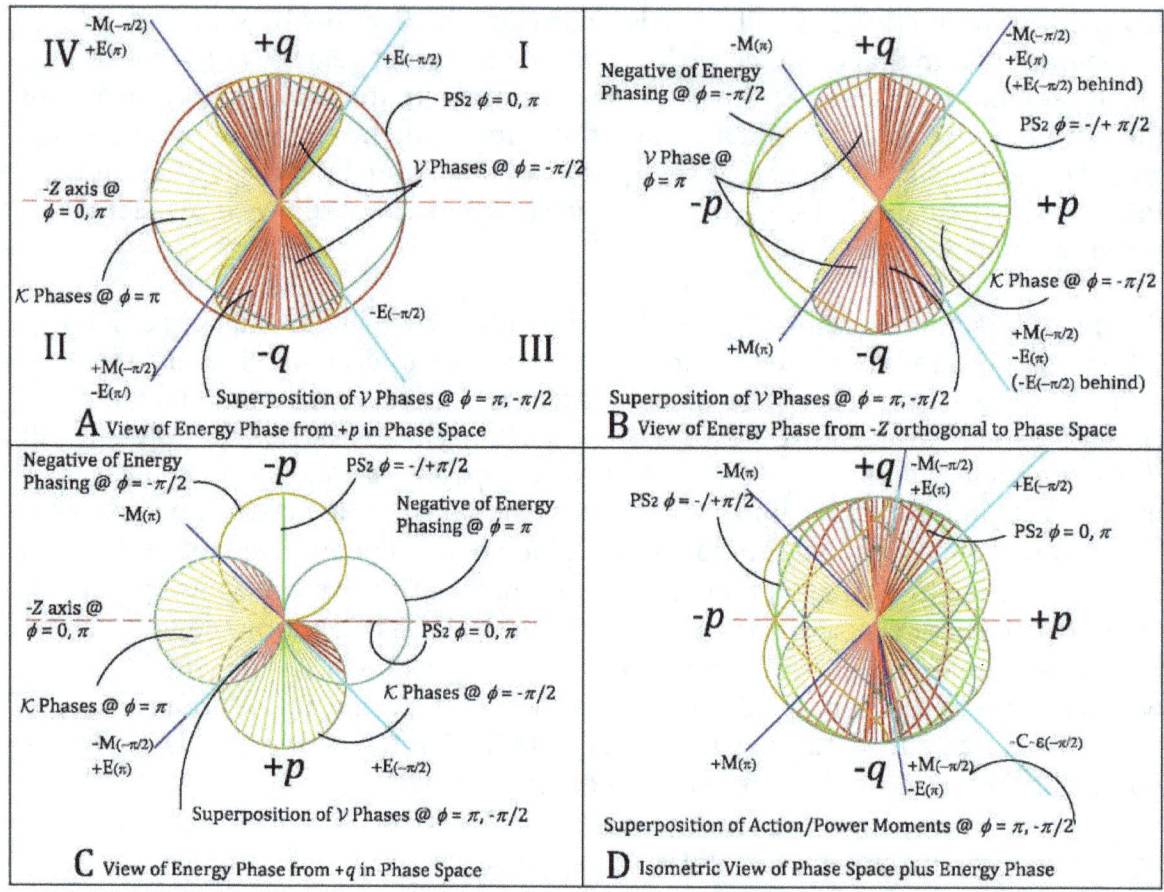

Superposition of Energy Phase Space and conventional Phase Space
for Cyclical Path Oscillation (Phase Space Rotates as ϕ about $+q/-q$ Axis)

Figure 24

It bears emphasis that the action/power moments are contiguous with, in the plane of, disk ϕ, but remain fixed over time at an angular displacement of $\pi/4$ from both the θ / q-axis and the plane of circle C_{+p}. While they rotate with θ, it is the rotation of ϕ about its axis in the zp plane that actually advances the moments and thereby the rotation of θ and not the other way around. This advance constitutes a tangential, in terms of our phase graphs, longitudinal momentum component of the otherwise transverse oscillation of the O_{psi} between $+q$ and $-q$, which generates the angular momentum of θ.

Simple Harmonic Motion and Rotational Oscillation or Spin Space
Accounting for the Generation of Spin and Charge

If we were to sum up all the paths over one cycle of PS3, by summing up the paths around the q-axis, it would look something like the views in Figure 25. The similarity of 25-A with Figure 22 is clear. The concentration of the potential energy components about the +/- q poles and the q-axis and of the kinetic energy around the $+p$ circle is instructive. The persistence of the action/power moments is also clear. This indicates the invariant, maximum value of the action and power of the oscillation over time and the energy fields indicate the invariance of both $\mathcal{V}$ and $\mathcal{K}$ and hence the existence of an invariant Lagrangian of 0, as

$$\mathcal{L} = \mathcal{K} - |\mathcal{V}| = |p_0 \dot{q}_0| - |\dot{p}_0 q_0| = 0 \qquad (3.1)$$

The Hamiltonian is obviously invariant. In the earlier case of (2.14) we had determined a Hamiltonian of 1 and therefore an average $\mathcal{V}$ and $\mathcal{K}$ each of ½. In the case of PS3, the energies are doubled due to the presence of the inverse components of the phase space paths as seen in 23-C, thus $\mathcal{V}$ and $\mathcal{K}$ each equal 1.

$$\mathcal{H} = \mathcal{K} + |\mathcal{V}| = |p_0 \dot{q}_0| + |\dot{p}_0 q_0| = 2 p_0 q_0 \omega_0 \qquad (3.2)$$

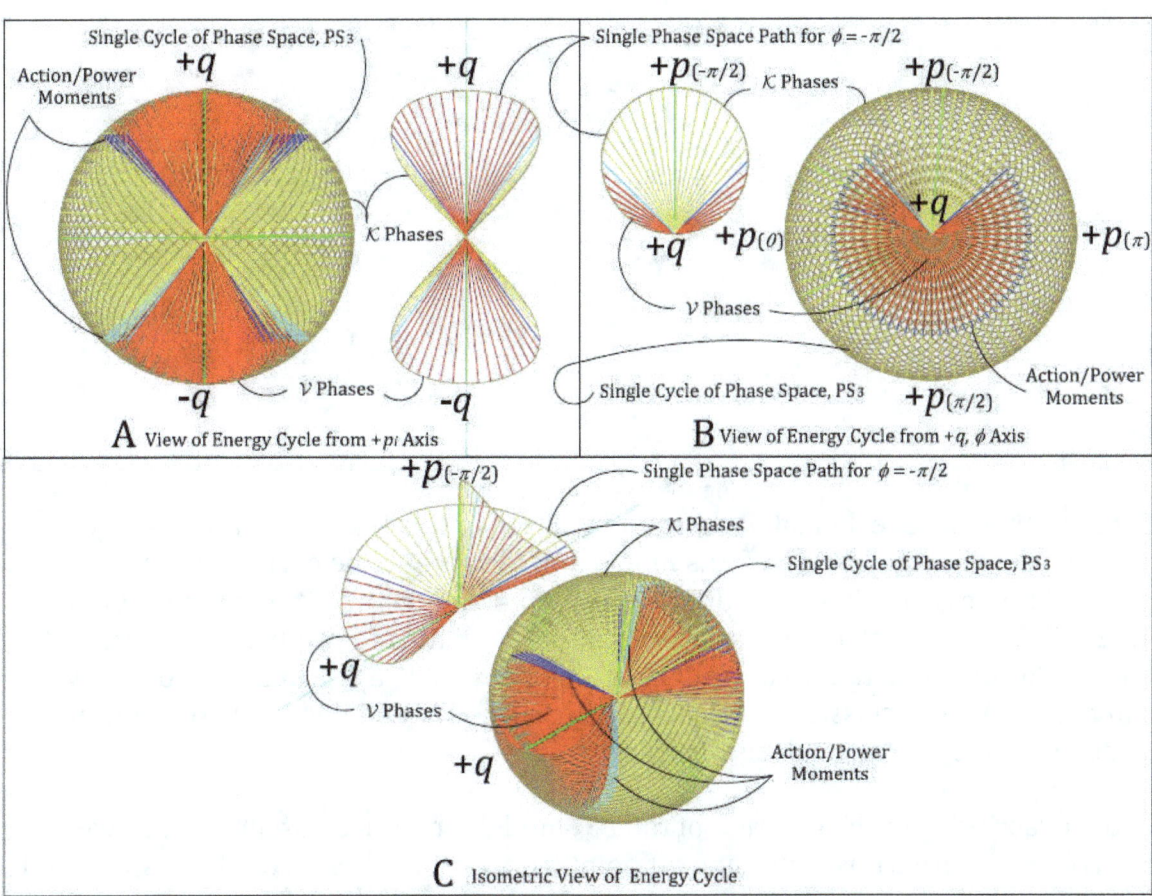

Superposition of Phase Spaces over Whole Energy Cycle
for Cyclical Path Oscillation (Phase Space Rotates as ϕ about $+q/-q$ Axis)

Figure 25

The action/power moments as shown in Figure 24-C have an orthogonal projection along any $+p$ co-radial in the zp plane of $1/\sqrt{2}$ and as in 24-A, along the $+/-q$-axis of $1/\sqrt{2}$. With respect to its initial position in C_{+p}, each moment, $+/-M_i$ and $+/-E$, has an orthogonal distance to its $+p_i(-\pi/4)$ radial of $\sqrt{3}/2$ and a distance from the PS_3 center to the projection along the radial of $1/2$. Thus, the sine of the angles ε and μ between each moment and the equilibrium radial at $+p_i(-\pi/4)$ of $+/-K$ is $\sqrt{3}/2$ and the cosine is $1/2$. This is depicted in the following Spin Diagram 1 of Figure 26.

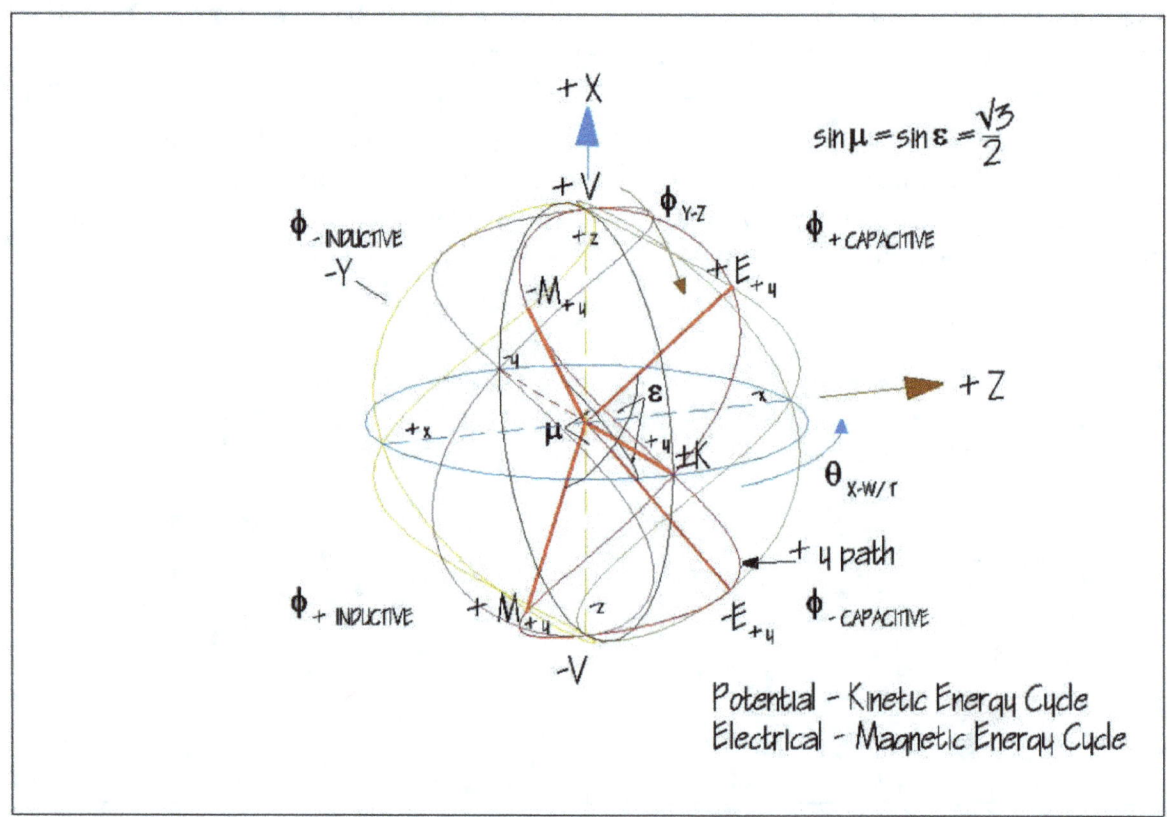

Figure 26 – Spin Diagram 1

Those with a more technical background will see in this last paragraph a suggestion of a quantum interpretation of this model and they would be correct. The action in the quantum case is Planck's reduced quantum of action or $\hbar$ (h-bar) as previewed in (2.13). We must ask then, just what is it that oscillates to produce a quantum PS_3. The reader might be tempted to suggest a lepton or perhaps a quark field, but we will pursue a more classical approach, and to do so we will need to take another aside, this time into an analysis of stress.

In the description of the creation of the PS_3 model for a linear oscillation, O_l, we referred to Figure 21 as a jumping off point from our discussion of phase space and the phase graphs in the interests of continuity by showing that PS_3 was a natural extension of that discussion. We will now look at another way of developing that model for a cyclical oscillation, O_c, from a discussion of stress.

Aside #2
A reading or review of Aside #2 is recommended at this point.
End of Aside #2

Based on the aside comments concerning stress, strain and elasticity, we can state that the theory of general relativity models space as an elastic medium, albeit coupled with time so that the stress tensor has an additional dimension, that of time. In that theory, particles of matter and energy couple with spacetime in a manner that warps or distorts, i.e. strains that spacetime, resulting in its curvature. Such curvature in turn directs the motion or travel of these particles along geodesics or energy conserving paths through spacetime that we recognize as gravity. A gravitational field therefore consists of the curvilinear strain of spacetime, stress-induced by large aggregates of mass and energy, upon which individual particles of mass and energy are constrained to follow. Such spacetime is elastic in the sense that it becomes distorted from rectilinearity by a transiting celestial body as the body moves through it, but returns to its general undistorted configuration after the body moves on.

Some authors would say that it is not space that distorts due to the presence of mass/energy, but time. In fact, in general relativity space and time are not considered as separate properties, but rather interchangeable components of a 4-dimensional spacetime. This is a topic that we could spend volumes on, but it suffices to point out that time is a means of accounting for changes in the configuration of particles or bodies of mass and energy observed to be *in space*. It is my opinion that time is simply a measure of distortions, changes or motion in or of space. If there were no motion or other change in space, there would be no time. If time is therefore a comparison of two different rates of change in space, the most fundamental of such changes in terms of its scope is the expansion rate of the universe. This appears to be exponential, due to the perceived acceleration of universal expansion. By comparison the speed of light as a calibrator of time is important but of secondary primacy, since it is a measure of the rate of change of position of electromagnetic waves occurring *in an expanding universe* of apparently finite age, and therefore would appear to be conditioned by any changing stress and density of that expansion, perhaps invariant over time, perhaps not. If time itself expands linearly along with each of three spatial dimensions due to cosmic expansion, then the speed of light, c, will be invariant by definition. However, this does not speak to whether all the processes that occur now in one second, would have occurred 13 billion years ago in one second. Perhaps some would, perhaps some wouldn't.

Our interest here is not primarily with time, however, but stress within the context of Simple Harmonic Motion and Uniform Circular Motion. We can envision an instance of static stress, such as a loaded beam, in which the stress is not a function of time, but discussions of SHM necessarily involve time or at least frequency. We want to return to our discussion of PS_3, 3 dimensional phase space and the phase sphere, and examine how it might relate to an oscillation of stress and surface force without the need of a body force, either as an initiating factor or an ongoing component of SHM.

We will start by imaging a single point in an inertially dense spacetime. The time component I am referring to in this "spacetime" is not some "unseen" dimension, but rather the same 3 dimensional space that we saw a moment ago, *all of it*, however changed by whatever has transpired within it, including by an ongoing expansion. What I mean by an "inertially dense" spacetime is a three dimensional modeling space, a virtual 3-D blackboard so to speak, that is everywhere inert, that is it will not change unless we, or other things within it, including expansion, do something to make it change. Such change, once engendered, can be represented by an additional, fourth dimension where we keep track of the changes. We can call that dimension a record or a history of change or we can call it time, but it does not exist "out there" in some other realm of multitudinous, corresponding 3-D spaces. That fourth dimension is all in our head, but that's alright, since so is the 3-D blackboard. Once we put a point in spacetime, it stays there unless and until we erase it. Oh, and this spacetime is the same thing as the one all around us that we call the universe.

This single point is so far out in extra galactic space that there is no other body, no point of light visible to us since we have no Hubble space telescope with us. Even our bodies are invisible. Just our minds and the single point that we can see. We want so see what happens to this point if we allow a condition of isotropic expansion of the space around it. The entire region around the point is moving away from it, which is another instance of inertia. There are after all two types of inertia, position inertia, q, in which a body at rest at some position stays at that position vis-à-vis our position, and momentum inertia, p, in which a body in motion along some trajectory stays in motion along that trajectory vis-à-vis our trajectory or position. Simple harmonic motion is a process that relates the ordered interplay of these two types of inertia so that we have the emergence of a concept of frequency, ω, and its inverted notion, time, t, as in how many <u>times</u> does the SHM occur while something else is occurring.

To my knowledge, inertia is not generally characterized as being of two types such as this, though this is consistent with Newton's first law. We can always find a trajectory and velocity parallel to a moving body that will put us in its rest frame, a rest frame that would otherwise be a moving frame to us. Rest inertia in this sense is always relative to a co-moving frame of reference of other bodies, but this does not preclude a rest frame defined by isotropic red-shift in an expanding cosmos.

It should be remembered that the rate of expansion, as given by the Hubble rate, H_0, at approximately 73 kilometers per megaparsec per second, is actually quite small. This indicates the universe is expanding, straining, at approximately $2.37... \times 10^{-18}$ meters per meter per second, which is approximately 88.6 times smaller than the reduce Compton wavelength of a neutron, $\lambda_{C,n}$, a virtually undetectable length on the human scale. The meter of course does not represent a natural length scale. If this strain, which as a ratio of relative length change per time is not length scale dependent, is figured with respect to $\lambda_{C,n}$, the length augmentation per second from expansion relative to that scale is 4.97... x 10^{-34} meter per $\lambda_{C,n}$ per second. This is approximately 30.78 times the generally accepted, theoretical smallest length scale of 1.61... x 10^{-35} meters, the Planck length. But the second does not represent a natural time scale either. Assuming the speed of light in a vacuum to be invariant, the time taken for light to transit a distance equivalent to $\lambda_{C,n}$ is the inverse of the angular frequency, obtained as

$$t_{C,n} = \frac{\lambda_{C,n}}{c} = \omega_{C,n}^{-1} = 7.00...x10^{-25} \text{ seconds}. \tag{3.3}$$

This indicates that the expansion strain rate with respect to a neutron scale of length and time, $t_{C,n}$, is the dimensionless strain number per time using a theoretical neutron radian, θ_n, as the time scale,

$$H_0 \omega_0^{-1} = 1.65...x10^{-42} / t_{C,n} \tag{3.4}$$

The strain distortion length per meter at this time scale is

$$\lambda_{C,n} H_0 \omega_0^{-1} = 3.48...x10^{-58} \text{ meter}/\theta_n \tag{3.5}$$

(3.4) is, once again, 30.78 times the theoretical smallest time scale, the Planck time at 5.39... x10^{-44} seconds. This ratio of times and lengths will be examined in the Verification section. However, (3.5) is 2.15... x 10^{-23} smaller than the Planck length. No experimental device currently available or to this writer's knowledge, even conceivable can penetrate to the Planck scale, let alone this much smaller scale.

Nuclear and other subatomic particle interactions, presumably occurring at or near the speed of light, operate on a scale that dwarfs this expansion rate. The length of time it would take light, whose velocity is supposed to be scale invariant, to transit the strain distance per meter is 7.90... x 10^{-27} seconds, during which time the expansion strain would be a mere 1.87... x 10^{-44} meters per meter, nine orders of magnitude less than the Planck length. The length of time it would take light to transit *this* distance is approximately 6.24... x10^{-53} seconds, again nine orders of magnitude less than the Planck time. Thus any random variation in position or conjugate momentum in the phase space of a nucleon dwarfs in scale any achievable determination of an absolute position in space and time.

Still, we would like to find a way to gauge the expansion locally, if indirectly. We have our position-inertia as discussed at the location of our point, and we can assume, instead of a second point, a translucent, momentum-inertial sphere around the central initial point at a scalable distance that we call r_0, moving out at a yet to be determined expansion rate. It is important that while we think of the way this and additional spheres will respond to changes in themselves and their positions, between each other and in their environment, we must understand that essentially they are not "things" at all. They simply represent loci of dynamic stress equilibrium, strain movements and their relationships in a small section of a single spacetime continuum in response to its overall expansion.

Our position-inertia is what we normally call mass, m, while our momentum-inertia is our old friend from phase space, conjugate momentum, p, which at r_0 is p_0. In terms of a position in spacetime, the mass is a measure of the inertial constant, ה, times the square root of the Gaussian curvature of the phase sphere PS$_3$ which we will designate as K, which happens to also be the angular wave number, κ, represented by the sphere or

$$\kappa = \sqrt{K} = \sqrt{\frac{1}{r_0^2}} = \frac{1}{r_0} \tag{3.6}$$

Thus the unit mass, m_0, is

$$m_0 = ה\kappa_0 \tag{3.7}$$

the linear inertial density over the distance r_0 is

$$\lambda_0 = \frac{ה\kappa_0}{r_0} = ה\kappa_0^2 \tag{3.8}$$

The time rate of change in the unit momentum at the spherical shell per unit of surface area, is

$$\frac{\dot{p}_0}{A_0} = \frac{\tau_0}{A_0} = f_{t0} \tag{3.9}$$

As shown, this is equal to the stress at the surface of the sphere. It is important that we understand what this means. If we are standing at a point in space at which every point around us is moving away from us at some finite speed, with the understanding that the same is happening for every other point, we have to ask ourselves what determines this speed. If the whole of the space around us were to be infinitely inert, it would mean that it could not move at all and there would be no expansion. If the whole of that space had zero inertia, it would mean that whatever and whenever some agency put it into motion, the expansion would be instantaneous. The fact that it has a finite speed that appears to be increasing over time, indicates that such space has an inertial component. Like a massive flywheel that requires decreasing energy input per radian of motion over time in order to accelerate, such a space, due to its inertial property would be expected to accelerate its expansion over time. With expanding spacetime such inertial acceleration constitutes a stress force field.

For an isotropic force field moving out from a central point which has no dimensional component, that is a radius of zero and a vanishing surface area, the stress at the point is infinite. This is physically untenable, so instead of a point we are compelled to assume that some geometric domain intervenes between the isotropic outward expansion force and the center point of the sphere. With expansion we should expect the generation of a spherical loci at radius, r_0, at which the expansion tension stress equals the reactive stress, establishing the elastic potential energy density of the interior of the sphere.

Once equalized, given only a tension component of the stress, we would expect a dilatation or increase in r_0 related to the expansion rate. As discussed in the Aside #2 on stress and strain analysis, however, there are other components of stress including lateral compression, shear and torsion to consider. The net effect of this would be the production of torsional, i.e. limited rotational oscillation and the creation of a central angular acceleration of the sphere that works counter to any dilatation at its surface. Since the spherical discreteness we are examining is a locus of stress equilibrium in an otherwise continuous manifold, and not a discrete body in isolation in a true vacuum, we would not expect full rotation of the sphere, which would create unsustainable axial strain and stress at the poles of the rotation. Rather we would expect axial oscillation in keeping with our model of PS$_3$. Once initiated, with increasing expansion stress, the increasing rotational strains are anticipated to accelerate tangentially and therefore centrally, resulting in a net differential central force radiating isotropically and responsible ultimately for coupling with a gravitational field.

If we let the sphere expand, we will once again have no way to gauge the expansion around it, and if we keep it fixed, we are in the same pickle. (At this point we have no measuring rods and no clocks.) We think about it for a few minutes and realize isotropic expansion implies expansion about all points, so that if we put some additional, similar spheres around the single one and keep them all sized at r_0, allowing them all to touch, we might eventually be able to notice some variation in their motion with expansion. We can get twelve such spheres around the initial sphere, as in Photo 1, and find it interesting that the only way we can get them all in place is to create a lattice of four axes about the first sphere. We notice that each axis is normal to an arrangement of the central sphere with six spheres forming a hexagonal disk around it, and that each such arrangement has three sets of two opposing spheres in common with three other disks, making a system of 4 intersecting disks around a central sphere of equal size.

Cuboctahedral Lattice
Photo 1

Tetrahedral Aperture
Photo 2

Octahedral Aperture
Photo 3

We find that each and every sphere on the periphery of this arrangement has two other spheres in direct contact with it in the form of an equilateral triangle and in fact each sphere belongs to two such groups on opposite sides. We can transpose any three by rolling them simultaneously about the six in which they are nested, by a turn of 1/3 π. This rearranges the axes, bending three of them at the center around the first and affects the symmetry, but in the end returns the thirteen spheres to the same tight density they had originally. Turning them back restores the 4-axial symmetry.

Now, if we allow the space around them to expand isotropically, after a period of time they each end up a distance αr_0 apart from the central sphere and from each in the adjacent pair on either side on the periphery. If we draw lines connecting the centers of each sphere, we notice the emergence of some recognizable geometry. The lines joining the twelve, periphery spheres form a cuboctahedron, which is the polyhedron formed by drawing a line between the midpoints of the 4 edges of each face of a cube. This creates a diamond on the face of each of the 6 cubic faces and an equilateral triangle where each of the 8 cubic vertices had been. This means that the twelve spheres in their pre-expansion position form a similar rectilinear grid, a cube with a sphere centered at each of its edge midpoints, for which the previously mentioned 4 axes are the cubic diagonals. The 8 vertices of this cube are positioned in the regions outside the three spheres about each corner and the centers of each sphere coincide with one of the midpoints of the cubic edges. The cubic edge measures $2\sqrt{2}r_0$ in this initial configuration.

What we have is the superposition core of an interlaced 3-axis rectilinear or hexahedral lattice centered through the cubic faces and a 4-axis alternating double tetrahedral plus octahedral lattice centered along the diagonals. The first lattice actually embeds a 3-axis octahedral lattice through the center of each sphere and a second 3-axis stretched rectangular lattice with a $\pi/4$ twist between adjacent axes through the periphery spherical centers and around the central sphere. The second lattice embodies the obvious local tetrahedral lattice along the diagonals. I have called it local, because if this close packing of spheres is carried out indefinitely, each sphere will have tetrahedral packing in each direction at each of its diagonals, but the diagonal and therefore lattice axis will run through an octahedral cell before entering another tetrahedral cell and then another sphere.

Looking at Photos 2 and 3, we do some quick checks and find that the cosine of each angle of the equilateral triangles of $\pi/3$ is ½ and the sine is $\frac{\sqrt{3}}{2}$ for an area of $\sqrt{3}r_0^2$ spanned by the three spherical centers, while the spherical cross-sectional area is $\frac{\pi}{2}r_0^2$; a difference for the tetrahedral "free space" aperture of $0.16125\ldots r_0^2$. The computation for the diamond on the six cubic faces is a bit simpler since each side is $2r_0$ for an area of $4r_0^2$ minus the spherical cross section of πr_0^2 for a difference for the octahedral "free space" aperture of $0.85840\ldots r_0^2$. This means that the octahedral aperture is $5.32330\ldots$ times larger than that of the tetrahedral aperture. It also means that the inertial density of the spheres is less about the customary cubic surface axes than about the diagonal axes, where we include only the area within the corresponding cuboctahedral surface, by a factor of

$$\frac{\pi}{4} \bigg/ \frac{\pi/2}{\sqrt{3}} = \frac{\sqrt{3}}{2} = 0.86602\ldots \tag{3.10}$$

If we define the density as the ratio of the cross-sectional areas of the adjacent spheres to the corresponding aperture we have

$$\frac{4\pi}{0.85840...} \Big/ \frac{3\pi}{0.16125...} = 0.25046... \tag{3.11}$$

Cuboctahedral Lattice showing both hexahedral and octahedral components

Lattice extension along 3-axes surfaces develops octahedron with stability of vertices and density of surfaces

Lattice extension along 4 axes diagonals develops cube with instability at vertices and porous surfaces

Photo 4

Photo 5
Cuboctahedral Lattice

Photo 6

Photo 4 shows additional elements to the cuboctahedron suggestive of an extended lattice. These configurations are made with small magnetic spheres, which have magnetic dipoles analogous to a neutron or proton magnetic moment. Photo 5 shows the stability of extension along the hexahedral 3-axes and density of the octahedral, diagonal 4-axes. Photo 6 shows an extension along the 4-axes with instability evident at the cubic vertices and a much greater porosity of the cubic surfaces. The internal structure of both Photo 5 and Photo 6, as in an indefinitely extended lattice, is the same cuboctahedral lattice. It is only at the boundaries, principally of the hexahedral component and in particular of the vertices, that the instabilities and therefore the differences emerge.

It is worth noting in the context of this lattice geometry that a cuboctahedral edge length is equal to the diameter of the central sphere, so that it is also equal to the radius of a sphere, R, circumscribed through the vertices. The distance from the center of the central sphere to the midpoint of the cuboctahedral edges is $\sqrt{3}/2\, R$ and from that center to the cubic faces is $\sqrt{1/2}\, R$. The first of these coefficients is the sine of the angles ε and μ mentioned earlier in connection with Figure 26 and is significant to quantum mechanics as the magnitude, S, of the spin angular momentum of all fermions, for which the quantum spin number is $s = \tfrac{1}{2}$ and S is

$$S = \sqrt{s(s+1)}\hbar = \sqrt{\tfrac{1}{2}(\tfrac{1}{2}+1)}\hbar = \sqrt{3}/2\, \hbar \tag{3.12}$$

where $\hbar$ has been previously discussed as an invariant action. In terms of our discussion of PS_3 and SHM then, if $R = q_0$,

$$S = \sqrt{3}/2\, S_0 = \sqrt{3}/2\, \hbar c \,. \tag{3.13}$$

The value $\sqrt{\frac{1}{2}}$ is the coefficient of q_0, p_0, τ_0, and c_0, at the moments +/-E and +/-M, when the action and power each reach a maximum ½ in PS$_2$ and an invariant ½ each as they rotate in PS$_3$.

No other Platonic or Archimedean polyhedral lattice embodies these significant, fundamental geometric coefficients. The takeaway from this is that isotropic expansion performed globally results in the emergence of the above-described lattice without any exogenous constraints. Of interest is the fact that this analysis suggests that the principal axes affected by expansion are the 4 axes of the diagonals. Assuming an inertial spacetime, this means that the principal tension axes in the locale under investigation, and thus on the surface of the central sphere, are these four diagonals and not the three cubic face axes. Thus the shear and shearing rotational/torsion stresses that would be expected by elastic stress-strain analysis to accompany an expansion tension are then about these axes and in the plane of each of the eight triple spheres delineating the cubic vertices. The inertia is more concentrated about these axes in response to such stress. These are free to rotate about the diagonals with less inertial resistance from other spherical domains than are the four spheres defining each cubic face about the three cubic face axes.

It is of interest that there is a two-dimensional correspondence to this three dimensional emergent phenomena. Rayleigh-Benard convection is the designation for the formation of a hexagonal lattice pattern of cells in a shallow layer of fluid subjected to gravity and a temperature, i.e. energy, gradient from below, in conjunction with the viscosity and thermal properties of the fluid. In the case of our current thought experiment, gravity and the properties of the fluid are replaced by the inertial properties of spacetime and the energy gradient is due to its isotropic expansion. The resulting planar convection cells are replaced by the stress/strain oscillations of spacetime. The form each emergent phenomenon takes is due to geometric constraint.

An essential aspect of this inertial spacetime is the concept of continuity, that is that the points *right next to* any given point cannot be transposed. While we can see that a lattice can emerge from stretching such continuity, it does not follow that such continuity can emerge from a collection of points not inherently so constrained, i.e. points free to move in the manner of a random walk vis-à-vis other points. Assuming an inertial, and within certain limits, elastic continuum, we would expect that with continued expansion and the geometric constraints of PS$_3$ and the above emergent lattice, that on some scale determined by the inertial density of the continuum, instances of torsion would occur.

In Figure 27, we have an example of such torsion operating in two orthogonal directions. There are two sets of co-ordinates shown with capital letters for reference purposes in keeping with the prior development of PS$_3$, one for the diagonal and one for the surface axes. The lower case co-ordinates indicate points in the field that are moved by the stress/strain relationship. In graphic #1, we have defined a square with edges of $\sqrt{2}$ in keeping with the cuboctahedral description and our prior development of the moments E and M in PS$_2$ and PS$_3$. The torsional stress potential of #1 results with expansion in the defined strain of #2 shown by the co-ordinate pairs, followed by a second torsion strain in #3, which sets up a recoil stress potential indicated by the axial vector ϕ in #4. Upon recoil, the potential ϕ becomes the active axial vector ϕ, and initiates the angular momentum vector θ in #5. This last step gives the Spin Diagram seen in Figure 26, which is advanced about θ by half a rotation from #5.

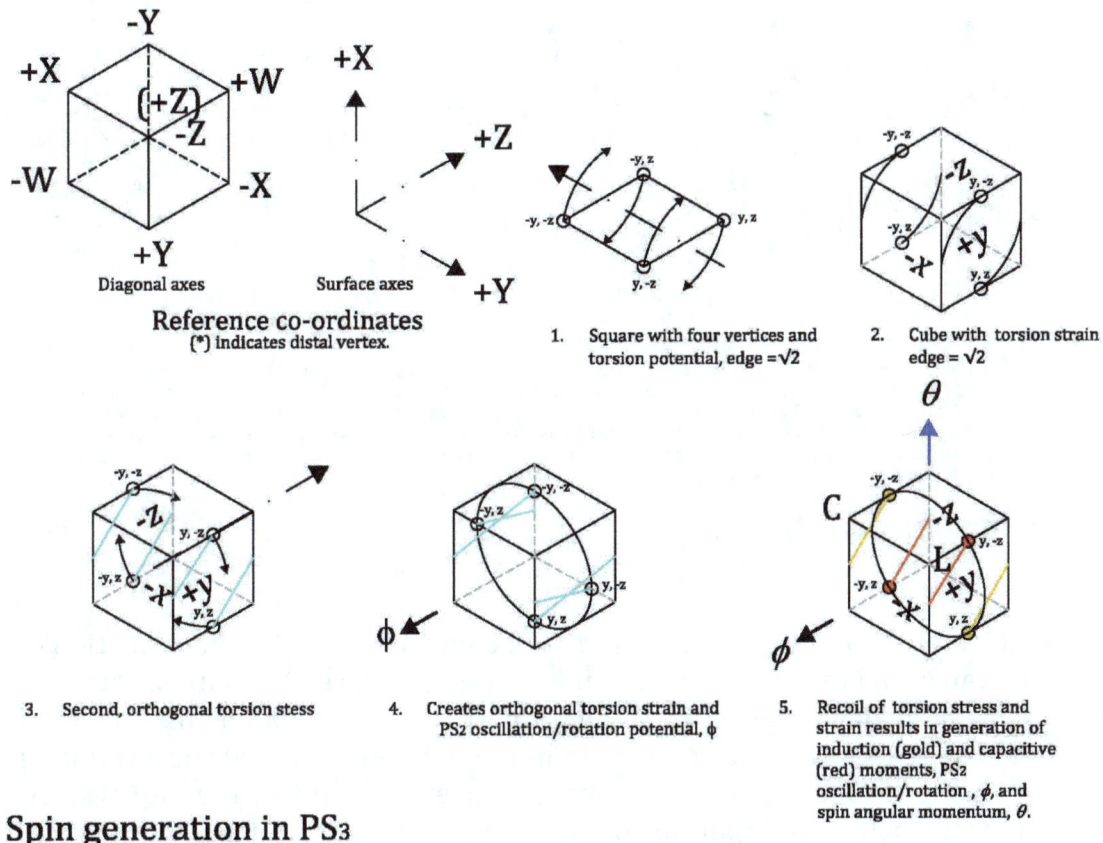

Spin generation in PS$_3$

Figure 27 – Initial Strain State

In #5, the various y,z strains are colored to reflect their functional states at the point in time at which they constitute the +/- E and +/- M moments as shown. Figure 28 shows a continuation of the rotation of θ over one cycle at ½ π stages or phases. As we will see, this is the resonant state of PS$_3$ which is recognized as the neutron. Each of the four phases consists of a ½ π rotation about ϕ followed by a ½ π rotation about θ according to the axial vectors shown. Each of these rotational sequences is equivalent to a right hand twist at each of the upper diagonals as shown. Such a sequence of twists, as well as the corresponding sequence of double rotations, rotates the permanently displaced +/- x faces of the cube about the X axis, while the +/-y and the +/-z faces alternately oscillate about their initial positions along axes Z and Y.

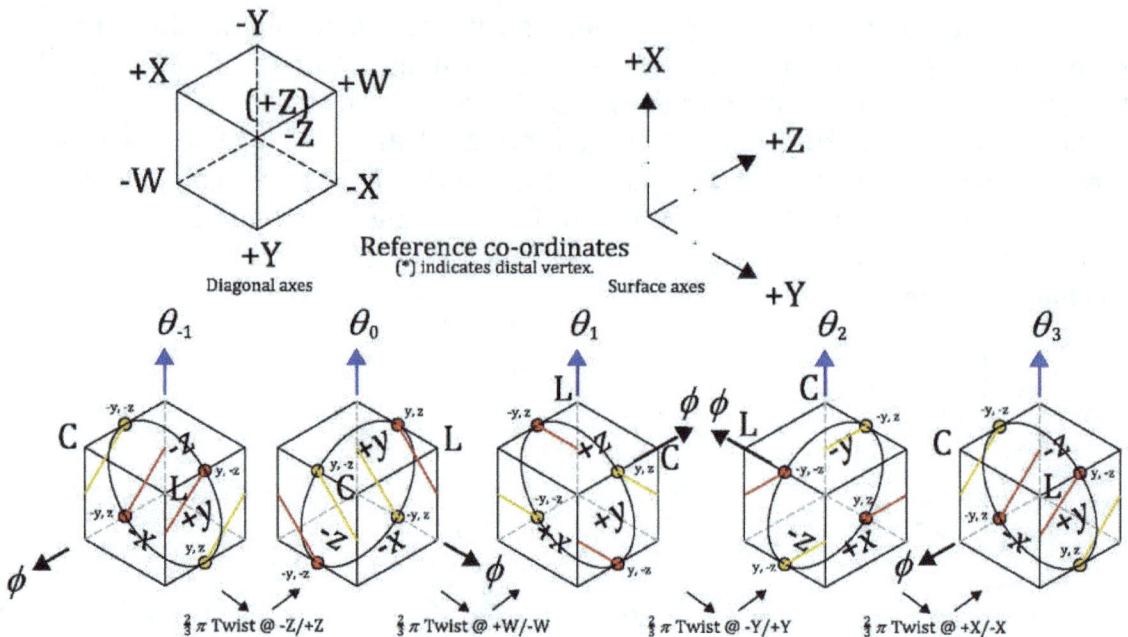

Combination of PS2 rotation, ϕ, and PS3 spin, θ, each over $\frac{1}{2}\pi$ is equivalent to a twist at diagonal axes of $\frac{2}{3}\pi$ as shown. The result is an oscillation of the torsion stress and strains, here shown with representations for each phase. The capacitive, C, and inductive, L, torques are found by crossing the two equilibrium or unstrained positions into the capacitive moments for the capacitive torque and the two inductive moments into their equilibrium or unstrained positions for the inductive torque. This particular sequence represents the resonant condition which is recognized in quantum phenomenology as the neutron. The quarks of the standard theory are the nodes/antinodes of the oscillation.

Resonant spin state, the neutron

Figure 28 – Resonant Strain State

The twisting motion of the triplets about the diagonals constitutes an oscillation of the tetra-octahedral lattice and the principle stress/strain relationship in the continuum responsible for generating the property of spin. Why it is this and not the hexahedral lattice that is primary can be found in the notion of the conservation of angular momentum. Such conservation directs the energy flow in a torsional stress into whatever structure incorporates the smallest moment of inertia. The moment of inertia, I, of the cuboctahedron of Photo 1 is less about the diagonal axis through the tetrahedral aperture (TA) as shown in Photo 2 than about the surface axis through the octahedral aperture (OA) in Photo 3. Assuming uniform density of the spheres, the sum of the distance from each spherical center to the associated axis gives a measure of I as

$$I_{TA} = 6(1) + 2\left(3\left(\tfrac{1}{\sqrt{3}}\right)\right) = 6\left(1 + \tfrac{1}{\sqrt{3}}\right) = 9.46410\ldots \tag{3.14}$$

where 1 is the distance from the central sphere to each of the 6 adjacent spheres in the central plane and $\tfrac{1}{\sqrt{3}}$ is the distance from each of the triple spheres about the pair of TAs to the corresponding axis, and

$$I_{OA} = 4(1) + 2\left(4\left(\tfrac{1}{\sqrt{2}}\right)\right) = 4\left(1 + \sqrt{2}\right) = 9.65685\ldots \tag{3.15}$$

where 1 is the distance from the central sphere to each of the 4 adjacent spheres in the central plane and $\tfrac{1}{\sqrt{2}}$ is the distance from each of the quadruple spheres about the pair of *OA*s to the corresponding axis. This gives a slight advantage to torsional oscillation about the diagonals.

As a description of this oscillation, we can rotate diagonal -Z with its two pairs of triplets by a rotation of 2/3 π ccw. This moves the cubic face +y, onto which we can draw an arrow pointing up, to the top with a rotation ccw of ½ π of the arrow to point left. We repeat the same rotation diagonal +W, which moves +y back to its original position, but now rotated ccw π from its original orientation with the arrow pointing down. At diagonal –Y we repeat the procedure, which rotates +y to the bottom of the cube, presumably in the same orientation as when it was on top. (You can peek if you want and make sure, but the arrow does point left.) Finally we focus on diagonal +X and rotate the axis as before. +y returns to its original position with the arrow pointing up. No co-ordinates have been harmed in the making of this movie. No entanglement of an arbitrary x, y and z, so that we have to untangle them every even number of rotations. The net effect is a sustained angular momentum of the system about the top face. An oscillation over 4 axes creates a rotation in three without an actual complete bodily rotation of the center sphere. In fact, when we analyze its motion, it very accurately embodies the motion of the PS_2 disk in the PS_3 sphere. In short, the expansion of an inertially dense spacetime overtime naturally and necessarily leads to the emergence of a discrete unit, i.e. a quantum, oscillation with angular momentum of ½ S_0. It is such oscillation that is described by PS_3.

There is a more work needed to nail this down. If we cross the action/power moments of PS_3, with respect to their positions prior to displacement on the equilibrium circle C_{+p} into those positions for the +/-M and from those positions into the moments for +/-E, we obtain two axial vectors or torques, one for the two capacitive moments, $C\varepsilon$, and one for the two inductive moments, $L\mu$. The angles for this crossing, μ for the inductive and ε for the capacitive moments, are $\pi/3$.

These torques necessarily rotate synchronically with the moments in ϕ with the inductive torque leading the capacitive torque by ½ π, both inclined away from the angular momentum vector. These apply a maximum torque to each of the θ axes of the rotating PS_2 disk, mitigating the recoil of θ to align with ϕ, and advancing the rotation of the restorative force of SHM. They also generate a magnetic moment antiparallel to the angular momentum or spin vector. Thus

$$\left[\left(-\mathrm{M}\times p_{-3\pi/4}\right)+\left(+\mathrm{M}\times p_{+\pi/4}\right)\right]=\mathbf{L}_\mu \qquad (3.16)$$

$$\left[\left(-\tau_0\times -q_0\right)+\left(+\tau_0\times +q_0\right)\right]\sin\mu = 2\left(i\tfrac{\sqrt{3}}{2}\tau_0 q_0\right) \qquad (3.17)$$

$$\left[\left(p_{-\pi/4}\times +\mathrm{E}\right)+\left(p_{+3\pi/4}\times -\mathrm{E}\right)\right]=\mathbf{C}_\varepsilon \qquad (3.18)$$

$$\left[\left(-q_0\times +\tau_0\right)+\left(+q_0\times -\tau_0\right)\right]\sin\varepsilon = 2\left(-i\tfrac{\sqrt{3}}{2}\tau_0 q_0\right) \qquad (3.19)$$

While the two components in each torque can be represented as one axial vector, in terms of the magnitude of the torque, they operate separately on each half of the rotations ϕ and θ. These are shown in Figure 29 for the resonant state, the neutron.

The upper figure shows the four figure eight paths of points +/- y, +/-z on the PS$_2$ disk prior to the torsional strain. PS$_2$, which is centered on the axial vector ϕ, is positioned so that the physical point +z is displaced to the point +q of PS$_3$ in Figure 23, one of the two functional points, +/-V, of maximum potential energy for *+z and all other points on the circle of PS$_2$*. This is designated as the time $\theta(\frac{\pi}{2})$, where the time $\theta(0)$ is represented by the position of the axial vector ϕ parallel to the +Y axis.

The moments +/-E and +/-M, represented by the black "X", are located at time $\theta(\frac{\pi}{4})$, and $-M_{+z}$ indicates the position of +z when it is at the moment of maximum power on the discharge or induction leg of its cycle traveling in the –X direction. The angles ε and μ are shown. It is important to note that the selection of the points +/-z and +/-y are arbitrary and that the moments E and M exist for all points on the circle C$_{+p}$ that includes these four points.

The lower figure shows the oscillation at time $\theta(\frac{\pi}{2}+\frac{\pi}{4})$ when +z has reached -M and the other points +y, -z, and -y have reached the other moments shown. The two torques generated by these moments are shown at $\mathbf{L}_\mu$ and $\mathbf{C}_e$. Most significantly, the effects of these torques on what we will call the nodes and antinodes of ϕ and θ are shown by the small red, ε, and gold, μ, vectors along the θ rotational path. The inset on the lower left of the figure enlarges these and also shows the condition of ϕ at maximum potential energy point +/- V. All are as viewing the sphere from the exterior toward its center. Note that at the positions of +/-V and +/-W, ε and μ are orthogonal and that their dot product therefore vanishes and their cross product is maximum. At K$_0$ and K$_\pi$, they are parallel, the cross product vanishes and the dot product is maximum. We will see the effect of this on spin and charge in a few minutes.

Overtime, as the continuum continues to expand, its mechanical impedance decreases and the inductive torque advances and drops generally antiparallel to the angular momentum vector, reversing the direction of the magnetic moment vector to align with the spin and transmitting a portion of its power and energy as the electron, decreasing its frequency slightly in the process. In the event of a retarding of the inductive torque, which generally will not occur in the context of expansion, the capacitive torque drops generally anti-parallel to the spin vector, creating the antiproton and transmitting the positron.

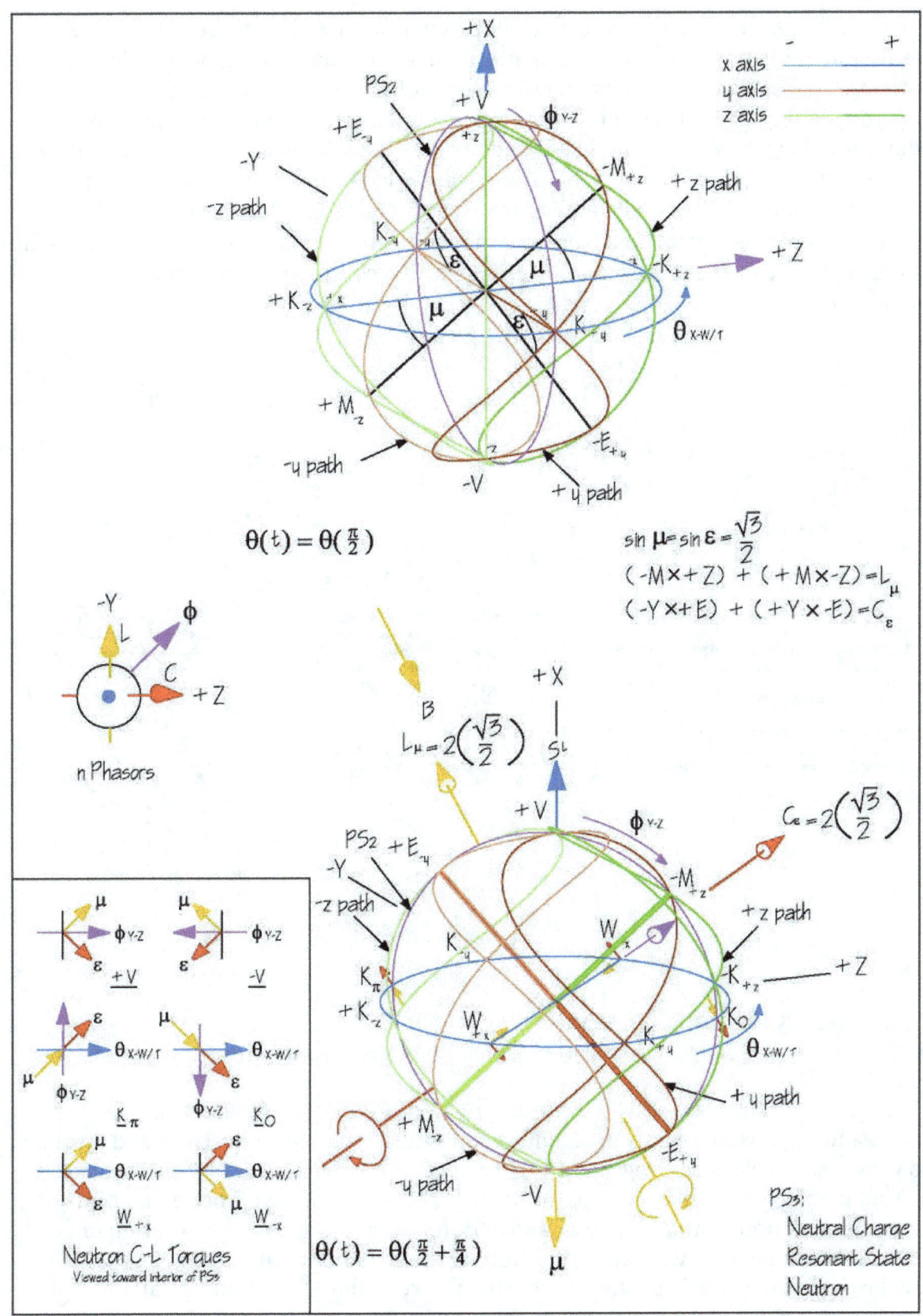

Figure 29 - Spin Diagram 2, Neutron

Figure 30 indicates the strain relationships associated with beta-decay and the emission of the electron. The drop in mechanical impedance and inertial density arising from spacetime expansion effectively pulls at the nodes of rotation ϕ, increasing the strain. These nodes are designated by the W points on the lower diagram of Figure 29. As we will see shortly, this is particularly significant at the W_{-x} node. Over time, this advances the ϕ axial vector in θ without a corresponding rotation of ϕ itself. This is seen in Figure 30 as indicated at step 2. Recoil of the strain as indicated by the four point pairs results in a rotation reversal of ϕ and a flip in the spin vector θ. A portion of the energy of PS$_3$ continues on as the emitted electron state.

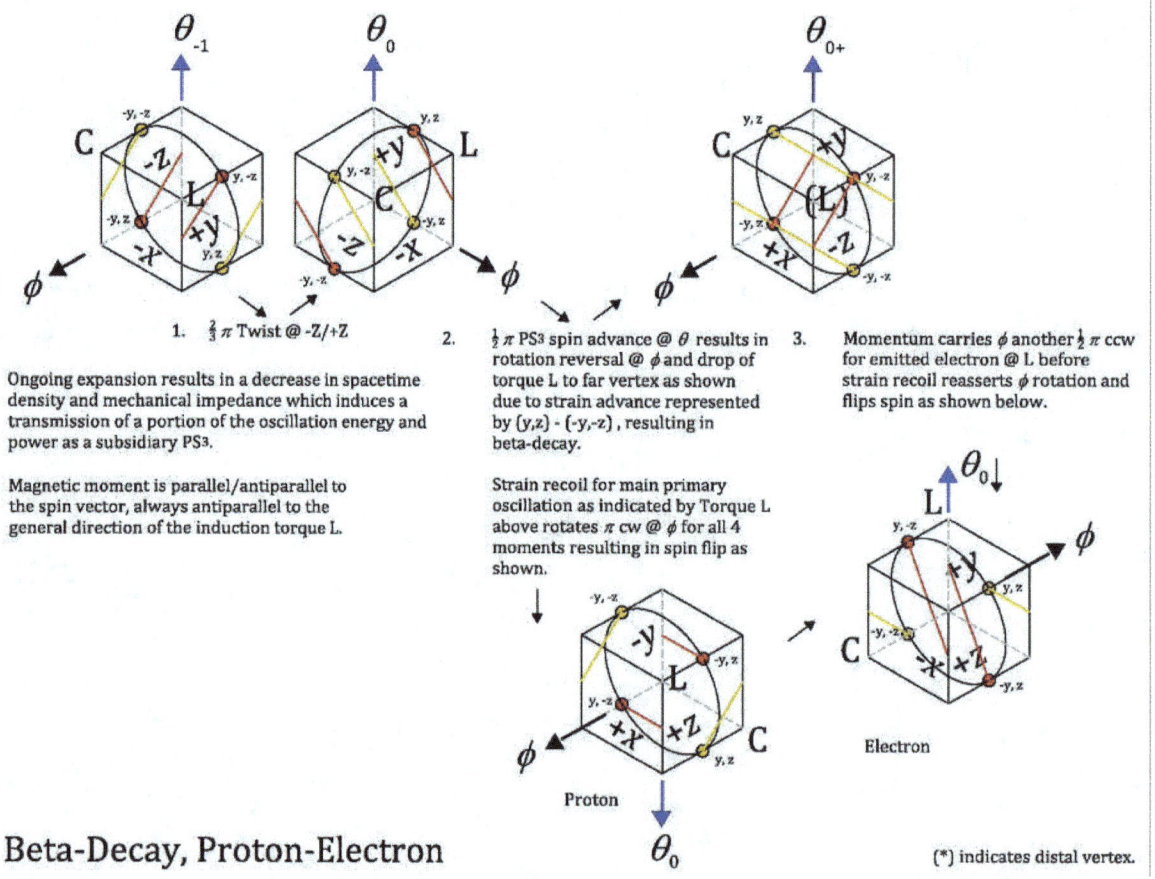

Figure 30 – Strain States of ordinary Matter at Beta-Decay

Let's do a quick review of what is occurring in this quantum oscillation. The energy of cosmic expansion results in an emergent oscillation of torsion stress and strain on a very small scale according to well-defined geometric constraints. The physical strain displacement and recoil paths incorporate well-defined functional points of maximum concentration of power and action, which act as torques rotating in synchronous manner with and conditioning the path nodes and antinodes, which we will refer to collectively as the nodes. The nature of that conditioning can be understood through analysis of the vector operations of ε and μ at the nodes. The results of this analysis can be read in the spin and charge tables, Figures 38-39.

We start in all cases by calculating the product of the torques $\mathbf{L}_\mu$ and $\mathbf{C}_\varepsilon$ and the PS_3 radials at the points shown to establish the magnitude of the vectors ε and μ at each of those points. As the torques lie along the diagonals of a cuboctahedron and the nodal points lie along its surface axes, the angle in all cases is $0.9553\ldots$ with a sine of $\sqrt{\frac{2}{3}}$. As a result, the magnitude of ε and μ in all cases is the magnitude of the torques for each half of the cycle, $\frac{\sqrt{3}}{2}$, times the sine or $\frac{1}{\sqrt{2}}$. The angle between ε and μ and the tangent vectors for each of the rotations θ and ϕ is in all cases $\frac{\pi}{4}$ with a sine and cosine of $\frac{1}{\sqrt{2}}$ while the angle between ε and μ at the points +/-V and +/-W is an invariant $\frac{\pi}{2}$ and at the points K_0 and K_π is either 0 or π. As stated, this means their dot product vanishes at V and W and the cross product vanishes at K.

The maximum angular momentum of PS_3 is determined by the points K where the kinetic energy is greatest. As seen in the charge table, the various dot products in all cases equal +/- ½ as a coefficient of the angular momentum. This is in keeping with the observed phenomena that all fermions have $\pm\frac{1}{2}$ spin angular momentum. In the case of the neutron, all the values of the dot products are positive, indicating a potential for the interacting torques and rotational dynamics to augment each other. It is thus an unstable condition, even though it represents a resonant state of PS_3. By analogy, a bridge that begins to oscillate at resonant frequency is inherently unstable.

With respect to the various cross products in the table, the radial sense designation used is a filled dot for a centripetally directed vector and a hollow dot for a counter-centripetal vector product. It bears mentioning that while ε and μ are equal at the given nodes, by their natures the differentials of the inductive or kinetic vector, μ, are always increasing and those of the capacitive or potential vector, ε, are always decreasing. For this reason, in the charge tables μ is always crossed into ε with the direction of the vector product given by the right hand rule.

Column 7 gives the charge at each node of the ϕ rotation, W_{+x} and W_{-x}, (+/-W). Notice that while these two points are the nodes of ϕ, they are circulating on the θ or C_{+p} circle, as indicated by the row headings. In the case of the neutron, the $\mu\times$ and the $\varepsilon\times$ are antiparallel and cancel, indicating a neutral charge for the neutron PS_3. Note also that the radial sense for $\mu\times\varepsilon$ at W_{-x} is centrifugal, indicating a potential for emission of energy from this node, while that at W_{+x} is centripetal, indicating a corresponding potential for retention of energy at this node.

Figure 30 gives a representation of the strain events at the point of beta decay for ordinary matter. Figure 31 gives the four phases of the rotational cycles of the electron and proton, with the relative positions of the C and L torques shown. Note that all the cubic representations of the strain states of PS_3 are out of phase from the corresponding Spin Diagrams, by $\frac{\pi}{4}$ as each shows from Figure 27 #1, initial, undisplaced positions at the vertices of the square as the points strained to the (gold/red) power/action moments of the Strain States. The Spin Diagrams indicate positions from the normal vectors of an XYZ co-ordinate system displaced to those same moments.

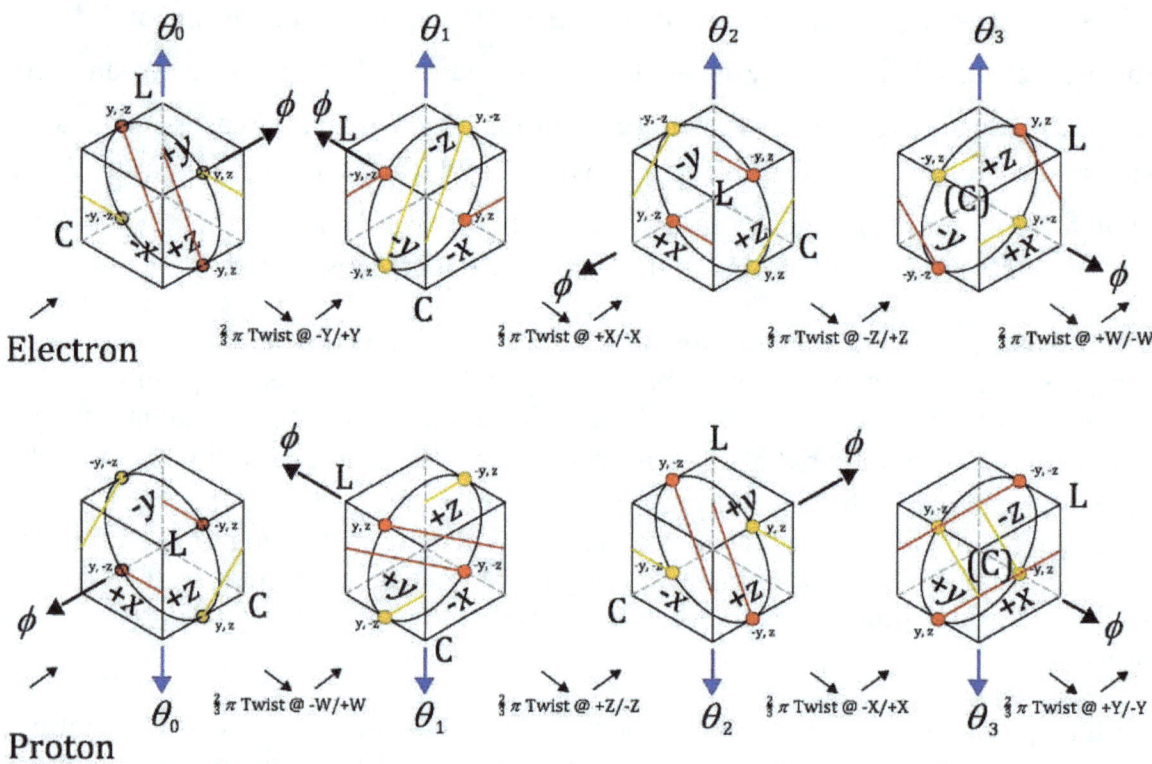

Proton & Electron Rotational Oscillation

Figure 31 – Inductive Strain States, Electron and Proton

The following Spin Diagrams for the proton and electron show the dynamics of Figure 31 in greater detail at a given moment in time, with the same nodal detail for the torques and their local vectors μ and ε as seen with the neutron in Figure 29. With respect to the charge table for ordinary matter, for the proton, the dot products of these vectors are of opposite sense at each of the nodal points with respect to ϕ and θ, and are negative with respect to their common product, indicating the antiparallel nature of their effect. The spin magnitude remains ½ but the stability of the proton over the neutron is indicated by the opposite senses for μ and ε.

With respect to the cross products and charge, the condition at W_{-x} is centrifugal for both $\mu\times$ and $\varepsilon\times$ as it is for $\mu\times\varepsilon$ with the net result of a charge of +1. At W_{+x} the charge is similarly +1, but is is centripetally directed.

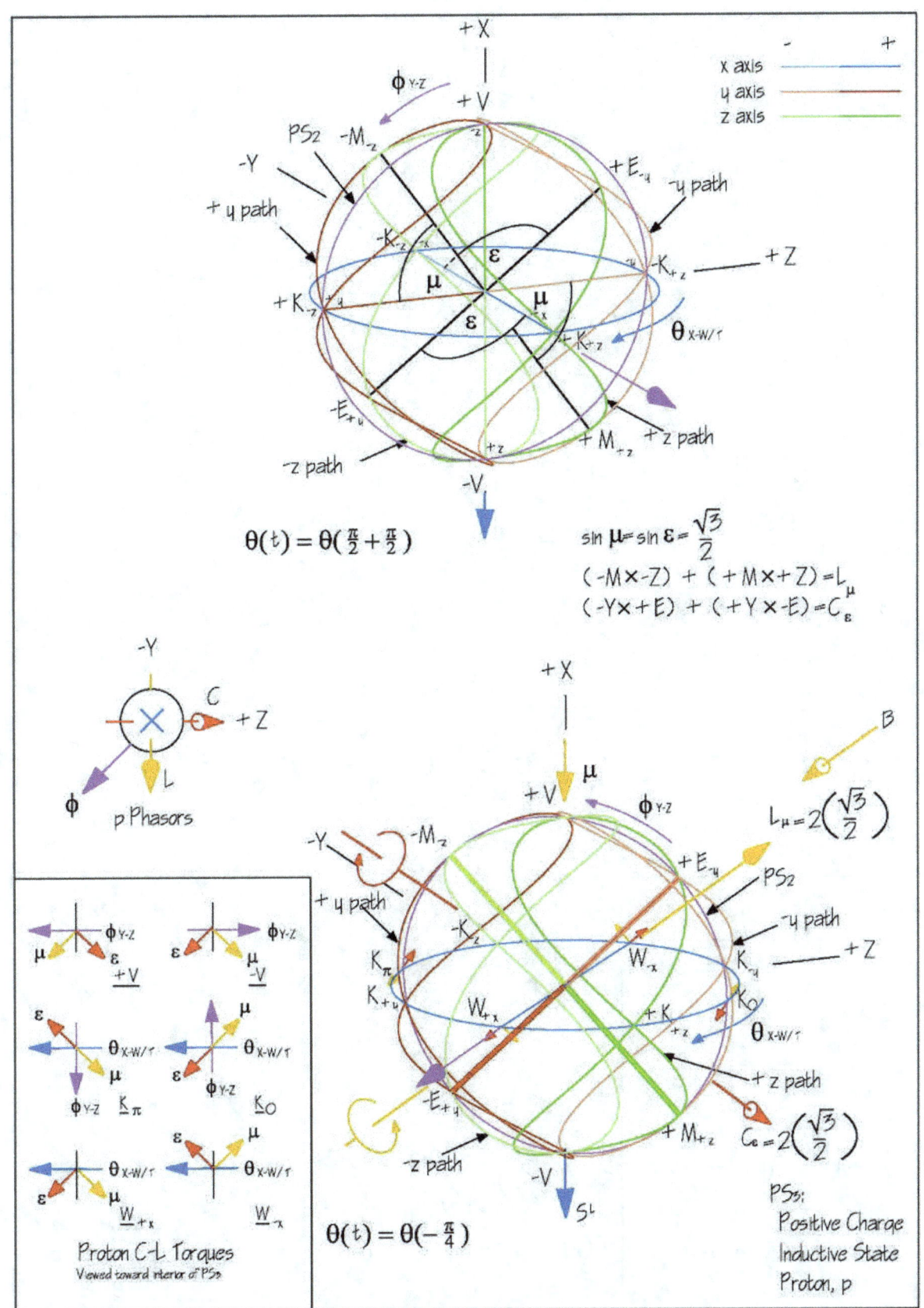

Figure 32 – Spin Diagram 3, Proton

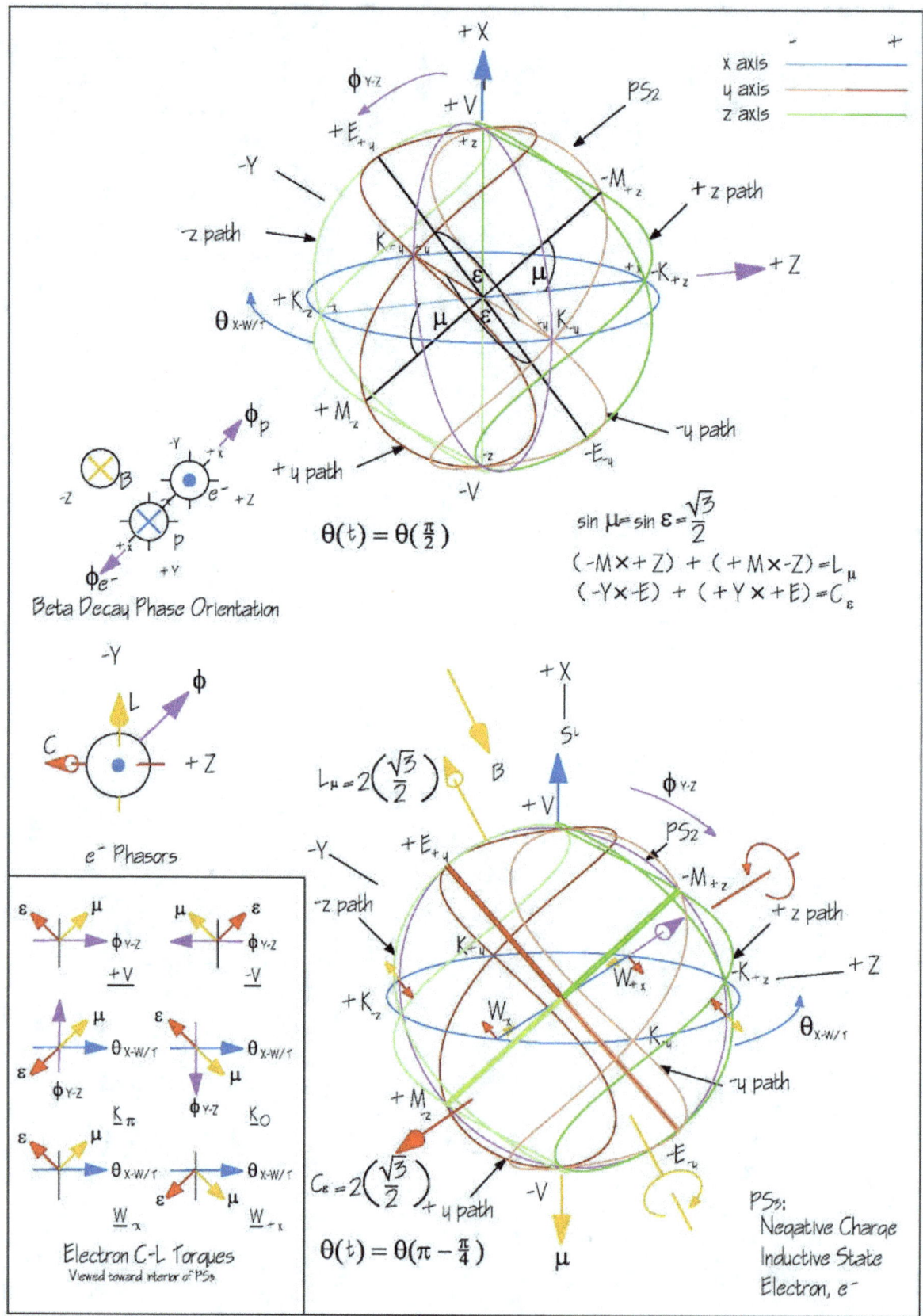

Figure 33 – Spin Diagram 4, Electron

In the case of Figure 33 for the electron, while the dot products of μ and ε remain negative due to the antiparallel nature of the vectors, the dot products of $\mu \cdot$ are all positive due to the inductive nature of the proton-electron strain states; the inductive torque is generally parallel to the direction of motion of all the nodal points of ϕ and θ in the electron. The $\varepsilon \cdot$ products are each antiparallel to that motion and therefore negative.

With respect to the cross products, as with the proton, at the K points $\mu \times \varepsilon$ vanishes and their individual products with ϕ and θ are antiparallel and therefore cancel. At W_{-x} the cross products are centripetal for both $\mu \times$ and $\varepsilon \times$ giving a net charge of 1, but antiparallel to $\mu \times \varepsilon$ for a negative charge as shown.

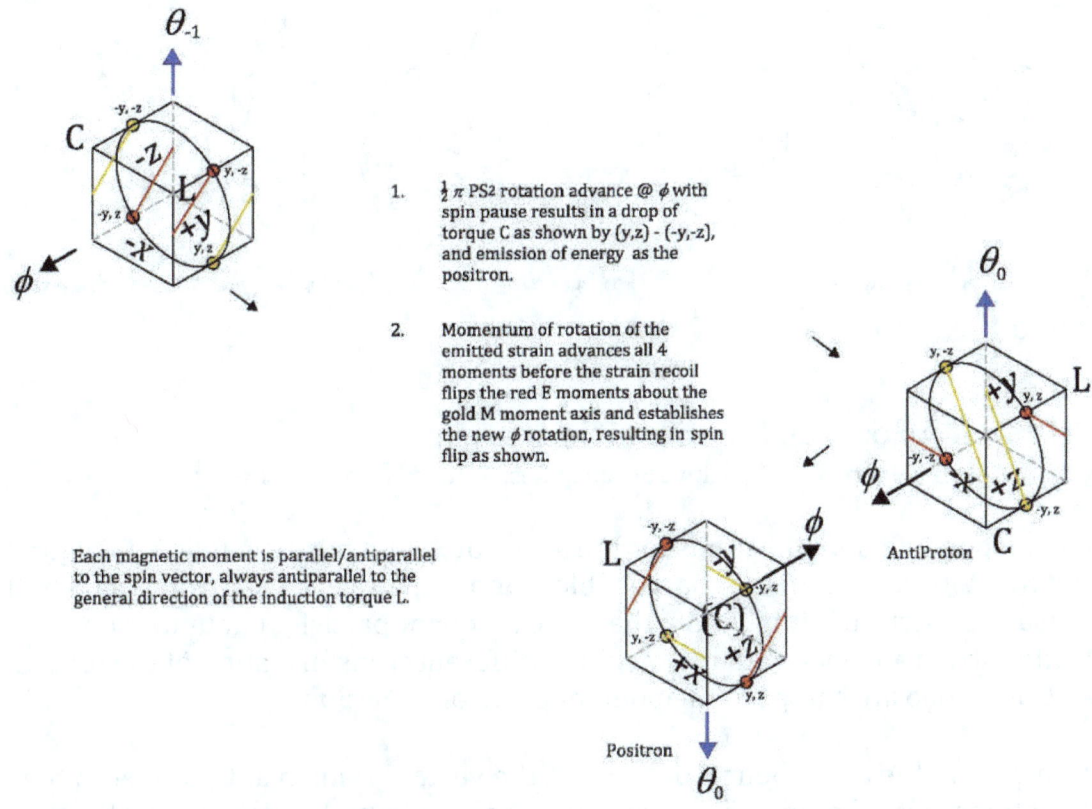

Beta-Decay, AntiProton-Positron

Figure 34 – Strain States of Anti Matter at Beta-Decay

Figure 34 is a representation of the strain states at the point of beta decay for anti matter. In this case, as can be seen there is an advance of the ϕ rotational strain without a corresponding advance of θ, which emits the positron with a spin reversal. This is equivalent to a retardation in spin and indicates a general capacitive state, which is why it is rare in the context of an inductive, expansionary state of the cosmos. Figure 35 represents the strain states for the antiproton and positron and shows the symmetry of the relationship with ordinary matter.

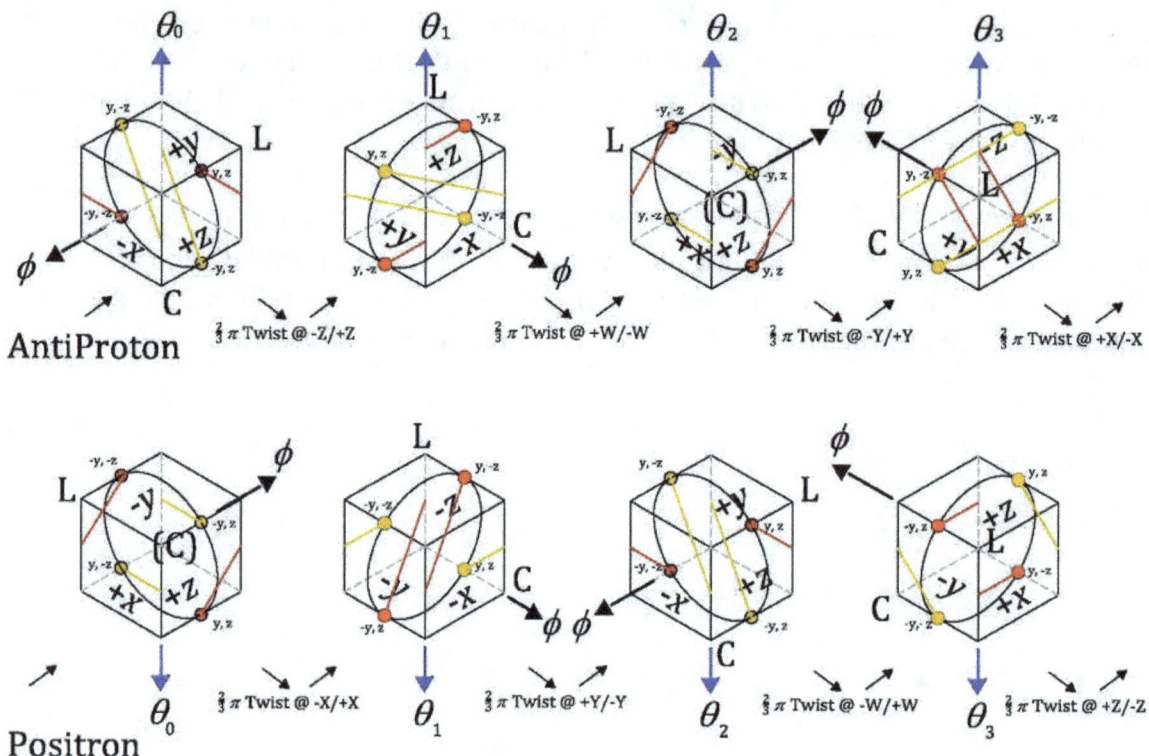

AntiProton & Positron Rotational Oscillation

Figure 35 – Capacitive Strain States, Anti Proton and Positron

The corresponding Spin Diagrams for the antiproton and positron follow in Figure 36 and 37. With respect to the charge tables for antimatter, the neutron is said to be its own antiparticle, but it is simply the same resonant particle that undergoes a capacitive strain advance with the resulting different transformation of the torques C and L and opposite charge at the point of positron emission.

With respect to the dot products of μ and ε, the vectors remain antiparallel so the sense of their mutual products at K remains negative in both particles. For the dot products with ϕ and θ, the senses are all opposite those of the proton and electron. This means for the positron that μ is antiparallel and ε is parallel to ϕ and θ at all nodal points, indicating a capacitive strain, phase state.

With respect to the cross products, the condition at W_{-x} is centrifugal for both $\mu \times$ and $\varepsilon \times$, indicating the point of transmission, but for $\mu \times \varepsilon$ is centripetal and antiparallel for a charge of -1. For the positron, at W_{-x} all are centripetal for a charge of +1.

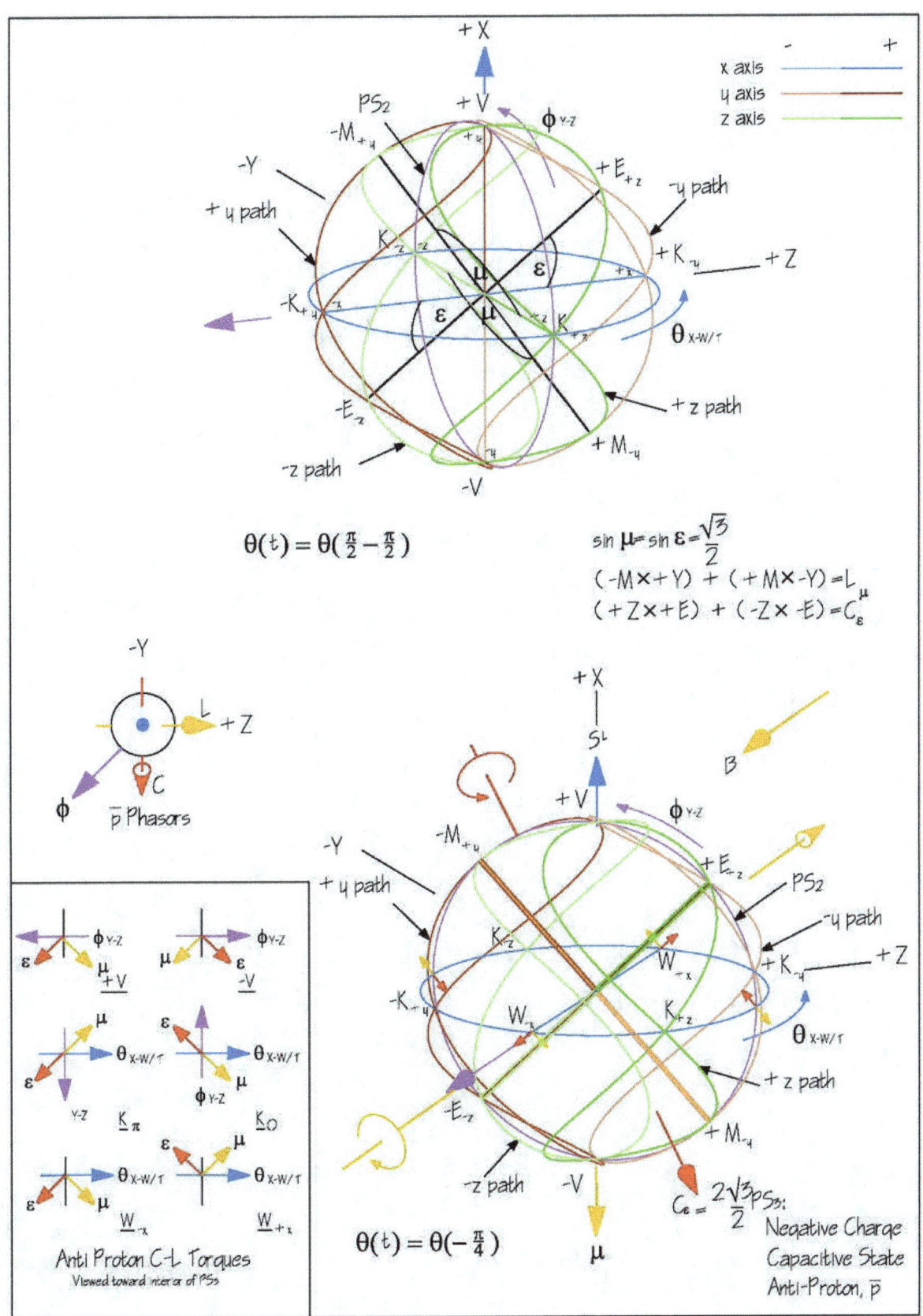

Figure 36 – Spin Diagram 5, the AntiProton

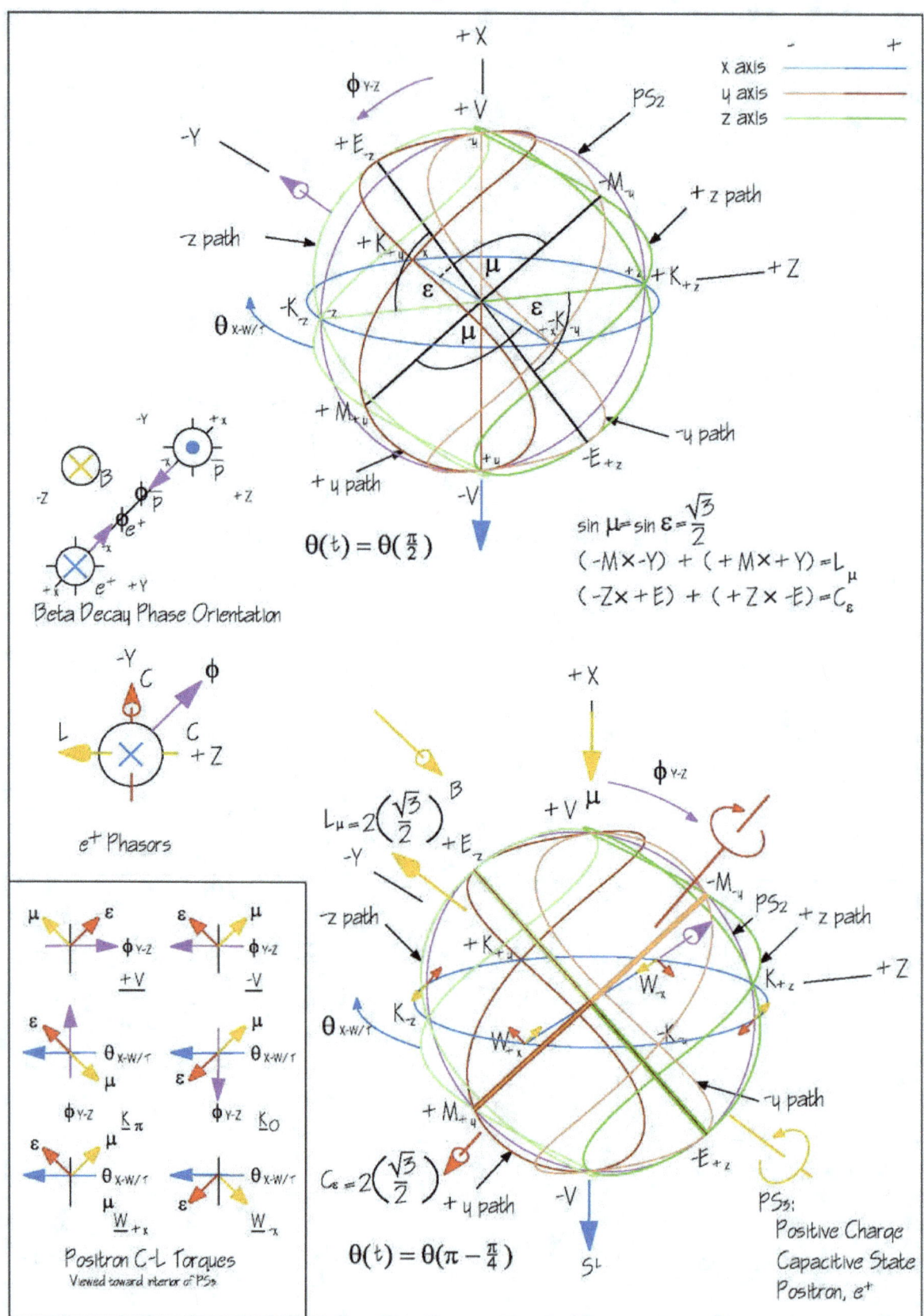

Figure 37 – Spin Diagram 6, the Positron

		1	2	3	4	5	6	7	8
	Diagram, Node/Antinode, Rotation	$\mu \cdot \varepsilon$ ε,μ $=\frac{1}{\sqrt{2}}$	$\mu \cdot$ Into θ,ϕ	$\varepsilon \cdot$ Into θ,ϕ	$S = \mu \times \varepsilon$ $\mu,\varepsilon = \frac{1}{\sqrt{2}}$	$\mu\times$ Into θ,ϕ	$\varepsilon\times$ Into θ,ϕ	$q = \frac{S}{|S|}(\mu\times + \varepsilon\times)$	Total Charge q_{W-}, q_{W+} $q_{V\pm}, q_{V\mp}$
Diagram 2 - Neutron	$W_{+x}\text{-}\theta$	0	$+\frac{1}{2}$	$+\frac{1}{2}$	$\bullet\frac{1}{2}$	$\bullet\frac{1}{2}$	$\circ\frac{1}{2}$	$\bullet 1\left(\bullet\frac{1}{2}+\circ\frac{1}{2}\right)=0$	
	$W_{-x}\text{-}\theta$	0	$+\frac{1}{2}$	$+\frac{1}{2}$	$\circ\frac{1}{2}$	$\circ\frac{1}{2}$	$\bullet\frac{1}{2}$	$\circ 1\left(\circ\frac{1}{2}+\bullet\frac{1}{2}\right)=0$	0
	$K_o\text{-}\theta$	$+\frac{1}{2}$	$+\frac{1}{2}$	$+\frac{1}{2}$	0	$\circ\frac{1}{2}$	$\circ\frac{1}{2}$	0	
	$K_\pi\text{-}\theta$	$+\frac{1}{2}$	$+\frac{1}{2}$	$+\frac{1}{2}$	0	$\bullet\frac{1}{2}$	$\bullet\frac{1}{2}$	0	
	$K_o\text{-}\phi$	$+\frac{1}{2}$	$+\frac{1}{2}$	$+\frac{1}{2}$	0	$\bullet\frac{1}{2}$	$\bullet\frac{1}{2}$	0	
	$K_\pi\text{-}\phi$	$+\frac{1}{2}$	$+\frac{1}{2}$	$+\frac{1}{2}$	0	$\circ\frac{1}{2}$	$\circ\frac{1}{2}$	0	
	$+V\text{-}\phi$	0	$+\frac{1}{2}$	$+\frac{1}{2}$	$\bullet\frac{1}{2}$	$\bullet\frac{1}{2}$	$\circ\frac{1}{2}$	$\bullet 1\left(\bullet\frac{1}{2}+\circ\frac{1}{2}\right)=0$	
	$-V\text{-}\phi$	0	$+\frac{1}{2}$	$+\frac{1}{2}$	$\circ\frac{1}{2}$	$\circ\frac{1}{2}$	$\bullet\frac{1}{2}$	$\circ 1\left(\circ\frac{1}{2}+\bullet\frac{1}{2}\right)=0$	[0]
Diagram 3 - Proton	$W_{+x}\text{-}\theta$	0	$-\frac{1}{2}$	$+\frac{1}{2}$	$\bullet\frac{1}{2}$	$\bullet\frac{1}{2}$	$\bullet\frac{1}{2}$	$\bullet 1\left(\bullet\frac{1}{2}+\bullet\frac{1}{2}\right)=+1$	
	$W_{-x}\text{-}\theta$	0	$-\frac{1}{2}$	$+\frac{1}{2}$	$\circ\frac{1}{2}$	$\circ\frac{1}{2}$	$\circ\frac{1}{2}$	$\circ 1\left(\circ\frac{1}{2}+\circ\frac{1}{2}\right)=+1$	+1
	$K_o\text{-}\theta$	$-\frac{1}{2}$	$-\frac{1}{2}$	$+\frac{1}{2}$	0	$\circ\frac{1}{2}$	$\bullet\frac{1}{2}$	0	
	$K_\pi\text{-}\theta$	$-\frac{1}{2}$	$-\frac{1}{2}$	$+\frac{1}{2}$	0	$\bullet\frac{1}{2}$	$\circ\frac{1}{2}$	0	
	$K_o\text{-}\phi$	$-\frac{1}{2}$	$+\frac{1}{2}$	$-\frac{1}{2}$	0	$\circ\frac{1}{2}$	$\bullet\frac{1}{2}$	0	
	$K_\pi\text{-}\phi$	$-\frac{1}{2}$	$+\frac{1}{2}$	$-\frac{1}{2}$	0	$\bullet\frac{1}{2}$	$\circ\frac{1}{2}$	0	
	$+V\text{-}\phi$	0	$+\frac{1}{2}$	$-\frac{1}{2}$	$\circ\frac{1}{2}$	$\bullet\frac{1}{2}$	$\bullet\frac{1}{2}$	$\circ 1\left(\bullet\frac{1}{2}+\bullet\frac{1}{2}\right)=-1$	
	$-V\text{-}\phi$	0	$+\frac{1}{2}$	$-\frac{1}{2}$	$\bullet\frac{1}{2}$	$\circ\frac{1}{2}$	$\circ\frac{1}{2}$	$\bullet 1\left(\circ\frac{1}{2}+\circ\frac{1}{2}\right)=-1$	$[-1]=-i1$
Diagram 4 - Electron	$W_{+x}\text{-}\theta$	0	$+\frac{1}{2}$	$-\frac{1}{2}$	$\bullet\frac{1}{2}$	$\circ\frac{1}{2}$	$\circ\frac{1}{2}$	$\bullet 1\left(\circ\frac{1}{2}+\circ\frac{1}{2}\right)=-1$	
	$W_{-x}\text{-}\theta$	0	$+\frac{1}{2}$	$-\frac{1}{2}$	$\circ\frac{1}{2}$	$\bullet\frac{1}{2}$	$\bullet\frac{1}{2}$	$\circ 1\left(\bullet\frac{1}{2}+\bullet\frac{1}{2}\right)=-1$	-1
	$K_o\text{-}\theta$	$-\frac{1}{2}$	$+\frac{1}{2}$	$-\frac{1}{2}$	0	$\circ\frac{1}{2}$	$\bullet\frac{1}{2}$	0	
	$K_\pi\text{-}\theta$	$-\frac{1}{2}$	$+\frac{1}{2}$	$-\frac{1}{2}$	0	$\bullet\frac{1}{2}$	$\circ\frac{1}{2}$	0	
	$K_o\text{-}\phi$	$-\frac{1}{2}$	$+\frac{1}{2}$	$-\frac{1}{2}$	0	$\bullet\frac{1}{2}$	$\circ\frac{1}{2}$	0	
	$K_\pi\text{-}\phi$	$-\frac{1}{2}$	$+\frac{1}{2}$	$-\frac{1}{2}$	0	$\circ\frac{1}{2}$	$\bullet\frac{1}{2}$	0	
	$+V\text{-}\phi$	0	$+\frac{1}{2}$	$-\frac{1}{2}$	$\circ\frac{1}{2}$	$\bullet\frac{1}{2}$	$\bullet\frac{1}{2}$	$\circ 1\left(\bullet\frac{1}{2}+\bullet\frac{1}{2}\right)=-1$	$[-1]=-i1$
	$-V\text{-}\phi$	0	$+\frac{1}{2}$	$-\frac{1}{2}$	$\bullet\frac{1}{2}$	$\circ\frac{1}{2}$	$\circ\frac{1}{2}$	$\bullet 1\left(\circ\frac{1}{2}+\circ\frac{1}{2}\right)=-1$	

Figure 38 – Charge and Spin Table for Ordinary Matter

		1	2	3	4	5	6	7	8
	Diagram, Node/Antinode, Rotation	$\mu \cdot \varepsilon$ ε,μ $= \frac{1}{\sqrt{2}}$	$\mu \cdot$ Into θ, ϕ	$\varepsilon \cdot$ Into θ, ϕ	$S = \mu \times \varepsilon$ $\mu, \varepsilon = \frac{1}{\sqrt{2}}$	$\mu \times$ Into θ, ϕ	$\varepsilon \times$ Into θ, ϕ	$q = \frac{S}{|S|}(\mu \times + \varepsilon \times)$	Total Charge q_{W-}, q_{W+} $q_{V\pm}, q_{V\mp}$
Diagram 2 – Neutron	$W_{+x}\text{-}\theta$	0	$+\frac{1}{2}$	$+\frac{1}{2}$	$\bullet\frac{1}{2}$	$\bullet\frac{1}{2}$	$\circ\frac{1}{2}$	$\bullet 1\left(\bullet\frac{1}{2}+\circ\frac{1}{2}\right)=0$	
	$W_{-x}\text{-}\theta$	0	$+\frac{1}{2}$	$+\frac{1}{2}$	$\circ\frac{1}{2}$	$\circ\frac{1}{2}$	$\bullet\frac{1}{2}$	$\circ 1\left(\circ\frac{1}{2}+\bullet\frac{1}{2}\right)=0$	0
	$K_o\text{-}\theta$	$+\frac{1}{2}$	$+\frac{1}{2}$	$+\frac{1}{2}$	0	$\circ\frac{1}{2}$	$\circ\frac{1}{2}$	0	
	$K_\pi\text{-}\theta$	$+\frac{1}{2}$	$+\frac{1}{2}$	$+\frac{1}{2}$	0	$\bullet\frac{1}{2}$	$\bullet\frac{1}{2}$	0	
	$K_o\text{-}\phi$	$+\frac{1}{2}$	$+\frac{1}{2}$	$+\frac{1}{2}$	0	$\bullet\frac{1}{2}$	$\bullet\frac{1}{2}$	0	
	$K_\pi\text{-}\phi$	$+\frac{1}{2}$	$+\frac{1}{2}$	$+\frac{1}{2}$	0	$\circ\frac{1}{2}$	$\circ\frac{1}{2}$	0	
	$+V\text{-}\phi$	0	$+\frac{1}{2}$	$+\frac{1}{2}$	$\bullet\frac{1}{2}$	$\bullet\frac{1}{2}$	$\circ\frac{1}{2}$	$\bullet 1\left(\bullet\frac{1}{2}+\circ\frac{1}{2}\right)=0$	
	$-V\text{-}\phi$	0	$+\frac{1}{2}$	$+\frac{1}{2}$	$\circ\frac{1}{2}$	$\circ\frac{1}{2}$	$\bullet\frac{1}{2}$	$\circ 1\left(\circ\frac{1}{2}+\bullet\frac{1}{2}\right)=0$	[0]
Diagram 5 – Anti Proton	$W_{+x}\text{-}\theta$	0	$+\frac{1}{2}$	$-\frac{1}{2}$	$\circ\frac{1}{2}$	$\bullet\frac{1}{2}$	$\bullet\frac{1}{2}$	$\circ 1\left(\bullet\frac{1}{2}+\bullet\frac{1}{2}\right)=-1$	
	$W_{-x}\text{-}\theta$	0	$+\frac{1}{2}$	$-\frac{1}{2}$	$\bullet\frac{1}{2}$	$\circ\frac{1}{2}$	$\circ\frac{1}{2}$	$\circ 1\left(\circ\frac{1}{2}+\circ\frac{1}{2}\right)=-1$	-1
	$K_o\text{-}\theta$	$-\frac{1}{2}$	$+\frac{1}{2}$	$-\frac{1}{2}$	0	$\circ\frac{1}{2}$	$\bullet\frac{1}{2}$	0	
	$K_\pi\text{-}\theta$	$-\frac{1}{2}$	$+\frac{1}{2}$	$-\frac{1}{2}$	0	$\bullet\frac{1}{2}$	$\circ\frac{1}{2}$	0	
	$K_o\text{-}\phi$	$-\frac{1}{2}$	$-\frac{1}{2}$	$+\frac{1}{2}$	0	$\circ\frac{1}{2}$	$\bullet\frac{1}{2}$	0	
	$K_\pi\text{-}\phi$	$-\frac{1}{2}$	$-\frac{1}{2}$	$+\frac{1}{2}$	0	$\bullet\frac{1}{2}$	$\circ\frac{1}{2}$	0	
	$+V\text{-}\phi$	0	$-\frac{1}{2}$	$+\frac{1}{2}$	$\bullet\frac{1}{2}$	$\bullet\frac{1}{2}$	$\bullet\frac{1}{2}$	$\bullet 1\left(\bullet\frac{1}{2}+\bullet\frac{1}{2}\right)=+1$	
	$-V\text{-}\phi$	0	$-\frac{1}{2}$	$+\frac{1}{2}$	$\circ\frac{1}{2}$	$\circ\frac{1}{2}$	$\circ\frac{1}{2}$	$\circ 1\left(\circ\frac{1}{2}+\circ\frac{1}{2}\right)=+1$	$[+1]=+i1$
Diagram 6 – Positron	$W_{+x}\text{-}\theta$	0	$-\frac{1}{2}$	$+\frac{1}{2}$	$\circ\frac{1}{2}$	$\circ\frac{1}{2}$	$\circ\frac{1}{2}$	$\circ 1\left(\circ\frac{1}{2}+\circ\frac{1}{2}\right)=+1$	
	$W_{-x}\text{-}\theta$	0	$-\frac{1}{2}$	$+\frac{1}{2}$	$\bullet\frac{1}{2}$	$\bullet\frac{1}{2}$	$\bullet\frac{1}{2}$	$\bullet 1\left(\bullet\frac{1}{2}+\bullet\frac{1}{2}\right)=+1$	+1
	$K_o\text{-}\theta$	$-\frac{1}{2}$	$-\frac{1}{2}$	$+\frac{1}{2}$	0	$\circ\frac{1}{2}$	$\bullet\frac{1}{2}$	0	
	$K_\pi\text{-}\theta$	$-\frac{1}{2}$	$-\frac{1}{2}$	$+\frac{1}{2}$	0	$\bullet\frac{1}{2}$	$\circ\frac{1}{2}$	0	
	$K_o\text{-}\phi$	$-\frac{1}{2}$	$-\frac{1}{2}$	$+\frac{1}{2}$	0	$\bullet\frac{1}{2}$	$\circ\frac{1}{2}$	0	
	$K_\pi\text{-}\phi$	$-\frac{1}{2}$	$-\frac{1}{2}$	$+\frac{1}{2}$	0	$\circ\frac{1}{2}$	$\bullet\frac{1}{2}$	0	
	$+V\text{-}\phi$	0	$-\frac{1}{2}$	$+\frac{1}{2}$	$\bullet\frac{1}{2}$	$\bullet\frac{1}{2}$	$\bullet\frac{1}{2}$	$\bullet 1\left(\bullet\frac{1}{2}+\bullet\frac{1}{2}\right)=+1$	
	$-V\text{-}\phi$	0	$-\frac{1}{2}$	$+\frac{1}{2}$	$\circ\frac{1}{2}$	$\circ\frac{1}{2}$	$\circ\frac{1}{2}$	$\circ 1\left(\circ\frac{1}{2}+\circ\frac{1}{2}\right)=+1$	$[+1]=+i1$

Figure 39 – Charge and Spin Table for Anti Matter

		1	2	3	4	5	6	7	8
	Diagram, Node/Antinode, Rotation	$\mu \cdot \varepsilon$ $\mu, \varepsilon = \sqrt{\frac{2}{3}}$	$\mu \cdot$ Into θ, ϕ	$\varepsilon \cdot$ Into θ, ϕ	$S = \mu \times \varepsilon$ $\mu, \varepsilon = \sqrt{\frac{2}{3}}$	$\mu \times$ Into θ, ϕ	$\varepsilon \times$ Into θ, ϕ	$q = \frac{S}{|S|}[\mathrm{T}(\mu \times + \varepsilon \times)]$ $\mathrm{T} = \frac{\sqrt{3}}{2}$	Total Charge q_{W-}, q_{W+} $q_{V\pm}, q_{V\mp}$
Diagram 2 - Neutron	$W_{+x}\text{-}\theta$	0	$+\frac{1}{\sqrt{3}}$	$+\frac{1}{\sqrt{3}}$	$\bullet\frac{2}{3}$	$\bullet\frac{1}{\sqrt{3}}$	$\circ\frac{1}{\sqrt{3}}$	$\bullet 1\left(\bullet\frac{1}{2}+\circ\frac{1}{2}\right)=0$	
	$W_{-x}\text{-}\theta$	0	$+\frac{1}{\sqrt{3}}$	$+\frac{1}{\sqrt{3}}$	$\frac{2}{3}$	$\circ\frac{1}{\sqrt{3}}$	$\bullet\frac{1}{\sqrt{3}}$	$\circ 1\left(\circ\frac{1}{2}+\bullet\frac{1}{2}\right)=0$	0
	$K_o\text{-}\theta$	$+\frac{2}{3}$	$+\frac{1}{\sqrt{3}}$	$+\frac{1}{\sqrt{3}}$	0	$\frac{1}{\sqrt{3}}$	$\frac{1}{\sqrt{3}}$	0	
	$K_\pi\text{-}\theta$	$+\frac{2}{3}$	$+\frac{1}{\sqrt{3}}$	$+\frac{1}{\sqrt{3}}$	0	$\frac{1}{\sqrt{3}}$	$\frac{1}{\sqrt{3}}$	0	
	$K_o\text{-}\phi$	$+\frac{2}{3}$	$+\frac{1}{\sqrt{3}}$	$+\frac{1}{\sqrt{3}}$	0	$\frac{1}{\sqrt{3}}$	$\frac{1}{\sqrt{3}}$	0	
	$K_\pi\text{-}\phi$	$+\frac{2}{3}$	$+\frac{1}{\sqrt{3}}$	$+\frac{1}{\sqrt{3}}$	0	$\circ\frac{1}{\sqrt{3}}$	$\bullet\frac{1}{\sqrt{3}}$	0	
	$+V\text{-}\phi$	0	$+\frac{1}{\sqrt{3}}$	$+\frac{1}{\sqrt{3}}$	$\bullet\frac{2}{3}$	$\bullet\frac{1}{\sqrt{3}}$	$\circ\frac{1}{\sqrt{3}}$	$\bullet 1\left(\bullet\frac{1}{2}+\circ\frac{1}{2}\right)=0$	
	$-V\text{-}\phi$	0	$+\frac{1}{\sqrt{3}}$	$+\frac{1}{\sqrt{3}}$	$\frac{2}{3}$	$\frac{1}{\sqrt{3}}$	$\frac{1}{\sqrt{3}}$	$\circ 1\left(\circ\frac{1}{2}+\bullet\frac{1}{2}\right)=0$	[0]
Diagram 3 - Proton	$W_{+x}\text{-}\theta$	0	$-\frac{1}{\sqrt{3}}$	$+\frac{1}{\sqrt{3}}$	$\bullet\frac{2}{3}$	$\bullet\frac{1}{\sqrt{3}}$	$\frac{1}{\sqrt{3}}$	$\bullet 1\left(\bullet\frac{1}{2}+\bullet\frac{1}{2}\right)=+1$	
	$W_{-x}\text{-}\theta$	0	$-\frac{1}{\sqrt{3}}$	$+\frac{1}{\sqrt{3}}$	$\frac{2}{3}$	$\circ\frac{1}{\sqrt{3}}$	$\frac{1}{\sqrt{3}}$	$\circ 1\left(\circ\frac{1}{2}+\circ\frac{1}{2}\right)=+1$	+1
	$K_o\text{-}\theta$	$-\frac{2}{3}$	$-\frac{1}{\sqrt{3}}$	$+\frac{1}{\sqrt{3}}$	0	$\circ\frac{1}{\sqrt{3}}$	$\bullet\frac{1}{\sqrt{3}}$	0	
	$K_\pi\text{-}\theta$	$-\frac{2}{3}$	$-\frac{1}{\sqrt{3}}$	$+\frac{1}{\sqrt{3}}$	0	$\circ\frac{1}{\sqrt{3}}$	$\bullet\frac{1}{\sqrt{3}}$	0	
	$K_o\text{-}\phi$	$-\frac{2}{3}$	$+\frac{1}{\sqrt{3}}$	$-\frac{1}{\sqrt{3}}$	0	$\circ\frac{1}{\sqrt{3}}$	$\bullet\frac{1}{\sqrt{3}}$	0	
	$K_\pi\text{-}\phi$	$-\frac{2}{3}$	$+\frac{1}{\sqrt{3}}$	$-\frac{1}{\sqrt{3}}$	0	$\circ\frac{1}{\sqrt{3}}$	$\frac{1}{\sqrt{3}}$	0	
	$+V\text{-}\phi$	0	$+\frac{1}{\sqrt{3}}$	$-\frac{1}{\sqrt{3}}$	$\bullet\frac{2}{3}$	$\circ\frac{1}{\sqrt{3}}$	$\bullet\frac{1}{\sqrt{3}}$	$\circ 1\left(\circ\frac{1}{2}+\bullet\frac{1}{2}\right)=-1$	
	$-V\text{-}\phi$	0	$+\frac{1}{\sqrt{3}}$	$-\frac{1}{\sqrt{3}}$	$\bullet\frac{2}{3}$	$\circ\frac{1}{\sqrt{3}}$	$\frac{1}{\sqrt{3}}$	$\bullet 1\left(\bullet\frac{1}{2}+\circ\frac{1}{2}\right)=-1$	$[-1]=-i1$
Diagram 4 - Electron	$W_{+x}\text{-}\theta$	0	$+\frac{1}{\sqrt{3}}$	$-\frac{1}{\sqrt{3}}$	$\bullet\frac{2}{3}$	$\circ\frac{1}{\sqrt{3}}$	$\bullet\frac{1}{\sqrt{3}}$	$\bullet 1\left(\bullet\frac{1}{2}+\circ\frac{1}{2}\right)=-1$	
	$W_{-x}\text{-}\theta$	0	$+\frac{1}{\sqrt{3}}$	$-\frac{1}{\sqrt{3}}$	$\frac{2}{3}$	$\bullet\frac{1}{\sqrt{3}}$	$\bullet\frac{1}{\sqrt{3}}$	$\circ 1\left(\circ\frac{1}{2}+\bullet\frac{1}{2}\right)=-1$	-1
	$K_o\text{-}\theta$	$-\frac{2}{3}$	$+\frac{1}{\sqrt{3}}$	$-\frac{1}{\sqrt{3}}$	0	$\bullet\frac{1}{\sqrt{3}}$	$\bullet\frac{1}{\sqrt{3}}$	0	
	$K_\pi\text{-}\theta$	$-\frac{2}{3}$	$+\frac{1}{\sqrt{3}}$	$-\frac{1}{\sqrt{3}}$	0	$\bullet\frac{1}{\sqrt{3}}$	$\bullet\frac{1}{\sqrt{3}}$	0	
	$K_o\text{-}\phi$	$-\frac{2}{3}$	$+\frac{1}{\sqrt{3}}$	$-\frac{1}{\sqrt{3}}$	0	$\bullet\frac{1}{\sqrt{3}}$	$\bullet\frac{1}{\sqrt{3}}$	0	
	$K_\pi\text{-}\phi$	$-\frac{2}{3}$	$+\frac{1}{\sqrt{3}}$	$-\frac{1}{\sqrt{3}}$	0	$\frac{1}{\sqrt{3}}$	$\bullet\frac{1}{\sqrt{3}}$	0	
	$+V\text{-}\phi$	0	$+\frac{1}{\sqrt{3}}$	$-\frac{1}{\sqrt{3}}$	$\frac{2}{3}$	$\bullet\frac{1}{\sqrt{3}}$	$\circ\frac{1}{\sqrt{3}}$	$\circ 1\left(\bullet\frac{1}{2}+\circ\frac{1}{2}\right)=-1$	$[-1]=-i1$
	$-V\text{-}\phi$	0	$+\frac{1}{\sqrt{3}}$	$-\frac{1}{\sqrt{3}}$	$\bullet\frac{2}{3}$	$\frac{1}{\sqrt{3}}$	$\frac{1}{\sqrt{3}}$	$\bullet 1\left(\circ\frac{1}{2}+\bullet\frac{1}{2}\right)=-1$	

Figure 40 – Charge and Spin Table for Ordinary Matter for C & L = 1

		1	2	3	4	5	6	7	8
Diagram, Node/Antinode, Rotation		$\mu \cdot \varepsilon$ $\mu, \varepsilon = \sqrt{\frac{2}{3}}$	$\mu \cdot$ Into θ, ϕ	$\varepsilon \cdot$ Into θ, ϕ	$S = \mu \times \varepsilon$ $\mu, \varepsilon = \sqrt{\frac{2}{3}}$	$\mu \times$ Into θ, ϕ	$\varepsilon \times$ Into θ, ϕ	$q = \frac{S}{|S|}(\mu \times + \varepsilon \times)$ $T = \frac{\sqrt{3}}{2}$	Total Charge q_{W-}, q_{W+} $q_{V\pm}, q_{V\mp}$
Diagram 2 – Neutron	W_{+x}-θ	0	$+\frac{1}{\sqrt{3}}$	$+\frac{1}{\sqrt{3}}$	$\bullet\frac{2}{3}$	$\bullet\frac{1}{\sqrt{3}}$	$\circ\frac{1}{\sqrt{3}}$	$\bullet 1\left(\bullet\frac{1}{2}+\circ\frac{1}{2}\right)=0$	
	W_{-x}-θ	0	$+\frac{1}{\sqrt{3}}$	$+\frac{1}{\sqrt{3}}$	$\circ\frac{2}{3}$	$\circ\frac{1}{\sqrt{3}}$	$\bullet\frac{1}{\sqrt{3}}$	$\circ 1\left(\circ\frac{1}{2}+\bullet\frac{1}{2}\right)=0$	0
	K_o-θ	$+\frac{2}{3}$	$+\frac{1}{\sqrt{3}}$	$+\frac{1}{\sqrt{3}}$	0	$\circ\frac{1}{\sqrt{3}}$	$\circ\frac{1}{\sqrt{3}}$	0	
	K_π-θ	$+\frac{2}{3}$	$+\frac{1}{\sqrt{3}}$	$+\frac{1}{\sqrt{3}}$	0	$\bullet\frac{1}{\sqrt{3}}$	$\bullet\frac{1}{\sqrt{3}}$	0	
	K_o-ϕ	$+\frac{2}{3}$	$+\frac{1}{\sqrt{3}}$	$+\frac{1}{\sqrt{3}}$	0	$\bullet\frac{1}{\sqrt{3}}$	$\bullet\frac{1}{\sqrt{3}}$	0	
	K_π-ϕ	$+\frac{2}{3}$	$+\frac{1}{\sqrt{3}}$	$+\frac{1}{\sqrt{3}}$	0	$\circ\frac{1}{\sqrt{3}}$	$\circ\frac{1}{\sqrt{3}}$	0	
	$+V$-ϕ	0	$+\frac{1}{\sqrt{3}}$	$+\frac{1}{\sqrt{3}}$	$\bullet\frac{2}{3}$	$\bullet\frac{1}{\sqrt{3}}$	$\circ\frac{1}{\sqrt{3}}$	$\bullet 1\left(\bullet\frac{1}{2}+\circ\frac{1}{2}\right)=0$	
	$-V$-ϕ	0	$+\frac{1}{\sqrt{3}}$	$+\frac{1}{\sqrt{3}}$	$\circ\frac{2}{3}$	$\circ\frac{1}{\sqrt{3}}$	$\bullet\frac{1}{\sqrt{3}}$	$\circ 1\left(\circ\frac{1}{2}+\bullet\frac{1}{2}\right)=0$	[0]
Diagram 5 – Anti Proton	W_{+x}-θ	0	$+\frac{1}{\sqrt{3}}$	$-\frac{1}{\sqrt{3}}$	$\circ\frac{2}{3}$	$\bullet\frac{1}{\sqrt{3}}$	$\bullet\frac{1}{\sqrt{3}}$	$\circ 1\left(\bullet\frac{1}{2}+\bullet\frac{1}{2}\right)=-1$	
	W_{-x}-θ	0	$+\frac{1}{\sqrt{3}}$	$-\frac{1}{\sqrt{3}}$	$\bullet\frac{2}{3}$	$\circ\frac{1}{\sqrt{3}}$	$\circ\frac{1}{\sqrt{3}}$	$\circ 1\left(\circ\frac{1}{2}+\circ\frac{1}{2}\right)=-1$	-1
	K_o-θ	$-\frac{2}{3}$	$+\frac{1}{\sqrt{3}}$	$-\frac{1}{\sqrt{3}}$	0	$\frac{1}{\sqrt{3}}$	$\bullet\frac{1}{\sqrt{3}}$	0	
	K_π-θ	$-\frac{2}{3}$	$+\frac{1}{\sqrt{3}}$	$-\frac{1}{\sqrt{3}}$	0	$\circ\frac{1}{\sqrt{3}}$	$\circ\frac{1}{\sqrt{3}}$	0	
	K_o-ϕ	$-\frac{2}{3}$	$-\frac{1}{\sqrt{3}}$	$+\frac{1}{\sqrt{3}}$	0	$\frac{1}{\sqrt{3}}$	$\bullet\frac{1}{\sqrt{3}}$	0	
	K_π-ϕ	$-\frac{2}{3}$	$-\frac{1}{\sqrt{3}}$	$+\frac{1}{\sqrt{3}}$	0	$\frac{1}{\sqrt{3}}$	$\frac{1}{\sqrt{3}}$	0	
	$+V$-ϕ	0	$-\frac{1}{\sqrt{3}}$	$+\frac{1}{\sqrt{3}}$	$\bullet\frac{2}{3}$	$\bullet\frac{1}{\sqrt{3}}$	$\bullet\frac{1}{\sqrt{3}}$	$\bullet 1\left(\frac{1}{2}+\frac{1}{2}\right)=+1$	
	$-V$-ϕ	0	$-\frac{1}{\sqrt{3}}$	$+\frac{1}{\sqrt{3}}$	$\circ\frac{2}{3}$	$\circ\frac{1}{\sqrt{3}}$	$\circ\frac{1}{\sqrt{3}}$	$\circ 1\left(\circ\frac{1}{2}+\circ\frac{1}{2}\right)=+1$	$[+1]=+i1$
Diagram 6 – Positron	W_{+x}-θ	0	$-\frac{1}{\sqrt{3}}$	$+\frac{1}{\sqrt{3}}$	$\frac{2}{3}$	$\frac{1}{\sqrt{3}}$	$\frac{1}{\sqrt{3}}$	$\circ 1\left(\frac{1}{2}+\frac{1}{2}\right)=+1$	
	W_{-x}-θ	0	$-\frac{1}{\sqrt{3}}$	$+\frac{1}{\sqrt{3}}$	$\bullet\frac{2}{3}$	$\bullet\frac{1}{\sqrt{3}}$	$\bullet\frac{1}{\sqrt{3}}$	$\bullet 1\left(\frac{1}{2}+\frac{1}{2}\right)=+1$	+1
	K_o-θ	$-\frac{2}{3}$	$-\frac{1}{\sqrt{3}}$	$+\frac{1}{\sqrt{3}}$	0	$\frac{1}{\sqrt{3}}$	$\bullet\frac{1}{\sqrt{3}}$	0	
	K_π-θ	$-\frac{2}{3}$	$-\frac{1}{\sqrt{3}}$	$+\frac{1}{\sqrt{3}}$	0	$\bullet\frac{1}{\sqrt{3}}$	$\circ\frac{1}{\sqrt{3}}$	0	
	K_o-ϕ	$-\frac{2}{3}$	$-\frac{1}{\sqrt{3}}$	$+\frac{1}{\sqrt{3}}$	0	$\frac{1}{\sqrt{3}}$	$\bullet\frac{1}{\sqrt{3}}$	0	
	K_π-ϕ	$-\frac{2}{3}$	$-\frac{1}{\sqrt{3}}$	$+\frac{1}{\sqrt{3}}$	0	$\frac{1}{\sqrt{3}}$	$\bullet\frac{1}{\sqrt{3}}$	0	
	$+V$-ϕ	0	$-\frac{1}{\sqrt{3}}$	$+\frac{1}{\sqrt{3}}$	$\bullet\frac{2}{3}$	$\bullet\frac{1}{\sqrt{3}}$	$\bullet\frac{1}{\sqrt{3}}$	$\bullet 1\left(\bullet\frac{1}{2}+\bullet\frac{1}{2}\right)=+1$	$[+1]=+i1$
	$-V$-ϕ	0	$-\frac{1}{\sqrt{3}}$	$+\frac{1}{\sqrt{3}}$	$\circ\frac{2}{3}$	$\circ\frac{1}{\sqrt{3}}$	$\circ\frac{1}{\sqrt{3}}$	$\circ 1\left(\circ\frac{1}{2}+\circ\frac{1}{2}\right)=+1$	

Figure 41 – Charge and Spin Table for Anti Matter for C & L =1

Figures 40 and 41 give the same information as Figures 38 and 39 for the dot and cross products of μ and ε, however using a value of 1 for the torques C and L instead of $\frac{\sqrt{3}}{2}$. This is done to give an indication of the tie-in of this analysis with the current quark phenomenology of the standard model. I will leave it to others to work out the details. The correct value of C and L is worked back into the analysis in Column 7 of these tables to arrive at the correct charge values as indicated at the Column heading.

Figure 42 gives a comparison of the strain cycle states for each of the states delineated above. It shows the strain sequence over $\theta_{(0,1,2,3)}$ for the position (y, z) as taken from the Figures 28, 31 & 35. For the neutron, proton, and antiproton, the same sequence occurs for the three remaining positions, $(y,-z)$, $(-y,-z)$, and $(-y, z)$ with appropriate transposition of the sequence numbers, rotating clockwise around the axial vector $\boldsymbol{\theta}$ as viewed from the top of the cube. The strain is therefore rotationally symmetric for each of the four strains. In the case of the electron and positron states, the (y, z) and the $(-y,-z$, not shown) strains are half-rotationally symmetric about $\boldsymbol{\theta}$, as are the $(y,-z)$ and the $(-y, z$, not shown) strains. This asymmetry is an indication of two properties of these states; a significant elongation from a generally spherical to a pronounced prolate form of the strain along the x or

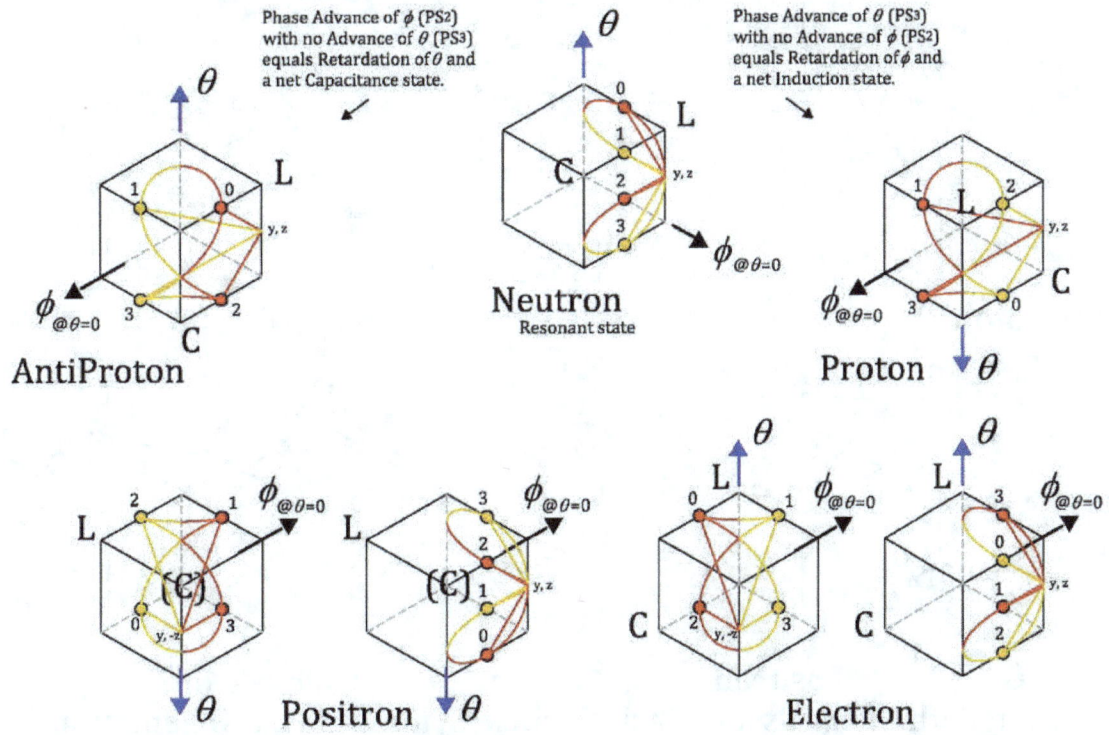

Figure 42 – Comparison of Strain States

generally (*y, z*)-(-*y,*-*z*) axis, and a propensity for ***θ*** to flip about the generally (*y, -z*)-(-*y, z*) and X axis plane. Such a flip transmits a traveling torsion wave that is recognized in the standard phenomenology as a photon or boson, a force carrier.

In Figure 43, the same particles are shown so that all the ***θ*** vectors are parallel. This gives a better view of the symmetry between ordinary and anti matter, the first of which is advanced in the rotation of ***θ*** and the second of which is retarded. Notice that the same advanced/retarded symmetry holds for the emitted states, the leptons of the standard phenomenology, electron and positron. For the positron, two ***θ*** flips are shown for the position (*y, -z*). One is an observational flip, which means a snapshot view of its position shown in Figure 42 taken from the backside and upside down and shows the same strain state as in Figure 42. The physical spin flip shown represents an actual flip of the spin due to a recoil of the strain as shown. Note the difference in the initial position of (*y, -z*) in the two depictions. A similar condition applies to the electron as well. These physical flips indicate a reduction in stress/strain of the state to that of the basic or ground energy state with the emission of a photon.

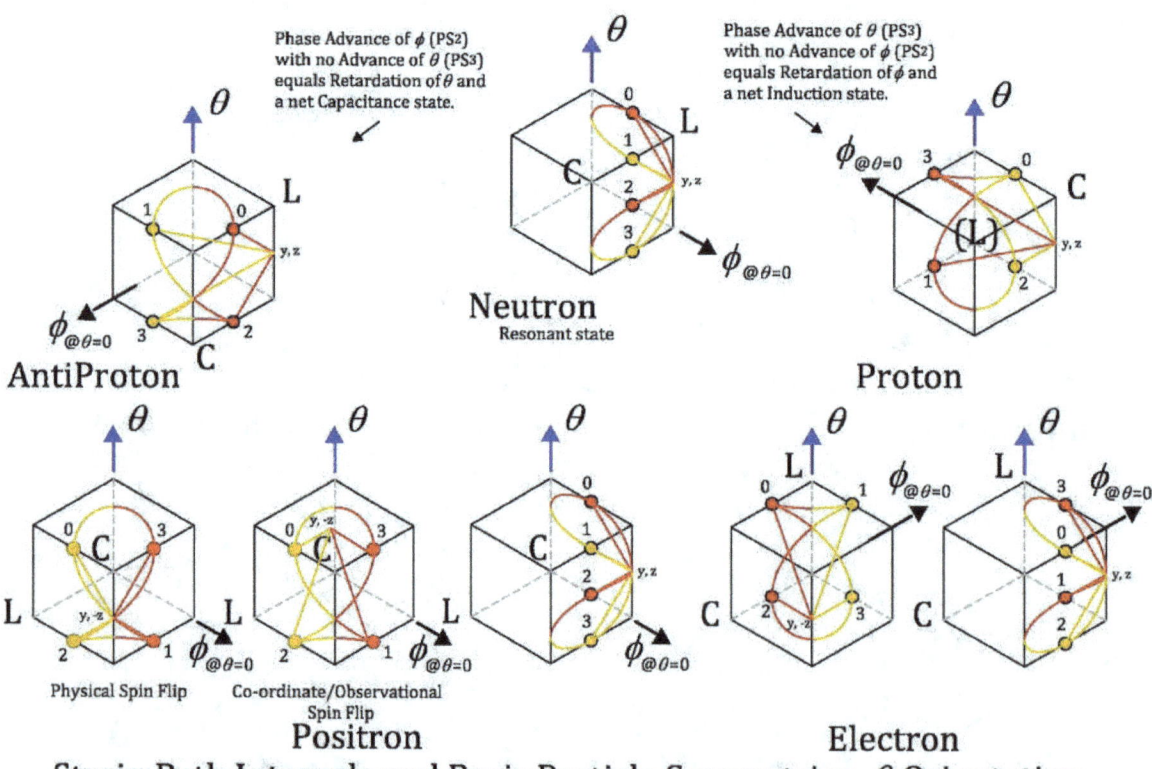

Figure 43 – Comparison of Strain States with Parallel Spin

The following photos are of toy models representing the strain phase and spin states of the neutron, proton and antiproton based on the above analysis and diagrams. The smaller red and gold (or yellow) nubs represent the E and M moments and their relationship to the long brown ϕ axial vector along the rotating *x* axis. The longer red and gold vectors represent the C and L torques all of which circulate about the blue spin angular momentum vectors. The orange vectors parallel or antiparallel to the spin vectors represent the effective magnetic moments of the quantum states.

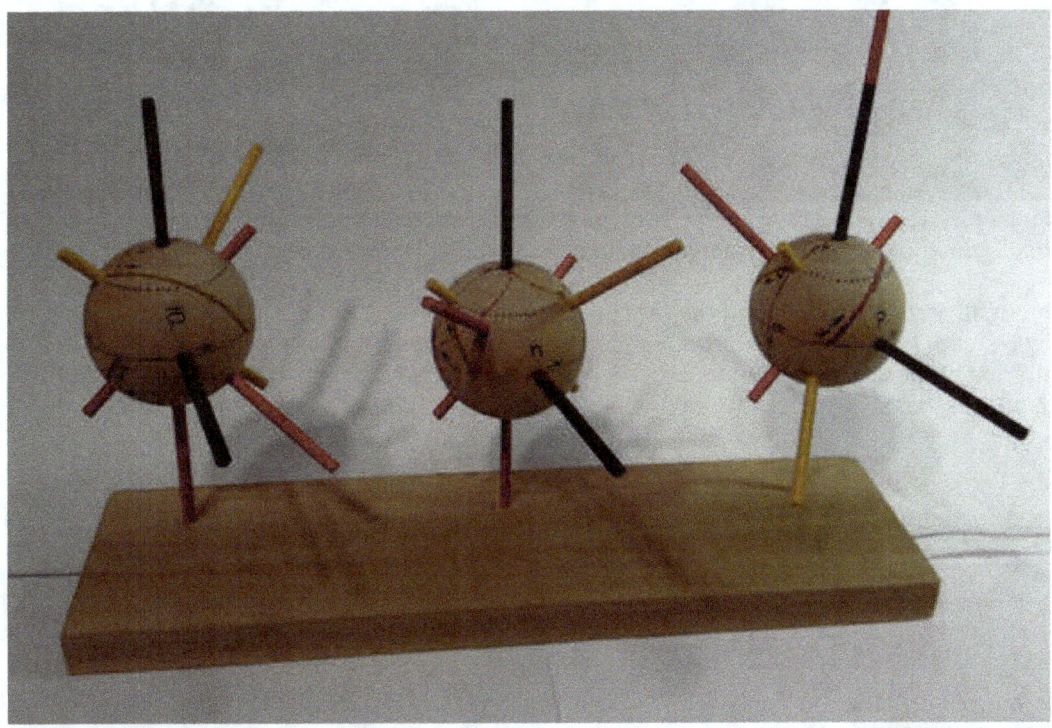

PS$_3$ Anti-Proton   PS$_3$ Neutron   PS$_3$ Proton

Toy Models of 3-D Phase Space for the neutron, proton and anti-proton. The 2 capacitive moments (small red vectors) and 2 inductive moments (small yellow vectors) are in the plane of the 2-D phase space whose rotation is indicated by the brown axial vector representing the principal restorative force of SHM. These moments in turn generate the capacitive (long red axial vector) and inductive (long yellow axial vector) torques that prevent energy dispersion and generate spin angular momentum (blue axial vectors). The magnetic dipole moments (orange vectors) are determined by the inductive torque and roughly anti-parallel to it. The proton results from an advance of PS$_2$ about the angular momentum and the resulting flip in the inductive torque. The anti-proton results from a retardation of PS$_2$ about the angular momentum and the resulting flip of the capacitive torque.

Photo 7

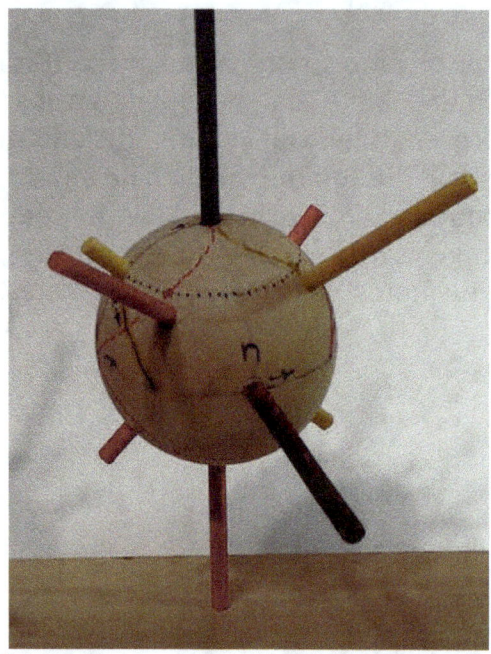
Neutron state at capacitive torque showing displacement of inductive moment behind
Photo 8

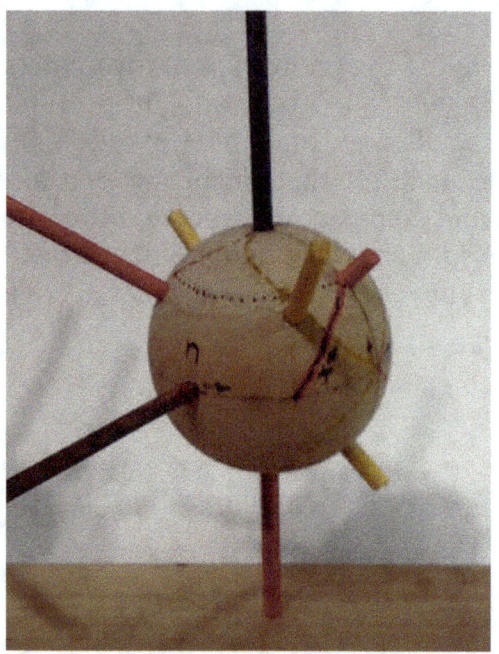
Neutron state at inductive torque showing displacement of capacitive moment behind
Photo 9

The small arrow to the right of the "n" shows direction of the PS_2 disk axis revolution in PS_3. The arrow beneath and pointed toward the red capacitive moment in the right photo shows the direction of rotation of the PS_2 disk about its center. The top portion of the figure 8 path crossing at that point shows the oscillatory path etched by that point of the disk on the PS_3 spherical shell as the disk rotates. The neutron state is a resonant state.

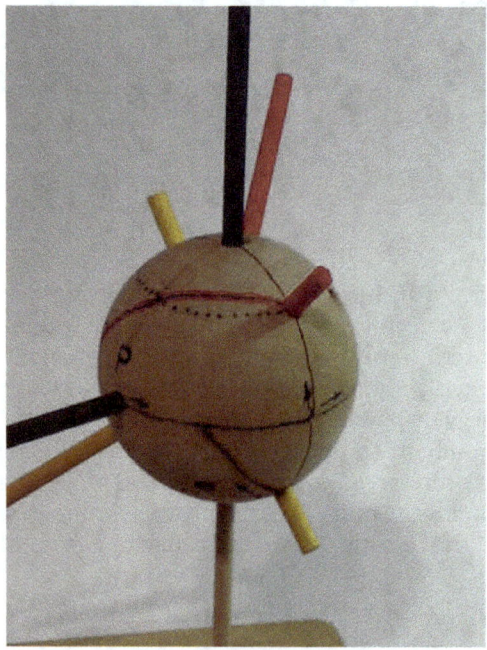
Proton state showing the advance of the capacitive and inductive moments and their displacement paths
Photo 10

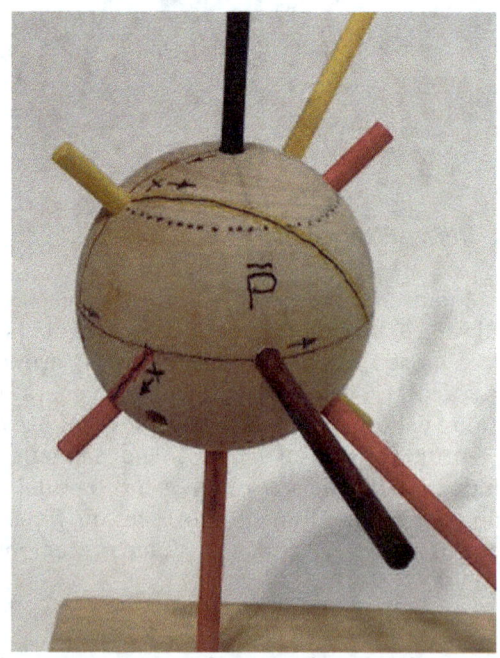
Anti-proton state showing the retardation of the capacitive and inductive moments and their displacement paths
Photo 11

Verification

Our discussion on with Simple Harmonic Motion started with its representations in classical phase space and sinusoidal functions, took an aside into some basic notions of calculus and isotropic differential change, developed an extension of phase space to three dimensions to incorporate potential and kinetic energy, took another aside into the stress-strain analysis of three-dimensional materials and mediums and the difference between body and surface forces, examined some properties of an expanding spacetime fabric (STF), including the inevitable emergence of geometric constraints due to the stress-strain relationship of inertial continuity, to finally arrive at the statement that the basic particles of matter that we observe everywhere about us necessarily arise in such inertially continuous spacetime in the process of expansion as microscopic individual or quantum instances of SHM. This may be reasonable, but if it is true it should be accurate. We should be able to predict some of the readily observed properties of such matter and in a way that current models cannot.

Specifically, as functions of an expanding spacetime we should be able to:

1) derive a gravitational quantum, predict Newton's gravitational constant and his law of gravity, in a manner consistent with general relativity, and
2) derive a model of beta decay tied to the expansion, predict the neutron and electron mass ratios, predict the expansion rate of the cosmos, account for the missing mass of beta decay, and predict the value of elementary charge.

1) Quantum Gravity

Current attempts at a unification of theoretical physics rest in part on the dimensional analysis of Newton's gravitational constant, G_N, by Max Planck and its assumed applicability at a theoretically smallest limiting scale, where we have displacement, r_0, mass, m_0, and time, t_0, and τ_0 is a fundamental unit of a presumed body force, the subscript naught again indicating a fundamental unit, i.e. equal to one in some natural scale:

$$G_N = \frac{r_0^2}{m_0^2}(\tau_0) = \frac{r_0^2}{m_0^2}\left(m_0 \frac{r_0}{t_0^2}\right) = \frac{r_0^3}{m_0 t_0^2} \tag{4.1}$$

It bears emphasizing that if we are to anticipate a quantum theory of gravity, then the function of Newton's constant in the context of Newton's gravitational law is to produce the force between two massive bodies by converting their respective masses and the distance of separation of their centers of mass to some fundamental units of each property and multiplying this by a fundamental unit or quantum of force as in the second term above.

Since c, the speed of light, and $\hbar$, Planck's quantum of action, are assumed to be invariant at any scale, their expression in terms of the units of some as yet undefined natural length scale, r_0, commensurate with τ_0 is

$$c = \frac{r_0}{t_0} \quad \therefore t_0 = \frac{r_0}{c}, \text{ and}$$

$$\hbar = \frac{m_0 r_0^2}{t_0} \quad \therefore m_0 = \frac{\hbar}{cr_0}$$

(4.2)

Substituting these conclusions for time and mass back into (4.1), and assuming the same natural units for the length scale, we have the following expression for G_N,

$$G_N = \frac{r_0^3}{m_0 t_0^2} = \frac{c^3}{\hbar} r_0^2 = \frac{c^3}{\hbar} A_0$$

(4.3)

Since G_N, c, and $\hbar$ have reasonably well determined values, we can rearrange and solve to evaluate the Planck scale, here in SI units, via the Planck area, A_P,

$$A_0 = r_0^2 = \frac{G_N \hbar}{c^3} = 2.6116...\text{x}10^{-70} \text{ meter}^2 = A_P$$

(4.4)

and its square root, the Planck length, l_P, where

$$r_0 = 1.616...\text{x}10^{-35} \text{ meter} = l_P \, .$$

(4.5)

The remainder of the Planck scale values are easily determined using (4.2).

$$t_0 = 5.3908...\text{x}10^{-44} \text{ second}$$

$$m_0 = 2.1766...\text{x}10^{-8} \text{ kilograms}$$

(4.6)

A_P is generally deemed to be a low end cutoff scale for definable physical phenomena, but this shows that if it had any other value, lesser or greater, the invariance of one or more of the three familiar constants would be challenged. In addition, in the above discussion concerning cosmic expansion and its implications for the nuclear scale, we have already encountered meaningful distance and time intervals far below even the Planck scale values.

In (4.1), τ_0 is presumed to be a body force of mass times acceleration and when evaluated using the Planck scale values, due to that extremely short time scale indicats a truly astronomical acceleration and force of

$$\tau_0 = 1.2104...\text{x}10^{44} \text{ newton} \, .$$

(4.7)

Hence the need for a big bang, of whatever unknown cause, to overcome gravity at this scale . . .

Gravitational Quantum

However, in keeping with the analysis of general relativity, τ_0 can, and in this writers opinion should, be considered a surface or stress force, the product of a fundamental cross-sectional area, A_0, and a fundamental unit of tension stress, f_0, as previously discussed and recapitulated here as

$$\tau_0 = A_0 f_0$$

(4.8)

and a differential stress force at that. Rearranging this to indicate stress as a function of force and cross-section and taking the total derivative gives

$$f_0 = \frac{\tau_0}{A_0}$$

$$df_0 = \frac{\partial f_0}{\partial \tau_0}d\tau_0 + \frac{\partial f_0}{\partial A_0}dA_0 = \frac{1}{A_0}d\tau_0 - \frac{\tau_0}{A_0^2}dA_0 \qquad (4.9)$$

The unit or fundamental mode subscript designations remain, because we are examining changes in those fundamentals over time. Separating and inverting this function we have two differential equations, the first as a differential change in force as a function of a change in tension stress or

$$d\tau_0 = A_0 df_0 \qquad (4.10)$$

and the second as a differential change in cross-section as a function of that change in stress as

$$dA_0 = -\frac{A_0^2}{\tau_0}df_0 = -\frac{\tau_0}{f_0^2}df_0 = -\frac{A_0}{f_0}df_0 = -A_0 d\ln f_0 \qquad (4.11)$$

From our comments on isotropic stress, T_0, at (6.11) in the asides we know that an isotropic stress resulting in an isotropic dilatation is related to the stress at each face by a factor of 6. In addition, if that isotropic stress is operating along the diagonals of the cube or along the cuboctahedral 4-axis octahedral component, it will be equiangular to all three faces or at a factor of $\sqrt{3}$ for a total factor of $6\sqrt{3}$, denoted γ_3, to give

$$df_0 = \frac{dT_0}{6\sqrt{3}} \qquad (4.12)$$

making the differential stress force

$$d\tau_0 = \frac{A_0}{6\sqrt{3}}dT_0 = \frac{A_0}{\gamma_3}dT_0 \; . \qquad (4.13)$$

Newton's Gravitational Constant

If we now use this differential stress force in Planck's dimensional analysis instead of the body force, we have

$$G_N = \frac{r_0^2}{m_0^2}(d\tau_0) = \frac{r_0^4}{\hbar^2/c^2}\left(\frac{r_0^2}{6\sqrt{3}}dT_0\right) = \frac{c^2}{\gamma_3 \hbar^2}r_0^6 dT_0 = \frac{r_0^6}{\gamma_3 \bar{\lambda}^2}dT_0 \qquad (4.14)$$

From Aside #1, since dT_0 equals 1, and we know the values of the other invariants, we can rearrange again and solve for the fundamental length scale, applicable to the current expansion extent of the cosmos and get

$$r_0 = \left(\gamma_3 \frac{G_N \hbar^2}{c^2}\frac{1}{dT_0}\right)^{\frac{1}{6}} = \left(\gamma_3 G_N \bar{\lambda}^2 \frac{1}{dT_0}\right)^{\frac{1}{6}} = 2.1002\ldots x10^{-16} \text{ meter} = \bar{\lambda}_{C,n} \qquad (4.15)$$

As indicated, this evaluates as the reduced Compton wavelength of the neutron. This indicates that the neutron scale, the resonant state of a quantum PS_3, is the fundamental physical scale. Applying this result to (4.2) we get the neutron mass as the fundamental mass, (fundamental need not mean smallest in regards to mass), and to (4.11) and (4.12), we find that

$$dA_0 = A_P \qquad (4.16)$$

the Planck area, and a gravitational quantum, G_0, for use in (4.14) and in Newton's Law is

$$G_0 = d\tau_0 = \gamma_3^{-1} r_0^2 dT_0 = \gamma_3^{-1} \lambda_{C,n}^2 dT_0 = 4.2443...\times 10^{-33} \; Newton \qquad (4.17)$$

Using (4.14) and the current CODATA values for $\lambda_{C,n}, \hbar$ and c gives a theoretical value for G_N of $6.6731971...\times 10^{-11}$ m³ / kg-s², within the standard uncertainty of the current CODATA value of $6.67384(80) \times 10^{-11}$, in conformance with our PS3 model. This does not indicate that it is the reduced Compton wavelength that produces gravity, rather it is the change in central force as a result of in a change in the expansion stress that determines both $\lambda_{C,n}$ and G_0.

Evaluation of the tension stress <u>force</u> at the boundary of a unit PS3 based on this scale is

$$\tau_n = \tau_0 = \frac{\hbar}{c} \frac{c^2}{\lambda_{C,n}^2} = \frac{\hbar}{c} \omega_n^2 = \frac{m_n c^2}{\lambda_{C,n}} = 7.1676...\times 10^5 \; Newton \qquad (4.18)$$

The ratio of tension stress force to differential force or gravitational quantum, which is also the ratio of the tension <u>stress</u> to the differential stress is

$$\frac{\tau_0}{d\tau_0} = \frac{\tau_0}{G_0} = \frac{T_0}{dT_0} = 1.6887...\times 10^{38}. \qquad (4.19)$$

Inverting this gives the differential of the natural log of the expansion stress

$$d\ln T_0 = \frac{dT_0}{T_0} = 5.9214...\times 10^{-39} \qquad (4.20)$$

This is also the general scale of the ratio of the gravitational force to the strong force, this latter interaction being what τ_0 represents. It will prove of interest that the square root of (4.19) is equal to the ratio of the neutron reduced Compton wavelength and the Planck length as

$$\frac{r_0}{l_P} = \sqrt{\frac{T_0}{dT_0}} = \sqrt{d\ln T_0^{-1}} = 1.29952...\times 10^{19}. \qquad (4.21)$$

With respect to gravity, the rotational oscillation of the PS3 results in a centripetally directed tension stress force <u>differential</u> in response to expansion tension force that accounts for gravity, as found in (4.17) above. This is found in the above Spin Diagrams and the Figure 40 and 41 Charge and Spin Tables at the centripetally directed cross products for the W_{+x} and $+V$ nodal points for the neutron, the W_{+x} and $-V$ nodal points for the proton and electron, and the W_{-x} and $+V$ nodal points for the antiproton and positron.

Quantum Newtonian Law of Gravity

A quantum version of Newton's law can thus be shown from the above as

$$F_G = \frac{n_{M1} n_{M2}}{n_{r_0}^2} \left(\frac{A_0}{\gamma_3} dT_0 \right) = \frac{n_{M1} n_{M2}}{n_{r_0}^2} G_0 \qquad (4.22)$$

where the mass and distance properties are expressed in units of the fundamental neutron scale. Substituting

$$n_{Ma} = \frac{M_a}{m_0} = M_a \frac{r_0}{\frac{\hbar}{c}} \tag{4.23}$$

$$n_{r_0} = \frac{d}{r_0} \tag{4.24}$$

gives Newton's law

$$F_G = \frac{M_1 M_2}{d^2}\left(\frac{r_0^4}{\left(\frac{\hbar}{c}\right)^2} \frac{A_0}{\gamma_3} dT_0\right) = \frac{M_1 M_2}{d^2} G_N \tag{4.25}$$

Since the interaction depicted here is mediated directly by the spacetime fabric (STF) and not by any messenger particles or gravitons and since it has been operable with expansion since the initial generation of the rest mass particles, it is not a case of "action at a distance".

Since all the forces in this discussion of PS₃ are stress forces of the STF, the individual PS₃ or rest mass quanta, as well as any transmitted photon, naturally couple with the STF in keeping with the field equation of general relativity. In addition, such generation of quanta from a classical phase space and field is straightforward and requires no other constructs such as Minkowski time for interpretation.

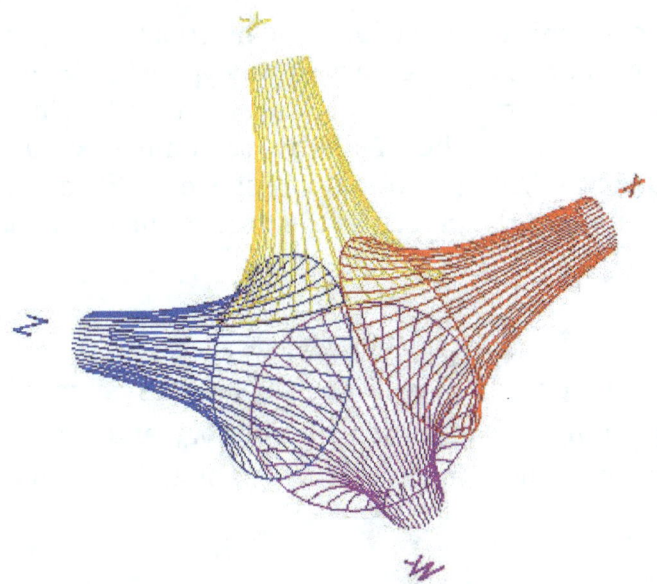

Figure 44 – One Half of an Inversphere

We can, however, offer the following 4 dimensional interpretation of a 2-sphere, the surface of a PS₃, in the context of the above development. Each of the four axes of the cubic diagonals can be defined as the central axis of a pair of pseudo-spheres, one on each side of the central 2-sphere, so that they intersect each other orthogonally at their rims as seen in Figure 44, which shows the top half of such an arrangement. A pseudo-sphere has constant negative curvature, and the rim intersections will be found to coincide with the intersections of the corresponding three cubic surface axes and the surface of a sphere of curvature related to the positive of the pseudo-spheres; if the pseudo-sphere curvatures are -1, that of the sphere is $+\frac{\sqrt{3}}{2}$. We can call this contraption of eight pseudo-spheres an inversphere. Sequential oscillatory twists of the four axes as in Figure 28 and in the development of the cuboctahedral lattice produces the rotation of PS$_2$ in PS$_3$.

2) Quantum PS$_3$ Oscillation States, Cosmic Expansion and Beta Decay

With expansion, the interstitial areas of the lattice respond primarily to tension, and resulting tension strain leads to a decrease in ambient inertial density. The mechanical impedance of the interstitial area decreases as well, and a portion of the energy of the quantum oscillation is transmitted in the process of beta decay as a lepton and corresponding neutrino. Therefore, we would expect the expansion rate, i.e. the Hubble rate, to be linearly coupled with that decrease in linear density and mechanical impedance and thereby beta decay. Photonic messenger particles, in turn, are generated by the activity of the emitted electron.

Now for a mathematical development of this claim, we review the following:

Classical Wave Mechanics

The mechanisms of harmonic motion of an ideal inertial/elastic, continuous medium give rise to discrete phenomena in the form of wave phasing, θ, expressed as a wave period, semi or quarter period, or here as radian. Such discreteness can be quantized in terms of distance as the angular wave number, κ, and in terms of time as the angular frequency, ω, of the motion. The speed of the motion, c, in either standing or traveling form, is then given as the ratio of the frequency to wave number as

$$c = \frac{\omega}{\kappa} = \frac{\frac{\partial \theta}{\partial t}}{\frac{\partial \theta}{\partial r}} = \frac{\partial r}{\partial t}. \tag{4.26}$$

Such ideal wave bearing continuum will typically have a resonant frequency, ω_0, and hence a corresponding resonant wave number, κ_0, and we can thereby designate natural distance and time units based upon these resonance values as

$$r_0 = \kappa_0^{-1} \tag{4.27}$$

and

$$t_0 = \omega_0^{-1} \tag{4.28}$$

and (4.26) can be restated variously as

$$c = \frac{\omega_0}{\kappa_0} = \frac{\frac{\partial \theta}{\partial t_0}}{\frac{\partial \theta}{\partial r_0}} = \frac{\partial r_0}{\partial t_0} = \frac{r_0}{t_0} = r_0 \omega_0 \tag{4.29}$$

Note the correspondence of (4.26) with the first line of (4.2) and that as long as the distance and time variables remain coupled by the phase variable, they vary proportionally or are co-variant with respect to any change in θ.

While the relationship given by (4.26) is descriptive of the phenomena of quantization, the dynamics of the wave is explained by the properties of the underlying continuum substrate. According to classical wave mechanics, in this case of an ideal stretched string, the wave speed squared is directly related to the tension force through the string and indirectly related to its inertial or mass density as

$$c^2 = \frac{\tau_0}{\lambda_0}. \tag{4.30}$$

Thus an increase in the tension force or a decrease in inertial density necessarily results in an increase in the wave speed. Coupling (4.29) and (4.30) gives the basic wave equation

$$-\kappa_0^2 = \frac{\partial^2 \theta}{\partial r_0^2} = \frac{1}{c^2}\frac{\partial^2 \theta}{\partial t_0^2} = \frac{1}{\frac{\tau_0}{\lambda_0}}\frac{\partial^2 \theta}{\partial t_0^2} = -\frac{1}{\frac{\tau_0}{\lambda_0}}\omega_0^2 \tag{4.31}$$

In the last term, the dynamic properties of the wave are found in the ratio of force to density, which determines the wave speed and thereby produces the observed quantization found in the displacement and time derivatives. An increase in the wave speed over time will result in a decrease in the wave number, i.e. in a red shift, if the time standard given by t_0 is held fixed. However, if time and displacement standards are co-variant, then the nominal wave speed will remain invariant, even though that speed in absolute terms, i.e. measured against some universal time scale, is increasing.

Note that if the medium, in this case the string, is stretched, it indicates a decrease in inertial density throughout its extent, along with a possible net force increase. We can apply this same reasoning to an arbitrary one dimensional tension component of a three dimensional space under isotropic extension or expansion.

As expressed in the second line of (4.2), if the speed of light is invariant at any scale, then Planck's constant simply points to a more fundamental relationship in which the product of particle mass and length scale, as given by the particle's reduced Compton wavelength, is invariant. Particle mass, m_q then, is an inverse measure of the particle wavelength, and can also be expressed as the particle angular wave number, κ_q. Thus

$$\frac{\hbar}{c} = m_q r_q = m_q \lambdabar_{C,q} = \frac{1}{\lambdabar_{C,q}}\lambdabar_{C,q} = \kappa_q \lambdabar_{C,q} = ת \tag{4.32}$$

where the final term is a constant of inertia, as introduced earlier and designated tav, ת, which is equal to 1 in a natural system as seen in the fourth term, and is simply a proportionality factor relating conventional measures of mass to distance in the second and third term. While there are cases in which the transverse wave speed of a medium may be different from its longitudinal speed, in cases where they are the same, the greater the wave number, the greater will be the curvature of the wave form. If the wave medium is the spacetime fabric, the greater the curvature, the more particle mass, i.e. inertia, will be incorporated by the wave action. The value in breaking down Planck's constant into the inertial constant and the speed of light, is that it allows us to remove the time dimension from the fundamental invariant, giving us a dynamic component or moment that is invariant in any reference time-frame.

Combining (4.32) with (4.26) and (4.30) gives the quantum wave equation

$$\lambda_q = \hbar \kappa_q^2 = \frac{1}{c^2}\hbar \omega_q^2 = \frac{1}{c^2}\tau_q \tag{4.33}$$

where λ_q is the inertial density of the quantum waveform and τ_q is the wave force of that quantum.

Applying (4.32) to (4.3) with (4.4) gives

$$G_N = \frac{r_0^3}{m_0 t_0^2} = \frac{c^3}{\hbar}r_0^2 = \frac{c^2}{\hbar}A_P \tag{4.34}$$

and to (4.14) gives

$$G_N = \frac{r_0^2}{m_0^2}(d\tau_0) = \frac{r_0^4}{\hbar^2}\left(\frac{r_0^2 dT_0}{6\sqrt{3}}\right) = \frac{r_0^4}{\gamma_3 \hbar^2}A_0 dT_0 \tag{4.35}$$

Substituting (4.35) and (4.11) and (4.12) into (4.34) gives

$$\frac{r_0^4}{\gamma_3 \hbar^2}A_0 dT_0 = \frac{c^2}{\hbar}\left(-\frac{A_0}{\gamma_3}d\ln T_0\right) \tag{4.36}$$

Here the difference in sense between the two terms indicates the centripetal direction of the left term and the counter centripetal of the right. Rearranging so that the space dimensions are on the left and the properties with time dimension, apart from the stress differential, on the right gives

$$\frac{A_0^2}{\hbar}dT_0 = c^2(-d\ln T_0) \tag{4.37}$$

while some rearrangement

$$\frac{r_0^4}{\hbar} = c^2\left(\frac{-d\ln T_0}{dT_0}\right) = -c^2\frac{1}{T_0} \tag{4.38}$$

and inversion gives a three dimensional equivalent of the wave speed equation of (4.30)

$$\frac{\hbar}{r_0^4} = -\frac{1}{c^2}T_0. \tag{4.39}$$

This is descriptive of the spacetime substrate and not of any particular quantum wave.

Returning to (4.37), assuming that the inertial and space parameters are invariant over time, and that the stress derivative remains at unity, so that the left hand term of the equation is invariant, if we assume that the log of the tension increases with expansion, that is, over cosmic time, then going back in time indicates an increase in the differential log of the expansion and a corresponding decrease in the square of the wave speed, if the invariance of the left hand term is to be preserved. In other words, expansion results in a decrease in inertial density of the spacetime substrate and a corresponding increase in the wave speed. If the decrease in the differential is not balanced by the increase in wave speed with expansion, then the stress differential on the left must change accordingly, with a corresponding effect on the invariance of the gravitational quantum.

We are assuming that the speed of a traveling wave, i.e. of electromagnetic radiation, is the same as that of a discrete, standing oscillation, a rest mass waveform. Such oscillation is equivalent to an electromagnetic wave circling a center of spin at a distance of the quantum's reduced Compton wavelength. If we think of such wavelength as the arm of a quantum clock whose tip travels always at the speed of light, then extending the length of the arm results in a decrease in its angular velocity. By contrast, if we think of time as a measure of the clocks angular velocity as is customary, then time must slow down as the arm extends if the end speed is to remain invariant.

In general relativity, time is said to dilate, but in a different manner. Increased inertial density, as in a gravitational sink, causes our quantum clock to contract its arm instead of extend it, in keeping with the inertial constant, with a corresponding decrease in angular velocity. The end of the clock arm, then, slows down as measured from some global perspective. Rising out of that sink causes the clock arm to lengthen and the angular velocity to increase.

If the speed of the arm tip is to remain constant with decreasing density and an extension of the arm, either the angular velocity must decrease or the time unit must extend to t_{0e}, to account for an increase in circumferential travel per unit of initial time, t_{0i}. Inverting the usual expression for velocity, the time standard must vary with the length standard if c is to remain invariant as

$$c^{-1} = \frac{t_{0e}}{r_{0e}} = \frac{t_{0i}}{r_{0i}}$$

$$\therefore t_{0e} = \frac{r_{0e}}{r_{0i}} t_{0i}$$

(4.40)

Similarly in terms of angular frequency

$$c = r_{0e}\omega_{0e} = r_{0i}\omega_{0i}$$

$$\therefore \omega_{0e}^{-1} = \frac{r_{0e}}{r_{0i}\omega_{0i}}$$

(4.41)

Beta Decay as a Function of Expansion

We are now ready to tackle our claim concerning the coupling of beta decay to the Hubble rate. Expansion of the STF does not indicate an equal linear decrease in density either inside, locally outside or remotely outside the fundamental quantum oscillation. The region remotely outside any oscillation is primarily under tension stress and attendant strain, and with extension suffers a decrease in linear density and related mechanical impedance, Z, where impedance, which essentially relates units of time to units of mass of a wave bearing medium, is defined as follows, using the customary theoretical unit values

$$\lambda_0 c = \frac{\tau_0}{c} = Z_0 \qquad (4.42)$$

The region about the periphery of the oscillation participates in the oscillation and exhibits a combination of tension and shear stress/strain and corresponding density fluctuations similar to what we might find in the ergosphere of an extreme Kerr quantum black hole, which is what the neutron is. The region between the nodes of the oscillation remains at the same density, unless a change of external impedance allows transmission of a small portion of its energy and therefore a change in inertial density.

Neutron/Electron Mass Ratio

In order for the energy of beta decay, which we will quantify as the mass of the electron, to be transmitted from the neutron waveform, the density and impedance at its boundary must decrease sufficient to permit that mass-energy to pass. The electron mass, m_e, is determined according to geometric constraints of the neutron oscillation and is approximately 0.000543867 ... of the neutron mass, m_0.

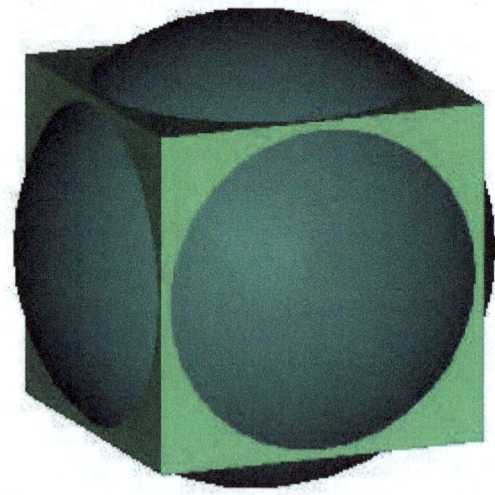

Figure 45

Figure 45 shows a concentric sphere and cube which have equal individual surface areas. Assuming a sphere of radius r_0 and area A_s and a corresponding cube with area A_c and edge length l_c, we have

$$A_s = 4\pi r_0^2 = 4\pi = 6l_c^2 = A_c \qquad (4.43)$$

$$l_c = \sqrt{\tfrac{2}{3}\pi} \qquad (4.44)$$

Tension stress on the surface of each would be equal, though the sphere represents isotropic stress while the cube represents a breakdown of the orthogonal components of such stress in keeping with the above development. The points mid way along each cubic edge are loci of closest stress/strain equivalence between cube and sphere. They are also the points of optimal shear stress and strain in the PS3 rotational oscillation, as evidenced by the action/power moments. Such stress force operates in an oscillatory manner toward a leading adjacent vertex, directed by the two resultant torques, C and L, aligned with two of the cubic diagonals, toward one or the other of the two vertices beyond the leading adjacent one.

Over time the length of the moments vary as δr_0, in the context of an expanding STF, generally in an increasing direction. The edge of the cube represents a limit for the increase in the moments, which is reflected by an increase in C and L and their orthogonal vector representations ε and μ of the Spin Diagrams. The result is an increase in the cross-product along the $W_{+x} - W_{-x}$ axis for ϕ, an advance of the moments and a transmission of energy and power at that W_{-x} node as beta decay, where δr_0^2 represents the relative energy and therefore mass of the transmitted oscillation.

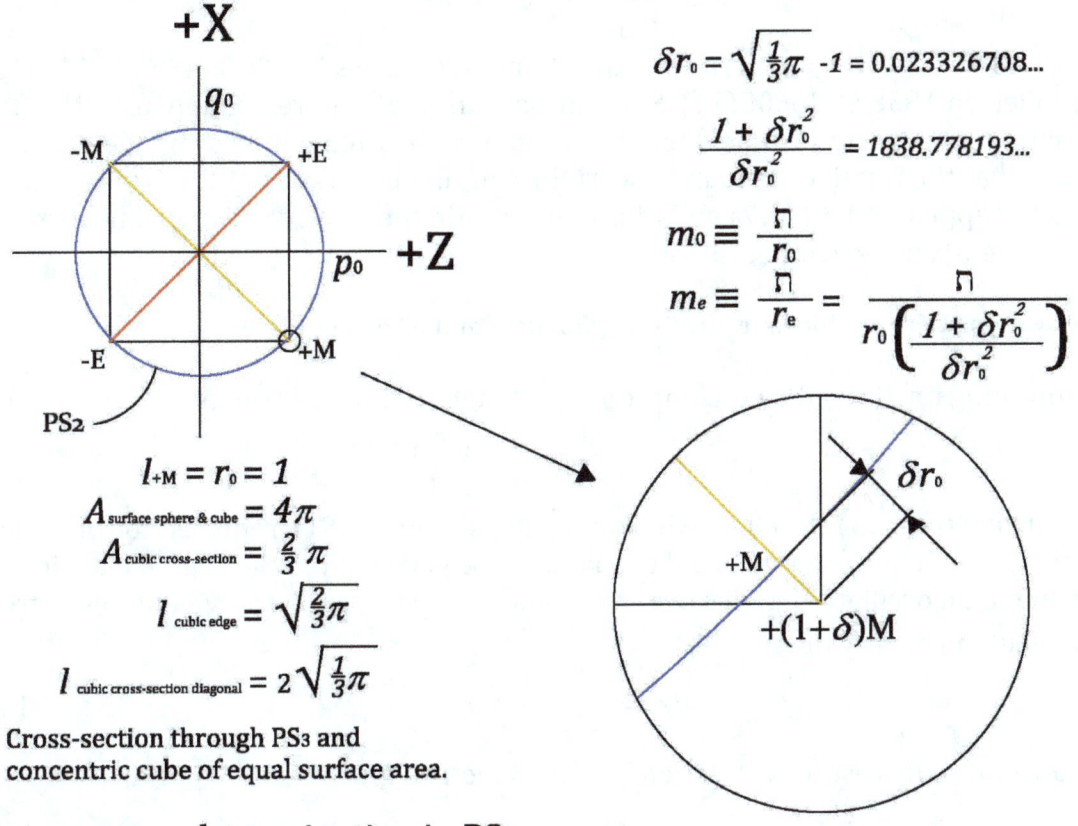

Electron mass determination in PS3

Figure 46

Figure 46 shows a cross-section through this structure at the X-Z plane of the developed PS₃ and aligned with PS₂ so that the moments +/-E and +/-M are aligned with the four half diagonals of the cubic cross-section. This indicates the moments rotating in alignment with the mid points of two of the four edges of the upper and lower cubic faces. Each half diagonal length is therefore $\sqrt{\frac{\pi}{3}}$ to the parallel moment length (not strength) of r_0 of 1.

The square of the differential δr_0 is reflected in the cross-product of the differential values of ε and μ as

$$\delta r_0 = \sqrt{\tfrac{\pi}{3}} - 1 = 0.023326708\ldots \tag{4.45}$$

$$\delta r_0^2 = 0.0005441353061\ldots \tag{4.46}$$

The ratio of the differential stress to the augmented total according to the resulting strain is

$$\frac{\delta r_0^2}{1 + \delta r_0^2} = 0.0005438393841\ldots \tag{4.47}$$

which when inverted is

$$\frac{1 + \delta r_0^2}{\delta r_0^2} = 1838.778193\ldots \tag{4.48}$$

The 2010 CODATA ratio of the electron to neutron mass is 0.00054386734461(32) or inverted 1838.6836605(11). Since mass computation presumably uses Newton's gravitational constant somewhere in the standardization of mass and weight, and given that the relative standard uncertainty of that constant at 1.2 x 10⁻⁴ is relatively large, it appears that (4.47) and (4.48) are within the relative standard uncertainty of the neutron-electron mass ratio.

Derivation of the Hubble Rate, the Expansion Rate of the Cosmos

Continuing on, the reduced Compton wavelength of the electron is

$$\lambdabar_{C,e} = r_e = \frac{\hbar}{m_e} = \frac{m_0 r_0}{m_e} \tag{4.49}$$

According to (4.33) the change in inertial density of the STF required for beta-decay is the loss of mass/energy equal to that of the electron from the region exterior to the neutron oscillation nodes over a distance r_e, to be replaced by beta-decay from the neutron energy or

$$d\lambda_0 = \frac{\hbar}{r_e^2} = \frac{1}{c^2} \hbar \omega_e^2 = \frac{1}{c^2} \tau_e \tag{4.50}$$

where ω_e is the rest mass frequency of the electron given by

$$\omega_e = \frac{c}{r_e} \tag{4.51}$$

and τ_e is the wave force of the electron rest mass. The differential density is the decrease in inertia over the distance of a wavelength required to generate a waveform of such mass.

With separation of one of the wave speed components in (4.50), a change in the linear inertial density over time is equal to a change in the impedance over distance as

$$\frac{d\lambda_0}{dt} = \frac{\tau_e}{c}\frac{1}{dr} = \frac{dZ_0}{dr} \quad (4.52)$$

Since the values of the inertial constant as Planck's constant over the speed of light and the electron reduced Compton are well determined, we can solve for $d\lambda_0$ and get

$$d\lambda_0 = 2.3589...x10^{-18} \, kg/m \quad (4.53)$$

Since the change in linear inertial density is a linear change, we might expect this expression to reflect the Hubble rate, which instead of a velocity per megaparsec of recession of galaxies, can be viewed as a dimensionless linear strain of space and therefore of time, and in fact (4.53) is a very close approximation. Converting kilometers per megaparsec to a dimensionless strain over a second, assuming a Hubble, H_0, of 73[1] km per second per mps gives a spacetime strain of $2.3657...x10^{-18}$ per second. This indicates that the Hubble rate is capable of generating the force required for beta-decay. However, we would like something more precise and dimensionally correct.

Returning to (4.50), we can decompose the wave speed invariants

$$\frac{\hbar}{r_e^2} = \frac{1}{(r_e\omega_e)(r_0\omega_0)}\hbar\omega_e^2 \quad (4.54)$$

then rearrange and multiply through by r_e to get

$$m_e = \frac{\hbar}{r_e} = \frac{\hbar\omega_e}{r_e}\frac{\omega_e}{\omega_0}\frac{r_e}{r_0\omega_e} = dZ_0\left(\frac{\omega_e}{\omega_0}\right)H_0 \quad (4.55)$$

where the change in impedance is stated as the quotient of the change in expansion force and the wave speed and the Hubble rate is shown as the spacetime length and simultaneous time strain for each second, as in (4.40) and (4.41),

$$H_0 = \frac{r_e}{r_0\omega_e} = \frac{r_e\,t_0}{r_0\,\theta_e} = \frac{\omega_0\,t_0}{\omega_e\,\theta_e} = 2.36838922...x10^{-18}\,s. \quad (4.56)$$

Transferring the frequency ratios to the mass side of the equation and substituting from (4.41) the resonant mass of the neutron as a function of the product of the expansion rate and the concurrent change in mechanical impedance is

---

[1] A study by Ron Eastman, Brian Schmidt and Robert Kirshner in 1994 and quoted in Kirshner's book, The Extravagant Universe, found an H_0 of 73 km/s/mps +/- 8 km.

$$m_0 = \hbar\kappa_0 = \frac{\hbar}{r_0} = \frac{\hbar}{r_e}\frac{r_e}{r_0} = dZ_0 H_0 \ . \qquad (4.57)$$

Evaluation of (4.56) in conventional astronomical terms is 73.082 kilometers per megaparsec for each second of current time. That is, a unit of space and co-variant time are currently extended/dilated at this rate. The implication is that space and time are currently expanding logarithmically, therefore at an accelerating pace and such expansion drives the resonant frequency as indicated by the conjugate of the frequency, wave number hence mass.

Thus, if the Hubble rate of expansion is roughly 73 kilometers per second per mpc, this indicates that every local section of space is moving away from every other at approximately 2.37 x10⁻¹⁸ meters per second per meter of separation. However, we would expect this expansion to show up primarily in the large voids between galactic filaments and clusters and not in these galactic environs or filaments of baryonic matter due to the counter effects of gravity and electromagnetism. It follows conventionally that inversion of this number would give us the approximate time since all the matter was at the same locale and that the universe has been expanding, or 4.22 x10¹⁷ seconds, which is roughly 13.4 billion years.

However, as this number represents an expansion via a compounded augmentation of the scale of spacetime itself, and not simply an extension of matter within that spacetime, the following equation for the doubling of spacetime applies, giving us the Hubble time, τ_H as

$$\tau_H = \frac{\ln 2}{H_0} = 2.92666...x10^{17} s, \qquad (4.58)$$

This indicates that space is doubling at a current rate of every 9.280 billion years, measured in terms of today's seconds. If we assume that the wavelength of the cosmic background radiation at approximately 5mm embodies that augmentation, while harkening back to a period of primal beta decay as indicated by the Compton wavelength over 2π of an electron, this represents a doubling of some 30 times, or

$$\frac{\ln\left(\frac{.005}{2\pi}\right)}{r_e}}{\ln 2} = \frac{\ln 2.060...x10^9}{\ln 2} = 30.94...doublings \qquad (4.59)$$

a lifetime in terms of today's measure of time of roughly 288 billion years. If we extrapolate back on the same basis for the expansion over the scale of r_0 to r_e, prior to beta-decay where it may or may not be applicable, we have an additional doubling of 10.84 times or

$$\frac{\ln(1830.6842...)}{\ln 2} = 10.84.... \qquad (4.60)$$

or a total doubling of the Hubble time of 41.78 or 393.47 billion years in current time as

$$(2.927...x10^{17})(41.78...) = 1.2227...x10^{19} s \qquad (4.61)$$

Finally, if we envision that a current expansion factor can be derived by a comparison of the Planck length and the neutron Compton wavelength, keeping in mind that we can multiply both terms by the speed of light without affecting their ratio and express the quotient as a coefficient of expansion in light seconds, given as

$$\kappa_{exp} = \frac{r_0}{l_p} = \frac{2.10019...x10^{-16}m}{1.61612...x10^{-35}m} = 1.29952...x10^{19}\,ls \qquad (4.62)$$

We have a close agreement with (4.61) at 412 billion years.

In another vein, we can multiply this figure as with (4.58) to get the extent of doubling, in terms of current time standards over the most recent doubling period, as 285 billion years or

$$C_x = \ln 2(\kappa_{exp}) = 9.00758...x10^{18}\,ls. \qquad (4.63)$$

Dividing by (4.58) we get the number of doublings since the initial factor established by beta-decay and get

$$\frac{C_x}{\tau_H} = \frac{9.00758...x10^{18}\,ls}{2.92666...x10^{17}\,s} = 30.77...doublings \qquad (4.64)$$

compared to (4.59).

With respect to the period before beta-decay or the last scattering of the standard model cosmology, it is not clear from this extant modeling that rest mass quanta emerged from an initial big bang. Rather it appears likely that such matter emerges in an ongoing manner from galactic inertial centers, i.e. black holes which can be gravitational field sources as well as sinks, and their connecting filaments in response to the tension stress of expansion of the surrounding, relatively mass free voids, as evidenced by the observance of episodic gamma ray bursts of unknown origin.

The Missing Mass of Beta Decay

We are not quite through with our investigation. While the ratio of neutron-electron mass as developed here is compelling, there is still a matter of the missing mass of beta decay. According to the CODATA ratios, the difference between the neutron-electron mass ratio and proton-electron mass ratio is

$$\frac{m_n}{m_e} - \frac{m_p}{m_e} = 1838.6836... - 1836.1526... = 2.5310... \qquad (4.65)$$

Since the relative mass of the electron in this case is 1, there is a relative mass or equivalent energy of 1.530... that is unaccounted for. If it is assumed that mass is a property that is somehow bound up in the confines of a discrete particle, this is a puzzlement. However, if it is understood to be a measure of the resistance of stress to a straight line force, i.e. a measure of redirection of oscillatory energy and therefore of curvature of spacetime strain, the problem vanishes.

Consider the function

$$W(n) = \ln_0 e_n^n \qquad (4.66)$$

which is related to the Lambert W function, where n can be any real number, though we will only be considering the integers. The significant feature of this function is that it generates a system of natural logs, $\ln_n$, and corresponding exponential bases, e_n, that can be used as normalizing factors, so that

$$\ln_n e_n = 1, \; \ln_n e_{-n} = -1 \;. \qquad (4.67)$$

At $n(0)$, this is simply the natural log and exponential base, and

$$W(0) = \ln_0 e_0^0 = 0 \qquad (4.68)$$

In the following Figure 47 we have graphed the significant portion of the natural log and exponential functions. Note the functions mirror each other along the line $y = x$, as do their derivatives. We can define the exponential base, e_0, on both x and y axes by the point on each function at which the lines (blue) whose slopes represent the derivatives intersect each other and the origin of the system. The only other instances of such intersection would be when the functions reach negative infinity along both axes, which of course they never do in the context of Euclidean space. They do on the Riemannian complex sphere, however.

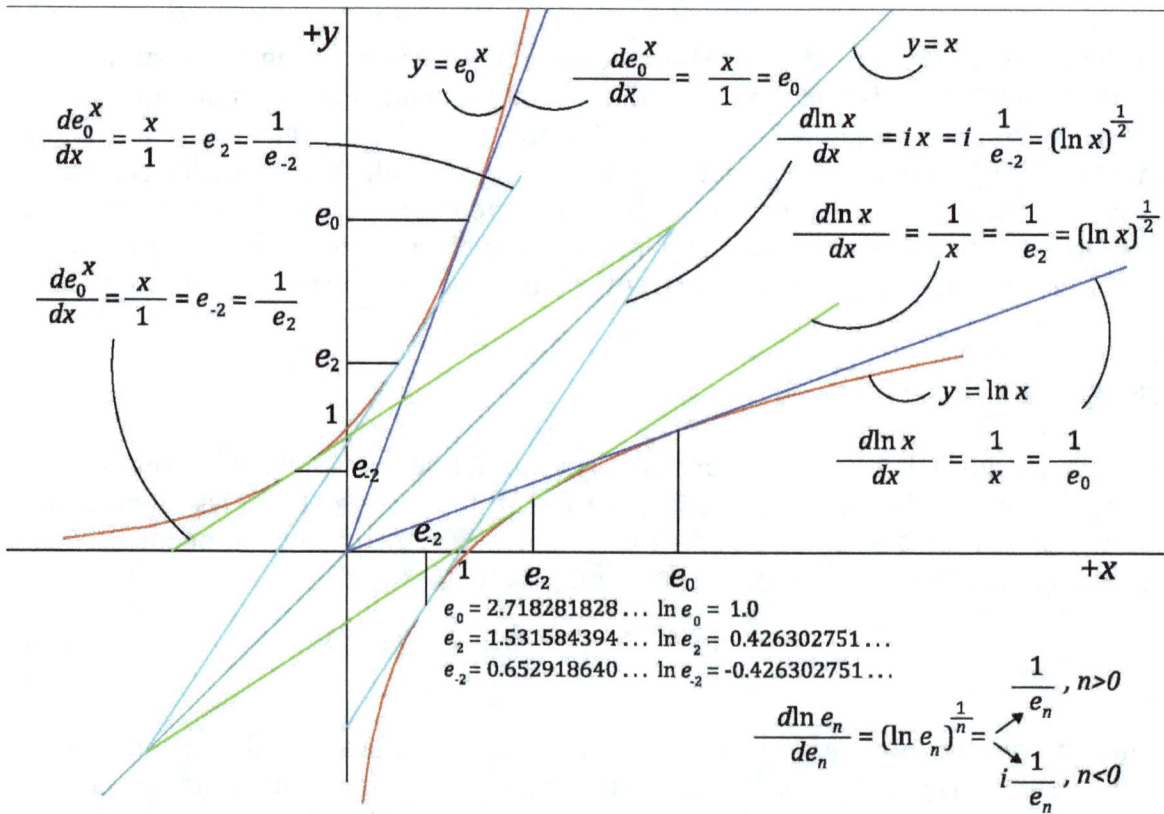

System of exponent bases e_n, shown for e_0 and $e_{\pm 2}$ where $e_{-2} = e_2^{-1}$

Figure 47

The whole system of e_n, for $n>0$, occurs in the range $1 < x < e_0$, and as n increases, the slope values converge while their intersection moves toward the negative infinities of both x and y. At the point $x = 1$, the derivatives of both functions equal 1 and their slopes are parallel. In terms of the Riemannian sphere, the lines actually form to great circles about the spheres equator. From that point, n decreases from negative infinity toward 0 at the y asymptote. Thus $n<0$, occurs in the range $0 < x < 1$. The x and y axes then are the doubles for the slopes e_0 and e_0^{-1}, corresponding to the doubles for the rest of range of n.

For the range $e_0 < x < +\infty$, the slopes of the two derivatives diverge as x increases, and there are no real subscript functions of e_n. Note that for the range $n<0$, however, according to the derivative of the natural log with respect to a change in x, the slope has imaginary sense, which generally indicates a rotation of some manner or another.

The following Figure 48 table shows the results of this function for the first three integers, and an assumption of results carried to infinity. The second Figure 48 table shows related results with the introduction of imaginary sense to the various function.

$f(n)$ \ n	0	1	2	3	...	∞
e_n	2.718281828..	1.763222834..	1.531584394..	1.419024454..		1
$e_{\_n} = e_n^{-1}$	0.367879441..	0.567143291..	0.652918640..	0.704709490..		1
e_n^n	1	1.763222834..	2.345750756..	2.857390779..		∞
$e_{\_n}^n = e_n^{-n}$	1	0.567143291..	0.426302751..	0.349969632..		0
$\ln_0 e_n$	1	0.567143291..	0.426302751..	0.349969632..		0
$\ln_0 e_{\_n}$	-1	-0.567143291..	-0.426302751..	-0.349969632..		0
$\ln_n e_n$	1	1	1	1		1
$\ln_n e_{\_n}$	-1	-1	-1	-1		-1
$\ln_0 e_n^n$ = $W(n)$	0	0.567143291..	0.852605502..	1.049908893..		∞
$\ln_0 e_{\_n}^n$ = $W(-n)$	0	-0.567143291..	-0.852605502..	-1.049908893..		$-\infty$
$\ln_n e_n^n$	0	1	2	3		∞
$\ln_n e_{\_n}^n$	0	-1	-2	-3		$-\infty$

Figure 48

n	f(n)	n ln₀ eₙ	in ln₀ eₙ	n ln₀ ieₙ	in ln₀ ieₙ
1		0.5671...	i0.5671...	0.5671...+iπ/2(=+1 iπ/2)	-π/2 + i0.5671...
2		0.8526...	i0.8526...	0.8526...+iπ(=+2 iπ/2)	-π + i0.8526...
3		1.0499...	i1.0499...	1.0499...+i3π/2(=+3 iπ/2)	-3π/2 + i1.0499...
4		1.2021...	i1.2021...	1.2021...+i2π(=+4 iπ/2)	-2π + i1.2021...
5		1.3067...	i1.3067...	1.3067...+i5π/2(=+5 iπ/2)	-5π/2 + i1.3067...
6		1.4324...	i1.4324...	1.4324...+i3π(=+6 iπ/2)	-3π + i1.4324...

Figure 49

The function in (4.66) finds form in the following equation, where the negative sense in the subscript has the same meaning it does in the superscript exponent, that is it represents inversion.

$$\ln_0 e_n = e_n^{-n} = e_{-n}^n \tag{4.69}$$

Thus for n = 2, we have the following, where it is understood that e_2 is a normalizing coefficient for any variable x, in particular for an instant unit variable property x_0

$$\ln_0 e_2 (e_2)^2 = \ln_0 e_2 (e_{-2})^{-2} = 1 \tag{4.70}$$

and with the variable, x_0, we have

$$(\ln_0 e_2 x_0)^{\frac{1}{2}} = e_2^{-1} = e_{-2} = \frac{d \ln e_2 x_0}{d e_2 x_0} = \frac{1}{1.53158...} \tag{4.71}$$

$$(\ln_0 e_{-2} x_0)^{\frac{1}{2}} = ie_{-2}^{-1} = ie_2 = i\frac{d e_{-2} x_0}{d \ln e_{-2} x_0} = i\frac{1}{1.53158...} \tag{4.72}$$

In the final table, it is clear that the integers, n, are the count of the rotations of ½ π and of the powers and hence the number of orders of i, both indications of a degree of orthogonal structure.

We are interested here specifically in the factor e_2. As a review of Figure 47 hopefully makes clear, the value in the subscript exponential bases is in determining a coefficient of proportionality between two related differentials, one of which is a function of the nth-root of the logarithm to the others linear function. We refer back to the start of our discussion of SHM at (1.2) and the relation between an oscillator's frequency as a function of the square root of the quotient of the pendulum length over the motivating gravitational acceleration. Again in the above development of the neutron scale for quantum gravity at (4.21) we have an expression of the change in the linear scale of r_0 as the square root of the change in the natural log of the expansion stress scale, f_0. We can model quantum mass as a linear function of space by the reduced angular wavelength, r_0, or time by the frequency, ω_0. Using the inertial constant and/or the speed of light we have

$$m_0 = f(r) = \hbar r_0^{-1} \tag{4.73}$$

$$m_0 = f(\omega) = \hbar c \omega_0 = \hbar \omega_0 \tag{4.74}$$

Stress is modeled as a function of the square of both of these

$$f_0 = f(\omega^2, r^2) = n\omega_0^2 r_0^{-2} \qquad (4.75)$$

We will use r_0 for our discussion, since we previously discussed beta decay as a function of its increase. Thus a change in stress with expansion leads to a increase in r_0, where a preliminary decrease in mass of the fundamental oscillation, the neutron, is equal to the mass of the emitted electron or positron as developed above or

$$m_e = \Delta m_0 = m_0 \left(\frac{\delta r_0^2}{1 + \delta r_0^2} \right) \qquad (4.76)$$

The change in stress/energy density of the oscillation is

$$dE_1 = df_0 = f'(A_0^{-2}) = -\frac{n\omega_0^2}{A_0^2} dA_0 = -\frac{\tau_0}{A_0^2} dA_0 = -\frac{f_0}{A_0} dA_0 \qquad (4.77)$$

where it is clear that a change in the log of the stress is inversely equal to a change in the in the log of the cross-section,

$$d \ln f_0 = \frac{df_0}{f_0} = -\frac{dA_0}{A_0} = -d \ln A_0 . \qquad (4.78)$$

Obviously, since both stress and cross-section are unit values,

$$\ln f_0 = \ln A_0 \qquad (4.79)$$

Thus for a logarithmic expansion of the cosmos, in accordance with (4.78) and using (4.72), we substitute f_0 for A_0 which is the square of $r_0 = x$ and get

$$(\ln f_0)^{\frac{1}{2}} = ie_{-2} = i\frac{df_0}{d \ln f_0} = i\frac{df_0}{d \ln f_0} = i0.65291... = \frac{1.0}{i1.53158...} \qquad (4.80)$$

The imaginary sense assigned to the natural log differential is an indication of transverse motion and other energy associated with the change in stress, resulting in the change to that of the reduced Compton wavelength of the proton.

The Hamiltonian or total energy of the system resulting from beta decay is therefore the energy of the neutron, less the rest mass energy of the electron due to the change in stress, less the change in spin energy due to the natural log of the stress, to equal the mass or rest mass energy of the proton.

$$E_0 - E_e(df_0) - E_0(d \operatorname{nl} f_0) - E_p = 0 \qquad (4.81)$$

In terms of mass

$$m_0 - 1m_e - 1.53158...\Delta m_0 - m_p = 0 \qquad (4.82)$$

This is accurate to a factor of 2.16 x 10⁻⁷.

Evaluation of Elementary Charge

An final observation is in order, this about charge. We discussed it as a function of the fundamental oscillation of PS$_3$, but did not relate it to experimental data and the SI fundamental charge, the coulomb, C. The coulomb, or ampere per second, is equivalent in mechanical dimensions to one kilogram-meter per second, a measure of momentum. A fundamental unit or elementary charge, e_0, is established as

$$e_0 = 1.60217653(14) \times 10^{-19} \text{ Coulomb} \tag{4.83}$$

As a measure of momentum, in connection with our development of PS₃ and the transmission of momentum with beta decay at W$_{-x}$, the fundamental unit of conjugate momentum, using angular frequency, is reasonably close to (4.83)

$$p_0 = \hbar\omega_0 = 5.02130...\times 10^{-19} \, kg \cdot m/s \tag{4.84}$$

Charge is related to each of the two rotational nodes, W$_{-x}$ and W$_{+x}$, indicating the need to apply semi-periodic frequency, which we can do by dividing (4.84) by π. In addition, the charge generation is conditioned by the product of the momentum and the mechanical impedance of the STF (not to be confused with the electromagnetically derived characteristic impedance of the vacuum), which is

$$Z_0 = \hbar\omega_0\kappa_0 = 0.002390877...kg/s$$

$$\zeta = \left(\frac{1+Z_0}{\pi}\right) = 0.319070926... \tag{4.85}$$

$$\zeta^2 = \left(\frac{1+Z_0}{\pi}\right)^2 = 0.101806256...$$

where we define the total factor, ζ, and its square for later use. Thus we would anticipate an elementary charge of

$$e_0 = p_0\zeta = \hbar\omega_0\zeta = 1.602152647...\times 10^{-19} \, kg \cdot m/s \tag{4.86}$$

This varies from the established value by a factor of 1.000015...which once again is in the same order of magnitude as the relative uncertainty for the gravitational constant.

Fine Structure Constant

Further development, using the familiar identity for the inverse of the fine structure constant, α, a dimensionless number and therefore the ratio of two like-property magnitudes, as

$$\alpha^{-1} \equiv \hbar c \frac{4\pi\varepsilon_0}{e_0^2} = 137.0359989... \tag{4.87}$$

and the permeability, μ_0, and permittivity, ε_0, relationship, where μ_0 is in units of inductance per meter or henrys per meter which reduces to units of force per current squared or newton's per ampere squared, and ε_0 is in units of capacitance per meter or farads per meter which reduces to ampere squared per newton over the speed of light in vacuo squared, so that

$$\varepsilon_0 = \frac{1}{c^2\mu_0} \tag{4.88}$$

and with rearrangement in (4.87) gives the following

$$e_0^2 = -\hbar c^2(\alpha 4\pi\varepsilon_0) = -\hbar\frac{\alpha 4\pi}{\mu_0} = -\hbar\frac{\alpha}{10^{-7}} = -\hbar(\hbar\omega_0^2)\zeta^2 \tag{4.89}$$

It is noted that the value of μ_0 is set by convention in relating charge, q, (of which elementary charge, e_0, is an effective quantum) and current, $i = dq/dt$, resulting in the exactness of the denominator of the next to last term. Since the negative sense of the right terms above can be attributed to the current, therefore charge, squared, it can be incorporated therein, canceling such sense in the charge squared term. This suggests the transparent presence of a current squared argument in (4.89), for which the fine structure constant is a coefficient, since from Ampere's Law for one ampere² of current, where the denominator on the right is in newton, we have

$$\mu_0 = 2\pi \left(2 \times 10^{-7} N\right) \frac{d}{L} i_0^{-2} \tag{4.90}$$

2 x 10⁻⁷ newton is the force generated for each meter length of two conductors of infinite length and negligible cross-section and one meter apart in a vacuum with one ampere of constant current flowing in each conductor. The d and L obviously cancel and the i_0^2 component and therefore the force is positive or negative depending on whether the currents are parallel and attractive or antiparallel and repulsive.

Inserting this into (4.89) with some rearrangement gives the ratio of elementary charge squared to current squared as the product of the modified fine structure constant, α' as shown, and the inertial constant. If the fine structure constant is dimensionless and its denominator is a force from the above, then α' is an inverse force, which in terms of our PS₃ development is the inertial constant times a frequency squared and k is an unknown proportionality factor for the frequency as

$$\frac{e_0^2}{i_0^2} = \frac{\alpha}{10^{-7}} \sqcap = \alpha' \sqcap = \frac{k^2}{\sqcap \omega_e^2} \sqcap = \frac{k^2}{\omega_e^2} \tag{4.91}$$

If the force in the last term is the base transverse wave force of the electron as in the above development, then k is an angular measure per unit of elementary charge as,

$$k = \omega_e \frac{e_0}{i_0} = \left(\frac{\theta_e}{s}\right) \frac{e_0}{\frac{ne_0}{s}} = 124.3840198\ldots \frac{\theta_e}{e_0} \tag{4.92}$$

Using this value with (4.91) gives

$$\alpha = \frac{k^2 \left(10^{-7}\right)}{\sqcap \omega_e^2} = \frac{k^2 \left(10^{-7} N\right)}{0.212013671\ldots N} = 0.007297352\ldots \frac{\theta^2}{e_0^2} \tag{4.93}$$

With another look at (4.89), we get the following relationships between the fundamental wave force and α'

$$\alpha' = \tau_0 \zeta^2 = \sqcap \omega_0^2 \zeta^2 = \zeta \omega_0 e_0 \tag{4.94}$$

Special Relativity and Muon & Tau Families

Concerning the compatibility of this model and special relativity, I have written about this extensively elsewhere. Suffice it to say that the PS₃ model is one of constrained stress/strain in the STF, which acts as discrete units of rest mass with derived properties. Each discrete state, remains a wave form and in response to interaction with other states is free to translate and rotate in space according to the ambient energies. It will therefore contract its characteristic strain radius in response to acceleration in keeping with the Fitzgerald-Lorentz length contraction, resulting in an increase in spin energy/mass according to the definition of the inertial constant

$$♩ \equiv mr \equiv \frac{m}{\kappa} \qquad (4.95)$$

As to the two other families of leptons, the muon and tau, and their theoretical related hadrons, based on their short lifetime and granted my limited knowledge of the experimental background for their theoretical introduction, it is my perception that they are simply the basic PS_3 states we have discussed, altered by relativistic dynamics and interaction. We would expect these states to behave in a generally ordered fashion under constraints of high energy collision and those defined by geometry and mathematics. The evolution of a catalogue of such short lived phenomenology, while useful, does not indicate the need or wisdom of elevating that phenomenology to ontology. I would grant the status of "fundamental rest mass particle" only to common, stable, relatively long-lived states, including the neutron, of course.

Conclusions

We can only talk about nature and the physical world because we have a conscious experience of it. That sensory experience is mediated by minds conditioned to look for patterns to help navigate that world. The degree to which such navigation is so facilitated is largely a measure of how well those patterns can be found and understood as interconnected parts of the complete fabric of experience. The process of such discovery is necessarily marked by periods of progress and temporary setback involving mental effort without recompense, but over the long haul it is a satisfactory venture.

It should be no surprise therefore that these minds feel the need to partition what is essentially an experience of a whole into a physical world and a mental world and perhaps some others in an attempt at understanding. Experience is experimental. What works is real, and what doesn't is, well, just in the mind; imaginary or perhaps just more complex.

We observe, we think with the hope of understanding, we create models in our heads of what we think is going on behind the scenes of our observation, we put the models to the test, and we observe again. When the model is accurate, we take notice, and when it is both accurate and precise, we call it true. The model merges with the field of observation and we quickly forget that the model is just a model. Without diminishing in one iota the relevance of such models, it bears remembering that the reality of the model is in our head. But then so is the conscious experience by way of the senses.

The problem with successful models is that they may be successful in their predictive power without sufficient understanding of the fundamentals on which their success is based. Axiomatic amnesia. It is all too easy to confuse logically necessary results of the way the model is framed with the assumptions made about the necessity of what is going on behind the curtain. Then when a dead end is encountered, it is too often deemed to be a problem with the superstructure of the model rather than a problem with its foundation.

No one has ever seen the microscopic world directly. The invention of that apparatus gave us a view of patterns unimaginable before. The microbiological world in particular teemed with living one-celled entities, quantum life so to speak, and it was natural to assume that the division of inert physical stuff continued down to some fundamental level as well. And so it does. It is just that each of those bacteria and other microbial beings don't create the biosphere. The biosphere is a set of ambient conditions that allow such beings to thrive. In a similar manner it is not the quantum particles of physics that create the phenomenal physical world, rather it is the principal of change of inertial continuity in space, which necessarily means over time, that creates the particles. If there is a Higgs field, this is what it is.

It is this writer's position that the center stage quantum world can only be understood against its classical backdrop of an expanding cosmos. I can't see behind the stage anymore than anyone else. What I do know is that if we are to understand what is going on back there, we must be logical. I know that there must be some ontological regime that operates more or less the same across the whole of the cosmos, if we are to make sense of it, and that it is more likely that such regime is to be physically found universally instead of replicated exactly at each microscopic locale across its extents. That is, the mass of a neutron here and 10 billion light years from here (after doing all the de rigueur relativistic computations) must be the same now not because of some initial investment at the big bang, but due to an operative condition at *both locales right now*. This sounds like a function of space itself.

The instant development of PS_3 as a model for understanding, for accounting for momentum and displacement and kinetic and potential energy and action and power and then spin angular momentum and charge within the context of Simple Harmonic Motion is a mental construct. I don't know for a fact if the void that space appears to be among the multitude of stars is a true vacuum or the only thing that really is. What I do know is that if we treat it as such, as an inertial – elastic wave bearing continuum, which appears to be fully consistent with the structure of PS_3, we can correctly account for or predict, without any extraneous factors:

Newton's gravitational constant as a quantum effect
Quantum spin invariance and magnitude
Quantum charge invariance and magnitude
The ratio of neutron/electron mass
The missing mass of beta decay and therefore
The ratios of neutron/proton and proton/electron
The nature of ordinary and anti matter
Relationship to the fine structure constant
All this as a function of an exponential Hubble rate

I don't know about the tauon and muon families, but there is more with respect to the consistency of this model with special and general relativity, including a description of the neutron as an extreme Kerr quantum black hole.

I believe this is a step up the ladder from the existing understanding. If it furthers the discussion, it deserves to be vetted.

You be the judge.

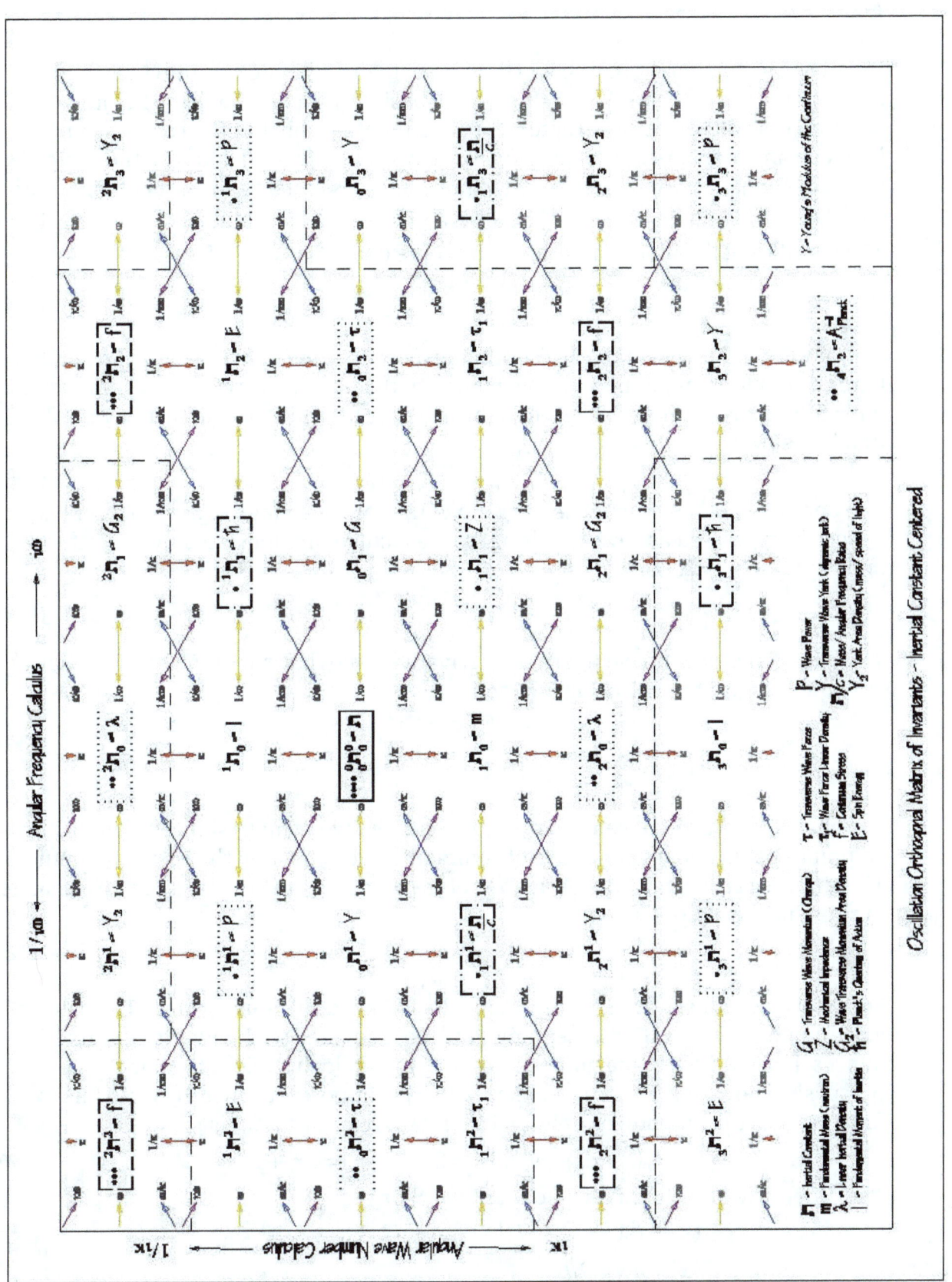

Matrix of PS₃ Functions and Invariants

Aside #1

Before we go further, it may be worthwhile to take a brief detour into the fundamentals of calculus. While this is very basic, it is hoped it will be helpful. A function is a mathematical structure, $F(x$, etc.$)$ that establishes the relationship between an independent variable(s) numerical value, (x, etc.) and a mathematical result of computation, (y), using that value. Thus, y is equal to some function of x or

$$y = F(x) \tag{5.1}$$

The specifics of the function are given by some equation, such as

$$y = 2x + 3, \text{ or} \tag{5.2}$$
$$y_n = x^n$$

where x is generally held to vary continuously over some interval, $a \leq x \leq b$. In this latter example, y_n might be a line segment length if $n = 1$, an area if $n = 2$, or a volume if $n = 3$. The n is not normally used in this manner for the y, of course, but we will find it useful. Thus

$$F_n(x) = y_n = x^n \tag{5.3}$$

The derivative of the function is the ratio of the rate of change in the function $F(x)$, generally expressed as $F'(x)$, resulting from a change in x, indicated by the letter d and expressed as dx. Thus

$$F'_n(x) = \frac{dy_n}{dx} = ? \tag{5.4}$$

The question is how do we figure the derivative and what does it mean? Let's say n is 1, and we have the following function

$$F_1(x) = 2x^1 \tag{5.5}$$
$$y_1 = 2x^1$$

x might be the weight of some variable commodity we want to ship, 2 might be the shipping cost in dollars per ounce of commodity, and y is then the cost of shipping. The shipping rate is then the derivative of the shipping cost with respect to a change in shipping weight or

$$F'_1(x) = \frac{dy_1}{dx^1} = 2. \tag{5.6}$$

Some folks would say at this point, "Why do you need calculus for this? The value of y per unit of x gives us the same thing." Indeed, provided we know x is in units of ounces. If n had been some number larger than 1, calculus might be more persuasive, however. The way we derive (5.6) is important. Since $F(x)$ is a linear function of x, the function of x plus a little bit more or $F(x + dx)$ is just

$$F_1(x + dx) = y_1 + dy_1 = 2(x + dx)^1 = 2x^1 + 2dx^1 \tag{5.7}$$

The difference or differential amount of $dF(x)$ or y is

$$dF_1(x) = F_1(x + dx) - F_1(x) = y_1 + dy_1 - y_1 = 2x^1 + 2dx^1 - 2x^1$$
$$dF_1(x) = dy_1 = 2dx^1 \tag{5.8}$$

and the derivative is the rate at which a differential amount of y is generated for every differential amount of x,

$$\therefore F_1'(x) = \frac{dy_1}{dx^1} = 2\frac{dx^1}{dx^1} = 2 \tag{5.9}$$

Here we get into the ambiguities of language. The derivative asks, "what is the cost of shipping per ounce of commodity equal to?" and the answer generally given by the last term in the equation is "it equals 2 dollars." Everyone knows what is being asked and everyone understands the answer. The problem is that it is not good grammar. As the good math and science student will remember being told, "The dimensional units on each side of the equation have to be the same."

The correct answer to the question is " it equals 2 dollars *per ounce of commodity*." Looking at the next to the last term of (5.9) does not help. Obviously the *dx* terms cancel out, but before that operation, the term reads "it equals 2 dollar*ounce per ounce", or for the purist, ""2 dollar*very small weight per very small weight". What type of unit is a "dollar*weight" anyway?

What is happening is that the nimble human mind understands the implicit context even if it is only subconsciously aware of it and fills in the gaps. Usually. (5.9) really becomes

$$F_1'(x) = \frac{dy_1}{dx^1} = \frac{2y_0}{x_0} \tag{5.10}$$

where y_0 is, of course, a unit of *y*, in this case a dollar, and x_0 is a unit of *x*, in this case an ounce. We are assuming in this discussion that the differential amount, *dx* can be much smaller than 1, as will *dy* as a result, but the derivative is still expressed in units of both *as a limiting rate*. If *dy* is contextually in fact $2y_0$, then *dx* becomes x_0.

If *x* is some function of time, *F(t)*, then so is *y*, since
$$F(t) = x = ct$$

$$F'(t) = \frac{dx_1}{dt^1} = \frac{cx_0}{t_0} \tag{5.11}$$

$$\therefore dx_1 = \frac{cx_0}{t_0} dt^1$$

then

$$F_1'(F(t)) = \frac{dy_1}{dt^1}\left(\frac{t_0}{cx_0}\right) = \frac{2y_0}{x_0}$$

$$= \frac{dy_1}{dt^1} = \frac{cx_0}{t_0}\left(\frac{2y_0}{x_0}\right) = c\left(\frac{2y_0}{t_0}\right) \tag{5.12}$$

In the above example, *c* is the number of ounces, say 480, of commodity *x* being produced in a unit of time, say a day, so (5.12) becomes 480 ounces per day times 2 dollars shipping cost per ounce or 960 dollars in shipping costs per day, i.e. the rate of change in y per *one unit change* in t.

The take away here is that while we are accustomed to thinking of the derivative as expressing the ratios of exceedingly small quantities, at the limit that ratio is between two finite numbers, the consequent or denominator of which is always expressed as one unit.

A second observation about calculus is that it is generally stated with respect to an anisotropic differential. That is, one end of a linear mapping in a Cartesian system representing an independent variable is held fixed with respect to the appropriate axis as is its function on the second axis. This is reflected in the math as in the following, where the subscript after the independent variable indicates the number of differentials per variable,

$$F_2(x)_1 = y_2 = x^2$$

$$F_2(x+dx)_1 = y_2 + dy_2 = (x+1dx)^2 = x^2 + 2xdx + dx^2$$

$$F_2(dx)_1 = dy_2 = 2xdx + dx^2 \tag{5.13}$$

$$F_2'(x)_1 = \frac{dy_2}{dx} = \frac{2xdx + dx^2}{dx}$$

and finally

$$F_2'(x)_1 = \frac{dy_2}{dx} = 2x + dx \tag{5.14}$$

This is all well and good if one is only interested in an interpretation in which the derivative is the slope of the curve given by $F(x)$ in the above. If dx is much less than unity it can be discarded in most cases in the example of (5.14). It can become important in integration of some such functions, however.

$F(x)$ might also represent the area of a physical space in which the relationship of the perimeter, $4x$, to the area, y_2, is of interest. We then might need to think of the differential as isotropic or bilateral, as in the following, where the differential occurs at both ends of the linear variable, thus

$$F_2(x+2dx)_2 = y_2 + dy_2 = (dx + x + dx)^2 = (x+2dx)^2 = x^2 + 4xdx + 4dx^2 \tag{5.15}$$

and

$$F_2'(x)_2 = \frac{dy_2}{dx} = \frac{4xdx^1 + 4dx^2}{dx} = 4x + 4dx^1 \tag{5.16}$$

For a volume function or

$$F_3(x)_2 = y_3 = x^3 \tag{5.17}$$

$$F_3'(x)_2 = \frac{dy_3}{dx} = \frac{6x^2 dx^1 + 12xdx^2 + 8dx^3}{dx} = 6x^2 + 12xdx^1 + 8dx^2 \tag{5.18}$$

The magnitude of the resultant of addition of two orthogonal unit vectors, v_{01} and v_{02}, is $\sqrt{2}v_0$, and that of three orthogonal unit vectors is $\sqrt{3}v_0$. Therefore in a condition in which an isotropic volume change, dy_3, is represented by an orthonormal unit vector on each of six cubic faces, the ratio of the magnitude of a vector equidistant from all six, i.e. along one of the cubic diagonals, to each of the normal unit vectors will be $6\sqrt{3}$. Okay. Returning to Phase Space 3.

Aside #2

In non-technical contexts, the terms "stress" and "force" are often used interchangeably, though the distinction is generally understood even then. In informal settings "force" is generally used to convey the idea of an external agent operating on a separate system to produce some change. "Stress" is used to convey an internal change in some system resulting generally from the operation of an external agent, though there is the possibility of internally generated stress. As discussed earlier, this common understanding finds technical expression in the distinction between a body force, F_b, and a surface or stress force, $\tau_{t,s}$. Both have the properties of mass times displacement over time-squared. Displacement indicates the physical change occurring and time-squared indicates the acceleration with which the change occurs. Mass represents a property of the agent or the system that, at least in a classical, non-relativistic context, is not affected by the change or its rapidity, the property of inertia, a property to stay the same whatever else is happening. A surface or stress force has the additional definition as a stress integrated over the surface or cross-sectional area of its operation.

In a discussion of the interaction of separate agents, i.e. bodies, where the transfer of momentum and energy is conceptually concentrated to a point, we can generally do pretty well without reference to the concept of stress. When we get into a discussion of more diffuse arrangements of energy, such as in the behavior of fluids or the concept of gravitational or electromagnetic fields, we often need to use the concept of stress, that is a force operating over an area and the related concept of an energy/inertial density, the energy, mass or analogous property such as charge contained within a linear, planar, volume or higher dimensional boundary.

In various modeling, such as the ideal gas law and as expressed in Gauss's divergence theorem, the energy content, E, of a defined volume, V, is equal to the tension stress, f_t, general described as a pressure, at the surface or boundary of that volume, the stress being the tension force, τ_t, per unit area, A_0. Here we will confine that volume to a unit cube volume, V_0, so that

$$\frac{E}{V_0} = \frac{\tau_t}{A_0} = f_t \qquad (6.1)$$

The energy might be the kinetic energy of an ideal gas or an electric charge and the stress, the corresponding pressure or electric flux at a boundary surface.

In mechanical terms, the energy might be the elastic potential energy of a beam or other structural material, giving a potential energy density, V_1, for the left hand term above, and the stress, in accordance with Hooke's law, is the expression of Young's modulus of elasticity, Y, and the strain, ε_t, i.e. the stretching or extension that occurs in the volume as a linear response to the stress or

$$V_1 = \frac{V}{V_0} = \tfrac{1}{2} f_t \varepsilon_t = \tfrac{1}{2} Y \varepsilon_t^2 = \frac{f_t^2}{2Y} \quad (6.2)$$

where the strain is defined as a relative change in the length of the material, here designated by a unit displacement, and is therefore a dimensionless number,

$$\varepsilon_t \equiv \frac{\Delta q_0}{q_0} . \quad (6.3)$$

For convenience, we will state that

$$V_0 = A_0 q_0 = q_0^3 . \quad (6.4)$$

Young's modulus

$$Y = \frac{f_t}{\varepsilon_t} \quad (6.5)$$

is a stress potential, an inherent property of the material, with the dimensions of stress, as shown here, the linear ratio of stress to strain. This calls for some mathematical clarification, since Y is not a function of ε_t or f_t, and does not become undefined if ε_t is zero. Rather f_t is a function of ε_t and Y or ε_t is a function of f_t and Y. Y is a ratio such that if ε_t equals 1, theoretically f_t equals Y.

From (6.2) it is clear that if there is no strain, Y is equal to the elastic potential energy density, V_1. The one half coefficient of Y is a function of the work, W, done to extend the structural member the distance Δq_0, as

$$W = \tfrac{1}{2} \tau_t \Delta q_0 \quad (6.6)$$

In general, an extension in an isotropic elastic solid material along one dimension, x, results in a decrease or negative lateral extension along the other two dimensions, A, where A can refer to either lateral dimension, y or z. The lateral extension, ε_A, is then

$$\varepsilon_A \equiv \frac{\Delta A}{A} \quad (6.7)$$

and is inversely related to the longitudinal extension by a ratio known as Poisson's ratio, σ, which in an ideal isotropic solid will be 1/3, or

$$\sigma = -\frac{\varepsilon_A}{\varepsilon_t} = \frac{1}{3} . \quad (6.8)$$

The discussion will be facilitated if we use the following index convention to designate the arbitrary principal axes, where the first of the indices indicates the direction of a unit vector normal to a unit cross-section and the second indicates the direction of the surface force and therefore stress vector. Thus for a tension force, the indices are the same, and the complete set of elastic tension stress equations for a unit cube are

$$\sum_{t=0}^{6} \varepsilon_t = \begin{bmatrix} +\tfrac{1}{Y} f_{xx} & -\tfrac{\sigma}{Y} f_{yy} & -\tfrac{\sigma}{Y} f_{zz} \\ -\tfrac{\sigma}{Y} f_{xx} & +\tfrac{1}{Y} f_{yy} & -\tfrac{\sigma}{Y} f_{zz} \\ -\tfrac{\sigma}{Y} f_{xx} & -\tfrac{\sigma}{Y} f_{yy} & +\tfrac{1}{Y} f_{zz} \end{bmatrix} + \begin{bmatrix} +\tfrac{1}{Y} f_{-x-x} & -\tfrac{\sigma}{Y} f_{-y-y} & -\tfrac{\sigma}{Y} f_{-z-z} \\ -\tfrac{\sigma}{Y} f_{-x-x} & +\tfrac{1}{Y} f_{-y-y} & -\tfrac{\sigma}{Y} f_{-z-z} \\ -\tfrac{\sigma}{Y} f_{-x-x} & -\tfrac{\sigma}{Y} f_{-y-y} & +\tfrac{1}{Y} f_{-z-z} \end{bmatrix} \quad (6.9)$$

For a condition of isotropic stress in an isotropic material this becomes

$$\sum_{t=0}^{6} \varepsilon_t = \begin{bmatrix} +\frac{1}{3Y} f_{xx} \\ +\frac{1}{3Y} f_{yy} \\ +\frac{1}{3Y} f_{zz} \end{bmatrix} + \begin{bmatrix} +\frac{1}{3Y} f_{-x-x} \\ +\frac{1}{3Y} f_{-y-y} \\ +\frac{1}{3Y} f_{-z-z} \end{bmatrix} \quad (6.10)$$

The scalar value or magnitude of the total isotropic stress, T, is then

$$\sum_{t=0}^{6} \varepsilon_t = 6 \left| \frac{1}{3Y} f_t \right| = 2 \frac{f_t}{Y} = \frac{T}{Y} . \quad (6.11)$$

As all the stress vectors are directed out from the surface of the unit cube, this results in a net increase in its volume or a volume strain called a dilatation, ε_Δ or

$$\varepsilon_\Delta \equiv \frac{\Delta V}{V} \quad (6.12)$$

The dilatation is inversely related to a hydrostatic or mechanically analogous pressure, p_h, a negative tension, at the surface of the volume by a volume or bulk modulus, B, itself analogous to Young's modulus,

$$p_h = -B\varepsilon_\Delta . \quad (6.13)$$

In fact, in an ideal isotropic material, we find that

$$Y = 3B(1-2\sigma)$$
$$\therefore Y = B \quad (6.14)$$

As a result, the pressure, bulk modulus and dilatation are analogously related to the potential energy density of a material as

$$V_1 = -\tfrac{1}{2} p_h \varepsilon_\Delta = \tfrac{1}{2} B \varepsilon_\Delta^2 = \frac{p_h^2}{2B} \quad (6.15)$$

There are two other related types of stress in an elastic material, shear stress and the related concept of torsion. While tension stress operates normal or perpendicular to the surfaces of the unit cube, shear stress operates parallel to its edges. It should not be too surprising, therefore, if the first tension stress in (6.10) above, f_{xx}, also contributed to the shear stress in the four adjacent cubic faces, in fact, once to each edge of each adjacent face. However, since each edge is shared by two surfaces, in an isotropic condition the net is one fourth of the total shear force in each direction per face. If we think of the normal tension vector, f_{xx}, as equally divided as extension vectors to each of the adjacent surface mutual boundary edges, we have

$$|f_{xx}| = |f_{yx} + f_{zx} + f_{-yx} + f_{-zx}| . \quad (6.16)$$

Shear is not just the force across the appropriate cross sectional area, however. If this same condition applied to all six faces of the cube, there might be a dilatation or increase in volume, but there would be no shearing distortion, which is a relative flattening of the cross sectional surface by an increase in one of the diagonals vis-à-vis the other. The shear, once again a dimensionless value, is measured as the ratio of the displacement of one cubic edge, e, to the length of the adjacent, orthogonal edges, $e_\perp$, in the same plane, in this case the yz plane. Therefore the shear strain, ε_s, is

$$\varepsilon_s \equiv \frac{\Delta e}{e_\perp} = \tan\alpha \qquad (6.17)$$

where α is obviously the arctan of ε_s.

The total shear on a unit cube in the yz plane then is

$$\varepsilon_{yz} = \varepsilon_{zy} = \tfrac{1}{2}\left(\frac{\Delta e_y}{e_{0z}} + \frac{\Delta e_z}{e_{0y}}\right) \qquad (6.18)$$

The shear strain and stress, f_s, have a shear modulus or modulus of rigidity, μ, given as

$$f_s = 2\mu\varepsilon_s \qquad (6.19)$$

and an analogous relationship to Young's modulus as

$$Y = 2\mu(1+\sigma). \qquad (6.20)$$

The corresponding potential energy density in terms of shear along any dimension is

$$V_1 = f_s\varepsilon_s = 2\mu\varepsilon_s^2 = \frac{f_s^2}{2\mu} \qquad (6.21)$$

If the shear stresses along all edges of a given cross-section are not equal, there will be a rotation, ϕ_{yz}, here shown about the yz plane, in the direction of the net angular increase given by

$$\phi_{yz} = \tfrac{1}{2}\left(\frac{\Delta e_z}{e_{0y}} - \frac{\Delta e_y}{e_{0z}}\right) \qquad (6.22)$$

For a unit cube, assuming no rotational shear strain, the total strain at the surface of the cube is the following symmetric matrix, where the diagonals represent the tension strain as in (6.10)

$$E \equiv \begin{bmatrix} \varepsilon_{xx} & \varepsilon_{xy} & \varepsilon_{xz} \\ \varepsilon_{yx} & \varepsilon_{yy} & \varepsilon_{yz} \\ \varepsilon_{zx} & \varepsilon_{zy} & \varepsilon_{zz} \end{bmatrix} + \begin{bmatrix} \varepsilon_{-xx} & \varepsilon_{-xy} & \varepsilon_{-xz} \\ \varepsilon_{-yx} & \varepsilon_{-yy} & \varepsilon_{-yz} \\ \varepsilon_{-zx} & \varepsilon_{-zy} & \varepsilon_{-zz} \end{bmatrix} \qquad (6.23)$$

The following antisymmetric matrix represents the rotational components of the shear as

$$\Phi = \begin{bmatrix} 0 & \phi_{xy} & \phi_{xz} \\ \phi_{yx} & 0 & \phi_{yz} \\ \phi_{zx} & \phi_{zy} & 0 \end{bmatrix} + \begin{bmatrix} 0 & \phi_{-xy} & \phi_{-xz} \\ \phi_{-yx} & 0 & \phi_{-yz} \\ \phi_{-zx} & \phi_{-zy} & 0 \end{bmatrix} \qquad (6.24)$$

Whether a material responds to a shear stress by flattening across its corresponding cross-section or rotating about the cross-sectional axis is a function of the configuration of the shear stress about the axis and the torsional rigidity of the material, its resistance to torque or twisting.

In the event of a shear stress and strain uniformly distributed in one angular direction about the cross-section, we have an instance of torsion or twisting of the elastic medium in response to a torque. This is generally modeled on a rigid tube or rod whose length largely exceeds its cross-section. For a solid rod, the torque, M, is given as the product of the radius of the rod, r, and the circumferential shearing stress, f_ϕ, and the cross sectional area of the rod, $A = 2\pi r^2$ or

$$dM = 2\pi r^2 f_\phi dr \qquad (6.25)$$

The shearing strain is in keeping with (6.17) where $\Delta\varepsilon = r\phi$ is the circumferential displacement or rotation of the cross-section about the longitudinal axis and $\varepsilon_\perp = l$ is the length of that axis and the rod exhibiting a reacting torque to the applied torque. In terms of the SHM of a pendulum, the pendulum length and the length of the torsion rod are analogous as are the displacement of the plumb bob and the torsion shear. In fact, the torsion pendulum is common and is responsible for determining Newton's gravitational constant. Thus we have

$$f_\phi = \mu \varepsilon_\phi$$
$$\text{where } \varepsilon_\phi = \frac{r\phi}{l} \qquad (6.26)$$

Thus the torque for a solid rod is

$$M = \frac{\pi}{2} \mu \frac{r^4}{l} \phi . \qquad (6.27)$$

The elastic potential energy density per unit length of the rod is

$$V_1 = \tfrac{1}{2}\frac{M\phi}{l} = \frac{\pi}{4} \mu \frac{r^4}{l^2} \phi^2 = \frac{M^2}{\pi \mu r^4} \qquad (6.28)$$

In summary, stress potential, quantified as a stress modulus, is the property of a three dimensional elastic material or medium that distributes a change in one or more spatial dimensions, and therefore its potential energy density, to all three dimensions in the form of various stress forces and that distributes a change in stress in one or more spatial dimensions, and therefore its potential energy density, to all three dimensions in the form of various physical strains. It is the property by which such a medium stretches, compresses, bends, shears, and rotates in response to a change in configuration within its extent or at its boundary. The elasticity of such a medium is a measure of the degree to which the original physical configuration is restored with a cessation or diminution of the initial stress and strain. The plasticity of such a medium is a measure of the degree to which the original physical configuration fails to be restored with a cessation or diminution of the initial stress and strain. In general an elastic medium will become plastic beyond a certain limit at which the stress to strain ratios, as given by (6.5), cease to be linear, and the strain or deformation becomes permanent. In addition, physical elastic media generally have the propensity for hysteresis in varying degrees, in which an oscillating medium during its kinetic energy phases exhibits greater kinetic energy as it moves away from its equilibrium condition toward either point of maximum displacement than on the returns back towards the equilibrium position. Thus it results in a loss of energy over time, in contrast to SHM in which the ideal system is defined as being closed.

Stress and strain are perhaps best described mathematically using the stress, **F**, and strain, **E**, tensors, so that the elastic potential energy density of a material is

$$V_1 = \tfrac{1}{2} \mathbf{F} : \mathbf{E} \tag{6.29}$$

Where in an anisotropic condition or for one half of a unit cube the double dot product expands as

$$V_1 = \tfrac{1}{2}\left[f_{xx}\varepsilon_{xx} + f_{yy}\varepsilon_{yy} + f_{zz}\varepsilon_{zz} + 2\left(f_{yz}\varepsilon_{yz} + f_{zx}\varepsilon_{zx} + f_{xy}\varepsilon_{xy}\right) \right] \tag{6.30}$$

Aside #3 (taken from an earlier work-in-progress copy of this development)

In a manner that relates to the cuboctahedral lattice, we can examine the effect of an isotropic strain, along with the corresponding stress, on a unit volume of space. In light of previous comments, we can imagine the center of this cube as one center of expansion and the other as extra dimensional, represented by the indefinite extension of the four diagonals through the eight vertices. We will integrate the differential components of the cube to compare the work done on each boundary component to the change in the corresponding core, in this case a volume. We are interested in the relative contributions of each component as an order of differentiation over time to the initial unit volume, V, and not to the changing magnitude of the volume itself. That is, from (5.18) in Aside #1, we have $6x^2 dx$ differential surfaces, $12x dx^2$ differential edges and $8dx^3$ differential vertices. We substitute the following boundary place-hold identities for Surface, Edge and vertices (Corner), $1^2 S \equiv x^2$, $1^1 E \equiv x^1$, and $1^0 C \equiv x^0$ so as to maintain proper integration. It will be helpful if we assign a "normal" boundary strain vector to each of these components, which in each case will be in the direction in which the boundary is increasing. Thus

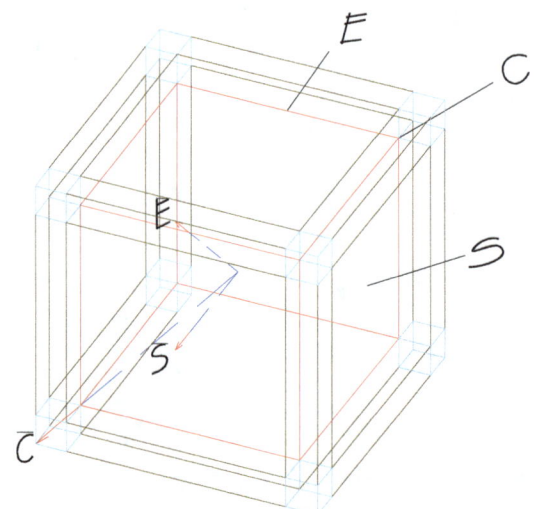

Cubic Expansion

$$|\mathbf{S}| = \left|\sqrt{\tfrac{1}{2}}\,\mathbf{E}\right| = \left|\sqrt{\tfrac{1}{3}}\,\mathbf{C}\right| \qquad (7.1)$$

$$|\mathbf{E}| = \left|\sqrt{2}\,\mathbf{S}\right| = \left|\sqrt{\tfrac{2}{3}}\,\mathbf{C}\right| \qquad (7.2)$$

$$|\mathbf{C}| = \left|\sqrt{3}\,\mathbf{S}\right| = \left|\sqrt{\tfrac{3}{2}}\,\mathbf{E}\right| \qquad (7.3)$$

In the following discussion, no assumption is made about the universal configuration or number of dimensions of the space in which the unit cube is embedded. We are only interested, at least initially, in the local geometry, which is assumed to be flat and therefore Euclidean. Thus it is background-independent. As to a fourth spatial dimension, we will see that change in or motion of such dimension is interchangeable with a dimension of time in a three spatial dimension context.

In this case the integration will be simultaneous on each order, as indicated by the pre-subscript n, in $\int {}_n dx^n$ so that we have

$$\int_V dV = 6x^2 \int_0^a {}_1 dx^1 + 12x^1 \int_0^a {}_2 dx^2 + 8x^0 \int_0^a {}_3 dx^3 \qquad (7.4)$$

$$\int_V dV = 6S \int_0^a dx + 12E \left(\int_0^a dx \right)\left(\int_0^a dx \right) + 8C \left(\int_0^a dx \right)\left(\int_0^a dx \right)\left(\int_0^a dx \right) \qquad (7.5)$$

$$\Delta V = 6aS + 12a^2 E + 8a^3 C \qquad (7.6)$$

Solving for the following ratios, all at unity, where the designations S, E and C are unit names, their dimensional quantities being absorbed in the numerical coefficients of a^n, i.e. 6 square units times a, 12 length units times a^2, etc., gives the value of a for each equivalence. The ratios have been stated with the highest order in the consequent or denominator so they are decreasing from infinity as dx increases, until unity is reached as stated. We have (showing the negative for the sake of symmetry)

$$\frac{S}{E+C} = \frac{6a}{12a^2 + 8a^3} = \frac{1}{2a + \tfrac{4}{3}a^2} = 1 \therefore a = -\tfrac{3}{4} \pm \tfrac{1}{4}\sqrt{21} = 0.39564...,-1.89564... \qquad (7.7)$$

$$\frac{S}{E} = \frac{6a}{12a^2} = \frac{\tfrac{1}{2}}{a} = 1 \therefore a = \tfrac{1}{2} = 0.5 \qquad (7.8)$$

$$\frac{E}{C} = \frac{12a^2}{8a^3} = \frac{\tfrac{2}{3}}{a} = 1 \therefore a = \tfrac{2}{3} = 0.66666... \qquad (7.9)$$

$$\frac{S}{C} = \frac{6a}{8a^3} = \frac{\tfrac{3}{4}}{a^2} = 1 \therefore a = \pm \tfrac{\sqrt{3}}{2} = \pm 0.86602... \qquad (7.10)$$

$$\frac{E}{S+C} = \frac{12a^2}{6a + 8a^3} = 1 \therefore a = \tfrac{3}{4} \pm i\tfrac{1}{4}\sqrt{3} = \tfrac{\sqrt{3}}{2} e^{\pm i\tfrac{\pi}{6}} = 0.86602...e^{\pm i\tfrac{\pi}{6}} \qquad (7.11)$$

$$\frac{S+E}{C} = \frac{6a + 12a^2}{8a^3} = 1 \therefore a = \tfrac{3}{4} \pm \tfrac{1}{4}\sqrt{21} = 1.89564...,-0.39564... \qquad (7.12)$$

If we think of the cube as embedded in an isotropic elastic continuum, which is of some inertial density and under tension, dx represents the work done in displacing or distorting the medium, and by virtue of Gauss' theorem, the integration of that work represents the energy of the distortion. By way of reference, in an ideal elastic medium, the stress operating on the locale is a function of the strain and the elastic modulus as

$$F = \frac{Y\mathbf{E} - 3\sigma \bar{P}\mathbf{1}}{1 + \sigma} \qquad (7.13)$$

where **F** is the stress tensor, **E** is the strain tensor, Y is Young's modulus of elasticity, σ is Poisson's ratio or the negative ratio of lateral to axial or shear to tension strain, $\bar{P}$ is the mean pressure in the medium, and **1** is the idemfactor or unit tensor. Assuming a value of σ of -1/3 for an ideal isotropic 3 dimensional medium we have

$$\mathbf{F} = \frac{3}{2}(Y\mathbf{E} + \bar{P}\mathbf{1}). \tag{7.14}$$

The vector fundamental tension stress component is

$$\mathbf{f} = Y\mathbf{e} \tag{7.15}$$

and is related to the energy distribution by Gauss' theorem for the radial strain

$$E_r = \int_V \nabla \cdot \mathbf{e}_r \, dv = \oint_S \mathbf{e}_r \cdot d\mathbf{S} \tag{7.16}$$

and Stokes' theorem for the angular or tangential strain

$$E_t = \int_S \nabla \times \mathbf{e}_t \cdot d\mathbf{S} = \oint_r \mathbf{e}_t \cdot d\mathbf{r}_t \tag{7.17}$$

These boundary order ratios, then, are inflection points indicating the energy contributions and potential energy gradient changes over time among the boundary components. In an ideal static, kinematic case the change in the ratios with an increase in dx would have no functional effect on the components, if dx has the same magnitude for each of them as it increases. This would amount to a simple change of scale. The real solutions above would appear to reflect this static condition. However, in a dynamic condition, we might imagine that as each ratio decreases below unity and past the inflection point, the magnitude of the consequent exceeds and affects the antecedent or numerator, whose magnitude then becomes a partial function of the consequent. This would appear to be the case for the complex solutions in particular, which correspond with an angular gradient potential of the boundary vectors from that of the antecedent to the direction of that of the consequent.

These evaluations were done with Maple. It is significant that if we convert (7.11) to complex polar notation as in the last term, the modulus is equal to the value for a in (7.10). It is important that we understand that the ratios represent the point at which the change in volume due to the sum totals of all component orders in the antecedent and consequent are equal. It is not the point at which one single component of a given S, E, or C times its appropriate $\int_n dx^n$ is equal to another, since this happens for all at the point where $a = 1$.

In these evaluations, the S component of the strain and hence of the work predominates until (7.7) is reached. At this point, the stress will begin to shift from a predominance of tension to that of shear, meaning there will be a potential for the surface and edge strains to oscillate. As the edges and vertices ring each of the surfaces, the system remains basically stable, however. At the point of (7.8) the edges assume dominance over the surfaces and a gradient is produced for the bulk strain and the tension stress in the direction of the edges. Once again, the 2:1 symmetry of edges to surface maintains stability. At (7.9) the vertices contribute more work than the edges and the strain gradient shifts in their direction. Thus there is a vector potential from the surfaces to the edges to the vertices. Once more the symmetry between vertices and edges maintains stability.

Jumping to (7.12), at this point the strain contributed by the vertices dominates both of the other components combined and the related stress is greatest at these locations. This would result in a dissipation of the energy altogether, were it not for the unusual and unique condition created by (7.10) and (7.11). The point at which the strains of the vertices come to equal those of the surfaces is also the point at which their combined strain comes to equal that of the edges, as given by the modulus of the latter's ratio. We can assume that the imaginary component of this ratio indicates a rotational component of $\pi/6$ or 30°, and since the vertices are assuming a predominance over the surfaces at this point, having already exceeded the edge strain, and as there is an imbalance in the number of vertices to surfaces, a necessary break in symmetry ensues.

We can imagine a rotational potential of the surface strain in the direction of the vertices, which by virtue of the asymmetry between S and C, of 3 degrees of rotational freedom and 4 possible rotational axes, results in an eventual rotational strain about one pair of the axes. This is simultaneous with a shift of the Es in the direction of S + C and a dragging of the strains at each of the two axial C poles. This then leads to a rotation of the axial Cs in the direction of one of the three E pairs extending from those two vertices. The equation of (7.11) gives this rotational relationship. The nature of the ambiguous sense in the argument is indicative of the equation of a rotation and its complex conjugate, when viewed from both senses of its axis, i.e. by rotating it about the real axis, where ± means plus <u>and</u> minus and not plus <u>or</u> minus, if we adjust the Euler identity to

$$e^{\pm i\theta} = \sin\theta \pm i\cos\theta. \tag{7.18}$$

One end of the axis of strain then can be shown as indicated by the "symmetry breaking" in (7.21).

$$12a^2 E = \left(6aS + 8a^3 C\right) \tag{7.19}$$

$$12\left(\tfrac{\sqrt{3}}{2}e^{\pm i\tfrac{\pi}{6}}\right)^2 E = 6\left(\tfrac{\sqrt{3}}{2}e^{\pm i\tfrac{\pi}{6}}\right)S + 8\left(\tfrac{\sqrt{3}}{2}e^{\pm i\tfrac{\pi}{6}}\right)^3 C \tag{7.20}$$

$$e^{-i\tfrac{\pi}{3}}E = \tfrac{1}{\sqrt{3}}\left(e^{+i\tfrac{\pi}{6}}S + e^{+i\tfrac{\pi}{2}}C\right) \tag{7.21}$$

Thus, the strain vector E, rotated in some direction $\tfrac{\pi}{3}$, is equal to $\tfrac{1}{\sqrt{3}}$ of the S and C strains rotated $\tfrac{2\pi}{3}$ in the opposite direction, presumably in the same plane. In fact, this states that C rotates $\tfrac{\pi}{2}$ while S rotates $\tfrac{\pi}{6}$. We can see specifically how these rotations occur in Spin Diagrams 1 and 2. We can also see there how a rotation back in time of $\tfrac{\pi}{3}$ equals one forward in time by $\tfrac{2\pi}{3}$ and vice-versa, if their plane of rotation, ϕ, is itself rotating at a constant rate with respect to an orthogonal plane, θ, that is where the two axes intersect at the centers of rotation. However, it is shown there that this corresponds with a rotation of θ, back $\frac{\pi}{4}$ and forward $\frac{3\pi}{4}$, indicating a variability in the strain velocity.

It should be understood that this cubic structure is simply an expression of the orthogonal tendency for stress equalization and energy conservation. The condition found at (7.10) and (7.11), then becomes a stable dynamic condition of rotational oscillation or spin, within certain parameters of inertial density and mechanical impedance. If the isotropic tension in this situation was sufficient to increase the strain indefinitely, if the medium was to lose its elasticity and become plastic or even rupture, any tendency to oscillate would be overcome by the transfer of energy via strain to the vertices. Local energy would not be conserved, but be drawn away by the strain.

It is essential to extrapolate this scenario to a hypercube, H, to achieve a full understanding. We will skip the integrals but show the results for the corollary of (7.6) as

$$\Delta H = 8aV + 24a^2 S + 32a^3 E + 16a^4 C$$
$$= 1aV + 3a^2 S + 4a^3 E + 2a^4 C \tag{7.22}$$

There are 25 combinations with corresponding non-ordered permutations or sub-combinations, for the 4-cube; 7 involving all 4 parameters, 12 permutations involving all sub-combinations of 3, and 6 one to one relationships. With the 3-cube, there are 2 single real positive solutions at (7.8) and (7.9), one instance of a complex solution at (7.11), one correspondence between a real and a complex solution at (7.10) and (7.11) where the real value of a in one is equal to the complex modulus in the other, and one instance of a correspondence of solutions with sense inversion, (7.7) and (7.12), that is their solutions have the same magnitude, but of opposite sense. As might be expected, the 4-cube shows significantly more of these symmetries. It should be noted that while an attempt has been made to analyze the ratios qualitatively so that all are represented as decreasing with respect to an increasing dx, they have not all been checked quantitatively, and some may be increasing as shown. In fact, (7.35) and (7.37) are found to be increasing at the point represented by the first positive solution and decreasing at the second. For (7.32) it is worth stating that for every value of the ratio $0.75 < \left(\frac{S}{V+E}\right) < +\infty$, the modulus is ½ and the argument ranges from 0 to ½ π.

It is important to remember that a given component in the 3-cube is identical to the same component in the 4-cube, but the relationships between them are different. An edge still is bounded by 2 vertices, but there are 4 edges intersecting at each vertex of the 4-cube. A line segment in an *x-y* plane is qualitatively no different than one in the *z-x* or for that matter *z-w* plane. In fact a point in 3-space also has a location in n-space, at least in Euclidean n-space. In the following, it is also important to remember that *a* is not the value of the corresponding ratio, but rather the value found in both antecedent and consequent when the ratio equals 1. The evaluations are based on the following identities in (7.23),

$$V \equiv 1a, S \equiv 3a^2, E \equiv 4a^3, C = 2a^4 \tag{7.23}$$

$$\frac{V}{S}, a = \tfrac{1}{3} \tag{7.24}$$

$$\frac{V}{E}, a = \pm \tfrac{1}{2} \tag{7.25}$$

$$\frac{V}{C}, a = \tfrac{1}{\sqrt[3]{2}}, -\tfrac{1}{2}\left(\tfrac{1}{\sqrt[3]{2}}\right) \ldots \pm i \tfrac{\sqrt{3}}{2}\left(\tfrac{1}{\sqrt[3]{2}}\right) = \tfrac{1}{\sqrt[3]{2}} e^{\pm i \tfrac{2\pi}{3}} = 0.79370\ldots e^{\pm i \tfrac{2\pi}{3}} \tag{7.26}$$

$$\frac{S}{E}, a = 0, \tfrac{3}{4} \tag{7.27}$$

$$\frac{S}{C}, a = 0, \pm \sqrt{\tfrac{3}{2}} \tag{7.28}$$

$$\frac{E}{C}, a = 0, 0, 2 \tag{7.29}$$

$$\frac{V}{S+E}, a = -1, \tfrac{1}{4} \tag{7.30}$$

$$\frac{V+S}{E}, a = -\tfrac{1}{4}, 1 \tag{7.31}$$

$$\frac{S}{V+E}, a = \tfrac{3}{8} \pm i \tfrac{1}{8}\sqrt{7} = \tfrac{1}{2} e^{\pm i 0.722734248\ldots} \tag{7.32}$$

$$\frac{V}{S+C}, a = 0.31290\ldots, -0.15645\ldots + i1.25436\ldots = 1.26408\ldots e^{\pm i 1.694883228\ldots} \tag{7.33}$$

$$\frac{V+S}{C}, a = -1, -0.36602\ldots, 1.36602\ldots \tag{7.34}$$

$$\frac{V+C}{S}, a = -1.36602\ldots, 0.36602\ldots, 1 \tag{7.35}$$

$$\frac{V}{E+C}, -1.85463\ldots, -0.59696\ldots, 0.45160\ldots \tag{7.36}$$

$$\frac{V+C}{E}, a = -0.45160\ldots, 0.59696\ldots, 1.85463\ldots \tag{7.37}$$

$$\frac{V+E}{C}, a = 2.1120\ldots, -0.05604\ldots \pm i 0.48331\ldots = 0.48655\ldots e^{\pm i 1.686235431\ldots} \tag{7.38}$$

$$\frac{S}{E+C}, a = -2.58113\ldots, 0, 0.58113\ldots \tag{7.39}$$

$$\frac{S+E}{C}, a = -0.58113\ldots, 0, 2.58113\ldots \tag{7.40}$$

$$\frac{E}{S+C}, a = 0, 1 \pm i \tfrac{1}{\sqrt{2}} = \sqrt{\tfrac{3}{2}} e^{\pm i 0.615479709\ldots} \tag{7.41}$$

$$\frac{V}{S+E+C}, a = 0.24415\ldots, -1.12207\ldots \pm i0.88817\ldots = 1.43105\ldots e^{\pm i 2.472026458\ldots} \quad (7.42)$$

$$\frac{E}{V+S+C}, a = -0.24415\ldots, 1.12207\ldots \pm i0.88817\ldots = 1.43105\ldots e^{\pm i 0.669566197\ldots} \quad (7.43)$$

$$\frac{V+E+C}{S}, a = -2.63993\ldots, 0.31996\ldots \pm i0.29498\ldots = 0.43519\ldots e^{\pm i 0.744798022\ldots} \quad (7.44)$$

$$\frac{V+S+E}{C}, a = 2.63993\ldots, -0.31996\ldots \pm i0.29498\ldots = 0.43519\ldots e^{\pm i 2.396794631\ldots} \quad (7.45)$$

$$\frac{V+S}{E+C}, a = -2.51702\ldots, -0.25673\ldots, 0.77375\ldots \quad (7.46)$$

$$\frac{E+S}{V+C}, a = -0.77375\ldots, 0.25673\ldots, 2.51702\ldots \quad (7.47)$$

$$\frac{V+E}{S+C}, a = 1, \tfrac{1}{2} \pm i\tfrac{1}{2} = \tfrac{1}{\sqrt{2}} e^{\pm i \tfrac{\pi}{4}} \quad (7.48)$$

Once again using Maple, there are a total of 10 couplings involving complex solutions, of which one is exclusively complex and one other has only a zero for the third and real solution. Only one single real positive solution is given. There are, however, 7 corresponding pairs of solutions involving sense inversion, 5 real and 2 complex. Note that all cases of sense inversion involve a combination of one or more components in either the antecedent and/or consequent and the sense change is associated with a transposition of one or two components in each pair. These do not appear to have any special relationship to the conditions of the 3-cube, at first glance, and we have not investigated them further.

There are several, however, that appear to have a direct relationship to some of the ratios of the 3-cube. Two conditions of correspondence are found between a real positive solution and the complex modulus of a complex solution with a positive real component. (7.28) $\left(\frac{S}{C}\right)$ and (7.41) $\left(\frac{E}{S+C}\right)$ are directly related to (7.10) and (7.11) respectively, the real solution and the modulus of the complex of the second two being equal to the product of the first and $\sqrt{2}^{-1}$. The argument of (7.41) is the angle at the center of a cube between a radial normal to an edge of the cube and one extended along a diagonal to a vertex. (7.25) $\left(\frac{V}{E}\right)$ and (7.32) $\left(\frac{V}{E+S}\right)$ are related to (7.8) $\left(\frac{S}{E}\right)_3$ with a common value for their real solutions and the modulus of the complex one. The cosine of the argument of (7.32) is equal to the solution of (7.27) $\left(\frac{S}{E}\right)_4$, which is the same ratio coupling as (7.8). This pairing (7.32) in turn has a modulus equal to the real and imaginary components of an additional complex solution in (7.48) $\left(\frac{V+E}{S+C}\right)$. This latter solution has an argument of $\pi/4$ or 45° which appears to be an extremely stable condition, as found in a sine wave model as the point of maximum power of the wave, where the product of the transverse wave force and transverse wave speed are maximum. It is also the angle of the strain vector E discussed above for the 3-cube, with respect to the plane normal to the spin angular momentum vector as shown in the spin diagrams. In the model developed here, this condition is found to be invariant and rotates about the oscillation's angular momentum vector.

Finally, (7.41) $\left(\frac{E}{S+C}\right)$, (7.48) $\left(\frac{V+E}{S+C}\right)$, and (7.26) $\left(\frac{V}{C}\right)$ are found to be related in a most profound way in the mechanism of the oscillation herein described. The imaginary component of (7.41) equals the modulus of (7.48). Note that (7.26) represents a $\frac{2\pi}{3}$ rotation due to the interplay between the volume and vertex components of strain and a modulus of that strain of $\frac{1}{\sqrt[3]{2}}$. Using the equation for (7.26) or

$$aV = 2a^4 C \tag{7.49}$$

$$\tfrac{1}{\sqrt[3]{2}} e^{\pm i \frac{2\pi}{3}} V = 2 \left(\tfrac{1}{\sqrt[3]{2}} e^{\pm i \frac{2\pi}{3}} \right)^4 C \tag{7.50}$$

tells us that a rotational oscillation of the 4-volume (boundary) strain V of modulus $\frac{1}{\sqrt[3]{2}}$ by $\frac{2\pi}{3}$ is equal to 4 axial rotations about the vertices of the same modulus and argument, where the 2 in the consequent indicates simultaneous rotations of opposite sense at each end of an axis. The oscillation of V is fourth dimensional, and therefore beyond our direct sensory ken, however, the 4 vertices are not, and we can envision the above consequent, the expression in 3 dimension of this four dimensional rotation, as a sequence of 4, $\frac{2\pi}{3}$ rotations about the 4 diagonals of a 3-cube. This sequence leaves the cube unchanged and avoids the entanglement condition, i.e. the continuity of Euclidean 3-coordinates of the cube are not twisted by the sequence. This condition of limits on the twistability of the continuum strain is a necessary consequence of its inertial/elastic properties. As the rotation of V is continuous, we would imagine that the sequence of 4 rotations is continuous, i.e. the strain rotates from one reference diagonal to another about one of the three surface axes of the 3-cube. We can also envision this as one diagonal axis rotating $\frac{2\pi}{3}$, followed by a 2π rotation of the same sense about one of the adjacent 3-cube surface axes. We can also treat it as a sequence of 4 orthogonal permutations.

We can show this configuration simply. If we align a hyperbolic surface of revolution about the y axis of the curve

$$xy = \tfrac{1}{2}, \text{ for } x \leq \tfrac{1}{\sqrt{2}} \tag{7.51}$$

at each of the eight vertices of a cube so that each of them is at the angle of the argument given by (7.41) as just described, and so that the rims or circles of their bases intersect at the centers of each of the six surfaces of the cube, the following will be found concerning this geometry, which we will call an inversphere. We can also, as an alternative, create a similar construct using a pseudosphere in place of the above surface of revolution. Given a constant negative curvature of -1 for each pseudosphere, the resulting inversphere would have a constant negative curvature. With respect to the inversphere:

1. Each surface of revolution, which we might call a hyper-axis or h-axis and which can be represented by a complex plane, with the imaginary dimension parallel to the circumference of the revolution and the real along the diagonal axis, will have a curvature of negative 1 at the rim, remaining negative while decreasing, that is, moving toward zero or flatness, with distance along the asymptote. Here the left four of Figure 4 are shown, their designations corresponding with the axes of Figure 3 below.

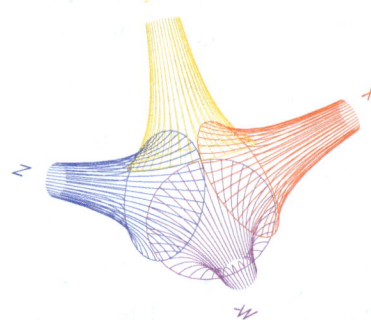

Figure 1

2. The rims will have a radius of $\frac{1}{\sqrt{2}}$. The area of the circle formed by the rim is therefore $\frac{\pi}{2}$, and its complex representation is $\frac{1}{\sqrt{2}}e^{i\theta}$ corresponding with $\left(\frac{V+E}{S+C}\right)$.
3. The rims will intersect orthogonally with each other at the cubic surface centers, so that there are three h-axes adjacent to a given h-axis along the cubic edges which we will refer to as the proximal axes.
4. The rims from h-axes located diagonally across the cubic <u>surface</u> from each other will be parallel or tangential at the same point at which they intersect with their proximal axes. We will call the corresponding parallel axes the distal axes. One set of mutually distal axes can be called the positive h-axes.
5. Each h-axis has a spatial inversion or anti-axis which is proximal to the distal axes of that h-axis. The set of their spatial inversions can be called the negative h-axes.
6. Each rim intersection is a $\frac{2\pi}{3}$ rotation from the others about the cubic diagonal, associating it with $\left(\frac{V}{C}\right)$.
7. The distance between cube surface centers describes an octahedron of edge length $\sqrt{\frac{3}{2}}$. The surface area of the octahedron is therefore $3\sqrt{3}$ and the volume is $\frac{\sqrt{3}}{2}$. The radial normal to the octahedron face is ½.
8. The cube will have an edge measure of $\sqrt{3}$. The surface area of the cube is 18 and the volume is $3\sqrt{3}$.

9. The concentric sphere intersecting at the rim intersections will have a radius of $\frac{\sqrt{3}}{2}$. The surface area is 3π and the volume is $\frac{\sqrt{3}}{2}\pi$.
10. We can think of this arrangement as the expression of a 4-cube in a 3-space, where the orthogonality condition of the 4-D space is met by the rim intersections, the center of each component of sphere, cube, octahedron and h-axis intersections being a common system center.
11. This configuration can be reduced to a 3-space orthogonal system simply by collapsing the cube along the W hyper-axis, as in the figure at left below, resulting in the co-ordinate system at right.

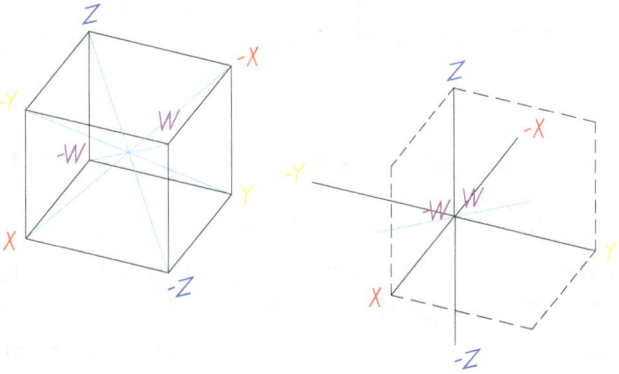

Figure 2

The condition of (7.8) $\left(\frac{S}{E}\right)_3$, (7.25) $\left(\frac{V}{E}\right)$ and (7.32) $\left(\frac{S}{V+E}\right)$ is represented by (7.48) $\left(\frac{V+E}{S+C}\right)$ at each h-axis. Thus the orthogonal projections of the argument of (7.48), described as extending from the system center to each cubic edge midpoint, are equal to the modulus of (7.41) $\left(\frac{E}{S+C}\right)_4$, and the argument of (7.26) $\left(\frac{V}{C}\right)$ is the rotation of that axis between proximal intersections and cubic surface centers.

In terms of $a = \int dx$ we are only interested in positive or increasing real values, although in the context of complex values, some negative real components as in (7.26) are of interest. A deeper analysis would no doubt find significance in all of the couplings, but we are only interested in the general manner in which the 4-cube and the 3-cube couplings might interact. In this regards it is important to remember that in the case of the 4-cube, the volume is a boundary that is increasing while in the case of the 3-cube, it is the base space, held constant, upon which the boundary changes are taking place.

From the perspective of a rotational oscillation, as found in a torsion pendulum or a jump rope oscillation, of interest are those couplings of two boundary parameters, V + E and S+ C, which have an intervening parameter, S and E respectively. More interestingly, in both these cases, V + E for the 4-cube and S + C for both 4-cube and 3-cube, the two-parameter components also have a ratio between themselves whose solution is ($\pm$) real and equal to the modulus of the companion ratio. (7.48) gives the special case of V + E with S + C. Unlike the other three rotational oscillator couplings, it has a positive real solution in addition to its complex solution. It also has the two parameter component ratios in common with the other two oscillators of the 4-cube. The remaining couplings with complex solutions all have intervals between their real and complex moduli solutions, for most exceeding 1, which mitigates against oscillation, with one exception. (7.26) $\left(\frac{V}{C}\right)$ has a real solution that equals its modulus, thereby indicating rotational oscillation. In addition, the cosine of its argument is equal to the modulus and real solution for $\left(\frac{S}{V+E}\right)$ and $\left(\frac{V}{E}\right)$ at $\frac{1}{2}$ and its sine, to the 3-cube modulus and real solution for $\left(\frac{E}{S+C}\right)$ and $\left(\frac{S}{C}\right)$ at $\frac{\sqrt{3}}{2}$, and to the 4-cube modulus and real for $\left(\frac{E}{S+C}\right)$ and $\left(\frac{S}{C}\right)$ at $\frac{1}{\sqrt{2}}\frac{\sqrt{3}}{\sqrt{2}} = \frac{\sqrt{3}}{2}$. Thus the rotational parameters of the other rotational oscillation or spin couplings, can be found in the simple ratio of $\left(\frac{V}{C}\right)$.

Within the context of the 4-cube, the first value that arises is (7.42) $\left(\frac{V}{S+E+C}\right)$ followed closely by (7.30) $\left(\frac{V}{S+E}\right)$. This simply shows that the vertex component adds very little at this juncture, although it does have a rotational element, but the negative real component indicates a significant rotation which would seem out of synch with the small real strain. A similar comment could be made about (7.33) $\left(\frac{V}{S+C}\right)$ which is next in the real order, though the potential rotation is much less. This is followed by (7.24) $\left(\frac{V}{S}\right)$ which has no rotational component. It is significant in that it is the value of Poisson's ratio in an ideal isotropic elastic solid, relating the axial to lateral strain and thereby, tension to shear stress.

Next is (7.35) $\left(\frac{V+C}{S}\right)$ with no rotational component, followed by (7.44) $\left(\frac{V+E+C}{S}\right)$, which has a rotational component. The real solution and therefore the strain is negative, however, and is out of scale with the modulus of the complex solution, which would mitigate against rotational oscillation. This modulus and the positive solution of (7.36) $\left(\frac{V}{E+C}\right)$ are the first values to exceed any of the solutions for the 3-cube. The next ratios (7.25) $\left(\frac{V}{E}\right)$ and (7.32) $\left(\frac{S}{V+E}\right)$ involve the first of the oscillatory groups. The real solution of the first and modulus of the second are equal to each other and to that of (7.8) $\left(\frac{S}{E}\right)_3$, while the cosine of the argument of $\left(\frac{S}{V+E}\right)$ coincides with the real solution of (7.27) $\left(\frac{S}{E}\right)_4$. Thus we might associate an actual oscillation of the 4-cube with the potential $\left(\frac{S}{E}\right)_3$ of the 3-cube. This is followed by (7.39) $\left(\frac{S}{E+C}\right)$ which has a real solution and is the 4-cube corollary of the first ratio of the 3-cube. It is of no special interest other than being, along with (7.9) $\left(\frac{E}{C}\right)_3$ a precursor for the next coupling, which is (7.48) $\frac{V+E}{S+C}$, perhaps the most important of the whole assemblage. Together, $\left(\frac{S}{E+C}\right)$ and $\left(\frac{E}{C}\right)_3$ indicate a growing predominance of E and C over S and then C over E, or shear stress over tension, followed eventually by torsion over shear.

The argument of $\left(\frac{V+E}{S+C}\right)$ represents the power of the strain oscillation, first in the oscillatory twisting of the hyper-axes at $\left(\frac{V}{C}\right)$, then subsequently with the rotational oscillation of the 3-cube itself. Given the above description of the inversphere, the modulus of this solution represents the radius of and in the plane of the rim of the h-axis at the point at which its curvature is -1. The argument is the power phase of an oscillation which can be found as a phase constant in the eventual rotational oscillation of the 3-cube. This is followed by (7.46) $\left(\frac{V+S}{E+C}\right)$, which adds no new oscillatory components, but does show the gaining dominance of the higher order boundary components, E and C. This culminates in a new oscillatory condition at (7.26) $\left(\frac{V}{C}\right)$.

Note that the real value and the modulus of $\left(\frac{V}{C}\right)$ is slightly more than the values of $\left(\frac{V+E}{S+C}\right), \left(\frac{V}{E}\right), \left(\frac{S}{E}\right)_4$ and slightly less than the values of $\left(\frac{E}{S+C}, \frac{S}{C}\right)_3$ at oscillation. We can interpret the condition at $\left(\frac{V}{C}\right)$ as an oscillation about each of the 8 vertices. Each oscillation involves a twisting or torsion ultimately of $\frac{2\pi}{3}$ in each direction about each h-axis. The proximal axes will twist counter to the instant rotation sense of a given h-axis as will the anti-axis, all as viewed from the exterior of the system. The distal axes will twist with the same sense as the given h-axis, thus the directional sense of these axes corresponds with their rotational sense vis-à-vis the other axes. The strain on the enclosed sphere at maximum twist will be of a simultaneous lengthening along each <u>cubic</u> axis and flattening in the plane of said axis and the <u>cubic</u> axis from which the strain occurred and at which it is at a minimum, ideally zero, as indicated in the figure below. The two pairs of distal axes on each surface create two countervailing torques, which in this oscillatory condition are in equilibrium.

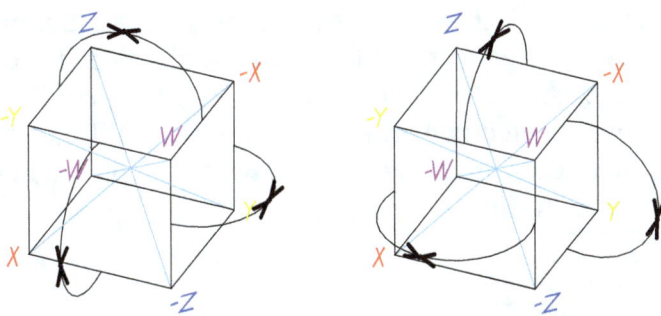

Figure 3

This initial symmetrical condition of $\frac{2\pi}{3}$ rotational oscillation of each of the four diagonals is broken upon *dx* reaching the oscillatory threshold given by $\left(\frac{E}{S+C}, \frac{S}{C}\right)_3$ at $\frac{\sqrt{3}}{2}$. This results in a permanent rotation of $\frac{\pi}{3}$ of one pair of the E vectors as indicated by (7.21), thence the whole system <u>strain</u> continues to <u>oscillate</u>, while the <u>stresses rotate</u> and generate an angular momentum vector. (7.26) indicates the rotation of the stresses in time among and about the four diagonals, which represent the four orthogonal axes of H. The oscillation of the 3-cube is supported and driven by the 4-stress which is concentrated in one transforming axis. (7.48) $\left(\frac{V+E}{S+C}\right)$ represents the power moments or positions of maximum conversion of kinetic to potential energy and vice versa.

Finally, (7.41) $\left(\frac{E}{S+C}\right)_4$ represents, in addition to the diagonals, a capacitive and an inductive torque that is co-linear with two of the diagonals and is the product of crossing into the power moments from their positions of equilibrium strain and rotates with them about the angular momentum vector, all described later. The modulus and the solution to $\left(\frac{S}{C}\right)_4$ at $\sqrt{\frac{3}{2}}$ represents the radial length from the center of the inversphere and 3-cube to the midpoint of the cubic edge. The solution $a = \sqrt{\frac{3}{2}} e^{\pm i 0.61547\ldots}$ in this case indicates a rotation of this vector into the diagonal or one of the h-axis or of **E** into **C**. Solving for $\left(\frac{E}{S+C}\right)_4$

$$4a^3 E = 3a^2 S + 2a^4 C$$

$$4\left(\sqrt{\tfrac{3}{2}} e^{\pm i 0.61547\ldots}\right)^3 E = 3\left(\sqrt{\tfrac{3}{2}} e^{\pm i 0.61547\ldots}\right)^2 S + 2\left(\sqrt{\tfrac{3}{2}} e^{\pm i 0.61547\ldots}\right)^4 C \qquad (7.52)$$

after reduction and some parsing gives

$$2\left(\sqrt{\tfrac{3}{2}} e^{-i 0.61547\ldots}\right) E = \left(\sqrt{\tfrac{3}{2}} e^{i \frac{\pi}{2}}\right)^2 S + \tfrac{3}{2} e^{+i 1.23095\ldots} C. \qquad (7.53)$$

Here as with the companion relationship for the 3-cube, we have "broken symmetry" with the rotational senses, and see that rotation of two edge strains into an adjacent corner is equal to two orthogonal rotations of a surface strain and a flip of a vertex strain from one h-axis to a proximal axis. The moduli in this case correspond to the metrics of the inversphere, where $\frac{3}{2}$ is the distance from the cubic center to the cubic vertex. We will see that this represents an instance of beta decay, where the surface and vertex rotations indicate the flip of the electrical phase torques from one pair of vertices to one of three proximal pairs. In the case of the inductive torque, we have an electron emission along with a flip of the magnetic moment, and in the case of the capacitive torque, we find a positron emission, without the magnetic moment flip.

Equal Area Cube and Sphere

The position indicated as the midpoint on the cubic edge is of special interest. If we analyze a concentric cube and a sphere of equal surface area, and presumably of equal total surface stress, we will find that the radial to the midpoint as represented by $\left(\frac{S}{C}\right)_4$ exceeds the spherical radius (and the path of the rotational oscillation strain) at that point by a factor of

$$\delta r = \frac{\sqrt{2\pi}}{3}\left(\sqrt{\frac{3}{2}} - \frac{3}{\sqrt{2\pi}}\right) = \sqrt{\frac{\pi}{3}} - 1 = 0.023326708\ldots. \qquad (7.54)$$

This indicates that the rotational path of the strain constricts the diagonals and restricts the operation given by $\left(\frac{E}{S+C}\right)_4$. Thus this differential must be overcome by the increase in stress of that operation. If we assume that the differential given by (7.54) is one component of the cross sectional area on which an orthogonal stress is operating, then the square of that value gives a differential stress required for the diagonal to flip of

$$\delta r^2 = 0.0005441353061\ldots. \qquad (7.55)$$

The ratio of differential stress to the augmented total is then

$$\frac{\delta r^2}{1+\delta r^2} = \frac{0.0005441353061}{1.0005441353061} = 0.0005438393841\ldots \qquad (7.56)$$

which when inverted is

$$\frac{1+\delta r^2}{\delta r^2} = 1838.778193 \qquad (7.57)$$

It bears noting that the 2002 CODATA ratio of the electron to neutron mass is 0.00054386734481(38), or within $2.796\ldots\times 10^{-8}$ of the value of (7.56). Thus the ratio of the differential stress needed to produce beta decay and the stress of fundamental oscillation correlates significantly with the ratio of the mass-energy of the product of that decay, the electron, and that of the fundamental oscillation, the neutron.

With reference to beta decay, one additional observation concerns the weak mixing angle yielded by the final measured asymmetry stated in the September, 2005 issue of Physics Today by Bertram Schwarzschild in "Tiny Mirror Asymmetry in Electron

Scattering Confirms the Inconstancy of the Weak Coupling Constant" as $\sin^2 \theta_W = 0.2397 \pm 0.0013$. If we consider the surface area of a sphere in steradians as 4π, the portion spanned by each cubic edge in conjuction with the above development is one twelfth that or an area of $\pi/3$. A linear component of that measure would therefore be $\sqrt{\frac{\pi}{3}}$ and would correspond generally and perhaps in some statistical manner with the distance from a cubic surface vector to a vertex vector as in the interplay between S and C at $\left(\frac{E}{S+C}, \frac{S}{C}\right)_{3,4}$. The arc distance between the mid-point of that arc and each of the three parameters E, S, and C is then $\frac{1}{2}\sqrt{\frac{\pi}{3}}$. We then have the following, which is stated phenomenologically and without causal analysis

$$\sin^2\left(\tfrac{1}{2}\sqrt{\tfrac{\pi}{3}}\right) = 0.239735827. \qquad (7.58)$$

Bibliography and Other Resources

Astronomy and Astrophysics 338, 856-862 (1998), "Magnetically supported tori in active galactic nuclei", Lovelace, Romanova, and Biermann.

Exploring Black Holes, Taylor and Wheeler, Addison Wesley Longman, Inc, New York, 2000.

The Extravagant Universe, Kirshner, Princeton University Press, Princeton, NJ, 2002.

The Feynman Lectures on Physics :Commemorative Issue, Feynman, Leighton, Sands, Volume I, Addison-Wesley Publishing Company, Inc., Reading, Massachusetts 1963.

Fundamentals of Physics, Fifth Edition, Halliday, Resnick, Walker, John Wiley & Sons, Inc. New York, 1997.

Gravitation, Misner, Thorne, and Wheeler, W.H. Freeman and Company, New York, 1973.

Mathematical Methods for Physicists, Fifth Edition, Arfken and Weber, Harcourt Academic Press, New York, 2001.

Physics of Waves, Elmore and Heald, Dover Publications, Inc., New York, 1985. This was the primary source for wave, elasticity and tensor equations.

The Six Core Theories of Modern Physics, Stevens, The MIT Press, Cambridge, Massachusetts, 1995.

Three Roads to Quantum Gravity, Smolin, Basic Books, New York, 2001.

Visual Complex Analysis, Needham, Oxford University Press, Oxford, England, 1997

The Theoretical Minimum: What You Need to Know to Start Doing Physics, Susskind and Hrabovsky, Basic Books, New York, 2013

National Institute of Standards and Technology, These are the **2002 CODATA recommended values** of the fundamental physical constants, the latest CODATA values available. For additional information, including the bibliographic citation of the source article for the 1998 CODATA values, see P. J. Mohr and B. N. Taylor, "The 2002 CODATA Recommended Values of the Fundamental Physical Constants, Web Version 4.0," available at physics.nist.gov/constants. This database was developed by J. Baker, M. Douma, and S. Kotochigova. (National Institute of Standards and Technology, Gaithersburg, MD 20899, 9 December 2003).

Table of Nuclides, Nuclear Data Evaluation Lab., Korea Atomic Energy Research Institute (c) 2000-2002, http://yoyo.cc.monash.edu.au/~simcam/ton/index.html

R.R.Kinsey, et al., *The NUDAT/PCNUDAT Program for Nuclear Data*, paper submitted to the 9 th International Symposium of Capture-Gamma ray Spectroscopy and Related Topics, Budapest, Hungary, Octover 1996. Data extracted from NUDAT database (Jan. 14/1999)

Schwarzschild, Bertram, "Tiny Mirror Asymmetry in Electron Scattering Confirms the Inconstancy of the Weak Coupling Constant", Physics Today, September, 2005

Wapstra, A. H. and Bos, K., "The 1983 atomic-mass evaluation. I. Atomic mass table," Nucl. Phys. A 432, 1-54, 1985, quoted at http://hyperphysics.phy-astr.gsu.edu/hbase/nucene/nucbin2.html

Rest Mass Quantization as a Function of Spacetime Exponential Expansion Stress

—:—

Compactification of Time,
Geometrization of Quantum Mass and Gravity, and
the Fundamental Quantum Metric

September 19, 2012

Martin Gibson
P.O. Box 2358
Southern Pines, NC 28388
910-585-1234
martin@uniservent.org

Copyright © Martin Gibson 2021 All Rights Reserved

Rest Mass Quantization as a Function of Spacetime Exponential Expansion Stress

Compactification of Time, Geometrization of Quantum Mass and Gravity, and the

Fundamental Quantum Metric

H. Martin Gibson

P.O. Box 2358, Southern Pines, NC, 28388

martin@uniservent.com

Abstract

This analysis provides a geometric model, capable of being visualized in three spatial dimensions, of rest mass quantization as an emergent property of a classical spacetime continuum by way of a fundamental, locally discrete rotational oscillation that is a function of the exponential expansion stress of that spacetime. A non-Minkowski spacetime is developed in which time is modeled as a local, compactified dimension exhibiting Lorentz covariance and in which fundamental quantum rest mass, m_0, and spin energy, E_0, is a measure of the angular wave number, κ_0, and angular frequency, ω_0, of a resonant oscillation. Quantum gravity, dG_0, arises naturally as the quantum differential of the transverse wave force of this oscillation with respect to a change in spacetime expansion stress, dT_0. The Planck area, dA_0, is shown to be the differential of a fundamental unit area with respect to that change in expansion stress. The strong interaction is the operation of that wave force between two or more quanta within a shared, local wave force domain. This quantum state is expressed as a modification of a chargeless extreme Kerr metric with an oscillation at resonant frequency of the ϕ coordinates imposed by continuity conditions which prevent coordinate entanglement. Such oscillation results in a rotation of the wave phase at the same frequency. It thereby describes a physical spinor, constituting the quantum magnetic field and the property of ½ spin, and isospin in the presence of other quanta. The ergosphere of this quantum metric is the wave force domain of the strong interaction. From a universal bookkeeper reference frame, the fundamental quantum scale is the neutron scale given by the neutron reduced Compton wavelength. Finally, the analysis indicates that cosmic expansion is accelerating exponentially from a condition of maximum density, is presumably cyclical and that in terms of the current time scale, it is approximately 285 billion years into the current expansion cycle. General relativity requires the following refinement in this model; spacetime acquires the property of inertial density as a potential energy density independent of any energy or rest mass quanta, has an exponential expansion rate, and admits torsion that prevents the orientation entanglement condition.

1 – Kinematics and the Geometrization of Time

"Mechanics . . . is generally regarded as consisting of kinematics and dynamics. Kinematics . . . is the science that deals with the motions of bodies or particles without any regard to the causes of these motions. Studying the positions of bodies as a function of time, kinematics can be conceived as a space-time geometry of motions, the fundamental notions of which are the concepts of length and time. By contrast, dynamics, . . . is the science that studies the motions of bodies as the result of causative interactions. As it is the task of dynamics to explain the motions described by kinematics, dynamics requires concepts additional to those used in kinematics, for "to explain" goes beyond "to describe"." [1]

To take up the task set forth by Max Jammer, we might look for explanation of dynamics in a greater understanding of those "concepts additional", chief of which is mass; in particular we might seek "to explain" mass through a more detailed description of the kinematic concepts of length and time. We would seek to find a definition of mass as a measure of length and/or time. In order to properly undertake such an investigation, we must first examine the concepts of length and time.

Length is a concept used to quantify the apparent spatial separation of entities, where entity might be any distinction within the field of observation, including the two ends of a rod. It is of interest that the magnitude of time is also referred to as a length. We easily conflate measures of separation in time and in space with one term, length, and to contrast them as a ratio, speed. However, there is no more than a conventional preference for the ordering of that relationship, as a mile in four minutes and a four minute mile despite a numerical difference indicate the same physical change, the race speed or

$$c_{race} = \frac{1 \text{ mile}}{4 \text{ minutes}} = \frac{4 \text{ minutes}}{1 \text{ mile}} = \frac{1 \text{ space or time interval}}{1 \text{ time or space interval}}. \quad (0.1)$$

In a similar fashion, we can state a number of times per time or of lengths per length, i.e. a frequency in time or space, as

$$f_t = \frac{4 \text{ flashes}}{1 \text{ second}}, \text{ or } f_l = \frac{3 \text{ feet}}{1 \text{ yard}}. \tag{0.2}$$

A length of spatial or temporal separation can be termed an interval between entities or events, as in general relativity. A single entity can have multiple events, as with a flashing beacon, and a single event can have multiple entities, as with a "big bang", as well as multiple perceptions of the event. This does not mean that time and distance are the same qualities by virtue of the use of this common reference term, but it suggests we might equate them mathematically with some universally acknowledged gauge. Thus the speed of light in vacuo, held to be a maximum, is used to gauge a length of time, converting it to a length of distance. We might also use as our gauge some minimum, for example the Hubble rate, approximately 7.87×10^{-27} times smaller than the speed of light.

The use of the same term for a separation by time and by space can be misleading. Spatial length is a primary concept, understood by common experience. In simplest manner, its magnitude is determined by holding two objects in proximity, one of which is a standard and the other of which is a test object. We might also consider temporal length as a primary concept, however, we tend to define time <u>quantitatively</u> in terms of a primary spatial length component of an otherwise cyclic or periodic concept, as a comparison of the length rate of change along the circumference of a clock face contemporaneous with some other change.

Taking a hint from the nomenclature of simple arithmetic, we state that a velocity is some translational displacement divided by the number of <u>times</u> some cyclic distance is transited at a constant rate, i.e. the number of times a clock hand tip transits a circumferential distance on the clock face designated as a unit standard interval. In the final analysis velocity is a comparison of two physical lengths, where the customary practical human standard is gauged to correspond with the tangential distance the earth rotates at the equator during (approximately) 1/86,400th of its diurnal cycle, i.e. a second.

The reader may object that it is not the length transited, but the angular speed that marks out time, pointing to the cyclical property with which it is customarily endowed. For a fixed reference frame, all 60 second analog clocks move ideally at the same angular rate, resulting in a varied velocity at hand tip that is a function of the hand length. We might envision that this velocity is limited by the speed of light, and for an ideal clock we stipulate that the length of time taken for light to travel from the center of the clock face to the end of the hand, be it hour, minute, second, nanosecond or yoctosecond, is equal to the length of time for the tip of the hand to travel the same distance tangentially about the face for one radian. Thus its angular frequency, ω, will be inversely related and gauged to the length of its arm, r, or abstractly to an angular wave length, λ, and consequently directly related by the angular wave number, κ, by the constant velocity, c, given by the familiar relationships

$$\omega = \frac{d\theta}{dt} = \frac{c}{r} = \frac{c}{\lambda} = c\kappa. \qquad (0.3)$$

Some rearrangement and integration of the angular measure, using a normalized value for the speed, $c = 1$, gives

$$r\int_0^1 d\theta = c\int_0^{(\omega^{-1})} dt, \quad \therefore |r| = |t| \qquad (0.4)$$

If we treat r as a 3-vector, $\mathbf{r}$, (calling the clock arm $\mathbf{r}$), its origin at the center and its extension point at the circumference of the clock, it is clear that $d\theta$ is orthogonal to $\mathbf{r}$. The unit integral of $d\theta$, along with the orthogonal sense, i, is thus an operator that transforms $\mathbf{r}$ orthogonally into an instant tangent vector, $c\mathbf{t}$, that carries the tip of $\mathbf{r}$ with it, rotating $\mathbf{r}$ about its origin as

$$\mathbf{r} = ic\mathbf{t} \tag{0.5}$$

for which the scalar form, leaving the i for emphasis, is

$$r = ict. \tag{0.6}$$

Such orthogonality is what a dimensional relationship between space and time demands. The c is simply a reminder that r and t are normalized, and can be left out by using the ought subscript to indicate unit values in

$$r_0 \equiv it_0. \tag{0.7}$$

Since r is radial and t is tangential, it is immediately apparent that in addition to being orthogonal to a spatial length, r, time is locally cyclical. After a period of 2π it will return to its starting point and continue to cycle at the invariant rate or angular frequency

$$\omega_0 = \frac{d\theta}{dt} = \frac{c}{r_0}. \tag{0.8}$$

We can rotate and translate r_0 to any direction and place in three dimensional space, and t_0 will remain extended orthogonally from the instant point of r_0, as in Time Scale 1.

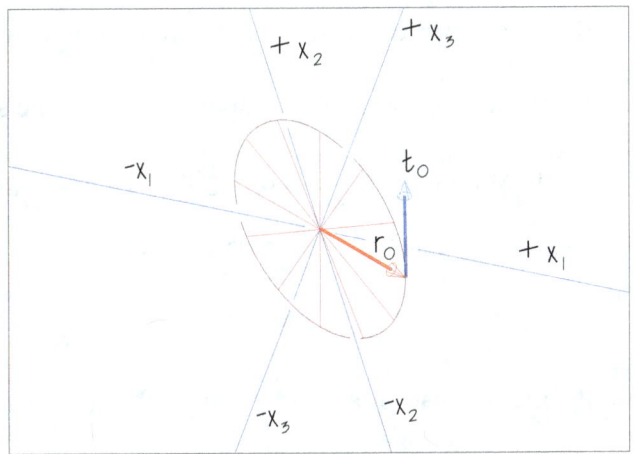

Clock Face Fixed with Rotating Hand
Time Scale 1
Figure 1

We can equate the instant r_0 to a unit base vector along an instant spatial dimension x_1, for which x_2 and x_3 are the remaining instant orthogonal dimensions. Since we are limited to three spatial dimensions, $x_{i=1,2,3}$, in most graphic representations the addition of an orthogonal linear dimension of time, $t = x_0$, involves representational difficulty. If we shift the origin of t_0 to the origin of the vector r_0, so that t_0 is co-linear with another unit vector along x_2, call it ir_0, we have a 2 dimensional graphic representation of spacetime by substituting the dimension x_0 for x_2. In a 3 dimensional depiction, we can make the equation of $x_0 = x_3$, representing space as a two dimensional plane, x_1-x_2. Both methods are used in discussions of general relativity, with the familiar warping of spacetime represented by a curving funnel in the 3-D depiction. These representations essentially depict time as a linear dimension substituted for one of the suppressed spatial ones.

While such representation has its time tested merits, it yet depends upon the explicit relationship of equation (0.5), which in turn retains the implicit relationship of equation

(0.8). We would hope to find a representation of spacetime which can depict time explicitly as orthogonal to all three dimensions of space, without the suppression of one or two spatial dimensions. In such case, time is thought of as a compactified dimension resident on some local scale, r_0, at each locus of 3-D space.

For such a registration of time, instead of a hand moving about a clock face, we might imagine the entire transparent face rotating about some center. Any spot on the circumference at a distance of r_0 from the center represents the origin of a tangent unit time vector t_0, its direction either clockwise or counterclockwise depending on which side of the face one is viewing. The clock face, i.e. time itself, then is moving orthogonal to two spatial dimensions, say x_1 and x_2, as shown in Time Scale 2. Note that the face is moving orthogonal to any instant r_0 superimposed upon it and to any arbitrary x_1 and x_2 coordinates centered on the origin of r_0.

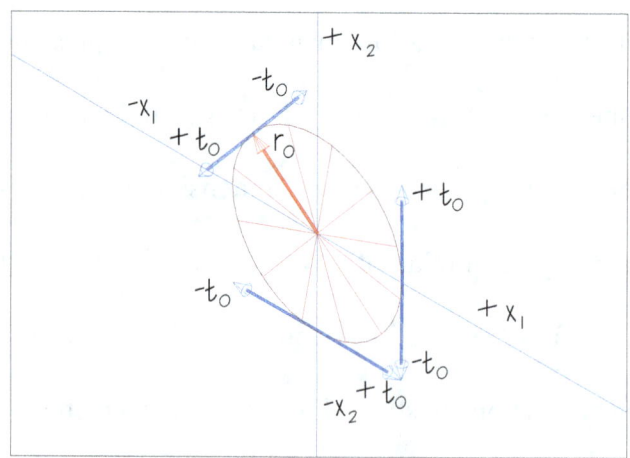

Clock Face Rotates with Hand
Time Scale 2
Figure 2

Since we have stipulated above that the clock hand can be rotated or translated without changing the relationship of equation (0.5), the same can be said for a rotation or translation both in 2-D and in 3-D space of the whole clock face. Sticking to the 2-D case, in x_1-x_2, we can designate a pair of differential vectors, *dt*, pointing clockwise and counterclockwise, at each possible location of the point of an r_0 about the clock face, so that the sum of all *dt* forms two superimposed circles about the instant center of the clock face. The dimension of time then forms a circle of radius r_0 about each point in x_1-x_2. This can be related to a polar coordinate system, in which the arm of the clock face, r_0, is a norm and the x_1-x_2 plane is sectioned as the θ coordinate about its origin.

For a 3-D space, in x_1-x_2-x_3, we can once again designate a differential vector pair, *dt*, at each possible location of the point of an r_0 about the clock face and at each possible orientation of the clock face within the 3-space, so that *dt* can point anywhere in a tangential plane and so that the sum of all *dt* form a sphere about the instant center of the clock face. Thus the dimension of time, *t*, is orthogonal to all three spatial dimensions, x_i, of any arbitrary spatial orientation at the points $x_i = +/-1$.

Now we can simplify and make things a bit more definite as in Time Scale 3. For any clock face θ of radius r_0 in θ, an arbitrary x_1-x_2 plane, rotating about an axis, **θ**, aligned with the x_3 axis orthogonal to x_1-x_2, we can find a second clock face ϕ of equal r_0, concentric with, orthogonal to, and rotating with θ, i.e. spinning like a coin, while itself simultaneously rotating at the same frequency, $\omega_\phi = \omega_\theta$, about an arbitrary axis, **ϕ**, where ϕ rotates in θ and with θ. We can now choose a clock hand, r_0, its origin at the center of the concentric clock faces, initially at x_2 at one of the two radial intersections of ϕ and θ, and rotate it with ϕ about $\boldsymbol{\phi}$, so that

1. at $t(\theta) = 0$, $\boldsymbol{r}_0$ points to $(0,+1,0)$ and $\boldsymbol{t}_{0\phi}$ points to $(0,+1,+1)$;

2. at $t(\theta) = \pi/2$, $\boldsymbol{r}_0$ points to $(0, 0, +1)$ and $\boldsymbol{t}_{0\phi}$ points to $(+1, 0, +1)$;

3. at $t(\theta) = \pi$, $\boldsymbol{r}_0$ points to $(0,+1,0)$ and $\boldsymbol{t}_{0\phi}$ points to $(0,+1,-1)$;

4. at $t(\theta) = 3\pi/2$, $\boldsymbol{r}_0$ points to $(0,0,-1)$ and $\boldsymbol{t}_{0\phi}$ points to $(+1,0,-1)$; and finally

5. at $t(\theta) = 2\pi$, $\boldsymbol{r}_0$ points to $(0,+1,0)$ and $\boldsymbol{t}_{0\phi}$ points to $(0,+1,+1)$;.

There are an infinite number of $\boldsymbol{r}_0$ in ϕ, they each intersect with the clock face of θ twice and at the same location in θ with each cycle of ϕ about $\boldsymbol{\theta}$, and they each extend once to each of the extrema in the ϕ coordinate at $+/-\pi/2$, i.e. at $x_3 = +1,-1$. Thus the point of each $\boldsymbol{r}_0$ and the origin of its time vector $\boldsymbol{t}_0$, traces a figure eight oscillation about one half of the spherical shell formed by the sum of all time vectors $d\boldsymbol{t}$, while a wave phase rotates counterclockwise about $\boldsymbol{\theta}$. Note that his motion avoids the coordinate entanglement condition as depicted in <u>Gravitation</u> by Misner, et al., [2]. We can use this graphic depiction of time to great advantage later in our discussion.

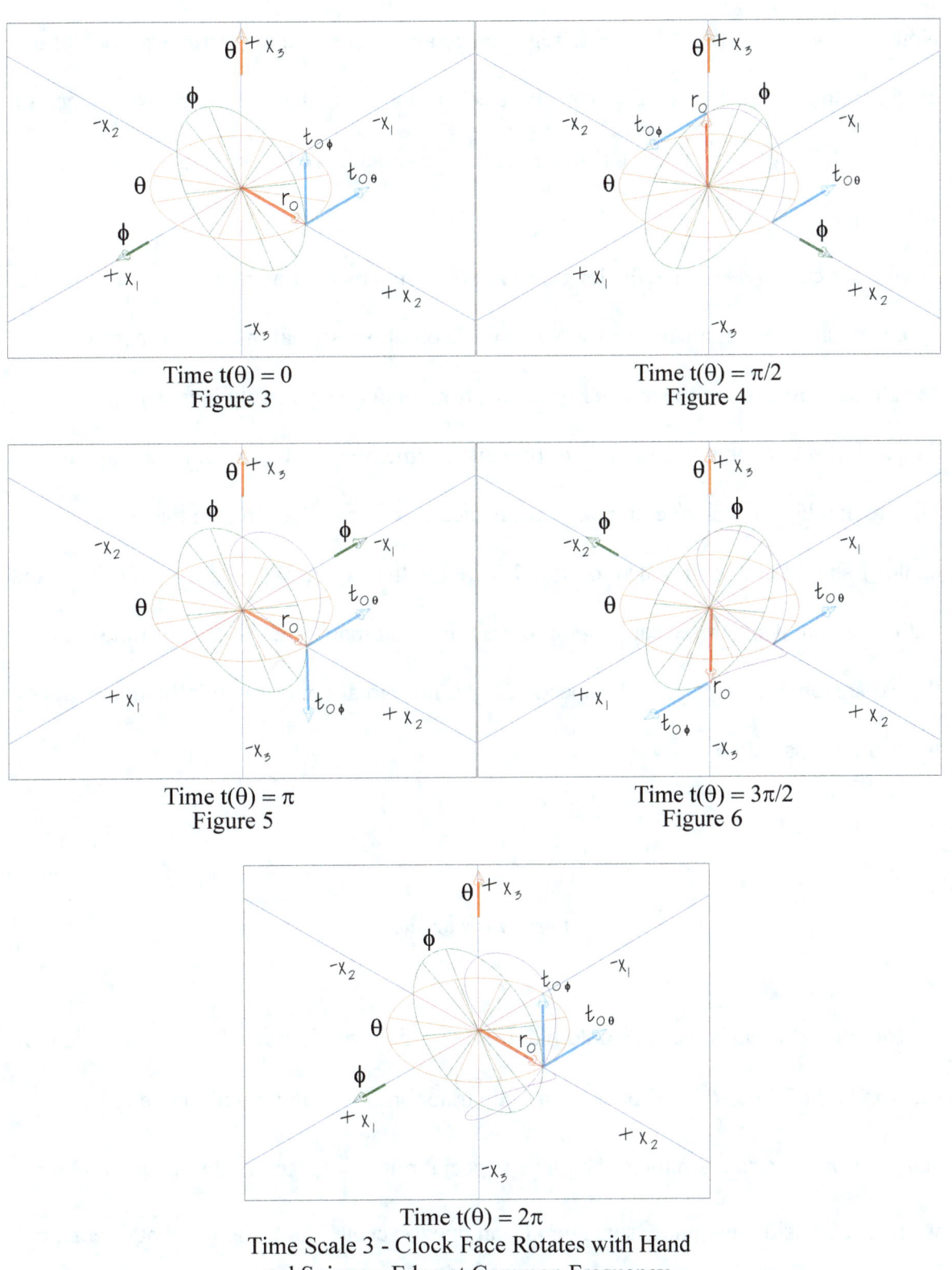

Time t(θ) = 0
Figure 3

Time t(θ) = π/2
Figure 4

Time t(θ) = π
Figure 5

Time t(θ) = 3π/2
Figure 6

Time t(θ) = 2π
Time Scale 3 - Clock Face Rotates with Hand
and Spins on Edge at Common Frequency
Figure 7

Note that the same instant of time is represented anywhere on the spherical surface of this clock, so that the surface constitutes a time co-ordinate singularity. We can keep track of the "length" of time by a count of the number of oscillations of a given r_0.

Finally we can envision that the length of r_0 is in some manner augmented or diminished by a very small amount continually with each oscillation, so that the time dimension is seen to be wound up in the manner of a kite string about a constantly increasing or decreasing spatial unit sphere. It is important to remember, however that there are an infinite number of such dt continually connected in spherical fashion, so the string analogy should not be stretched too far. It is really the expanse of 3-space both about and within such unit sphere, expanding or contracting, that marks the passage of time. It is the expansion of this space at the speed of light, not radialy but tangentially, that gauges time in this spacetime.

Lorentz Covariance

To complete this analysis, we would like to see if this formulation is Lorentz covariant, if the standard of time, t_0, will undergo a scale transformation along with the length standard, r_0, according to the principles of special relativity. Returning to equation (0.6), we might envision that under some condition due to acceleration, such as that of cosmic expansion, r_0 contracts to $r_0^o < r_0$. We divide that equation into its contracted version,

$$\frac{r_0^o}{r_0} = \frac{ict_0^o}{ict_0} = \frac{t_0^o}{t_0} \qquad (0.9)$$

and find the unit time standard varies according to the ratio of the unit lengths, as

$$t_0^o = \frac{r_0^o}{r_0} t_0. \qquad (0.10)$$

In special relativity as represented by Charles Stevens [3], time intervals transform according to

$$t' = \gamma(1-\beta)t \qquad (0.11)$$

where t is the interval in reference frame F and t' is the same interval viewed in reference frame M moving relative to F at velocity, v, as a fraction of the speed of light, c, giving the ratio identity β, which cannot be greater than 1, as

$$\beta \equiv \frac{v}{c} \qquad (0.12)$$

and the value of γ, which cannot be less than 1, as

$$\gamma \equiv \frac{1}{\sqrt{1-\beta^2}}. \qquad (0.13)$$

One minus β approaches 0 faster than the inverse of equation (0.13), so the combined factor never exceeds 1 and approaches 0 at the limit. If a relationship can be established between the time dimensions in equations (0.10) and (0.11), then we might expect a relationship between the factors on the right sides. We can do this by viewing a unit standard, t_0, from F and from M.

The spatial interval transformation, in which we have aligned r with an arbitrary x_i axis, is

$$r' = \gamma(r - vt) \qquad (0.14)$$

Substituting from equation (0.6) for t, we have

$$r' = \gamma\left(r - v\frac{r}{c}\right) = \gamma(1-\beta)r \qquad (0.15)$$

which is symmetric with equation (0.11).

Rearranging gives an expression of a proper time, τ, and a proper length, σ, which are invariants of M. We are not using Minkowski space for our 4-vector and r is simply ct, so that multiplying equation (0.16) through by c gives us equation (0.17). This proper length will be shown to be related to r_0^o.

$$\tau \equiv \frac{t'}{\gamma} = (1-\beta)t \qquad (0.16)$$

$$\sigma \equiv \frac{r'}{\gamma} = (1-\beta)r . \qquad (0.17)$$

In the Chart 1 graphic representation of a Lorentz transformation we have aligned the spatial axis, r, of a stationary reference frame, F, with the direction of travel of a moving frame, M, making it a pure transformation or boost. This is expressed for the time dimension by equation (0.11) and for the space dimension by equation (0.15). In each equation, the unprimed coordinate with respect to F is modified by the two related factors, $(1-\beta)$ and γ, to arrive at the primed co-ordinate with respect to M.

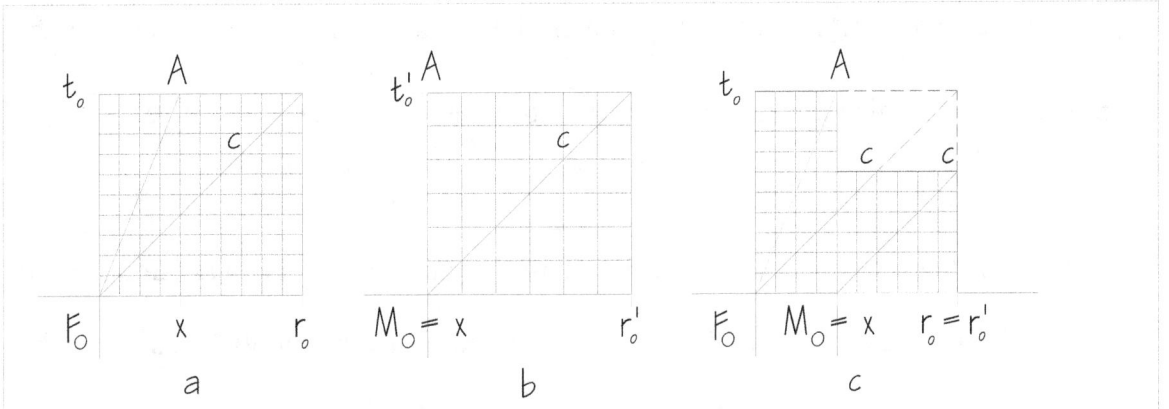

Chart 1 - (1-β) Component of Boost
Figure 8

While the customary analysis is for an arbitrary *x*, or in this case *r* and *t* in *F*, we will apply the same to a space and time unit length in *F*, r_0 and t_0, orthogonally aligned. This is shown in Chart 1.a. The path of *M*, moving at a velocity *v*, or a distance of *x* in time t_0, is drawn as the sloped line, and terminates at point $A = F(x, t_0)$. The unit of spacetime has been marked off in decimal fractions. The limit of relative frame velocity, *c*, is shown with its inverted slope of 1 t_0 per 1 r_0.

In Chart 1.b, the operation of (1-β) on *F* indicates the effect of the motion of *M*, which transforms the unit spacetime from that shown in 1.a. Assume that both *F* and *M* begin to receive a periodic signal from beyond the left edge of their respective charts when those charts are coincident at $r = r' = 0$. They both know that the signal flashes are spaced one-tenth of t_0 apart. As shown, *x* and therefore β happens to be 0.4, resulting in a (1-β) of 0.6. After one t_0, *F* counts ten flash intervals, but *M* has by that time moved four intervals to the right and only counts six intervals. As a result, for *F* the perceived time elapsed before the first signal reaches r_0, therefore the distance from 0 to r_0 is ten-tenths or unity, while for M that time and distance is six-tenths of t_0 and r_0 respectively.

Note that the path of M observed from F in Charts 1.a and 1.c, the diagonal through space and time, is perceived by M in his own view of this spacetime, as simply a path through time, shown by the vertical line, $M_O - t'_0$. Note also that the shortening of the time scale is required if c is to remain normalized and invariant.

This is not the time dilation and space contraction of relativity, however. If the signal had been coming from the right, during the time t_0, M would have counted fourteen intervals to a count of ten for F, or a factor of $(1+\beta)$. This is simply an instance of the Doppler effect, a frequency shift.

As can be seen in Chart 1.c, the gauge or scale factor of the spacetime is the same in both frames, as indicated by the identical grid intervals. The unit time and distance scales of the spacetime for each are not themselves modified by this observed modification, and we will disregard it in the remainder of the discussion. It is of interest, though, that the product of these two factors equals the square of the inverse of the other factor, γ, or

$$(1-\beta)(1+\beta) = (1-\beta^2) = \sqrt{1-\beta^2}^2 = \gamma^{-2}. \tag{0.18}$$

It is the factor γ that we are primarily interested in, as it embodies the change in the scale of spacetime reflected in a measured interval through the Fitzgerald-Lorentz length contraction,

$$\gamma \, \Delta r = \Delta r' \qquad (0.19)$$

and through time dilation,

$$\gamma \, \Delta t = \Delta t'. \qquad (0.20)$$

These in turn are related to a change in the proper time, τ, and proper length, σ, as in the identity terms of equations (0.16) and (0.17) as

$$\gamma \, dt = dt' = \gamma \, d\tau \qquad (0.21)$$

$$\gamma \, dr = dr' = \gamma \, d\sigma \qquad (0.22)$$

Following this line of thought, we substitute the unit standards for the unprimed interval coordinates in equations (0.16) and (0.17) and their contractions for the primed to arrive at an expression of a unit proper time, τ_0, and a unit proper length, σ_0, where each is the representation of the unit standards of M in F,

$$\gamma \, \tau_0 \equiv t_0^{\,o} = \gamma (1 - \beta) t_0 \qquad (0.23)$$

$$\gamma \, \sigma_0 \equiv r_0^{\,o} = \gamma (1 - \beta) r_0 \qquad (0.24)$$

Some care is in order here. While the length contraction is often interpreted as a property by which a moving body shrinks absolutely in proportion to its velocity with respect to a stationary frame, and while this may in some instances be true, its fundamental statement is that the unit standard by which a length, l, is measured in a moving frame is smaller than the unit standard in the stationary frame with respect to which it is deemed to be moving and from which it is held to be shorter.

In a similar manner, time dilation is deemed to indicate that a given duration of time in a moving frame is measured as moving slower from a stationary frame; thus the usual depiction of the space traveler who returns to earth after 50 years of near speed of light travel, having aged only a couple of earth years. As in the last paragraph, equation (0.20) states the same physical condition as equation (0.19), that the unit standard of time in a moving frame is smaller than the unit standard in a stationary frame, thus an interval of time is measured as greater, i.e. longer as is a length, in the moving frame, but this does not necessarily mean slower.

If our clocks in both the moving and the stationary frame are defined as having hands of a length measured by equation (0.19), and the speed of the end of the hand is the speed of light, c, then the moving frame will have a longer arm and its angular velocity will necessarily be less than that of the stationary frame, and the clock in M will rotate at a slower rate than in F. This is the general interpretation of time dilation. On the other hand, if the length of the hand in M is set to the unit length standard, smaller in M than it is in F, then the speed of light constraint for the speed of the hand tip will result in an increased angular speed and the clock in M will spin faster. In such case time will still be measured as greater, i.e. longer in M than in F, as a count of the number of clock cycles would indicate, in keeping with equation (0.20), since γ in this case is a measure of the relative angular frequencies of M and F. This is so even though the length of the clock hand path in keeping with equation (0.8) is the same, or

$$r_0 \omega_0 = r_0^o \omega_0^o = c \qquad (0.25)$$

since

$$\frac{r_0}{r_0^o} = \frac{\omega_0^o}{\omega_0} = \frac{c}{r_0^o \omega_0} = \gamma . \qquad (0.26)$$

With this in mind, we can combine equations (0.20) and (0.19) as we did in equation (0.9), converting from incremental to differential values, and get the equivalent of equation (0.10), where this last case explicitly shows the equivalence of the differential length ratio and γ,

$$dt' = \frac{dr'}{dr} dt = \gamma \, dt . \qquad (0.27)$$

We have a temporary conundrum, however, as $\gamma \geq 1$, but the unit length ratios in equation (0.10) and again if inverted from equation (0.26) is less than 1. The problem arises from the nature of a unit standard. If it is fixed, any change in an interval, differential or incremental, will vary directly, proportional to the standard. If the standard itself varies, then the numerical value of a fixed interval will vary indirectly to the change in the standard.

Given a fixed interval, $l \equiv l'$, which is related nominally by γ as measured from frames M over F, equation (0.27) measures the identical interval, $dt \equiv dt'$ from two different physical standards. Equation (0.10) relates two unit standards, $t_0 > t_0^o$, that vary proportionally to the two other unit standards, $r_0 > r_0^o$, all related by c. Thus

$$\frac{l}{r_0}\gamma = \frac{l'}{r_0^o} \tag{0.28}$$

$$\therefore \gamma = \frac{r_0}{r_0^o} \tag{0.29}$$

We return now to the charts to see how this might be represented graphically. Chart 2 shows an enlargement of the top portion of Chart 1.a in the neighborhood of the time t_0 in F. We are analyzing only the effects of the factor γ on the two reference frames and disregarding the Doppler effect of $(1-\beta)$. Point A represents the intersection of the line of motion of M in F and the time coordinate in F for time t_0, designated as $F(x,t_0)$. In reference frame M, based on the above discussion and equations (0.19) and (0.20), this same point would be measured as $M(\gamma x, \gamma t_0)$, which as drawn for $\beta = 0.4$, so that $\gamma = 1.0910...$, would be $(0.4364..., 1.0910...)$. Finally, based on these same two equations this point is expressed as the intersection of x' and t_0', as shown in the square brackets or $M(x', t_0')$.

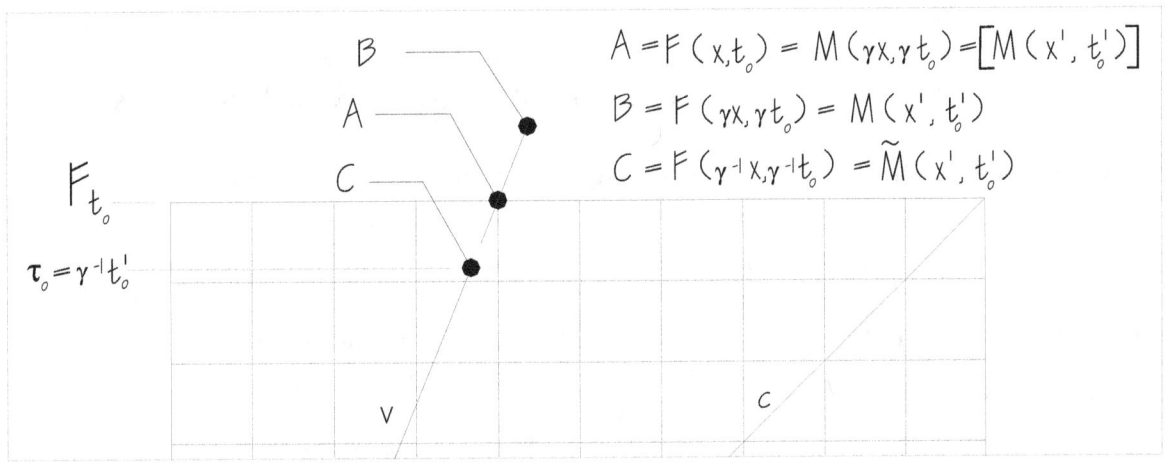

Chart 2 - γ Component of Lorentz Transformation
Figure 9

Point *B* shows the co-ordinates in *F* corresponding to the numerical values of $M(\gamma x, \gamma t_0)$, and therefore represents an expansion of the line for *v* by the value of *γ*. Thus it expresses *M* in terms of *F* and is numerically equal to the value $M(x', t_0')$.

Point *C* shows the numerical value in terms of *F* for the inverse of $M(x', t_0')$ or $\tilde{M}(x', t_0')$. Thus if we were to designate $M(x', t_0')$ in *M* as $M(0.4, 1.0)$, $\tilde{M}(x', t_0')$ would be $F(0.366..., 0.916...)$. The time component of *C* then represents the *proper time*, τ_0, the naught subscript used to indicate its specificity to a unit time standard, t_0', of *M*, when measured from *F*, and in keeping with the concept of time dilation, it is longer in *M* than in *F*. Thus for a value in *M* of $t_0' = 1$, *F* will perceive an elapsed time in *M* of $\tau_0 = 0.916...$. Once again, while generally interpreted as a slowing of time in *M*, this "lengthening" of time can be attributed to a shortening of the time standard.

This is all very interesting, but it would be more illustrative if we could find an essential depiction of the relationship of F and M involving γ and τ_0. For instance, the length of v from F_0 to the three points, A, B and C, embodies the factor of γ, but that factor does not arise naturally, or at least readily, from an analysis of the charting of v.

The problem lies in the dual utility of the chart itself. On the one hand it represents a Cartesian background for the plotting of two related bits of data, location in time and in space. From this perspective, the right hand end of the speed of light curve, c, at the upper right corner of the chart, represents the time elapsed in F during the displacement of a light wave or photon by one unit, or $F(r_0, t_0)$. On the other hand, it is a 2-D chart of spacetime itself, where the speed of light determines the unit speed for the passage of a stationary reference frame through time or of a displacement through space with no passage of time. This second usage means that in time t_0, the limit of travel of a spacetime vector in the unit spacetime is a circle, or in our chart, a quarter circle, described by the unit spacetime vector, R_0, where

$$R_0 = \left(r^2 + t^2\right)^{\frac{1}{2}}. \tag{0.30}$$

We will use the designation R_0 for both the vector and the circle described using it as a radius, dependent on context. When R_0 is orthogonal to the time axis it is a pure space vector and equals r_0 and when orthogonal to the space axis is a pure time vector and equals t_0. It should be mentioned that in a 4-D spacetime, R_0 is an invariant 4-vector, but that it is not the same 4-vector residing in Minkowski space, as generally used in relativity, as the time vector is not subtracted, but rather is added to the three space vectors, as in equation (0.50).

Drawing this condition on the unit spacetime for F gives us Chart 3.a, and we notice immediately that the velocity curve used for the moving frame M terminates at A, beyond the limit imposed by c; that is, it violates one of the basic assumptions of relativity. To correct this, in Chart 3.b we draw the velocity curve, v^o, through the intersection of R_0 and x_1 at A^0, as shown in close-up in Chart 4, and find on closer inspection that this corresponds with the time value of τ_0 for x_1. In fact, for any value of $0 < x < r_0$, this condition will be found to hold, which means that the secant of the angle between t_0 and v^o equals γ, or

$$\frac{\overline{OB^o}}{t_0} = \frac{\overline{OA^o}}{\tau_0} = \gamma. \tag{0.31}$$

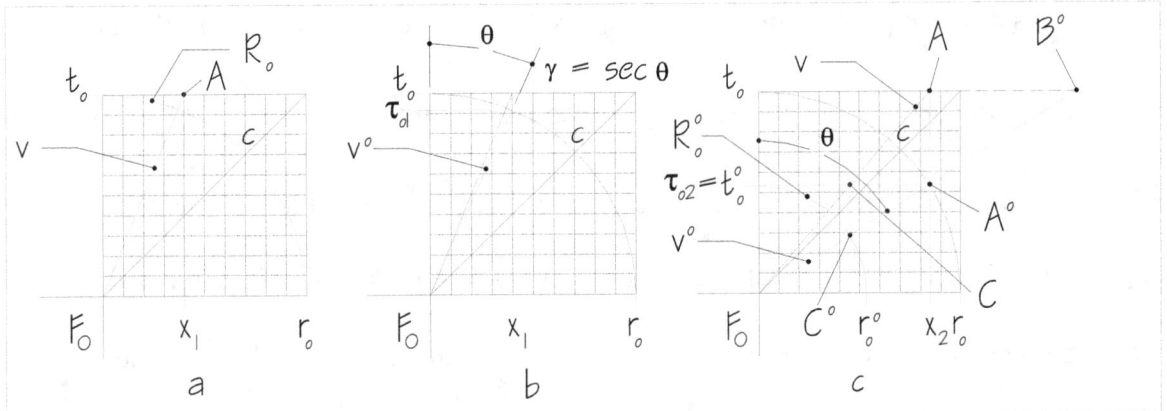

Chart 3 - Contraction of γ Component of Lorentz Transformation
Figure 10

Chart 3.c, a condition at a much higher velocity, shows more clearly the relationship of v^o and τ_0. We construct a second circle for R_0^o such that $|R_0^o| = |r_0^o| = |t_0^o|$, where

$$t_0^o \equiv \tau_0 \tag{0.32}$$

The orthogonal projection of the intersection of R_0 and v^o onto t_0 intersects at τ_0 and intersects the curve v at point C, while an orthogonal projection from C onto r_0 intersects R_0^o and v^o at the same point, C^o. Thus we have the similar triangles, $B^o A A^o \sim A^o C C^o$, and R_0^o represents a contraction in the moving frame of the unit spacetime vector R_0.

Chart 4 is a close-up view of the top portion of Chart 3.b, showing the contraction of v into v^o. We will call this reference frame F^o. A^o represents the same physical condition as A in F. The displacement x remains as the same percentage of r_0, but the time scale at that point is now the proper time, τ_0. In square brackets, that same point in M^o is numerically identical to $M(\gamma x, \gamma t_0)$ at A, however, the unit time standard, $t_0^o = 1$ and the related x^o, are in M^o instead of F. Thus the unit standard is local to M^o instead of the expression of a stationary spatial background. If we can think of the time dimension M_t^o as being inclined along the slope of v^o instead of orthogonal as in F, the same spacetime numerical values hold in both reference frames.

In keeping with this observation, C^o is proportionally the same to A^o as C is to A. In M^o, the unit time and space standards apply as indicated by R_0^o, so that $M^o(x^o, t_0^o)$ is numerically equal to $F(x, t_0)$ shown at A, or in this case, (0.4, 1.0). For the proper time, with all values in units of F, we have the following identity,

$$\tau_0 \equiv t_0^o \equiv \gamma^{-1} t_0 = \gamma F_t(C^o) \tag{0.33}$$

Point B^o indicates the nature of time dilation as conventionally figured. At point A^o, M^o has traveled the same length of time as F, as given by $R_0 = t_0$, but to F this is registered as the proper time τ_0. By the time M^o reaches B^o, which F registers as 1.0, F has moved on in time to γt_0. The length of v^o, $\overline{OB^o}$, is equal to γ.

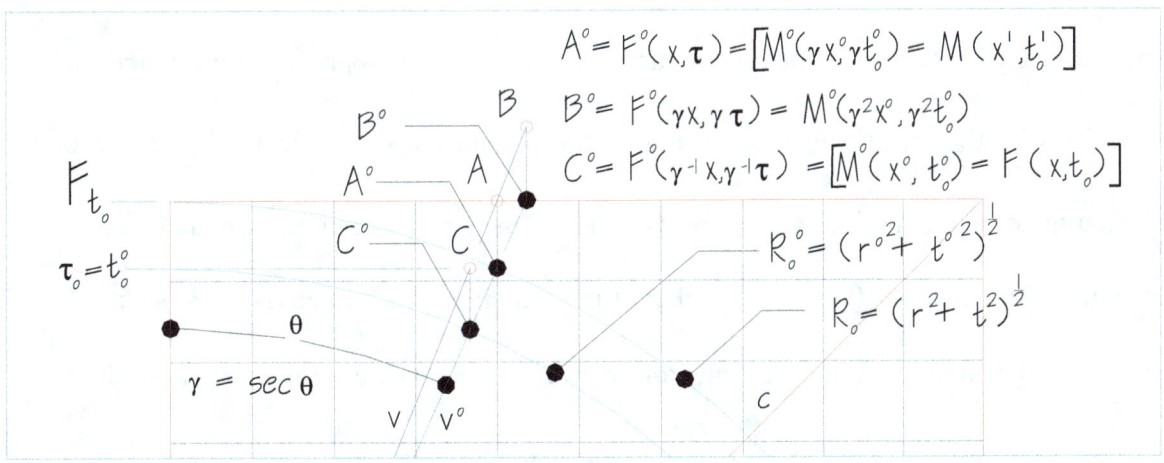

Chart 4 - Closeup of Contraction
Figure 11

Perhaps the most significant aspect of this representation is that the secant of the angle of $t_0 \, O \, v^o$ establishes γ, underscoring the geometric nature of time. Expanding on the relationships of equation (0.31), we have

$$\frac{\overline{OB^o}}{t_0} = \frac{\overline{OA^o}}{\tau_0} = \frac{R_0}{R_0^o} = \frac{t_0}{\tau_0} = \frac{\overline{OB^o}}{\overline{OA^o}} = \frac{\overline{OB}}{\overline{OA}} = \frac{\overline{OA}}{\overline{OC}} = \gamma \qquad (0.34)$$

We have examined the Lorentz transformation with respect to time and proper time, but a similar analysis could be made with respect to space and a proper distance as modified from the conventional Minkowski representation as noted earlier. The case of time is more germane to our present discussion, as will be seen.

The above charts suggest that spacetime curvature is as much a matter of curvature of time as it is of space. Chart 3 indicates that as a moving reference frame approaches the speed of light and v approaches co-linearity with c, v^o and γ approach infinity and co-linearity with the space axis, r, and the time and distance scales indicated by R_0^o become exceedingly small, and in the same proportion. This is precisely what we analyzed initially with equation (0.5) and a cyclical time dimension. If we envision M as an accelerating reference frame starting from rest at F_O and accelerating to c within the first unit of spacetime, we would find that the contracted velocity curve, v^o, and the collinear contracted time dimension would curve, and that under certain constraints, would arc like a quarter circle. Its constantly shortening time standard, t_0^o, then is aligned with the arc of γ, and its space standard, r_0^o, correspondingly shortened, rotates with and orthogonally to it.

A physical instance of this shortening of both r_0^o and t_0^o can be found by examining the nature of the deBroglie wavelength of a massive particle. We assume that the reduced Compton wavelength, $\lambdabar_C$ is indicative of the rest state of such a particle, and is determined by dividing the reduced Planck's constant, $\hbar$, by the product of the speed of light and the particles rest mass, m,

$$\lambdabar_C = \frac{\hbar}{mc}. \qquad (0.35)$$

The reduced deBroglie wavelength, $\lambdabar_{dB}$, is the quotient of $\hbar$ and the particle's relativistic momentum, p, at velocity, v, given as

$$\lambdabar_{dB} = \frac{\hbar}{p} = \frac{\hbar}{\gamma m v} \qquad (0.36)$$

where the factor γ is the same as used in the development above. Combining and some rearrangement gives us the ratio of these reduced wavelengths as

$$\frac{\lambdabar_C}{\lambdabar_{dB}} = \gamma \frac{v}{c} = \gamma \beta = \gamma x. \qquad (0.37)$$

In Chart 4, this is represented by the tangent of angle θ between the time axis in F and v^o and gives the ratio of the particle velocity in F, where $A^o(x) = A(x)$, and the contracted unit standard, r_0^o. Thus

$$\frac{\lambdabar_C}{\lambdabar_{dB}} = \frac{x}{r_0^{\,o}} = \frac{x}{t_0^{\,o}} = \frac{x}{R_0^{\,o}} \tag{0.38}$$

and as we approach the limit of the speed of light and x approaches $r_0 = 1$, we have

$$\lambdabar_C = \frac{r_0}{r_0^{\,o}} \lambdabar_{dB} = \gamma \, \lambdabar_{dB} \tag{0.39}$$

Thus, in the extreme case

$$\text{if } \lambdabar_C = r_0, \text{ then } \lambdabar_{dB} = r_0^{\,o}. \tag{0.40}$$

Since the frequency and wavelength are related as

$$\lambdabar_C \omega_C = \lambdabar_{dB} \omega_{dB} = c \tag{0.41}$$

rearrangement gives, in the extreme case

$$\gamma = \frac{\lambdabar_C}{\lambdabar_{dB}} = \frac{\omega_{dB}}{\omega_C} = \frac{t_C}{t_{dB}} = \frac{r_0}{r_0^{\,o}} = \frac{\omega_0^{\,o}}{\omega_0} = \frac{t_0}{t_0^{\,o}}, \tag{0.42}$$

therefore we also have

$$\text{if } t_C = t_0, \text{ then } t_{dB} = t_0^{\,o}. \tag{0.43}$$

Considering a normalized frequency, that is, where the angular displacement, $\theta = \theta_0$, always equals 1 and the time consequent varies according to the particular F from which it is observed, we can integrate equation (0.25) for any time $t = qt_0 = q^o t_0^{\,o}$

$$r_0 \omega_0 \int_0^t dt = r_0^{\,o} \omega_0^{\,o} \int_0^t dt \tag{0.44}$$

$$r_0 \omega_0 (qt_0) = r_0^{\,o} \omega_0^{\,o} (qt_0) = r_0^{\,o} \omega_0^{\,o} (q^o t_0^{\,o}) \tag{0.45}$$

$$qr_0 = q\gamma r_0^{\,o} = q^o r_0^{\,o} \tag{0.46}$$

and finally

$$r_0 = \gamma r_0^o = \frac{q^o}{q} r_0^o \tag{0.47}$$

therefore

$$t_0 = \gamma t_0^o = \frac{q^o}{q} t_0^o \tag{0.48}$$

showing that γ is simply the frequency ratio of the unit standards of space and time between a moving and a stationary reference frame.

It is worth noting the case when $x = R_0^o = r_0^o = \frac{1}{\sqrt{2}} r_0$, that is, when r equals t at the intersection of the curve of c and R_0. If we consider a massive particle as some manner of physical stationary waveform, i.e. a bound, rotating wave, a ratio of r and t of unity represents the point at which the translational displacement of the particle in space begins to outrun the transverse wave displacement, i.e. its displacement in time. It is the point at which the contracted velocity, v^o, equals the speed of light. Prior to that point the waveform would conceivably flatten in space in the form of an oblate spheroid. From that point on, the waveform must contract in all dimensions so that its transverse motion remains in phase with its translation, conforming to the deBroglie wavelength.

It follows immediately that from any reference frame F in 4-D spacetime, for a moving frame M, a unit standard can be given for space by r_0^o and for time by t_0^o, both related to a 4-vector (of additive components), R_0^o, as

$$R_0^o = \tfrac{1}{\sqrt{2}} \left(\left(r_0^o\right)^2 + \left(t_0^o\right)^2 \right)^{\frac{1}{2}} = \frac{R_0}{\gamma} \tag{0.49}$$

where x_{0i} are 3 orthonormal bases, symmetric with respect to r_0,

$$R_0 = \tfrac{1}{\sqrt{2}}\left(r_0^2 + t_0^2\right)^{\frac{1}{2}} = \tfrac{1}{\sqrt{2}}\left(\left(\tfrac{1}{\sqrt{3}}x_{01}\right)^2 + \left(\tfrac{1}{\sqrt{3}}x_{02}\right)^2 + \left(\tfrac{1}{\sqrt{3}}x_{03}\right)^2 + t_0^2\right)^{\frac{1}{2}}. \qquad (0.50)$$

If we shift the origin of t_0^o in Chart 3.c from the origin of r_0^o to its point, we have the configuration shown in Time Scale 1 and 2. From there we can extrapolate to the 3-D form shown in Time Scale 3 for the expression of a 3 dimensional clock.

A statement is in order concerning the "relativity" of the reference frames F and M, and that of the spacetime scales R_0 and R_0^o. Assuming that F resides in an expanding 3-manifold, if that residence is isotropic with respect to cosmological red shift, then we can state that the local position of F is stationary with respect to space and in an extremal position of change with respect to time. Otherwise, F would experience a blue shift in the direction of its travel with respect to space. In similar fashion, F could experience such an anisotropy while rotating about a center, perhaps galactic or supergalactic, that is itself stationary or isotropic with respect to cosmological red shift. Thus at every point in spacetime, assuming an isotropic expansion, there exists a potential F for which R_0 is a local maximum, though R_0 at all points need not be identical. For any moving reference frame M at that same point, $R_0^o < R_0$.

By this analysis, we can envision a 4-D spacetime with Lorentz covariance in which the time dimension is modeled as a compactified rotational dimension orthogonal to the three space dimensions, as developed earlier. Having taken this side-trip into an investigation of spatial and temporal length, we can now look at the concept of mass.

2 – Geometrization of Mass in Classical and Quantum Theories

In his book, Concepts of Mass in Contemporary Physics and Philosophy, Max Jammer delineates three types of mass; inertial, active gravitational (corresponding to a source) and passive gravitational (corresponding to a test particle), and concludes that the jury is still out as to whether these represent distinct concepts of mass. Looking at the related concept of inertia, we can readily see that it can be quantified in terms of length and time by the concepts of linear and angular displacement and their derivatives. For simplicity, we limit our thought experiments to analysis in one spatial dimension, unless stated otherwise.

Inertial Mass

Inertia is a resistance to any change in the momentum of a body:

1. An absolute or infinite inertia would indicate immobility or a displacement of $dx = 0$ from the reference frame of that body or a change in velocity of $dv = 0$ from any arbitrary reference frame, resulting from interaction with another body.

2. An absolute lack of or zero inertia would indicate an instantaneous displacement of a, an undefined or relative infinite displacement or change in velocity due to a finite displacement with zero passage of time resulting from such interaction.

3. A finite displacement of a body, *a*, over a finite time duration resulting from its interaction is an inverse measure of its finite inertia, i.e. of its inertial mass.

While "body" has been historically conceived as a classical entity, substitution of the term "particle" understood in quantum terms, should not change the meaning of "inertia". A free body or particle is classically conceptualized as moving within and through a flat, three dimensional space, said space of itself and in the absence of any field potential or other bodies or particles of either mass or energy, constituting both a phenomenological and an ontological void. By the above definition and our expansion of it, however, a space upon which we can superimpose a metric, in and of itself exhibits the property of inertia, since it has a definite resistance to change and in the case of physical space, appears to have a finite, albeit accelerating, expansive momentum as evidenced by cosmological red-shift. By virtue of such property, space even without quantum fields can not be said to be either a phenomenological or an ontological void. Within such space, time can be seen as the path of its inertial change.

In the interest of gaining a geometric, descriptive explanation of mass, inertial or gravitational, we will investigate inertial mass first in a classical target-test body. In general relativity, gravitational field source mass is geometrized in <u>direct</u> relationship to length, and we can find a direct relationship between mass and length in the aggregation of bodies or particles. As in the case of stellar configurations, the product of the volume of the body and its average volume inertial density computes the mass of the body, so that for a given density, the reduced circumference of the body gives a geometrized approximation of its mass. For inertial test body *a* the magnitude of its mass, m_a, is indirectly proportional to the displacement, x_a, over the time interval of an interaction, t_a, under a given impulse, J, from another body or source, which results in a final velocity for *a* of v_a,

$$m_a = \frac{J}{v_a} = \tfrac{1}{2} J \frac{t_a}{x_a}. \tag{1.1}$$

The definition of impulse is the integral of force with respect to time which is equal to a change in momentum, ΔP,

$$J = \int_{t_i}^{t_f} F(t)\, dt = \Delta P. \tag{1.2}$$

While in general the force, hence the acceleration, will vary over the duration of the impulse, for ease of illustration, we will use a constant force and acceleration, i.e., the average over the duration. In this case *a* is accelerated from an initial velocity, v_{ai}, to a final velocity, v_{af}, over the time interval $t_\Delta = t_f - t_i = t_a - t_0 = t_a$. The time subscripts indicate that at time $t_i = t_0$, $v_{ai} = 0$. Starting at the end of such interaction, at time $t_f = t_a$, the final velocity of *a* will be reached at $v_{af} = v_a$, and it will continue on at that velocity as viewed from its original reference frame, *F*.

We assume that the source of the impulse and the test body exist in a steady state in their respective rest frames and in isolation from each other and any other interactions both before and after their collision, but that during their interaction they each undergo an acceleration and an exchange of momentum and energy. Thus the acceleration for *a* is

$$a_a = \frac{v_{af} - v_{ai}}{t_a} = \frac{2x_a}{t_a^2} \tag{1.3}$$

and the force is

$$F = m_a \frac{2x_a}{t_a^2} \tag{1.4}$$

Since the time interval for the acceleration of the body and the time interval found in the statement of its velocity is the same as the interaction interval, t_f, we have the following time independent parameter of the interaction

$$\mathsf{n} = \int_{t_i}^{t_f} J(t)\, dt = \int_{t_i}^{t_f} F(t)\, t_f\, dt = \tfrac{1}{2} F t_f^2 = m\, x_f \tag{1.5}$$

where the letter n (tav) is an inertial constant of the interaction, of mass-length dimensions. Equation (1.1) can then be expressed in a time independent scalar form where mass is the inverse measure of the space interval of the interaction,

$$m_a = \frac{\mathsf{n}}{x_a}. \tag{1.6}$$

We can postulate a second condition, with J unchanged, for a body b, for which

$$m_b > m_a. \tag{1.7}$$

Therefore, we have

$$m_b = \frac{J}{v_b} = \tfrac{1}{2} J \frac{t_b}{x_b} = \frac{\mathsf{n}}{x_b} \tag{1.8}$$

and the following inequality is apparent

$$v_b < v_a. \tag{1.9}$$

This suggests that if x_b is equal to or less than x_a,

$$t_b \geq t_a \tag{1.10}$$

and/or if t_b is equal to or greater than t_a,

$$x_b \leq x_a, \tag{1.11}$$

but that the time intervals cannot be equal if $x_b = x_a$. However, inequality (1.9) would also hold as long as

$$\frac{\Delta x}{x_a} < \frac{\Delta t}{t_a}. \tag{1.12}$$

In any case, the inverse velocity will be greater for v_b, so that if п is invariant, the mass of the test body is an inverse measure of the displacement and a direct measure of the inverse velocity of the interaction, and a geometrized mass should reflect that kinematic relationship. If the test body is "tethered" in some manner with respect to its initial linear reference frame, so that it is free to move along some circular path into an alternate dimension, that velocity becomes an angular measure instead of a linear one, and mass becomes a measure of the angular frequency.

It is of interest that if we consider a source for our impulse above from a classical body, A, of mass, M_A, where

$$M_A \gg m_a, \tag{1.13}$$

moving with an initial velocity of V_A, prior to the interaction with a, we find the interaction conforms to the following equation

$$v_a = \frac{2M_A}{(M_A + m_a)} V_A. \tag{1.14}$$

Therefore, at the extreme, where m_a is negligible,

$$v_a \approx 2V_A \tag{1.15}$$

and the final velocity of the test body is principally a function of the source velocity and not of the source mass. We would expect that a similar relationship would hold, if instead of representing a source in an elastic collision, M_A represented a gravitational source. If gravitational and inertial mass are equivalent, then $M_A V_A$ represents the impulse generated by that source, and the final velocity of a test particle a is limited by equation (1.15). Thus if V_A is limited by the speed of light, c, then v_a will be limited to $2c$. While this appears to be a violation of the postulates of relativity, when we examine the properties of an extreme Kerr metric later, we will find some justification, since 2 is the coefficient for the tangential or angular component of the metric at the horizon.

With respect to a quantum interaction, we see that equation (1.5) is related to the action, S, of the interaction, using Maupertuis' principle, by

$$\begin{aligned} S &= \int_{x_i}^{x_f} J(x) \cdot dx = \frac{2m}{t_f} \int_{x_i}^{x_f} x_f \cdot dx = \frac{2m x_f}{t_f} \cdot \frac{x_f}{2} \\ &= m x_f \cdot \frac{2 x_f}{2 t_f} = \hbar \cdot \frac{v_f}{2} = \hbar \cdot c = \hbar \omega \cdot x_f = \hbar \end{aligned} \tag{1.16}$$

As $S = \hbar$ is an invariant of the quantum interaction, and m and x are inversely related, so t must be inversely related to m as well (and directly related to x as demonstrated in the previous section). Inverse time is the expression of a rate or in this case unit frequency of interaction, so that mass is the dynamic representation of the kinematics of that unit or angular frequency of the interaction, which varies in proportion to the mass and with respect to the above relationship of a and b, as

$$\omega_b > \omega_a. \qquad (1.17)$$

Equation (1.16) indicates that the ratio of mass to frequency is equal to the ratio of the inertial constant and one half the "final" velocity of the particle. If S and $\sqcap$ are both invariants, then so must be v_f, and with some substitution and rearrangement we have

$$m = \frac{\sqcap}{\tfrac{1}{2}v_f}\omega = \frac{\sqcap}{c}\omega, \qquad (1.18)$$

where

$$c = \frac{v_f}{2} = \frac{2x_f}{2t_f} = \frac{x_f}{t_f} \qquad (1.19)$$

From this we have the following expressions for the impulse,

$$J = \Delta P = 2mc = 2\sqcap\omega \qquad (1.20)$$

which states mass as frequency in a quantum interpretation, since by multiplying through by $\tfrac{1}{2}c$, (differentiating with respect to time and integrating with respect to displacement), we have the mass-energy relationship

$$E = \tfrac{1}{2}Jc = mc^2 = \sqcap c\omega = \hbar\omega = \frac{\hbar c}{\lambda}. \qquad (1.21)$$

It follows that

$$\sqcap = \frac{\hbar}{c}. \qquad (1.22)$$

Returning to equation (1.1) and substituting from equation (1.20) gives the following relationship between the length of the interaction and m_a, which is as equation (1.6),

$$m_a = \frac{Jt_a}{2x_a} = \frac{\hbar\omega_a}{c} = \frac{\hbar}{\lambda_a} = \hbar\kappa_a. \qquad (1.23)$$

We find that for individual quantum mass, i.e. that of the neutron, electron, proton, muon, etc., x_q is equal to the Compton reduced wavelength, $\lambdabar_{C,q}$, for that quantum, as given by

$$x_q = \lambdabar_{C,q} = \frac{\hbar}{m_q}. \qquad (1.24)$$

Quantum analysis assumes the two fundamental invariants, $\hbar$ and c, to which we have now added an inertial constant, $\hbar$. Some simple numerical analysis applied to the variables of mass, displacement, and time in conjunction with the equations for impulse, the inertial constant, interaction terminal velocity and action will help to clarify the geometric relationship of mass, length and time.

In the following table, Row A gives our initial, normalized condition for the variables valued in brackets in the left-hand column. In the remaining rows of this table we have substituted a new body of the given mass, and assumed different space and time values according to various impulse assumptions. The column on the right states whether the set of assumptions in the variables column violates any of the assumed invariants just stated.

Rows B and C maintain the same impulse and have the same v_f, but the space and time intervals differ and neither maintains the velocity Row A. Row D maintains that velocity, but violates the action and the inertial constant condition. It also departs from the initial value of the impulse. The stipulation of a set value for the impulse was a convenience for purposes of development of our argument, but it is not a necessary or even anticipated condition of a physical interaction. Row E is constrained to maintain that impulse, thus maintaining the velocity found in B and C, but results in a violation of all three invariants and is not a quantum solution.

Only Row F and the related G, while necessarily departing from the initial impulse, avoid a violation of the three invariants. What F and G show at a glance, assuming a quantum context, is that quantum inertial test mass is an inverse measure of space and time, the latter two of which are identically gauged in keeping with the development of the previous section on kinematics in which we saw that $r_0^o = t_0^o = R_0^o$.

In Row F, if the space and time standards are assumed to be smaller by the inverse of the factor γ due to a contraction in a moving frame after impulse, the increase in the mass is found to be by the factor γ, showing that Row F is consistent with the postulates of special relativity. Row G, on the other hand, shows an increase in the space and time standards in keeping with a change in γ and a corresponding decrease in mass as we might find in a moving frame that has decreased its velocity from a prior greater differential with respect to some rest frame.

As a source, mass is a direct measure of the impulse as shown by the second column of Rows F and G. Further, since the space and time intervals are identical, and we might assume symmetrical, i.e. interchangeable, it is apparent that the impulse has the same dimensional form as the spin energy of the quantum. Again, using the angular frequency in computing the final velocity, we have the same form for the inertial constant and that velocity, so that in natural units,

$$c^2 = |\eta c| = |\hbar| \qquad (1.25)$$

and mass and frequency measure the same physical condition, interaction per time interval, i.e. the smaller the interaction time, the greater the mass and frequency, as

$$\frac{m}{\omega} = \frac{\hbar}{c^2} = \frac{\eta}{c} = 1. \qquad (1.26)$$

	$F(m, x_f, t_f)$ $= 2mx_f t_f^{-2}$	J $= \int_{ti}^{tf} F(t)\,dt$ $= Ft_f$	η $= \int_{ti}^{tf} J(t)\,dt$ $= mx_f$	$c =$ $\frac{v_f}{2} = \frac{x_f}{t_f}$ $= x_f \omega_f$	$S(=\hbar)$ $= \int_{xi}^{xf} J(x)\,dx$ $= \eta\frac{v_f}{2}$	Violations of η, c, S
A	$(1,1,1)$	$2\frac{1(1)}{1^2}(1) = 2$	$1(1) = 1$	$1 = 1(1)$	$1(1) = 1$	
B	$(2,\tfrac{1}{2},1)$	$2\frac{2(\tfrac{1}{2})}{1^2}(1) = 2$	$2(\tfrac{1}{2}) = 1$	$\tfrac{1}{2} = \tfrac{1}{2}(1)$	$1(\tfrac{1}{2}) = \tfrac{1}{2}$	c, S
C	$(2,1,2)$	$2\frac{2(1)}{2^2}(2) = 2$	$2(1) = 2$	$\tfrac{1}{2} = 1(\tfrac{1}{2})$	$2(\tfrac{1}{2}) = 1$	η, c
D	$(2,1,1)$	$2\frac{2(1)}{1^2}(1) = 4$	$2(1) = 2$	$1 = 1(1)$	$2(1) = 2$	η, S
E	$\left(2, \tfrac{1}{\sqrt{2}}, \sqrt{2}\right)$	$2\frac{2\left(\tfrac{1}{\sqrt{2}}\right)}{\sqrt{2}^2}\sqrt{2} = 2$	$2\left(\tfrac{1}{\sqrt{2}}\right) = \sqrt{2}$	$\tfrac{1}{2} = \tfrac{1}{\sqrt{2}}\left(\tfrac{1}{\sqrt{2}}\right)$	$\sqrt{2}\left(\tfrac{1}{2}\right) = \tfrac{1}{\sqrt{2}}$	η, c, S
F	$(2,\tfrac{1}{2},\tfrac{1}{2})$	$2\frac{2(\tfrac{1}{2})}{\tfrac{1}{2}^2}(\tfrac{1}{2}) = 4$	$2(\tfrac{1}{2}) = 1$	$1 = \tfrac{1}{2}(2)$	$1(1) = 1$	None

G	$(\frac{1}{2}, 2, 2)$	$2\frac{\frac{1}{2}(2)}{2^2}(2) = 1$	$\frac{1}{2}(2) = 1$	$1 = 2(\frac{1}{2})$	$1(1) = 1$	None

Table 1 – Numerical Analysis of Invariant Violation of Certain Variable Assumptions

The symmetries are yet more pronounced since the speed of light, written in terms of the properties of a wave, can be stated as the ratio of the angular frequency and angular wave number, κ,

$$c = \frac{\omega}{\kappa} \qquad (1.27)$$

which when substituted into equation (1.26) gives us the symmetrical statement for the inertial constant,

$$\sqcap = \frac{m}{\kappa}. \qquad (1.28)$$

A couple of words are in order concerning frequency, which are no doubt obvious to reader. First, since it is an expression of the ratio between a count of the number of units or radian contemporaneous with a unit of time, in keeping with the comments concerning equation (0.1), it is equal to a count of one radian per fraction of some larger unit of time. A base or unit frequency would be an extremal, normalized frequency of one radian or other briefest instance of change per one smallest standard of time, $t_0^o = r_0^o = R_0^o$. Thus an interaction of the greatest frequency and therefore greatest energy per equation (1.21) will be the one of shortest duration. Second, while such normalized frequency might be a conventional angular frequency of one radian per unit of time, it might equally be one unit of space per unit of time as

$$c = r_0\omega_0 = r_0 \frac{1}{t_0} = \frac{r_0}{t_0}. \tag{1.29}$$

If we state with respect to Rows F and G that

$$x_f = r_0 \tag{1.30}$$

then the displacement x_f resulting from impulse J, can be a reference to a rotational tangent vector at the circumference of the previously depicted rotating clock, instead of the customary translational displacement vector at or from a point-like particle. Such impulse, under the constraints of an invariant c, results in a contraction of the clock, and a decrease in r_0 and t_0 in keeping with γ, and mass is correspondingly measured as increased. Such impulse could be the result of an inelastic collision with a photon-like source or the acceleration arising from some field potential. It is important to note in regards to a field gradient, that the increase in mass, as with the impulse, can be continual and not in discrete steps and still adhere to equation (1.21), since the frequency can increase continually, while the action, $S = \hbar$, remains invariant at any frequency.

To make graphic sense of this in terms of an inertial spacetime continuum, for modeling purposes we can think of an elastic collision between two bodies of equal mass and spherical shape, a and a', constrained so that their motion oscillates in simple harmonic motion. Next we consider a, instead of being a body or particle, to be a 3 dimensional continuum, non-particulate in composition, isotropic but for a boundary plane at the point of impact from a', where a' is moving normal to and in the direction of a. Instead of the mass quantity of body a, continuum a has a linear inertial density, λ_a. That density is subject to variability and elastic strain in addition to being inertial, so that as a' meets a, the inertial density immediately in front of the line of travel of a' increases, slowing its velocity, and the continuum around the area of impact of a' is strained and curved inward.

If we had assumed an <u>inelastic</u> collision, at some point the momentum of a' would be absorbed by a, which would then remain in a distorted condition, marked by a finite degree of strain and curvature of the continuum around the area of impact, and the impulse would continue to diminish indefinitely into the interior of a.

Given a fixed initial momentum of a', the greater the inertial density and therefore the mass of a, the smaller will be the penetration of a' and the radius of the strain at the area of the impact. We can envision that there are two instances of curving, one as a generally deformed hemispherical area around the area of interaction of a and a' and the other along the sides of the generally toric deformation of the initial plane of the interaction. For simplicity we will assume that the radii are of equal magnitude, though necessarily of different sense.

With an elastic collision, at some point all the momentum of A will be transferred to a, but in the case of a continuum a, at such point all the kinetic energy of A is transferred to the elastic potential energy of the stress and strain at the deformation of a. As the restorative forces in the plane of the interaction of aa' exceed the compression force of a' on a, a force which is normal to the interaction plane, a will rebound and begin to work on a', which will travel in the opposite direction, eventually to be expelled from the plane of the initial impact.

We imagine this interaction now with another half continuum mirrored opposite a, so that the total system of the aa' interaction constitutes a resonant oscillation of a localized, ellipsoidal section of the combined continuum along the center axis aa', in which $m_a = m_{a'} = m_0$.

Using equation (1.6), we can state the linear inertial density, λ_0, of the continuum at the system as follows,

$$\lambda_0 = \frac{m_0}{r_0} = \frac{\hbar}{r_0^2} = \hbar \kappa_0^2. \qquad (1.31)$$

This indicates that the linear inertial density is equal to the inertial constant times the curvature of the deformation or strain, as shown in the last term. Assuming an isotropic Gaussian curvature, k, given by

$$k = \kappa_0^2 = \frac{1}{r_0^2}, \qquad (1.32)$$

this means that mass is a measure of <u>linear</u> curvature, given by the angular wave number, κ, once again indirectly related to the length scale, as

$$m_0 = \frac{\hbar}{r_0} = \hbar \kappa_0. \qquad (1.33)$$

Now we stipulate that instead of a linear oscillation, we have a torsional oscillation of a small section of a generally rigid continuum about an axis, θ, so that at resonant frequency a wavelength of characteristic angular wave number, κ_0, develops. As a torque vector, θ flips in direction and oscillates in intensity with each change in the direction of the torsion. We next rotate the axis of torsion about a center of wavelength, such that the transverse momentum of θ carries the two nodes into a newly defined axis, ϕ, in the general helical path of the oscillation. Restorative forces prevent rotation of ϕ in θ beyond a point and the original oscillation of θ continues, now aligned as ϕ, between its original nodal poles and causing those displaced nodes to rotate about θ without entangling or twisting the continuum beyond a general range of one half π. As a result, the torque vector for ϕ does not oscillate as did θ. It rotates about θ, so that θ ceases to flip its direction and becomes a sustained angular momentum vector according to the rotation. Note that θ remains the primary torque, with ϕ as a derivative, so that the interaction between the two is non-commutative and there is a tendency over time for ϕ to realign as θ.

The motion can be crudely emulated by imagining a basketball held out in your two hands so that the label is facing up and away from you, i.e. it would be readable by someone facing you and looking down on it. The label represents the ultimate $+\theta$ direction and you can start the initial oscillation by moving your hands back and forth so as to rotate the ball about an axis through the center of the ball and label. When the ball is rotated so that your right hand is closest to your body, start the return counterclockwise cycle, but initiate a second rotation so that you start a new motion by 1) rotating the right hand over, left hand under and twisting 90 degrees ccw so that someone on your left can tilt their head to the right and read the label, 2) returning the hands to their first position, but twisting the ball back so that the label is facing you upside down, 3) rotating the left hand over, right hand under and twisting 90 degrees ccw again so that someone on your right can tilt their head to the left and read the label, and 4) returning the hands to their original position, but twisting the ball over so that the label is readable by someone at a distance in front of you, 5) continuing on to step number 1. Continue the process so that the motion between each step is smooth and the label remains ideally in the same plane or in a wide cone of even angle. Each hand will trace a figure 8 with each cycle, crossing at 45 degrees to the plane and in the direction of rotation with each half cycle. The label and its antipode represent the original boundaries of θ, now of ϕ, and rotate while avoiding entanglement. The a) left and right hands and the b) top and bottom of the ball when at step 2 oscillate between the top and bottom position, 90 degrees out of phase, while the motion of the hands between position 1 and 3 clearly shows, in the context of a continuum, the torsion involved in both a and b.

Such rotational oscillation mimics the rotation of our three dimensional clock developed in the previous section on kinematics. While locally constrained by the stress of an expanding spacetime, the wave phasing which manifests as spin is free to transform translationally and rotationally subject to field and particle interactions and perturbations.

The torsion forms a sinusoidal wave rotating about θ, with a sustained displacement of the initial θ nodes. In this context, the moments of maximum capacitive and inductive power of the wave generate a capacitive, C_ε, and an inductive, L_μ, torque (crossed from a position of zero displacement into the most immediate direction of increasing force, thus backwards in time for L_μ) that rotate with θ, 90 degrees out of phase with each other, C_ε generally parallel and L_μ generally anti-parallel. These torques interact with the nodes of ϕ to prevent realignment with θ. The rotation of the inductive torque generates an effective magnetic moment anti-parallel to the angular momentum vector of θ.

In the presence of an isotropic expansion, the characteristic inertial density and related mechanical impedance, given by

$$Z_0 = \lambda_0 c = \eta \kappa_0 \omega_0, \qquad (1.34)$$

decrease over time, advancing the inductive moment and dropping the frequency. This results in the flip of the inductive torque to a generally parallel position behind the capacitive torque, which also flips the effective magnetic moment to parallel. A power transmission results anti-parallel to the inductive torque and the spin vector, along with a generation of charge, the latter of which is a measure of transverse wave momentum; we recognize this as beta decay. The decreased frequency is that of the proton, and that of the emitted wave is that of the electron. The rest frequency of the latter is a function of geometry and the expansion rate.

With a retarding of the capacitive moment, the capacitive torque flips anti-parallel, the oscillation becomes an anti-proton and emits a positron. The expansion of spacetime is more conducive to the inductive advance, resulting in a predominance of the proton-electron system.

Thus a geometrization of massive-particle mass involves the representation of quanta as three dimensional clocks and indicates that particle mass is a measure of the frequency of the clock. As the above continuum is a representation of 3-D space, its quantization represents an oscillation of a local section of space that is made discrete by the boundaries of its harmonic oscillation. If that oscillation is seen to be at resonant frequency, ω_0, then we have the following relationship to the wave speed in such continuum

$$\kappa_0 = \frac{\omega_0}{c} \tag{1.35}$$

which when substituted into equation (1.31) gives the following wave equation, where τ_0 is the tension force in the continuum,

$$\mathsf{n}\kappa_0^2 = \lambda_0 = \frac{1}{c^2}\tau_0 = \frac{\mathsf{n}\omega_0^2}{c^2} \tag{1.36}$$

where by canceling the inertial constants and substituting conventional derivatives for the Euler versions we have the dimensionally equivalent wave equation

$$\frac{\partial^2 \psi}{\partial x_i^2} = \frac{1}{c^2}\frac{\partial^2 \psi}{\partial t^2} \tag{1.37}$$

Finally, integrating equation (1.36) with respect to the wave number shows the basic wave nature of mass – energy equivalence as

$$-im_0 = \frac{\mathsf{n}\kappa_0^2}{i\kappa_0} = \frac{1}{c^2}\frac{\mathsf{n}\omega_0^2}{i\kappa_0} = -i\frac{E_0}{c^2} \tag{1.38}$$

From this discussion we can state some basic quantum dynamic properties of interest in terms of the inertial constant:

Interaction impulse = transverse wave momentum	$\Delta p = mc = \hbar\omega$
Force – stress force = transverse wave force	$\tau = \hbar\omega^2$
Action = spin angular momentum	$S = \hbar = \hbar c = \hbar\frac{\omega}{\kappa}$
Rest Mass	$m = \hbar\frac{\omega}{c} = \hbar\kappa$
Spin Energy	$E = mc^2 = \hbar\omega = \hbar c\omega$

For the etymologically inclined, the word *mass* is from the German *massieren* meaning *to knead* dough, and evokes the notion of folding and stretching the dough with the heel of the hand, turning it 90°, and repeating the process. This action forms gluten, allowing the dough to catch the gas of the leavening agent and expand. The symmetry of this scenario, with its orthogonal folding and rotation of dough, and the torsional rotation of the 3-D clock developed herein as an analogue of the fundamental rest mass rotational oscillation of spacetime is inescapable.

Gravitational Mass (Source)

In general relativity, gravitational source mass is converted from conventional units related to a force, M_{kg}, to units of length, r, as M_l, by the conversion factor of G_N/c^2, where G_N is Newton's gravitational constant and c^2 is the speed of light in a vacuum squared, both of which are taken as free parameters, as

$$r = M_l = \frac{G_N}{c^2}M_{kg} = \left(7.424 \times 10^{-28} \tfrac{m}{kg}\right)M_{kg}. \qquad (1.39)$$

here evaluated using the CODATA SI values.[4] This procedure facilitates computation, as when used in a metric, so that if M_l is the geometrized mass of an extreme Kerr black hole, the reduced circumference at the horizon is $r_h = M_l$.

It bears noting that the relationship between the two measures of mass is <u>direct</u> and appears to be classical, so that we can state a differential form

$$dr = dM_l = \frac{G_N}{c^2} dM_{kg} \qquad (1.40)$$

We should acknowledge, however, the obvious and logical possibility that M_{kg} is an aggregation of some basic quantized units of mass of one or more magnitudes. Consideration of this equation using the smallest of rest-mass quanta, the electron, gives a linear measure of its mass in orders of magnitude of 10^{-58} meters. For the proton and neutron, the figure is slightly larger at 10^{-54} meters. However, all of these are much smaller than the Planck length of 10^{-35} the reputed smallest of determinable physical scales, raising possible theoretical questions about the use of equation (1.39) in determining a geometrized mass for individual quanta.

As previously discussed at equation (1.24), the mass of an individual quantum, a neutron, proton, electron, tau, or muon is related to its reduced Compton wavelength by the following, where the r_q is the reduced circumference and the norm of a polar coordinate system centered on q.

$$\lambda_{C,q} = \frac{\hbar}{c} \frac{1}{m_q} = \frac{\daleth}{m_q} = r_q. \qquad (1.41)$$

In the SI system, $\daleth$ (tav), evaluates as

$$\Pi \equiv \frac{\hbar}{c} = m_q r_q = 3.5176 \times 10^{-43} \, kg \cdot m \tag{1.42}$$

In summary, in contrast to the direct relationship in the geometrization of mass in the classical application of general relativity, in quantum theory conventional mass is <u>indirectly</u> related to length. If we consider a relativistic quantum qualitatively, we know that the deBroglie wavelength decreases as the relativistic mass and the particles momentum increases, indicating once again the inverse relationship of mass and length.

3 – Derivation of Newton's Gravitational Law with Quantum and General Relativistic Principles

We would like to derive Newton's Gravitational Law from quantum principles, while in keeping with the principles of general relativity. The quantum principle we are interested in is that of fundamental discrete units or quantities of rest mass. This means that we seek to express the gravitational force, F, of his law as a product of 1) the number, n_a, of some as yet unknown fundamental discrete units of mass, m_0, in two aggregate bodies of mass, M_a, 2) the curvature of space, k, expressed as the inverse square of the massive bodies separation in numbers, n_r, of some as yet unknown minimum unit of length specifically of a reduced circumference, r_0, and 3) a fundamental discrete unit or quantum differential of gravitational force, dG_0, as

$$F_{m_1 m_2 k} = n_{M1} n_{M2} n_r^{-2} dG_0 \tag{2.1}$$

We state Newton's Law, where G_N is Newton's gravitational constant, conventionally considered a free parameter, as

$$F_{M_1M_2k} = \frac{M_1M_2}{R^2}G_N = M_1M_2kG_N. \qquad (2.2)$$

Assuming a 3-space that is isotropic with respect to a source mass, M_1, here we have made use of the observation that the inverse square component of the distance of separation, R, of M_1 and M_2 is the reduced circumference of the spacetime around M_1, making the inverse of the square of R the measure of the Gaussian curvature, k, at the location of M_2, using

$$k = \frac{1}{R^2}. \qquad (2.3)$$

The left hand side of equation (2.2) represents a force. Some reflection will tell us that if it is to be related to general relativity, the right hand side must represent the product of a 4-stress, T, and an area, A, or in keeping with the last paragraph, an inverse curvature. Thus this equation is dimensionally equivalent to

$$F = TA = Tk^{-1}. \qquad (2.4)$$

Some rearrangement gives us a scalar form

$$k = F^{-1}T \qquad (2.5)$$

where the curvature of spacetime given by the left term is related to the stress-energy density of the right by the inverse force. This is thus related to the field equation of general relativity, customarily expressed in tensor form as

$$G_{\alpha\beta} = -8\pi G_N T_{\alpha\beta} \tag{2.6}$$

where the Einstein curvature tensor on the left is similarly related to the stress-energy tensor on the right by the geometrically based numerical coefficient and Newton's gravitational constant, which we will see contains a force differential.

Analyzing G_N dimensionally, we know it has the dimensions of distance, r, cubed divided by the product of a mass, m, and time, t, squared. If it in fact conceals a force differential, extracting that force in the third term shows G_N to be the product of that force and the inverse square of a linear inertial density, λ, as

$$G_N = \frac{r^3}{mt^2} = \frac{r^2}{m^2}\frac{mr}{t^2} = \lambda^{-2} dF \tag{2.7}$$

The inverse inertial densities in Newton's constant then convert the product of the masses on the right side of Einstein's field equation (2.6), of which there are two, one in the force differential and one in the stress tensor, to the product of two distances. The result, however, does not give the dimensions of curvature. With respect to equation (2.5), which has an inverse force on the right, the G_N as shown in equation (2.7) has a direct force. Using the fundamental identity of space and time as shown in equation (0.7), we can make the following dimensional substitutions into equation (2.7),

$$G_N = \frac{r^3}{mt^2} = \frac{(it)^3}{m(-ir)t} = \frac{t^2}{mr} = dF^{-1} \tag{2.8}$$

53

which converts G_N to an inverse force and equation (2.6) assumes the dimensional form of equation (2.5).

Expressed as a force, gravity is centrally directed toward the bodies of mass and within the context of a flat spacetime, assumed to be isotropic about each. The curvature in such conditions is considered generally spherical, so that some rearrangement of equation (2.4) in the absence of any rotation of the two bodies about each other, results in a centripetal gravitational tension stress, f_3, where the subscript indicates the dimensional order of the stress

$$f_3 = \frac{F}{A} = Fk. \tag{2.9}$$

The stress in the case of general relativity is a 4-stress, however, so that we are looking for a formulation that makes explicit the relationship between a 3-stress and a 4-stress, T_4

$$f_3 \equiv \frac{F}{A} \sim T \equiv T_4. \tag{2.10}$$

Additionally, we are interested in the 4-stress associated with an accelerating expansion of space, so we take a closer look at the geometry of stress, specifically of isotropic expansion stress. We examine the case of energy density - stress in an n-manifold that is expanding in response to its expanding n+1-core. First, in Stress Diagram 1 we examine the differential area of a 2-sphere on a 3-ball, such as an expanding balloon. We imagine that the balloon is expanding due to a differential pressure normal to the balloon surface, so that there is a 3-stress (tension), dT_3, orthogonal to the balloon's surface, the 2-sphere. We look at a differential square on the surface of the balloon and see that the sum of the

2-stress (transverse or shear), df_2, in the balloon surface at that locus should be equal to the orthogonal tension stress, or

$$dT_3 = \gamma_2 df_2 \qquad (2.11)$$

where γ_2 is a geometric factor summing the shear stress.

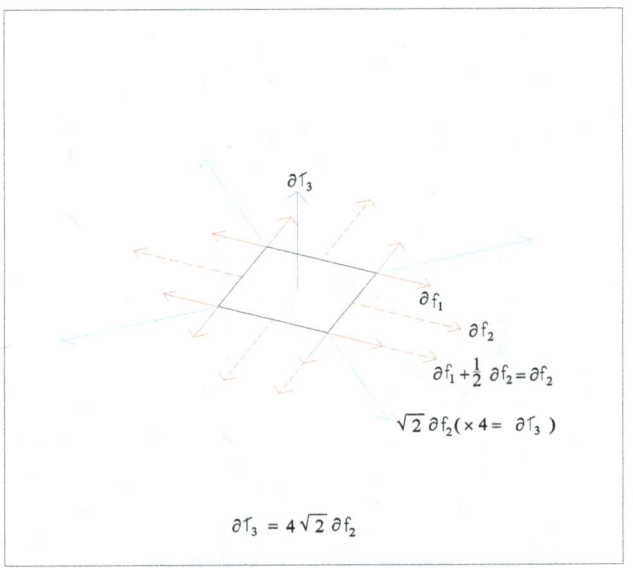

Stress Diagram 1
Figure 12

It is the displacement of the vertices of the square that defines the change, so instead of a normal unit vector to each mid-edge of the differential square, we stipulate a ½ vector at each vertex, along with a ½ extension or shear vector from the adjacent edge, giving a total of 8, 1 vectors at the four vertices. With a total of 4 resultants of the vector pairs at each vertex, we have

$$\gamma_2 = 4\left(\sqrt{1^2 + 1^2}\right) = 4\sqrt{2} \qquad (2.12)$$

and equation (2.11) becomes

$$dT_3 = 4\sqrt{2} df_2 \qquad (2.13)$$

Extending this approach with analogous elasticity conditions to a 3-space on a 4-core, in Stress Diagram 2 we have a 4-stress, (which necessarily cannot be shown) normal to and equal to an isotropic 3-stress, as

$$dT_4 = \gamma_3 df_3. \tag{2.14}$$

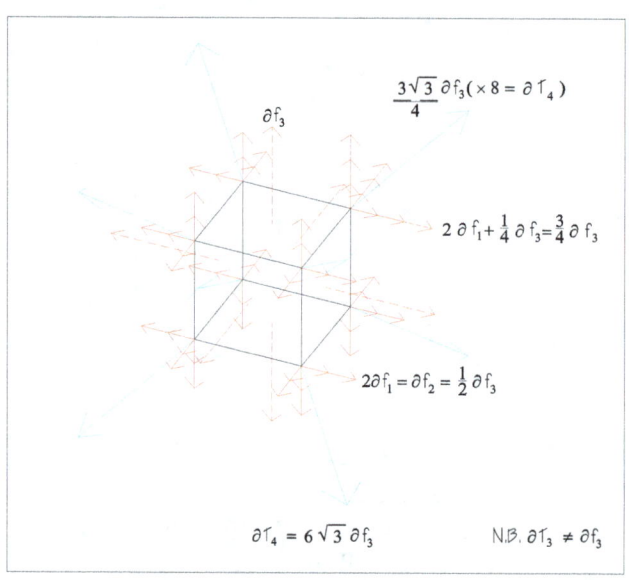

Stress Diagram 2
Figure 13

This time we consider a differential cube, and instead of the customary assignment of an orthonormal tension stress vector to the center of each of the faces of the cube, we assign a quarter of each normal vector to the corners of each face, collinear with and equal to the shear vectors of the two adjacent surfaces. This is equivalent to a Poisson's ratio of 1/3. The sum of these ¼ tension vectors and the two parallel ¼ shear vectors is a ¾ vector, so that there are 3, ¾ orthogonal stress vectors at each vertex. The resultant of the three orthogonal components at each corner then, aligned with the cubic diagonal, is the total stress contributed to each of the 8 vertices by an isotropic stress, so that the geometric factor relating the stresses in equation (2.14) is

$$\gamma_3 = 8\left(\sqrt{\left(\tfrac{3}{4}\right)^2 + \left(\tfrac{3}{4}\right)^2 + \left(\tfrac{3}{4}\right)^2}\right) = 8\frac{3\sqrt{3}}{4} = 6\sqrt{3} \qquad (2.15)$$

and equation (2.14) becomes

$$dT_4 = 6\sqrt{3}\, df_3 \qquad (2.16)$$

Next we examine a scalar expression of the equation (2.10) in light of this adjustment, where we specify that A_0 is a fundamental quantum unit area,

$$\gamma_3^{-1} T = \frac{F}{A_0}, \qquad (2.17)$$

with the derivatives for an invariant T being

$$\gamma_3^{-1} dT = \frac{\partial T}{\partial F} dF - \frac{\partial T}{\partial A} dA = \frac{1}{A_0} dF - \frac{F}{A_0^2} dA = 0. \qquad (2.18)$$

Separating and inverting this function we have the two following differential equations, the first of which is straight forward,

$$dF = \gamma_3^{-1} A_0 dT \equiv \left(\gamma_3^{-1} \kappa_0^{-2} dT\right) \qquad (2.19)$$

and the second one expressing various parsings of interest, especially those in which the stress force is removed from the equation,

$$dA = -\gamma_3^{-1} \frac{A_0^2}{F} dT = -\frac{F}{\gamma_3^{-1} T^2} dT = -\frac{A_0}{T} dT$$
$$= -A_0 d\ln T \equiv \left(-\kappa_0^{-2} d\ln T\right) \qquad (2.20)$$

According to the above specifications a quantum formulation for Newton's Law, as previously stated, would be

$$F_{m_1 m_2 k} = n_{M1} n_{M2} n_r^{-2} dG_0. \qquad (2.21)$$

An aggregate mass is the product of the number of quanta in that aggregate times the fundamental unit of mass or with rearrangement

$$n_{Ma} = \frac{M_a}{m_0} \qquad (2.22)$$

and the reduced circumference of the separation of the two bodies of mass is the product of the number of unit lengths in that separation and the minimum or quantum unit length, or

$$n_r = \frac{R}{r_0}. \qquad (2.23)$$

Substituting equation (2.22) and equation (2.23) into equation (2.21), noting the dimensional equivalence of the bracketed term to equation (2.7), gives

$$F_{M_1 M_2 k} = \frac{M_1 M_2}{R^2} \left(\frac{r_0^2}{m_0^2} dG_0 \right) \qquad (2.24)$$

Assuming that the gravitational quantum is equivalent to the formulation from equation (2.19) and substituting from its middle term, gives the following, in which the stress differential is normalized in its relationship to dG_0 as $dT_0 = 1$,

$$F_{M_1 M_2 k} = \frac{M_1 M_2}{R^2} \left(\gamma_3^{-1} \frac{r_0^4}{m_0^2} dT_0 \right) = \frac{M_1 M_2}{R^2} G_N. \qquad (2.25)$$

In keeping with earlier development, we restate the relationship between the above postulated quantum mass, m_0, and length, r_0, the latter stated as the reduced Compton wavelength,

$$m_0 = \frac{\hbar}{c}\lambdabar_{C,0}^{-1} = \frac{\hbar}{c}r_0^{-1} = \frac{\sqcap}{r_0}. \tag{2.26}$$

We substitute from equation (2.26) into the bracketed term of equations (2.24) and (2.25), and get

$$G_N = \frac{dG_0}{\lambdabar_0^2} = \frac{r_0^4}{\sqcap^2}dG_0 = \gamma_3^{-1}\frac{r_0^6}{\sqcap^2}dT_0, \tag{2.27}$$

which in a natural system simply equals γ_3^{-1}.

After some rearrangement, we have

$$r_0 = \left(\gamma_3 \sqcap^2 G_N dT_0^{-1}\right)^{1/6}. \tag{2.28}$$

Since with respect to dG_0, dT_0 equals 1, and as we know the other invariants in the right hand term, we can solve for r_0, and find that in the SI system it equals the reduced Compton wavelength of the neutron or

$$r_{0n} = 2.100246...\times 10^{-16}\,m \cong 2.10019...\times 10^{-16}\,m = \lambdabar_{C,n} \tag{2.29}$$

within the standard uncertainty for G_N. The "n" in the subscript "$0n$" is redundant and simply emphasizes the neutron scale as the fundamental, quantum scale. All other values for the fundamental properties incorporate and can be computed from this value. Therefore, the fundamental gravitational mass is the neutron mass or

$$m_0 = m_n = 1.67492...\times 10^{-27}\,kg. \tag{2.30}$$

This is not stating that the neutron is the only particle responsible for the generation of quantum gravity. In the full development of this model, the proton is seen to be a neutron which has undergone a frequency drop due to the transmission of a portion of its power as the electron in the process of beta decay. It is the derivative of the wave force of the fundamental oscillation with respect to stress that is the gravitational quantum.

The gravitational quantum then is variously

$$dG_0 = \gamma_3^{-1}\kappa_{0n}^{-2}dT_0 = \gamma_3^{-1}A_{0n}dT_0 = \gamma_3^{-1}T_0 dA = 4.244...x10^{-33} N, \qquad (2.31)$$

where the last algebraic term makes use of equation (2.20).

Some rearrangement gives

$$T_0 = \gamma_3 \frac{dG_0}{dA} = \frac{A_{0n}}{dA}dT_0 \qquad (2.32)$$

With this development, we can get the spin energy density-stress, T_0, of the neutron, which we assume to be a quantum waveform, where E_{0n} is the spin energy of the neutron and

$$\tau_0 = \hbar\omega_0^2 \qquad (2.33)$$

is the transverse wave force of the oscillation,

$$\gamma_3^{-1}T_0 = \frac{E_{0n}}{r_{0n}^3} = \frac{m_n c^2}{r_{0n}^3} = \frac{\hbar\omega_0^2}{r_{0n}^2} = \frac{\tau_0}{r_{0n}^2} = 1.625...x10^{37} N/m^2. \qquad (2.34)$$

Substituting this into equation (2.31) for the gravitational quantum and rearranging, we get the following expression and value for the differential of the unit area, $dA_0 = dA$, which we find is equal to the Planck area,

$$dA_0 = \gamma_3 T_0^{-1} dG_0 = A_{Pl} = 2.6116...x10^{-70} m^2 \qquad (2.35)$$

This analysis indicates that Newton's gravitational constant contains a quantum differential, and that the neutron scale is the fundamental scale of an expanded spacetime. It also indicates a relationship to the Planck scale, and we would like to determine more of that relationship next.

4 – Analysis of the Relationship between the Neutron and the Planck Scale

If we use the conventional geometrization factor from general relativity for mass, G_N/c^2, for the neutron we get a length measure of a hypothetical quantum black hole horizon as

$$r_{hn} = m_{l,n} = \frac{G_N}{c^2} m_n = 1.243...x10^{-54} m. \qquad (3.1)$$

Comparing this with the neutron reduced Compton, we get the dimensionless number

$$\frac{m_{l,n}}{\lambdabar_{C,n}} = \frac{r_{hn}}{\lambdabar_{C,n}} = 5.92...x10^{-39}. \qquad (3.2)$$

Squaring equation (3.1) to get the inverse curvature of a hypothetical quantum inertial sink at that scale gives

$$r_{hn}^2 = 1.545...x10^{-108} \qquad (3.3)$$

which is related to the Planck area by the same ratio or

$$\frac{r_{hn}^2}{A_{Pl}} = 5.92...\text{x}10^{-39}. \tag{3.4}$$

It bears noting that this is in the range of the factor separating the gravitational and the strong interactions.

Using the structure for Newton's constant developed above, we analyze the conventional geometrization factor, where we make use of the classical wave relationship,

$$\tau_0 = \lambda_0 c^2 \tag{3.5}$$

in which τ_0 is the linear tension force and in this case the transverse wave force in a wave bearing medium, λ_0 is the linear inertial density of that medium and c is its speed of wave propagation. We find that the conventional conversion factor is equal to the differential of the natural log of the expansion stress divided by the linear inertial density,

$$\frac{G_N}{c^2} = \left(\frac{dG_0}{d\lambda_0^2}\right)\frac{1}{c^2} = \frac{1}{\lambda_0}\left(\gamma_3^{-1}\frac{r_{0n}^2}{\lambda_0 c^2}\right)dT_0 = \frac{1}{\lambda_0}\left(\gamma_3^{-1}\frac{r_{0n}^2}{\tau_0}\right)dT_0 = \frac{1}{\lambda_0 T_0}dT_0 \tag{3.6}$$

$$\frac{G_N}{c^2} = \frac{1}{\lambda_0 T_0}dT_0 = \frac{d\ln T_0}{\lambda_0} = \frac{\hbar_{C,n}}{m_n}d\ln T_0 \tag{3.7}$$

Using CODATA values for the neutron mass and reduced Compton to determine λ_0, we can solve for $d\ln T_0$ and get the factor found in equations (3.2) and (3.4)

$$d\ln T_0 = T_0^{-1}dT_0 = \gamma_3^{-1}\frac{r_{0n}^3}{m_n c^2}dT_0 = 5.92146...\text{x}10^{-39} \tag{3.8}$$

Inverting and multiplying through by $dT_0 = 1$ gives the value of T_0,

$$T_0 = \gamma_3 \frac{m_n c^2}{A_{0n} r_{0n}} = \gamma_3 \frac{\lambda_0 c^2}{A_{0n}} = 1.6888...\text{x}10^{38} \, N/m^2 \tag{3.9}$$

from which we can get the transverse quantum wave force of the neutron

$$\tau_{0n} = \gamma_3^{-1} T_0 A_{0n} = 7.1676...\times 10^5 \, N \tag{3.10}$$

With the gravitational quantum as the differential of the quantum transverse wave force with respect to differential stress, $\tau'(T)$, we have the ratio of that differential and the wave force itself, $\tau(T)$, or equation (2.31) over equation (3.10)

$$d \ln T_0 = \frac{\tau'(T)}{\tau(T)} = \frac{d\tau_0}{\tau_{0n}} = \frac{dG_0}{\tau_{0n}} = 5.92146...\times 10^{-39} \tag{3.11}$$

which is the ratio of the gravitational and the strong interactions.

Rearranging equation (3.10) and taking the derivative of inverse curvature with respect to the isotropic stress results in an evaluation equal to the Planck area,

$$dA_0 = -\gamma_3 \frac{\tau_{0n}}{T_0^2} dT_0 = -A_{0n} d \ln T_0 = -A_{Pl}. \tag{3.12}$$

once again indicating that the Planck area represents a differential of expansion stress. To verify this statement, we substitute equations (2.33), (3.5), and (3.9) for the expansion force and stress into the second term here and find

$$\begin{aligned} dA_0 &= -\gamma_3 \frac{\hbar \omega_{0n}^2}{\gamma_3^2 \lambda_0^2 c^4 A_0^{-2}} dT_0 = \left(-\gamma_3^{-1} \frac{A_0}{\lambda_0^2} dT_0 \right) \frac{\hbar \omega_{0n}^2 r_{0n}^2}{c^4} \\ &= -G_N \frac{\hbar c^2}{c^4} = -G_N \frac{\hbar}{c^3} \end{aligned} \tag{3.13}$$

From this analysis of the differential nature of the Planck area and the endnote comments,[i] which suggest expansion along a hyperbolic manifold, from equation (3.12) we can show the Planck length as a differential value, as

$$dr_0 = |dA_0|^{\frac{1}{2}} = r_{0n}\sqrt{d\ln T_0} = r_{Pl} = 1.6161...\times 10^{-35}\,m. \qquad (3.14)$$

5 – Cosmological Implications

Basic to our discussion is the assumption that spacetime is expanding relative to our local frame of reference. This means that over time a local fixed unit length standard becomes an ever decreasing proportion of some linear measure of the cosmic extent. If we project backwards in time, we can assume that at some point that measure of cosmic extent was equal to the current local length standard or unity.

The current concept of a big bang start of cosmic spacetime expansion implies an initial condition of maximum inertial density, possibly infinite, which decreases with the expansion of space from an extremely small volume, possibly zero, i.e. from a singularity. Instead of emergence from a singularity, the space component of spacetime can be modeled as a boundary on the next higher dimensional manifold itself under expansion, analogous to a circle drawn on the surface of an expanding balloon. Alternately, we might imagine a spherical balloon of fixed size with a circular wave emanating from one spot, widening in diameter as it approaches an equator before shrinking again as it nears the antipode. An analogous inertial spacetime oscillates on a cosmic scale between a maximum density and rarification, between a maximum compression and maximum extension. The fact that the expansion appears to be accelerating indicates that the expansion rate is best understood exponentially. We can then take the condition of maximum density as unity instead of as a singularity, and gauge any expansion with respect to that unity for A_0 and r_0 as inversely related to the associated increase in stress T_0 due to expansion according to equations (3.12) and (3.14).

The current expansion factor, κ_{exp}, the ratio of the current fundamental neutron scale to the Planck length, is equal to the inverse square root of the differential natural log of the expansion stress,

$$\kappa_{exp} = \frac{r_{0n}}{dr_0} = \sqrt{d \ln T_0}^{-1} = 1.29952...x10^{19} \qquad (4.1)$$

As this expansion is at an exponential rate, in terms of doubling from an initial condition of maximum density equal to the linear inertial density of the neutron scale, λ_0, with time and space normalized, in terms of the whole or an arbitrary unit standard, cosmic expansion, C_x, is

$$C_x = \ln 2 (\kappa_{exp}) = 9.00764...x10^{18} \text{ light seconds } = 2.8544...x10^{11} \text{ light years} \qquad (4.2)$$

Note that the last term would indicate, if interpreted as a straight line increase at the speed of light, an expansion age of the cosmos of 285.44 billion years.

An exponential expansion rate, X_e, derived in the full development of this model and shown to equate to a predicted Hubble rate of 72.791 km/mps/s and supported by independent studies as 73 km/mps/s +/- 8 km [7] and 72 km/mps/s [8], shows the change in unit scale per second as

$$X_e = H_0 = \frac{\Delta r_0}{r_0} \Big/ \text{second} = 2.35896...\times 10^{-18}/\text{s} \qquad (4.3)$$

If we interpret this as a straight line expansion rate from an initial singularity, inverting would give the age of the cosmos in current units as

$$X_e^{-1} = 13.433 \text{ billion years} \qquad (4.4)$$

However, if the Hubble rate is exponential or compounding, the following gives the Hubble time, τ_H, as a time in current units for a doubling in spatial linear extent, or

$$\tau_H = \ln 2 X_e^{-1} = 9.311 \text{ billion years} \qquad (4.5)$$

The product of the expansion rate and the expansion factor is the number of doublings or

$$X_e \kappa_{\exp} = 30.655... \text{ doublings} = 285.43 \text{ billion years} \qquad (4.6)$$

Following this logic, if the wavelength of the cosmic microwave background is approximately $3.3mm$ and indicates an expansion along with spacetime from a primal epoch of beta decay as gauged by the electron Compton wavelength, $\lambda_{C,e}$, dividing the natural log of such expansion by the natural log of 2 gives the number of doublings based on those parameters or

$$\ln\left(\frac{.0033}{\lambda_{C,e}}\right)(\ln 2)^{-1} = \frac{\ln 1.360...\times 10^9}{\ln 2} = 30.34... \text{ doublings} = 282.5 \text{ billion years} \qquad (4.7)$$

in very close agreement with equation (4.6).

This observation indicates that r_0, related to the reduced electron Compton wavelength, $\lambdabar_{C,e}$, by the ratio of the neutron to electron Compton wavelength of $0.000543...$, remains stable as spacetime and the CMB expands and indicates that such quanta did not have a geometry of the Planck scale at an early epoch, which instead of starting from a singularity with all the physical dilemma that entails, started expansion from a maximum finite density. The Planck length, then, is the ratio of the neutron reduced Compton and the cosmic extension from an initial compact condition of maximum density, and continues to decrease with expansion.

Alternately, but not contradictory, if we think of the cosmic extent of 3-space as a fixed unit, what appears mathematically from a local perspective as expansion is from the universal perspective a process of regional and local concentration of inertial density. With respect to our analogy of the fixed balloon above, the linear (and area) density of the balloon in the absence of a wave is invariant over the surface of the sphere, but a wave moving over its surface creates a density differential at the wave front, increasing as it approaches a pole and antipole and decreasing as it approaches an equator. From the reference frame of the traveling wave front approaching the poles, the stress related to the wave front, T_0, increases and r_0, as a related unit standard which in the case of the balloon we might give as the distance perpendicular to the given polar diameter, decreases over time. The ratio of r_0 with respect to the balloon's extent, B_x, its radius at the equator, represents a decreasing differential length, dr_0, and can be expressed as the cosine of the angle of declination of the wave front.

The wave front in this analogy represents the current local quantum scale given by r_{0n}. If we were to rotate the balloon about the given polar axis at the same frequency as the wave's movement over its face, each point in the wave front would mimic the action of our 3-D clock. From either of the above perspectives, the energy per cosmic extent is invariant and cosmological red-shift is apparent, and in neither case is the Planck scale a fixed discrete scale.

Black Hole Metrics

Assuming that the above and supportive analysis does indicate that the neutron is a quantum inertial sink, but not a quantum black "hole", then a maximum linear inertial density is given by

$$\lambda_0 = \frac{m_n}{r_{0n}} = 7.975...x10^{-12} \, kg/m \tag{4.8}$$

This would seem quite small, but for its bulk implications. For a volume density, we would figure the number of hypothetical fundamental rest mass quanta per volume of such quanta, tightly packed. Using a packing system of one sphere with twelve contacting identical spheres, and disregarding any expansive effects of spin, charge, etc., we can compute the theoretical maximum density and find that it equals roughly

$$2.2549...x10^{46} \, quanta/m^3 \tag{4.9}$$

Inverting the neutron mass gives the number of such quanta per kilogram or

$$5.9704...x10^{26} \, quanta/kg \tag{4.10}$$

for a maximum theoretical density of

$$3.7768...x10^{19} \, kg/m^3 \tag{4.11}$$

or a density per sphere of one meter radius

$$\rho_{sphere} = 1.5820...\text{x}10^{20} \, kg \, / \, meter \, sphere \qquad (4.12)$$

From this we can find a threshold black hole mass, $M_{kg,TBH}$ for an aggregation of quanta by using the following for a flat Euclidean space, where r_{Max} is the reduced circumference of a celestial body of maximum density,

$$r_{Max} = \left(\frac{M_{kg,TBH}}{\rho_{sphere}} \right)^{\frac{1}{3}} \qquad (4.13)$$

Assuming $r_h = M_l$ as with an extreme Kerr spacetime

$$\frac{G_N}{c^2} M_{kg,TBH} = M_l = r_h = r_{Max} = \left(\frac{M_{kg,TBH}}{\rho_{sphere}} \right)^{\frac{1}{3}} \qquad (4.14)$$

for an extreme Kerr horizon gives

$$M_{kg,TBH} = \left(\frac{c^6}{G_N^3 \rho_{sphere}} \right)^{\frac{1}{2}} = 3.930 \text{x} 10^{30} \, kg \qquad (4.15)$$

which using the above density gives us the evaluations in the following table or approximately two solar masses for the threshold.

Here in column 3, from equation (4.13) we compute the r_{Max} for various celestial bodies, Earth, Sun, Milky Way galactic BH and Virgo cluster BH, and include the theoretical threshold size black hole and the Universe, as listed in column 1. "Flat Spacetime" does not specify that the pertinent body has no curvature effect on the surrounding spacetime, but rather that the curvature of individual quanta, i.e. quantum gravity, is not effected by the aggregate mass and remains the same as for an individual quantum in isolation in flat spacetime, i.e. there is no assumed collapse of each quantum waveform toward a quantum singularity, though there may be a state similar to a Bose-Einstein condensate. The fourth column gives the reduced circumference at the horizon of an extreme, charge free, Kerr black hole according to the conventional interpretation of general relativity. The fifth column simply makes explicit whether the third column figure resides within the fourth. This indicates that the rest mass quanta inside a black hole horizon could congregate at maximum density without precipitating a singularity.

	Mass in kg	Radius, r_{Max} in m, Density = ρ_{Sphere} Flat Spacetime	Mass in meters $M_l = \frac{G_N}{c^2} M_{kg} = r_h$	Is r within $M_l = r_h$ at Horizon?
Earth	*5.9742×10^{24}	33.55	4.44×10^{-3}	No
Sun	*1.989×10^{30}	2.325×10^3	1.477×10^3	No
Kerr BH threshold	3.930×10^{30}	2.913×10^3	2.913×10^3	At Horizon
Milky Way	*5.2×10^{36}	3.20×10^5	3.86×10^9	Yes
Virgo cluster	*6×10^{39}	3.36×10^6	4.45×10^{12}	Yes
Universe	1.67×10^{53} = 10^{80} nucleon	1.02×10^{11}	1.24×10^{26} = 13.1×10^9 light yrs	Yes

*[5] Figures from Exploring Black Holes, by Taylor and Wheeler, Addison Wesley Longman, 2000

Table 2 - Chart of Various Celestial Mass Geometrizations

Of interest is the fact that the universe appears to be within its own horizon, which conventionally would tend to imply that its constituents should be contracting, and that there are black holes within black holes. Also the mass in meters being equal to the reputed age of the universe times the speed of light seems a bit serendipitous unless of course that mass, i.e. the number of currently theorized nucleons, was estimated using the above geometrization equation. But this figure is not the currently estimated extent of the universe, which has a lower end range of 78 billion light years or 24 Gpc according to a study by Cornish, et al.[6] Finally it is noted that the hypothetical mass of the known universe at maximum density and a radius of 102 million kilometers, would fit inside the earth's solar orbit in flat spacetime.

6 – The Quantum Metric

We turn now to the metric, specifically a chargeless extreme Kerr metric in the equatorial plane (the ϕ coordinates are suppressed), in which the angular momentum parameter, a, is equal to the horizon reduced circumference and the geometrized mass, or $a = r_h = M_l$. The timelike metric at the horizon is

$$d\tau^2 = \left(1 - \frac{2M_l}{r_h}\right) dt^2 + \frac{4M_l a}{r_h} dt d\theta - \frac{dr^2}{\left(1 - \frac{2M_l}{r_h} + \frac{a^2}{r_h^2}\right)} - \left(1 + \frac{a^2}{r_h^2} + \frac{2M_l a^2}{r_h^3}\right) r_h^2 d\theta^2 \quad (5.1)$$

Substituting for $a = M_l$ gives

$$d\tau^2 = \left(1 - \frac{2M_l}{r_h}\right) dt^2 + \frac{4M_l^2}{r_h} dt d\theta - \frac{dr^2}{\left(1 - \frac{M_l}{r_h}\right)^2} - \left(r_h^2 + M_l^2 + \frac{2M_l^3}{r_h}\right) d\theta^2 \quad (5.2)$$

We make the following observation concerning the dr^2 term. While the conventional interpretation is that the term goes to infinity as the denominator approaches zero, and any infalling test particle transits the horizon, the math can also be interpreted in terms of a limit for radial motion. A mathematical conflation is at work in the formulation, since the differentials are deemed to approach zero in the limit, but are effectively treated as dimensional units, i.e. equal to one of some infinitesimal scale. This is necessary since the product of a non zero co-efficient and a zero differential at the limit would be zero. This is warranted since we find a similar non-zero differential without a coefficient on the left side of the equation.

This is contradicted, however, if the metric component represented by the differential has a natural limit where it is necessarily zero. Thus if the horizon in an extreme Kerr spacetime represents that limit, dr equals zero at the limit of that horizon coincident with the term in the denominator, the coefficient and the differential cancel. The result is simply -1 as shown below, which when factored gives an imaginary or orthogonal sense, i.e. it rotates any differential change into tangency. The horizon, then, is effectively a physical asymptote. Thus at the event horizon, where $r = r_h = M$, this simplifies to

$$d\tau^2 = -dt^2 + 4r_h dt d\theta - (2r)^2 d\theta^2 - dr^2 = (idt - i2r_h d\theta)^2 + (idr)^2 \qquad (5.3)$$

This can be factored as a complex number and its conjugate

$$d\tau^2 = \left[(idt - i2r_h d\theta) + i(idr)\right]\left[(idt - i2r_h d\theta) - i(idr)\right] \qquad (5.4)$$

or can be simplified as follows,

$$d\tau^2 = \left[(idt - i2r_h d\theta) - dr_h\right]\left[(idt - i2r_h d\theta) + dr_h\right] \qquad (5.5)$$

where r_h is the reduced circumference at the horizon and $dr_h = 0$ is a zero vector with respect to the radial, giving a proper time of

$$d\tau = \pm i\left(dt - 2r_h d\theta\right) \tag{5.6}$$

If we assume that for bookkeeper time the differential is in the plane of the horizon, and time as developed earlier flows with the rotational motion of the ergosphere, so that

$$dt = r_h d\theta \tag{5.7}$$

then the proper time is found to flow orthogonally to that rotational motion, into the negative and positive ϕ coordinates, since

$$d\tau = \mp i dt \tag{5.8}$$

This will be significant in our statement of the quantum metric.

From this perspective, at the static limit and the start of the ergosphere, where $r = 2M_l$, pure radial motion is no longer possible, and a rotational component or frame dragging element is injected into the equation so that at the event horizon, all motion is rotational as indicated by the "imaginary" or orthogonal senses. Note that if we consider spacetime as an inertio-elastic continuum, frame dragging is simply the wave strain associated with a rotational waveform, be it macrocosmic or quantum. Instead of gravitational collapse, this argues that any incremental matter or light accruing to the inertial sink is smeared out and bound at the horizon in a state resembling a Bose-Einstein condensate.

We now get to the meat of the matter with an expression of the quantum metric. The dynamics of the quantum waveform is not extremely complicated, but it does involve some rather lengthy, non-standard analysis using methods of complex classical wave physics extended to 4 dimensions, and is beyond the scope of the present discussion. We will simply state that its kinematics prevent the orientation entanglement condition.

With reference to Quantum Inertial Sink Diagram 1, the timelike quantum metric is given as a modified chargeless extreme Kerr metric. The modification is in the ϕ coordinates as shown here, where the quantum mass has been explicitly geometrized as r_{0n},

$$d\tau^2 = \left(1 - \frac{2r_{0n}}{r_{0n}}\right)dt^2 + \frac{4r_{0n}^2}{r_{0n}}dtd\theta - \frac{dr^2}{\left(1 - \frac{r_{0n}}{r_{0n}}\right)^2} - R^2 d\theta^2 \mp \left\{\left(e^{\pm i(\omega_0 t \mp \theta)} Ld\phi\right)^2\right\} \quad (5.9)$$

The caveat stated above concerning the limit of radial motion represented by r_{0n} remains. In the last term, the complex exponential is defined as

$$e^{\pm i(\omega_0 t \mp \theta)} = \text{Re}\left(e^{\pm i(\omega_0 t \mp \theta)}\right) = \cos_{ccw}(\omega_0 t + \theta) \text{ or } \cos_{cw}(\omega_0 t + \theta)$$
$$= \cos(\omega_0(+t) - \theta) \text{ or } \cos(\omega_0(-t) + \theta) \quad (5.10)$$

Either the real or the imaginary part could of course be used. The *ccw* term indicates rotation in the upper hemisphere according to the right hand rule, while the *cw* term indicates clockwise rotation in the bottom hemisphere according to the left hand rule, when viewed from the exterior of the corresponding rotational pole.

The plus and minus curly bracket has the following definition and indicates a flipping of the sign of the $d\phi$ vector, with every π rotation of θ, plus being parallel and minus being anti-parallel with respect to the RHR spin axial vector. It thus performs a function similar to a mathematical spin matrix.

$$\pm\{a\} \equiv \frac{\cos(\omega_0 t - \theta)}{|\cos(\omega_0 t - \theta)|} a, \quad \mp\{a\} \equiv -\frac{\cos(\omega_0 t - \theta)}{|\cos(\omega_0 t - \theta)|} a \qquad (5.11)$$

Obviously, θ and ϕ rotate at the same frequency, with the axis of the ϕ rotation rotating in the equatorial plane. This motion avoids the orientation entanglement condition and is necessitated by the assumed continuity condition of a classical spacetime continuum and the density property postulated in this development. When analyzed it is apparent that the motion is that of a transverse wave traveling in tight orbit around the spin axis, its amplitudes inclined toward the poles, analogous to a gravitationally bound, electromagnetic wave, and in fact constitutes the magnetic field of the quantum.

This diagram is a cross-section through the spin axis and shows the relationship of the static limit, the ergosphere, and the horizon. The ergosphere is the domain of the strong interaction. The transverse or ϕ differential is limited in its motion toward the spin poles to the point on the static limit where $L = 1$.

The metric simplifies at the horizon with no radial motion as

$$d\tau^2 = -dt^2 + 4r_{0n} dt d\theta - R^2 d\theta^2 \mp \{\cos^2(\omega_0 t - \theta) L^2 d\phi^2\} \qquad (5.12)$$

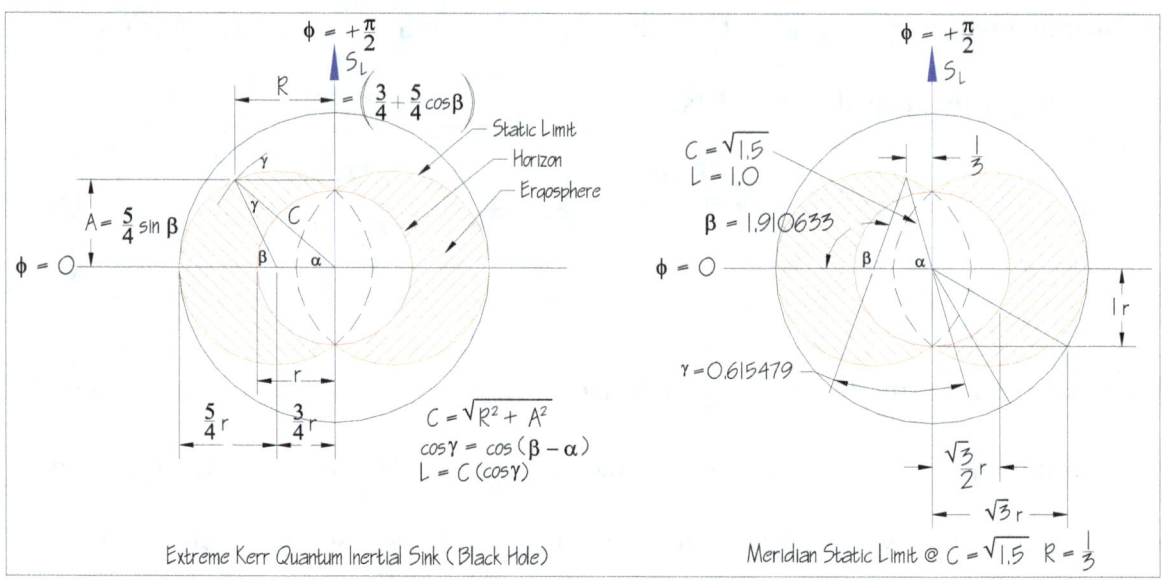

Quantum Inertial Sink 1
Figure 14

From this diagram we have the following coefficient component for ϕ along the meridians at the static limit

$$L = \tfrac{4}{5}r_{0n} + \tfrac{3}{5}R = \tfrac{4}{5}r_{0n} + \tfrac{3}{5}\left(\tfrac{3}{4}+\tfrac{5}{4}\cos\beta\right)r_{0n} = \left(\tfrac{5}{4}+\tfrac{3}{4}\cos\beta\right)r_{0n} \tag{5.13}$$

Substituting this in equation (5.12) simplifies at the horizon along the equatorial plane of a fixed spin axis where $\cos\beta = 1$, as

$$d\tau^2 = \left(idt - i2r_{0n}d\theta\right)^2 \mp \left\{\cos^2\left(\omega_0 t - \theta\right)\left(2r_{0n}\right)^2 d\phi^2\right\} \tag{5.14}$$

The corresponding spacelike metrics is

$$d\sigma^2 = -\left(idt - i2r_{0n}d\theta\right)^2 \pm \left\{\cos^2\left(\omega_0 t - \theta\right)\left(2r_{0n}\right)^2 d\phi^2\right\} \tag{5.15}$$

giving the fundamental symmetry

$$d\sigma^2 \equiv -d\tau^2 \tag{5.16}$$

and for the proper time and space, indicating the orthogonal nature of space and time,

$$d\sigma \equiv id\tau. \tag{5.17}$$

This can be represented by the following anti-symmetric orthonormal matrix at r_0,

		Direction of ortho normal vector dx_i with respect to		
		X Axis	Y Axis	Z Axis
Vector dx_i originating at	X = +1	0	$+rd\theta$	$+r\sin\omega t d\phi$
	Y = +1	$-rd\theta$	0	$-r\cos\omega t d\phi$
	Z = +1	$-r\sin\omega t d\phi$	$+r\cos\omega t d\phi$	0
	X = -1	0	$-rd\theta$	$-r\sin\omega t d\phi$
	Y = -1	$+rd\theta$	0	$+r\cos\omega t d\phi$
	Z = -1	$+r\sin\omega t d\phi$	$-r\cos\omega t d\phi$	0

Table 3 - Quantum Anti-Symmetric Orthonormal Matrix at r_0

In the presence of an anti-parallel external magnetic field as shown in Quantum Inertial Spin Diagram 2, the quantum spin axis inclines toward the equatorial plane and precesses about its initial position. The resulting coefficients of ½ spin can be seen here. Note also that the Heisenberg "observational" uncertainty is limited by the inverse curvature of the horizon to

$$r_0^2 c = m_{I0} r_0 c = \bar{r} c = \hbar. \tag{5.18}$$

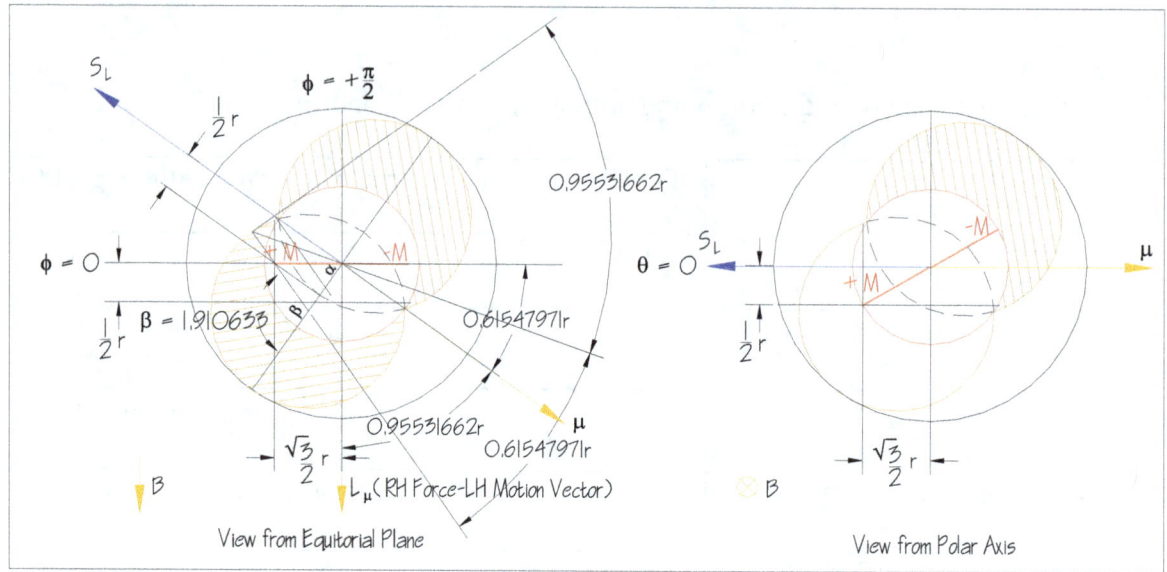

Quantum Inertial Sink 2
Figure 15

Full Model Development

This model is elsewhere more fully developed and presented as the 3-D representation of a classical 4-D oscillation. Expansion acts as an EMF that drives the fundamental frequency, both by mechanical analogy and as the actual mechanical or piezoelectric basis for electro-magnetism. The rest-mass quantum is thus a small simple harmonic oscillator, with a potential-kinetic, capacitive-inductive energy cycle, in a general inductive mode during expansion, of which the waveform of ordinary matter is the result. During universal contraction, a capacitive mode ensues, resulting in a predominance of anti-matter.

Over a short period of time, in particular in the absence of confinement at nuclear density, expansion leads to a drop in mechanical impedance, resulting in a transmission of energy and power at the boundary of a neutral or resonant quantum. The result is beta decay, which is tuned to the expansion rate for any isolated neutral quantum and generates the electromagnetic interaction, which is properly considered an intra-action of the spacetime continuum. The rest-mass ratios between the neutron, electron and proton and the "missing" mass of beta decay arise naturally in this analysis. Finally, quark phenomenology of fractional charge is shown to be the property of the nodes and antinodes of the quantum waveform.

Endnote

[i] A derivative taken on a flat rectilinear area,

$$\frac{\pm d\mathrm{A}}{dr} = \frac{(\mathrm{A} \pm d\mathrm{A}) - \mathrm{A}}{dr} = \frac{(r \pm dr)^2 - r^2}{dr} = \frac{\pm 2rdr + dr^2}{dr}$$

gives a differential area of

$$\pm d\mathrm{A} = \pm 2rdr + dr^2.$$

Now consider a hyperbolic surface, specifically the derivative of the inverse curvature of a pseudosphere, which is of constant negative curvature, for simplicity $k = -1$, where r_i is the interior radius and r_e is the exterior radius, and we have the function, where either r_e or r_i, could be used as the variable

$$k_{r_e}^{-1} = k^{-1}(r_e) = -r_e^{-1} r_e = -r_i r_e$$

The curvature is conserved, therefore the differential is zero or

$$dk_{r_e}^{-1} = (-r_i - dr_i)(r_e - dr_e) - (-r_i r_e) = r_i dr_e - r_e dr_i + dr_i dr_e = 0$$

The senses of the radii and their differentials indicate a direction toward (+) or away from (-) the exterior of the pseudosphere. Note that the differentials are of the same sense. Thus the above equation indicates a change toward the mouth or rim of the pseudosphere, as r_i is increasing and r_e is decreasing. At the point of normalization, where $r_i = r_e = r_0 = 1$,
we have

$$dr_e - dr_i = -dr_i dr_e.$$

Therefore $dr_i = -x^{-1}, dr_e = -x \therefore dr_i dr_e = 1$ and after a sense inversion we have the solution

$$x - x^{-1} = 1$$
$$x = \sqrt{\tfrac{5}{4}} + \tfrac{1}{2} = 1.618033... = \Phi$$

the well known coefficient of conservative evolution of a system.
Note that the product $(x)(x^{-1})1 = 1$ is conserved.

At the point where $r_i = \Phi^{-1}$ and $r_e = \Phi$,
we have, where the differential senses are explicit,

$$-r_i(-dr_e) + r_e(-dr_i) = -(-dr_i)(-dr_e)$$

and we can normalize the differentials at $k_{r_e}^{-1}(\Phi)$ as

$$|dr_i| = |dr_e| = |dr_o|$$

giving

$$\frac{dr_i}{dr_e} = \frac{dr_e}{dr_i} = 1,$$

therefore

$$dr_i^2 = dr_e^2 = dr_0^2 = d\mathrm{A}_0.$$

Then the invariant inverse curvature is equal to the square of normalized differentials

$$k_{r_e}^{-1}(\Phi) = -r_i r_e = -dr_i dr_e = -dr_0^2 = -d\mathrm{A}_0 = -1.$$

However, for any such conservative hyperbolic system of any invariant finite curvature, we can state the following,

$$dr_i = r_i \Phi^{-1}, \quad dr_e = r_e \Phi,$$

so that
$$dr_i dr_e = |r_i r_e|$$
and we have the following relationship between the inverse curvature function and its differential components
$$k_{r_e}^{-1} = k_{r_e}^{-1} + dk_{r_e}^{-1} = r_i dr_e - r_e dr_i = -r_i r_e \Phi + r_e r_i \Phi^{-1} = -dr_i dr_e = -r_i r_e.$$
Finally, with some substitution, for the function and its derivative, as
$$dr_i = \frac{r_i}{\Phi^2 r_e} dr_e \text{ and } r_i = \frac{-k_{r_e}^{-1}}{r_e}$$
we have, with rearrangement and simplification
$$k_{r_e}^{-1} + dk_{r_e}^{-1} = \frac{-k_{r_e}^{-1}}{r_e} dr_e - \frac{-k_{r_e}^{-1}}{\Phi^2 r_e} dr_e = -\left(1 - \Phi^{-2}\right) k_{r_e}^{-1} d\ln r_e = -\Phi^{-1} k_{r_e}^{-1} d\ln r_e.$$
The symmetrical condition for $k_{r_i}^{-1}$ is
$$k_{r_i}^{-1} + dk_{r_i}^{-1} = \frac{-k_{r_i}^{-1} \Phi^2}{r_i} dr_i - \frac{-k_{r_i}^{-1}}{r_i} dr_i = -\left(\Phi^2 - 1\right) k_{r_i}^{-1} d\ln r_i = -\Phi k_{r_i}^{-1} d\ln r_i,$$
and obviously
$$d\ln r_i = \Phi^{-1}, \ d\ln r_e = \Phi.$$
Since
$$-dr_i dr_e = -\Phi^{-1} k_{r_e}^{-1} d\ln r_e = -\Phi k_{r_i}^{-1} d\ln r_i$$
we have
$$dk_{r_i r_e}^{-1} = \Phi^{-1} k_{r_e}^{-1} d\ln r_e - \Phi k_{r_i}^{-1} d\ln r_i = 0,$$
and finally
$$k^{-1} = d\mathrm{A}_0 = -dr_0^2 = -dr_i dr_e = -\Phi^{-1} k_{r_e}^{-1} d\ln r_e = -\Phi k_{r_i}^{-1} d\ln r_i.$$

Bibliography

Citations

[1] M. Jammer, <u>Concepts of Mass in Contemporary Physics and Philosophy</u>, Princeton University Press, Princeton, NJ (2000).

[2] C. W. Misner, K. S. Thorne, and J. A. Wheeler, <u>Gravitation</u>, W.H. Freeman and Company, New York (1973).

[3] C. F. Stevens, <u>The Six Core Theories of Modern Physics</u>, The MIT Press, Cambridge, MA (1995).

[4] National Institute of Standards and Technology, These are the **2002 CODATA recommended values** of the fundamental physical constants, the latest CODATA values available. For additional information, including the bibliographic citation of the source article for the 1998 CODATA values, see P. J. Mohr and B. N. Taylor, "The 2002 CODATA Recommended Values of the Fundamental Physical Constants, Web Version 4.0," available at physics.nist.gov/constants. This database was developed by J. Baker, M. Douma, and S. Kotochigova. (National Institute of Standards and Technology, Gaithersburg, MD 20899, 9 December 2003).

[5] E. F. Taylor and J. A. Wheeler, <u>Exploring Black Holes</u>, Addison Wesley Longman, Inc, New York (2000).

[6] N. J. Cornish, et al, "Constraining the Topology of the Universe", Physical Review Letters, Volume 92, Number 201302 (2004).

[7] a study by R. Eastman, B. Schmidt and R. P. Kirshner in 1994 quoted in <u>The Extravagant Universe</u>, R. P. Kirshner, Princeton University Press, Princeton, NJ (2002).

[8] W. L. Freedman, et al, Astrophysical Journal, 533, 47-72 (2001).

General References

<u>The Feynman Lectures on Physics</u> :Commemorative Issue, Feynman, Leighton, Sands, Volume I, Addison-Wesley Publishing Company, Inc., Reading, Massachusetts 1963.

<u>Fundamentals of Physics</u>, Fifth Edition, Halliday, Resnick, Walker, John Wiley & Sons, Inc. New York, 1997.

<u>Mathematical Methods for Physicists</u>, Fifth Edition, Arfken and Weber, Harcourt Academic Press, New York, 2001.

Physics of Waves, Elmore and Heald, Dover Publications, Inc., New York, 1985.
This was the primary source for wave, elasticity and tensor equations.

Visual Complex Analysis, Needham, Oxford University Press, Oxford, 1997.

A Dimensional Analysis of the Dimensionless Fine Structure Constant

—:—

By Martin Gibson

December 14, 2010

Martin Gibson
P.O. Box 2358
Southern Pines, NC 28388
910-585-1234
martin@uniservent.org

Copyright © Martin Gibson 2021 All Rights Reserved

A Dimensional Analysis of the Dimensionless Fine Structure Constant
By Mart Gibson

We can remove some of the mystery and the mysticism surrounding the fine structure constant, α, by employing the following dimensional and functional analysis. Clearly, the dimensionless context of α is due to the fact that it is a ratio of two measurements/computations of like dimensionality. It is therefore itself a factor in the coefficient of one of those dimensional structures, which we might surmise includes the dimension of time as in an impulse or force.

Using the CODATA definition of the fine structure constant, α, with fundamental or quantum charge, e_0, the reduced Planck's constant of action, $\hbar$, light speed in vacuo, c, and the permittivity of the vacuum, ε_0, we have,

$$\alpha = \frac{e_0^2}{\hbar c 4\pi \varepsilon_0} = \frac{c e_0^2 \mu_0}{\hbar 4\pi}, \qquad (1.1)$$

where the permeability constant, μ_0 is expressed (with dimensions shown in curly brackets) as

$$\mu_0 = \frac{1}{c^2 \varepsilon_0} = q \left\{ \frac{N}{A^2} \right\}. \qquad (1.2)$$

μ_0 is defined both quantitatively, as a number q, and qualitatively (dimensionally) in terms of the magnetic force in newton$\{N\}$ induced by 2 moving charges or electrical currents in ampere$\{A\}$. We note that the ampere is one of the fundamental units of the International System of Units (SI) along with the three units of mass, length and time, i.e. kilogram$\{kg\}$, meter$\{m\}$, and second$\{s\}$, which define the Newton as

$$N = m \cdot l / t^2 \{kg \cdot m/s^2\}. \qquad (1.3)$$

Thus in dimensional terms, μ_0 is the ratio, here equal to q, of a force to the square of a unit current, (technically the cross product of the field strength of one unit current operating on a parallel unit current at a distance of one unit of length.) Here the naught subscripts indicate a normalized or unit value of the respective variables, or as in the case of μ_0 or ε_0, a universal constant.

Specifically, for two parallel wires, a and b, of indefinite length, one meter apart, d_0, in vacuo, each carrying a unit of current, i_0, of one ampere or coulomb$\{C\}$ per second, a magnetic force, F_{ba}, of 2×10^{-7} N is generated on one of the wires by the other for each meter length, l_0, of the two, toward each other or positive if the currents are parallel and away from each other or negative if they are anti-parallel; thus expressed,

$$\frac{F_{ba}}{l_0} = \frac{\mu_0 i_{0a} i_{0b}}{2\pi d_0} = \frac{2 \times 10^{-7}}{1} \left\{ \frac{N}{m} \right\}. \tag{1.4}$$

With transposition for the value of F_{ba} as stated, we have the defined value of μ_0 as,

$$\mu_0 = \frac{2\pi F_{ba}}{i_0^2} = \frac{4\pi \times 10^{-7}}{\left(n_{A0} e_0 / t_0\right)^2} \left\{ \frac{N}{(C/s)^2} \right\} = \frac{4\pi \times 10^{-7}}{1^2} \left\{ \frac{N}{A^2} \right\} \therefore q = 4\pi \times 10^{-7}. \tag{1.5}$$

In fact it is the value of μ_0 that defines $i_{0a} = i_{0b} = 1\,A = 1\,C/s$ and $l_0 = d_0 = 1\,m$. We note that l_0 and d_0 are orthogonal to each other. Since a coulomb is deemed to be made up of a finite number, n_{A0}, of fundamental unit charges, e_0, for each ampere or

$$n_{A0} e_0 = 1\,C \tag{1.6}$$

an ampere can be written as in the divisor of the middle term of (1.5). Substituting this into (1.1), gives

$$\alpha = \frac{c e_0^2 \mu_0}{\hbar 4\pi} = \frac{e_0^2 10^{-7}}{\frac{\hbar}{c} n_{A0}^2 e_0^2 / t_0^2} \left\{ \frac{C^2 N}{\frac{kg \cdot m^2 s^{-1}}{ms^{-1}} (C^2)/s^2} \right\} = \frac{10^{-7}}{\Pi\, n_{A0}^2 / t_0^2} \left\{ \frac{N}{kg \cdot m/s^2} \right\} \tag{1.7}$$

In the last term, the constants of action and the speed of light are reduced and the unit charge squared terms are canceled. As a change in the relative placement of two charges produces a force, and a force is a momentum differential, i.e. an impulse, per unit of time, a fundamental quantum or unit of charge can be viewed as a quantum of momentum or a unit impulse, potential or kinetic, i.e. static or moving. In the next to last term, the number, n_{A0}, is the number of such impulses in an ampere of current, or when combined with one of the time dimensions, an expression of the frequency of such impulses. Canceling the fundamental charges leaves a measure of force in the antecedent, and in the consequent a frequency squared times Planck's quantum of action divided by the speed of light, which resolves dimensionally to a measure of mass-length.

In the last term, we introduce the inertial constant, Π(tav), as a time independent, fundamental mass–length unit, where for any rest mass quantum, m_q,

$$m_q = \frac{\hbar}{c^2} \omega_q = \frac{\hbar}{c} \kappa_q = \Pi \kappa_q \quad \therefore \frac{\hbar}{c} = \Pi. \tag{1.8}$$

Omega and kappa are angular wave frequency and number, respectively, indicating that mass is essentially a proxy for wave number. The inertial constant derivation is the time integral of a quantum impulse, $\Pi = J(t) \int_0^1 dt = m_q \lambdabar_{C,q}$, which is invariant.

Equation (1.7) expresses the fine structure constant as a dimensionless number. It is dimensionless in the sense that a ratio of two like qualities is dimensionless, yet such dimensionless number can also be seen as a coefficient, whole or partial, in this case as a quantifier of the consequent (divisor) of such ratio. [It is the factor required, in product with the inertial constant times the square of its frequency found in two 1 ampere currents as figured above, to produce an induced force of 10^{-7} Newton.]

Some rearrangement of (1.7) gives

$$n_{A0}^{2} e_{0}^{2}/t_{0}^{2} \left\{ \frac{C^2}{s^2} \right\} = \frac{e_{0}^{2} 10^{-7}}{\alpha \eta} \left\{ \frac{NC^2}{kg \cdot m} \right\} = i_{0}^{2} \{A^2\} \qquad (1.9)$$

Here we have an expression of the current at one ampere, squared. Once again canceling the fundamental unit charges gives

$$(n_{A0}/t_{0})^{2} \left\{ \frac{\#^2}{s^2} \right\} = \frac{10^{-7}}{\alpha \eta} \left\{ \frac{N}{kg \cdot m} \right\} = \frac{i_{0}^{2}}{e_{0}^{2}} \left\{ \frac{A^2}{\text{quantum charge}^2} \right\} = i_{e}^{2} = 3.895644... \times 10^{37} \left\{ \left(\frac{\text{charge impulses}}{s} \right)^{2} \right\} (1.10)$$

where it is apparent that the left hand term is the square of a frequency, f_{A0}^{2}, perhaps periodic, semi-periodic, or some other duration. This frequency, in fact angular, gives the number of instances of the inertial moment or constant, η, a time-independent quantum of (wave) momentum and force, found in $10^{-7}/\alpha$ N, or

$$\eta (n_{A0}/t_{0})^{2} \{kg \cdot m/s^2\} = \eta f_{A0}^{2} \{N\} = \frac{10^{-7}}{\alpha} \{N\} = \alpha' \{N\} = 137.035999... \times 10^{-7} \{N\} (1.11)$$

It is immediately clear that the value of α is dictated by the dividend at 10^{-7}, since the presumed invariant is their quotient, α', and a change in the dividend necessitates a corresponding change in the divisor, α, the mysterious fine structure constant. In practice, 10^{-7} sets the length of a meter in terms of c in (1.4) and (1.5), so that a nominal change in the dividend would result in a nominal change in the speed of light and in the various wavelengths found in the fine structure series, and therefore in α as well. What would change the value of α' is a change in the value of a unit of time, t_0, so that a nominal lengthening of a time unit (to include more n_{A0} per second) would nominally increase the force on the right. It is the duration of the second that determines the frequency, which determines α', and given a nominal 10^{-7}, determines α.

If we take a dimensional look at (1.2) in the context of (1.4), it appears that μ_0, the permeability of the vacuum, is converting the two current flows into a force component. It is equally correct to think that each current flow constitutes a force that interacts to produce a magnetic field force between them, their cross product given by F_{ba}, which gives μ_0 the dimensions of an inverse force. If current is expressed in units of force, then charge becomes a count of momenta or impulse as the time integral of the current, and the fundamental unit of charge becomes a fundamental unit of some type of momentum as discussed above.

This discussion is facilitated by the conceptualization of rest mass quanta as local rotational oscillations of an inertially dense spatial continuum, rather than the current Standard Model view as point particles that exhibit wave properties under certain circumstances. While a complete exposition of the mechanics of such oscillation is beyond the present scope of this monograph, (it can be provided) we can get a rough clue to this oscillation by visualizing a disk with its edge oscillating transversely about a stationary center, resulting in a wave phase rotation about that center.

Such quantum oscillations are instances of simple harmonic motion which can be expected to exhibit certain fundamental wave characteristics such as transverse wave force and transverse wave momentum, represented by a fundamental or resonant angular frequency, ω_0. This is not a body force or momentum of a particle, but rather a wave force and momentum of the density field oscillation, of which there are two simultaneous and opposed instances, one for each half of the cycle. In the quantum context, using the inertial constant as a quantum of mass-length, the fundamental transverse wave force and wave momentum are represented by

$$\mathrm{n}\omega_0^2, \text{ fundamental quantum transverse wave force} \tag{1.12}$$

$$\mathrm{n}\omega_0, \text{ fundamental quantum transverse wave momentum} \tag{1.13}$$

which travel around with the oscillation, with force leading momentum by $\pi/2$. If we assume for the sake of argument, that the neutron represents this fundamental oscillation, then using the mass of the neutron in light of (1.8) gives an expression and an evaluation of the wave force and momentum respectively as

$$\mathrm{n}\omega_0^2 = \frac{\hbar}{c}\omega_n^2 = m_n c \omega_n = 716,766.8351 \text{ N} \tag{1.14}$$

$$\mathrm{n}\omega_0 = \frac{\hbar}{c}\omega_n = m_n c = 5.02130545...\text{x}10^{-19} \text{ kg}\cdot\text{m/s} \tag{1.15}$$

The above is a tremendous amount of force, especially for a single quantum, but it pales in comparison to the stress found by figuring the small cross sectional area upon which the stresses operate, which are on the order of 10^{37} pascals.

Note the following ratio, in which the π converts the angular frequency to a semi-periodic frequency

$$\frac{\mathrm{n}\omega_0/\pi}{e_0} = \frac{1.59833...\text{x}10^{-19}}{1.60217...\text{x}10^{-19}} = 0.997599940... = 1 - 0.002400060... \tag{1.16}$$

where both terms are figured in units of momentum. That is, electron charge <u>is</u> neutron wave momentum observed as a result of beta decay.

In the above referenced wave conceptualization, the fundamental frequency is driven by cosmic expansion, and the fundamental wave force itself is a function of the expansion stress, an isotropic stress that pervades space, isotropic that is, except at the "surface" of the oscillation where the cross product effects of microscopic torsion make themselves felt as particle spin.

The mechanical impedance, Z_0 of such inertial space, (not to be confused with the SI impedance of the vacuum), is the quotient of the fundamental characteristic wave force and the speed of wave propagation, evaluated here and compared with the last term of (1.16)

$$Z_0 = \frac{\hbar \omega_0^2}{c} = \hbar \omega_0 \kappa_0 = 0.002390877.... \quad (1.17)$$

Thus

$$\frac{\hbar \omega_0 (1+Z_0)/\pi}{e_0} = \frac{\hbar \omega_0 (1+\hbar \omega_0 \kappa_0)/\pi}{e_0} = .999985079... = 1 - .000014921..., \quad (1.18)$$

where the subtrahend of the last term represents the difference of the antecedent on the left from a theoretically precise equation with e_0. As the value of antecedent depends on a measured observation of the neutron mass, which presumably uses Newton's constant, G, for its evaluation in terms of some mass standard, and as the relative uncertainty of that constant is 10^{-4} and (1.18) is precise to approximately 10^{-5}, these results recommend pursuit of this line of reasoning.

Combining (1.18), (1.11) and (1.4) we have an expression relating fundamental charge and electromagnetic induced force in terms of the inertial constant and the resonant frequency and wavelength of the vacuum, and in light of the above uncertainty,

$$2\hbar \left(\frac{\pi}{\hbar \omega_0 (1+\hbar \omega_0 \kappa_0)} \right)^2 \simeq \frac{2\hbar}{e_0^2} = 2\hbar f_{A0}^2 = \frac{F_{a0b0}}{\alpha} = \frac{2 \times 10^{-7}}{\alpha} N. \quad (1.19)$$

With respect to the Standard Model, it has always seemed incomprehensible to me that such separate and diverse ontologies as that of the electron and proton could fit together so precisely with the help of the equally incomprehensibly elusive neutrino, to produce a neutron. It's as if an archaeologist, finding one large piece of pottery with a piece of its rim missing, next found a much smaller piece nearby that filled the missing void, but for a similar sized piece still lost, then concluded that his partial reconstruction represented the first time the pieces were ever joined! Add to this that his dig found billions and billions of similarly fitting parts. If the electron and proton (and neutrino) fit together so well, it is because they started out together as a unit. In fact what is happening, is that the expansion of the cosmos leads to a drop in the inertial density of space, and a related drop in its impedance. This produces a discontinuity at the boundary, i.e. node of the oscillation, and results in the transmission of power and energy which is registered as beta decay. The results are precisely determined by geometry, accounting for the observed neutron/electron, proton/electron, and neutron/proton mass ratios. Classical linear wave analysis, taken to three and four dimensions explicates the basics of quantum dynamics, in which it is seen that the lepton and quark phenomenology of the Standard Model can be understood as the nodal/antinodal structure of a multi-dimensional bound wave system.

There is much, much more to this analysis for the sufficiently curious party. Quantum gravity is a natural outcome of its development.

A Classical Complex 4-Wave Foundation

of the

Cosmic-Quantum Mechanism

-

Fundamental Rest Mass Quanta as
Simple Harmonic Oscillations of
the Spacetime Continuum
at Resonant Frequency and Wave Number,
Driven by Cosmic Expansion

By Mart Gibson

October 14, 2006
Revisions and Addition of the Section on The Quantum Metric
as of February 1, 2007

Mart Gibson
P.O. Box 2358
Southern Pines, NC 28388
910-692-7462
martin@uniservent.org

Copyright © Martin Gibson 2021 All Rights Reserved

A Classical Complex 4-Wave Foundation of the Cosmic-Quantum Mechanism

Mart Gibson

Abstract

A model of a fundamental ½ spin quantum, specifically the neutron, is developed as a simple harmonic oscillation of an expanding 3-space of variable inertial density and resonant frequency in an underlying 4-continuum. The oscillation is shown to be an extreme Kerr quantum inertial sink, aka black hole, for which the quantum metric is given. The uncertainty principle is examined in light thereof, as well as the inappropriateness of the factor G_N/c^2 in converting mass measure in kilograms to measure in meters for an individual quantum. The correct quantum conversion factor is developed as the inertial constant, $\beth(tav) = \hbar/c$, which multiplied by the inverse of the quantum mass in kilograms gives the correct mass in meters. This shows the neutron mass to be a measure of curvature as the reduced circumference at its horizon, equal to its Compton wavelength over 2π. Within the static limit, the ergosphere of the oscillation is shown to be the domain of the strong interaction. Expansion provides a mechanical analogue of an EMF which drives the neutral quantum. Absent inertial confinement, a differential decrease in inertial density creates a discontinuity, inducing a decrease in frequency to that of the proton, with transmission of the electron. Quantum gravity arises as the derivative of the wave force with respect to the expansion tension stress, equal to the inverse curvature divided by a geometric factor of $6\sqrt{3}$ and the Planck area as the derivative of the fundamental cross-sectional scale and inverse curvature with respect to a change in stress. An exponential Hubble rate is coupled with the differential wave force and thereby beta decay. The nature of matter and anti-matter as inductive and capacitive states, respectively, is a straightforward consequence of this analysis. A quantum physical mechanism, with animation, modeling the above is developed along with the derivation of the inertial constant, $\beth$. An orthogonal matrix of the wave symmetries, functions, invariants, and their couplings is examined, clearly showing the relationship of the electromagnetic and gravitational interactions in a unified field.

A Classical Complex 4-Wave Foundation of the Cosmic-Quantum Mechanism

Table of Contents

Abstract	
1 – Background and Fundamentals	
Motivation	1
The Quantum Metric	6
Wave Bearing Continuum	18
Geometry and Topology	28
2 – The Fundamental Inertial Quantum as a Simple Harmonic Oscillator	
Isotropic Expansion and the Generation of Rotational Oscillation	41
Dynamic Functions	60
Quantum Rotational Oscillation or Spin	72
Generation of Charge	84
3 – Cosmic Expansion as the Driving EMF of a Quantum State	
The Reactance States and Electron-Positron Generation	91
Cosmic Expansion Rate and Expansion Force	99
4 – Strong Interactions	103
5 – Cosmic Considerations and Speculations	107
6 – Evaluations	
Observed Values	117
Theoretical Values	118
Appendix A – Direct Product and Inverse Square Law	121
Gravitation	121
Electrostatics	122
Electrodynamics	122
Appendix B – Wave Transmission at a Discontinuity	125
Appendix C – Vector Orthogonality	125
Appendix D – Exponentiation	126
Bibliography	144

Diagrams, Tables, and Figures

Table of Black Hole Mass and Horizon	13
Quantum Inertial Sink Diagram 1	14
Quantum Inertial Sink Diagram 2	17
Figure 1 – Electro Magnetic Wave	18
Figure 2 – Electro Magnetic Wave Source	26
Table of Dedekind Cuts	32
Manifold Charts	34
Table of 0-Monad Orthogonality Potentials	36
Chart of 0-Monad Orthogonality	37
Table of Boundary Parameters	38
Table of n-Cube Breakdown	38
Table of n-Sphere Breakdown	39
Diagram of Cubic Expansion	41
Figure 3 – Hyper-Axis Intersection	48
Figure 4 – 4 to 3 Space Contraction	49
Figure 5 – Hyper-Axis Oscillation	52
Equal Area Cube and Sphere	53
Wave Diagram 1 – Double Rotation, $\phi(\theta)$	55
Wave Diagram 2 – Kinematic Functions of η	55
Spin Diagram 1 – Spin Energy Cycle	56
Table of 3-Space Orthogonal Spin Permutations	57
Table of 4-Space Orthogonal Spin Permutations	57
Diagram of 4-Space Orthogonal Spin Permutations	58
Table of 4 to 3-Space Contractions (3)	59
Wave Diagram 3 – Simple Harmonic Oscillation of ϕ	61
Wave Diagram 4 – Dynamic Functions of $\phi(t)$	63
Matrix of Invariants	67
Spin Chart 1 – $t = 0$, RHV/RHR	68
Spin Chart 2 – $t = n2\pi$, RHV/RHR	70
Spin Chart 3 – $t = 0$, LHV/LHR	71
Spin Chart 4 – $t = n2\pi$, LHV/LHR	71
Spin Diagram 2 – Neutron	75
Spin Diagram 3 – Proton	78
Spin Diagram 4 – Electron	80
Spin Diagram 5 – Anti-Proton	81
Spin Diagram 6 – Positron	82
Spin Table 1 – Inductive State, Ordinary Matter	85
Spin Table 2 – Capacitive State, Anti Matter	86
2-D Representation of a 3-Torus	113
Diagram of Cosmic - Planck Scale	115
Diagram of Invariance of the Fundamental Length and Cosmic Extent	115

1 – Background and Fundamentals

Motivation

Based on a presumed unity on some ontological level, an understanding of the fundamental phenomena that inform the physical world would appear to hinge on the ability to link the classical realm of large aggregates of matter and the quantum world of individual particle interactions. It requires that we find an expression of the very large, gravity, as a quantum effect and of the very small, individual particles of mass and energy, as a cosmic effect. Currently, the first of these attempts focuses on the Planck scale, held to comprise the fundamental, discrete units of space, time and mass, and the second, on the high energy conditions held to dominate at a point or locus of cosmic inception. We will try a fresh approach which derives the phenomenology of quantum effects, including gravity, from the wave bearing ontology of a cosmic continuum, a spacetime which we will discuss briefly in qualitative terms.

We would expect to find the unification of the large and the small amenable to mathematical expression in an equation uniting the basic invariants of each. We might expect to find a solution to the following, in which the familiar energy-mass equation of relativity and energy-frequency equation of quantum mechanics are joined. Thus, the inherent energy of a particle, E, equal to the mass, m, times the square of the speed of light, c^2, is equated qualitatively with Planck's quantum of action, $\hbar$, times the angular frequency of a particle, ω. To elucidate this procedure, we look for a constant that couples the two expressions and find a candidate in the inertial constant, ת (tav), which will be subsequently derived, and which we introduce provisionally now as

$$ת = \frac{\hbar}{c}. \tag{1.1}$$

There is nothing new in this coupling of h-bar and the speed of light, but it has, to this writer's knowledge, never been identified, by any name or symbol, as an invariant of significance in its own right. We will subsequently see the wisdom in doing this.

The presence of ω is an indication that a quantum particle is some manner of oscillation, and since it appears to have a discrete value over some interval of time, we will assume that it is an instance of simple harmonic motion, i.e. that there are no harmonic overtones. It is further presumed that in the context of such periodic phenomena, as in the case of a traveling wave, the following equation for the wave velocity applies, in which κ is the angular wave number, hereinafter simply referred to as "wave number", and $\partial\theta$ is the change in the wave phase commensurate with a change in time, for ω, or in space, for κ, along the length of the wave propagation;

$$c = \frac{\omega}{\kappa} = \frac{\partial\theta}{\partial t}\frac{\partial x}{\partial\theta} = \frac{\partial x}{\partial t} \tag{1.2}$$

This applies even if we envision the oscillation to be more or less fixed at a locus in space. That is, if there is an actual motion on some scale associated with the oscillation and not just a periodic phenomena emanating from a point in space, if it is an actual standing wave of some sort, then it will have a wave number just as would a traveling wave. Assuming the following equivalence, yields the rest mass as the product of the inertial constant and the wave number,

$$E = mc^2 = \hbar\omega \tag{1.3}$$

$$m = \frac{1}{c^2}\hbar\omega = \daleth\kappa. \tag{1.4}$$

Thus, on a quantum level, mass is a measure of the wave number of an oscillation.

We will next assume that the wave nature of an individual particle is an indication of its basic structure and not simply a statistical artifact, i.e. that such particles are not points, but have some inherent size as exhibited by a wave amplitude and length; that the Compton wavelength over 2π, lambda-bar with a subscript C, $\bar{\lambda}_C$, or reduced Compton wavelength, expresses such, and is the modulus, $\hat{r}$, of some manner of quantum complex wave, which we will investigate, and we might state

$$\bar{\lambda}_{C,part} = \hat{r}_{part}. \tag{1.5}$$

Finally, as the curvature, κ, of a linear path is the inverse of the radius of the path at any point, we might expect the wave number to also be a measure of that curvature as

$$\kappa_{C,part} = \frac{1}{\hat{r}_{part}} = \frac{1}{\bar{\lambda}_{C,part}}. \tag{1.6}$$

A review of the 2002 CODATA values for particle mass and $\bar{\lambda}_C$ for the neutron, proton, electron, muon, and tau, confirm (1.1) and (1.4) and show that in all cases

$$m_{part}\bar{\lambda}_{C,part} = \frac{m_{part}}{\kappa_{C,part}} = \frac{\hbar}{c} = \daleth = 3.51767...\times 10^{-43}\, kg \cdot m. \tag{1.7}$$

Thus rest mass is a measure of the wave number of a particle and, for a fundamental oscillation, of some element of spacetime curvature. Specifically, for a two dimensional curvature, herein assumed to be isotropic, we might look for $\kappa^2_{C,part}$ in a fundamental.

Assuming a physical meaning of this relationship as to the nature of an actual oscillation indicates that three dimensional space is a wave bearing continuum of inertial-elastic properties, from which the mass, m_0, of a <u>fundamental</u> oscillation at resonant frequency, ω_0, is derived, where

$$\frac{\daleth}{c} = \frac{m_0}{\omega_0} = \frac{m_0}{c\kappa_0} \tag{1.8}$$

is an invariant of the system. That is, a particle is a sustained, confined oscillation of a local volume strain of the medium, which derives its mass from the inertial properties of that medium as indicated by its oscillatory frequency and wave number.

We will return to this derivation in a moment, but first we should provide some motivation from the large scale world for pursuing this line of reasoning. In classical Newtonian dynamics, the strength or magnitude of gravitational attractive force, F_g, is directly proportional to the product of the mass, M_a, of two interacting bodies and inversely proportional to the square of the distance, d, separating their centers of mass times an empirically determined gravitational constant, G_N. This finds mathematical expression in Newton's law of universal gravitational attraction generally stated as

$$F_g = \frac{M_1 M_2}{d^2} G_N \qquad (1.9)$$

We would like to find some natural derivation of G_N, arising from a quantum geometrical analysis and independent of empirical determination. Based on the observation that the greatest example of gravitational force, as found in an inertial sink, i.e. a black hole, appears to be a neutron star that has exceeded a certain critical mass, and that the neutron is the more massive of the two atomic nucleons, the other being the proton, we will forward the provisional postulate that this particle serves a principle role in the operation of gravity on the quantum scale.

We will assume for a minute that (1.9) is operational at that scale, so that the maximum quantum gravitational force between two nucleon would be anticipated between two neutron of mass m_n, in contact, which we will take to mean at a distance of separation of their centers of oscillation of twice their $\lambdabar_C$. In this regards it is further provisionally assumed that, on a quantum level, the force attributed to gravity is a manifestation of the centripetal force associated with the spin angular momentum of the particle, $\hbar$, and its oscillatory transverse wave force, so that it is operating at a distance $\hat{\mathbf{r}}$ (using spherical co-ordinates), from the "surface" of one to the center of the adjacent oscillation. In keeping with general relativity, which treats gravity as a function of curvature, we state the curvature of such surface, assumed herein to be isotropic, as $|\hat{\mathbf{r}}|^{-2}$. Thus (1.9) becomes

$$F_{(n\cdot n)g} = \frac{m_{n1} m_{n2}}{\lambdabar_{C,n}^2} G_N = \frac{m_{n1} m_{n2}}{|\hat{\mathbf{r}}_n|^2} G_N = m_{n1} m_{n2} \kappa_{C,n}^2 G_N \qquad (1.10)$$

where the n in the suffixes is for *neutron*. Referring to the CODATA source again, for all values on the right, we find that this evaluates to

$$F_{(n\cdot n)g} = 4.24425...\times 10^{-33} N \cong \frac{1}{6\sqrt{3}} \lambdabar_{C,n}^2 = \frac{1}{6\sqrt{3}\kappa_{C,n}^2}. \qquad (1.11)$$

within a factor of

$$\frac{\frac{1}{6\sqrt{3}} \lambdabar_{C,n}^2}{F_{(n\cdot n)g}} - 1 = 0.000014648... \qquad (1.12)$$

of the CODATA value of $\lambdabar_{C,n}^2$ divided by $6\sqrt{3}$, which is within the relative standard uncertainty for G_N at 0.00015. A very close approximation to this factor will crop up again with respect to other constants, indicating that any measurement process that includes G_N somewhere in its mix could involve this discrepancy. The numerical co-efficient will be derived in a moment. The very close agreement of this number with the magnitude of the neutrons's reduced Compton wavelength is not found if the same procedure is used for the proton, electron, muon, and tau. This suggests a fundamental tie-in with the geometries, i.e. the curvature of the neutron. It will be noted that while the magnitude of $F_{(n\cdot n)g}$ (times $6\sqrt{3}$) is equal to $\lambdabar_{C,n}^2$, it evaluates in SI units of force or mdt^{-2} as Newton, whereas $\lambdabar_{C,n}^2$ is in units of length squared, d^2, or area. This hints at the derivative nature of $F_{(n\cdot n)g}$, which judging by the units involved is a change in force, dF, per change in stress, dT, where stress is defined as a force operating on a cross-sectional or surface area of a volume, the sine of the angle of incidence of said force to the plane of said area varying anywhere from 0, in which the stress is a pure shear stress, to 1, in which the stress is a pure tension stress, with any combination possible between these two extremes. Thus, with a little basic calculus, we have the scalar differential form

$$\frac{F}{A} = T \therefore F = AT, \; dF = AdT \tag{1.13}$$

The derivative form, with the dimensional units expressed in SI terms, is

$$\frac{F_{(n \cdot n)g}}{1 \text{ unit of tension}} = \frac{dF}{dT} = \frac{d(kg \cdot m/s^2)}{d\left(\frac{kg \cdot m/s^2}{m^2}\right)} = \frac{\alpha(kg \cdot m/s^2)}{1\left(\frac{kg \cdot m/s^2}{m^2}\right)} = \alpha m^2 \tag{1.14}$$

α in this case is a number, corresponding to the very small number in (1.11) and is, in mathematical terms, the tangent or slope of the curve, $F = AT$ at the point (T,F), and if A is constant, it is the (linear) curve. It bears acknowledging what is generally implicit, that while the derivative as stated in the second term of (1.14) is taken at the limit, where both dF and dT are exceedingly small, the quantity α is the number of units of force, in this case Newton, per one unit of Pascal or Newton per square meter. Thus the derivative is always normalized or reduced to a unit value of the independent variable as the tangent in the following example is normalized as

$$\tan\frac{\pi}{3} = \frac{\sin\frac{\pi}{3}}{\cos\frac{\pi}{3}} = \frac{.86602...}{.5} = \frac{\sqrt{3}}{1} \tag{1.15}$$

$F_{(n \cdot n)g}$ then can be seen as a differential change in a stress force, and as a quantum of gravitational force, G_q as

$$G_q = F_{(n \cdot n)g} = \alpha m^2 (1 \, pascal) = \alpha \, Newton \tag{1.16}$$

Newton's equation can then be stated in a quantum form as the direct product of the number of quanta (primarily nucleon) in two bodies of mass and inversely to the square of the distance separating them in quantum units times the quantum of gravity as

$$F_g = \frac{n_{M1} n_{M2}}{n_{\hat{r}_n}^2} G_q \tag{1.17}$$

In the full development of this line of reasoning, it will be shown that the proton is simply the neutron which has transmitted a portion of its energy in a wave form that we know as the electron, with a similar analysis for the anti-proton and positron. The number of quanta then is the number of fundamental oscillators or nucleon in each body of mass.

From (1.10) we can see that

$$G_N = \frac{|\hat{\mathbf{r}}_n|^2}{m_n^2} G_q = \frac{G_q}{\lambda_0^2} \tag{1.18}$$

where lambda, (not to be confused with the Compton wavelength) is the linear inertial density of a wave bearing medium or

$$\lambda_0 = \frac{m_n}{|\hat{\mathbf{r}}_n|} \tag{1.19}$$

Substituting the following for aggregate mass, M_a and distance, d, (1.17) becomes (1.9),

$$n_{Ma} \equiv \frac{M_a}{m_n} \tag{1.20}$$

$$n_{\hat{r}_n} \equiv \frac{d}{|\hat{r}_n|}, \tag{1.21}$$

$$F_g = \frac{M_1 M_2}{d^2}\left(\frac{|\hat{r}_n|^2}{m_n^2}\right)G_q = \frac{M_1 M_2}{d^2}G_N. \tag{1.22}$$

Therefore, the possibility of deriving Newton's equation from first principles exists, if we can find some quantum mechanism responsible for the phenomenological fact that

$$G_q = \frac{1}{6\sqrt{3}}\lambda_{C,n}^2 = \frac{1}{6\sqrt{3}\kappa_{C,n}^2}. \tag{1.23}$$

We have a notational decision to make concerning (1.23), that is whether to express the gravitational quantum as a derivative or as a differential. The magnitude will be unchanged in either case, however the units will differ. For use as a component in Newton's constant, G_N, it will be necessary to treat it as a differential, dG_q, in order for the units to correspond to the Newtonian convention of treating gravity as a force. Thus,

$$dG_q = dF = \frac{\lambda_{C,n}^2}{6\sqrt{3}}dT = \frac{|\hat{r}_n|^2}{6\sqrt{3}}dT = \alpha \; kg \cdot m/s^2 \tag{1.24}$$

and Newton, as convention has it, becomes

$$G_N = \frac{|\hat{r}_n|^2}{m_n^2}dG_q = \beta\alpha \; m^3/kg \cdot s^2 \tag{1.25}$$

For purposes of the development of this model and in keeping with the conventions of general relativity, we will find it best to treat it as a derivative, which expresses the quantum gravitational derivative as inverse curvature

$$G_q = \frac{dF}{dT} = \frac{\lambda_{C,n}^2}{6\sqrt{3}} = \frac{|\hat{r}_n|^2}{6\sqrt{3}} = \frac{1}{6\sqrt{3}\kappa_{C,n}^2} = \alpha \; m^2. \tag{1.26}$$

As developed herein, we will find that the force differential is in fact a differential of transverse wave force, $d\tau_0$, and the stress differential is of spacetime expansion stress, dT_0, while the cross product of the moduli expressed as orthogonal tangent vectors gives the inverse curvature of spacetime at the surface of the fundamental quanta. Thus,

$$G_q = \frac{d\tau_0}{dT_0} = \frac{i\hat{r}_n \times j\hat{r}_n}{6\sqrt{3}} = \frac{k|\hat{r}_n|^2}{6\sqrt{3}} = \alpha \; m^2 \tag{1.27}$$

where a greater curvature is indicated by a smaller cross product, positive or spherical in this case. We shall see that it also has a negative or hyperbolic counterpart. In the following discussion of the metric, we will find it convenient at times to refer to curvature by its inverse, r^2, instead of by its direct measure, κ^2.

The Quantum Metric

With respect to the gravitational modeling of general relativity, the local curvature (at a point) of spacetime can be expressed by a metric which is itself a function of the energy density-stress, **T**, of the almost local (nearby) assemblage of matter-energy particle-fields. This energy density-stress describes how the local spacetime will curve while the description of the curvature, **G**, describes how any local quantum particle-fields will move within that spacetime. Thus we have the field equation for general relativity,

$$\mathbf{G} = 8\pi \mathbf{T}. \tag{1.28}$$

Note the functional similarity of this form with (1.27).

From **G** we can extract a 4 dimensional spacetime metric describing this motion in terms for time, τ, or space, σ, as the square of a vector. For a flat locus of spacetime, for a differential timelike change we have, as a function of one dimension of time and three of space

$$d\tau^2 = dt^2 - \left[dx_1^2 + dx_2^2 + dx_3^2 \right] \tag{1.29}$$

and for a differential spacelike change

$$d\sigma^2 = -dt^2 + \left[dx_1^2 + dx_2^2 + dx_3^2 \right] \tag{1.30}$$

For an isotropic condition, the three space terms in brackets could be expressed as

$$dX^2 = \left(\sqrt{3} dx_i \right)^2 = \left[dx_1^2 + dx_2^2 + dx_3^2 \right] \tag{1.31}$$

It is usual to express both timelike and spacelike versions as the square of units of differential length, where the square exponent indicates a squaring of the differential and not a second order differential. Thus the dynamic description of the energy density-stress on the right of (1.28), which involves mass, must be converted to express mass as units of length as well. This indicates that the local curvature as delineated by **G** is a property not simply of a spacetime, but rather of spacetimemass or, if you prefer, spacetimedensity. This is done by use of the following conversion factor,

$$M_l = \frac{G_N}{c^2} M_{kg} = \left(7.424 \times 10^{-28} \tfrac{m}{kg} \right) M_{kg} \tag{1.32}$$

where $G_N c^{-2}$, in units of length over mass, is a reasonable choice for representing mass as a length. This can then be used in the Schwarzschild metric which describes the vicinity of a non-rotating black hole or gravitational sink, its horizon being the boundary from within which, if crossed by a quantum particle, there is no return. While not conventionally recognized, we might surmise that within the horizon the gravitational field strength exceeds the electromagnetic and weak field strengths. The metrics, in polar co-ordinates, where r is the reduced circumference, or circumference around the sink divided by 2π at a point in the vicinity of the sink, are

$$d\tau^2 = \left(1 - \frac{2M_l}{r} \right) dt^2 - \frac{dr^2}{\left(1 - \frac{2M_l}{r} \right)} - r^2 \left(d\theta^2 + \sin^2\theta d\phi^2 \right) \text{ and} \tag{1.33}$$

$$d\sigma^2 = -\left(1-\frac{2M_l}{r}\right)dt^2 + \frac{dr^2}{\left(1-\frac{2M_l}{r}\right)} + r^2\left(d\theta^2 + \sin^2\theta d\phi^2\right). \tag{1.34}$$

The squares root of the timelike and spacelike metrics are the proper time and distance as would be measured at the location of the point under observation and is invariant with respect to any other frame of reference. The information on the right is the bookkeeper or global space and time components of the same event.

The horizon is determined by the case in which $r = 2M_l$, and indicates that at that point, the bookkeeper time, dt, stretches toward infinity, and the radial change of the second term, dr, comes to a stop. The third term is in the plane of the horizon, where the curvature is given (the smaller the number, the greater the curvature) by

$$r^2 = (2M_l)^2 = \frac{1}{\kappa^2} \tag{1.35}$$

Thus the reduced circumference, r, is a local measure of linear curvature, κ, which is known as the quotient of the some standard r_{std} and r, by $\kappa = r_{std}/r$. In this regards it is worth noting that the maximum curvature allowed according to a conventional interpretation of general relativity is given by the Planck area or

$$r^2 = A_{Pl} = \frac{G_N}{c^2}\frac{\hbar}{c} = \frac{G_N}{c^2}\mathsf{n} = 2.612 \times 10^{-70} m^2 \tag{1.36}$$

where we note the use of the inertial constant from (1.1) in the next to the last term. The square root of this parameter gives the Planck length or

$$r_{Pl} = \sqrt{A_{Pl}} = 1.616 \times 10^{-35} m \tag{1.37}$$

The mass in kilograms associated with such curvature according to the Schwarzschild metric, using (1.35) and (1.32), is then

$$M_{kg} = \frac{r_{Pl}}{2}\frac{c^2}{G_N} = 1.088 \times 10^{-8} kg. \tag{1.38}$$

We must mention one other type of metric, for a rotating black hole of neutral charge, which is the Kerr metric. Since it involves angular momentum, it is not spherically symmetrical as the Schwarzschild metric and we use the co-ordinates for the equatorial plane of the system to simplify things. Thus for the timelike metric

$$d\tau^2 = \left(1-\frac{2M_l}{r}\right)dt^2 + \frac{4M_l a}{r}dtd\theta - \frac{dr^2}{\left(1-\frac{2M_l}{r}+\frac{a^2}{r^2}\right)} - \left(1+\frac{a^2}{r^2}+\frac{2M_l a^2}{r^3}\right)r^2 d\theta^2 \tag{1.39}$$

where $a = J/M_l$ is the angular momentum parameter, and J is the angular momentum of the rotating black hole. At the extremum, $J = M_l^2$, so that $a = M_l$, and (1.39) becomes

$$d\tau^2 = \left(1-\frac{2M_l}{r}\right)dt^2 + \frac{4M_l^2}{r}dtd\theta - \frac{dr^2}{\left(1-\frac{M_l}{r}\right)^2} - \left(r^2 + M_l^2 + \frac{2M_l^3}{r}\right)d\theta^2 \tag{1.40}$$

The solution at $r = 2M_l$, while resulting in the vanishing of the dt^2 term, as before, does not have the same effect for the radial term, and we might surmise that the horizon is given by $r = M_l$, resulting in the vanishing of the denominator for the radial term, dr^2. The point at $r = 2M_l$, then constitutes the static limit and the region between the two, $r < r_{ergosphere} < 2r$, is termed the ergosphere, shown in Diagram 1 on page 14. Particles within this domain, including light, are swept along in the direction of rotation. The radial condition at the horizon indicates that it would take an infinite amount of radial change from a bookkeeper or global perspective for any differential radial change to have an effect on the metric. We might interpret this as an indication that from within the horizon over any time duration nothing can emerge. Hence a black hole is portrayed as a domain from which there is no escape; its horizon often as a portal to an infinite abyss.

We might take a different tack however. If it is understood that the horizon is a limit for dr which approaches zero as the denominator approaches zero, the terms in the coefficient cancel and we have for the radial term simply $-1dr^2$, which when factored to give us the proper time results in an imaginary, i.e. an exclusively orthogonal or tangential component to the radial differential, or at the limit $dr_0 \equiv (dr = 0)$.

At the horizon, for the condition $r = M_l$, (1.40) becomes

$$d\tau^2 = -dt^2 + 4rdtd\theta - (2r)^2 d\theta^2 - dr^2 = (idt - i2rd\theta)^2 + (idr)^2 \tag{1.41}$$

$$d\tau^2 = \left[(idt - i2rd\theta) + i(idr)\right]\left[(idt - i2rd\theta) - i(idr)\right] \tag{1.42}$$

$$d\tau^2 = \left[(idt - i2rd\theta) - dr_0\right]\left[(idt - i2rd\theta) + dr_0\right] \tag{1.43}$$

and by this somewhat circuitous logic, we see that the proper time is

$$d\tau = \pm i(dt - 2rd\theta). \tag{1.44}$$

As the radial differential vanishes, instead of a black hole masking a possible singularity as perhaps conventionally considered, we have an internal domain within this horizon that prevents all radial motion or penetration, but does not preclude rotation, (or we might surmise, a possible fourth dimensional component.) All motion, including frame dragging, then is directed about the surface of the sphere as indicated by the imaginary sense, principally within the ergosphere. Thus the ergosphere admits both transverse and radial strain of spacetime, (but none purely radial), the latter of which makes dr only appear to stretch toward infinity as it approaches the horizon where $r_0 \equiv r + dr_0 = M_l$.

The curvature at the limit, r_0^2, can be seen as the gravitational limit, so that any incremental addition to the mass by incoming quanta simply augments $M_l = r_0$. Accordingly, in light of the Kerr metric, analysis of the Planck length at (1.37) as before gives us a mass in kilograms, known as the Planck mass, of

$$M_{Pl} = r_{Pl}\frac{c^2}{G_N} = \frac{\hbar}{cr_{Pl}} = \frac{\hbar}{r_{Pl}} = 2.176 \times 10^{-8} kg. \tag{1.45}$$

Since r_{Pl} is an instance of r_0, or the reduced circumference at the horizon, which is itself a measure of both space and time or spacetime with respect to a certain critical mass, M_0, then and although (1.32) might well apply to aggregates of mass quanta and energy, for the curvature of spacetime associated with individual quanta we might expect the term involving the inertial constant to apply from (1.7) and (1.45) for a conversion factor for quantum mass$_{kilograms}$ to mass$_{length}$ where

$$M_{Pl} r_{Pl} = M_0 r_0 = \daleth = 3.5176 \times 10^{-43} kg \cdot m \tag{1.46}$$

of

$$r_0 = M_{0l} = \frac{\daleth}{M_{0kg}} \tag{1.47}$$

With some transposition we see that the inertial constant converts quantum curvature of spacetime$_{mass}$ to quantum mass$_{kilograms}$,

$$M_{0kg} = \frac{\daleth}{\sqrt{r_0^2}} = \daleth \sqrt{\kappa_0^2} \tag{1.48}$$

If we analyze the horizon for a fundamental quantum, specifically for the neutron as analyzed above in connection with a gravitational quantum, using the Kerr metric and the conventional conversion factor (1.32), we get a reduced circumference of

$$r_{0n} = M_{ln} = \frac{G_N}{c^2} m_n = 1.243 \times 10^{-54} \tag{1.49}$$

and a curvature indicated by

$$r_{0n}^2 = 1.545 \times 10^{-108}. \tag{1.50}$$

(1.49) is smaller than the Compton wavelength over 2π, and (1.50), smaller than the Planck area, both by a factor of 5.92×10^{-39}. If the Planck scale does represent some absolute scale limit and therefore curvature limit, it is apparent that the neutron does not represent a candidate for a quantum inertial sink. We might look for some significance in this factor, and return to (1.24) for use in an analysis of (1.32) in determining a length measure of mass. Using that gravitational quantum differential and the therein derived form of Newton's constant, G_N, of (1.25)

$$G_N = \frac{dG_q}{\lambda_0^2} \tag{1.51}$$

we can state

$$\frac{G_N}{c^2} = \left(\frac{dG_q}{\lambda_0^2} \right) \frac{1}{c^2} = \frac{1}{\lambda_0} \left(\frac{\lambdabar_{C,n}^2}{6\sqrt{3}\lambda_0 c^2} \right) dT_0 = \frac{1}{\lambda_0} \left(\frac{|\hat{r}_n|^2}{6\sqrt{3}\tau_0} \right) dT_0 = \frac{1}{\lambda_0 T_0} dT_0 \tag{1.52}$$

We posit that the Compton wavelength over 2π is the reduced circumference of the neutron at its "surface" or horizon, $\hat{r}_n$, for use in the following metrics, and that $|\hat{r}_n|^2$ is a measure of the (inverse) curvature at that locus. The term in brackets in the third and fourth parsing is, as we shall see, the inverse isotropic expansion stress, where λ_0 is a presumed fundamental linear inertial density of spacetime and τ_0 is the corresponding fundamental quantum transverse wave force, according to the classical wave relationship

$$\tau_0 = \lambda_0 c^2. \tag{1.53}$$

(1.52) then can be stated as

$$\frac{G_N}{c^2} = \frac{1}{\lambda_0 T_0} dT_0 = \frac{\hat{r}_n}{m_n} d\ln T_0 \tag{1.54}$$

Here the differential of the natural log of the expansion stress is a dimensionless number, equal in the SI system to the following, as determined by the CODATA values of the neutron mass, neutron Compton wavelength over 2π, and the speed of light,

$$d\ln T_0 = T_0^{-1} dT = \frac{\lambdabar_{C,n}^3}{6\sqrt{3} m_n c^2} dT = 5.92146 \times 10^{-39} \left(N/m^2\right)^{-1} \tag{1.55}$$

and the units and magnitude of (1.54) are as found in (1.32). In a natural system in which the unit length scale is gauged to $\hat{r}_n$, and in which the inertial constant, $\daleth = \frac{\hbar}{c}$, as well as $\hbar$ and c themselves are all set to one and a unit of time is equal to a unit of length, the differential of the natural log of the stress as well as the stress itself is equal to one, while the log itself vanishes.

If we examine (1.32) then as a logarithmic function of the change in isotropic expansion stress, we have, in SI, where $Z_0 = 2.3908 \times 10^{-3} kg/s$ is the characteristic mechanical impedance of spacetime, not the same as the CODATA impedance of the vacuum,

$$\frac{G_N}{c^2} \frac{1}{d\ln T_0} = \frac{1}{(\lambda_0 c)^2} \frac{dG_q}{d\ln T_0} = \frac{1}{Z_0^2} \frac{dG_q}{d\ln T_0} = \frac{\hat{r}_n}{m_n} = 1.2539 \times 10^{11} \frac{m}{kg} \tag{1.56}$$

Thus as isotropic expansion stress increases, $d\ln T_0$ decreases, so that the inertial density must decrease proportionally for the left term of (1.54) to remain constant. But we might anticipate this local change if energy, i.e. inertial density per <u>universal</u> volume, is to be conserved as the universal volume increases. Our model here is of a spacetime which has an inherent universal inertial density irrespective of the presence of quanta, rest mass or otherwise, i.e. a spacetimemass. This requires an adjustment of perspective from the conventional modeling, as there is no true physical void or vacuum in this model.

Since established models do not appear to incorporate the existence and therefore the significance of T_0, at least in the form found herein, which evaluates to $1.6888 \times 10^{38} N/m^2$, it is understandable why the gap between the gravitational and other interactions appears so intractable. It also states that the value of a <u>quantum</u> mass expressed in terms of length is undervalued by that same magnitude.

Now, when we apply (1.32) using the modification of (1.56) to determine the spacetime curvature at the periphery of a neutron for use in the Kerr metric, we have,

$$M_l = \frac{\hat{r}_n}{m_n} m_n = \hat{r}_n = \lambdabar_{C,n} = \frac{1}{\kappa_n} = 2.1002 \times 10^{-16} m. \tag{1.57}$$

It is significant here, that while aggregates of rest mass quanta can be effectively represented by units of length <u>directly</u> proportional to the aggregate mass, for individual quanta, as judged by the ratios of rest mass and Compton wavelength for various particles, mass is <u>inversely</u> proportional to length in keeping with (1.7). The reduced circumference per quantum is <u>inversely</u> proportional to mass, but then again that is what spacetime curvature tells us, since a flat spacetime in the absence of quanta has an "infinite" radius.

At the point at which two neutron are in contact as judged by $\hat{r}_n = \lambda_{C,n}$, using one as the center of mass and the other as the test particle, the extreme Kerr metric in timelike form, suppressing one dimension, is

$$d\tau^2 = \left(1 - \frac{2|\hat{r}_n|}{\hat{r}_n}\right)dt^2 + \frac{4|\hat{r}_n|^2}{\hat{r}_n}dtd\theta - \frac{dr^2}{\left(1 - \frac{|\hat{r}_n|}{\hat{r}_n}\right)^2} - \left(|\hat{r}_n|^2 + |\hat{r}_n|^2 + \frac{2|\hat{r}_n|^3}{\hat{r}_n}\right)d\theta^2 \quad (1.58)$$

$$d\tau^2 = -dt^2 + 4|\hat{r}_n|dtd\theta - 4|\hat{r}_n|^2 d\theta^2 \quad (1.59)$$

and for the proper time,

$$d\tau = \pm i\left(dt - 2\hat{r}_n d\theta\right). \quad (1.60)$$

As with (1.40)-(1.44), dr^2 at the horizon vanishes, in this case expressing the stationary or "rest" nature of rest mass quanta. Per below, proper and bookkeeper time have the same magnitude at the horizon, indicating a property of universal cosmic bookkeeper time (but not external to spacetime and not necessarily of constant unit value). They are obviously orthogonal in some manner as indicated by the imaginary sense, and we might consider the possibility of the condition in which time is simply a measure of the transverse motion or

$$dt = \hat{r}_n d\theta \text{ or} \quad (1.61)$$
$$dt - 2\hat{r}_n d\theta = -dt \text{ so that} \quad (1.62)$$
$$d\tau = \mp idt \quad (1.63)$$

which applies at the horizon. We can think of the proper time as registered in the (curved) plane of that surface, and idt as being fourth dimensional or normal to the horizon and hence in the direction of isotropic expansion.

If $2\hat{r}_n d\theta$ is in the plane or surface of the static limit, in the equatorial plane, and we suppose that

$$dt = 2\hat{r}_n d\theta, \text{ then} \quad (1.64)$$
$$d\tau = \pm i\left(dt - 2\hat{r}_n d\theta\right) = 0 \quad (1.65)$$

and proper time effectively stops as with a lightlike vector. This suggests that the static limit of a rest quantum is a circulating light wave.

The curvature at the horizon is indicated by

$$r_0^2 = |\hat{r}_n|^2 = 4.4108 \times 10^{-32} m^2 \quad (1.66)$$

If we contrast this with the quantum gravitational differential, itself an expression of fundamental curvature, we have

$$\frac{dG_q}{r_0^2} = dG_q \kappa_0^2 = \frac{1}{6\sqrt{3}} dT_0. \tag{1.67}$$

The isotropic expansion stress, a 4-stress, is orthogonal to all three ortho-normal components of a 3-space, and can be modeled as constrained to the four cubic diagonals at an angle of 0.615479 to the cubic edge and 0.955316 to the cubic surface normal vector. $1/\sqrt{3}$ is the sine of 0.615479 and the cosine of 0.955316. Each ortho-normal component of this stress in that 3-space is 1/6 of the total 4-stress. Combined, these factors give us the coefficient on the right.

We can see this gravitational, i.e. geometric, condition of the static limit as an operation of the strong force. For two neutron in contact, the center of mass of each sits ideally on the others static limit. The center of a proton, with slightly larger Compton, would sit right outside the limit, while an electron would be removed by a factor of almost 2,000.

Assuming that the above and supportive analysis does indicate a quantum inertial sink, and that the neutron is such, then a maximum linear inertial density is given by the inverse of (1.56). This would seem quite small, but for its bulk implications. For a volume density, we would figure the number of hypothetical fundamental rest mass quanta per volume of such quanta, tightly packed. Using a system of one sphere with twelve contacting identical spheres, and disregarding any expansive dynamic effects of spin, charge, etc., we can compute the density and find that it equals roughly 2.2549×10^{46} $quanta/m^3$, while inverting the neutron mass gives the number of such quanta per kilogram or 5.9704×10^{26} $quanta/kg$, or a maximum theoretical density of 3.7768×10^{19} kg/m^3 or $\rho_{sphere} = 1.5820 \times 10^{20}$ $kg/meter\ sphere$, where ρ_{sphere} is the volume density in kilograms for a sphere with a radius of 1 meter.

From this we can find a threshold inertial sink size for an aggregation of quanta by using the following for a flat Euclidean space,

$$r = \left(\frac{M_{kg}}{\rho_{sphere}}\right)^{\frac{1}{3}} \tag{1.68}$$

Assuming $r = M_l$ as with (1.32)

$$\frac{G_N}{c^2} M_{kg} = \left(\frac{M_{kg}}{\rho_{sphere}}\right)^{\frac{1}{3}} \tag{1.69}$$

for an extreme Kerr spacetime horizon gives

$$M_{kg} = \left(\frac{c^6}{G_N^3 \rho_{sphere}}\right)^{\frac{1}{2}} = 3.930 \times 10^{30} kg \tag{1.70}$$

which using the above density gives us the evaluations in the following table or approximately two solar masses for the threshold.

Here in column 3, from (1.68) we compute the radius for various celestial bodies, Earth, Sun, Milky Way galactic BH and Virgo cluster BH, and include the theoretical threshold size black hole and the Universe, as listed in column 1. "Flat Spacetime" does not specify that the pertinent body has no curvature effect on the surrounding spacetime, but rather that the curvature of individual quanta, i.e. quantum gravity, is not effected by the aggregate mass and remains the same as for an individual quantum in isolation in flat spacetime, i.e. there is no assumed collapse of the quantum waveform toward a singularity. The fourth column gives the reduced circumference at the horizon of an extreme, charge free, Kerr black hole according to the conventional interpretation of general relativity. The fifth column simply makes explicit whether the third column figure resides within the fourth. This indicates the possibility that the rest mass quanta inside a black hole horizon could congregate at maximum density without precipitating a singularity.

	Mass in kg	Radius, r in m, Density = ρ_{Sphere} Flat Spacetime	Mass in meters $M_l = \frac{G_N}{c^2} M_{kg} = r_0$	Is r within $M_l = r_0$ at Horizon?
Earth	5.9742×10^{24}	33.55	4.44×10^{-3}	No
Sun	1.989×10^{30}	2.325×10^3	1.477×10^3	No
Kerr BH threshhold	3.930×10^{30}	2.913×10^3	2.913×10^3	At Horizon
Milky Way	5.2×10^{36}	3.20×10^5	3.86×10^9	Yes
Virgo cluster	6×10^{39}	3.36×10^6	4.45×10^{12}	Yes
Universe	$1.67 \times 10^{53} =$ 10^{80} nucleon	1.02×10^{11}	$1.24 \times 10^{26} =$ 13.1×10^9 light yrs	Yes

Of interest is the fact that the universe appears to be within its own horizon, which conventionally would tend to imply that its constituents should be contracting, and that there are black holes within black holes. Also the mass in meters being equal to the reputed age of the universe times the speed of light seems a bit serendipitous unless of course that mass, i.e. the number of currently theorized nucleons, was estimated using the above mass$_{length}$ equation. But this figure is not the currently theorized (observed) extent of the universe, which is in the 150 billion light year range. Finally it is noted the the hypothetical mass of the known universe at maximum density and a radius of 102 million kilometers, would fit inside the earth's orbit in flat spacetime.

Returning to the Kerr metric as the basis for a metric for a quantum inertial sink, what is left out in this metric is the remaining parameter, ϕ, which covers the conditions over the surface of the horizon other than the equator. The following Cartesian ortho-normal matrix is taken from Section 2, subsection Isotropic Expansion and the Generation of Rotational Oscillation in the development which follows, where the trace components vanish as $dr \to 0$, and where $\theta = \omega t$,

$$\begin{pmatrix} 0 & \hat{r}_n d\theta & \hat{r}_n \sin\omega t d\phi \\ -\hat{r}_n d\theta & 0 & -\hat{r}_n \cos\omega t d\phi \\ -\hat{r}_n \sin\omega t d\phi & \hat{r}_n \cos\omega t d\phi & 0 \end{pmatrix} + \begin{pmatrix} -0 & -\hat{r}_n d\theta & -\hat{r}_n \sin\omega t d\phi \\ \hat{r}_n d\theta & -0 & \hat{r}_n \cos\omega t d\phi \\ \hat{r}_n \sin\omega t d\phi & -\hat{r}_n \cos\omega t d\phi & -0 \end{pmatrix} \quad (1.71)$$

With respect to the uncertainty principle, for a given spin axis, S_L, and equatorial plane, as indicated by the $\hat{r}_n d\theta$ components, indicating the quantum orientation, without a determination of θ, we can not state of the transverse wave momentum at the $\hat{r}_n (\sin/\cos)\theta d\phi$ components. If we could determine the momentum vectors from all the above components at a moment in time, we still could not determine the spin (tangential and therefore axial) vectors of $S_L = \hbar$, which could be pointed toward $+x_3$ or $-x_1$.

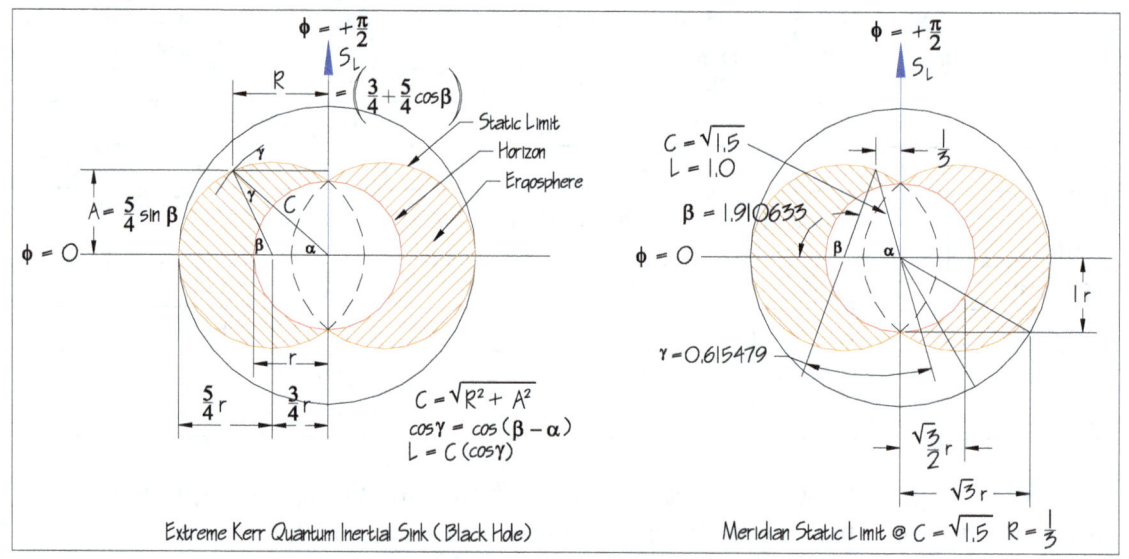

Quantum Inertial Sink Diagram 1

Based on the above diagram, we might state our quantum inertial sink metric, solved for the horizon, where $\hat{r}_n \equiv M_l = a = r$, $R = \left(\frac{3}{4} + \frac{5}{4}\cos\beta\right)\hat{r}_n$, the normal distance from the spin axis to the static limit, and $L = C\cos\gamma$, the tangent reduced circumference, C, at any point along a constant meridian of the static limit projected on to the tangent plane of the static limit, and $\pm e^{\pm i\omega_0 t}$ represents one of eight orthogonal, azimuthal clocks located along the meridians at $\theta = \left(0, \frac{\pi}{2}, \pi, \frac{3\pi}{2},\right)$ and $\phi > 0$ for $ccw(+\cos\omega t, +i\sin\omega t, -\cos\omega t, -i\sin\omega t)$ and at $\theta = \left(0, \frac{\pi}{2}, \pi, \frac{3\pi}{2},\right)$ and $\phi < 0$ for $cw(+\cos\omega t, +i\sin\omega t, -\cos\omega t, -i\sin\omega t)$, all viewed from outside the sphere.

Per the right hand rule for angular momentum, clocks in the "northern" hemisphere turn counterclockwise, while those in the southern turn clockwise. The sense of the trig functions indicates parallel (+) and anti-parallel (-) with respect to the angular momentum vector or north. At $\phi = 0$, the oscillation is binary digital and therefore has no chirality, though its intensity is sinusoidal, while at $\phi = \pm\frac{\pi}{2}$, the four functions merge through parallel transport along their respective meridians and are senseless with respect to north and south. As a result, the trig function could also be written

$$\pm\left\{e^{\pm i(\omega_0 t \mp \theta)}\right\} = \text{Re}\left(e^{\pm i(\omega_0 t \mp \theta)}\right) = \cos_{ccw}(\omega_0 t + \theta) \text{ or } \cos_{cw}(\omega_0 t + \theta) \quad (1.72)$$
$$= \cos(\omega_0(+t) - \theta) \text{ or } \cos(\omega_0(-t) + \theta)$$

where

$$\pm\{a\} = \frac{\cos(\omega_0 t - \theta)}{|\cos(\omega_0 t - \theta)|} a, \mp\{a\} = -\frac{\cos(\omega_0 t - \theta)}{|\cos(\omega_0 t - \theta)|} a \quad (1.73)$$

The metric at the quantum horizon and static limit with respect to the equatorial plane of its spin angular momentum axis then is

$$d\tau^2 = \left(1 - \frac{2|\hat{r}_n|}{\hat{r}_n}\right)dt^2 + \frac{4|\hat{r}_n|^2}{\hat{r}_n} dt d\theta - \frac{dr^2}{\left(1 - \frac{|\hat{r}_n|}{\hat{r}_n}\right)^2} - R^2 d\theta^2 \mp \left\{\left(e^{\pm i(\omega_0 t \mp \theta)} L d\phi\right)^2\right\} \quad (1.74)$$

$$d\tau^2 = -dt^2 + 4|\hat{r}_n| dt d\theta - R^2 d\theta^2 \mp \left\{\cos^2(\omega_0 t - \theta) L^2 d\phi^2\right\} \quad (1.75)$$

Of interest is the fact that
$$L = \tfrac{4}{5}\hat{r}_n + \tfrac{3}{5}R = \tfrac{4}{5}\hat{r}_n + \tfrac{3}{5}\left(\tfrac{3}{4} + \tfrac{5}{4}\cos\beta\right)\hat{r}_n = \left(\tfrac{5}{4} + \tfrac{3}{4}\cos\beta\right)\hat{r}_n \quad (1.76)$$

Thus for a fixed spin axis from the equatorial plane, (1.75) becomes the following quantum timelike metric for the condition $\hat{r}_n \equiv M_l = a = r$

$$d\tau^2 = \left(idt - i2|\hat{r}_n|d\theta\right)^2 \mp \left\{\cos^2(\omega_0 t - \theta)\left(2|\hat{r}_n|\right)^2 d\phi^2\right\} \quad (1.77)$$

and we might imagine the spacelike metric to be

$$d\sigma^2 = -\left(idt - i2|\hat{r}_n|d\theta\right)^2 \pm \left\{\cos^2(\omega_0 t - \theta)\left(2|\hat{r}_n|\right)^2 d\phi^2\right\} \quad (1.78)$$

so that on this fundamental level an essential symmetry exists as

$$d\sigma^2 \equiv -d\tau^2 \quad (1.79)$$

and for the proper time and space, indicating that time is orthogonal to any measure of space,

$$d\sigma \equiv id\tau. \quad (1.80)$$

The condition at $L = 1\hat{r}_n$ gives angles of $\gamma = .615479708...$ and $\beta = 1.910633236...$ at both poles. This value of γ is pivotal as the angle of cubic center-to-vertex diagonal to the diagonal across a cubic surface, and thereby to isotropic stress. The value of β appears to be the path integral over one quadrant or one fourth cycle of the figure eight shown later at Spin Diagram 1 – Spin Energy Cycle on page 56. As $L = 1\hat{r}_n$ equals the horizon's reduced circumference, its value squared equals the horizon inverse curvature, representing an effective limit to the approach of the static limit toward the horizon. It is also the sinusoidal limit of the azimuthal clocks, which will be cotemporaneous at the ring $R = \tfrac{1}{3}\hat{r}_n$ where $C = \sqrt{1.5}$.

The metrics indicate that the time, t, and spin position, given by θ, are inextricably coupled in the rotation of the quantum and therefore the frequency, while their coupling in the coefficient of $\mp/\pm\{\cos(\omega_0 t - \theta)\}$ maintains that same frequency in the meridian differential about a second axis, $\Phi = (\omega_0 t \mp \frac{\pi}{2})$, orthogonal to and rotating about S_L. Thus the system evidences an instance of double rotation which avoids the entanglement condition, where the S_L components rotate through $\theta = 0$, while the Φ components oscillate through $\phi = 0$, via $\phi = \pm\frac{\pi}{2}$.

The metrics further indicate that the static limit, in the global or bookkeeper frame of reference, does not appear to arc along the meridian to the angular momentum pole at the horizon as in the diagram, but rather appears to have a uniform reduced circumference about a spherical shell. Continuity is maintained through the spin axis, i.e. the azimuthal clocks merge via parallel transport. From the global perspective for both θ and ϕ, however, the proper distance and time are modified by a factor of ½ from a modulus of $2\hat{r}_n$. As a result, with respect to θ and ϕ, the static limit is deemed to be equal to the horizon at $\hat{r}_n$, and the path integral of the Φ components, $\phi = 0 \rightarrow +\frac{\pi}{2} \rightarrow 0 \rightarrow -\frac{\pi}{2} \rightarrow 0$ appears a figure 8 as in Spin Diagram 1. The value of the path integral is invariant as given by

$$\oint_{\theta=2\pi} d\phi = 4\beta = 4(1.910633...) = 7.642532... \quad (1.81)$$

From the global perspective, the halving of the moduli of θ and ϕ effectively collapses the quantum horizon to a singularity, though it is a false singularity, as it is the static limit, via the metric, that reduces to the horizon. As $\hat{r}_n$ is the limit of radial change, and $\hat{r}_n^2$ the curvature limit, this also establishes the angular momentum limit, shown first in length then SI units

$$S_L = |\hat{r}_n|^2 c = M_l |\hat{r}_n| c = \frac{n}{|\hat{r}_n|}|\hat{r}_n| c = nc = \hbar \quad (1.82)$$

giving us another basis for the uncertainty principle. In light of the above this limit also shows the geometric basis of gravity as curvature as with (1.27), reiterated here,

$$G_q = \frac{d\tau_0}{dT_0} = \frac{i\hat{r}_n \times j\hat{r}_n}{6\sqrt{3}} = \frac{k|\hat{r}_n|^2}{6\sqrt{3}} \quad (1.83)$$

where k is a unit vector normal to the horizon tangent plane. The cross product can be of negative sense and indicate negative curvature, which in conjunction with the complex axes of the inversphere as indicated in the following development, and associated with the value of γ above, suggests a wave node to a 4-D, hyperbolic component.

The quantum metric of (1.75) has observational problems that manifests as the uncertainty principle, as well. Any test particle approaching the static limit has its own quantum metric and quantum gravitational and magnetic fields that interact with the initial quantum as it comes within its vicinity. In keeping with the development herein, all rest mass quanta have a magnetic induction torque, L_μ, at an angle of 0.955316 anti-parallel to the spin angular momentum vector. In the presence of an external magnetic field, B, the torque vector tends to align parallel with the field. This alignment cants the spin vector toward the same angle with respect to the initial position as shown in the following Quantum Inertial Sink Diagram 2.

Associated with the torque and orthogonal to it is a pair of magnetic power moments, $\pm M$, arising from the wave characteristics of the quantum oscillation. These power moments are an expression of the wave restorative action as it returns from a point of maximum amplitude toward its equilibrium strain position in the equatorial plane of Quantum Inertial Sink Diagram 1, while simultaneously twisting about the spin axis. It is this pair of power moments that produce the inductive torque, L_μ. If the external magnetic field, produced by a quantum or aggregate source, is of sufficient strength, $\pm M$ will be pulled into the equatorial plane and the spin, S_L, and, in the case of the neutron, anti-parallel effective magnetic moment, μ, will align as shown here, and precess about L_μ. The projection of $\pm M$ into the plane of S_L/μ - L_μ is as shown in the polar axis view, and contributes the characteristic coefficients of ½ spin to rest mass particles. As the approach of a test particle-field or an applied external field will always tend to align L_μ as shown, the spin vector, whose poles are the loci of wave amplitude, will respond accordingly, creating the phenomena of isotropic spin. Thus the application of the quantum metric is limited by this aspect of particle interaction.

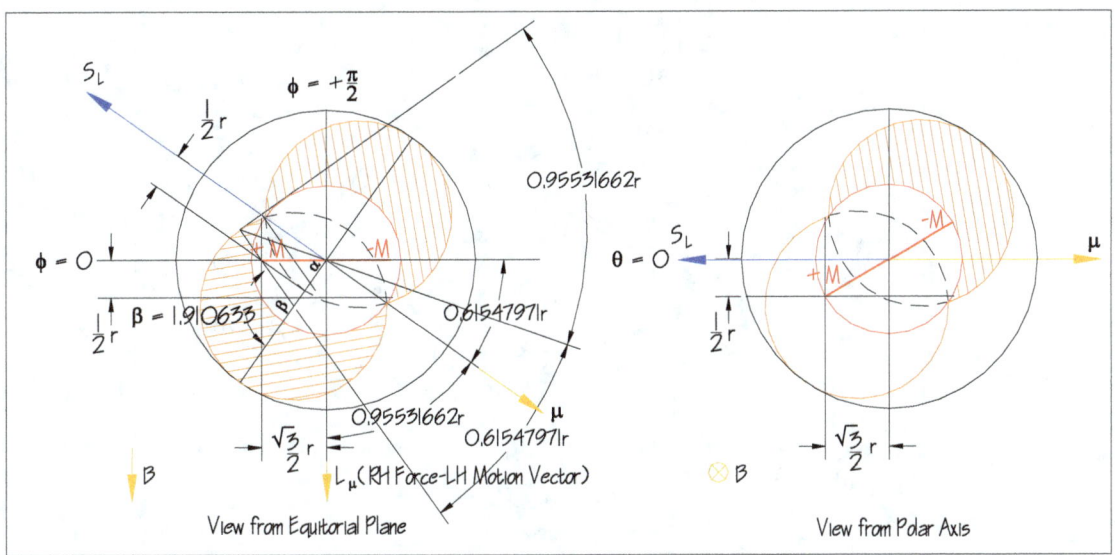

Quantum Inertial Sink Diagram 2

Although there may be epistemological limits on the application of a quantum metric, in principle this analysis shows a fundamental quantum, the neutron, to be an inertial sink, whose curvature and static limit, ostensibly a physical expression of confined rotational oscillation, is responsible for both quantum gravity and the strong interaction, and through multiples of such quanta in celestial aggregations, for larger scale curvature of spacetime, or as we have attempted to show, curvature of spacetimemass or spacetimedensity. The cosmological implications of this analysis, in particular with respect to the significance of the Planck scale, will be taken up in the final section.

Wave Bearing Continuum

Above we stated that the oscillatory nature of quantum particles indicates that three dimensional space is a wave bearing continuum of inertial-elastic properties, and should next provide some justification for that statement. Regardless of whether space, absent any oscillation, is defined as a void or, along with time, as part of a kinematic stage or protean backdrop of a four dimensional spacetime as in general relativity, it is generally recognized as permitting the translation of electromagnetic oscillations. While these oscillations are variously modeled as photons, radiation or rays, energy, messenger particles, and self propagating electromagnetic or EM waves, they are all viewed as traveling through space, hence in some manner space is allowing or permitting the penetration of such oscillations; therefore, it transfers stress, and we might surmise, strain. In terms of the EM wave model, they are recognized as transverse waves in which the electric fields, E, vary sinusoidally in phase with, hence at the same frequency as and orthogonally to the magnetic fields, B, clockwise when viewed from the direction of wave travel. Thus the cross product of E into B gives

$$E \times B \Rightarrow \text{direction of wave travel} \tag{1.84}$$

Various properties of EM wave propagation are shown here.

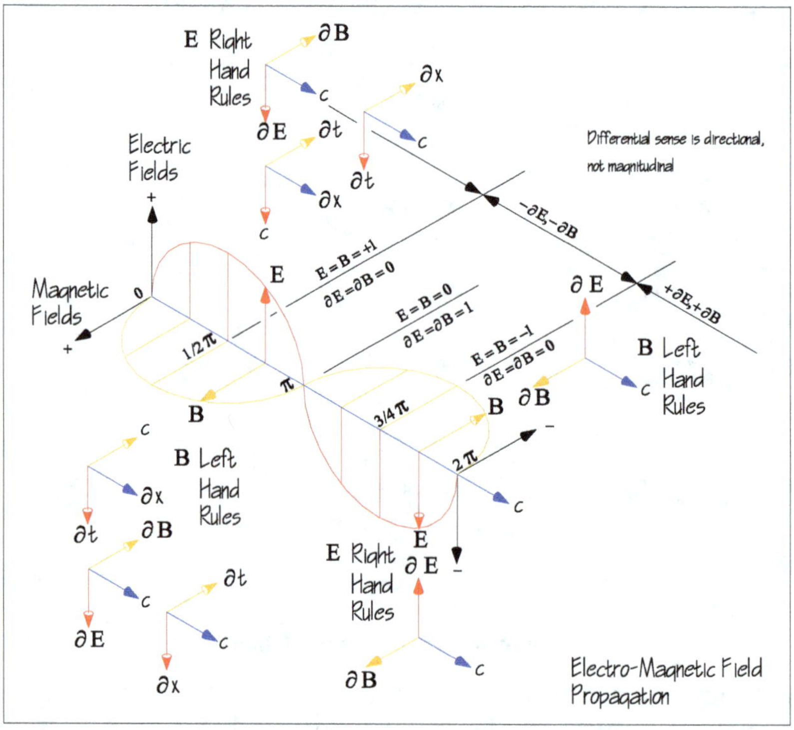

Figure 1 - Electro Magnetic Wave

It is established that for an induced electric field, using Faraday's law of induction, the change in the electric field over space is related to the change in the magnetic field over time by the scalar equation

$$\frac{\partial E}{\partial x} = -\frac{\partial B}{\partial t} = \left[\frac{-i\partial B}{i\partial t}\right]. \tag{1.85}$$

Similarly, for an induced magnetic field, using Maxwell's law of induction, the change in the magnetic field over space is related to the change in the electric field over time by

$$\frac{\partial B}{\partial x} = -\mu_0 \varepsilon_0 \frac{\partial E}{\partial t} = \left[\frac{1}{c^2} \frac{-i\partial E}{i\partial t}\right] \quad (1.86)$$

We can shed some light on this condition and on the nature of the fundamental ½ spin particles by using the orthogonal sense of the bracketed terms to develop a vector form of these equations. While vector division is not normally defined, if we stipulate an orthogonal condition, as in the above wave, we can define vector division between orthogonal vectors, which by virtue of a ½ π rotation indicated by $\pm i$ can become parallel or anti-parallel vectors. In fact we will find that the operation of the cross product is equivalent to division involving a vector with imaginary sense in either the dividend or the divisor. This is done not to confer any notational advantage, but to indicate the underlying rotational symmetry of the system. Thus, for the right hand rule, we can state

$$\mathbf{a}(i \div)\mathbf{b} = \mathbf{b}(\div i)\mathbf{a} \equiv \mathbf{a}(\times)\mathbf{b}. \quad (1.87)$$

The right hand rule reverse, which is equivalent to a left hand rule is

$$\mathbf{a}(-i \div)\mathbf{b} = \mathbf{b}(\div -i)\mathbf{a} \equiv -\mathbf{a}(\times)\mathbf{b} = \mathbf{b}(\times)\mathbf{a}. \quad (1.88)$$

indicating that the right hand rule is equal to the left hand rule reverse.

With some rearrangement of (1.85) and (1.86) we have

$$i\partial E = \frac{\partial x}{i\partial t} \partial B = c\partial B \text{ and} \quad (1.89)$$

$$i\partial B = \left(\frac{i\partial t}{\partial x}\right)^2 \frac{\partial x}{i\partial t} \partial E = \frac{i\partial t}{\partial x} \partial E = \frac{1}{c} \partial E. \quad (1.90)$$

We might write this in vector terms as follows shortly, with the understanding that the i represents a counterclockwise rotation of ∂E in the plane of E-B, when viewed from the direction of wave travel, and c is a scaling or normalizing factor equal in magnitude to the speed of wave travel. In other words, if the unit of distance x was equal to the unit of time t, c would equal 1. Thus, given the right hand rule, from

$$\partial x \equiv ic\partial t \quad (1.91)$$

we have

$$\frac{\partial \mathbf{x}}{i\partial \mathbf{t}} \equiv \partial \mathbf{t} \times_R \partial \mathbf{x} = \frac{i\partial \mathbf{x}}{-\partial \mathbf{t}} \equiv \partial \mathbf{x} \times_R -\partial \mathbf{t} = \mathbf{c} \quad (1.92)$$

indicating the inherent orthogonality between $\mathbf{x}$ and $\mathbf{t}$. $\mathbf{t}$ can be modeled as a vector anywhere in the plane, t, orthogonal to $\mathbf{x}$, so that $i\mathbf{t}$ represents a rotation of $\mathbf{t}$ into $\mathbf{x}$ or a collapse of the t plane onto $\mathbf{x}$. This appears to indicate that time and space are commutative, in the sense that they are interchangeable if we can think of time as constituting an orthogonal plane about a given spatial dimension, x. This extends to the left hand rule as well as

$$\frac{-i\partial \mathbf{x}}{\partial \mathbf{t}} \equiv \partial \mathbf{x} \times_L \partial \mathbf{t} = \frac{\partial \mathbf{x}}{-i(-\partial \mathbf{t})} \equiv -\partial \mathbf{t} \times_L \partial \mathbf{x} = \mathbf{c} \quad (1.93)$$

The direction of $\mathbf{c}$ is therefore determined by the direction of $\partial \mathbf{x}$, which is logical, and not by $\partial \mathbf{t}$ which is orthogonal to all of 3-space. The inversion symmetries are

$$\frac{-\partial \mathbf{x}}{i\partial t} \equiv \partial \mathbf{t} \times_R -\partial \mathbf{x} = \frac{i(-\partial \mathbf{x})}{-\partial t} \equiv -\partial \mathbf{x} \times_R -\partial \mathbf{t} = -\mathbf{c} \text{ and} \qquad (1.94)$$

$$\frac{-\partial \mathbf{x}}{-i(-\partial t)} \equiv -\partial \mathbf{t} \times_L -\partial \mathbf{x} = \frac{-i(-\partial \mathbf{x})}{\partial t} \equiv -\partial \mathbf{x} \times_L \partial \mathbf{t} = -\mathbf{c} \qquad (1.95)$$

In (1.92) to (1.95), we have a fixed orthogonal relationship between $\partial \mathbf{t}$, $\partial \mathbf{x}$, and $\mathbf{c}$, which we might call a native right hand relationship as given by the first crossed term in (1.92), even though we have a symmetry of right and left hand cross products. Note that there is no $\partial \mathbf{t} \times_L \partial \mathbf{x}$. With a native left hand relationship, $\partial \mathbf{t}$ and $\partial \mathbf{x}$ are transposed, as found in the Left Hand Rules section with the magnetic fields in the EM wave diagram above, and we have

$$\partial \mathbf{x} \equiv -ic\partial t \qquad (1.96)$$

$$\frac{\partial \mathbf{x}}{-i\partial t} \equiv \partial \mathbf{t} \times_L \partial \mathbf{x} = \frac{-i\partial \mathbf{x}}{-\partial t} \equiv \partial \mathbf{x} \times_L -\partial \mathbf{t} = \mathbf{c} \qquad (1.97)$$

$$\frac{i\partial \mathbf{x}}{\partial t} \equiv \partial \mathbf{x} \times_R \partial \mathbf{t} = \frac{\partial \mathbf{x}}{i(-\partial t)} \equiv -\partial \mathbf{t} \times_R \partial \mathbf{x} = \mathbf{c} \qquad (1.98)$$

$$\frac{-\partial \mathbf{x}}{-i\partial t} \equiv \partial \mathbf{t} \times_L -\partial \mathbf{x} = \frac{-i(-\partial \mathbf{x})}{-\partial t} \equiv -\partial \mathbf{x} \times_L -\partial \mathbf{t} = -\mathbf{c} \qquad (1.99)$$

$$\frac{-\partial \mathbf{x}}{i(-\partial t)} \equiv -\partial \mathbf{t} \times_R -\partial \mathbf{x} = \frac{i(-\partial \mathbf{x})}{\partial t} \equiv -\partial \mathbf{x} \times_R \partial \mathbf{t} = -\mathbf{c} \qquad (1.100)$$

In both the left and right hand native relationships, a positive $\mathbf{c}$ results from a rotation of one vector parallel into the other, while the negative $\mathbf{c}$ results from a rotation of that vector anti-parallel into the other. Notice that an i in both the dividend and the divisor is not defined as a cross product, though it might be depending on the context. The sense of the imaginary designation, then indicates by convention whether the crossing is to the left, clockwise, or to the right, counterclockwise, and its position shows it as an operator crossing the following vector into the other component of the quotient. Thus we have the following identity that is tacit in (1.91) through (1.100),

$$\mathbf{c} = \frac{\partial \mathbf{x}}{i\partial t} \equiv \partial \mathbf{t} \times_R \partial \mathbf{x} \equiv \frac{i\partial t}{\partial \mathbf{x}} = \frac{1}{\mathbf{c}} \qquad (1.101)$$

With this in mind, returning to (1.89) we can state with reference to the diagram,

$$\frac{i\partial \mathbf{E}}{\partial \mathbf{B}} \equiv \mathbf{c} \equiv \frac{\partial \mathbf{x}}{i\partial t} = \frac{i\partial \mathbf{x}}{\partial t}. \qquad (1.102)$$

for the transverse wave speed and direction as well as the longitudinal phase speed and direction. A similar identity holds for (1.90), if we multiply through by -1, as

$$\frac{-i\partial \mathbf{B}}{\partial \mathbf{E}} \equiv \frac{1}{\mathbf{c}} \equiv \frac{-i\partial t}{\partial \mathbf{x}} \qquad (1.103)$$

This follows the left hand rule, as indicated by the fact that the translational sense of $\mathbf{c}$ in (1.103) is unaffected by the rotational sense change as in comparing (1.97) and (1.92). In right hand form this is

$$\frac{\partial \mathbf{B}}{i\partial \mathbf{E}} \equiv \frac{1}{\mathbf{c}} \equiv \frac{\partial t}{i\partial \mathbf{x}} = \frac{i\partial t}{\partial \mathbf{x}} \qquad (1.104)$$

which is simply the inverse of (1.102). In more familiar form, using the right hand rule in the cross, both (1.102) and (1.104) are
$$\partial \mathbf{E} \times \partial \mathbf{B} = \mathbf{c} = \partial \mathbf{t} \times \partial \mathbf{x} = \partial \mathbf{x} \times \partial \mathbf{t} \tag{1.105}$$
and we can see that by virtue of (1.92) and (1.98) that crossing involving time is commutative.

Clearly in Maxwell the product of the permeability, μ_0, and the permittivity, ε_0, constants as c^2 inverts the time and space differentials and relates a change in the electric fields to a change in space and a change in the magnetic fields to a change in time, precisely as in Faraday. It further shows, in keeping with (1.92), that
$$c^2 = \left(\frac{\partial \mathbf{x}}{i\partial \mathbf{t}}\right) \cdot \left(\frac{\partial \mathbf{x}}{i\partial \mathbf{t}}\right) = \left(\frac{\partial \mathbf{x}}{i\partial \mathbf{t}}\right) \cdot \left(\frac{i\partial \mathbf{x}}{\partial \mathbf{t}}\right) \tag{1.106}$$
as a scalar, where the second term is a parallel dot product and the third is an anti-parallel dot product, this latter case only if $\partial \mathbf{t}$ and $\partial \mathbf{x}$ represent the same vectors in each quotient. If they do not as in the crossing diagrams representing the transverse wave speed in the EM wave diagram above, we would have
$$-c^2 = \left(\frac{\partial \mathbf{x}}{i\partial \mathbf{t}}\right) \times \left(\frac{\partial \mathbf{x}}{i\partial \mathbf{t}}\right) \neq \left(\frac{\partial \mathbf{x}}{i\partial \mathbf{t}}\right) \times \left(\frac{i\partial \mathbf{x}}{\partial \mathbf{t}}\right) = c^2 \tag{1.107}$$

The use of identities indicates that as normalized vectors, where the time and distance scale are equal, $\partial \mathbf{E} \leftrightarrow \partial \mathbf{t}$, are interchangeable as are $\partial \mathbf{B} \leftrightarrow \partial \mathbf{x}$. Thus
$$\partial \mathbf{E} \times \partial \mathbf{B} = \partial \mathbf{E} \times \partial \mathbf{x} = \mathbf{c} \text{ and} \tag{1.108}$$
$$\partial \mathbf{t} \times \partial \mathbf{x} = \partial \mathbf{t} \times \partial \mathbf{B} = \mathbf{c}. \tag{1.109}$$

Also equivalent, applying the RHR to the magnetic field vectos are $\partial \mathbf{E} \leftrightarrow \partial \mathbf{x}$ and $\partial \mathbf{B} \leftrightarrow \partial \mathbf{t}$ so that
$$\partial \mathbf{E} \times \partial \mathbf{B} = \partial \mathbf{E} \times \partial \mathbf{t} = \mathbf{c} \text{ and} \tag{1.110}$$
$$\partial \mathbf{x} \times \partial \mathbf{t} = \partial \mathbf{x} \times \partial \mathbf{B} = \mathbf{c}. \tag{1.111}$$

This would be so much nonsense, with every vector shown as an orthogonal identity with every other, were it not for the fact that what is being elucidated is the simple orthogonal geometry of spacetime itself, and particularly of time with respect to all of 3-space. If we rotate the EM wave in the diagram ¼ π clockwise about the direction of travel, $\mathbf{c}$, the amplitudes of E and B can be viewed as the real and imaginary amplitudes corresponding to a complex modulus, $R = \sqrt{2}E = \sqrt{2}B$, which represents the amplitude of a sinusoidal traveling wave with a transverse displacement in the R-c plane. The wave, then, is analogous to a traveling wave on an ideal stretched string, but instead of a displacement of a medium of finite and much smaller cross-section relative to the amplitude, we have a sinusoidal strain in a medium in which continuity extends the cross-section indefinitely, far exceeding the magnitude of the amplitude. The handedness of the wave would appear to be a function of the matter particles and fields that interact with the wave and not of the wave itself.

This condition is shown in the above diagram of the EM wave. The E, right hand rules group of three cross products at the top of the diagram shows the differential vectors which apply at the end of the positive electric field vector shown in the second quadrant phase, between ½ π and π. If we had used a vector between 0 and ½ π, these three would be rotated one π turn about the blue vectors, **c** and $\partial \mathbf{x}$. The differentials reverse direction at the antinodes at ½ π and 1½ π, instantaneously vanishing, and reach their maximum, indicated by 1, at the nodes. The cross product at the top and the one at four o'clock from it are identical, but the situation would be unchanged if we exchanged the ∂t and $\partial \mathbf{x}$ in the second one in keeping with (1.92) and (1.98). The third product, directly beneath the first, has been rotated so that **c** is shown as the transverse wave velocity vector. Thus we have two directions of wave speed, transverse for the physical motion, and longitudinal for the wave phase as shown by the blue **c**s. An analogous situation is shown for the magnetic fields at B, left hand rules.

Recognizing that a four minute/mile and ¼ mile/minute represents the same velocity,

$$\mathbf{c} = \frac{1}{\mathbf{c}} \tag{1.112}$$

$$\therefore \mathbf{c}^2 = \mathbf{c} \times \mathbf{c} = 1 = \mathbf{c}, \text{ a vector or} \tag{1.113}$$

$$c^2 = \mathbf{c} \cdot \mathbf{c} = 1 = c, \text{ a scalar} \tag{1.114}$$

depending on whether the squaring indicates an orthogonal condition, i.e. a cross product, as when taking the product of two sides of a square, or a dot product, in which case the product is, conventionally, no longer a vector. We can make it a vector once more by applying the gradient, so that

$$\nabla c = \mathbf{c} \tag{1.115}$$

Actually, implicit in (1.113) is that the cross product is as in (1.107) and therefore we have,

$$\mathbf{c}^2 = i\mathbf{c} \times \mathbf{c} = 1 = k\mathbf{c} \tag{1.116}$$

where the k indicates that the product, **c**, is orthogonal to both of the other two **c**'s. Alternatively we can express this as

$$i\mathbf{c} \times j\mathbf{c} = k\mathbf{c} \tag{1.117}$$

giving the second **c** in the cross product its own orthogonal sense, or we could use subscripts

$$\mathbf{c}_i \times \mathbf{c}_j = \mathbf{c}_k \text{ or } \mathbf{c}_1 \times \mathbf{c}_2 = \mathbf{c}_3. \tag{1.118}$$

In the case of the dot product, then, we simply have

$$\mathbf{c}_1 \cdot \mathbf{c}_1 = c \tag{1.119}$$

and the gradient becomes

$$\nabla_1 (\mathbf{c}_1 \cdot \mathbf{c}_1) = \mathbf{c}_1. \tag{1.120}$$

In the case of a normalized c, (or **c**), t is simply another instance of x_i, where

$$x_i = (x, y, \text{ or } z) = (x_i, x_j, \text{ or } x_k) = (x_1, x_2, \text{ or } x_3) \tag{1.121}$$

and any two orthogonal x are interchangeable, and we can go with the established convention and call t, x_0, or

$$x_i = (t, x, y, \text{ or } z) = (x_i, x_j, x_k \text{ or } x_l) = (x_0, x_1, x_2, \text{ or } x_3) \tag{1.122}$$

In the second of these scenarios, we placed the time dimension at the end of the sequence or shifted the subscripts to the left, depending on your perspective, showing that time and space really are interchangeable. A velocity or any other kinematic derivative is simply the rate of change in one dimension with respect to a change in another. In the above treatment, i, by itself and without the other subscripts, is considered a generic orthogonal sense or operator, independent of any formal complex notation. It is simply directed at an angle of $\pi/2$ with respect to the other term in a binary operation. If it is constrained in a plane, then it itself can be either + or -, conventionally counterclockwise or clockwise, as directly viewed, the mirror image or "view from behind" being reversed. (This indicates that in addition to an intrinsic degree of freedom, $+\pi/2$ or $-\pi/2$, and an extrinsic infinitely variable degree of freedom determined by whatever constrains the plane, there is an observational degree of freedom set by the view sense.) If there is no planar constraint, then the extrinsic degree becomes intrinsic, and i indicates any direction in a plane which is orthogonal from the original direction. Whether it is $+\pi/2$ or $-\pi/2$ is determined by observation or some other constraint, i.e. whether a right-hand or left-hand rule is applied. With this in mind, it bears noting that while $-iE$ and iB represent rotations within the E-B plane, $+/- ix$, and in some contexts, $+/- it$, represents a rotation into that plane from the x axis.

From this development we can combine (1.102) and (1.104), for the transverse wave speeds, keeping in mind (1.112), to get

$$\left(\frac{i\partial \mathbf{E}}{\partial \mathbf{B}}\right) \times \left(\frac{-i\partial \mathbf{B}}{\partial \mathbf{E}}\right) = \left(\frac{\partial \mathbf{x}}{i\partial \mathbf{t}}\right) \times \left(\frac{i\partial \mathbf{t}}{\partial \mathbf{x}}\right) = \mathbf{c} \times \frac{1}{\mathbf{c}} = \mathbf{c}^2 = \mathbf{1}. \quad (1.123)$$

Thus the rotational change in the electric field as a function of a change in the magnetic field times the rotational change in the magnetic field as a function of the change in the electric field is an invariant vector, $\mathbf{1}$. From (1.113) and from (1.114) and (1.120) there are two versions of this scenario, which we will explore in a moment.

The left hand term of (1.123) indicates that the induced changes in the electric and magnetic fields in some manner cancel over time and distance along the path of the wave's travel until, at the point of what we would recognize as a node, the value of both $\mathbf{E}$ and $\mathbf{B}$ is 0. The transverse changes in the fields, however, do not stop at that point, since the 0 point is a relative zero, a point of equilibrium, and the functions are continuous through such point, $\partial \mathbf{E}$ and $\partial \mathbf{B}$ being at a maximum with respect to $\partial \mathbf{t}$ or $\partial \mathbf{x}$.

We can imagine this as two blades of a pair of scissors that start in a position orthogonal to each other and are brought together toward a point 45° from each. As they come together, due to a common pivot point outside the edge of each blade, their open edges each become shorter, until they vanish at the end of their travel. The use of the scissors metaphor is not accidental. Though our treatment here uses vectors, the wave can also be modeled as a tensor field.

Thus the E and B fields can be modeled to represent the shear stress components and corresponding strains of a stress and a strain tensor, where $\mathbf{c}$ in the direction of travel, represents the tension components. The wave form of both fields is sinusoidal, at $\pi/2$

rotation from each other as indicated above. At the point of maximum shear strain and stress, corresponding with the amplitude of the wave, the transverse wave momentum instantaneously vanishes, just as the transverse, shear stresses instantaneously vanish at the node. **c** then, while invariant with respect to the speed of the wave phase, has a transverse component that oscillates and reaches a maximum of 1 as both **E** and **B** reach 0.

As we emerge through the antinodes at ½ π and 1½ π, (1.123) becomes

$$\left(\frac{-i\partial \mathbf{E}}{-\partial \mathbf{B}}\right) \times \left(\frac{+i\partial \mathbf{B}}{-\partial \mathbf{E}}\right) = \left(\frac{-i\partial \mathbf{x}}{\partial \mathbf{t}}\right) \times \left(\frac{i\partial \mathbf{t}}{\partial \mathbf{x}}\right) = \mathbf{c}^2 = 1 \qquad (1.124)$$

leaving the product unchanged, although the field directions have all reversed, having multiplied both terms of each quotient by -1. This differential sense change is set by the amplitude of the wave, which is a dynamic function of other variables. Of course, (1.124) reverts to (1.123) through a cancellation of the senses, and the changes in the differential senses are always simultaneous, so that **c** remains invariant, that is under the following of the two above mentioned conditions.

If we recast (1.123) in light of (1.120), we have the following normalized gradient in the direction of wave travel, which describes the radial propagation of an electromagnetic wave from some source,

$$\nabla_x\left(\left(\frac{i\partial \mathbf{E}}{\partial \mathbf{B}}\right) \cdot \left(\frac{-i\partial \mathbf{B}}{\partial \mathbf{E}}\right)\right) = \nabla_x\left(\left(\frac{\partial \mathbf{x}}{i\partial \mathbf{t}}\right) \cdot \left(\frac{i\partial \mathbf{t}}{\partial \mathbf{x}}\right)\right) = \nabla_x \mathbf{c}^2 = \mathbf{c}^2 = 1. \qquad (1.125)$$

Note that the second term is algebraically self normalizing.

To arrive at a dynamic expression for the wave, we need some scaling factor that will indicate its energy and mass, i.e. its angular frequency and wave number. Using (1.2) and (1.4) we have

$$E = \hbar \mathbf{\kappa} \cdot \nabla_x \mathbf{c}^2 = \hbar c \omega = \hbar \omega . \qquad (1.126)$$

We would like now to see if there might be another application of (1.123) using the cross product. We assume, for purposes that we will later make clear, that the wave radiates from some locus, so that at that source, the directions of propagation can be resolved in terms of three co-ordinates, x, y, and z as shown in the following subscripts. We would expect, therefore, that (1.123) would take the form

$$\mathbf{c}_x \times \mathbf{c}_{-z} = \mathbf{c}_y \qquad (1.127)$$

where we choose subscripts which best serve our long term purpose.

In keeping with the orthogonal nature of the electric and magnetic field orientation, and their cross product which results in a third orthogonal vector, (1.127) requires that <u>either</u> the electric <u>or</u> the magnetic fields and their differentials be common to both vectors $\mathbf{c}_x$ and $\mathbf{c}_{-z}$, unless the wave can be modeled as a 4-D wave. Thus we will make the field orientations (not that of the differentials) explicit with the appropriate subscripts. Since there is no evidence that magnetic monopoles exist, we will assume that at the source locus a magnetic dipole does exist. This implies that any electrical fields radiate at and from the center of the dipole perhaps as a ring of quantum charge. All we are interested in are two such fields orthogonal to each other. One such configuration that might satisfy this relationship is

$$\left(\frac{i\partial \mathbf{E}_z}{\partial \mathbf{B}_{-y}}\right) \times \left(\frac{-i\partial \mathbf{B}_{-y}}{\partial \mathbf{E}_x}\right) = \left(\frac{\partial \mathbf{x}_{-y}}{i\partial t_z}\right) \times \left(\frac{i\partial t_x}{\partial \mathbf{x}_{-y}}\right) = \mathbf{c}_x \times \mathbf{c}_{-z} = \mathbf{c}_y \quad (1.128)$$

We would imagine, given the dipole nature of B and our innate sense of the justice of symmetry, that another such simultaneous configuration exists so that

$$\left(\frac{i\partial \mathbf{E}_{-z}}{\partial \mathbf{B}_y}\right) \times \left(\frac{-i\partial \mathbf{B}_y}{\partial \mathbf{E}_{-x}}\right) = \left(\frac{\partial \mathbf{x}_y}{i\partial t_{-z}}\right) \times \left(\frac{i\partial t_{-x}}{\partial \mathbf{x}_y}\right) = \mathbf{c}_x \times \mathbf{c}_{-z} = \mathbf{c}_y \quad (1.129)$$

But wait! This is anything but symmetrical! Since the same right hand rule for the cross product has been used for both (1.128) and (1.129), the velocity cross products and their final cross product are the same in both cases. We will explore this further, but our intention at this point is simply to show that the same ontology which supports a classical wave mechanism found in a propagated electromagnetic wave can be modeled as the support for the quantum source mechanism, and we can model both as a spacetime continuum with inertial-elastic wave bearing properties.

Returning to (1.129), if we assume symmetry at the source, we must assume that instead of the right hand rule for the cross product, we should use the left hand rule for the field polar opposites of those in (1.128). Doing this we have

$$\left(\frac{-i\partial \mathbf{E}_{-z}}{\partial \mathbf{B}_y}\right) \times_L \left(\frac{i\partial \mathbf{B}_y}{\partial \mathbf{E}_{-x}}\right) = \left(\frac{i\partial \mathbf{x}_y}{\partial t_{-z}}\right) \times_L \left(\frac{\partial t_{-x}}{i\partial \mathbf{x}_y}\right) = \mathbf{c}_{-x} \times_L \mathbf{c}_z = \mathbf{c}_{-y}. \quad (1.130)$$

This implies a basic symmetry at the quantum source which is broken as a result of propagation of the EM wave, resulting in a preponderant phenomenology of right hand electrical, left hand magnetic, versions of such.

What this shows us about the source, by adding (1.128) and (1.130), referring to Figure 2, is:

1) the change in electrical fields as a function of a change in the magnetic field over time ∂t results in a radial tension stress along a path of potential wave propagation, $\mathbf{c}_{+x}$ - $\mathbf{c}_{-x}$ which in an equilibrium condition are balanced, preventing propagation,
2) the change in the magnetic fields as a function of a change in the electrical fields over the distance ∂x results in the velocity vectors $\mathbf{c}_{+z}$ - $\mathbf{c}_{-z}$ and a rotation of the magnetic dipole, B_{+y}- B_{-y}, making it an axial vector, in this case shown as a left hand vector, and which we will call ϕ,
3) the cross product of the velocity vectors in (1) and (2) creates the vectors $\mathbf{c}_{+y}$ - $\mathbf{c}_{-y}$ and results in a rotation about a second axial vector, this time shown as right hand by the blue arrow, S_L, which we will call θ, at E_{+z}- E_{-z}, which remains, absent any perturbations, fixed in space over time, whereas ϕ remains fixed in time, i.e. phasing, over space as it rotates about θ.

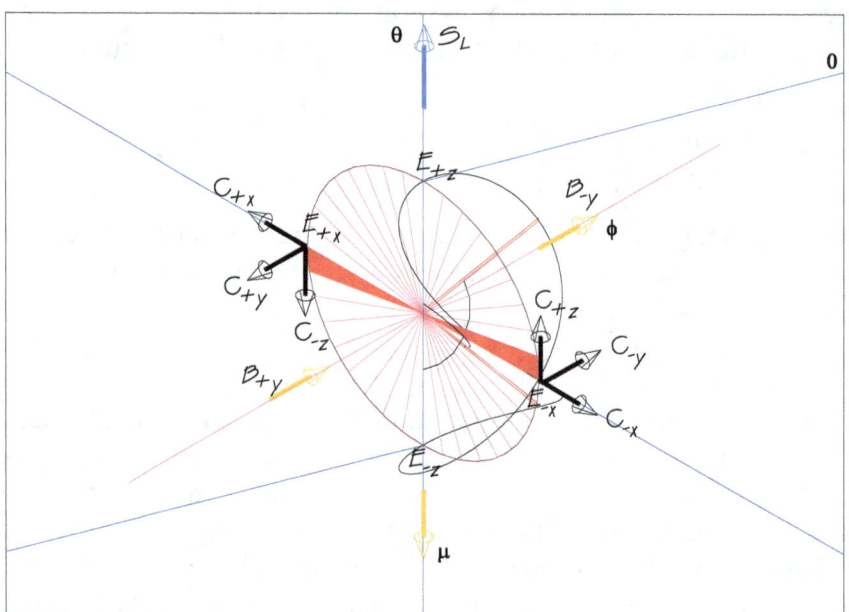

Figure 2 – Electro Magnetic Wave Source

θ is a spin angular momentum vector while the yellow axial vector pointing to the bottom of the figure, mirrored to θ and shown as a left hand axial vector, μ, is the effective magnetic moment of the source. The angle between ϕ and the line E_{+z}-0, and the analogous one involving E_{-z} are angular strains constituting the electric fields, which oscillate in place over time, while the angle between μ and ϕ on either side of the figure is a permanent orthogonal distortion which rotates as it moves about θ. The figure 8 path that appears to be bent around the central disk on the right hand side of the figure with its center at E_{-x} is the oscillatory path of the <u>strain</u> which that point takes as ϕ rotates about θ. There are an infinite number of such, one corresponding to each point over the circumferential distance of ϕ and with each corresponding to two points on the circumferential path of θ over one 2π cycle. Thus the stress rotates about θ while the strain oscillates along the figure 8 path.

We will investigate this model in greater depth. The opposition of the spin and magnetic moments indicates that it is an electron, neutron or an anti-proton, but we will see that it in fact represents a neutron. There are obviously some steps between this form and the generation of electromagnetic radiation, in particular involving beta decay and the generation of an electron or positron. The significance now is that if we apply the same logic to (1.128) and (1.130) as we did to (1.125), we have the standing wave corollary of the energy equation found in (1.126) for a rest mass particle, in which factors of $\frac{1}{\sqrt{2}}^2$ related to the root mean square for the maximum power of a wave and the two instances of instantaneous spin energy, $\mathbf{E}_{+y}$ and $\mathbf{E}_{-y}$, equal unity giving

$$E = 2\left|\hbar\boldsymbol{\kappa}_{\pm y} \cdot \left(\tfrac{1}{\sqrt{2}}\mathbf{c}_{\pm x} \times_{R/L} \tfrac{1}{\sqrt{2}}\mathbf{c}_{\mp z}\right)\right|$$
$$= \left|\hbar\boldsymbol{\kappa}_{+y} \cdot \left(\tfrac{1}{\sqrt{2}}\mathbf{c}_{+x} \times_R \tfrac{1}{\sqrt{2}}\mathbf{c}_{-z}\right) - \hbar\boldsymbol{\kappa}_{-y} \cdot \left(\tfrac{1}{\sqrt{2}}\mathbf{c}_{-x} \times_L \tfrac{1}{\sqrt{2}}\mathbf{c}_{+z}\right)\right|. \qquad (1.131)$$
$$= mc^2 = \hbar c\omega = \hbar\kappa^{-1}\omega^2 = \hbar\omega$$

As we will see, the spin angular momentum is modified by these power and other geometric factors, resulting in a coefficient of $\sqrt{3}/2$. The above figure implies that the oscillation pictured is an instance of simple harmonic motion. An animation of this motion is available. We will examine the kinematics and dynamics of such motion, using a model of standing wave motion on a one dimensional string.

Before we examine this oscillation more closely, a few words about continuity and isotropic expansion are in order. It should be stated that what we are talking about is not a process of inflation as currently generally conceived. As modeled here it is not a hyper rapid and extensive early epoch phenomena, but rather an ongoing process, that while slow from our frame of reference, is accelerating in an exponential manner. It would not be surprising if that exponential manner is in fact a complex exponential expansion and thus an indication of cyclical expansion and contraction.

Geometry and Topology.

Geometry deals with the subject of the shape of space and objects within that space, while topology is concerned primarily with the subject of continuity of a space and is intimately connected with the subject of set theory. These disciplines are vast, and a thorough elaboration far beyond the expertise and inclination of this writer, however, we will touch on a few matters that have some bearing on the subject under development, without undue concern for rigor.

At the heart of set theory and topology lies a paradox that receives little apparent attention and that is the nature of a point. In geometric analysis, it is readily understood that a point is dimensionless and therefore has no size. What it does have is <u>location</u>, so it is always described by its relationship to something else, never mind that such something else is itself thought of as a point or collection of points. This last item is the departure for set theory and topology which is content to state that a point is understood by the neighborhood or locus of other points around it.

If a point has no size, there is little point trying to add up all the contiguous points in a space to create a neighborhood, since there are as many points between one point and the point next to it as there are seconds in eternity. There is no manner of determining a scale from scratch, so to speak, in this madness. It is sufficient to draw a closed loop on a supposed continuum and state, "This is my neighborhood...", and draw another loop connected to that one, and state, "this is an adjoining neighborhood", etc.

The continuum is what we start with, and we partition it by defining a portion of it as separate from the rest of it, as we with all due seriousness define a portion of the good earth to ourselves and our heirs in perpetuity and build a house. The line separating us from our neighbor has no width, but we will both be aware when it is crossed, because it does have length. It has a dimensionless starting and ending point, and the line can be seen or at least envisioned by virtue of the fact that they and many other points along the continuum in between are separated by a whole lot of infinities of dimensionless points, that is, a whole lot of nothing.

It is easy enough to imagine that the widthless circumferences we use to describe our sets of mathematical neighborhoods can be shrunk to a point, and this is in fact the method that is used to distinguish the genus or different orders of topological spaces that make up an n-space or space of n dimensions. If any and all closed looped lines or analogous structures drawn on an n-space can be shrunk to a point, that n-space is considered a genus 0, and is held to be equivalent to a sphere. If some of those loops run the risk of being caught up on a donut hole or the hole forming a coffee cup handle or the n-space equivalent, they are a genus 1 n-space. If they can get caught on two different holes they are a genus 2 n-space. This is apparently pretty straight forward, and at the time of this writing a gentlemen from Russia was about to be awarded a pot full of money for proving that the genus 0 3-space is the simplest of such and equivalent to the genus 0 of the 2 and other numbered spaces. The problem is that this method of distinguishing a genus is based on a fallacy, since *you can never shrink a loop down to a point*. A loop has length and every one agrees that a length has dimension, in fact it is the very essence of dimension, but it can never be shrunk to a dimensionless entity, any more than, going in the other direction, any number of planes can be stacked to form a solid. They are simply different types of things.

To be sure, we can use calculus to integrate an area and arrive at a volume, but that is because of the underlying assumption of continuity on which the calculus is based. Without the basis of continuity, the functions and the derivatives which they father would fail. At whatever scale we try to reduce it to, a loop remains a loop and never will arrive at a point. Which is not to say that holes don't exist. Holes are null sets, just as each and every point to which a loop might approach as a limit is a null set, and one thing that both set theorists and geometers will agree on is that points and null sets do exists. But you, that is a dimensioned entity, will never get there, since there's no there, there to get to. The loop at the limit does not become the point, it surrounds the null set point and retains its loopness. Points, however, do have location, as we have said, within a given continuum, be it line, plane, volume, etc., and so does a hole, and that location is where a loop drawn around it, within a given continuum, is the smallest. In the sense of having location, therefore, neither hole nor point is a null set and both are defined within a continuum and have structure. It is in order to discuss that definition, the structure of a point and a null set, that we entered on this digression.

Poor Gödel and his ordered infinities! There are at least twice as many rational numbers as irrational, since any irrational number whose binary digits we may be counting, that is calling out or naming, has one rational number on either side of it at each digit along its span; so much for denumerable infinities. Infinity, at least in the sense of divisibility of a continuum, is the rule and not the exception, and counting is simply a way of drawing temporary loops on the expanse of it. It is just that any number of null sets added together do not make a continuum. And a continuum is both whole, that is one, and infinite at the same time, all the while being the home of every null set and conceivable collection thereof. Pretty simple when you think about it, or perhaps that is, if you stop trying to think about it.

Actually, it is not *quite* that simple. The reason the point can never be reached is that a spherical surface, a 2-sphere as it is generally called, approaches Euclidean flatness with increasing refinement of scale and paradoxically flatness has no scale. Even the curvature of the loop, by itself, gives no concept of scale. At some point the continued shrinking of the loop and the shrinking adjacent field of reference, our tunneling scope of observation, if they co-vary or assume a common velocity of shrinkage, takes on the quality of a state, a static condition, instead of that of change and motion.

The only way the point can be reached is for us to define it as being there. We must state, "Enough of this tedium. I order you to loose the one dimensional quality of a 1-sphere (a circle), and become a dimensionless 0-sphere (a point)!" And it happens. But something else, very strange, happens. We can never place another point right next to the first point, since we can always zoom in towards the first point at a scale that finds the second receding into the distance. We can continue to zero in on the point, for a length of time equal to that already spent on this shrinking endeavor; for a day, a week, an eternity, but we never appear to get any closer to the point. At any scale it continues to appear just the same, just a point. It will never become a disk or a hole. We can make it change into cross hairs if we like, but they will not change in appearance as we approach, always of some length, but forever of immeasurably small width and inscalability. But then the reason for establishing, for pinning down a point is not possession, but rather station and partition of an otherwise protean and irrational continuum, and this can be done to a reasonable precision and accuracy within a few orders of magnitude.

It is curvature that grants scalability. Even the most rigorous Cartesian logic assumes an arc of continuity across the vast Pythagorean expanse between the x and y co-ordinates in staking out an ordinate and abscissa. And while nature has given us the asymptote, she has also given us the tangent and Herr Riemann to show us that the two can be in fact the same thing in curved space. Lines do touch, and just as they may be made to intersect at one point in our crosshairs, they can be made as a circle and a line, and thereby, two circles, in the world of 1-spheres embedded in a universe of 2-spheres, to intersect at just one point. . .or in some cases two.

It is relatively easy to think of a 2-sphere, the surface of a 3-ball or three-dimensional volume, as two dimensional, and their denizens as one-dimensional combinations of lines and their circumscribed 2-balls. We might consider a figure eight and its enclosed 2-balls and speculate on whether it consisted of two separate disks united at a point of tangency or a single 1-sphere that somehow had one portion inverted. There is no way on close inspection to tell an up or down, in or out side of a 2-space, which includes the 2-sphere and any number of other 2-manifolds, such as the torus, and including the Klein bottle in which the sides are continuous, so there is no way of determining if the 8 is conjoined or self-crossed.

In fact, within that 2-space, or 2-manifold, which is generally speaking a space without a boundary, the idea of an in and out is meaningless. It is only from our three dimensional perspective that the notion of an in and out arises. What does have meaning is whether or not there is a boundary to the space. A 2-sphere has no boundary, since the sides, the in and out surfaces of the "skin" that we recognize as separating the interior of the 3-ball from the exterior, do not exist for the 2-denizen and there are no edges to bump into, as for example we would find in the case of a 2-ball or disk which is bounded by a circle or 1-sphere. The boundary of a space, generally speaking, is one dimension less than the space itself, so that a 1-line segment is bounded only by its two 0-endpoints, a 2-disk by a surrounding 1-circle, a 3-ball by the surrounding 2-sphere. An n-space is not bounded by the next higher dimension, though it may be closed or open to it.

It is here, perhaps that the system gets sticky. For topology, the ordering of spaces, in addition to the reference to n-dimensions and genus number and boundary condition, allows for some fuzziness by asking whether the space is closed or open. This is not the same thing as whether or not it has a hole in it or even a boundary. It has to do with whether or not, when you strip the boundary off a ball or interior, you take any of the ball with you. It is a matter of whether the n-boundary belongs to the n+1-ball. If it is a closed space apple, it means when you peel it you take some of the apple with you. If it is an open space banana, you get only the 2-peel and leave the 3-fruit intact. With fruit this is understandable enough, but with n-space manifolds it can get tricky.

This is just another example of the open or closed interval question, which gets back to the questionable ontology of a point. It really is an issue of whether the boundary is approached asymptotically or tangentially. Can you get right up next to the boundary point and leave the point without leaving the point next to it? If it is a Euclidean space and you are approaching the open boundary asymptotically, you can never get quite close enough to avoid leaving some of the ball behind with the rind, and yet Euclidean space says that, by fiat, you can. If we are dealing with a curved space continuum which is truly connected, then any cut we make will be tangential <u>at</u> some point, and the point will remain with both the ball and the boundary space, because the n-boundary of an n+1-enclosure has no dimension in the direction normal to the cut. An n-space which is a boundary is not like a peel or a skin of an n+1-fruit of any variety. It has no thickness. In general, however, this does not appear to be topologically admitted. If you take the n-space away from the n+1-enclosure, that enclosure remains intact.

We might imagine that an n-space might be closed with respect to adjacent n+1-dimensional spaces and open with respect to an n-1-dimensional boundary. Thus an open disk or 2-ball might be morphed into a spherical bowl with a small opening at one end. Any denizen of the 2-ball, turned almost 2-sphere, as he approached this opening would be whisked along the edge, but never quite get to it. The openness implies that the edge can never quite be reached. If this edge (not the opening itself) is closed, he might reach it but move along it tangentially without recognizing it as any more limitation than the interior or exterior that only we 3-denizens can see, and that limits his motion tangentially to the almost 2-sphere itself.

It gets stickier. The notion of topology says that any space of the same dimension and genus number is in some sense equivalent. Thus with a 2-ball (a flat disk) and a 2-sphere (surface of a 3-ball), though the first is with and the second is without boundary. The geometry of the space, whether it is concave or flat, doesn't matter. However, it does not say that you can inflate a flat disk and get a 3-ball. That is adding a dimension out of compressed air. Nor does it let you remove the boundary of the aforementioned morphed disk turned almost sphere to close the opening in the jar, no matter how small it lets you make it.

And yet it states the equivalence of a coffee cup and a donut, and lets you shape one side of the latter into the concave vessel of the former. The reason is that the torus/coffee cup/donut was always the 2-dimensional boundary for the 3-dimensional volume of stuff out of which each shape was made, while the 2-ball disk was never a boundary, at least an enclosing boundary, of any 3-d thing. Perhaps physically a drawing on paper, but conceptually without any interior, it has no 3-d substance out of which to construct the 3-d interior for a 2-d surface to be the boundary of. It <u>is</u> the interior, of a 1-sphere.

Yet all is not lost. If we can shrink a closed loop (closed in terms of set theory as well as geometrically) to a point on the surface of a sphere by approaching it tangentially, we can surely enlarge it to that point's antipode and cover the sphere, and if we can cover a sphere that is already there, surely we can trace out the same contours of empty space and create one from a closed loop boundary of a closed flat disk. The question of whether or not the gap surrounding the antipode can be sealed off depends on whether or not the boundary approaching it is closed to it or open. If the 2-disk itself is closed, that is if it "owns" its boundary, and the boundary is itself closed on both sides, that is if it is truly only 1-dimensional, the gap can close, since there is no "fuzziness" of an asymptotic approach to prevent closure. The shrinking loop reaches tangency with all the tangent lines forming the plane of tangency at the antipode and in fact becomes self-tangent. We might even think of a point, instead of as null set, as a circle which is self-tangent.

We must digress further in our topological digression from the subject of the fundamental interactions with some comments about open and closed-ness. In topology and set theory[1], the concept of the real number line, in fact the whole concept of a set, is based on the use of the Dedekind cut, which stipulates that a space (including a 1-space or line) can be cut with a parenthesis, which represents an open or asymptotic approach to the cut, or a bracket, which represents a closed or tangential approach to the cut. The cut itself, unless it is a pre-existent boundary of the space under the knife, must be approached from both sides, so a scissors must be used, giving us four possible types of cuts, represented by

I.	Closed – Open	Tangential - Asymptotic	L---](---R
II.	Open – Closed	Asymptotic - Tangential	L---)[---R
III.	Open – Open	Asymptotic - Asymptotic	L---)(---R
IV.	Closed – Closed	Tangential – Tangential	L---][---R

where the L and R serves only to emphasize the left and right hand section of the line or a left and right hand set. Unfortunately, of the four possible types, set theory only allows the first three, the first two of which are held to represent the cut at a rational number, and the third of which is held to represent the cut at an irrational number. The set of all such cuts forms the real number line and it is a feature of this line, that, by definition, if we make a cut so that x is a number in L or an element of the L set, and y is a number in R or an element of the R set, where the largest number in L will be less than the smallest number in R, then x will be less than y and either L will have a largest element or R will have a smallest element. That is, either L or R will be closed, but not both, and a number z can belong to either L or R, but not both.

This no-doubt arose in such manner because the first set theorists were accountants and a penny could either belong in one account or the other, but not both, and it could not be cut. If they had been dress makers or carpenters, they would have realized that when you cut a piece of fabric or mark off a piece of lumber, the line of demarcation belongs to each side of the partitioned space, unraveled thread and sawdust notwithstanding; or

---

[1] An Introduction to the Elements of Mathematics, John N. Fujii, John Wiley & Sons, New York, 1961

better yet, had they been surveyors with real live property owners to remind them of the fact that, at the very least, the line did not belong exclusively to the other side.

On a fundamental level, numbers are used for counting or for measuring; counting is generally conceived as being a quantum or integer affair, and while use of a tape measure can be thought of as counting the units and fractions thereof, the notion of continuity is essential to the application of linear measure. If the distance being measured exceeds the length of the tape or measuring rod, a position mark is made on the field and that same mark is held to be the end of the previous extension and the start of the next. That mark *is* in the last element in the set of the first measure and the first element in the set of the next.

In the real and physical world of, from this writer's perspective, undeniable continuity, the world of conservation of energy and momentum and mass, only the fourth of the above cuts takes place. To be sure, there will always be questions of precisely where the cut took place, but there is never any supposition that if one side is neatly cleaved the other is out there in never-never land, perpetually waiting for the asymptotic axe to fall. One is tempted to point to Schrödinger and Heisenberg, but theirs are epistemological issues and not ontological, as indicated by EPR and quantum entanglement.

With the above in mind, we can stipulate that a 2-ball can be charted into a 2-sphere, and if a closed loop without boundary can be made to vanish, then it can certainly be made to reappear, this time with a vengeance. By squeezing the pode and the antipode of our 2-sphere together, at the center of its 3-ball, we can create a horn torus, a torus without any hole, except for the one single point of self tangency at the center. Now, that closed point can be expanded as a closed loop toward the center circle of the toric annulus in the equatorial plane and we have a donut. From here we can form a coffee cup, and now that the genii are out of the bottle, as many holes as we desire, with no lack of continuity. And now that we have created a 3-ball of genus-n out of a 2-ball which presumably we could have created out of a 1-ball, why not fold the 3-ball-n into a 4-ball-n, etc. It is all just a continuum, after all. But let us concentrate on the horn torus for a while, which, at the risk of disturbing Decartes' spirit, will herein be referred to as a monad.

This manifold is a hybrid between a sphere and a torus, in the sense that it forms a torus for which the hole is but a single point. Since a single point has no dimension, it is effectively the null set, $\varnothing$, and as such represents the defining hole of a torus; however, as a single point maintains linear connectedness and closure between neighboring points, and here neighboring hemi-folds, this manifold is topologically equivalent to a sphere.

Of added interest is that this manifold is invertible through the center in the manner of the Möbius strip, and maintains continuity between a lower section and an upper section of the opposite lateral half. If we envision a vertical cross-section through the center of the manifold, we have a figure 8 laying on its side or the symbol for infinity, ∞, (whimsically enough, if we view the manifold from above and consider the center point to be a hole, we have the symbol for 0, while viewing it rotated 90 degrees from the exterior side, we have the symbol for 1.) We can make the ∞ by tracing two circles intersecting at a point or with one stroke in the familiar fashion, crossing at the center. If we trace the figure in this latter manner, but keep our marker to one side of the line, we will see that it in fact inverts, on the inside of one of the circles, transferring to the outside of the adjacent one as it crosses. The central crossing point thus represents a point of inflection for the space.

The central point also maintains connectedness between the upper and lower exteriors of the manifold along the tangent line, which is the central axis of a tangent sheaf or bundle of circle arcs of radius greater than that of the monad's annulus, tangent to the annular surface at the central point. Necessarily, the full circle elements of the sheaf form the set of all elements exterior to the 2-monad, whose interior we will call a 3-core to distinguish it from a 3-ball or 3-donut. The 3-core, then, can be seen as the set of all circles tangent to the central point and the central tangent axis with a radius smaller than that of the monadic annulus. Significantly, the central point also maintains 3-core diametric connectedness.

It follows that the central point is simultaneously an element of the exterior and of the interior of the 2-monad while being the central element of the 2-monad itself. In fact from the perspective of the 2-monad, it is central to all three sets. The cross section of the 2-monad forms the aforementioned figure 8, which is itself a 1-monad and it follows that the exterior, interior or 2-core, and the space itself bear the same relationship to the central point as those of the 2-monad. The two apparent disks or 2-balls of the interior are in fact continuous through the center diameter of the 1-monad. We might infer from this that the same relationship between exterior, interior and the space itself holds for all n-monads.

The following diagram of manifold connectivity is illustrative of the fact that this manifold is topologically equivalent to the other three. The charts are two dimensional Euclidean representations of the corresponding manifolds, and can be thought of as differential areas, which with integration over the manifold surface undergo the necessary transformation or stretching required to tile the surface.

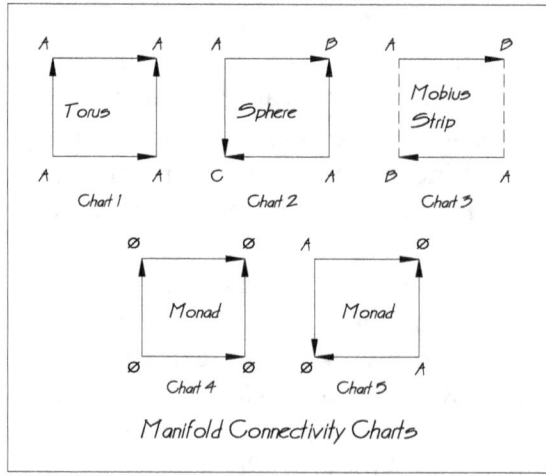

Manifold Charts

i.   In Chart 1, the Torus, the left and right sides join in common orientation as indicated by the direction arrows to form a tube, which then joins top to bottom to form the familiar donut shape. The repetitive corner designation, in this case A, indicates a common point in the torus manifold. We could also start by joining the top and bottom then join the sides, giving us a different orientation.

ii.   In Chart 2, the Sphere, corners A can be thought of as joining at an equator, while the top and right sides and the bottom and left sides come together at the B and C poles respectively.

iii.   In Chart 3, the Möbius strip, the top and bottom sides are twisted one half turn before joining to form the usual shape, while the dotted sides indicate unjoined edges.

iv.   Charts 4 and 5 represent two equally valid means of composing the Monad. The first, in which the common point, $\emptyset$, is at the center, is equivalent to the Torus, while the second, in which the B and C poles are drawn together at the center, $\emptyset$, is equivalent to the Sphere and, some mental gymnastics will show, to the Möbius strip. However, instead of twisting the strip to join A and B, one common corner is joined first as when joining A to A in Chart 2, followed by a twisting through the center of the ring to joint the other two corners. The opposite sides are not joined right to left or top to bottom, however, but rather the co-terminal edges are joined as with the Sphere, top to right and bottom to left, in the case where the A's are joined first.

As this particular manifold is topologically equivalent to both a sphere and a torus and in addition the Möbius strip, and granting the closure and opening procedure of above, the Klein bottle, it can be seen as a parent of all of these in 2-space and we might conjecture in n-space and to hold the place generally reserved for the sphere. With respect to the mapping of the complex and projective plane onto the sphere, with infinity at the top pole and zero at its antipode, with the monad we have that same manifold with the infinity and zero point joined at the center of things, the unit sphere occupying the equatorial cross-sectional plane, and the negative z axis, which is the 2-space itself, forming the apple skin and the positive z axis vanishing in the core.

If we recognize that the imaginary sense i is simply another way of representing a half π rotation, and make of it a motive operator, we have a manifold that rotates about its central axis or horn, counter clockwise by convention, and about its annulus, upward through the horn center and downward at the periphery. That is, if we imagine the 2-monad as divided into octants, we can assign a velocity potential for each of the four octants in the top half to rotate counterclockwise viewed from above, and simultaneously outward and down on the surface of the form toward the octant below, while a simultaneous velocity potential exists for each of the lower octants to rotate in the same direction on the lower half but inward through the center and to the diagonally opposite octant on the upper half. This last motion amounts to an inversion of the 2-space surface through the center in the manner of the Möbius strip.

It remains to consider the nature of a 0-monad or monadic point and the 1-core. We can consider the 0-monad as the central point of an n-monad itself. Unlike a Euclidean point, which is generally held to be dimensionless and directionless, the 0-monad maintains an orientation or direction. This direction, which we might think of as a differential vector, is the 1-core and is tangent to the corresponding 1-monad or 2-monad of which it is an element. It is, however, for an isolated point indeterminate in its orientation, having with respect to a 3-core, 6 potential orientations or three degrees of freedom as constrained or bounded by the 8 cubic vertices which are the parameters of the 0-monad. The velocity potentials, therefore can be assigned to vectors extending from one to another of the vertices.

This is not a stipulation of a lattice but rather of an orthogonal potential. As an element in an n-monad, the 0-monad is closed on the 1-core and open on what would be the 2-core. That is, it is closed on the tangent line, and open on any line normal to that tangent line and point. The exception is when it is at the position of the point of central tangency, in which case it is closed on all sets, the n-monad, its n+1-cores, and its exterior, and hence is a point of continuity in all n-space. At the central point, in connection with the dynamics just indicated, it has 2 extrinsic degrees of freedom, annular rotation, either up or down, and chirality, or left or right handed, clockwise or counterclockwise rotation. Of these, only the contrast of chirality to annular rotation, hence 1 degree of freedom, is intrinsic.

If we consider the 8 vertices of the 0-monad, with regard to the various possible flow patterns or velocity potential vectors connecting them, we can see that the above scenario involving the use of i as a motive operator can be mapped onto the 0-monad, so that we have the following table of possible flows for a right hand, counter-clockwise, $+i$ convention. The related rotations indicate a rotation of the unit vector as seen from the positive axis normal to the rotation, resulting in a transformation of a given vertex to a vertex indicated as an out vector direction. Thus a rotation of the $+x$ unit vector **i** ccw about and seen from the $+z$ axis to the position of the $+y$ unit vector **j** is referred to as $+i$. The cw rotation of **i** about $+z$ to the $-y$ axis is $-i$. Rotation of **j** ccw about $+x$ to **k** at $+z$ is $+j$; **j** cw about $+x$ is $-j$; **k** ccw about $+y$ to **i** at $+x$ is $+k$; cw to $-x$ is $-k$. Where two rotations, within semi-colons, are indicated for each vertex, they are not equivalent and are non-commutative. Hence, $+i$ and $-k$ are the two ½ π rotations capable of taking vertex 1 to the position of vertex 2. The criteria here have been to have an equal number of vectors pointing to and from and total for each vertex. Other scenarios are possible, such as a counter rotational flow for the top and bottom which reverses over time.

Vertex Number	Relative Octant in x,y,z	In Vector from	Out Vector to	Related Rotations viewed from +z, +x &+y axis
1	+x,+y,+z	4;7*	2;5	$+i,-k; +k,-j$
2	-x,+y,+z	1;8*	3;6	$+i,+j; -k,-j$
3	-x,-y,+z	2;5*	4;7	$+i,+k; -k,+j$
4	+x,-y,+z	3;6*	1;8	$+i,-j; +k,+j$
5	+x,+y,-z	8;1	6;3*	$+i,+k; -l = -i +\delta'$
6	-x,+y,-z	5;2	7;4*	$+i,-j; -l = -i +\delta'$
7	-x,-y,-z	6;3	8;1*	$+i,-k; -l = -i +\delta'$
8	+x,-y,-z	7;4	5;2*	$+i,+j; -l = -i +\delta'$

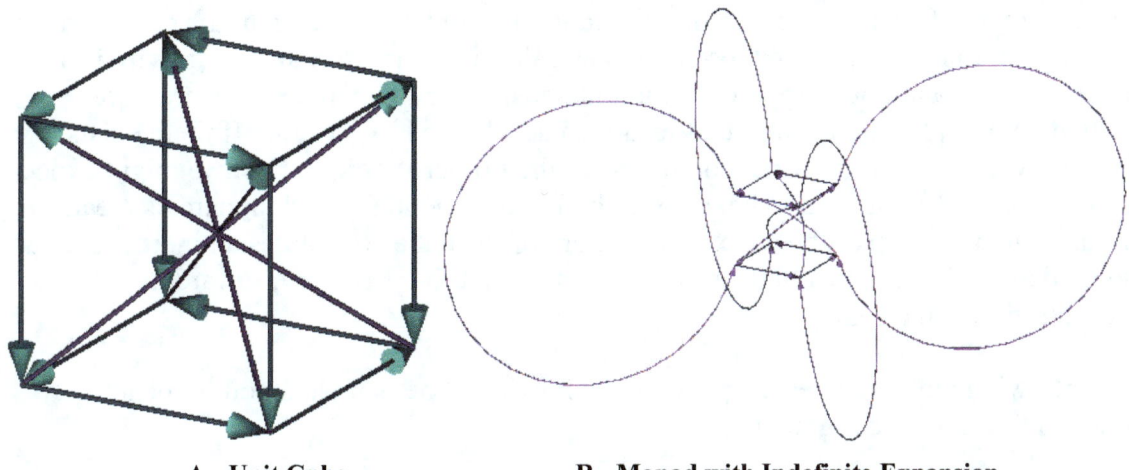

A - Unit Cube   **B - Monad with Indefinite Expansion**

Monad Orthogonality

The asterisks indicate a diagonal vector or flow from the lower to the upper octants, which can also be interpreted as a clockwise rotation out of topological 3-space into geometric 4-space or rather, since it is actually in n-space, the ½ π rotation of the image of an n-form about an n-1 boundary, coupled with the orthogonal translation, δ', through n+1 space to another boundary of the n-form. Note the rotation is the same for each of the four vertices, though the translation is to a different octant in each case. This is consistent with similar constructions in lower n-space, as a square can be constructed by rotating a line segment orthogonally at one end and sweeping it to the diametrically opposite end, and a cube, by rotating a square similarly and translating it to the other side. By "sweeping" we are indicating not simply the translation of, for example, a square, but its extrusion, after copy and rotation, to span and thereby define a 2-cube and its interior. The same can be done with that interior to define a 3-cube and its 4-block, as we might call it.

The figures above show a unit cube constructed according to this table on the left, A, with expansion along one plane of the system shown on the right, B. The result, when coupled with the arrows interpreted as velocity potentials is a horn torus with potential rotation ccw about the upper and lower surface and through the top and into the bottom about the annulus. B can be morphed into a sphere by lengthening the inner diagonals as a spindle torus and into a conventional torus by spreading the top and bottom squares.

While A is presented as a unit form, it can be thought of as a differential form with potential structure, but no inherent size. Size or scale is conferred on it by dropping a unit of inertial density into the center, which then begins to circulate up and out toward and about the upper vertices, at which point angular momentum begins the expansion shown in B, so that instead of going down the four edges, inertia leads to an expansion in the direction of a top surface normal vector and out about the annulus. This is shown as three dimensional, being all that is graphically possible, but it can comprise as many dimensions as are needed to describe it, (di-mension being etymologically to divide or separate with the mind.)

It is the nature of such a configuration that regions more removed from the center will have less inertial density than those more central. Since, as we shall see, gravity is a quantum effect and not a property of inertial density per se, there exists a pressure gradient and Laplacian outward that results in accelerated expansion. If there exists a third derivative over time corresponding with the property of jerk, then we might expect such accelerated expansion to be, not steady, but exponential. Such dynamics create an inertial font at the top of the cube. The bottom of the cube constitutes an inertial sink as the vertices direct the inertial density down and about the periphery, toward a concentration at the center.

The following table outlines the parameters, in which line is understood to mean a bounded line or a line segment.

n-space, element	# in boundary	n-block	surfaces	edges	vertices
0-cube, point	2	1-block, line	0	0	2
1-cube, line	4	2-block, square	0	4	4
2-cube, square	6	3-block, cube	6	12	8
3-cube, cube	8	4-block, hypercube	24	32	16

It may be of some help to provide the algebraic support for these figures. Using the fundamental theorem of calculus, in which the differential dx is seen as the point boundary of a line segment, of which there are two, we have the following table for n-cubes, which we can think of as origin centered.

Space	Core = n+1-space		Boundary Order			
n	Open	Closed on Boundary	Volumes	Surfaces	Edges	Vertices
undif	$x^0=dx$		$x^3 dx^{n-2}$	$x^2 dx^{n-1}$	$x dx^n$	dx^{n+1}
0	x^1	$(x+2dx)^1 =$ $x+2dx$				$2dx$
1	x^2	$(x+2dx)^2 =$ $x^2+4xdx+4dx^2$			$4xdx$	$4dx^2$
2	x^3	$(x+2dx)^3 =$ $x^3+6x^2dx+12xdx^2+8dx^3$		$6x^2 dx$	$12xdx^2$	$8dx^3$
3	x^4	$(x+2dx)^4 =$ $x^4+8x^3dx+24x^2dx^2+32xdx^3+16dx^4$	$8x^3 dx$	$24x^2 dx^2$	$32xdx^3$	$16dx^4$
		Table of n-cube breakdown with n+1-core				

With respect to the n-sphere, in which the formulae have been converted to a function of diameter instead of radius, thus relating to the equations above by a factor of $\pi/2(n+1)$, we have the following table. With the exception of the 0-sphere which is effectively a one-half circumference, and this, subject to interpretation, the n-spheres are manifolds without boundary, that is, as manifolds they have no edge. The figures in the edge column do represent an edge of sorts as the extent of their curvature, as it can be seen that they are the formulae for a circle. With respect to the vertices, it can be seen that if we substitute a dr

for the *dx*, which has some justification at the limit, the formulae of the vertices is that of the n+1-space itself, indicating the single vertex is the center of the space.

Sphere	Ball = n+1-space, $x = 2r$ = the diameter		Boundary Order			
n	Open	Closed on Boundary	Volumes	Surfaces	Edges	Vertices
undif	$x^0 = dx$		$x^3 dx^{n-2}$	$x^2 dx^{n-1}$	$x dx^n$	dx^{n+1}
0	$(\pi/2)x^1$ πr^1	$(\pi/2)(x+2dx)^1 =$ $(\pi/2)x + \pi dx$				πdx πdr
1	$(\pi/4)x^2$ πr^2	$(\pi/4)(x+2dx)^2 =$ $(\pi/4)x^2 + \pi x dx + \pi dx^2$			$\pi x dx$ $2\pi r dx$	πdx^2 πdr^2
2	$(\pi/6)x^3$ $4\pi r^3/3$	$(\pi/6)(x+2dx)^3 =$ $(\pi/6)x^3 + \pi x^2 dx + 2\pi x dx^2 + (4/3)\pi dx^3$		$\pi x^2 dx$ $4\pi r^2 dx$	$2\pi x dx^2$ $4\pi r dx^2$	$4\pi dx^3/3$ $4\pi dr^3/3$
3	$(\pi/8)x^4$ $2\pi r^4$	$(\pi/8)(x+2dx)^4 =$ $(\pi/8)x^4 + \pi x^3 dx + 3\pi x^2 dx^2 + 4\pi x dx^3 + 2\pi dx^4$	$\pi x^3 dx$ $8\pi r^3 dx$	$3\pi x^2 dx^2$ $12\pi r^2 dx^2$	$4\pi x dx^3$ $8\pi r dx^3$	$2\pi dx^4$ $2\pi dr^4$

Table of n-sphere manifold breakdown and n+1-balls

We have used a cube as a basic form to show the simplicity with which an orthogonal structure can be morphed into an essentially curved structure. Conversely it shows the orthogonal structure that, given any symmetric motion, lies hidden in the most curved of spaces. The horn torus has positive curvature on the outside extent of its annulus were it approaches the configuration of a sphere and negative curvature on the inner confines where is approaches the configuration of a pseudosphere, with a region of flatness in the regions of its upper and lower annular extremal planes. Such regions of flatness, like the crest of a wave, carry the orthogonal potential found at the center outward with expansion, and we might think to find orthogonality in a quantum structure albeit in some sense inverted from the above. Such crest, in this case, represents the transition from negative to positive curvature. If we try to imagine the above cube as a 4-cube, and the monad as of one additional dimension, we might imagine such flatness as the region of our observed universe, the crest of the expanding n-monad.

In the interest of symmetry, it might be imagined that such a structure as A has a conjoined twin that mirrors the dynamics above along the bottom surface with a font from a left hand rotation, so that the center plane becomes an inertial subduction region. We might also envision similar reflection at the top surface, with expansion outward along the central plane. Finally we might imagine a case in which the conjoined system oscillates between extension along axis and plane. We have large scale examples of general toric structure in the collimated jets of active galactic nuclei and the central planes of spiral galaxies, and it would not be surprising to find such on a cosmic level.

The current general cosmological assumption appears to be that spacetime emerged via the big bang from a not quite singular, but probably spherical point. It is also assumed to have been hot. The concept of a thermal birth seems an attempt to explain the presumably conserved energy of the universe initially confined to a very small space. If spacetime itself has inertial properties from which the mass, momentum, energy and power of quantum particles are derived through motion, including its oscillatory activity, then the inertial font attributed to a hot, big bang need not have been either hot or a bang. The total energy of the universe can be contained in a very small, but finite locus of extreme inertial density as a potential energy density as just described.

Unification of the present duality of the general relativistic description of spacetime and the standard model description of quantum matter and its interactions is attempted in a big bang that is a common source for the presumed separate ontologies of both. This holds a paradox, for if gravity is a quantum property, that along with the other interactions, held all together prior to the cosmic inception with a binding intensity that exceeded even the darkest black hole, some yet more energetic mechanism was required to upset a presumed equilibrium and lead to expansion. If matter curves spacetime and in turn conforms to the confines of that curvature, then we have a problem. This is a very short dog chasing an even shorter tail.

The perspective behind the development herein is that spacetime is an inertial continuum that conforms in some manner to the above description of the monad, oscillating cosmically over time, probably in the manner of a conjoined system. This gives it simply described regions of positive, negative and flat curvature over the cosmic extent, whose local dynamic properties are conditioned by the curvature configuration. As we appear to find ourselves in a generally flat section of this system, we can start our investigation of a quantum structure with an assumption of local Euclidean flat space and see what possible curved structure arises out of an expansive change in that space. We might, for example, find that it appears to have some relationship to A as found at the center of B above after it has been transformed, along the annular surface to a point of isotropy or general flatness.

Isotropic expansion is generally stated as an expansive condition that has no center. In fact, expansion of an n-space has two centers; one is the center of the (n+1)-core for which the n-space forms the cover, and the other is the local center in the n-space from which everything around it is seen to expand. In the customary inflating balloon analogy used to explain isotropic expansion, all points on the surface of the balloon are moving away from the center of the balloon, while any loop drawn on the surface of the balloon will be seen to expand from a point in its center. As in the table concerning the n-spheres above, one vertex at the center end of the radius and the other at any of the circumferential radial end points making the n-sphere's boundary constitute the centers of expansion. The same holds for the n-cube, and any similarly conforming n-space. Any radial will do. Any expanding space requires simply an extra spatial center, either completely geometric or weighted by some other property, i.e. center of mass, energy, etc. and one (and every) local center. It requires, then, a change in the scale or gauge of only one parameter, **r**, common to each of the n+1 dimensions to effect the isotropic change. The local quantum center is intimately connected to the universal center through an isotropic expansion strain and associated stress.

2 – The Fundamental Inertial Quantum as a Simple Harmonic Oscillator

Isotropic Expansion and the Generation of Rotational Oscillation

We can examine the cube for Space(2) in the above table with the intention of analyzing the effect of an isotropic strain in a condition of assumed uniform dilatation, along with the corresponding stress, on a unit volume of space. In light of the comments just made, we can imagine the center of this cube as one center of expansion and the other as extra dimensional, represented by the indefinite extension of the four diagonals through the eight vertices. We will integrate the differentials to compare the contribution made by each boundary order to the change in the corresponding core, in this case a volume. We are interested in the relative contributions of each order over time to the initial unit volume, V, and not to the changing magnitude of the volume itself. We substitute the following boundary placehold identities for <u>S</u>urface, <u>E</u>dge and vertices (<u>C</u>orner), $1^2 S \equiv x^2$, $1^1 E \equiv x^1$, and $1^0 C \equiv x^0$ so as to maintain proper integration. It will be helpful if we assign a "normal" boundary strain vector to each of these components, which in each case will be in the direction in which the boundary is increasing. Thus

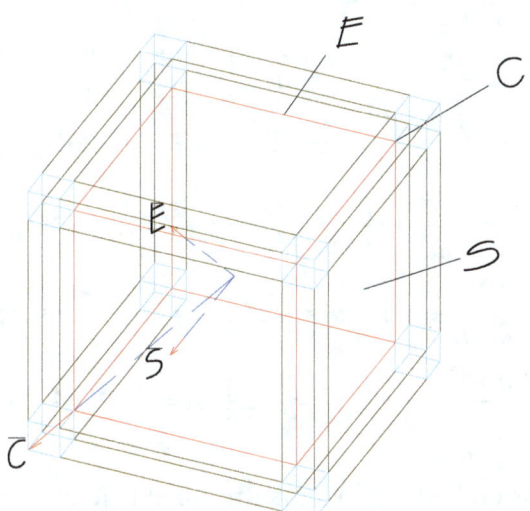

Cubic Expansion

$$|\mathbf{S}| = \left|\sqrt{\tfrac{1}{2}}\,\mathbf{E}\right| = \left|\sqrt{\tfrac{1}{3}}\,\mathbf{C}\right| \qquad (2.1)$$

$$|\mathbf{E}| = \left|\sqrt{2}\,\mathbf{S}\right| = \left|\sqrt{\tfrac{2}{3}}\,\mathbf{C}\right| \qquad (2.2)$$

$$|\mathbf{C}| = \left|\sqrt{3}\,\mathbf{S}\right| = \left|\sqrt{\tfrac{3}{2}}\,\mathbf{E}\right| \qquad (2.3)$$

In the following discussion, no assumption is made about the universal configuration or number of dimensions of the space in which the unit cube is embedded. We are only interested, at least initially, in the local geometry, which is assumed to be flat and therefore Euclidean. Thus it is background-independent. As to a fourth spatial dimension, we will see that change in or motion of such dimension is interchangeable with a dimension of time in a three spatial dimension context.

In this case the integration will be simultaneous on each order, as indicated by the pre-subscript n, in $\int {}_n dx^n$ so that we have

$$\int_V dV = 6x^2 \int_0^a {}_1 dx^1 + 12x^1 \int_0^a {}_2 dx^2 + 8x^0 \int_0^a {}_3 dx^3 \qquad (2.4)$$

$$\int_V dV = 6S \int_0^a dx + 12E \left(\int_0^a dx\right)\left(\int_0^a dx\right) + 8C \left(\int_0^a dx\right)\left(\int_0^a dx\right)\left(\int_0^a dx\right) \qquad (2.5)$$

$$\Delta V = 6aS + 12a^2 E + 8a^3 C \qquad (2.6)$$

Solving for the following ratios, all at unity, where the designations S, E and C are unit names, their dimensional quantities being absorbed in the numerical coefficients of a^n, i.e. 6 square units times a, 12 length units times a^2, etc., gives the value of a for each equivalence. The ratios have been stated with the highest order in the consequent or denominator so they are decreasing from infinity as dx increases, until unity is reached as stated. We have (showing the negative for the sake of symmetry)

$$\frac{S}{E+C} = \frac{6a}{12a^2 + 8a^3} = \frac{1}{2a + \tfrac{4}{3}a^2} = 1 \therefore a = -\tfrac{3}{4} \pm \tfrac{1}{4}\sqrt{21} = 0.39564..., -1.89564... \qquad (2.7)$$

$$\frac{S}{E} = \frac{6a}{12a^2} = \frac{\tfrac{1}{2}}{a} = 1 \therefore a = \tfrac{1}{2} = 0.5 \qquad (2.8)$$

$$\frac{E}{C} = \frac{12a^2}{8a^3} = \frac{\tfrac{2}{3}}{a} = 1 \therefore a = \tfrac{2}{3} = 0.66666... \qquad (2.9)$$

$$\frac{S}{C} = \frac{6a}{8a^3} = \frac{\tfrac{3}{4}}{a^2} = 1 \therefore a = \pm \tfrac{\sqrt{3}}{2} = \pm 0.86602... \qquad (2.10)$$

$$\frac{E}{S+C} = \frac{12a^2}{6a + 8a^3} = 1 \therefore a = \tfrac{3}{4} \pm i\tfrac{1}{4}\sqrt{3} = \tfrac{\sqrt{3}}{2} e^{\pm i \tfrac{\pi}{6}} = 0.86602...e^{\pm i \tfrac{\pi}{6}} \qquad (2.11)$$

$$\frac{S+E}{C} = \frac{6a + 12a^2}{8a^3} = 1 \therefore a = \tfrac{3}{4} \pm \tfrac{1}{4}\sqrt{21} = 1.89564..., -0.39564... \qquad (2.12)$$

If we think of the cube as embedded in an isotropic elastic continuum, which is of some inertial density and under tension, dx represents the work done in displacing or distorting the medium, and by virtue of Gauss' theorem, the integration of that work represents the energy of the distortion. By way of reference, in an ideal elastic medium, the stress operating on the locale is a function of the strain and the elastic modulus as

$$\mathbf{F} = \frac{Y\mathbf{E} - 3\sigma \bar{P}\mathbf{1}}{1 + \sigma} \qquad (2.13)$$

where $\mathbf{F}$ is the stress tensor, $\mathbf{E}$ is the strain tensor, Y is Young's modulus of elasticity, σ is Poisson's ratio or the negative ratio of lateral to axial or shear to tension strain, $\bar{P}$ is the mean pressure in the medium, and $\mathbf{1}$ is the idemfactor or unit tensor. Assuming a value of σ of $-1/3$ for an ideal isotropic 3 dimensional medium we have

$$\mathbf{F} = \frac{3}{2}\left(Y\mathbf{E} + \bar{P}\mathbf{1}\right). \qquad (2.14)$$

The vector fundamental tension stress component is

$$\mathbf{f} = Y\mathbf{e} \tag{2.15}$$

and is related to the energy distribution by Gauss' theorem for the radial strain

$$E_r = \int_V \nabla \cdot \mathbf{e}_r dv = \oint_S \mathbf{e}_r \cdot d\mathbf{S} \tag{2.16}$$

and Stokes' theorem for the angular or tangential strain

$$E_t = \int_S \nabla \times \mathbf{e}_t \cdot d\mathbf{S} = \oint_\mathbf{r} \mathbf{e}_t \cdot d\mathbf{r}_t \tag{2.17}$$

These boundary order ratios, then, are inflection points indicating the energy contributions and potential energy gradient changes over time among the boundary components. In an ideal static, kinematic case the change in the ratios with an increase in *dx* would have no functional effect on the components, if *dx* has the same magnitude for each of them as it increases. This would amount to a simple change of scale. The real solutions above would appear to reflect this static condition. However, in a dynamic condition, we might imagine that as each ratio decreases below unity and past the inflection point, the magnitude of the consequent exceeds and affects the antecedent or numerator, whose magnitude then becomes a partial function of the consequent. This would appear to be the case for the complex solutions in particular, which correspond with an angular gradient potential of the boundary vectors from that of the antecedent to the direction of that of the consequent.

These evaluations were done with Maple. It is significant that if we convert (2.11) to complex polar notation as in the last term, the modulus is equal to the value for *a* in (2.10). It is important that we understand that the ratios represent the point at which the change in volume due to the sum totals of all component orders in the antecedent and consequent are equal. It is not the point at which one single component of a given S, E, or C times its appropriate $\int_n dx^n$ is equal to another, since this happens for all at the point where $a = 1$.

In these evaluations, the S component of the strain and hence of the work predominates until (2.7) is reached. At this point, the stress will begin to shift from a predominance of tension to that of shear, meaning there will be a potential for the surface and edge strains to oscillate. As the edges and vertices ring each of the surfaces, the system remains basically stable, however. At the point of (2.8) the edges assume dominance over the surfaces and a gradient is produced for the bulk strain and the tension stress in the direction of the edges. Once again, the 2:1 symmetry of edges to surface maintains stability. At (2.9) the vertices contribute more work than the edges and the strain gradient shifts in their direction. Thus there is a vector potential from the surfaces to the edges to the vertices. Once more the symmetry between vertices and edges maintains stability.

Jumping to (2.12), at this point the strain contributed by the vertices dominates both of the other components combined and the related stress is greatest at these locations. This would result in a dissipation of the energy altogether, were it not for the unusual and unique condition created by (2.10) and (2.11). The point at which the strains of the vertices come to equal those of the surfaces is also the point at which their combined strain comes to equal that of the edges, as given by the modulus of the latter's ratio. We can assume that the imaginary component of this ratio indicates a rotational component of $\pi/6$ or 30°, and since the vertices are assuming a predominance over the surfaces at this point, having already exceeded the edge strain, and as there is an imbalance in the number of vertices to surfaces, a necessary break in symmetry ensues.

We can imagine a rotational potential of the surface strain in the direction of the vertices, which by virtue of the asymmetry between S and C, of 3 degrees of rotational freedom and 4 possible rotational axes, results in an eventual rotational strain about one pair of the axes. This is simultaneous with a shift of the Es in the direction of S + C and a dragging of the strains at each of the two axial C poles. This then leads to a rotation of the axial Cs in the direction of one of the three E pairs extending from those two vertices. The equation of (2.11) gives this rotational relationship. The nature of the ambiguous sense in the argument is indicative of the equation of a rotation and its complex conjugate, when viewed from both senses of its axis, i.e. by rotating it about the real axis, where ± means plus and minus and not plus or minus, if we adjust the Euler identity to

$$e^{\pm i\theta} = \sin\theta \pm i\cos\theta. \tag{2.18}$$

One end of the axis of strain then can be shown as indicated by the "symmetry breaking" in (2.21).

$$12a^2 E = \left(6aS + 8a^3 C\right) \tag{2.19}$$

$$12\left(\frac{\sqrt{3}}{2}e^{\pm i\frac{\pi}{6}}\right)^2 E = 6\left(\frac{\sqrt{3}}{2}e^{\pm i\frac{\pi}{6}}\right)S + 8\left(\frac{\sqrt{3}}{2}e^{\pm i\frac{\pi}{6}}\right)^3 C \tag{2.20}$$

$$e^{-i\frac{\pi}{3}} E = \frac{1}{\sqrt{3}}\left(e^{+i\frac{\pi}{6}} S + e^{+i\frac{\pi}{2}} C\right) \tag{2.21}$$

Thus, the strain vector E, rotated in some direction $\frac{\pi}{3}$, is equal to $\frac{1}{\sqrt{3}}$ of the S and C strains rotated $\frac{2\pi}{3}$ in the opposite direction, presumably in the same plane. In fact, this states that C rotates $\frac{\pi}{2}$ while S rotates $\frac{\pi}{6}$. We can see specifically how these rotations occur in Spin Diagrams 1 and 2. We can also see there how a rotation back in time of $\frac{\pi}{3}$ equals one forward in time by $\frac{2\pi}{3}$ and vice-versa, if their plane of rotation, ϕ, is itself rotating at a constant rate with respect to an orthogonal plane, θ, that is where the two axes intersect at the centers of rotation. However, it is shown there that this corresponds with a rotation of θ, back $\frac{\pi}{4}$ and forward $\frac{3\pi}{4}$, indicating a variability in the strain velocity.

It should be understood that this cubic structure is simply an expression of the orthogonal tendency for stress equalization and energy conservation. The condition found at (2.10) and (2.11), then becomes a stable dynamic condition of rotational oscillation or spin, within certain parameters of inertial density and mechanical impedance. If the isotropic tension in this situation was sufficient to increase the strain indefinitely, if the medium was to lose its elasticity and become plastic or even rupture, any tendency to oscillate would be overcome by the transfer of energy via strain to the vertices. Local energy would not be conserved, but be drawn away by the strain.

It is essential to extrapolate this scenario to the hypercube, H, to achieve a full understanding. We will skip the integrals but show the results for the corollary of (2.6) as

$$\Delta H = 8aV + 24a^2S + 32a^3E + 16a^4C \qquad (2.22)$$
$$= 1aV + 3a^2S + 4a^3E + 2a^4C$$

There are 25 combinations with corresponding non-ordered permutations or sub-combinations, for the 4-cube; 7 involving all 4 parameters, 12 permutations involving all sub-combinations of 3, and 6 one to one relationships. With the 3-Space, there are 2 single real positive solutions at (2.8) and (2.9), one instance of a complex solution at (2.11), one correspondence between a real and a complex solution at (2.10) and (2.11) where the real value of a in one is equal to the complex modulus in the other, and one instance of a correspondence of solutions with sense inversion, (2.7) and (2.12), that is their solutions have the same magnitude, but of opposite sense. As might be expected, the 4-Space shows significantly more of these symmetries. It should be noted that while an attempt has been made to analyze the ratios qualitatively so that all are represented as decreasing with respect to an increasing dx, they have not all been checked quantitatively, and some may be increasing as shown. In fact, (2.35) and (2.37) are found to be increasing at the point represented by the first positive solution and decreasing at the second. For (2.32) it is worth stating that for every value of the ratio $0.75 < \left(\frac{S}{V+E}\right) < +\infty$, the modulus is ½ and the argument ranges from 0 to ½ π.

It is important to remember that a given component in the 3-cube is identical to the same component in the 4-cube, but the relationships between them are different. An edge still is bounded by 2 vertices, but there are 4 edges intersecting at each vertex of the 4-cube. A line segment in an x-y plane is qualitatively no different than one in the z-x or for that matter z-w plane. In fact a point in 3-space also has a location in n-space, at least in Euclidean n-space. In the following, it is also important to remember that a is not the value of the corresponding ratio, but rather the value found in both antecedent and consequent when the ratio equals 1. The evaluations are based on the following identities in (2.23),

$$V \equiv 1a, S \equiv 3a^2, E \equiv 4a^3, C = 2a^4 \qquad (2.23)$$

$$\frac{V}{S}, a = \tfrac{1}{3} \qquad (2.24)$$

$$\frac{V}{E}, a = \pm\tfrac{1}{2} \qquad (2.25)$$

$$\frac{V}{C}, a = \tfrac{1}{\sqrt[3]{2}}, -\tfrac{1}{2}\left(\tfrac{1}{\sqrt[3]{2}}\right) \ldots \pm i\tfrac{\sqrt{3}}{2}\left(\tfrac{1}{\sqrt[3]{2}}\right) = \tfrac{1}{\sqrt[3]{2}}e^{\pm i\frac{2\pi}{3}} = 0.79370\ldots e^{\pm i\frac{2\pi}{3}} \qquad (2.26)$$

$$\frac{S}{E}, a = 0, \tfrac{3}{4} \qquad (2.27)$$

$$\frac{S}{C}, a = 0, \pm\sqrt{\tfrac{3}{2}} \qquad (2.28)$$

$$\frac{E}{C}, a = 0, 0, 2 \tag{2.29}$$

$$\frac{V}{S+E}, a = -1, \tfrac{1}{4} \tag{2.30}$$

$$\frac{V+S}{E}, a = -\tfrac{1}{4}, 1 \tag{2.31}$$

$$\frac{S}{V+E}, a = \tfrac{3}{8} \pm i\tfrac{1}{8}\sqrt{7} = \tfrac{1}{2}e^{\pm i 0.722734248...} \tag{2.32}$$

$$\frac{V}{S+C}, a = 0.31290..., -0.15645... + i1.25436... = 1.26408...e^{\pm i 1.694883228...} \tag{2.33}$$

$$\frac{V+S}{C}, a = -1, -0.36602..., 1.36602... \tag{2.34}$$

$$\frac{V+C}{S}, a = -1.36602..., 0.36602..., 1 \tag{2.35}$$

$$\frac{V}{E+C}, -1.85463..., -0.59696..., 0.45160... \tag{2.36}$$

$$\frac{V+C}{E}, a = -0.45160..., 0.59696..., 1.85463... \tag{2.37}$$

$$\frac{V+E}{C}, a = 2.1120..., -0.05604... \pm i0.48331... = 0.48655...e^{\pm i 1.686235431...} \tag{2.38}$$

$$\frac{S}{E+C}, a = -2.58113..., 0, 0.58113... \tag{2.39}$$

$$\frac{S+E}{C}, a = -0.58113..., 0, 2.58113... \tag{2.40}$$

$$\frac{E}{S+C}, a = 0, 1 \pm i\tfrac{1}{\sqrt{2}} = \sqrt{\tfrac{3}{2}}e^{\pm i 0.615479709...} \tag{2.41}$$

$$\frac{V}{S+E+C}, a = 0.24415..., -1.12207... \pm i0.88817... = 1.43105...e^{\pm i 2.472026458...} \tag{2.42}$$

$$\frac{E}{V+S+C}, a = -0.24415..., 1.12207... \pm i0.88817... = 1.43105...e^{\pm i 0.669566197...} \tag{2.43}$$

$$\frac{V+E+C}{S}, a = -2.63993..., 0.31996... \pm i0.29498... = 0.43519...e^{\pm i 0.744798022...} \tag{2.44}$$

$$\frac{V+S+E}{C}, a = 2.63993..., -0.31996... \pm i0.29498... = 0.43519...e^{\pm i 2.396794631...} \tag{2.45}$$

$$\frac{V+S}{E+C}, a = -2.51702..., -0.25673..., 0.77375... \tag{2.46}$$

$$\frac{E+S}{V+C}, a = -0.77375..., 0.25673..., 2.51702... \tag{2.47}$$

$$\frac{V+E}{S+C}, a = 1, \tfrac{1}{2} \pm i\tfrac{1}{2} = \tfrac{1}{\sqrt{2}}e^{\pm i\tfrac{\pi}{4}} \tag{2.48}$$

Once again using Maple, there are a total of 10 couplings involving complex solutions, of which one is exclusively complex and one other has only a zero for the third and real solution. Only one single real positive solution is given. There are, however, 7 corresponding pairs of solutions involving sense inversion, 5 real and 2 complex. Note that all cases of sense inversion involve a combination of one or more components in either the antecedent and/or consequent and the sense change is associated with a transposition of one or two components in each pair. These do not appear to have any special relationship to the conditions of the 3-cube, at first glance, and we have not investigated them further.

There are several, however, that appear to have a direct relationship to some of the ratios of the 3-cube. Two conditions of correspondence are found between a real positive solution and the complex modulus of a complex solution with a positive real component. (2.28) $\left(\frac{S}{C}\right)$ and (2.41) $\left(\frac{E}{S+C}\right)$ are directly related to (2.10) and (2.11) respectively, the real solution and the modulus of the complex of the second two being equal to the product of the first and $\sqrt{2}^{-1}$. The argument of (2.41) is the angle at the center of a cube between a radial normal to an edge of the cube and one extended along a diagonal to a vertex. (2.25) $\left(\frac{V}{E}\right)$ and (2.32) $\left(\frac{V}{E+S}\right)$ are related to (2.8) $\left(\frac{S}{E}\right)_3$ with a common value for their real solutions and the modulus of the complex one. The cosine of the argument of (2.32) is equal to the solution of (2.27) $\left(\frac{S}{E}\right)_4$, which is the same ratio coupling as (2.8). This pairing (2.32) in turn has a modulus equal to the real and imaginary components of an additional complex solution in (2.48) $\left(\frac{V+E}{S+C}\right)$. This latter solution has an argument of $\pi/4$ or 45° which appears to be an extremely stable condition, as found in a sine wave model as the point of maximum power of the wave, where the product of the transverse wave force and transverse wave speed are maximum. It is also the angle of the strain vector E discussed above for the 3-cube, with respect to the plane normal to the spin angular momentum vector as shown in the spin diagrams. In the model developed here, this condition is found to be invariant and rotates about the oscillation's angular momentum vector.

Finally, (2.41) $\left(\frac{E}{S+C}\right)$, (2.48) $\left(\frac{V+E}{S+C}\right)$, and (2.26) $\left(\frac{V}{C}\right)$ are found to be related in a most profound way in the mechanism of the oscillation herein described. The imaginary component of (2.41) equals the modulus of (2.48). Note that (2.26) represents a $\frac{2\pi}{3}$ rotation due to the interplay between the volume and vertex components of strain and a modulus of that strain of $\frac{1}{\sqrt[3]{2}}$. Using the equation for (2.26) or

$$aV = 2a^4 C \tag{2.49}$$

$$\frac{1}{\sqrt[3]{2}} e^{\pm i \frac{2\pi}{3}} V = 2 \left(\frac{1}{\sqrt[3]{2}} e^{\pm i \frac{2\pi}{3}} \right)^4 C \tag{2.50}$$

tells us that a rotational oscillation of the 4-volume (boundary) strain V of modulus $\frac{1}{\sqrt[3]{2}}$ by $\frac{2\pi}{3}$ is equal to 4 axial rotations about the vertices of the same modulus and argument, where the 2 in the consequent indicates simultaneous rotations of opposite sense at each end of an axis. The oscillation of V is fourth dimensional, and therefore beyond our direct sensory ken, however, the 4 vertices are not, and we can envision the above consequent, the expression in 3 dimension of this four dimensional rotation, as a sequence of 4, $\frac{2\pi}{3}$ rotations about the 4 diagonals of a 3-cube. This sequence leaves the cube unchanged and avoids the entanglement condition, i.e. the continuity of Euclidean 3-coordinates of the cube are not twisted by the sequence. This condition of limits on the twistability of the continuum strain is a necessary consequence of its inertial/elastic properties. As the rotation of V is continuous, we would imagine that the sequence of 4 rotations is continuous, i.e. the strain rotates from one reference diagonal to another about one of the three surface axes of the 3-cube. We can also envision this as one diagonal axis rotating $\frac{2\pi}{3}$, followed by a 2π rotation of the same sense about one of the adjacent 3-cube surface axes. We can also treat it as a sequence of 4 orthogonal permutations.

We can show this configuration simply. If we align a hyperbolic surface of revolution about the y axis of the curve

$$xy = \tfrac{1}{2}, \text{ for } x \leq \tfrac{1}{\sqrt{2}} \qquad (2.51)$$

at each of the eight vertices of a cube so that each of them is at the angle of the argument given by (2.41) as just described, and so that the rims or circles of their bases intersect at the centers of each of the six surfaces of the cube, the following will be found concerning this geometry, which we will call an inversphere. We can also, as an alternative, create a similar construct using a pseudosphere in place of the above surface of revolution. Given a constant negative curvature of -1 for each pseudosphere, the resulting inversphere would have a constant negative curvature. This points to the development of the monad above. With respect to the inversphere:

1. Each surface of revolution, which we might call a hyper-axis or h-axis and which can be represented by a complex plane, with the imaginary dimension parallel to the circumference of the revolution and the real along the diagonal axis, will have a curvature of negative 1 at the rim, remaining negative while decreasing, that is, moving toward zero or flatness, with distance along the asymptote. Here the left four of Figure 4 are shown, their designations corresponding with the axes of Figure 3 below.

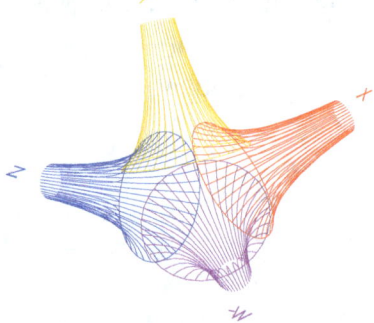

Figure 3

2. The rims will have a radius of $\frac{1}{\sqrt{2}}$. The area of the circle formed by the rim is therefore $\frac{\pi}{2}$, and its complex representation is $\frac{1}{\sqrt{2}}e^{i\theta}$ corresponding with $\left(\frac{V+E}{S+C}\right)$.
3. The rims will intersect orthogonally with each other at the cubic surface centers, so that there are three h-axes adjacent to a given h-axis along the cubic edges which we will refer to as the proximal axes.
4. The rims from h-axes located diagonally across the cubic <u>surface</u> from each other will be parallel or tangential at the same point at which they intersect with their proximal axes. We will call the corresponding parallel axes the distal axes. One set of mutually distal axes can be called the positive h-axes.
5. Each h-axis has a spatial inversion or anti-axis which is proximal to the distal axes of that h-axis. The set of their spatial inversions can be called the negative h-axes.
6. Each rim intersection is a $\frac{2\pi}{3}$ rotation from the others about the cubic diagonal, associating it with $\left(\frac{V}{C}\right)$.
7. The distance between cube surface centers describes an octahedron of edge length $\sqrt{\frac{3}{2}}$. The surface area of the octahedron is therefore $3\sqrt{3}$ and the volume is $\frac{\sqrt{3}}{2}$. The radial normal to the octahedron face is ½.
8. The cube will have an edge measure of $\sqrt{3}$. The surface area of the cube is 18 and the volume is $3\sqrt{3}$.
9. The concentric sphere intersecting at the rim intersections will have a radius of $\frac{\sqrt{3}}{2}$. The surface area is 3π and the volume is $\frac{\sqrt{3}}{2}\pi$.
10. We can think of this arrangement as the expression of a 4-cube in a 3-space, where the orthogonality condition of the 4-D space is met by the rim intersections, the center of each component of sphere, cube, octahedron and h-axis intersections being a common system center.
11. This configuration can be reduced to a 3-space orthogonal system simply by collapsing the cube along the W hyper-axis, as in the figure at left below, resulting in the co-ordinate system at right.

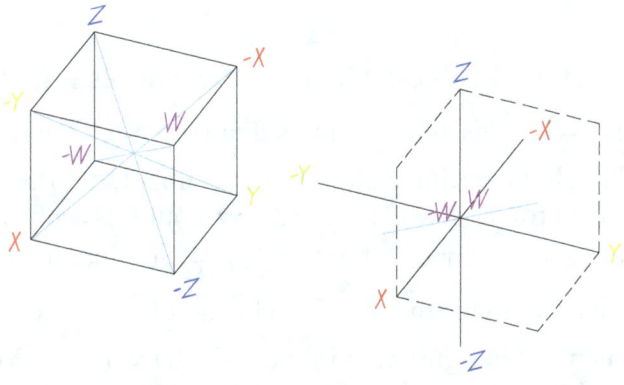

Figure 4

The condition of (2.8) $\left(\frac{S}{E}\right)_3$, (2.25) $\left(\frac{V}{E}\right)$ and (2.32) $\left(\frac{S}{V+E}\right)$ is represented by (2.48) $\left(\frac{V+E}{S+C}\right)$ at each h-axis. Thus the orthogonal projections of the argument of (2.48), described as extending from the system center to each cubic edge midpoint, are equal to the modulus of (2.41) $\left(\frac{E}{S+C}\right)_4$, and the argument of (2.26) $\left(\frac{V}{C}\right)$ is the rotation of that axis between proximal intersections and cubic surface centers.

In terms of $a = \int dx$ we are only interested in positive or increasing real values, although in the context of complex values, some negative real components as in (2.26) are of interest. A deeper analysis would no doubt find significance in all of the couplings, but we are only interested in the general manner in which the 4-cube and the 3-cube couplings might interact. In this regards it is important to remember that in the case of the 4-cube, the volume is a boundary that is increasing while in the case of the 3-cube, it is the base space, held constant, upon which the boundary changes are taking place.

From the perspective of a rotational oscillation, as found in a torsion pendulum or a jump rope oscillation, of interest are those couplings of two boundary parameters, V + E and S+ C, which have an intervening parameter, S and E respectively. More interestingly, in both these cases, V + E for the 4-cube and S + C for both 4-cube and 3-cube, the two-parameter components also have a ratio between themselves whose solution is ($\pm$) real and equal to the modulus of the companion ratio. (2.48) gives the special case of V + E with S + C. Unlike the other three rotational oscillator couplings, it has a positive real solution in addition to its complex solution. It also has the two parameter component ratios in common with the other two oscillators of the 4-cube. The remaining couplings with complex solutions all have intervals between their real and complex moduli solutions, for most exceeding 1, which mitigates against oscillation, with one exception. (2.26) $\left(\frac{V}{C}\right)$ has a real solution that equals its modulus, thereby indicating rotational oscillation. In addition, the cosine of its argument is equal to the modulus and real solution for $\left(\frac{S}{V+E}\right)$ and $\left(\frac{V}{E}\right)$ at $\frac{1}{2}$ and its sine, to the 3-cube modulus and real solution for $\left(\frac{E}{S+C}\right)$ and $\left(\frac{S}{C}\right)$ at $\frac{\sqrt{3}}{2}$, and to the 4-cube modulus and real for $\left(\frac{E}{S+C}\right)$ and $\left(\frac{S}{C}\right)$ at $\frac{1}{\sqrt{2}}\frac{\sqrt{3}}{\sqrt{2}} = \frac{\sqrt{3}}{2}$. Thus the rotational parameters of the other rotational oscillation or spin couplings, can be found in the simple ratio of $\left(\frac{V}{C}\right)$.

Within the context of the 4-cube, the first value that arises is (2.42) $\left(\frac{V}{S+E+C}\right)$ followed closely by (2.30) $\left(\frac{V}{S+E}\right)$. This simply shows that the vertex component adds very little at this juncture, although it does have a rotational element, but the negative real component indicates a significant rotation which would seem out of synch with the small real strain. A similar comment could be made about (2.33) $\left(\frac{V}{S+C}\right)$ which is next in the real order, though the potential rotation is much less. This is followed by (2.24) $\left(\frac{V}{S}\right)$ which has no rotational component. It is significant in that it is the value of Poisson's ratio in an ideal isotropic elastic solid, relating the axial to lateral strain and thereby, tension to shear stress.

Next is (2.35) $\left(\frac{V+C}{S}\right)$ with no rotational component, followed by (2.44) $\left(\frac{V+E+C}{S}\right)$, which has a rotational component. The real solution and therefore the strain is negative, however, and is out of scale with the modulus of the complex solution, which would mitigate against rotational oscillation. This modulus and the positive solution of (2.36) $\left(\frac{V}{E+C}\right)$ are the first values to exceed any of the solutions for the 3-cube. The next ratios (2.25) $\left(\frac{V}{E}\right)$ and (2.32) $\left(\frac{S}{V+E}\right)$ involve the first of the oscillatory groups. The real solution of the first and modulus of the second are equal to each other and to that of (2.8) $\left(\frac{S}{E}\right)_3$, while the cosine of the argument of $\left(\frac{S}{V+E}\right)$ coincides with the real solution of (2.27) $\left(\frac{S}{E}\right)_4$. Thus we might associate an actual oscillation of the 4-cube with the potential $\left(\frac{S}{E}\right)_3$ of the 3-cube. This is followed by (2.39) $\left(\frac{S}{E+C}\right)$ which has a real solution and is the 4-cube corollary of the first ratio of the 3-cube. It is of no special interest other than being, along with (2.9) $\left(\frac{E}{C}\right)_3$ a precursor for the next coupling, which is (2.48) $\frac{V+E}{S+C}$, perhaps the most important of the whole assemblage. Together, $\left(\frac{S}{E+C}\right)$ and $\left(\frac{E}{C}\right)_3$ indicate a growing predominance of E and C over S and then C over E, or shear stress over tension, followed eventually by torsion over shear.

The argument of $\left(\frac{V+E}{S+C}\right)$ represents the power of the strain oscillation, first in the oscillatory twisting of the hyper-axes at $\left(\frac{V}{C}\right)$, then subsequently with the rotational oscillation of the 3-cube itself. Given the above description of the inversphere, the modulus of this solution represents the radius of and in the plane of the rim of the h-axis at the point at which its curvature is -1. The argument is the power phase of an oscillation which can be found as a phase constant in the eventual rotational oscillation of the 3-cube. This is followed by (2.46) $\left(\frac{V+S}{E+C}\right)$, which adds no new oscillatory components, but does show the gaining dominance of the higher order boundary components, E and C. This culminates in a new oscillatory condition at (2.26) $\left(\frac{V}{C}\right)$.

Note that the real value and the modulus of $\left(\frac{V}{C}\right)$ is slightly more than the values of $\left(\frac{V+E}{S+C}\right), \left(\frac{V}{E}\right), \left(\frac{S}{E}\right)_4$ and slightly less than the values of $\left(\frac{E}{S+C}, \frac{S}{C}\right)_3$ at oscillation. We can interpret the condition at $\left(\frac{V}{C}\right)$ as an oscillation about each of the 8 vertices. Each oscillation involves a twisting or torsion ultimately of $\frac{2\pi}{3}$ in each direction about each h-axis. The proximal axes will twist counter to the instant rotation sense of a given h-axis as will the anti-axis, all as viewed from the exterior of the system. The distal axes will twist with the same sense as the given h-axis, thus the directional sense of these axes corresponds with their rotational sense vis-à-vis the other axes. The strain on the enclosed sphere at maximum twist will be of a simultaneous lengthening along each <u>cubic</u> axis and flattening in the plane of said axis and the <u>cubic</u> axis from which the strain occurred and at which it is at a minimum, ideally zero, as indicated in the figure below.

The two pairs of distal axes on each surface create two countervailing torques, which in this oscillatory condition are in equilibrium.

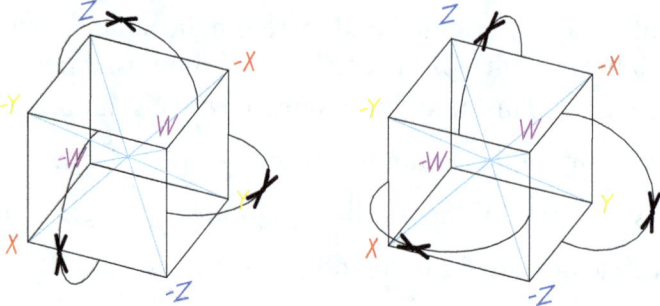

Figure 5

This initial symmetrical condition of $\frac{2\pi}{3}$ rotational oscillation of each of the four diagonals is broken upon dx reaching the oscillatory threshold given by $\left(\frac{E}{S+C}, \frac{S}{C}\right)_3$ at $\frac{\sqrt{3}}{2}$. This results in a permanent rotation of $\frac{\pi}{3}$ of one pair of the E vectors as indicated by (2.21), thence the whole system <u>strain</u> continues to <u>oscillate</u>, while the <u>stresses</u> <u>rotate</u> and generate an angular momentum vector. (2.26) indicates the rotation of the stresses in time among and about the four diagonals, which represent the four orthogonal axes of H. The oscillation of the 3-cube is supported and driven by the 4-stress which is concentrated in one transforming axis. (2.48) $\left(\frac{V+E}{S+C}\right)$ represents the power moments or positions of maximum conversion of kinetic to potential energy and vice versa.

Finally, (2.41) $\left(\frac{E}{S+C}\right)_4$ represents, in addition to the diagonals, a capacitive and an inductive torque that is co-linear with two of the diagonals and is the product of crossing into the power moments from their positions of equilibrium strain and rotates with them about the angular momentum vector, all described later. The modulus and the solution to $\left(\frac{S}{C}\right)_4$ at $\sqrt{\frac{3}{2}}$ represents the radial length from the center of the inversphere and 3-cube to the midpoint of the cubic edge. The solution $a = \sqrt{\frac{3}{2}} e^{\pm i 0.61547\ldots}$ in this case indicates a rotation of this vector into the diagonal or one of the h-axis or of **E** into **C**. Solving for $\left(\frac{E}{S+C}\right)_4$

$$4a^3 E = 3a^2 S + 2a^4 C$$

$$4\left(\sqrt{\tfrac{3}{2}} e^{\pm i 0.61547\ldots}\right)^3 E = 3\left(\sqrt{\tfrac{3}{2}} e^{\pm i 0.61547\ldots}\right)^2 S + 2\left(\sqrt{\tfrac{3}{2}} e^{\pm i 0.61547\ldots}\right)^4 C \qquad (2.52)$$

after reduction and some parsing gives

$$2\left(\sqrt{\tfrac{3}{2}} e^{-i 0.61547\ldots}\right) E = \left(\sqrt{\tfrac{3}{2}} e^{i \tfrac{\pi}{2}}\right)^2 S + \tfrac{3}{2} e^{+i 1.23095\ldots} C. \qquad (2.53)$$

Here as with the companion relationship for the 3-cube, we have "broken symmetry" with the rotational senses, and see that rotation of two edge strains into an adjacent corner is equal to two orthogonal rotations of a surface strain and a flip of a vertex strain from one h-axis to a proximal axis. The moduli in this case correspond to the metrics of the inversphere, where $\frac{3}{2}$ is the distance from the cubic center to the cubic vertex. We will see that this represents an instance of beta decay, where the surface and vertex rotations indicate the flip of the electrical phase torques from one pair of vertices to one of three proximal pairs. In the case of the inductive torque, we have an electron emission along with a flip of the magnetic moment, and in the case of the capacitive torque, we find a positron emission, without the magnetic moment flip.

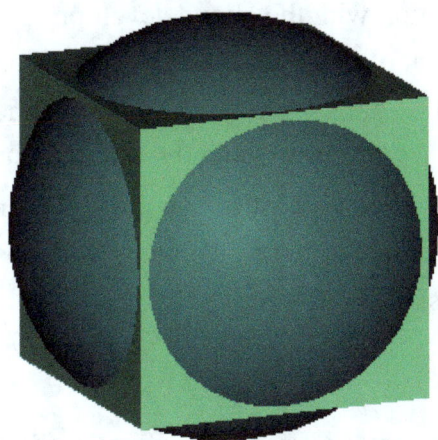

Equal Area Cube and Sphere

The position indicated as the midpoint on the cubic edge is of special interest. If we analyze a concentric cube and a sphere of equal surface area, and presumably of equal total surface stress, we will find that the radial to the midpoint as represented by $\left(\frac{S}{C}\right)_4$ exceeds the spherical radius (and the path of the rotational oscillation strain) at that point by a factor of

$$\delta r = \frac{\sqrt{2\pi}}{3}\left(\sqrt{\frac{3}{2}} - \frac{3}{\sqrt{2\pi}}\right) = \sqrt{\frac{\pi}{3}} - 1 = 0.023326708.... \qquad (2.54)$$

This indicates that the rotational path of the strain constricts the diagonals and restricts the operation given by $\left(\frac{E}{S+C}\right)_4$. Thus this differential must be overcome by the increase in stress of that operation. If we assume that the differential given by (2.54) is one component of the cross sectional area on which an orthogonal stress is operating, then the square of that value gives a differential stress required for the diagonal to flip of

$$\delta r^2 = 0.0005441353061.... \qquad (2.55)$$

The ratio of differential stress to the augmented total is then

$$\frac{\delta r^2}{1+\delta r^2} = \frac{0.0005441353061}{1.0005441353061} = 0.0005438393841... \qquad (2.56)$$

which when inverted is

$$\frac{1+\delta r^2}{\delta r^2} = 1838.778193 \qquad (2.57)$$

It bears noting that the 2002 CODATA ratio of the electron to neutron mass is 0.00054386734481(38), or within $2.796...\times 10^{-8}$ of the value of (2.56). Thus the ratio of the differential stress needed to produce beta decay and the stress of fundamental oscillation correlates significantly with the ratio of the mass-energy of the product of that decay, the electron, and that of the fundamental oscillation, the neutron.

With reference to beta decay, one additional observation concerns the weak mixing angle yielded by the final measured asymmetry stated in the September, 2005 issue of Physics Today by Bertram Schwarzschild in "Tiny Mirror Asymmetry in Electron Scattering Confirms the Inconstancy of the Weak Coupling Constant" as $\sin^2 \theta_W = 0.2397 \pm 0.0013$. If we consider the surface area of a sphere in steradians as 4π, the portion spanned by each cubic edge in conjuction with the above development is one twelfth that or an area of $\pi/3$. A linear component of that measure would therefore be $\sqrt{\frac{\pi}{3}}$ and would correspond generally and perhaps in some statistical manner with the distance from a cubic surface vector to a vertex vector as in the interplay between S and C at $\left(\frac{E}{S+C}, \frac{S}{C}\right)_{3,4}$. The arc distance between the mid-point of that arc and each of the three parameters E, S, and C is then $½ \sqrt{\frac{\pi}{3}}$. We then have the following, which is stated phenomenologically and without causal analysis

$$\sin^2\left(\tfrac{1}{2}\sqrt{\tfrac{\pi}{3}}\right) = 0.239735827. \tag{2.58}$$

This is the last ratio of interest, as it marks the final oscillatory condition for the 4-cube. We can show this development in the following orthogonal matrices. First the above rotational oscillation can be given by a 3-D strain spin matrix, $E_{\mu\nu}$, in which we assume a stationary spin angular momentum vector, $S_L (= \hbar)$, as our reference frame pointed in the $+z$ direction. We will give all six semi-axes, where $a/_{b^{-1}}$ is the direction of strain given as the double dot product of a into the b surface of the cube, as developed in <u>Physics of Waves</u>, Elmore and Heald, or

$$E_{\mu\nu} = \pm \mathbf{A} : \mathbf{B} = \pm \sum_{i=1}^{3}\sum_{j=1}^{3} a_{ij}b_{ij}$$
$$= \left(a_{xx}b_{xx} + a_{xy}b_{xy} + \ldots + a_{zz}b_{zz}\right) + \left(a_{-x-x}b_{-x-x} + a_{-x-y}b_{-x-y} + \ldots + a_{-z-z}b_{-z-z}\right) \tag{2.59}$$
$$= 2\left(a_{xx}b_{xx} + a_{xy}b_{xy} + \ldots + a_{zz}b_{zz}\right)$$

$$E_{\mu\nu} = \begin{pmatrix} x/_x & y/_x & z/_x \\ x/_y & y/_y & z/_y \\ x/_z & y/_z & z/_z \end{pmatrix} + \begin{pmatrix} x/_{-x} & y/_{-x} & z/_{-x} \\ x/_{-y} & y/_{-y} & z/_{-y} \\ x/_{-z} & y/_{-z} & z/_{-z} \end{pmatrix} \tag{2.60}$$

where a unit statement of a is

$$\begin{pmatrix} 1 & 1 & \sin\omega t \\ -1 & 1 & -\cos\omega t \\ -\sin\omega t & \cos\omega t & 1 \end{pmatrix} + \begin{pmatrix} -1 & -1 & -\sin\omega t \\ 1 & -1 & \cos\omega t \\ \sin\omega t & -\cos\omega t & -1 \end{pmatrix} \tag{2.61}$$

Note that plugging (2.61) into (2.60) for unit values of b gives

$$2\begin{pmatrix} 1 & 1 & \sin\omega t \\ -1 & 1 & -\cos\omega t \\ -\sin\omega t & \cos\omega t & 1 \end{pmatrix} \qquad (2.62)$$

The following Wave Diagrams show the conditions brought about by the above development.

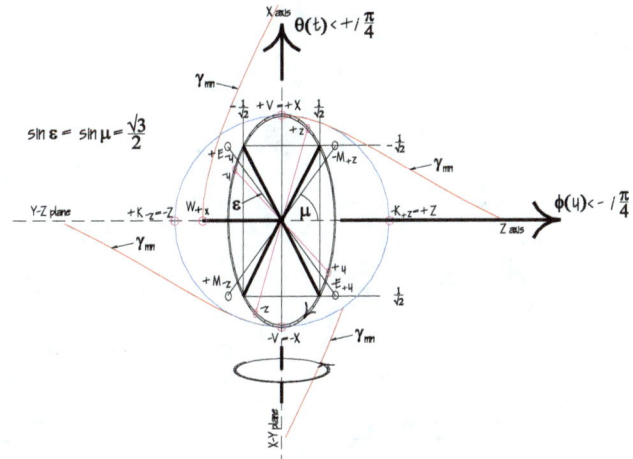

Wave Diagram 1 – Double Rotation, ϕ (θ)

The thin curved red lines represent the strains associated with $\left(\frac{V+E}{S+C}\right)$ where the argument of that coupling solution is found in the dark central cross representing the power moments, charge/potential (E) and induction/kinetic (M), represented by the ϕ vector, and $\left(\frac{V}{C}\right)$ which has rotated the x axis from its initial position, X. These power moments are analogous to the positions of maximum power of a wave on an ideal stretched string shown below.

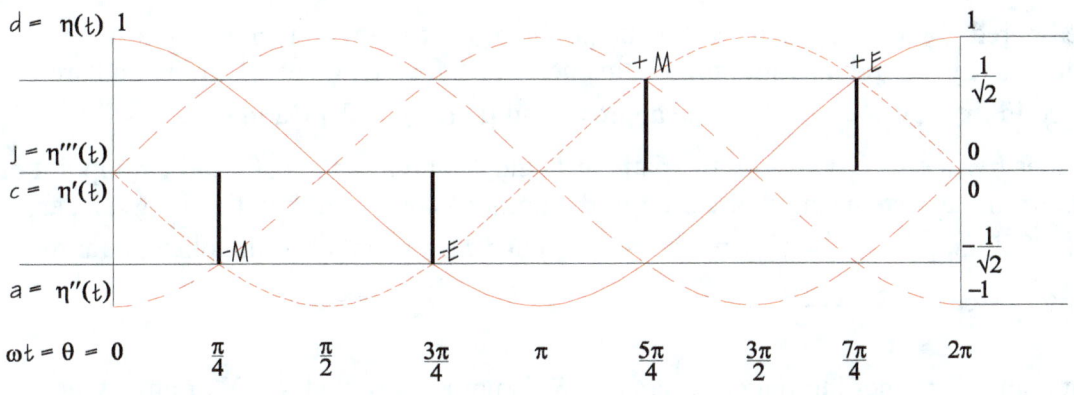

Wave Diagram 2 – Kinematic Functions of η(t)

The following Spin Diagram shows the strain path in red for point +y at $\theta = \omega t = n2\pi$ at which time the strain at that point is in a relative equilibrium position, (other than a ½ π twist) and transforming so that

$$x/y = -1, \; z/y = -\cos n2\pi = -1 \qquad (2.63)$$

Note that this recapitulates Figure 2 in the section, Wave Bearing Continuum.

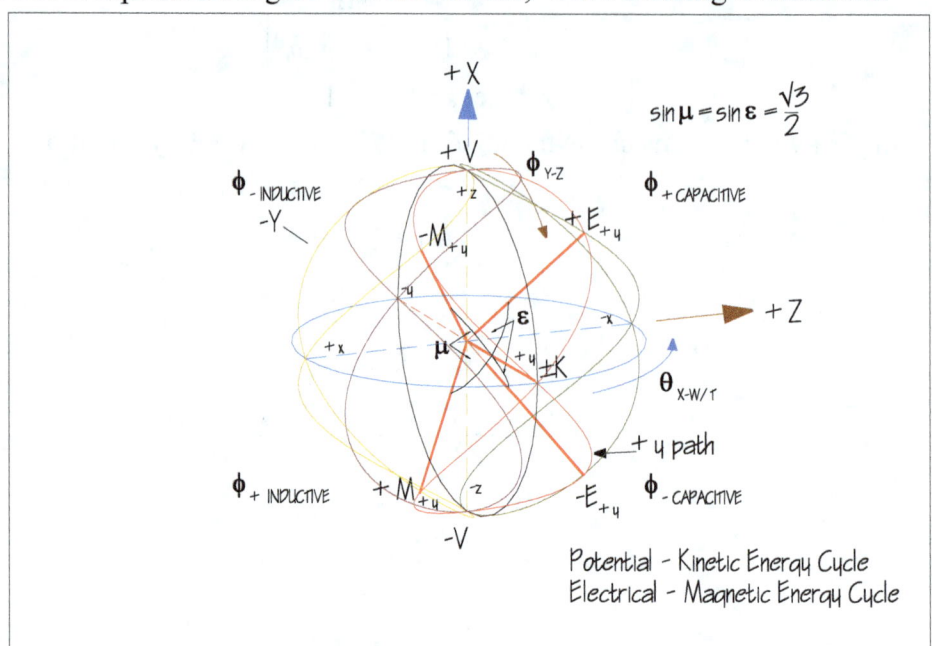

Spin Diagram 1 – Spin Energy Cycle

The initial condition consists of a symmetrical 4-oscillation given by $\left(\frac{V}{C}\right)$ at all h-axes. The initial co-ordinates are given by the upper case $+X$, $+Z$, and $-Y$. The oscillation about the $(+X, +Y, +Z)-(-X,-Y,-Z)$ h-axis, in which $+x$ oscillates from its equilibrium position through $+X$, between $+Y$ and $+Z$, is broken at $\left(\frac{E}{S+C}\right)_3, \left(\frac{S}{C}\right)_3$ as indicated by (2.21), when $+x$ is at $+Y$ by a clockwise rotation about the $+X$ axis to arrive at the position in the spin diagram. From this point the restorative forces begin the rotational oscillation in a counterclockwise sense about $+X$ and the spin vector, S_L.

The heavy red E and M moments show the points of maximum conversion of kinetic to potential energy, E, and potential to kinetic energy, M, for that locus of strain, and are separated in time by $\theta = \omega t = \frac{\pi}{2}$. The angles ε and μ are $\frac{\pi}{3}$, ((2.11) and (2.21) $\left(\frac{E}{S+C}\right)_3$).

The angles between the points on the path, at E and M and the plane, θ, at the midpoint of the line from the system origin or center to the point of maximum kinetic energy, K, are 0.615479709 (see (2.41) $\left(\frac{E}{S+C}\right)_4$) and between M and E moments and the plane, θ, at the center is $\frac{\pi}{4}$ (see (2.48) $\left(\frac{V+E}{S+C}\right)$).

Absent nuclear or other inertial confinement, S_L is not constrained to $+X$ or any other particular direction over time. The rotational oscillation continues at resonant frequency until the conditions found at $\left(\frac{E}{S+C}\right)_4, \left(\frac{S}{C}\right)_4$ precipitate a flip in the M moments and their corresponding inductive torques.

The symmetry between the 3-cube and the 4-cube can be represented by the following orthogonal matrices of space/time permutations involved in the above description when summed over one cycle. As there is inversion symmetry, only half the matrix is shown. The permutations indicate the physical distortion of the medium. The 0 time given below is at the start of an arbitrary cycle and corresponds with $\theta = \omega t = \pi$ in the diagram above (for reason having to do with the history of the generation of the graphics.)

This series of permutations can be carried out by a sequence of ccw $\frac{2\pi}{3}$ rotations about the cubic diagonals as with $\left(\frac{V}{C}\right)$ given by the combination of each diagonal's defining cube sense/axes, facing the cubic center. There is an obvious lack of symmetry in this matrix between all the axes, since +y and +z apparently oscillate over one π, while +x rotates around +z. Note that at any phase, when either +y or +z is at the equilibrium or mid point of its oscillation, the other axis, +z or +y, is at the extremum. This can be seen if we were to start our phases at ω^1. In this case, however, in the rotation column it appears that +y is rotating about +z, and that it is –z and –x that are at e and m respectively. The e and m in the table below merely indicate the condition at $\omega t = 0$, and they alternate with each $\pi/2$ change.

	Phase	rotation	m	e	ccw at
ω^0	$\omega t = 0$	+x	+y	+z	
ω^1	$\omega t = \pi/2$	+y	-z	-x	-x,-y,+z
ω^2	$\omega t = \pi$	-x	+y	-z	+x,-y,+z
ω^3	$\omega t = 3\pi/2$	-y	+z	-x	+x,+y,+z
ω^4	$\omega t = 2\pi$	+x	+y	+z	-x,+y,+z

In actuality, closer analysis shows that both +y and +z, as strains, oscillate between +X and –X, through their initial positions, +Y and +Z, while +x and –x rotate as <u>sustained strain and stress</u> points in the Y-Z plane, twisting so as to maintain continuity with their intital positions at +X and -X.

We can create a corresponding 4-D orthogonal permutation matrix, adding a fourth dimension, w, and representing it with the inversphere.

	Phase	e	m	e	m	ccw at cube
ω^0	$\omega t = 0$	-x	+y	+z	-w	
ω^1	$\omega t = \pi/2$	-w	-z	+x	+y	+x/-y edge
ω^2	$\omega t = \pi$	+x	+y	-z	-w	-x,+z,-y,+w face
ω^3	$\omega t = 3\pi/2$	-w	+z	+x	-y	-x/+y edge
ω^4	$\omega t = 2\pi$	-x	+y	+z	-w	+x,-z,+y,-w face
Results in a 4-D rotation ccw at z/-w edge						

The tie in between the 3-D and 4-D matrices follows, but first we should note a few properties. First, there is no sustained rotation about a diagonal axis, though there appears to be one analogous to that of the 3-D at one of the edges, and all permutations are shown to be oscillations between two ends of an axis through an intermediary axis. These oscillations can be generally described by a rotation of an axis not orthogonal to the other

4 which itself rotates about the center of the cube and through the plane bisecting the cube at the *+x, +y, -x, -y* vertices. The transformation between the ω^3 and $\omega^4 = \omega^0$ phase is analogous in three dimensions to a spatial inversion through the origin or center of the cube followed by a ccw 90° rotation at the bottom face.

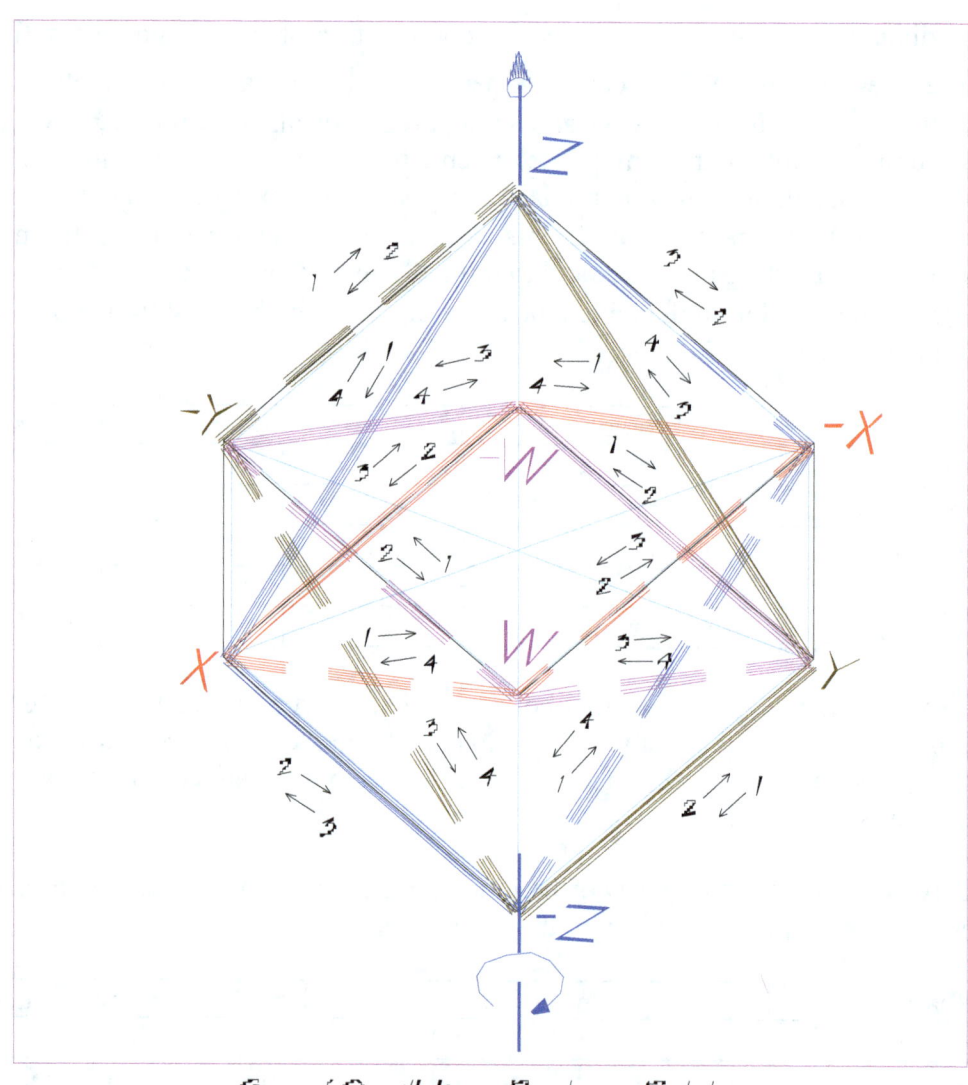

Sum of Oscillations Produces Rotation
1 - CCW at X/-Y Edge
2 - CCW at Top Face
3 - CCW at -X/Y Edge
4 - CCW at Bottom Face

As to the correspondence between the 3 and 4-D matrices, to remove the *w* dimension, we can start with the 4-D form and make the equations shown in the left column, abstracting their correspondence from the matrix.

	Phase	e	m	e	m
$\omega^0 = -w$	$\omega t = 0$	$-x$	$+y$	$+z$	
$\omega^1 = -w$	$\omega t = \pi/2$		$-z$	$+x$	$+y$
$\omega^2 = -w$	$\omega t = \pi$	$+x$	$+y$	$-z$	
$\omega^3 = -w$	$\omega t = 3\pi/2$		$+z$	$+x$	$-y$
$\omega^4 = -w$	$\omega t = 2\pi$	$-x$	$+y$	$+z$	

We then combine the first and last columns to get a table that is essentially the same as the 3-D form above with the important exception of the senses of the *xs* and the ambiguous nature of the first column.

	Phase	e&m	m	e
$\omega^0 = -w$	$\omega t = 0$	$-x$	$+y$	$+z$
$\omega^1 = -w$	$\omega t = \pi/2$	$+y$	$-z$	$+x$
$\omega^2 = -w$	$\omega t = \pi$	$+x$	$+y$	$-z$
$\omega^3 = -w$	$\omega t = 3\pi/2$	$-y$	$+z$	$+x$
$\omega^4 = -w$	$\omega t = 2\pi$	$-x$	$+y$	$+z$

We see, however, that this is associated with a contrary sense of the *w* in the frequency column. Inverting this sense is equivalent to a time reversal. Transposing this sense and converting the first oscillation denoted with an e&m to a rotation, we have the original. The underlying symmetry between the 3-D fundamental quantum <u>rotational</u> oscillation and a 4-D spatial oscillation becomes apparent.

	Phase	rotation	m	e
$\omega^0 = +w$	$\omega t = 0$	$+x$	$+y$	$+z$
$\omega^1 = +w$	$\omega t = \pi/2$	$+y$	$-z$	$-x$
$\omega^2 = +w$	$\omega t = \pi$	$-x$	$+y$	$-z$
$\omega^3 = +w$	$\omega t = 3\pi/2$	$-y$	$+z$	$-x$
$\omega^4 = +w$	$\omega t = 2\pi$	$+x$	$+y$	$+z$

Two items should be mentioned concerning this development with respect to the standard model. The first concerns the manner in which this description might be consistent with the quark model. It is obvious that there is an internal spin structure in the nature of the nodes, antinodes, and various moments and torques of the wave strain. We will see that an analysis of these features reveals a fractional charge, and that the phenomenology of quark confinement is the ontology of wave nodes and antinodes. The second concerns the other two flavors of particle families, generally centered around their leptons, the tau and muon. Since there is strong evidence that these last two particles mutate, specifically in the case of solar radiation, the assumption here is that they are relativistic products of beta decay from the same fundamental rotational oscillation, the neutron.

Dynamic Functions

To understand the dynamic functions of the above oscillation, we can start by examining the functions of an ideal string in "jump rope" oscillation, i.e. a standing wave of one half wavelength, shown below. The subscript noughts in the parameters indicate their fundamental characteristic values, which are properties of the resonance of the wave bearing medium. The oscillation in this case is of Simple Harmonic Motion, so there are no harmonic overtones traveling along the string, which is of some linear inertial density, λ_0, and some tension stress force, τ_0, given by

$$\tau_0 = f_0 \cdot A_0 \tag{2.64}$$

where f_0 is the tension stress and A_0 is the unit cross sectional area. This is related to the resonant angular frequency, ω_0, and commensurate angular wave number, κ_0, and velocity of the strain oscillation, c_0, as

$$\lambda_0 = \frac{1}{c_0^2}\tau_0, \text{ where } c_0 = \frac{\omega_0}{\kappa_0} = \frac{\partial \theta}{\partial t}\frac{\partial x}{\partial \theta} = \frac{\partial x}{\partial t} \tag{2.65}$$

When normalized,

$$c_0 = \frac{\partial x_0}{\partial t_0} = \frac{\partial t_0}{\partial x_0} = 1 \tag{2.66}$$

and in an isotropic space, for a unit value of x_0,

$$|x_0| = |\hat{\mathbf{r}}| = r_0 \tag{2.67}$$

Therefore, what may not be so obvious, but assuming the units of distance r, in A and in κ are the same, for $\theta = 1$,

$$\lambda_0 \omega_0^2 = \tau_0 \kappa_0^2 = f_0 \tag{2.68}$$

Also, of eventual interest, the mechanical impedance, Z_0, of the string is

$$Z_0 = \lambda_0 c_0 = \frac{\tau_0}{c_0} \tag{2.69}$$

and the power, P_0, transported by the wave, if it is a traveling wave, and hence retained if it is standing is

$$P_0 = \lambda_0 c_0^3 = Z_0 c_0^2 = \tau_0 c_0 \tag{2.70}$$

We can represent the oscillation with the Euler identity, using both the real and imaginary parts, as a complex standing wave, ϕ, where $\theta = \kappa x \pm \omega t$, and $\kappa x = 0$,

$$\phi = \eta + \varsigma = A(\cos\theta \pm i\sin\theta) = Ae^{i\theta} \tag{2.71}$$

so that for any time, t, and where A is a real amplitude, (not the cross-sectional area A_0 above) equal to the maximum radius of the string path and modulus of the complex polar form, r,

$$\begin{aligned}\eta &= A\cos\omega t \\ \varsigma &= iA\sin\omega t\end{aligned} \tag{2.72}$$

The ambiguity of the rotational sense is once again used, since in the case of a conservative or closed SHM where there is no damping or loss of energy of the system, time is cyclical and reverses with each half cycle of the oscillation. This also reflects the fact that what is clockwise on one side of the path is counterclockwise on the other, the latter of which is shown below.

We will stipulate that the phasing of ϕ and θ remain synchronized so that at time $t = 0$, the oscillation is at the top of the cross section on the right, at $\eta = A$. We can imagine that the cross section of the string, indicated by the small round circle at the end of the path radial, echos the path shown in blue below, maintaining the same orientation as the path, and if we imagine the path to represent the string cross section, that a density gradient resulting from the differential stretching of the string points along the radial from the small red circle through the center of the blue to its opposite side in the direction of increasing density. Thus, a two dimensional entity confined in awareness to the cross section and whose orientation is locked in any arbitrary y direction, parallel to η, would notice a rotation of the density gradient and might think that it is he that is rotating or spinning about with respect to the direction of the gradient. Thus to him the stress rotates while the strain oscillates in two orthogonal dimensions. A one dimensional being would simply feel the stress oscillate between two poles.

We can next think of the cross section of the string as reaching toward infinity, so that the displacement of the string is instead seen to be a lateral or transverse strain of the continuum in oscillation in two dimensions from and about its position of rest at the center of the path.

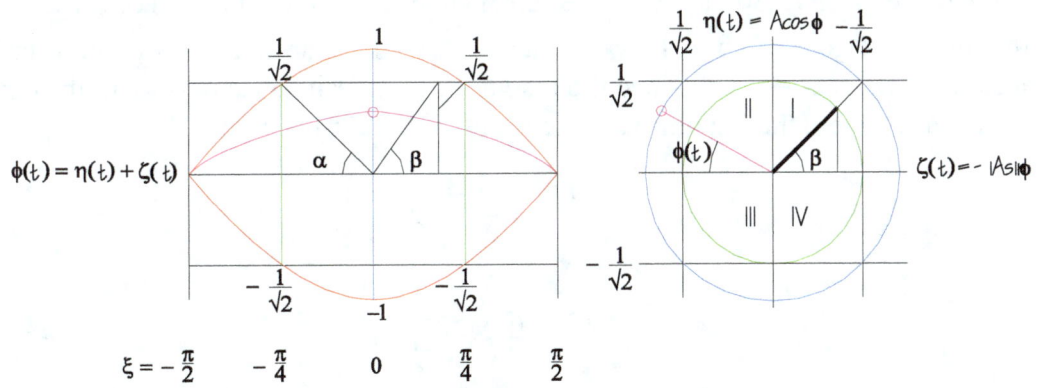

Wave Diagram 3 – Simple Harmonic Oscillation of ϕ

The usual kinematic functions of the oscillation can be given for ϕ, just as it was for η in Wave Diagram 2 above, assuming the ccw rotation as shown above, as

Displacement, r	$\phi(t) = Ae^{i\omega t}$	(2.73)
Velocity, c	$\phi'(t) = i\omega Ae^{i\omega t}$	(2.74)
Acceleration, a	$\phi''(t) = -\omega^2 Ae^{i\omega t}$	(2.75)
Jerk, j	$\phi'''(t) = -i\omega^3 Ae^{i\omega t}$	(2.76)

Due to our stipulation of conservation of energy, in a normalized system, where $d\omega = 0$ and $\omega = 1$, we have the following equivalence of the integral, $\Phi(t)$, and j

$$\Phi(t) = \phi'''(t) = -i\omega^3 A e^{i\omega t} = \frac{A e^{i\omega t}}{i\omega} \quad (2.77)$$

Their dynamic counterparts are:

Constant of Inertia, η(tav) $\quad {}_0^0\eta_0^0 \equiv m\phi(t) = mAe^{i\omega t} = \eta \quad (2.78)$

Transverse momentum, $G_0 \quad \eta_1 \equiv m\phi'(t) = m(i\omega A e^{i\omega t}) = i\eta\omega \quad (2.79)$

Transverse wave force, $\tau_0 \quad \eta_2 \equiv m\phi''(t) = m(-\omega^2 A e^{i\omega t}) = -\eta\omega^2 \quad (2.80)$

Transverse wave yank, $Y_0 \quad \eta_3 \equiv \eta^1 \equiv m\phi'''(t) = m(-i\omega^3 A e^{i\omega t}) = -i\eta\omega^3 = \frac{\eta}{i\omega} \quad (2.81)$

Some explanation is in order. Given a fixed ω_0 according to (2.65) and a fixed m, it is apparent that these functions are invariants of the system. (2.80) appears to be an expression of Hooke's law of force for simple harmonic motion generally given as

$$F = -(m\omega^2)\eta(t) = -k_{spring}\eta(t), \therefore -k_{spring} = -(m\omega^2) \quad (2.82)$$

where the spring constant, k_{spring}, includes a constant value for the mass, m. In this case, ϕ is substituted for η, with its magnitude a constant as a complex modulus, its instant direction given by the argument. Thus it is a scalar invariant, as is τ_0. Since the only difference between (2.80) and the other three functions are the powers of ω_0 and the $\frac{\pi}{2}$ directional change given by i, they are themselves all scalar invariants under rotation and translation. Since $A = r$, and since the displacement of r in ϕ_{yz} is normal to the direction of κ_x for any time t, the imaginary sense is entered, (2.78) becomes

$$\eta = mr = \frac{m}{\frac{\partial \theta}{\partial r}} = \frac{m}{i\kappa} = \frac{m_0}{i\kappa_0} \quad (2.83)$$

and

$$m_0 = i\eta\kappa_0 = \frac{\eta}{Ae^{i\kappa x}} \quad (2.84)$$

which we can substitute the last term back into (2.79), (2.80), and (2.81), resulting in the final term of each. For an arbitrary fixed reference frame, $Ae^{i\omega t}$ varies sinusoidally, but for a frame rotating at ω_0, with the complex modulus, is constant and at unity when divided by $Ae^{i\kappa x}$.

Thus mass is essentially the inverse wave modulus and the transverse wave number. (This latter should not be confused with what we might call the "amplitude wave number", $A^{-1} = r^{-1}$, which for the fundamental oscillation is the same as the wave number. Thus the neutron is an essentially spherical wave form, while the electron can be modeled in one of two alternatives as an extremely thin prolate spheroid wave form transmitted by the central oscillation at beta decay, whose major axis, $\pi\kappa_e^{-1}$, exceeds its minor axis, r_e, essentially by their ratios of $1.334775525...\times 10^6$. The neutron is then transformed into an oblate spheroid wave form of the proton, whose major axis exceeds its minor by a factor of 1.000000895.... This will be shown later)

The left hand identity of each is a compact and convenient notation for each invariant, $_0^0\pi_0^0$, where the subscript indicates complex differentiation and the superscript indicates complex integration, the right hand side with respect to time and the left hand side, not yet used, for the same calculus with respect to space. We can show these invariant functions graphically, and with the sense omitted, as

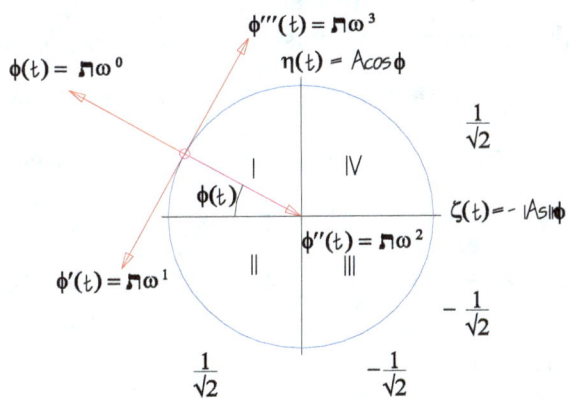

Wave Diagram 4 – Dynamic Functions of $\phi(t)$

The above discussion on the derivation of the wave form indicated that the rotational oscillation resulted after a $\frac{2\pi}{3}$ rotation about one of the h-axes with $\left(\frac{V}{C}\right)$, which results in an orthogonal rotation of all six cubic axes. This is reflected in the imaginary sense of (2.83). Thus we can differentiate with respect to space and get the following. As with the rotations with respect to time, the rotations of space about the h-axes with $\left(\frac{V}{C}\right)$ are cyclic, so that the first order of integration with respect to space is equal to the third order of differentiation.

Constant of Inertia, π $\qquad$ $_0^0\pi_0^0 \equiv m\phi(x) = \dfrac{\pi}{Ae^{i\kappa x}} Ae^{i\kappa x} = \pi$ $\qquad$ (2.85)

Oscillation mass, m_0 $\qquad$ $_1\pi \equiv m\phi'(x) = \dfrac{\pi}{Ae^{i\kappa x}}\left(i\kappa Ae^{i\kappa x}\right) = i\pi\kappa$ $\qquad$ (2.86)

Linear inertial density, λ_0 $_2\pi \equiv m\phi''(x) = \dfrac{\pi}{Ae^{i\kappa x}}\left(-\kappa^2 Ae^{i\kappa x}\right) = -\pi\kappa^2$ $\qquad$ (2.87)

Moment of inertia, I_0 $_3\pi \equiv {}^1\pi \equiv m\phi'''(x) = m\Phi(x) = \dfrac{\pi}{Ae^{i\kappa x}}\left(-i\omega^3 Ae^{i\kappa x}\right) = -i\pi\kappa^3 = \dfrac{\pi}{i\kappa}$ (2.88)

We can complete this picture by creating an orthogonal 4x4 matrix of the functions, using the inertial constant notation, in the second iteration of which we substitute integration for the third order of differentiation for both time and space, making .

$$\begin{bmatrix} {}_0\daleth_0 & {}_0\daleth_1 & {}_0\daleth_2 & {}_0\daleth_3 \\ {}_1\daleth_0 & {}_1\daleth_1 & {}_1\daleth_2 & {}_1\daleth_3 \\ {}_2\daleth_0 & {}_2\daleth_1 & {}_2\daleth_2 & {}_2\daleth_3 \\ {}_3\daleth_0 & {}_3\daleth_1 & {}_3\daleth_2 & {}_3\daleth_3 \end{bmatrix} = \begin{bmatrix} {}^1\daleth^1 & {}^1\daleth & {}^1\daleth_1 & {}^1\daleth_2 \\ \daleth^1 & {}^0_0\daleth^0_0 & \daleth_1 & \daleth_2 \\ {}_1\daleth^1 & {}_1\daleth & {}_1\daleth_1 & {}_1\daleth_2 \\ {}_2\daleth^1 & {}_2\daleth & {}_2\daleth_1 & {}_2\daleth_2 \end{bmatrix} \quad (2.89)$$

More conventionally, this second statement becomes the following, where we have reiterated the power function as conventionally stated on the right. It is of interest to note the orthogonal nature of the matrix by the sense of each function, and that the inertial constant is inherently an imaginary invariant, so that making its sense explicit would effectively rotate all the functions clockwise $\frac{\pi}{2}$, from their positions as follows.

$$\begin{array}{llll}
P_0 = -\daleth\kappa_0^{-1}\omega_0^{-1} & I_0 = -i\daleth\kappa_0^{-1} & \hbar = \daleth\kappa_0^{-1}\omega_0 & E_0 = i\daleth\kappa_0^{-1}\omega_0^2 \quad \overline{\left(P_0 = -\daleth\kappa_0^{-1}\omega_0^3\right)} \\
Y_0 = -i\daleth\omega_0^{-1} & \daleth\left(=\frac{m_0}{i\kappa_0}\right) & G_0 = i\daleth\omega_0 & \tau_0 = -\daleth\omega_0^2 \\
\frac{m_0}{\omega_0} = \daleth\kappa_0\omega_0^{-1} & m_0 = i\daleth\kappa_0 & Z_0 = -\daleth\kappa_0\omega_0 & \tau_1 = -i\daleth\kappa_0\omega_0^2 \\
Y_2 = i\daleth\kappa_0^2\omega_0^{-1} & \lambda_0 = -\daleth\kappa_0^2 & G_2 = -i\daleth\kappa_0^2\omega_0 & f_0 = \daleth\kappa_0^2\omega_0^2
\end{array} \quad (2.90)$$

The additional 9 functions are:

Mechanical Impedance, Z_0
$$_1\daleth_1 \equiv -\daleth\kappa_0\omega_0 \quad (2.91)$$

Transverse Momentum Surface Density, G_2
$$_2\daleth_1 \equiv -i\daleth\kappa_0^2\omega_0 \quad (2.92)$$

Planck's Quantum of Action, $\hbar$
$$^1\daleth_1 \equiv \daleth\kappa_0^{-1}\omega_0 = \daleth c_0 \quad (2.93)$$
(Spin Angular Momentum)

Linear Transverse Force Density, τ_1
$$_1\daleth_2 \equiv -i\daleth\kappa_0\omega_0^2 \quad (2.94)$$

Wave Stress, f_0
$$_2\daleth_2 \equiv {}^2\daleth^2 \equiv {}^2\daleth_2 \equiv {}_2\daleth^2 \equiv \daleth\kappa_0^2\omega_0^2 \quad (2.95)$$

Spin Energy, E_0
$$^1\daleth_2 \equiv i\daleth\kappa_0^{-1}\omega_0^2 \quad (2.96)$$

Mass Frequency Ratio, $\frac{m_0}{\omega_0}$
$$_1\daleth^1 \equiv \daleth\kappa_0\omega_0^{-1} = \frac{\daleth}{c_0} \quad (2.97)$$

Yank Surface Density, Y_2
$$_2\daleth^1 \equiv i\daleth\kappa_0^2\omega_0^{-1} \quad (2.98)$$

Wave Power, P_0
$$^1\daleth_3 \equiv {}^1\daleth^1 \equiv -\daleth\kappa_0^{-1}\omega_0^3 \quad (2.99)$$

(Yank Volume Density, Y_3)
$$_3\daleth_3 \equiv -\daleth\kappa_0^3\omega_0^3 \quad (2.100)$$

These derivative functions for a fundamental rotational oscillation, the neutron, are invariant functions of the resonant frequency, ω_0, and wave number, κ_0, of the continuum and not of any linear dimension of space or time. These latter two time and space parameters serve to gauge the interaction of the various functions and in fact to set the gauge, which is also the basis for the metric, for space and time itself. We can show this in greater detail in the orthogonal matrix that follows. The functions are instantaneous vectors, which together form a rotational tensor or spinor. The functions with asterisks, all of which are of real sense, are primary invariants of the system. Thus, while all are invariant with respect to the fundamental oscillation, with beta decay, for example, E and m for the central oscillation changes, though not for the system as a whole, which is conservative. The primary invariants with multiple asterisks can be more readily seen as properties of the continuum itself, with the single asterisks indicating invariants of oscillation, but all are rightly seen as spin potentials of the continuum. The primacy of the multiples is seen in (2.68) which is an expression of these four describing the necessary conditions for wave motion. Thus

$$\lambda_0 \equiv \mathsf{n}\kappa_0^2 = \frac{1}{\frac{\omega_0^2}{\kappa_0^2}}\mathsf{n}\omega_0^2 \equiv \frac{1}{\frac{\omega_0^2}{\kappa_0^2}}\tau_0 = \frac{f_0}{\omega_0^2}, \qquad (2.101)$$

while generally unrecognized is

$$\mathsf{n} \equiv \frac{1}{\kappa_0^2 \omega_0^2} f_0 \qquad (2.102)$$

or diagrammatically

$$\begin{array}{ccc} \mathsf{n} & \omega_0^2 & \tau_0 \\ \updownarrow \kappa_0^2 \quad \kappa_0^2\omega_0^2 \searrow\nearrow \frac{\omega_0^2}{\kappa_0^2} & & \updownarrow \kappa_0^2 \\ \lambda_0 & \omega_0^2 & f_0 \end{array} \qquad (2.103)$$

A similar condition with respect to the action, impedance, power and mass/frequency ratio is

$$\begin{array}{ccc} \hbar & \omega_0^2 & P_0 \\ \updownarrow \kappa_0^2 \quad \kappa_0^2\omega_0^2 \searrow\nearrow \frac{\omega_0^2}{\kappa_0^2} & & \updownarrow \kappa_0^2 \\ Z_0 & \omega_0^2 & \frac{\mathsf{n}}{c_0} \end{array} \qquad (2.104)$$

Among the secondary invariants, those whose magnitudes for a given particle vary from the fundamental, but whose relationships are still gauged as with the primary invariants, we have one involving the well known equation of Einstein,

$$\begin{array}{ccc} I_0 & \omega_0^2 & E_0 \\ \updownarrow \kappa_0^2 \quad \kappa_0^2\omega_0^2 \searrow\nearrow \frac{\omega_0^2}{\kappa_0^2} & & \updownarrow \kappa_0^2 \\ m_0 & \omega_0^2 & \tau_1 \end{array} \qquad (2.105)$$

In fact, the coupling between τ_0 and f_0 gauges the gravitational interaction, as the quantum of gravity is given

$$\tau_0 = -A_0 f_0 = -A_0 \frac{T_0}{6\sqrt{3}} = T_0 \frac{1}{6\sqrt{3}(-i\kappa)_0^2} \qquad (2.106)$$

$$G_q = \frac{d\tau_0}{dT_0} = -\frac{1}{6\sqrt{3}\kappa_0^2} \qquad (2.107)$$

where, by virtue of the spin mechanics of $\left(\frac{V}{C}\right)$, the isotropic 4-stress, T_0, operates on one pair of diagonal h-axis, at an angle of 0.615479..., relative to the six cubic faces (and normal to any three ortho-normal co-ordinates of a 3-space,) so that

$$T_0 = 6\sqrt{3}f_0 = 6\sqrt{3}\tau_0 A_0^{-1} \text{ and} \tag{2.108}$$

$$A_0 = 6\sqrt{3}\tau_0 T_0^{-1}. \tag{2.109}$$

We can state the relationship between the spin functions and that 4-stress as

$$\kappa\daleth_\omega = \daleth_{\mu\nu} = 4\pi\left(2T_{\mu\nu}\right) = 8\pi T_{\mu\nu} \tag{2.110}$$

Here $2T_{\mu\nu}$ is the 4-stress correspondence of the 3-strain component in (2.59)-(2.62), and 4π integrates the surface stress oscillation over a spherical surface over one cycle. Since $\daleth_{\mu\nu}$ is a geometrically defined set of functions as developed above, this is a background-independent quantum solution to the field equations of general relativity.

The following orthogonal matrix is the multi-faceted jewel of this rotational oscillatory system. In addition to the above functions, there are two others shown in the bottom two spots of the next to the rightmost column. Y, not to be confused with yank, Y, is the tension or Young's modulus of elasticity of the continuum, while the function under it is the inverse of the Planck area. Notice that it is gauged by the same general derivative, κ_0^2, with the stress f_0, as the stress is with the transverse, and in the case of the rotational oscillation, central wave force, τ_0. Thus the Planck area, from (2.109), is expressed as a derivative of area with respect to stress, similar to gravity at (1.23d) or

$$A_{Planck} = \frac{dA_0}{dT_0} = -\frac{6\sqrt{3}\tau_0}{\left(6\sqrt{3}\right)^2 f_0^2} = \frac{\daleth\omega_0^2}{6\sqrt{3}\daleth^2\kappa_0^4\omega_0^4} = \frac{\daleth}{6\sqrt{3}\daleth^2\kappa_0^6 c^2} = -\frac{G_q}{\lambda_0^2}\frac{\daleth}{c^2} = -G_N\frac{\hbar}{c^3}\frac{1}{dT_0} \tag{2.111}$$

Of similar interest is the relationship of the mechanical impedance with respect to Planck's quantum as

$$\hbar = 6\sqrt{3}Z_0 G_q. \tag{2.112}$$

Finally, the functions $\hbar$ and G are gauged by κ_0, where G_0 is related to the quantum of charge, e, by

$$G_0 = i\kappa_0 \hbar = i\daleth\omega_0 = \frac{\pi}{1.002406...}e. \tag{2.113}$$

thus

$$\frac{1.002406...}{\pi}\kappa_0 = -i\frac{e}{\hbar} = -i\frac{e}{\daleth c}. \tag{2.114}$$

The ratio of the gravitational quantum and the charge quantum, then is

$$\frac{e}{G_q} = i\kappa_0 Z_0 \frac{6\sqrt{3}(1.002406...)}{\pi}$$

$$= i\daleth\kappa_0^2\omega_0 \frac{6\sqrt{3}(1.002406...)}{\pi} \tag{2.115}$$

$$= G_2 \frac{6\sqrt{3}(1.002406...)}{\pi}$$

The invariant functions, then, are seen to be invariant differentials and the various instances of $i\kappa^n$ and $i\omega^n$, are their co-variant derivatives. Related to this is the fact

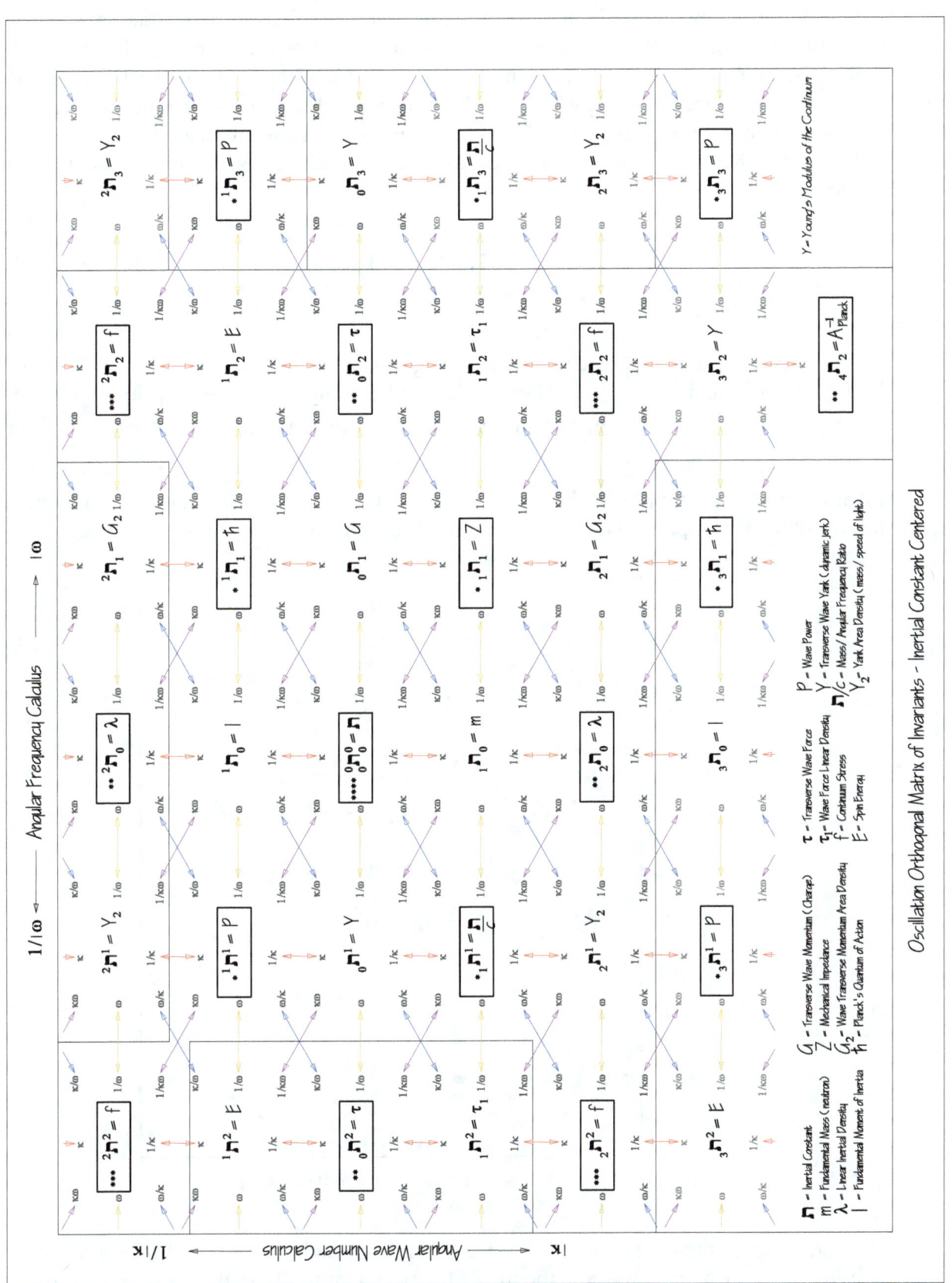

Matrix of Invariants

that, as indicated by the prefix noughts, the fundamental values of these functions constitute a system of natural dimensional unit values. Thus we have the following SI equivalents for the derivatives, imaginary sense omitted, along with the inertial constant, where θ is understood to be 1, which are the fundamental units of space and time, though they are obviously not quantum in the conventional sense of being discrete and indivisible. They represent, instead, the classical foundational parameters of the quantum world.

$$\daleth = 3.51767...\times 10^{-43} \ kg \cdot m/\theta \qquad (2.116)$$

$$x_0 = \kappa_0^{-1} = 2.10019...\times 10^{-16} \ m/\theta \qquad (2.117)$$

$$t_0 = \omega_0^{-1} = 7.00549...\times 10^{-25} \ s/\theta \qquad (2.118)$$

This matrix can also be represented to advantage by the following spin charts. The first, labeled, Right Hand View, Right Hand Rule, can be thought of as being prior to a strain and hence oscillation and is marked as at time $t = 0$. All time functions of $\daleth\omega_0^n$ are fixed in the same direction and therefore in phase and all space functions of $\daleth\kappa_0^n$ are arrayed together in sequential order with a counterclockwise rotation of the cycle starting at the left or 9:00. This represents a spin potential, analogous in some respects to a vector potential.

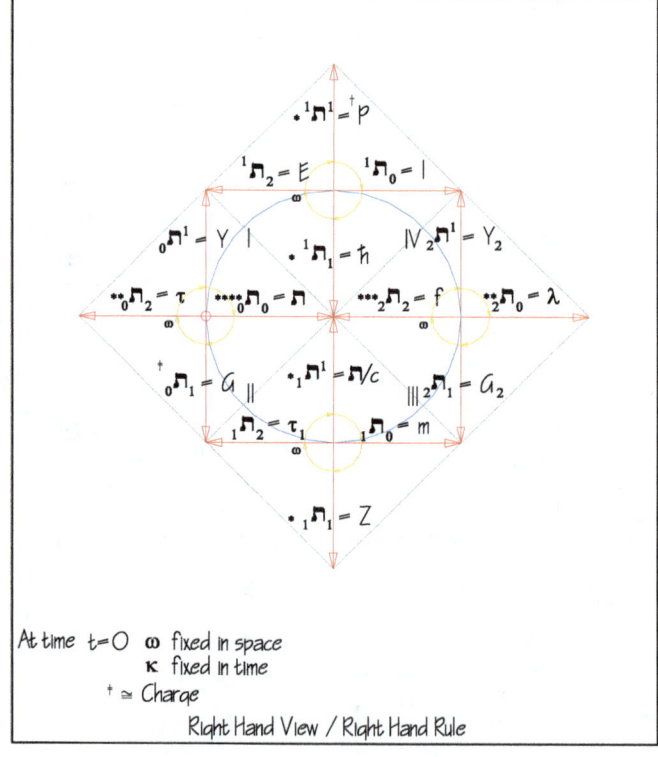

Spin Chart 1 – t = 0, RHV/RHR

We can imagine an X cut along the 45 degree lines through the center of the circle which is lying in an x-y plane. (Assume the top of the page is initially in the +x direction for reasons that will become clear.)

1. Pull the bottom arrow head of γ_c, which is pointing up toward +x at the center point, toward the –z direction, down out of the page, and pull the top arrow head of $\hbar$, which is pointing down at the center point, toward +z, so that the circle is now in the y-z plane, facing in the –x direction.
    a. Now, the vectors $\hbar$ will be pointing up and P down, Z up and γ_c down (and facing away from you) all having undergone a +i or $\pi/2$ ccw directional rotation about their group source or origin (viewed toward the center from the +y direction).
    b. This transformation performs an axial +i rotation of the remaining vectors of their groups, E and I for ^{1}ה and τ_I and m for $_1$ה, along with a simultaneous cw orbital rotation about the y axis.
    c. This also produces a -i axial rotation of τ-ה of $_0$ה and f-λ of $_2$ה and a simultaneous directional -i rotation of Y-G and Y_2-G_2 about the y axis.
   As a result, all the vectors will have undergone either an axial, axial/orbital or a directional rotation.
2. Simultaneous with (1), but second in a non-commutative order, in group $_0$ה pull the arrow head of ה which is pointing right at the center point, toward the -z direction (down out of the page) and in group $_2$ה on the right pull the arrow head of f, which is pointing left at the center point, toward the +z direction. The tendency or differential effect of this transformation, which is similar to (1), is to place the circle in the x-z plane, but this is in conflict with (1) which puts the circle in y-z. Since $\hbar$ is the spin angular momentum vector, we will give it a differential precedence, i.e. first in a non-commutative order, but after the full $\pi/2$ directional rotation of (1) and of $_0$ה and $_2$ה about Y-G and Y_2-G_2, we end up with
    a. The transformation of (1.a-c) rotated –i about the z axis (viewed once again toward the origin from the +z direction) into the x-z plane.
    b. Now, ה points outward toward +x and τ inward toward -x, and λ points inward toward +x and f points outward toward -x.
    c. This produces a +i axial rotation of Y-G and Y_2-G_2 along with a –i orbital rotation about the z = $\hbar$- γ_c axis.
3. The result is that all vectors of the groups undergo both a directional and an axial rotation, though the differential precedence or sequence varies. For the primary invariants, denoted by asterisks, of groups ^{1}ה and $_1$ה the non-commutative sequence is (1)directional-(2)axial, while for groups $_0$ה and $_2$ה the sequence is (1)axial-(2)directional. For the remaining components which meet at the corners of the square, there is also an orbital rotation along with the axial. The sequence is the same as for the primary invariants of their group. The result is Spin Chart 2, the diagram of time $t = \mathbf{n2\pi}$ which is in the x-z plane. Note that the primary invariants are all rotated in time with respect to diagram $t = \mathbf{0}$.

With an actual kinetic model, in the initial condition, the **RHV/RHR** label on each group square will be facing out, all right side up. In the final condition the model would be viewed on edge, with the right hand labels all facing to the right, but with the top and bottom square labels inverted and the left hand labels facing left, and similarly oriented.

To complete the symmetry of the picture we can also show a <u>L</u>eft <u>H</u>and <u>V</u>iew, <u>L</u>eft <u>H</u>and <u>R</u>ule in Spin Chart 3 that would be applied to the back of the **RHV/RHR** diagram. It can readily be seen that if you place the **LHV/LHR**, centered on the back of **RHV/RHR** and invert it 180 degrees, all the functions will line up back to back with each other. However, as we have stipulated that Spin Chart 3 follows the left hand rule, we will not invert the diagram, so that after the above operations are performed, the spin vectors are correct, so that $\hbar_{RHR} = -\hbar_{LHR}$ and their physical rotation representation is the same, i.e. $\hbar_{RHR}$ is ccw and $\hbar_{LHR}$ is cw when viewed from the vector arrow toward the center of the configuration.

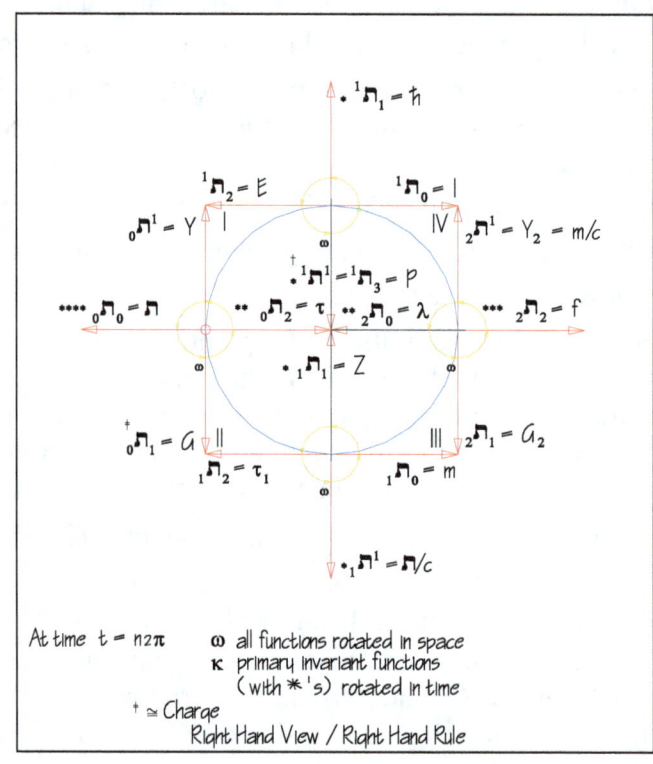

Spin Chart 2 – t = n2π, RHV/RHR

The result of this juxtaposition shows that Spin Chart 2 corresponds with the condition in Spin Diagram 1, at the +y position at the cross of the +y path. Spin Chart 4 corresponds with the similar position on the –y path. Both charts are viewed from the direction toward which θ is rotating at the +y and –y crossing. This corresponds with the condition found at Figure 2, in the section on EM waves.

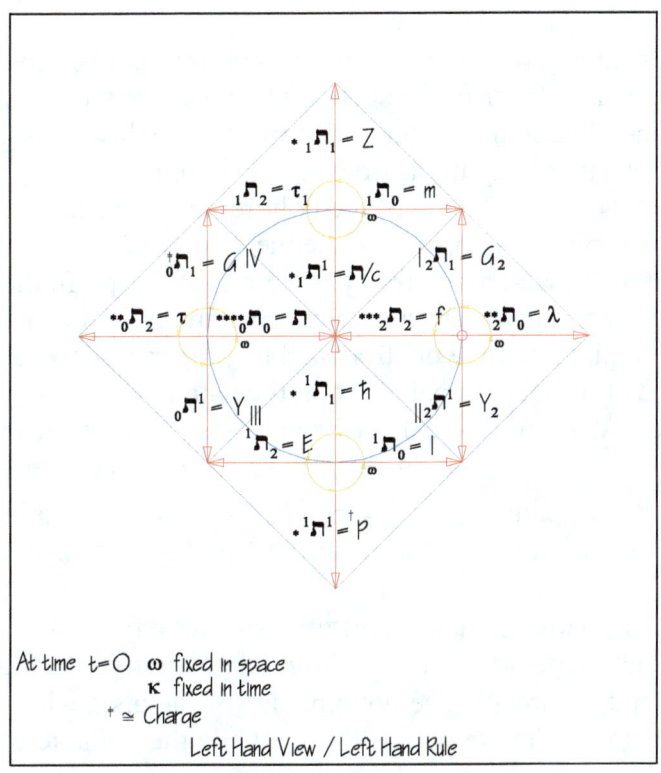

Spin Chart 3 – t = 0, LHV/LHR

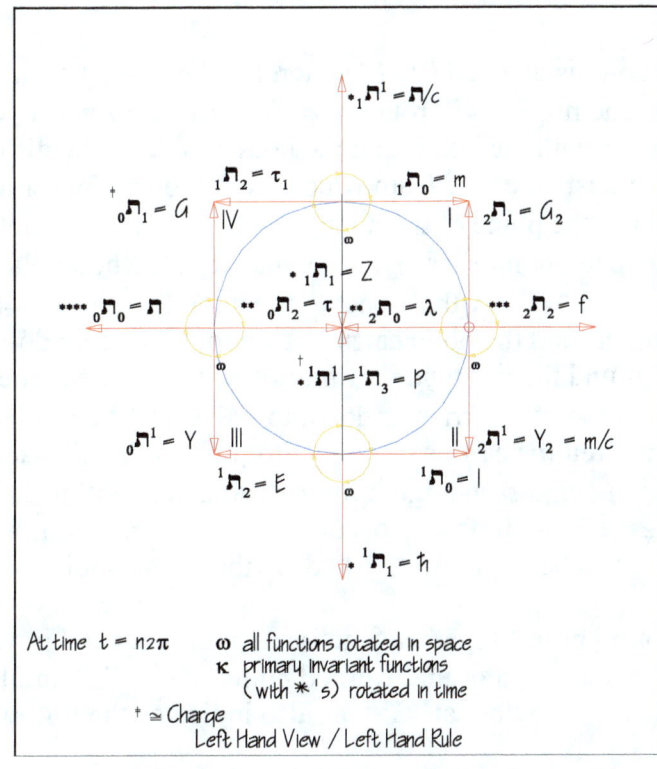

Spin Chart 4 – t = n2π, LHV/LHR

Quantum Rotational Oscillation or Spin

We can create a simple physical model of this condition, albeit in 2-D mode, using a flexible drumhead or a planar frame such as a crochet hoop over which we stretch a thin, flexible membrane. Pierce the center with a machine bolt with washers, head up and tightened so that it cannot slip in the opening, and preferably of greater length than the radius of the hoop. Fold the length of the bolt, representing the X axis, up against the bottom of the membrane or hoop and rotate the distal end around the hoop's circumference, allowing the bolt to slip and rotate freely in your loose grip. In the following, the lower case letters, x, y and z, refer to physical points originally in the undisturbed reference plane but now subject to displacement or strain and the upper case to the physical reference frame, X, Y and Z or to a functional point on the path of the strain, K, V, M, E, and W. In these graphics, the X, Y and Z axes have been rotated $+i$ from their customary positions, which amounts to a $\frac{2\pi}{3}$ ccw rotation about the XYZ diagonal. Thus $+V_{-z}$ indicates point $-z$ displaced to functional point $+V$. $+K_{-z}$ indicates $-z$ at K moving in the positive direction, toward $+V$. We will reference Wave Diagram 1 in this discussion.

The disk, ϕ, (+, originally up) and (−, originally down), formed by the bolt head and washer will now be roughly perpendicular to its original position in the plane of the membrane and hoop, θ, (+, up) and (−, down). The bottom edge of the disk, which we will call point $-z$, will be depressed to a point we will call $-V_{-z}$, while the point across the diameter at $+z$ will be elevated to $+V_{+z}$. The point $-y$ clockwise $\frac{\pi}{2}$ from $-V_{-z}$ on the edge of the disk and in the plane of the membrane, we will designate at $-iK_{-z} = K_{-y}$, and the diametrically opposed point $+y$ on the disk we will call at $+iK_{-z} = -iK_{+z}$.

As we move the bolt distal end $+i\theta_+$ (counterclockwise ½ π) around the hoop, the bolt head and adjacent membrane will rotate $-i\phi_+$ (clockwise ½ π)), as would be indicated by an axial vector pointing with the bolt along the bottom side of the distorted membrane. The sense subscripts indicate the side from which the rotation is observed. Point $-z$ will move up in the direction of the plane of the hoop, $-i\phi_+$, but $+i\theta_+$ to point $+K_{-z}$ in the plane of the hoop. Simultaneously, point $+z$ moves down toward the hoop plane, once again $-i\phi_+$, with respect to the disk and $+i\theta_+$ with respect to the hoop to point $-K_{+z}$. Another ccw ½ π rotation of the bolt around the hoop carries $-z$ up to $+V$ and $+z$ down to $-V$, once again with cw motion around the disk and ccw motion with respect to the hoop and $+V$. A third ccw ½ π rotation of the bolt carries $-z$ down to $-K_{-z}$ and $+z$ up to $+K_{+z}$ in the Y-Z plane. A fourth such rotation returns carries $-z$ and $+z$ to their initial displaced positions along the X axis. Note that both θ and ϕ undergo a full rotation, yet continuity is maintained with no twisting up of the medium, in fact, it occurs because the medium will not allow such indefinite distortion. The matrix in (2.61) describes this condition.

Note that this motion brings together functional point $-iK_{-z}$ and physical point $-z$ from functional point $-V_{-z}$ to a rest or undisturbed point $-Z = +K_{-z}$, simultaneously with $+z$ to $+Z$ and $+V_{+z}$ and $-iK_{+z}$ together at $-K_{+z}$. It also indicates the following conceptual spatial and time identities

$$-Y \equiv \left[K_{-y}(x) = -i_\theta K_{-z} = +i_\theta K_{+z} \right] \equiv \left[K_{-y}(t) = +K_{-y}, -K_{-y} \right] \quad (2.119)$$

and
$$+Y \equiv \left[K_{+y}(x) = -i_\theta K_{+z} = +i_\theta K_{-z} \right] \equiv \left[K_{+y}(t) = -K_{+y}, +K_{+y} \right]. \quad (2.120)$$

Clearly, $-V_\phi$ and $+V_\phi$ are two stable or observationally <u>static functional</u> points comprised of the constant <u>physical flux</u> of ϕ over one rotation of θ, while $-K_\phi$ and $+K_\phi$ form two iterations of a <u>moving functional</u> K_θ path around the $-V_\phi - +V_\phi$ axis in the θ plane comprised of the successive points of ϕ while at momentary <u>physical rest</u>. Thus, each point of ϕ coincides with a unique V once and a unique K twice with each cycle of θ. In terms of simple harmonic motion, the potential energy function of V, which we might surmise is also the energy of quantum static electrical charge or the electric field, is comprised of the continuum's sustained displacement-strain created by its oscillatory motion, while the kinetic energy function of K, the quantum manifestation of the energy of a magnetically induced current, is comprised of the sustained maximum momentum created by continuous change in rest position.

This is in keeping with Wave Diagram 2 in which the position of zero displacement corresponds with the instance of greatest momentum and zero acceleration and force. It is also what (2.101) would appear to indicate, since λ_0 is the inertial density, a function of position, $\phi''(x)$, and τ_0 is the tension force, a function of time, $\phi''(t)$, of the spacetime continuum, as

$$_2\Pi_2 \equiv \lambda_0 \omega^2 \equiv -\left(\Pi \kappa_0^2\right)\omega^2 = -\left(\Pi \omega_0^2\right)\kappa^2 \equiv \tau_0 \kappa^2 \equiv {_2\Pi_2} \quad (2.121)$$

Wave Diagram 1 assumes a point $\omega t_1 < \frac{1}{4}\pi$. At $t_0 = 0$, $+z$ was aligned with the X axis, and ϕ was aligned with the X-Y plane, normal to the plane of the paper. Rotation of θ brings $+x$ and the y-z orthogonal axes to the positions shown by the small red disks and radial lines within ϕ. The bold $\mathbf{X}$ superimposed on ϕ is the congregation of the $-M$, $-E$, $+M$, and $+E$ functional lines shown in Wave Diagram 2, and is the analogous condition for the quantum oscillation, in this case all being cotemporaneous in space as shown and rotating in time around the X axis. As will be shown in the spin diagrams that follow, the cross product of these vectors and their initial position in the Y-Z plane, operating through the stress-strain function, results in the creation of a wave guide at the points W_{+x} and W_{-x} (the latter not shown in the diagram) which is the boundary of the permanently rotated axis of wave travel. Such wave guide is seen to be earlier as the operation of $\left(\frac{V}{C}\right)$.

At the moment of the diagram, t_1, the ends of the $\mathbf{X}$ on the circumference of ϕ represent the displacement of points $+z'$, $+y'$, $-z'$, and $-y'$, each $\phi(\frac{1}{4}\pi - t_1)$ less than $+z$, $+y$, $-z$, and $-y$. At time $t_{(\frac{1}{4}\pi)}$, these last four points will have rotated that amount in ϕ and points $+x$ and $-x$ and the ends of the power cross, $\mathbf{X}$, will have rotated a corresponding amount in θ to the points $-M$, $-E$, $+M$, and $+E$ shown. As the wave speed for θ and ϕ is identical over the short term for a free quantum waveform, i.e. the neutron, and indefinitely under certain inertial conditions of atomic nuclear congregation, $+x$ and $-x$ constitute an effective boundary of the wave. Implicit in this model is that heat and friction is wholly a transfer of kinetic energy and momentum among separate quantum waveforms through translational, i.e. Brownian, motion and is not an operant condition of individual wave dynamics, thus there is no damping involved except as a boundary condition due to stress and strain at $+x$ and $-x$ as it might effect the fundamental eigenvalues, κ_0 and ω_0.

Spin Diagram 1 shows another view of this condition, this time with the path integral of an arbitrary point $+y$ over time, where the arbitrary y-z axes are indicated by the red and yellow broken lines at time, $t = 0$. Instead of the contemporaneous *M-E* power axes shown in Wave Diagram 1, however, we have the same positions in sequential order for $+y$, which is at rest at the functional point *K* at $t = 0$ at the instant of the diagram as indicated by the rotation arrows for θ and ϕ. As θ rotates $+2\pi$ as indicated at $+X$, and ϕ, -2π at $+x$, (instantaneously $+i$ at $+Z$ in the diagram), $+y$ will etch a figure 8 path, passing in sequence through 8 equal time segments of $\frac{1}{4}\pi$, through functional points $-E$, $-V$, $+M$, $+K$, $+E$, $+V$, $-M$, and back to $-K$. A continuum of other radials around the circumference of θ will do likewise, each $\partial\phi/\partial\theta$ from the adjacent ones. Four such paths of two orthogonal axes, of which $+y$ is one component, are shown.

The path of each radial is broken into four $\frac{1}{2}\pi$ phases, two capacitive and two inductive as shown, the capacitive phase always on the leading side in the direction of rotation of θ, and the inductive side always on the lagging side counter to the direction of rotation. Each such phase can further be seen as dominated by the kinetic energy of rotational oscillation between *K* and *E* for the capacitive and between *K* and *M* for the inductive phases and by the potential energy of displacement/strain between *E* and *V* and between *M* and *V*. The sine of the torsion strain angles, ε and μ, between *K-0-E* and *K-0-M* is $\frac{\sqrt{3}}{2}$ with a cosine of ½, indicating an angle of $1/3\,\pi$, while the sine and cosine of the angle between each of the power points and the plane of θ at the center is $\frac{1}{\sqrt{2}}$, indicating an angle of ¼ π. The sine and cosine of the angle between each plane of *K-0-E* and *K-0-M* and the plane of θ at the midpoint of the radial *0-K* is $\frac{\sqrt{2}}{\sqrt{3}}$ and $\frac{1}{\sqrt{3}}$, indicating an angle of .9553166, and an angle between the planes of angles ε and μ and the *X* axis of .6154797.

It bears emphasizing that the cross products, *K*×*E* and *K*×*M*, used in the following development and the corresponding diagrams represent the wave strain and are between the rest points with zero angular wave strain, *K*, and the points of strain at maximum instantaneous power, *M* and *E*. With respect to Spin Diagrams 1 and 2, this strain is balanced in resonance between the capacitive and the inductive phases. With the remaining diagrams, which involve a rotation of ϕ within θ and the *Y-Z* plane in the case of Diagrams 3 and 5 and an inversion of one axis in the case of Diagrams 4 and 6, one or the other of the phases predominates, indicating a sustained capacitive or inductive state of the wave. We would expect the predominant strain due to cosmic expansion to be a sustained inductive state or conversion of potential to kinetic energy, just as we might expect a capacitive state in association with a sustained condition of overall or local cosmic contraction.

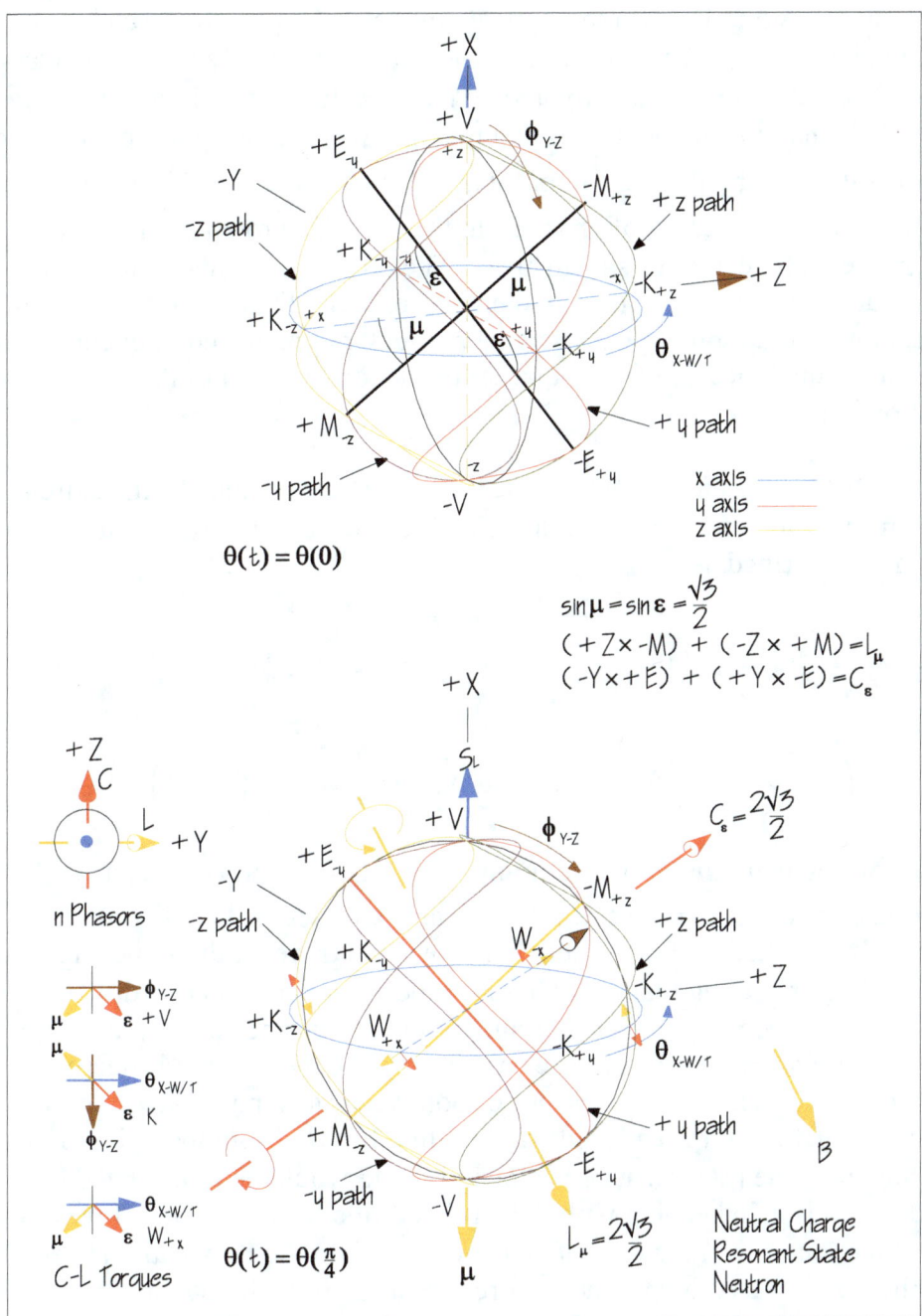

Spin Diagram 2 – Neutron

Spin Diagram 2, top figure, shows the oscillation at a point in time, $t(0) = -K$, for $+y$, shown as $-K_{+y}$ with the instantaneous power moments of $-M-+M$ and $-E-+E$. At time $t(0)$, these moments, shown by the dark cross, are not the active moments, the latter of which would be, as always, in the ϕ plane, if they were shown. Those shown are instead the positions to be reached by the strain/displacement of the y-z axes when they reach those physical points and assume those functional positions. The bottom figure shows this momentary assumption. It represents a point at $t(¼\pi)$ where $+y$ and $+z$ have rotated from the readers perspective $-¼\pi$ to $-E_{+y}$ and $-M_{+z}$ respectively in ϕ. Boundary nodes W_{+x} and W_{-x} and therefore ϕ have rotated $+¼\pi$ in θ. In the top figure, the tangential velocity, $\eta'(t)$, and thus momentum is at a maximum at the two poles of $-K_{+y}$ and $+K_{-y}$, to be followed at $t(½\pi)$ by the maximum for $\zeta'(t)$ at $-K_{+z}$ and $+K_{-z}$. The maxima, then, rotate with ϕ about θ, $½\pi$ out of phase from the W's. The rotation of θ creates a spin angular momentum vector, S_L, shown at $+X$. The power moments rotate with θ, at the same angular frequency and at the same phase position. Due to the steady state dynamics of the quantum oscillation, we can assign a particular dynamic component, i.e. force or momentum in equal measure or one quarter of the total to each of the power moments, as in the following.

The torsional strains, ε and μ, each crossed from the rest position to the extremum, create a torque on each side of the oscillation which is shown by the axial vectors C_ε and L_μ and which can be described as

$$(+\mathbf{Z}\times-\mathbf{M})+(-\mathbf{Z}\times+\mathbf{M})=\mathbf{L}_\mu \tag{2.122}$$

$$\left[\left(+\kappa_0^{-1}\times\tfrac{-1}{4}\mathsf{n}\omega_0^2\right)+\left(-\kappa_0^{-1}\times\tfrac{1}{4}\mathsf{n}\omega_0^2\right)\right]\sin\mu = i\tfrac{\sqrt{3}}{2}\kappa_0^{-1}\left(\tfrac{1}{2}\mathsf{n}\omega_0^2\right)=i\tfrac{\sqrt{3}}{4}\hbar\omega_0 \tag{2.123}$$

$$(+\mathbf{Y}\times-\mathbf{E})+(-\mathbf{Y}\times+\mathbf{E})=\mathbf{C}_\varepsilon \tag{2.124}$$

$$\left[\left(+\kappa_0^{-1}\times\tfrac{-1}{4}\mathsf{n}\omega_0^2\right)+\left(-\kappa_0^{-1}\times\tfrac{1}{4}\mathsf{n}\omega_0^2\right)\right]\sin\varepsilon = i\tfrac{\sqrt{3}}{2}\kappa_0^{-1}\left(\tfrac{1}{2}\mathsf{n}\omega_0^2\right)=i\tfrac{\sqrt{3}}{4}\hbar\omega_0 \tag{2.125}$$

The i's, as orthogonal sense, arise naturally in the crossing process, as in $j\times k = i$. These torques rotate in concert with the torsional strains and with θ. The torques are represented by the (active) h-axes detailed above and travel at the same angular velocity as θ and ϕ. It bears noting that while both C_ε and L_μ are shown according to the right hand rule, the force embodied in the strain at M is decreasing, while the momentum is increasing. At E we have the opposite case, where the momentum is decreasing while the force is increasing. The rotational sense of both vectors is in the direction of increasing force, forward in time for E and backward in time for M. From the point of view of motion, however, the rotation at L_μ is clockwise and would be represented by a left hand axial vector. This is in keeping with the left hand rule for induction. If we assume that L_μ is the direction of the magnetic field and cross it into the flow path between M and K on the spherical surface, the product, a force vector, will be in the direction of the oscillatory center.

The sum of (2.123) and (2.125) is a complex function of ϕ, in which we find the dimensions of spin energy, E_0, and in fact it is the spin energy of the system, which with complex integration with respect to a quarter cycle of time gives us the spin angular momentum as

$$S_L = i\tfrac{\sqrt{3}}{2}\mathsf{n}\omega_0^2\kappa_0^{-1}\int_{\omega t=0}^{\tfrac{\pi}{2}} e^{i\theta} = \tfrac{\sqrt{3}}{2}\int_t {}^1\mathsf{n}_2 = \tfrac{\sqrt{3}}{2}{}^1\mathsf{n}_1 = \tfrac{\sqrt{3}}{2}\hbar. \tag{2.126}$$

Had we started with the momenta of the moments being used, instead of the force in (2.122) and (2.124), we would have arrived at the action directly since dotting the momenta into the inverse wave number effectively integrates the momenta. Thus, instead of (2.123) and (2.125) we would have

$$S_{L\mu} = \left[\left(\tfrac{-1}{4}\hbar\omega_0 \cdot +\kappa_0^{-1}\right) + \left(\tfrac{1}{4}\hbar\omega_0^2 \cdot -\kappa_0^{-1}\right)\right]\cos\left(\tfrac{\pi}{2}-\mu\right) \quad (2.127)$$

$$= -\tfrac{\sqrt{3}}{2}\kappa_0^{-1}\left(\tfrac{1}{2}\hbar\omega_0\right) = -\tfrac{\sqrt{3}}{4}\hbar$$

$$S_{L\varepsilon} = \left[\left(\tfrac{-1}{4}\hbar\omega_0 \cdot +\kappa_0^{-1}\right) + \left(\tfrac{1}{4}\hbar\omega_0^2 \cdot -\kappa_0^{-1}\right)\right]\cos\left(\tfrac{\pi}{2}-\varepsilon\right) \quad (2.128)$$

$$= -\tfrac{\sqrt{3}}{2}\kappa_0^{-1}\left(\tfrac{1}{2}\hbar\omega_0\right) = -\tfrac{\sqrt{3}}{4}\hbar$$

$$S_L = S_{L\mu} + S_{L\varepsilon} = -\tfrac{\sqrt{3}}{2}\hbar \quad (2.129)$$

Vector μ is the effective magnetic moment vector of the oscillation which is the time averaged direction of L_μ. Vector B is the generalized direction of an assumed magnetic field. Were B of sufficient strength, L_μ would align with B, and S_L and μ would precess around B in the same direction as θ, which is in the direction indicated by L_μ as a left hand vector of motion. As shown in the phasor chart to the left of the diagram, in which the spin vector points toward the reader and the $+X$ direction, C leads L by ½ π.

Torques C_ε and L_μ in turn exert an equal shearing strain as shown by the small arrows on the ϕ boundary nodes W_{+x} and W_{-x}, and at $+V$ and $-V$ and again at the intersections of ϕ and θ, which we will call K_0 (between $-K_{+y}$ and $-K_{+z}$ in the top drawing) and K_π. The three charts labeled "C-L Torques" show the condition at $+V$, K_0 and W_{+x}, all as viewed from the exterior of the oscillation. The three antipodes of these, $-V$, K_π and W_{-x}, would be mirrored along the horizontal (with respect to the page) rotation vector of each chart, the rotation vectors being tangent differential components of the corresponding axial vectors. The six nodes/antinodes create a wave guide and boundary that works against the recoil of the strain from the x axis to the X axis.

The cross product of the small shear arrows, therefore, corresponding to L_μ and C_ε as indicated at W_{+x} and W_{-x}, form a charge vector (not shown) of relative magnitude 2/3 at each W, which is aligned with the ϕ rotation vector shown, its direction determined as indicated in the following section on charge generation. At the intersection of θ and ϕ the shear vectors cancel and the charge vanishes. In the case of resonance, as shown in Spin Diagram 2 for the neutral state or that of the neutron, the capacitive and inductive phases are balanced and there is no net charge force on the boundary nodes.

In an inductive state, that is, in a mode of discharge and current generation, the induction torque is predominant. This corresponds with a phase advance culminating in a $+i$ rotation of θ in the Y-Z plane, a long term physical increase in the rotational strain in the direction of spin. It is a release of kinetic energy over and above the normal oscillatory kinetic energy. This would be indicated by crossing the induction shear arrows into the capacitive shears at W_{+x} and W_{-x}, which in the case of Spin Diagram 2 would be anti-parallel to the ϕ rotation vector, and which would indicate a positive charge vector at the

W_{+x} boundary. With respect to L_μ as a left hand vector with respect to motion, the direction of motion of the μ shear vector will be reversed, which creates an unstable condition at W_{+x} and W_{-x} over time.

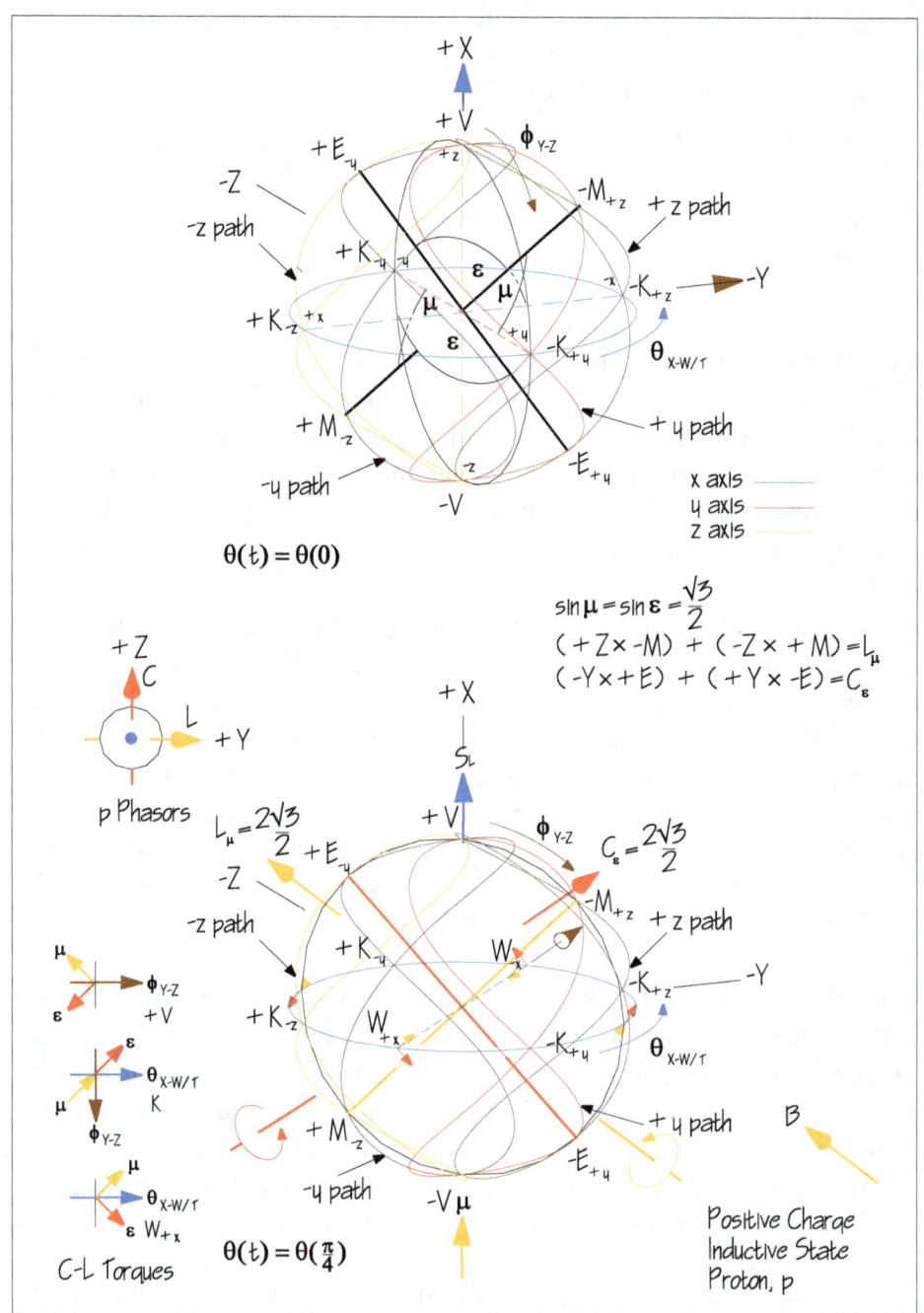

Spin Diagram 3 – Proton

The resulting condition, as shown in Spin Diagram 3, is that of the positive charge oscillation or the proton. The vantage point of the reader is shifted 90 degrees as indicated by the ϕ rotation vector which now aligns with the $-Y$ axis in the top figure. It is noted also that the power moments and their strain angles μ and ε have shifted $+i$ in Y-Z. In the bottom figure, the effect of this advance is to flip L_μ into the upper hemisphere where it continues to trail the capacitive vector by $-i$ and now trails W_{+x}. The result is a change in the shearing vectors at W_{+x} and W_{-x} such that the charge vector is now aligned with ϕ, with all the torques in general alignment with the rotation of θ, as indicated by S_L.

The phasor chart reflects this condition, with the capacitive vector continuing to lead the inductive, but with both in general alignment with the spin vector. The magnetic field vector B is now inverted from the position in Spin Diagram 2, to indicate that if activated as shown, L_μ would tend to align with it and S_L and μ to precess around B, once again in the direction indicated by L_μ. This indicates that if our perspective of observation is stationed by the direction of B in the transition between Diagram 2 and 3, the result will be a reversal of the direction of θ and a flip of the spin and capacitive vectors. This can be seen with a little effort by viewing Diagram 3 upside down and "looking up" at θ.

As might be imagined, this indicates an instance of beta decay. The generation of a charge vector at W_{+x} involves the transmission of wave energy past the boundary node in the form of a <u>reflected</u> negatively charged oscillation which constitutes the electron as shown in Spin Diagram 4. The predominant amount of the energy is transmitted from the medium as potential to the oscillation as kinetic with the resulting altered resonant frequency of the proton. The quantification of this process will be shown in a minute.

In Diagram 4, it should be noted that the reference grid shown is one half rotation from that shown in Diagram 3 to facilitate the viewing correspondences between the two. This is shown in the Phase Orientation chart at the left center of the diagram. Both diagrams share a common orientation for B. It is immediately apparent that the spin vector must flip from that of the proton as shown to provide continuity of the reflected wave. It is noted that the resulting oscillation involves an *inversion* of the y axis, while the x and z axes retain the same orientation with respect to their rest positions as does the neutral oscillation. Although Diagram 4 and related Diagram 6 are shown as spherical, the amplitude and inverse wave number for these oscillations are in fact not equal. While both amplitude and wave number and the covariant frequency are proportionally and equally smaller for the reflected oscillations, as differentials they have different physical interpretations. The amplitude is much smaller than that of the nuclear fundamental indicating a smaller size of the electron, while the equally smaller wave number that we will see is associated with the reflection would suggest a much elongated waveform with a proportionally reduced frequency. The inversion then indicates an inverted node, as the kinetic distal end of the reflected oscillation, which is encountered observationally as the electron orbital or cloud. The likelihood that this configuration involves an extension into a fourth-dimension should not be discounted, in which case the spherical form could be maintained.

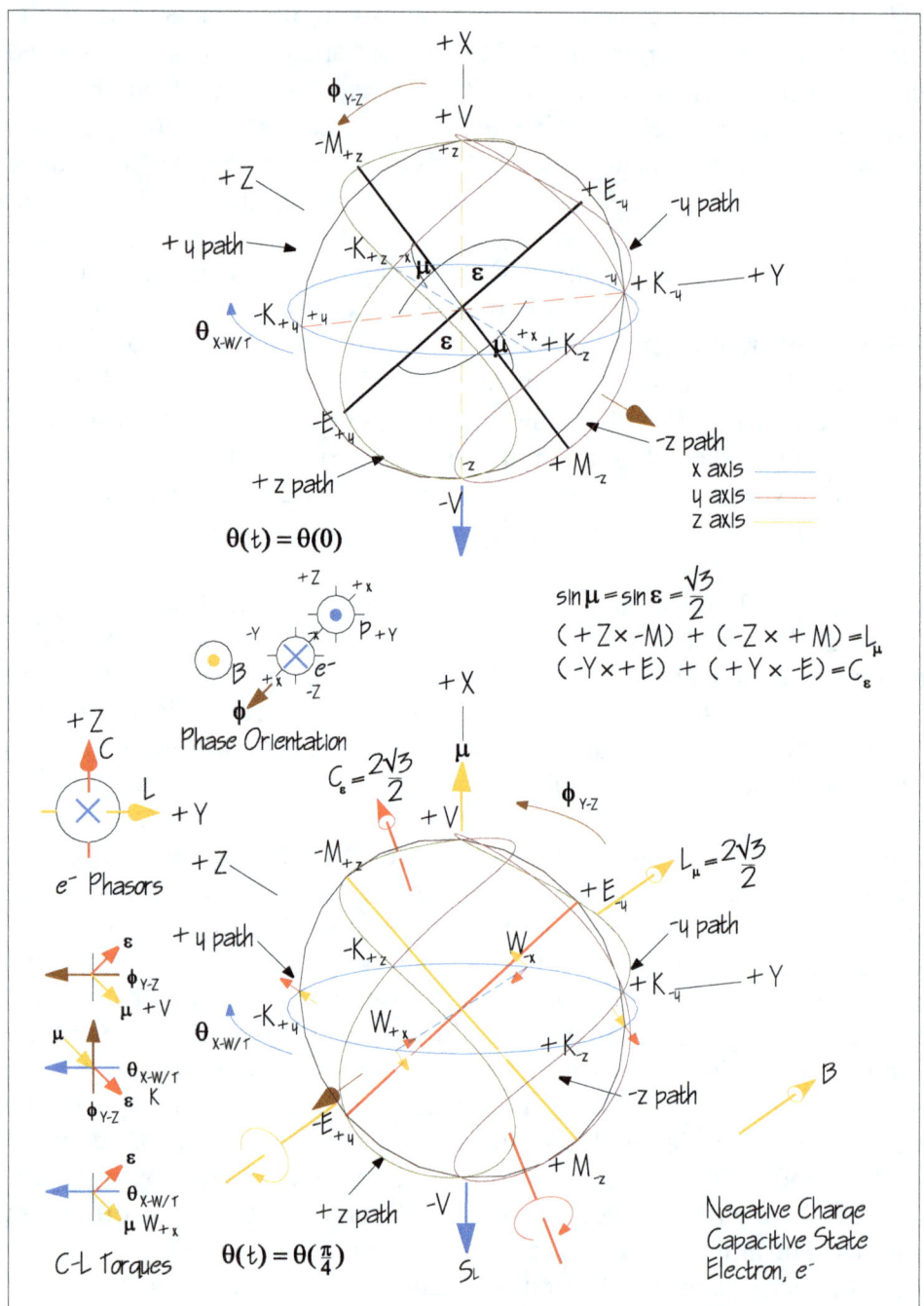

Spin Diagram 4 - Electron

The shearing vectors all point *against* the direction of spin indicating a negative or receptive charge, a capacitive or charging state of the oscillation. As before in the presence of a magnetic field, the alignment of L_μ with B results in a precession of S_L and μ.

The subject of reflected and transmitted oscillations with charges reversed, i.e. anti matter, is straightforward. Cosmic expansion is clearly the conversion of potential to kinetic energy, the energy of position to that of motion. This could not be more evident than in the notion of a "big bang". The natural state for an oscillation under this condition is that of conversion of potential to kinetic energy, of electrical field potential to magnetically induced current. It is this condition that accounts for the experienced predominance of matter over anti-matter.

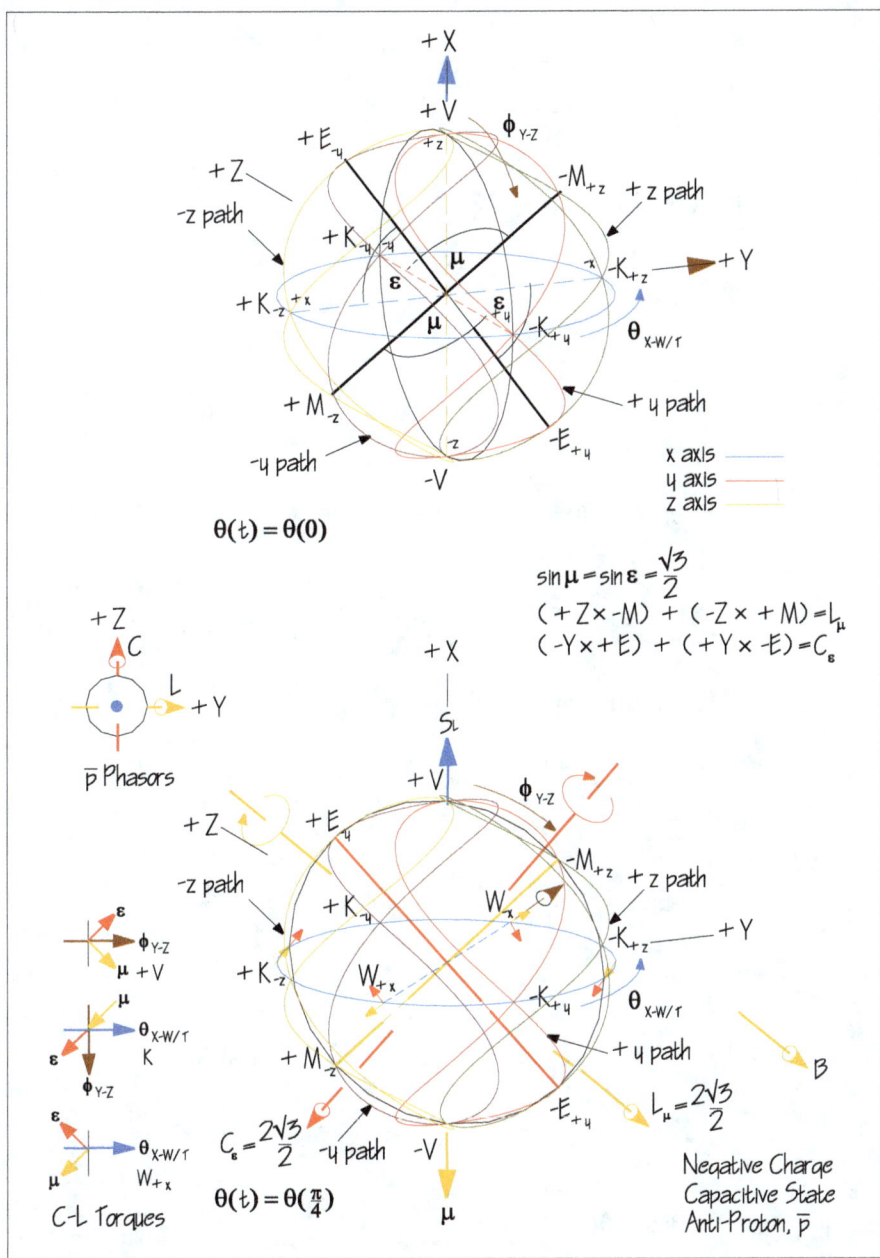

Spin Diagram 5 – Anti-Proton

Spin Diagram 5 shows the condition of a fundamental oscillation of the anti proton in the capacitive state resulting from the generation of a negative charge. Returning to Diagram 2 for a moment, if we cross the capacitive shear at W_{+x} and W_{-x} into the inductive, the resulting charge vector aligns parallel with the ϕ rotation vector, corresponding with a phase lag and eventual $-i$ rotation of θ in the Y-Z plane, a long term physical *decrease* in the rotational strain in the direction of spin. It is a restoration of kinetic energy to potential energy over and above that normal to the oscillatory cycle.

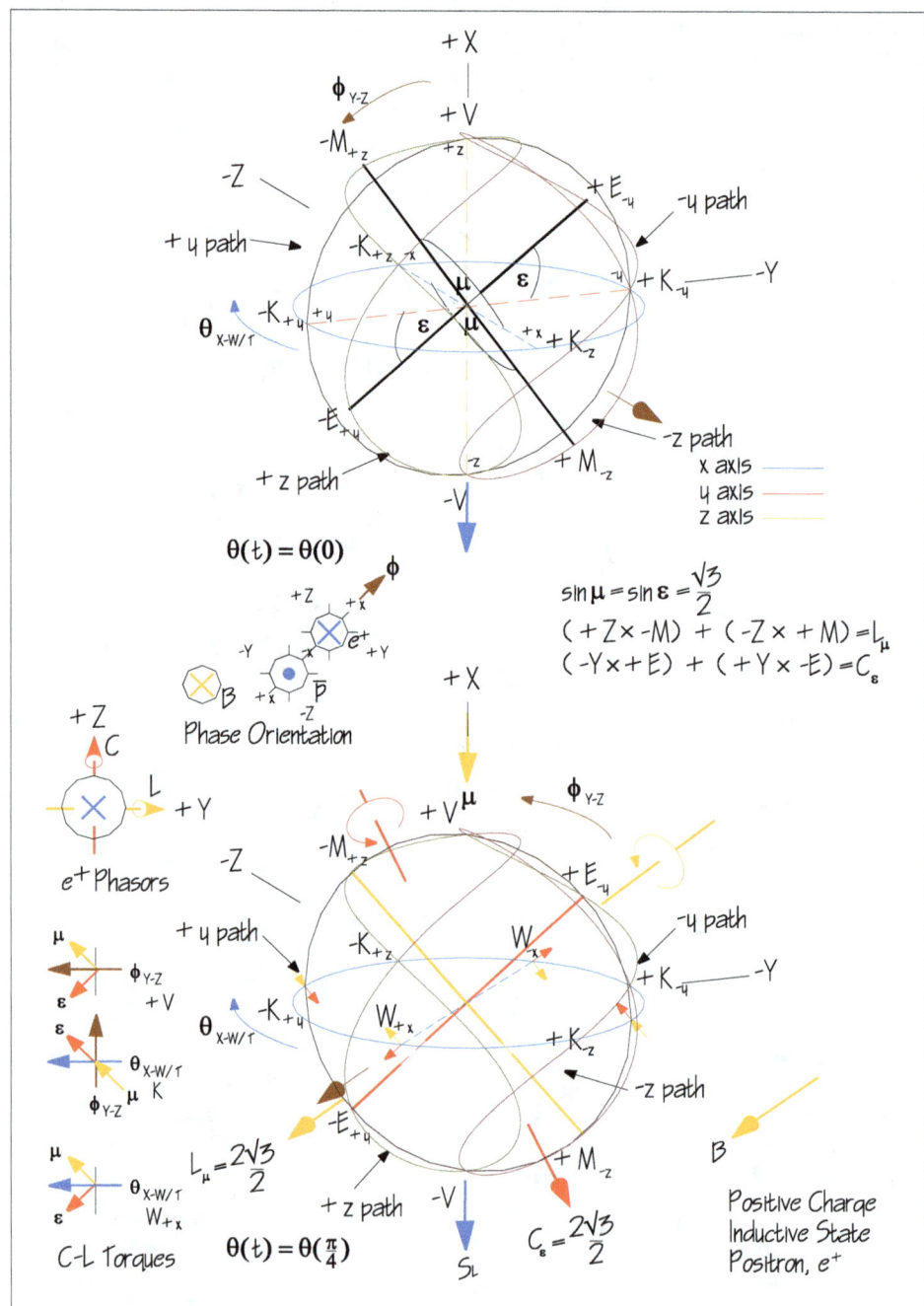

Spin Diagram 6 – Positron

Note that the reader orientation of Diagram 5 is rotated from that of Diagram 2, ½ π in the opposite direction from that of the proton and is thus out of phase from it by a value of π. The resulting lag results in a flip of the capacitive torque to general alignment with the effective magnetic moment, μ, and an applied magnetic field, B. The shearing arrows now all point counter to the direction of spin, θ, indicating a charging or capacitive mode.

Spin Diagram 6 shows the state of the positron, a wave reflected at the boundary node at W_{-x} in Diagram 2. It is analogous to the electron except for the reversal of the capacitive and inductive vectors, and hence the charge. As a result, it is in an inductive state and the shearing vectors all point in general alignment with the rotation of θ.

With respect to the nomenclature of "reflected" and "transmitted" waves, they have been chosen for historical reasons having to do with the transmission of power of a traveling wave at a discontinuity. This treatment is used subsequently within the context of a 4 dimensional continuum. We can think of the neutron, proton and anti-proton as 3-dimensional transmissions of a 4-D wave, where the electron and positron are reflected as the result of an impedance change.

Generation of Charge

The following tables outline the dynamics of the quanta shown in Spin Diagrams 2 – 5 with respect to the generation of charge. The tangential and centripetal acceleration attributed to each of the torques C_ε and L_μ at the six nodes is given by the dot product, in the case of tangential, or the cross product, in the case of centripetal, of shear vectors ε and μ and the rotation vectors, θ and ϕ.

In these tables, the small circles, filled and unfilled, preceding the values in columns 4-7 refer to the centripetal or counter-centripetal sense respectively, arising from the cross product operation. The specific choice of the options of θ and ϕ in columns 1-2 and 5-6 is determined by context with respect to the "C-L Torques" charts as shown for each node[2] in the diagram designation column. Columns 1-2, 4-6 give the values of the products without modification by the value of the torque involved, while column 7 is so modified.

A cursory glance at the tables and the diagrams will provide ample indication of a quantum structure that has corollaries with the quark and chromodynamic modeling of the standard theory. In particular, a physical basis for fractional electrical charge suggests itself and that for asymptotic freedom is more assertive. The nodes, while hardly rigid, as aspects of a discrete wave would be anticipated to register and transfer to any observational media the interactions of inelastic collision and scattering. Obviously, however, they are inseparable as aspects of an overall oscillation, and hence convey the quality of "confinement".

In the spin diagrams, the tangential rotational path of θ, as distinct from its axial vector representation, S_L, is the locus of sustained kinetic energy, tangential momentum and velocity and the electrical equivalence, current. The E moments transfer this velocity/current via the elastic strain of the medium to the static point of maximum potential or electrical amplitude at $+V$ and $-V$, thus charging the electrical field of the system. (The whole oscillation is, of course, free to rotate in space, but for any short term series of oscillations, these points can be considered static.) The potential or electrical charge energy of $+V$ and $-V$ present in the strain is converted by recoil or restoration of that strain to kinetic or velocity/current via the M moments. The vectors ε and μ both point in the direction of increasing strain, but their direction over time is not the same.

We can envision three general energy states of the oscillation, resonant, R, capacitive, C, or inductive, L. In the resonant state, ε points universally in the direction of the rotation tangent vectors, while μ points counter to those vectors, indicating the characteristic phase lead of ε and lag of μ for current with respect to potential. In columns 1 and 2, the dot product of these vectors and the rotation vectors cancel, indicating no net change or acceleration of the latter and a general condition of resonance, as indicated in column 3.

---

[2] As determined by context, the term *node* is used as a general term to include the category of *antinode*.

		1	2	3	4	5	6	7	8
	Diagram, Node/Antinode, Rotation	$\mu \cdot$ Into θ,ϕ	$\cdot \varepsilon$ From θ,ϕ	K-V State	$S = \mu \times \varepsilon$ $\mu, \varepsilon = \frac{\sqrt{2}}{\sqrt{3}}$	$\mu \times$ Into θ,ϕ	$\times \varepsilon$ From θ,ϕ	$q = \frac{S}{|S|}\left(\mathrm{T}(\mu\times + \times\varepsilon)\right)$ $\mathrm{T}_{\mu,\varepsilon} = \frac{\sqrt{3}}{4}$	Total Charge $q_{W-} - q_{W+}$ $q_{V\pm} - q_{V\mp}$
Diagram 2 - Neutron	W_{+x}-θ	$-\frac{1}{\sqrt{3}}$	$+\frac{1}{\sqrt{3}}$	R	$\circ \frac{2}{3}$	$\circ \frac{1}{\sqrt{3}}$	$\bullet \frac{1}{\sqrt{3}}$	$\circ \frac{1}{4} + \bullet \frac{1}{4} = 0$	
	W_{-x}-θ	$-\frac{1}{\sqrt{3}}$	$+\frac{1}{\sqrt{3}}$	R	$\bullet \frac{2}{3}$	$\bullet \frac{1}{\sqrt{3}}$	$\circ \frac{1}{\sqrt{3}}$	$\bullet \frac{1}{4} + \circ \frac{1}{4} = 0$	0
	K_o-θ	$-\frac{1}{\sqrt{3}}$	$+\frac{1}{\sqrt{3}}$	R	0	$\bullet \frac{1}{\sqrt{3}}$	$\bullet \frac{1}{\sqrt{3}}$	0	
	K_π-θ	$-\frac{1}{\sqrt{3}}$	$+\frac{1}{\sqrt{3}}$	R	0	$\circ \frac{1}{\sqrt{3}}$	$\circ \frac{1}{\sqrt{3}}$	0	
	K_o-ϕ	$-\frac{1}{\sqrt{3}}$	$+\frac{1}{\sqrt{3}}$	R	0	$\circ \frac{1}{\sqrt{3}}$	$\bullet \frac{1}{\sqrt{3}}$	0	
	K_π-ϕ	$-\frac{1}{\sqrt{3}}$	$+\frac{1}{\sqrt{3}}$	R	0	$\bullet \frac{1}{\sqrt{3}}$	$\circ \frac{1}{\sqrt{3}}$	0	
	$+V$-ϕ	$-\frac{1}{\sqrt{3}}$	$+\frac{1}{\sqrt{3}}$	R	$\circ \frac{2}{3}$	$\circ \frac{1}{\sqrt{3}}$	$\bullet \frac{1}{\sqrt{3}}$	$\circ \frac{1}{4} + \bullet \frac{1}{4} = 0$	
	$-V$-ϕ	$-\frac{1}{\sqrt{3}}$	$+\frac{1}{\sqrt{3}}$	R	$\bullet \frac{2}{3}$	$\bullet \frac{1}{\sqrt{3}}$	$\circ \frac{1}{\sqrt{3}}$	$\bullet \frac{1}{4} + \circ \frac{1}{4} = 0$	[0]
Diagram 3 - Proton	W_{+x}-θ	$+\frac{1}{\sqrt{3}}$	$+\frac{1}{\sqrt{3}}$	L	$\bullet \frac{2}{3}$	$\bullet \frac{1}{\sqrt{3}}$	$\bullet \frac{1}{\sqrt{3}}$	$\bullet 1\left(\bullet\frac{1}{4}+\bullet\frac{1}{4}\right) = +\frac{1}{2}$	
	W_{-x}-θ	$+\frac{1}{\sqrt{3}}$	$+\frac{1}{\sqrt{3}}$	L	$\circ \frac{2}{3}$	$\circ \frac{1}{\sqrt{3}}$	$\circ \frac{1}{\sqrt{3}}$	$\circ 1\left(\circ\frac{1}{4}+\circ\frac{1}{4}\right) = +\frac{1}{2}$	+1
	K_o-θ	$+\frac{1}{\sqrt{3}}$	$+\frac{1}{\sqrt{3}}$	L	0	$\bullet \frac{1}{\sqrt{3}}$	$\circ \frac{1}{\sqrt{3}}$	0	
	K_π-θ	$+\frac{1}{\sqrt{3}}$	$+\frac{1}{\sqrt{3}}$	L	0	$\circ \frac{1}{\sqrt{3}}$	$\bullet \frac{1}{\sqrt{3}}$	0	
	K_o-ϕ	$-\frac{1}{\sqrt{3}}$	$-\frac{1}{\sqrt{3}}$	C	0	$\bullet \frac{1}{\sqrt{3}}$	$\circ \frac{1}{\sqrt{3}}$	0	
	K_π-ϕ	$-\frac{1}{\sqrt{3}}$	$-\frac{1}{\sqrt{3}}$	C	0	$\circ \frac{1}{\sqrt{3}}$	$\bullet \frac{1}{\sqrt{3}}$	0	
	$+V$-ϕ	$-\frac{1}{\sqrt{3}}$	$-\frac{1}{\sqrt{3}}$	C	$\circ \frac{2}{3}$	$\bullet \frac{1}{\sqrt{3}}$	$\bullet \frac{1}{\sqrt{3}}$	$\circ 1\left(\bullet\frac{1}{4}+\bullet\frac{1}{4}\right) = -\frac{1}{2}$	
	$-V$-ϕ	$-\frac{1}{\sqrt{3}}$	$-\frac{1}{\sqrt{3}}$	C	$\bullet \frac{2}{3}$	$\circ \frac{1}{\sqrt{3}}$	$\circ \frac{1}{\sqrt{3}}$	$\bullet 1\left(\circ\frac{1}{4}+\circ\frac{1}{4}\right) = -\frac{1}{2}$	$[-1] = -i1$
Diagram 4 - Electron	W_{+x}-θ	$-\frac{1}{\sqrt{3}}$	$-\frac{1}{\sqrt{3}}$	C	$\circ \frac{2}{3}$	$\bullet \frac{1}{\sqrt{3}}$	$\bullet \frac{1}{\sqrt{3}}$	$\circ 1\left(\bullet\frac{1}{4}+\bullet\frac{1}{4}\right) = -\frac{1}{2}$	-1
	W_{-x}-θ	$-\frac{1}{\sqrt{3}}$	$-\frac{1}{\sqrt{3}}$	C	$\bullet \frac{2}{3}$	$\circ \frac{1}{\sqrt{3}}$	$\circ \frac{1}{\sqrt{3}}$	$\bullet 1\left(\circ\frac{1}{4}+\circ\frac{1}{4}\right) = -\frac{1}{2}$	
	K_o-θ	$-\frac{1}{\sqrt{3}}$	$-\frac{1}{\sqrt{3}}$	C	0	$\bullet \frac{1}{\sqrt{3}}$	$\circ \frac{1}{\sqrt{3}}$	0	
	K_π-θ	$-\frac{1}{\sqrt{3}}$	$-\frac{1}{\sqrt{3}}$	C	0	$\circ \frac{1}{\sqrt{3}}$	$\bullet \frac{1}{\sqrt{3}}$	0	
	K_o-ϕ	$-\frac{1}{\sqrt{3}}$	$-\frac{1}{\sqrt{3}}$	C	0	$\circ \frac{1}{\sqrt{3}}$	$\bullet \frac{1}{\sqrt{3}}$	0	
	K_π-ϕ	$-\frac{1}{\sqrt{3}}$	$-\frac{1}{\sqrt{3}}$	C	0	$\bullet \frac{1}{\sqrt{3}}$	$\circ \frac{1}{\sqrt{3}}$	0	
	$+V$-ϕ	$-\frac{1}{\sqrt{3}}$	$-\frac{1}{\sqrt{3}}$	C	$\circ \frac{2}{3}$	$\bullet \frac{1}{\sqrt{3}}$	$\bullet \frac{1}{\sqrt{3}}$	$\circ 1\left(\bullet\frac{1}{4}+\bullet\frac{1}{4}\right) = -\frac{1}{2}$	$[-1] = -i1$
	$-V$-ϕ	$-\frac{1}{\sqrt{3}}$	$-\frac{1}{\sqrt{3}}$	C	$\bullet \frac{2}{3}$	$\circ \frac{1}{\sqrt{3}}$	$\circ \frac{1}{\sqrt{3}}$	$\bullet 1\left(\circ\frac{1}{4}+\circ\frac{1}{4}\right) = -\frac{1}{2}$	

Spin Table 1 – Inductive State, Ordinary Matter

		1	2	3	4	5	6	7	8
	Diagram, Node/Antinode, Rotation	$\varepsilon \cdot$ Into θ,ϕ	$\cdot \mu$ From θ,ϕ	K-V State	$S = \varepsilon \times \mu$ $\mu,\varepsilon = \frac{\sqrt{2}}{\sqrt{3}}$	$\varepsilon \times$ Into θ,ϕ	$\times \mu$ From θ,ϕ	$q = \frac{S}{|S|}\left(\mathrm{T}(\mu \times + \times \varepsilon)\right)$ $\mathrm{T}_{\mu,\varepsilon} = \frac{\sqrt{3}}{4}$	Total Charge $q_{W-} - q_{W+}$ $q_{V\pm} - q_{V\mp}$
Diagram 2 – Anti Neutron	$W_{+x}-\theta$	$+\frac{1}{\sqrt{3}}$	$-\frac{1}{\sqrt{3}}$	R	$\bullet \frac{2}{3}$	$\circ \frac{1}{\sqrt{3}}$	$\bullet \frac{1}{\sqrt{3}}$	$\circ \frac{1}{4} + \bullet \frac{1}{4} = 0$	
	$W_{-x}-\theta$	$+\frac{1}{\sqrt{3}}$	$-\frac{1}{\sqrt{3}}$	R	$\circ \frac{2}{3}$	$\bullet \frac{1}{\sqrt{3}}$	$\circ \frac{1}{\sqrt{3}}$	$\bullet \frac{1}{4} + \circ \frac{1}{4} = 0$	0
	$K_o-\theta$	$+\frac{1}{\sqrt{3}}$	$-\frac{1}{\sqrt{3}}$	R	0	$\circ \frac{1}{\sqrt{3}}$	$\circ \frac{1}{\sqrt{3}}$	0	
	$K_\pi-\theta$	$+\frac{1}{\sqrt{3}}$	$-\frac{1}{\sqrt{3}}$	R	0	$\bullet \frac{1}{\sqrt{3}}$	$\bullet \frac{1}{\sqrt{3}}$	0	
	$K_o-\phi$	$+\frac{1}{\sqrt{3}}$	$-\frac{1}{\sqrt{3}}$	R	0	$\bullet \frac{1}{\sqrt{3}}$	$\bullet \frac{1}{\sqrt{3}}$	0	
	$K_\pi-\phi$	$+\frac{1}{\sqrt{3}}$	$-\frac{1}{\sqrt{3}}$	R	0	$\circ \frac{1}{\sqrt{3}}$	$\circ \frac{1}{\sqrt{3}}$	0	
	$+V-\phi$	$+\frac{1}{\sqrt{3}}$	$-\frac{1}{\sqrt{3}}$	R	$\bullet \frac{2}{3}$	$\circ \frac{1}{\sqrt{3}}$	$\bullet \frac{1}{\sqrt{3}}$	$\circ \frac{1}{4} + \bullet \frac{1}{4} = 0$	
	$-V-\phi$	$+\frac{1}{\sqrt{3}}$	$-\frac{1}{\sqrt{3}}$	R	$\circ \frac{2}{3}$	$\bullet \frac{1}{\sqrt{3}}$	$\circ \frac{1}{\sqrt{3}}$	$\bullet \frac{1}{4} + \circ \frac{1}{4} = 0$	[0]
Diagram 5 – Anti Proton	$W_{+x}-\theta$	$-\frac{1}{\sqrt{3}}$	$-\frac{1}{\sqrt{3}}$	C	$\circ \frac{2}{3}$	$\bullet \frac{1}{\sqrt{3}}$	$\bullet \frac{1}{\sqrt{3}}$	$\circ 1\left(\bullet \frac{1}{4} + \bullet \frac{1}{4}\right) = -\frac{1}{2}$	
	$W_{-x}-\theta$	$-\frac{1}{\sqrt{3}}$	$-\frac{1}{\sqrt{3}}$	C	$\bullet \frac{2}{3}$	$\circ \frac{1}{\sqrt{3}}$	$\circ \frac{1}{\sqrt{3}}$	$\bullet 1\left(\circ \frac{1}{4} + \circ \frac{1}{4}\right) = -\frac{1}{2}$	-1
	$K_o-\theta$	$-\frac{1}{\sqrt{3}}$	$-\frac{1}{\sqrt{3}}$	C	0	$\circ \frac{1}{\sqrt{3}}$	$\bullet \frac{1}{\sqrt{3}}$	0	
	$K_\pi-\theta$	$-\frac{1}{\sqrt{3}}$	$-\frac{1}{\sqrt{3}}$	C	0	$\bullet \frac{1}{\sqrt{3}}$	$\circ \frac{1}{\sqrt{3}}$	0	
	$K_o-\phi$	$+\frac{1}{\sqrt{3}}$	$+\frac{1}{\sqrt{3}}$	L	0	$\circ \frac{1}{\sqrt{3}}$	$\bullet \frac{1}{\sqrt{3}}$	0	
	$K_\pi-\phi$	$+\frac{1}{\sqrt{3}}$	$+\frac{1}{\sqrt{3}}$	L	0	$\bullet \frac{1}{\sqrt{3}}$	$\circ \frac{1}{\sqrt{3}}$	0	
	$+V-\phi$	$+\frac{1}{\sqrt{3}}$	$+\frac{1}{\sqrt{3}}$	L	$\bullet \frac{2}{3}$	$\bullet \frac{1}{\sqrt{3}}$	$\bullet \frac{1}{\sqrt{3}}$	$\bullet 1\left(\bullet \frac{1}{4} + \bullet \frac{1}{4}\right) = +\frac{1}{2}$	
	$-V-\phi$	$+\frac{1}{\sqrt{3}}$	$+\frac{1}{\sqrt{3}}$	L	$\circ \frac{2}{3}$	$\circ \frac{1}{\sqrt{3}}$	$\circ \frac{1}{\sqrt{3}}$	$\circ 1\left(\circ \frac{1}{4} + \circ \frac{1}{4}\right) = +\frac{1}{2}$	$[+1] = +i1$
Diagram 6 - Positron	$W_{+x}-\theta$	$+\frac{1}{\sqrt{3}}$	$+\frac{1}{\sqrt{3}}$	L	$\bullet \frac{2}{3}$	$\bullet \frac{1}{\sqrt{3}}$	$\bullet \frac{1}{\sqrt{3}}$	$\bullet 1\left(\bullet \frac{1}{4} + \bullet \frac{1}{4}\right) = +\frac{1}{2}$	+1
	$W_{-x}-\theta$	$+\frac{1}{\sqrt{3}}$	$+\frac{1}{\sqrt{3}}$	L	$\circ \frac{2}{3}$	$\circ \frac{1}{\sqrt{3}}$	$\circ \frac{1}{\sqrt{3}}$	$\circ 1\left(\circ \frac{1}{4} + \circ \frac{1}{4}\right) = +\frac{1}{2}$	
	$K_o-\theta$	$+\frac{1}{\sqrt{3}}$	$+\frac{1}{\sqrt{3}}$	L	0	$\circ \frac{1}{\sqrt{3}}$	$\bullet \frac{1}{\sqrt{3}}$	0	
	$K_\pi-\theta$	$+\frac{1}{\sqrt{3}}$	$+\frac{1}{\sqrt{3}}$	L	0	$\bullet \frac{1}{\sqrt{3}}$	$\circ \frac{1}{\sqrt{3}}$	0	
	$K_o-\phi$	$+\frac{1}{\sqrt{3}}$	$+\frac{1}{\sqrt{3}}$	L	0	$\bullet \frac{1}{\sqrt{3}}$	$\circ \frac{1}{\sqrt{3}}$	0	
	$K_\pi-\phi$	$+\frac{1}{\sqrt{3}}$	$+\frac{1}{\sqrt{3}}$	L	0	$\circ \frac{1}{\sqrt{3}}$	$\bullet \frac{1}{\sqrt{3}}$	0	
	$+V-\phi$	$+\frac{1}{\sqrt{3}}$	$+\frac{1}{\sqrt{3}}$	L	$\bullet \frac{2}{3}$	$\bullet \frac{1}{\sqrt{3}}$	$\bullet \frac{1}{\sqrt{3}}$	$\bullet 1\left(\bullet \frac{1}{4} + \bullet \frac{1}{4}\right) = +\frac{1}{2}$	$[+1] = +i1$
	$-V-\phi$	$+\frac{1}{\sqrt{3}}$	$+\frac{1}{\sqrt{3}}$	L	$\circ \frac{2}{3}$	$\circ \frac{1}{\sqrt{3}}$	$\circ \frac{1}{\sqrt{3}}$	$\circ 1\left(\circ \frac{1}{4} + \circ \frac{1}{4}\right) = +\frac{1}{2}$	

Spin Table 2 – Capacitive State, Anti Matter

Note that there is no physical difference between the neutron and the anti neutron, in which the capacitive and inductive strains are in equilibrium, though this oscillation appears in each table to reflect the underlying symmetry of the system. In Spin Table 1, it is embedded in an overall inductive state, as indicated by the vector operations at the top of each column, while in Spin Table 2 it is in an overall capacitive scheme.

In the case of the cross product operations, the results are shown with the novel sense to facilitate the algebra with respect to the spin surface. The crossing is done in the order outlined above for the inductive state of ordinary matter, that is, from μ to ε, θ or ϕ and from θ or ϕ to ε, and for the capacitive state of anti matter, from ε to μ, θ or ϕ and from θ or ϕ to μ. In this algebra, we have:

$$\circ 1 (\circ 1 + \circ 1) = +2 \qquad (2.130)$$

$$\bullet 1 (\bullet 1 + \bullet 1) = +2 \qquad (2.131)$$

$$\circ 1 (\bullet 1 + \bullet 1) = -2 \qquad (2.132)$$

$$\bullet 1 (\circ 1 + \circ 1) = -2 \qquad (2.133)$$

In the case of the neutron, the cross products at each of the W nodes cancel, resulting in no net change in the oscillatory state, potential/kinetic, electrical/magnetic. In the case of the proton, there is a net inductive change or positive charge along the axis of ϕ, and a corresponding capacitive change or negative charge along the axis of ϕ of the electron. In the case of the anti-proton, there is a net capacitive change or negative charge along ϕ, and a corresponding inductive change or positive charge along ϕ of the positron. Thus the sense of the charge is positive if $\mu \times \varepsilon$ is parallel to $(\mu \times + \times \varepsilon)$ and negative if they are anti-parallel, the positive charge representing an advance of the tangential momentum, or inductance, and the negative charge representing a retention of that momentum, capacitance. When $\mu \times$ is of opposite sense to $\times \varepsilon$, they cancel, there is no net charge and a state of resonance is maintained.

For the proton-electron pair and the inductive state, there is also a corresponding charge for each along the axial vector, S_L. Since these rotational and torque vectors are anti-parallel, indicating a negative charge for the pair, according to the C-L torque diagrams, the charge for both is $-i1$. They are complex, indicating their orthogonality. For the capacitive state, charge summation is from the same potential nodes, but the spins, relative to ε and μ, are reversed, leading to a reversal of charges or $+i1$ for the anti proton and positron.

The square brackets for these charges indicate that they pertain to the $+V$ and $-V$ potential nodes, which with respect to the ϕ oscillation are antinodes as are K_o and K_π. These former are the general loci of the wave boundary of θ, which has been rotated into the Y-Z plane and forms the wave boundary and nodes of ϕ at W_+ and W_-. Any transmission of momentum and energy through the boundary at W_+ and W_- would presumably involve the perpetual strain between the X and x axes and thereby the $+V$ and $-V$ nodes. The charges indicated at the latter, therefore, are not additional but are rather the complex expression of those at W_-. It should be noted that while we have assigned the charge to W_- for descriptive facility, the same charge can be applied to the opposite node. We would imagine, however, that with beta decay the reflected oscillation which becomes the electron occurs at one or the other end of the fundamental x axis, i.e. in the vicinity of $-V$ or $+V$, along the X axis. The direction of the axial vector ϕ is not necessarily a directional vector for electron reflection, but rather the indicator of ϕ spin for both particles at the time of transmission. This wave transmission/reflection results from a conversion of the potential energy density of the continuum to the kinetic energy of proton-electron oscillation through a decrease in the continuum impedance. It should not be ruled out that the electron wave in fact forms a spherical shell about the oscillation source, emanating from the two ends at $-V$ and $+V$, with its amplitude as developed at (3.34), and the shell as generally described in Spin Diagram 4.

K_o and K_π are antinodes for both ϕ and θ and not candidates for the longitudinal energy and momentum transmission of beta decay. This is indicated by the vanishing cross products, $\mu \times \varepsilon$ and $\varepsilon \times \mu$. The products of $\times \varepsilon$ and $\mu \times$ (and their anti matter counterparts), while not vanishing, cancel at each node, ϕ to θ for the resonant state, ε to μ for the proton and anti proton, and with respect to both for the electron and positron. This last condition is a reflection of the apparent lack of internal structure of these last two particles. The strain vectors ε and μ at all of the nodes are capacitive for the electron and inductive for the positron, resulting in an isotropic charge condition registering from particle interaction. In the case of the neutron they are balance between capacitance and inductance at each node. In that of the proton and anti proton one or the other predominates, and some indication of internal structure can be registered from collision and scattering.

Since a charge can be assigned to either node of ϕ or θ, we might surmise that the fundamental angular frequency, ω_0, associated with the tangential momentum be divided by π. Thus we can state a preliminary or raw quantification of the charge generated by the oscillation resulting in the process of beta decay, in light of (2.127) as a transfer of linear momentum or

$$e \approx \frac{1}{\pi} 4 \left| \left(\left(\frac{\sqrt{2}}{\sqrt{3}} \right) \frac{\sqrt{3}}{4} \hbar c \right) \times \frac{1}{\sqrt{2}} K_0 \right| = \frac{\pm i \hbar \omega_0}{\pi} = \pm i 2 \hbar v_0 \qquad (2.134)$$

where v_0 is the cyclic frequency of the oscillation. Conceptually, this reflects the fact that for each cycle of the fundamental oscillation, there are two phases of capacitance and two of inductance, two antinodes of maximum charge and two antinodes of maximum current. The ambiguous sense indicates the oscillation through both semi-cycles. Evaluation and comparison with CODATA observation is at (6.4) and (6.28). If we make this an equation, based on this data, we have

$$e = 1.002405818... \left(\frac{\pm i \hbar \omega_0}{\pi} \right) \approx (1+Z_0) \frac{\pm i \hbar \omega_0}{\pi} = (1+\hbar \kappa_0 \omega_0) \frac{\pm i \hbar \omega_0}{\pi} \qquad (2.135)$$

$$\approx 1.002390877... \left(\frac{\pm \hbar \omega_0}{\pi} \right) = \pm \frac{1}{\pi} \left(\hbar \omega_0 + \hbar^2 \omega_0^2 \kappa_0 \right) = \pm \frac{1}{\pi} \left(G_0 + G_0^2 \kappa_0 \right) \qquad (2.136)$$

where Z_0 is the mechanical impedance of the continuum and G_0 is the transverse wave momentum. Addition of this factor is off the observed value by a factor of 0.000014905…which we might compare with (1.12) at 0.000014648….

Pursuing this a bit further, we have

$$e^2 \approx \frac{(\pm i\mathsf{n}\omega_0)^2}{\pi^2} = -\frac{\mathsf{n}(\mathsf{n}\omega_0^2)}{\pi^2} = \mathsf{n}\frac{\lambda_0 c^2}{\pi^2} = \hbar\frac{Z_0}{\pi^2} = \frac{Z_0^2}{\pi^2 (i\kappa)_0^2}. \quad (2.137)$$

In the next to last term, the inertial density times c is the mechanical impedance of the continuum. In light of our prior discussion concerning the monad, or differential horn torus, it bears noting that if we transpose the π^2 term to the left hand side, we have the form of the equation for the surface area of the horn torus.

Further development, using the familiar identity for the fine structure constant

$$\alpha^{-1} = \hbar c \frac{4\pi\varepsilon_0}{e^2} \quad (2.138)$$

and the permeability, μ_0, and permittivity, ε_0, relationship

$$\varepsilon_0 = \frac{1}{c^2 \mu_0} \quad (2.139)$$

shows

$$e^2 = -\mathsf{n}c^2 (\alpha 4\pi\varepsilon_0) = -\mathsf{n}\frac{\alpha 4\pi}{\mu_0} = -\mathsf{n}\frac{\alpha}{10^{-7}} \approx -\mathsf{n}\frac{\lambda_0 c^2}{\pi^2}. \quad (2.140)$$

It is noted that the value of μ_0 is set by convention for computational facility in relating charge, q, (of which elementary charge, e, is an effective quantum) and current, $i = dq/dt$, resulting in the exactness of the denominator of the next to last term. Since the negative sense of the right terms above can be attributed to the current squared, it can be incorporated therein, canceling such sense in the charge squared term. With reference to Appendix A, this suggests the transparent presence of a current squared argument in (2.140), for which the fine structure constant is a coefficient, since for one ampere^2 of current, where the denominator on the right is in Newton,

$$\frac{4\pi}{\mu_0} = \frac{i^2}{10^{-7}}. \quad (2.141)$$

Thus (2.140) becomes

$$e^2 = \mathsf{n}\frac{\alpha}{10^{-7}} i^2 \quad (2.142)$$

Given the dimensions of n, and since 10^{-7} has the presumed units of force, the fine structure constant then, is dimensionless, yet may be the ratio of two forces. If e has the units of momentum, then this would necessarily be the case and current would have the units of force, as $F = dP_M/dt$. Since the value of 10^{-7} is conventional, we would like to convert it to some natural expression of force, presumably related to cosmic expansion, as

$$\left(\frac{e}{i}\right)^2 = \mathsf{n}\frac{\alpha}{10^{-7}} = \mathsf{n}\alpha' = \mathsf{n}\frac{k^2}{\mathsf{n}\omega_e^2}, \tag{2.143}$$

where we have transposed the current to give the square of the ratio of elementary charge to current and which will structure α' as

$$\alpha' = \frac{k^2}{\mathsf{n}\omega_e^2} = \frac{\alpha}{10^{-7}}. \tag{2.144}$$

In combination with (2.134), (in which it is assumed that the preliminary factor of π^2 is included in k^2) this gives

$$\left(\frac{e}{ki}\right)^2 = \frac{\mathsf{n}}{\mathsf{n}\omega_e^2} = \frac{1}{\omega_e^2} = \frac{1}{\partial\omega_0^2} \tag{2.145}$$

in which the coefficient k is a normalizing factor, so that inverting we have the frequency differential arising from cosmic expansion and responsible for the generation of charge

$$\partial\omega_0 = k\frac{i}{e} = \omega_e. \tag{2.146}$$

This suggests and another look reveals that (2.140) is in fact a differential equation in which

$$e^2 = \left(\frac{\partial i}{\partial\omega_0}\right)^2 = \mathsf{n}\alpha' = \mathsf{n}\left(7.297352568...x10^4\right) C^2/\theta^2, \text{ thus} \tag{2.147}$$

$$e = \frac{\partial i}{\partial\omega_0} = 270.1361239\sqrt{\mathsf{n}}\ C/\theta \tag{2.148}$$

3 - Cosmic Expansion as the Driving EMF of a Quantum State

The Reactance States and Electron-Positron Generation

In light of the above correspondence between the potential-kinetic and electrical-magnetic energy cycles of the quantum spin, the cosmic expansion can be seen to provide a driving electromotive force or emf, $\mathcal{E}$, which is necessarily tuned to the natural or resonant angular frequency of any local section of the 3-D cosmos. "Necessarily" indicates that the 3-D tension in the 3-space surface of an expanding 4-core is a function of the inertial density of that 4-core, expressed as either energy or mass density as

$$T = \Pi''(\kappa,\omega) = {}_2\Pi_2 = {}_4\Pi c^2, \tag{3.1}$$

where the volume inertial density, ρ_0, is

$$\rho_0 = -\lambda_0 \kappa_0^2 = \Pi \kappa_0^4 = {}_4\Pi . \tag{3.2}$$

While the 4-core appears to have elastic properties, in keeping with (2.101) the decrease in inertial density arising from expansion leads to a decrease in 3-D tension. At the boundaries of the oscillation, W_{+x} and W_{-x}, this results in a differential <u>increase</u> in the transverse wave speed and a differential increase in the amplitude A for ϕ, with a concomitant contraction of the W_{+x}-W_{-x} axis and differential increase in the wave number, κ_0, in keeping with the gravitational quantum of (2.107). If c is invariant, this indicates a corresponding increase in ω_0. We have a paradox of sorts due to (2.101), however, in that a decrease in the inertial density requires a decrease in the wave force and of ω_0 over time, so that ω_0 at some future time $t(p)$, becomes $\omega_0 \to \omega_p$, where $\omega_0 > \omega_p$ for an inductive cosmic state, i.e. that of general expansion.

While the continuum itself, in the field, and the oscillation in a state of resonance will retain the invariance of c, so that as functions of time $c(t)$,

$$c = c(0) = \frac{\omega_0}{\kappa_0} = c(p) = \frac{\omega_p}{\kappa_p} \tag{3.3}$$

we might surmise that within the boundary of the driven oscillation itself

$$c = \frac{\omega_0}{\kappa_0 \pm \delta\kappa} = \frac{\omega_0 \pm \delta\omega}{\kappa_0}, \tag{3.4}$$

where the delta indicates a variation in the frequency or wave number.

With respect to a driving emf in an RLC circuit, the amplitude of the emf, $\mathcal{E}$, is related to the current amplitude, I, by the impedance, Z, as

$$I = \frac{E}{Z} \tag{3.5}$$

where the impedance is

$$Z = \sqrt{R^2 + (X_L - X_C)^2} \qquad (3.6)$$

and R is the resistance, X_L is the inductive reactance and X_C is the capacitive reactance.

In a condition of resonance, represented by Spin Diagram 2,

$$X_L - X_C = 0 \quad (\text{Resonant State}) \qquad (3.7)$$

and the impedance is equal to R and to the mechanical impedance of (2.91),

$$Z = Z_0 = \bar{n}\omega_0\kappa_0 = \frac{\bar{n}\omega_0^2}{c} = \lambda_0 c = R \qquad (3.8)$$

Since we might reason that $E = \bar{n}\omega_0^2$, (3.5) in a resonant condition can be expressed as

$$I = c = \frac{\bar{n}\omega_0^2}{\bar{n}\omega_0\kappa_0} = \frac{\omega_0}{\kappa_0} \qquad (3.9)$$

where the current amplitude is equal to the speed of wave propagation.

The inductive reactance, X_L, is equal to the product of the driving frequency, $\omega_d = \omega_0$, and the inductance, L, while the capacitive reactance, X_C, is equal to the inverse product of the driving frequency, ω_0, and the capacitance, C, or

$$X_L \equiv \omega_0 L \text{ and} \qquad (3.10)$$

$$X_C \equiv \frac{1}{\omega_0 C}. \qquad (3.11)$$

The inductance then is the ratio of the wave momentum per wave velocity to the changed impedance of the expanding medium, arising from a change in resonant frequency squared, or changed force per wave velocity,

$$L = \frac{\bar{n}\omega_0 c^{-1}}{\bar{n}\omega_p^2 c^{-1}} = \frac{P_{M0} c^{-1}}{\tau_p c^{-1}}. \qquad (3.12)$$

The capacitance is the ratio of the fundamental mass as determined by the inertial density of the resonant and expanding medium to the impedance of the driving emf, or

$$C = \frac{\bar{n}\kappa_p}{\bar{n}\omega_0\kappa_0} = \frac{m_p}{Z_0} \qquad (3.13)$$

Therefore the inductive reactance, the ratio of the driving impedance to the decreased impedance of expansion, of the driving frequency squared to the decreased frequency squared, and of initial acceleration to subsequent acceleration, is

$$X_L = \frac{\hbar\omega_0^2 c^{-1}}{\hbar\omega_p^2 c^{-1}} = \frac{Z_0}{Z''(\omega)} = \frac{\omega_0^2}{\omega_p^2}. \quad (3.14)$$

The capacitive reactance, the ratio of the driving impedance to the decreased impedance due to a change in wave number or the ratio of initial to subsequent wave numbers, is

$$X_C = \frac{\hbar\omega_0 \kappa_0}{\hbar\omega_0 \kappa_p} = \frac{Z_0}{Z'(\kappa)} = \frac{\kappa_0}{\kappa_p} \quad (3.15)$$

For an inductive state indicated by the above,

$$X_L \equiv \omega_0 L > \frac{1}{\omega_0 C} \equiv X_C, \quad (3.16)$$

while for a capacitive state,

$$X_L \equiv \omega_0 L < \frac{1}{\omega_0 C} \equiv X_C \quad (3.17)$$

For an expanding section of the cosmos, we would expect an inductive state to predominate; in a contracting section, we would expect a capacitive state.

We might anticipate that for such an inductive state the following would be found for a half spin oscillation,

$$\frac{\sqrt{3}}{2} I = \frac{E}{\sqrt{Z_0^2 + (X_L - X_C)^2}}$$

$$\frac{\sqrt{3}}{2} c = \frac{\hbar\omega_0^2}{\sqrt{(\hbar\omega_0\kappa_0)^2 + \left(\frac{\omega_0^2}{\omega_p^2} - \frac{\kappa_0}{\kappa_p}\right)^2}} \quad (3.18)$$

Solving with the CODATA values of c, $\kappa_0 = \frac{m_n}{\hbar}$, and $\omega_0 = c\kappa_0$ per the evaluation section gives

$$X_L - X_C = 1.380373411...\times 10^{-3} \quad (3.19)$$

Some further algebra finds the following dimensionless ratios for the change in ω and κ of

$$R_m = \frac{\omega_0}{\omega_p} = \frac{\kappa_0}{\kappa_p} = \frac{\hbar\kappa_0}{\hbar\kappa_p} = \frac{m_n}{m_p} = 1.0013784732... \quad (3.20)$$

where the third ratio is that of the product of the inertial constant and the wave numbers of the driving oscillation and the driven resonant oscillation and the fourth is that of the mass of the neutron to that of the proton.

The deviation of this derived theoretical ratio from the CODATA observed ratio is

$$R_{m\ theoretical} - R_{m\ observed} = 1.0013784732... - 1.00137841870(58) = 5.45\times 10^{-8}. \quad (3.21)$$

While this is slightly outside the relative standard uncertainty indicated for the observed figure, when it is coupled with the broad uncertainty assigned to the gravitational constant, which no doubt enters systemically into the observed computation and its deviation from the theoretical value herein arrived at of 0.000015019, as per (1.12) the correlation is significant.

Of related interest is the following, where

$$\Delta_{\omega_0} = R_{m\,t} - 1 = \frac{\omega_0 - \omega_p}{\omega_p} = 0.0013784732... \quad (3.22)$$

so that the quotient of the mechanical impedance and the relative change in fundamental frequency, $\Delta_{\omega 0}$, times $R_{m\,t}$ is

$$\frac{Z_0}{R_{m\,t}(R_{m\,t}-1)} = \frac{\omega_p}{\Delta_{\omega_0}} \hbar \kappa_0 = \frac{\omega_p}{\Delta_{\omega_0}} m_0 = 1.732050837... \cong \sqrt{3} \quad (3.23)$$

within a deviation of 2.9×10^{-8}. An identical result applies to the corresponding operation of the wave number, κ. The appearance of $\sqrt{3}$ once again suggests a 4-D relationship.

The interpretation is that in the context of certain sufficient conditions of nuclear congregation and the confines of an inertial sink or font, the fundamental oscillation of frequency ω_0 and wave number κ_0 has an impedance matching the driving emf of cosmic expansion. These fundamentals decrease slowly, but exponentially over time with cosmic expansion in keeping with the driving frequency, ω_d, which, along with the mechanical impedance, Z_0, is a function of decreasing inertial density, λ_0, of an expanding cosmic 4-core.

Absent such congregation or confinement, the fundamental oscillation responds as a driven wave to the resonant frequency of the local continuum, ω_p, which, along with the mechanical impedance, Z_p, is a function of the exponentially decreasing inertial density, λ_p, of the expanding cosmic 3-surface. We can think of the differential between ω_d and ω_p as either a time or space gradient, as it is a spacetime gradient. The coefficient for the current amplitude of (3.18) arises from the rotational dynamics of the oscillation as outlined above, itself an expression of the 4-d orthogonality of those dynamics.

While this accounts for the energy of the proton, we still have that of the electron to include in the equation. We would anticipate that along with the transmitted energy of the proton, once again as a transitional state or conversion of the potential energy of inertial density to the dynamics of oscillation, there is a reflected component that accounts for the electron, as well as perhaps some other component that is accounted for by the neutrino in the standard model.

With reference to the Appendix – Wave Transmission at a Discontinuity, which is a one dimensional ideal string model, since our equations make use of the linear inertial density of the medium we can use the final impedance terms in the equations to solve for the complex amplitude reflection, $\check{R}_a$, and transmission, $\check{T}_a$, coefficients or ratios, and for the power reflection, R_p, and transmission, T_p, coefficients, which depend only on the properties of the medium. The complex amplitude reflection and transmission coefficients are applied to the complex modulus of the fundamental amplitude. The impedance for the driving cosmic core, Z_0, and expanding surface, Z_p, from above are

$$Z_0 = Z_1 = \hbar\omega_0\kappa_0 = 0.002390876881...kg\cdot\theta/s \qquad (3.24)$$

$$Z_p = Z_2 = \hbar\omega_p\kappa_p = 0.002384298968...kg\cdot\theta/s. \qquad (3.25)$$

and the above coefficients are

$$\check{R}_a = \frac{Z_1 - Z_2}{Z_1 + Z_2} \qquad (3.26)$$

$$\check{T}_a = \frac{2Z_1}{Z_1 + Z_2} \qquad (3.27)$$

$$R_p = \left(\frac{Z_1 - Z_2}{Z_1 + Z_2}\right)^2 \qquad (3.28)$$

$$T_p = \frac{4Z_1 Z_2}{(Z_1 + Z_2)^2} \qquad (3.29)$$

resulting in the following evaluations.

$$\check{R}_a = 0.001377522755... \qquad (3.30)$$

$$\check{T}_a = 1.001377522755... \qquad (3.31)$$

$$R_p = .000001897568941... \qquad (3.32)$$

$$T_p = 0.9999981024... \qquad (3.33)$$

Applying the dimensionless coefficient, $\check{R}_a$ to the amplitude = inverse wave number of the neutron, gives us an amplitude of the electron, A_e, as the reflected oscillation of

$$A_e = \check{R}_a \frac{1}{\kappa_0} = \check{R}_a \frac{\hbar}{m_n} = 2.893065241...\times 10^{-19} m. \qquad (3.34)$$

Since the proton has no established decay rate, and we have shown that its energy, therefore its frequency, is a direct function of a decreasing cosmic inertial density, we might anticipate that $\check{R}_a$, as with the other coefficients, is a differential or derivative of that decrease and of expansion. With respect to (2.145) through (2.148) we still do not have a value for (2.146) or k itself, though we can find it empirically by the ratio of neutron to electron mass, which according to the CODATA values is

$$\frac{m_n}{m_e} = 1838.6836598(13). \qquad (3.35)$$

We would like to find some mechanism, both on a quantum or intrinsic and cosmic or extrinsic level that would make the determination, since it is the value of this differential that determines k and α. We discussed this previously at (2.55) to (2.57). Recapitulating from that discussion, the variation of the unit area arising from a change in expansion stress over time is

$$\delta = \delta A = \delta r^2 = 0.0005441353061.... \tag{3.36}$$

The first delta is a coefficient which references the number magnitude without any applicable dimensions or units. The variance in the stress manifests itself in a fluctuation in the fundamental frequency and wave number of the oscillation over time within a limit that precludes transmission through the boundary. At the time such limit is exceeded and transmission occurs, therefore, we would anticipate the following ratio and range to hold, shown here for the wave number, but equally applicable to the frequency,

$$0.0005438393... < \frac{\delta \kappa}{\kappa_0 \pm \delta \kappa} < 0.0005444315.... \tag{3.37}$$

While in linear proportion to the variance, the change in wave number would tend to be negative in an expanding medium or toward a smaller number, that is to a lower ω and a proportionally lower κ, if c is invariant.

With reference to the energy of the system, we might imagine that while over time the kinetic and potential energy in resonance is balanced, at any instant the above variance may be in play so that the Lagrangian of the fundamental oscillation, where the kinetic and potential components cancel, is

$$L = (K_\phi - V_\phi) \pm \delta E = (0) \pm \delta E \tag{3.38}$$

The Hamiltonian or total energy of the system, then is

$$H = 2K_\phi - L = 2K_\phi \mp \delta E = K_\phi + V_\phi \mp \delta E \tag{3.39}$$

so that

$$\begin{aligned} H - \delta E &= 2K_\phi - \delta E = E_0 + E_p + \Delta E_{0-p} - \delta E_0 \\ &= E_0 + E_p + (\Delta_{0-p} - \delta) E_0 \\ &= \hbar \kappa_0 c^2 + \hbar \kappa_p c^2 + (\Delta_{0-p} - \delta) \hbar \kappa_0 c^2 \\ &= \hbar c \omega_0 + \hbar c \omega_p + (\Delta_{0-p} - \delta) \hbar c \omega_0 \end{aligned} \tag{3.40}$$

The variation on the left hand side represents the differential expansion energy, the first two terms on the right side are the neutron and proton mass-energy equivalence and the bracketed term is their difference. Since the amplitude and the inverse wave number appear to be equated in the case of the fundamental quantum oscillation, we might imagine an analogy with respect to the reflected wave, with the understanding that the smaller amplitude indicates a smaller wave number for the reflection. Thus substituting the reflection amplitude coefficient for the difference in wave number/angular frequency, and using the variance from (3.36), gives a difference of

$$\begin{aligned} \Delta_{0-p} - \delta &\simeq \check{R}_a - \delta = 0.001377522755... - 0.0005441353061... \\ &= 0.0008333874489... \end{aligned} \tag{3.41}$$

From another approach, with respect to (7.139) in the appendix on exponentiation, we might surmise that the acceleration normalizing factor represents a change in the Hamiltonian due to a dilative change in the time scale associated with beta decay,

$$ie_2 = \frac{\partial H}{\partial t_\beta} = i1.531584397... \tag{3.42}$$

so that canceling the δE terms from both sides, we have

$$H = \mathsf{n}\kappa_0 c^2 + \mathsf{n}\kappa_p c^2 + (1+ie_2)\delta' E_0 \text{ and} \tag{3.43}$$

$$H = \mathsf{n}c\omega_0 + \mathsf{n}c\omega_p + (1+ie_2)\delta' E_0 \tag{3.44}$$

where

$$\delta' = \frac{\check{R}_a}{(1+ie_2)} \tag{3.45}$$

With evaluation, we have

$$\delta' = 0.0005441346363... \tag{3.46}$$

$$ie_2\delta' = i0.0008333881188... \tag{3.47}$$

where apparently

$$ie_2\delta' = i(\Delta_{0-p} - \delta) \tag{3.48}$$

Note that δ is strictly a geometric derivation from (3.36) with no direct connection to the derivation of δ' from (3.30), and that their difference is

$$\Delta\delta = (\delta - \delta') = 6.7 \times 10^{-10}, \tag{3.49}$$

a significant correlation. A similar condition results where $ie_2\delta'$ is derived conceptually using (7.135) and (3.45) and independent of $\Delta_{0-p} - \delta$.

This suggests that the following are invariants of the system, where

$$\delta = \delta' \text{ and} \tag{3.50}$$

$$(1+ie_2) = \frac{\Delta_{0-p}}{\delta} = 2.531584397.... \tag{3.51}$$

From (3.39) and (3.40), where κ_e is the wave number of the reflected wave, we have the following equations in terms of potential (3.52) and kinetic (3.53) energy,

$$\mathsf{n}\kappa_0 c^2 - \Delta_{0-p}E = \mathsf{n}\kappa_0 c^2 - \mathsf{n}c^2(1+ie_2)\delta\kappa_0 = \mathsf{n}\kappa_0 c^2 - (1+ie_2)\mathsf{n}\kappa_e c^2 = \mathsf{n}\kappa_p c^2. \tag{3.52}$$

$$\mathsf{n}c\omega_0 - \Delta_{0-p}E = \mathsf{n}c\omega_0 - \mathsf{n}c(1+ie_2)\delta\omega_0 = \mathsf{n}c\omega_0 - (1+ie_2)\mathsf{n}c\omega_e = \mathsf{n}c\omega_p \tag{3.53}$$

Dividing through by the energy of the reflected wave, gives the coefficients of the driving, E_0, and resonant/transmitted, E_p, energies, in terms of their wave numbers and frequencies, and finally of rest masses with respect to the fundamental differentials

$$\frac{\kappa_0}{\kappa_e} - (1+ie_2) = \frac{\kappa_p}{\kappa_e} \tag{3.54}$$

$$\frac{\omega_0}{\omega_e} - (1+ie_2) = \frac{\omega_p}{\omega_e} \tag{3.55}$$

$$\frac{m_0}{m_e} - (1+ie_2) = \frac{m_p}{m_e} \tag{3.56}$$

By virtue of (2.83) and the CODATA values for the ratio of the neutron-electron mass, (3.35), and (3.51) expressed in terms of a fundamental unit value of δ, we have
$$1838.6836598... - 2.5315844... = 1836.152075.... \qquad (3.57)$$
where the CODATA value for the last term is 1836.15267261(85). Expressing this ratio in terms of the fundamental resonant energy given by κ_p, ω_p, and m_p, we have
$$\frac{\kappa_e}{\kappa_p} = \frac{\omega_e}{\omega_p} = \frac{m_e}{m_p} = 0.000544617199.... \qquad (3.58)$$

The deviation from the CODATA value for (3.58), then is
$$0.000544617199... - 0.00054461702173(25) = 1.773...\text{x}10^{-10}. \qquad (3.59)$$
Alternately, working backwards from the CODATA value for (3.58) gives us a theoretical value of
$$\frac{\kappa_e}{\kappa_0} = \frac{\omega_e}{\omega_0} = \frac{m_e}{m_0} = 0.0005438671685... \qquad (3.60)$$
and a similar deviation from the CODATA value. The theoretically derived value of the neutron to electron energy ratio, then compared with (3.35) is
$$\frac{\kappa_0}{\kappa_e} = \frac{\omega_0}{\omega_e} = \frac{m_0}{m_e} = 1838.684256.... \qquad (3.61)$$

Cosmic Expansion Rate and Expansion Force

With respect to the expansion force, we would expect equating of the variance operator with respect to a change in the local scales, t_l and x_l, and the differential operator with respect to cosmic time, t, and space, x, so that

$$\delta\kappa_0 = \partial\kappa_0 = \kappa_e, \text{ and} \tag{3.62}$$

$$\delta\omega_0 = \partial\omega_0 = \omega_e \tag{3.63}$$

A change in the fundamental mechanical impedance, using (2.101) then becomes

$$\partial Z_0 = -\mathsf{n}\kappa_e\omega_e = -\mathsf{n}\kappa_e^2 c = -\frac{1}{c}\mathsf{n}\omega_e^2 \tag{3.64}$$

$$-\mathsf{n}\frac{\partial\kappa_0^2}{\partial t} = -\frac{1}{c}\mathsf{n}\frac{\partial\omega_0^2}{\partial x} \tag{3.65}$$

$$-\frac{\partial\lambda_0}{\partial t} = -\frac{1}{c}\frac{\partial\tau_0}{\partial x} = -\frac{\partial Z_0}{\partial x} \tag{3.66}$$

In this final arrangement, a change in the fundamental linear inertial density over time arising from expansion is viewed as a conserved or unitary mass/energy distributed over an increasingly larger unit length, and is equal to the change in continuum impedance over a changing unit distance and scale. This can be seen graphically by referring to the Matrix of Invariants. As the inertial density differential is a linear function of that changing unit length, and as the time and distance scales are held to be invariant with respect to each other by virtue of c, then the time scale is increasing as well. In terms of a fixed scale of time, however, present, past or future, it is apparently an exponential function and in fact a measure of the cosmic expansion rate, X_e, and we should find

$$X_e = -\frac{1}{c}\mathsf{n}\frac{\omega_e^2}{\partial x} = -\left|\frac{1}{c^2}\mathsf{n}\omega_e^2\right| = -\left|\frac{1}{c^2}\partial\tau_0\right| = -\mathsf{n}\frac{\kappa_e^2}{\partial t} = -\frac{\partial\lambda_0}{\partial t}. \tag{3.67}$$

where the figures in absolute value brackets are converted from the second term by complex integration with respect to time.

Substituting the product of (3.60) and the CODATA based value of ω_0, gives an evaluation of

$$X_e = 2.35896879...x10^{-18}\Delta m/m/s \tag{3.68}$$

where the accelerating change in wave number is masked by the change in unit length.

This number, the rate of change in a meter unit of spacetime, per second, times the number of meters per megaparsec, gives the expansion rate in terms of the Hubble constant or

$$H_0 = X_e\left(3.08572x10^{22}m/mps\right) = 72,791.17172 m/mps/s \tag{3.69}$$

A study by Ron Eastman, Brian Schmidt and Robert Kirshner in 1994 and quoted in Kirshner's recent book, The Extravagant Universe, found an H_0 = 73 km/s/mps +/- 8km and an article in the *Astrophysical Journal*, 533, 47 - 72, (2001) by Freedman, W. L. et al.

gives the final results from the Hubble space telescope key project to measure the Hubble constant as H₀ = 72 km/s/mps. There are 3.08572×10^{22} meters per megaparsec.

Thus, if the Hubble rate of expansion is roughly 73 kilometers per second per megaparsec, and since there is no logical compulsion to think that we are at the current center of the universe, (except in the sense of a 3-space layer moving out from the center of a 4-core), this would tend to indicate that every local section of space, absent gravitational and electromagnetic constraints, is moving away from every other at approximately 2.35×10^{-18} meters per second per meter of separation. It follows logically that inversion of this number will give us the approximate time since all the matter was at the same locale, that the universe has been expanding, or 4.25×10^{17} seconds, which is roughly 13.5 billion years.

However, as this number might be deemed to represent an expansion via a compounded augmentation of the scale of spacetime itself, and not simply an extension of matter within that spacetime, we might surmise that this represents an exponential expansion, in which case the following equation for the doubling of spacetime applies, as

$$\tau_H = \ln 2 X_e^{-1} = 2.938 \times 10^{17} s = 9.311 \text{ billion years}, \qquad (3.70)$$

This indicates that spacetime is doubling at a current rate of every 9.311 billion years, measured in terms of today's seconds. If we assume that the wavelength of the cosmic background radiation at approximately 5mm embodies that augmentation, while harkening back to a period of primal beta decay as indicated by the Compton wavelength over 2π of an electron, this represents a doubling of some 30 times, or

$$\frac{\ln\left(\frac{.005}{2\pi}\right)}{\ln 2} = \frac{\ln 2.060....\times 10^9}{\ln 2} = 30.94...doublings \qquad (3.71)$$

a lifetime in terms of today's measure of time of roughly 288 billion years.

As

$$\ln 2 = 0.693147181... \qquad (3.72)$$

it is worth noting that this figure is effectively 70%, the factor of expansion attributed in current cosmological schemes to dark energy.

As this expansion rate as found in the Hubble constant is also found in the differential change in the linear inertial density and tension force operating on a fundamental quantum oscillation, it is confirmation of the fact that cosmic expansion provides the emf driving such oscillations.

Returning to (2.144), rearrangement gives

$$\partial n_2 \equiv n\omega_e^2 = n\partial\omega_0^2 = \frac{k^2 i^2}{\alpha'} = 0.212013542...N. \qquad (3.73)$$

Returning to (2.146), which we can rearrange and solve, using the CODATA value for elementary charge, gives

$$ki = e\partial\omega_0 = e\omega_e = 124.3839844...i \text{ (amperes)}. \qquad (3.74)$$

Thus the SI current per quantum differential change in frequency associated with beta decay is equal to the fundamental charge of the system. If we return to (2.147) and divide through by k to arrive at a normalized and naturalized value of elementary charge, e', we have,

$$\frac{e}{k} = e' = \frac{\partial i}{\partial \omega_0} = \frac{1}{\omega_e} = 1.288089085...x10^{-21} s/\theta \qquad (3.75)$$

which shows charge to be the inverse of the quantum differential change in angular frequency.

A final speculation regarding beta decay concerns the decay rate of the neutron, τ_n, which is reported in a paper by R.R.Kinsey, et al.,*The NUDAT/PCNUDAT Program for Nuclear Data* as being 624 seconds. Other recent sources report it in the 887 +/- 2 second range. If instead of the standard use of the *ln* 2 as the dividend used in computing a half life, in light of the above analysis of a 4-wave, we use the *ln* $\sqrt{2}$ to indicate the doubling of a 4-core, we have

$$\tau_n = \frac{\ln \sqrt{2}}{\frac{\omega_e}{\omega_0}} = 637.239... \text{ seconds}. \qquad (3.76)$$

4 – Strong Interactions

The above development shows the generation of quantum gravity and the electroweak interaction as a function of the spin dynamics of a fundamental oscillation. As functions of a single quantum, they are not so much "interactions" as intra-actions within the domain of a discrete, rotational oscillation of the spacetime continuum. With respect to interactions <u>between</u> separate oscillations, these functions are quantized by the fundamental characteristic dynamics of that continuum. The strong "force" that binds nucleons, by contrast, is truly a function between and among separate quanta, and can be understood to vary continually within a certain range. With respect to the domain of the strong interaction, we can imagine that it corresponds with the intersecting static limits of two adjacent quanta represented as extreme Kerr inertial sinks as in the Quantum Inertial Sink Diagrams 1 & 2 at the beginning of this development.

According to Boyle's Law, given adiabatic constraints, the energy per volume of a gas is equal to the pressure on the boundary of that volume. If we apply a similar logic to our discussion of spacetime in which expansion appears to be adiabatic, then density of the spin energy, E_0, within the boundary of a rotational oscillation will be equal to the tension at that boundary, held to be generally spherical, as

$$\frac{E_0}{V} = \frac{\tau_0}{A_0} = f_0 \text{ where} \qquad (4.1)$$

$$\frac{E_0}{V} = \frac{i\hbar \kappa_0^{-1} \omega_0^2}{\frac{4}{3}\pi (i\kappa)_0^{-3}} = -\frac{3\hbar \kappa_0^2 \omega_0^2}{4\pi} \qquad (4.2)$$

The tension, however, is also a function of the boundary configuration, as a given volume can be bound by a variable surface area, so that a decrease in boundary area, given no change in volume, must result in a decrease in the wave force if the energy is to be conserved. Disregarding the small difference between the spin energy of the neutron and proton, if two equal volumed quanta with the same energy density are brought in synchronic contact so that there is <u>no</u> tension <u>gradient</u> at their common boundary, the decrease in their total boundary area <u>with</u> <u>gradient</u> before and after conjunction results in a proportional decrease in the wave force of each. We can approximate this in general terms by looking at the difference between the combined surface area of two separate, equal spherical volumes and the surface area formed by two hemispheres connected by a cylinder of like radius about their combined volume. The following table shows this as,

Wave State	Volume	Total Surface Area	Difference	% Difference
2 Single Waves	$\frac{8\pi}{3}r^3$	$8\pi r^2$		
2 Conjoined Waves	$\frac{8\pi}{3}r^3$	$4\pi r^2 + \frac{8}{3}\pi r^2$	$\frac{4}{3}\pi r^2$	1/3 of a Single 1/6 of Combined

We might expect, then, a reduction in wave force of 1/6 to 1/3 of the quantum fundamental with the addition of each nucleon to an atomic nuclear congregation, which would result in an apparent reduction by a related amount in the mass/energy of the system. Hence this energy appears to be bound up in the nucleus, with the actual amount depending on the geometric configuration of the nucleus. In effect what occurs is a conversion of spin energy density, E_1, to the potential energy density of continuum mass, V_1, as a reduction in the wave number, and a net decrease in the wave tension, f_0, where

$$-\Delta E_1 = +\Delta V_1, \text{ where} \qquad (4.3)$$

$$E_1 = \frac{3f_0}{4\pi} = \frac{3\tau_0}{4\pi A_0} = \frac{3\tau_0}{4\pi}\kappa_0^2 \qquad (4.4)$$

In light of the above development of the nucleons as strain oscillations of spacetime, this shows the energy-stress force, $E_0\text{-}3\tau_0$, to be divided over the 4π steradians of the spherical volume. Therefore, the energy density of (4.4) is the energy per steradian of the oscillation. Taking the square root gives us the energy, and according to the matrix of invariants the other dynamic functions of spin, per radian of strain travel. Thus the ranges from the table above give the change of (4.3) per radian and relate that change to the primary parameters, ω_0 and κ_0, and thereby to energy and mass, and we have

$$\sqrt{\frac{1}{6}E_1} < \Delta E_1 < \sqrt{\frac{1}{3}E_1}, \qquad (4.5)$$

$$\sqrt{\frac{1}{6}\frac{3}{4\pi}f_0} < \Delta f_0 < \sqrt{\frac{1}{3}\frac{3}{4\pi}f_0}. \qquad (4.6)$$

This change in energy or binging energy is generally measured in million electron volts, Mev. In the Mev system, mass is converted to elementary charge and the time and distance dimensions are set to 1 or $1\,Mev = (mc^2/e)\times 10^6$. The binding energy per nucleon observed is in the 6 to 8.794… Mev range for all but the lightest elements, for which it is less. The upper end is the value for the binding energy of nickel, ^{62}Ni, as the most stable of elements. The only stable elements beneath the lower threshold are hydrogen, helium-3, and lithium, with helium-4 in-range at a stable 7.074 Mev.

The mass of the neutron is then 939.565…Mev, and the proton is 938.272…Mev. The energy density per steradian for their average is then

$$\frac{m_0 c^2}{eV} = \frac{3(938.918\ldots)}{4\pi} = 224.150\ldots Mev \qquad (4.7)$$

The range for the change in energy due to nuclear congregation as derived above is then

$$\sqrt{\frac{224.150\ldots}{6}}Mev < \Delta E_0 < \sqrt{\frac{224.150\ldots}{3}}Mev \qquad (4.8)$$

$$(6.112\ldots < \Delta E_0 < 8.643\ldots)\,Mev$$

Note that the square root operation is effectively the reverse of a cross product, in which the crossing of two orthogonal tangent vectors produces a radial or normal vectors. In that case the crossing of two forces can result in a third orthogonal force. Here, the square root decomposes a normal vector into two orthogonal tangent vectors. The resultant energy or force vectors, depending on the particular dynamics being analyzed, are in units of energy or force. The square root operates on Mev and not ev, since the basic units of energy density and stress are volume units and the unit Mev is in the same scale as the energy of the electron at .510... Mev or the difference between the proton and neutron at 1.293...Mev. Thus the SI stress force equivalent to energy in Mev remains in Newton or mass distance per time squared, and not the square root thereof.

As a refinement, with respect to nuclear congregation, twelve equal volume spheres will pack tightly around a thirteenth, so that we might imagine instead of distribution of the energy over 4π steradians, a maximum distribution over 12 equal sectors of $\pi/3$ steradians. The resulting range then becomes

$$\sqrt{\frac{234.729...}{6}} Mev < \Delta E_0 < \sqrt{\frac{234.729...}{3}} Mev, \qquad (4.9)$$

$$(6.254... < \Delta E_0 < 8.845...) Mev$$

the upper boundary of which is slightly over the observed upper limit to binding energy per nucleon. This is the energy per $\sqrt{\frac{\pi}{3}}$ radians. Note the tie in of this configuration with the comments on the weak mixing angle at (2.58). The top four elements in this regards, two nickel and two iron isotopes, taken from a study by Wapstra and Bos quoted at http://hyperphysics.phy-astr.gsu.edu/hbase/nucene/nucbin2.html, are

Nuclide	Mev per nucleon
^{62}Ni	8.79460 +/- 0.00003
^{58}Fe	8.79223 +/- 0.00003
^{56}Fe	8.79036 +/- 0.00003
^{60}Ni	8.78079 +/- 0.00003

Finally, with respect to the strong interaction and the quantum gravity differential, using the upper limit for comparison, we have a theoretical dimensionless ratio between the two

$$\frac{F_S}{F_{Gq}} = \frac{\sqrt{\frac{\pi}{3}}\sqrt{\frac{1}{4\pi}}\tau_0}{G_q dT} = \frac{\sqrt{\frac{1}{12}}\hbar\omega_0^2}{\frac{1}{6\sqrt{3}\kappa_0^2}dT} = \frac{3\hbar\kappa_0^2\omega_0^2}{dT} = 4.8750635...\times 10^{37} \, T/dT \qquad (4.10)$$

Comparing this with (4.2), we see that it is equal to the spin energy-stress density of the oscillation integrated over 4π steradians; that is, the volume with respect to E_0 and the surface area with respect to f_0.

5 - Cosmological Considerations and Speculations

Whether we see the decrease in inertial density with expansion as an elastic strain and concomitant stress <u>pulling</u> on the rotating wave front of a quantum oscillation or the densification of that front as an inertial body force <u>pushing</u> back on that front, the existence of forces that result in the restoration of any strain of the wave medium defines it as being elastic. Even an instance of inertial wave motion, as with an ideal jump rope, in which the kinetic energy is constrained, relies on the modulus of rigidity of the rope for the inertial constraint. As the modulus of rigidity is the shear modulus which can be expressed as a function of some tension modulus, it is clear that even an example of an ideal inertial wave depends on elastic properties for its expression. Attempting to make a jump rope out of bread dough will prove the point after only a few rotations. Apparently then, inertia and elasticity, or inertial density and tension/shear stress-strain are two phenomenalogical aspects of an underlying inertial continuum ontology.

Perhaps the real question is whether or not, in addition to being elastic, the spacetime continuum is capable of plasticity. Obviously, the elastic limits cannot be exceeded in the short term on the level of a quantum wave, or the wave would quickly dissipate, i.e. the medium would not sustain oscillation. Attempt to make a drumhead out of uncooked bread dough. On the larger scope of things, the vast areas of cosmic space devoid of galactic presence suggest regions of spacetime that have exceeded the limits of oscillatory strain and are in plastic flow. The whole issue of the missing dark matter which is deemed to hold sway at the peripheries of galaxies might itself vanish if we find that the tug between the galactic nuclear dynamics of gravity with increasing angular momentum and the isotropic expansion of the galactic environment are sufficient to exceed those elastic limits in the middle interstellar regions, leaving the outer regions to follow the dictates of angular momentum while the center region is under the domain of gravity and angular acceleration.

A terrestrial analogy of sorts can be found in the demolition of an old industrial brick chimney. When of sufficient height and toppled by explosion so that it pivots at its base, the potential for angular acceleration of the top can exceed that of gravitational interaction or free fall alone. If the shearing force arising from angular acceleration exceeds the bonding strength of the mortar, that acceleration can not be transmitted to the top portion of the chimney, which will break into two parts. The topmost part of the bottom portion is accelerated by angular forces greater than the gravitational ones and falls first with greatest velocity. The speed at impact for any differential length of the bottom section is proportional to its distance from the center of rotation, while the top portion is under the sway of gravity alone and hits the ground last, all portions at the same velocity.

The spacetime continuum itself is the cosmic mortar that keeps the quantum waves that comprise the stellar systems together. In a somewhat different scenario from our falling chimney, if a region of spacetime itself is under rotational stress to convey a common angular acceleration to the halo around a galactic core, we might imagine the rigidity of spacetime to be maintained to a yield boundary after which the shearing stresses would exceed the limit imposed by the shear modulus and a plastic state would ensue. Barred spiral galaxies would indicate an inner region of relative rigidity and elastic integrity of spacetime, with conventional spirals indicating a wider area of plasticity. As the tangential component of the angular acceleration increases, so does the centripetal component, and as the stress at the outer extents of the galactic arms increases to the yield limit, a state of plasticity ensues between those extents and the adjacent inner section. The new outer extent of the area within the galactic elastic circumference next accelerates to the tangential velocity productive of plastic flow, resulting in a differential shrinkage of the elastic circumference. The angular momentum already imparted to the outer regions at the inception of the yield would be maintained as a constant tangential velocity.

Apparently regions of maintained elasticity can exist within larger regions of plasticity and vice versa. If the elasticity, as the stress-strain relationship, is a function of inertial density, it would appear to be independent of time, and therefore reversible. Any geometrically defined process resulting in a densification or confinement of a plastic region would be capable of restoring elasticity.

With that caveat, we can now join the pieces of the spin function with the expansion stress. For an elastic medium of tension stress T_0, (where we call the stress a tension stress, although it is orthogonal to all of 3-space), with a strain of $\varepsilon = \frac{\Delta l}{l} = \frac{r_0}{1\ meter} = |\kappa_0^{-1}|$, the orthogonal modulus of elasticity, Y, is

$$Y = i\frac{T_0}{\varepsilon} = i\frac{T_0}{r_0} = 6\sqrt{3}(i\kappa_0)f_0 = 6\sqrt{3}\,_3\mathsf{n}_2$$

$$= i\frac{1.68877...x10^{38}\ N/m^2}{2.10019...x10^{-16}\ \Delta m/m} = i8.041025...x10^{53}\ N/m^2$$

(5.1)

The 4-D, either hyper-volume or spacetime, potential force density then is equal to the inverse of the Planck area, A_{Planck}, expressed as a derivative of area with respect to stress and not the differential area, dA,

$$F_1 = -\frac{Y}{r_0} = \frac{-T_0}{\varepsilon^2} = 6\sqrt{3}(i\kappa_0)^2 f_0 = \frac{f_0}{G_q} = 6\sqrt{3}\,_4\mathsf{n}_2$$

and (5.2)

$$= -\frac{1.68877...x10^{38}}{4.41082...x10^{-32}} = -3.82871...x10^{69}\ (N/m^2)/m^2 = -A_{Planck}^{-1}$$

$$A_{Planck} \equiv \frac{dA_0}{dT_0} = \frac{A_0}{T_0} = \frac{2.61184...x10^{-70}\ m^2}{1N/m^2}$$

(5.3)

The inertial notation, in which all orthogonal sense are tacit, points to the fact that although the modulus of elasticity is conventionally given the same dimensions as the stress, and the strain is a dimensionless number, the modulus in this case is actually an orthogonal stress derivative with respect to a change in extension or strain, which is a linear-stress/volume-force potential, as

$$f_0 = Yi\Delta r = \frac{1}{6\sqrt{3}}\frac{dT_0}{idr}i\Delta r = \frac{d\tau_0}{dA_0 \cdot idr}i\Delta r$$

(5.4)

which upon being integrated by $i\Delta r$, (dividing by $i\kappa$ in the Euler formalism), produces a stress, f, and (5.2) is then a 4-D force potential, which is integrated by the quantum gravitational constant, so that

$$f_0 = F_1 G_q = \frac{G_q}{A_{Planck}} \tag{5.5}$$

With respect to the rest of the parameters of the Planck scale, the square root of (5.2) gives us the inverse Planck length or what we can call the Planck wave number as

$$A_{Planck}^{-\frac{1}{2}} \equiv r_{Planck}^{-1} \equiv \kappa_{Planck} = 6.18765...\text{x}10^{34}\, m^{-1} \tag{5.6}$$

which with (2.84) gives the Planck mass, m_{Planck},

$$m_{Planck} \equiv \sqcap(i\kappa_{Planck}) = 2.17661...\text{x}10^{-8}\, kg \tag{5.7}$$

and with c gives us the inverse Planck time, or what we can call the Planck frequency as

$$t_{Planck}^{-1} \equiv \omega_{Planck} \equiv c(i\kappa_{Planck}) = 1.85501...\text{x}10^{43}\, s^{-1}. \tag{5.8}$$

Inverting (5.6) and (5.8), to express them as derivatives as in (5.3),

$$x_{Planck} \equiv \frac{dA_{Planck}^{\frac{1}{2}}}{dT_0} = \left|\frac{x_0}{\sqrt{T_0}}\right| = \left|\frac{x_0\sqrt{T_0}}{T_0}\right| = \frac{1.61612...\text{x}10^{-35}\, m}{1 N/m^2} \tag{5.9}$$

$$t_{Planck} \equiv \frac{c^{-1}dx_0}{dT_0} = \frac{dt_0}{dT_0} = \left|\frac{c^{-1}x_0}{\sqrt{T_0}}\right| = \left|\frac{c^{-1}x_0\sqrt{T_0}}{T_0}\right| = \frac{5.39080...\text{x}10^{-44}\, s}{1 N/m^2} \tag{5.10}$$

and doing the same with (5.7), we have

$$m_{Planck} \equiv \frac{\sqcap(i\kappa_{Planck})}{dT_0} = \left|m_0\sqrt{T}\right| = \left|\frac{m_0 T_0^{\frac{3}{2}}}{T_0}\right| = \frac{2.17661...\text{x}10^{-8}\, kg}{1 N/m^2}. \tag{5.11}$$

In terms of a present local section of spacetime, (5.3), (5.9), and (5.10) simply express relationships of the familiar invariants at those local conditions in natural units, as can be seen in the next to the last term of each, followed by the SI equivalents. It is the large magnitude of T_0 in SI units that makes the scale so small, which it is in terms of local natural units, x_0. If the value of x_{Planck} is extrapolated back to an initial condition of unity at the primal emanation, and T_0 retains its current SI value, then x_0 is expressed in units of x_{Planck} and is a measure of the extent of expansion of the cosmos, assuming maximum possible density at the primal emanation.

We can make one adjustment to these identities that will facilitate the final development by normalizing the value of c. We can do this by using the light second, l_{ls}, the distance light travels in one second, as our unit of length, so that the Planck length becomes

$$x_{Planck} \equiv \frac{dA_{Planck}^{\frac{1}{2}}}{dT_0} = \left|\frac{x_0\sqrt{T_0}}{T_0}\right| = \frac{5.39080...\text{x}10^{-44}\, l_{ls}}{1 N/m^2} \tag{5.12}$$

Conversely, we can use the light meter, t_{lm}, the time it takes light to travel one meter, as our unit of time. This is within an order of magnitude of one nanosecond, in which case,

$$t_{Planck} \equiv \frac{c^{-1}dx_0}{dT_0} = \frac{dt_0}{dT_0} = \left|\frac{c^{-1}x_0\sqrt{T_0}}{T_0}\right| = \frac{1.61612...\text{x}10^{-35}\, t_{lm}}{1 N/m^2}. \tag{5.13}$$

It must be understood, however, that if we do this for the Planck scale, we must do it for our current local scale. Thus the angular wave number in terms of the light second is

$$\kappa_0 = 1.42745...x10^{24} \; \theta/l_{ls} \tag{5.14}$$

and the angular frequency in terms of the light meter is

$$\omega_0 = 4.76146...x10^{15} \; \theta/t_{lm} \tag{5.15}$$

We also make the following geometric observation with respect to (5.4). An increase in the radius of a sphere or torus or a normal radial of a cube results in an increase by the same proportion squared to the surface area, subject to geometric configuration. Therefore, normally we would look for a change in the inertial density, λ_0, to be a linear function of ∂x through a change in the wave number κ_0, with a change in T_0 as a function of ∂A and a squaring of ∂x. Thus, given the following, where the bracketed figure in the second term appears to be the oscillatory spring constant,

$$\lambda_0 \omega_0^2 = -i\kappa_0 \left(-m_0 \omega_0^2\right) = -i\kappa_0 k_{spring} = \sqcap \kappa_0^2 \omega_0^2 \equiv \frac{T_0}{A_0} = f_0 \tag{5.16}$$

differentiating for the end terms, we have

$$\frac{\partial \left(\lambda_0 \omega_0^2\right)}{\partial \kappa_0} = \frac{\partial f_0}{\partial \kappa_0} = \kappa_0^2 \left(k_{spring}\right) \text{ and} \tag{5.17}$$

$$\frac{\partial f_0}{\partial A_0} = -\frac{T_0}{A_0^2} = -\frac{f_0}{A_0} \text{ to determine if} \tag{5.18}$$

$$\frac{\partial f_0}{\partial \kappa_0} = \kappa_0^2 \left(k_{spring}\right) \langle = \rangle A_{Planck}^{-1} = -\frac{6\sqrt{3} f_0}{A_0} = \frac{\partial T_0}{\partial A_0} \tag{5.19}$$

that is, that the inverse Planck area divided by the wave number squared is the oscillatory spring constant of spacetime. The problem with this is that the equality is valid only if mass varies with the wave number

$$k_{spring} = -6\sqrt{3} m_{\partial \kappa} \omega_0^2 \text{ where } m_{\partial \kappa} = m_0 \kappa_0 \tag{5.20}$$

so that

$$k_{spring} = -6\sqrt{3} \sqcap \kappa_0^2 \omega_0^2 = -T_0 \tag{5.21}$$

showing that the linear inertial density varies with the square of the augmentation of the length scale, in keeping with the identity, $\lambda_0 = \sqcap \kappa_0^2$, and points to its invariance vis a vis the other wave functions. Thus we are back to (5.16), which can be restated as

$$\frac{\partial \lambda_0}{\partial t_0^2} = \frac{\partial T_0}{\partial x_0^2} = f_0 \tag{5.22}$$

and see that a change in inertial density with a stretching of the length and, by virtue of $x/c = t$, time scale is an accelerating change equal to the change in force per change in area or dynamic stress, which we might find is exponential. We can think of this in physical terms as the continuous prevailing of the expansion force over inertia.

Returning to (5.12) and (5.13), a change in the scale of x or t is a function of the square root of the change in stress, an expression of the field strength, and for a normalized c is

$$x_{Planck} = A_{Planck}^{\frac{1}{2}} = \frac{x_0}{i\sqrt{T_0}} = \frac{1}{i\kappa_0\sqrt{k_{spring}}} = \frac{x_0}{i\sqrt{k_{spring}}} \qquad (5.23)$$

$$t_{Planck} = A_{Planck}^{\frac{1}{2}} = \frac{t_0}{i\sqrt{T_0}} = \frac{1}{i\omega_0\sqrt{k_{spring}}} = \frac{t_0}{i\sqrt{k_{spring}}} \qquad (5.24)$$

Inverting gives

$$\kappa_{Planck} = A_{Planck}^{-\frac{1}{2}} = i\kappa_0\sqrt{T_0} = i\kappa_0\sqrt{k_{spring}} \qquad (5.25)$$

$$\omega_{Planck} = A_{Planck}^{-\frac{1}{2}} = i\omega_0\sqrt{T_0} = i\omega_0\sqrt{k_{spring}} \qquad (5.26)$$

and mass is simply the product of κ_{Planck} and the inertial constant.

The inverse Planck parameters, (5.2), (5.6), and (5.8) and the mass term, (5.7), only have significance if we image that at some initial condition at primal emanation, but not necessarily a hot, big bang, they were all at a general condition of unity in the SI system. The wave number in (5.6) is in terms not of a meter <u>now</u>, but of the universal whole now or of a cosmic unity or extent, C_x. Thus we imagine the whole of the universe, at some primal epoch, collapsed to a sphere or horn torus of maximum density, in which the radii of all oscillations would ideally be in contact. If we think of the scale of those radii as being in terms of the current interpretation of the Planck length, then the scale of C_x would be one meter. If we think of that scale as being the current scale of x_0, at 10^{-16} meters, then C_x would be in the general neighborhood of 10^{11} meters, assuming a population of 10^{80} nucleon. Comparing this value with the current value for κ_0 we have the exponential factor for the change in κ_0 due to a change in inertial density from expansion

$$\kappa_{exp} = \frac{\kappa_{Planck}}{\kappa_0} = \sqrt{T_0} = \frac{6.18765...x10^{34}\,\theta/R_C}{4.76146...x10^{15}\,\theta/m} = 1.29952...x10^{19} \qquad (5.27)$$

indicating that the measure of unity has decreased, vis a vis the whole, according to the square root of the tension, T_0, from a universal unit, by the inverse factor of $7.69510...x10^{-20}$ to a current meter.

In similar fashion to (3.70) we can get a figure for the extent of cosmic expansion through a doubling of that extent, in terms of light seconds as

$$C_x = \ln 2(\kappa_{exp}) = 9.00764...x10^{18}\,s \qquad (5.28)$$

where we note the tie-in of this logarithmic change in the field with the discussion on beta decay at (3.42) and in the appendix on exponentiation at (7.139). This extent divided by the doubling rate in (3.70) gives us the number of times that extent has doubled or

$$\frac{C_x}{\tau_H} = \frac{9.00764...x10^{18}}{2.93834...x10^{17}} = 30.655... \qquad (5.29)$$

Approached another way, dividing the value of the expansion rate in (3.68) by the inverse of (5.27), where the expansion rate is the rate of decrease in the inertial density and thereby a measure of the linear scale, gives

$$\frac{X_e}{\kappa_{exp}^{-1}} = \frac{2.35896...x10^{-18}}{7.69510...x10^{-20}} = 30.655... \tag{5.30}$$

for a factor of a little over thirty times the increase from x_0 to 1 meter and an age of the cosmos of approximately 285 billion years. This compares very closely with (3.71) at 30.94 and 288 billion years which is based on the background microwave wavelength, and supports the assertion that the expansion of the cosmos is exponential and structured according to this development.

With respect to the Planck frequency, ω_{Planck}, the unit of time is once more universal and is in fact the time lapsed since the universal expansion began,

$$\omega_{exp} = \frac{\omega_{Planck}}{\omega_0} = \sqrt{T_0} = \frac{1.85501...x10^{43}\ \theta/T_C}{1.42745...x10^{24}\ \theta/s} = 1.29952...x10^{19} \tag{5.31}$$

What this says, of course, is that a second today is not what it was 285 billion years ago, but then again, neither is a meter, so c remains invariant.

The Planck mass, then is the product of (5.27) and m_0, the neutron mass, and gives its value at that early epoch, m_1. The remainder of the early epoch values for oscillation functions, as in the orthogonal matrix of invariants, can be found by substituting the values of ω_{Planck} and κ_{Planck} for the current values. At this scale the spin energy of a neutron would be, $1.95624...x10^9$ joules, the Planck energy. Based on this line of thinking, the potential energy density as in (4.4) at that epoch would be $1.10638...x10^{113}$ joules/cubic meter or $1.23101...x10^{96}$ kg/cubic meter. This assumes that the wave number is still the measure of mass and that the wave number of (5.6) is per today's meter, so that each oscillator is confined to a volume of roughly 10^{-105} cubic meters. As there are estimated to be roughly 10^{80} hadrons in the known universe, this indicates that at this scale the entire known compliment of matter would fit in a volume of approximately 10^{-25} cubic meters or roughly one cubic nanometer. The inertial density would be the same, regardless of whether there where any actual oscillations at that time, and though of great magnitude, is hardly infinite and does not suggest a singularity. The stress T_0 would be $4.81635...x10^{114}$ Newton/meter squared. The transverse wave force would be $1.21043...x10^{44}$ Newton.

If we idealize the cosmic structure as a horn torus, in which the center hole consists of a dense inertial locus, instead of as a sphere, then the horn becomes the center of the primal emanation or "big bang". Functional continuity is thereby maintained at and through the center, and expansion of a 3-D space from maximum density at that center and out around the annulus, with rotation about the axis, will lead to decreasing density with that strain, until such time as it reaches a half revolution of the annulus. From that locus the return toward the center will increase the density of the continuum as it is constrained back toward the horn in a cosmic inertial sink or "black hole". A similar effect would be produced by an oscillation through the horn, without the circumnavigation of the annulus, with some modification. This is more in keeping with the dictates of continuum logic, as circumnavigation of the annulus would appear to assume some type of ideal fluid, which has not been a part of our assumptions, i.e. no point neighborhoods that can flow past each other. Only stresses are assumed to be able to freely rotate. Finally, the structure might echo that of the fundamental quantum spin diagrams, expanding and contracting over cosmic time.

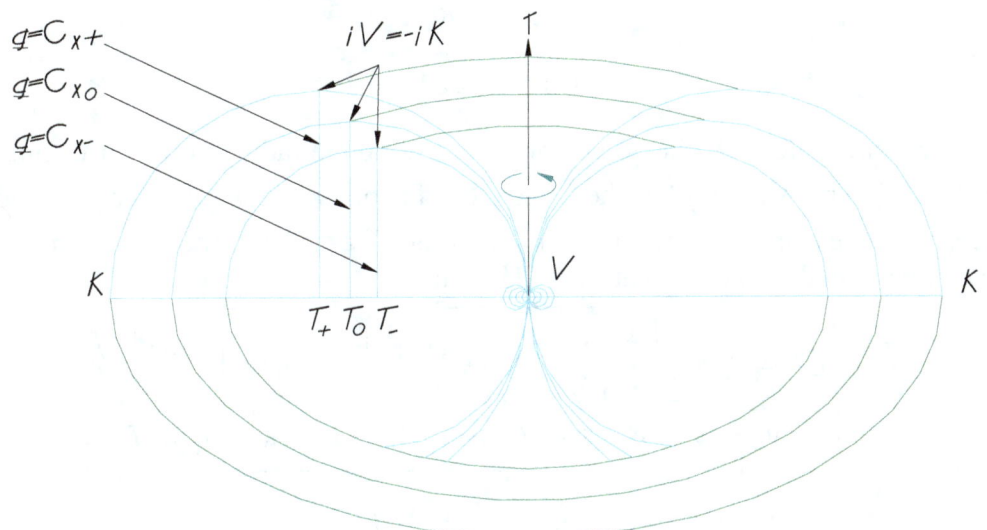

2-D Representatioon of a 3-Torus

With conceptual respect to the principles of expansion, the center of the torus comprises a locus, V, of maximum potential energy, and the extent of the annulus in the plane of the torus comprises a circumference, K, of maximum kinetic energy. This gives a linear potential energy density of

$$V_1 = \lambda_V c^2 = \tau_V \tag{5.32}$$

and a kinetic energy density of

$$K_1 = -\lambda_K c^2 = \tau_K, \tag{5.33}$$

both of which are equal to a tension force, τ, so that

$$V_1 = -K_1 g' \tag{5.34}$$

The negative sense reflects the relationship of Wave Diagram 2, in that the velocity/momentum and displacement are always of opposite sense. We might anticipate that τ_K vanishes at the extremum, since it represents a point of directional change, maximum momentum and zero acceleration, but this is so with respect to the central toric plane only, as it still has a centripetal component of angular acceleration about the polar axis. The horn torus represents an inversion of sorts of Wave Diagram 1, in which the antinodes of the central rotating x axis of that spin map to the toric central plane and the path of the nodes in the Y-Z plane map to the toric center. It is symmetrical to Spin Diagram 1 with the V poles brought together at the center of the sphere. The g' is a geometric factor that maps V_1 onto K_1. The points of maximum instantaneous power of the wave are found at $iV = -iK$ and the flat space of the torus.

The sum of the power at the locus of points K should be equal in magnitude to the power at point V, which we can express as

$$P = cE_1 = cV_1 = -c\sum K_1 \tag{5.35}$$

Assuming the invariance of c, combined with (5.32) & (5.33), this can be reduced to a contrast of the inertial densities, or

$$\pi\kappa_V^2 = -\pi\kappa_K^2 g', \text{ and} \tag{5.36}$$

$$g' = 2\pi(2\underline{g}) = -\frac{\kappa_V^2}{\kappa_K^2} \tag{5.37}$$

We might anticipate that g is a 4-vector representing the radius of the toric annulus, in which $\sqrt{3}x_0$ is a 3-vector, C_x is the cosmic expansion extent, a 4-vector, equal to g and to T, the cosmic time elapsed since a locus of stress at V has transformed to a point on the plane of the upper annulus extremus at, $iV = -iK$. Thus

$$\underline{g} = \underline{C}_x = \ln 2 \left(\frac{1}{2} \sqrt{\frac{x_0^2}{x_{Planck}^2} + \left(\sqrt{3T_0}\right)^2} \right) = \ln 2 |\kappa_{exp}| = \ln 2 |\sqrt{T_0}| \tag{5.38}$$

where the Planck length is an invariant equal to the initial $x_I = x_0$ at $T = 0$

$$\frac{x_0}{x_{Planck}} = \sqrt{\left(2\underline{C}_x(\ln 2)^{-1}\right)^2 - \left(\sqrt{3T_0}\right)^2} \tag{5.39}$$

A cosmic time vector, **T**, can be envisioned through the horn and in the case of the revolution of the annulus, remains locally tangential to the annulus as *T*. We might further imagine that **T** is an axial vector, imparting rotation to the torus as a whole. In the case of an oscillating structure, the vector simply reverses with the oscillation. We can assign a spatial axis and vector, **X**, which is parallel to **T**$_0$ at time zero at the center of the horn. It is immaterial which vector is stationary and which follows the annular tangent. Thus *T* constitutes a fourth spatial vector of motion, which we might call **W**, and which is locally indistinguishable from **X**, **Y** and **Z**, but which is normal to all three, as indicated previously. *T* is not to be confused with the *t* dimension of a standard model four-vector, as we are here interested in a time whose unit scale does not change with a change in spatial scale with expansion, and whose product with a cosmic frequency, Ω, will range between 0 and ½ π.

Any three-dimensional locus of points emerging from the horn undergoes an expansion which is perceived locally as isotropic. While the isotropic nature of the points equidistance from the horn center is apparent, the matter of those closer and further from that center require some scrutiny. The points at T_+ have a decreased inertial density from those at T_0, which is itself decreased from T_-, which indicates a gradient and an anisotropic condition. In addition, there would be no net motion of the locus from the center unless the expansion force, τ, bore the following relationship to *T*, or

$$\tau_{T_-} > \tau_{T_0} > \tau_{T_+} \tag{5.40}$$

as with the inertial densities

$$\rho_{T_-} > \rho_{T_0} > \rho_{T_+} \text{ and} \tag{5.41}$$

$$\lambda_{T_-} > \lambda_{T_0} > \lambda_{T_+} \tag{5.42}$$

and with the expansion stress, f,

$$f_{T_-} > f_{T_0} > f_{T_+} \tag{5.43}$$

Combining these inequalities with (2.121) we have the following expression for the expansion differentials, κ_e and ω_e, for $\rho_{T_+} - \rho_{T_0} = \rho_e$, $\lambda_{T_+} - \lambda_{T_0} = \lambda_e$ and $\tau_{T_+} - \tau_{T_0} = \tau_e$

$$_4\sqcap_2 \equiv \rho_e \omega_0^2 \equiv -(\sqcap \kappa_e^4)\omega_0^2 = -(\sqcap \omega_e^2 \kappa_e^2)\kappa_0^2 \equiv f_e \kappa_0^2. \quad (5.44)$$

$$_2\sqcap_2 \equiv \lambda_e \omega_0^2 \equiv -(\sqcap \kappa_e^2)\omega_0^2 = -(\sqcap \omega_e^2)\kappa_0^2 \equiv \tau_e \kappa_0^2. \quad (5.45)$$

Since

$$c \equiv \frac{\omega_0}{\kappa_0} \equiv \frac{\omega_p}{\kappa_p} \equiv \frac{\omega_e}{\kappa_e}, \quad (5.46)$$

we might surmise that this states that c is invariant, but in a special sense. In particular, it says that x_e and t_e are covariant. It does not, however, say that, for the following

$$\omega_0(T_0) = \omega_0(T_+) \text{ or} \quad (5.47)$$

$$\kappa_0(T_0) = \kappa_0(T_+), \quad (5.48)$$

where $\omega_p = \omega_0(T_+)$ and $\kappa_p = \kappa_0(T_+)$. Therefore, the expansion rate appears to be the operation of a complex exponential function, as might be expected from the above derivations, indicating its cyclical nature. We find here the oscillation of individual quanta as the function of an overall cosmic oscillation, where Ω is the cosmic frequency, iA_c is the imaginary part of the cosmic complex amplitude and $0 < \Omega T < \frac{\pi}{2}$, given by

$$m_0 = \frac{\sqcap}{A_c e^{i(\pm \Omega T)}} = \frac{\sqcap}{iA_c \sin(\pm \Omega T)} \quad (5.49)$$

where we have used the imaginary part of Euler and in which, owing to the close approximation of ΩT to $\pi/2$ we can use x_0 in lieu of C_x in the cosine,

$$\cos \Omega T \cong \cot \Omega T = \frac{x_{Planck}}{x_0} = \frac{\sqrt{T_0}}{T_0} \quad (5.50)$$

The left figure below shows the relationship in orders of magnitude of the Planck length, neutron scale, meter, and presumed cosmic extent in SI units at current cosmic time T_0. Also shown are the presumed cosmic extent and meter at the point of cosmic inception, T_1. At the right end of each parameter is the value in units of the Planck length.

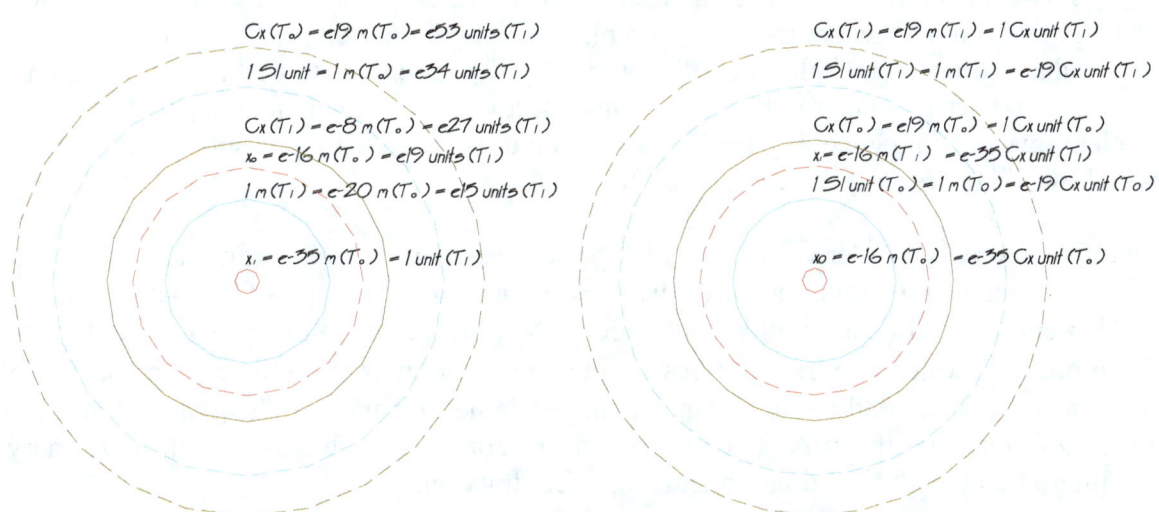

Cosmic - Planck Scale   **Invariance of the Fundamental Length and Cosmic Extent**

The above speculations concerning the form expansion might take need not be the only interpretation of the above analysis. It appears that these parameters might be interpreted in the context of a fixed 3-space without boundary in which the apparent expansion corresponds with an isotropic contraction of the oscillators toward the Planck length, in keeping with the development of the gravitational quantum and the hyperbolic nature of the inversphere, the ultimate fate of local groups being coalescence in inertial sinks. Red shift would come from this shrinkage.

In the right figure the invariance of the relationship among C_x, the meter, and x is shown at time T_1 and T_0. Note that the relationship is unchanged if 1 and 0 are transposed, as would be the case for cosmic expansion. In this figure, we are assuming that C_x is the fundamental unit and therefore remains unchanged, with all other values expressed in terms of it. Rotational oscillation of that extent, in conjunction with the inertia of its core, would induce the stress responsible for its quantum oscillations. x_0 has the value of the Planck length in terms of the invariant cosmic extent, and C_x and a meter, both at T_0, represent their apparent values at T_1, based on the assumption of expansion. By extention of this logic, the cosmic extent could oscillate between contraction in this manner and expansion as outlined before, with the invariance of the fundamental relationship intact throughout the oscillation.

Finally, in this regards, the condition in the right figure might be interpreted as representative of a hyperbolic 3-space without boundary of fixed extent, in which the apparent contraction or expansion is simply an artifact of negative curvature similar to an Escher print. It is not a necessary conclusion from the data that all hadronic and leptonic matter proceeds from one initial inertial font, such as a big bang. It is entirely feasible that such matter is generated by active galactic nuclei, galaxies being the largest discrete structures observed in the universe. The double rotation of the quanta is reflective of the conditions of helical stress-strain that would exist at the center of such galaxies, involving simultaneous toric annular and polar rotation, with black hole density and stress. The collimated, relativistic jets, with gamma frequency observed to issue from these loci suggest two such tori back to back at a sandwiched accretion disk, their central black holes in fact inertial fonts, generating hydrogen plasma. The lighter elements, including molecular hydrogen and helium of all isotopic configurations, with perhaps some lithium, would be expected to emerge from such high energy fonts, the heavier congregations of nuclear quanta being generally unable to withstand the relativistic pressure of generation which would necessarily exceed that of T_0.

One final word is in order with respect to special relativity. It is apparent that any instance of translational momentum increases the stress on and hence energy of an oscillation, fundamental or secondary, with a contraction of r, resulting in an increase in ω and therefore in κ, which by virtue of n results in an increase in or relativistic augmentation of m. This development, therefore, is in keeping with the Pythagorean theorem and the Lorentz framework, though not necessarily all the conclusions drawn from the customary SR interpretations of time dilation and length contraction.

6 – Evaluations

Observed Values

For evaluation purposes, the following, which are 2002 CODATA values, or computed directly from those values, are used with the exception of those denoted by †. Asterisks indicate observed values use in computing the theoretical values of this model.

Speed of electromagnetic wave propagation*
$$c = 299,792,458 \ m/s \qquad \text{*(6.1)}$$

Planck's Constant of Action*
$$\hbar = 1.05457168(18) x 10^{-34} \ kg \ m^2/s \qquad \text{*(6.2)}$$

Newton's Gravitational Constant
$$G_N = 6.6742(10) x 10^{-11} \ N \cdot m^2/kg^2 \qquad (6.3)$$

Elementary Charge
$$e = 1.60217653(14) x 10^{-19} \ C(kg \ m/s) \qquad (6.4)$$

Fine Structure Constant
$$\alpha = 7.97352568(24) x 10^{-3} \qquad (6.5)$$

Neutron Mass*
$$m_n = 1.67492728(29) x 10^{-27} \ kg \qquad \text{*(6.6)}$$

Neutron Compton Wavelength over 2π
$$\lambdabar_{C,n} = 2.100194157(14) x 10^{-16} \qquad (6.7)$$

Neutron-Electron Mass Ratio, Inverse
$$\frac{m_n}{m_e} = 1838.6836598(13), \quad \frac{m_e}{m_n} = 0.00054386734481(38) \qquad (6.8)$$

Neutron-Proton Mass Ratio, Inverse
$$\frac{m_n}{m_p} = 1.00137841870(58), \quad \frac{m_p}{m_n} = 0.99862347872(58) \qquad (6.9)$$

Proton-Electron Mass Ratio, Inverse*
$$\frac{m_p}{m_e} = 1836.15267261(85), \quad \frac{m_e}{m_p} = 0.00054461702173(25) \qquad \text{*(6.10)}$$

Planck Scale (Area)
$$A_P \equiv \frac{G_N \hbar}{c^3} = 2.611810291...x 10^{-70} \ m^2 \qquad (6.11)$$

Electron Angular Frequency, Inverse
$$\omega_e = \frac{c}{\lambdabar_{C,e}} = 7.7634407(52) x 10^{20} \ \theta/s, \quad \omega_e^{-1} = 1.28808867...x 10^{-21} s/\theta \qquad (6.12)$$

Hubble Constant†
$$H_0 = 7.2 x 10^4 \ \text{to} \ 7.3(8) x 10^4 \ m/s/mps \qquad (6.13)$$

Reduced Hubble Constant†
$$H_R = 2.333 x 10^{-18} \ \text{to} \ 2.368 x 10^{-18} \ \Delta m/s \qquad (6.14)$$

Theoretical Values

The computed values derived from this inertial theory as follows are based on the 2002 CODATA values for c, $\hbar$, m_n and the ratio of m_n to m_e or m_p to m_e.

Inertial Constant*
$$\eta = \frac{\hbar}{c} = 3.51767248(60) \times 10^{-43} \text{ kg m } (kg\, m/\theta) \tag{6.15}$$

Fundamental Wave Number
$$\kappa_0 = \frac{m_n}{\eta} = \lambdabar_{C,n}^{-1} = 4.76146453(81) \times 10^{15} \; \theta/m \tag{6.16}$$

Fundamental Angular Frequency
$$\omega_0 = c\kappa_0 = \frac{c}{\lambdabar_{C,n}} = 1.42745115(24) \times 10^{24} \; \theta/s \tag{6.17}$$

Fundamental Wave Transverse Momentum
$$\eta_1 \equiv P_M = -i5.02130565(85) \times 10^{-19} \text{ kg m/s} \tag{6.18}$$

Fundamental Wave Transverse Force
$$\eta_2 \equiv \tau_0 = -7.167668571... \times 10^5 \text{ N} \tag{6.19}$$

Fundamental Linear Inertial Density of the Spacetime Continuum
$$_2\eta \equiv \lambda_0 = -7.975106838... \times 10^{-12} \text{ kg/m} \tag{6.20}$$

Fundamental Mechanical Impedance of the Spacetime Continuum
$$_1\eta_1 \equiv Z_0 = 2.390876882... \times 10^{-3} \text{ N}/(m/s) \tag{6.21}$$

Fundamental Tension Stress/Volume Energy Density of the Spacetime Continuum
$$_2\eta_2 \equiv f_0 \equiv E_1 = 1.625021173... \times 10^{37} \text{ N}/m^2 \tag{6.22}$$

Fundamental Quantum Spin Energy
$$^1\eta_2 \equiv E = i1.505349565... \times 10^{-10} \text{ J } (N \cdot m) \tag{6.23}$$

Fundamental Quantum Power
$$^1\eta_3 \equiv P_{\omega_0} = 2.148812979... \times 10^{14} \text{ W} (J/s) \tag{6.24}$$

Fundamental Gravitational Quantum (Derivative form)
$$\frac{d\eta_2}{d\,_2\eta_2} \equiv G_q \equiv \frac{d\tau_0}{dT_0} \equiv \frac{1}{6\sqrt{3}\kappa_0^2} = 4.244309191... \times 10^{-33} \; m^2 \tag{6.25}$$

Newton's Gravitational Constant (Derived)
$$G_{Nq} \equiv \frac{dG_q}{\lambda_0^2} = 6.673197749... \times 10^{-11} \text{ N}/(kg/m)^2 \tag{6.26}$$

Planck Scale (Area, Derivative form)
$$\frac{d(\kappa_0^{-2})}{d\,_2\eta_2} \equiv A_P \equiv \frac{dA_0}{dT_0} \equiv -\frac{6\sqrt{3}\tau_0}{T_0^2} = -2.611848548... \times 10^{-70} \; m^2 \tag{6.27}$$

Fundamental Charge (Raw Derivation of SI Value)
$$e_{raw} = \frac{\eta \omega_0}{\pi} = 1.598331232... \times 10^{-19} \text{ kg m/s} \tag{6.28}$$

Natural Fundamental Charge
$$e' = \partial \omega_0^{-1} = \omega_e^{-1} = 1.288089086... \times 10^{-21} \text{ s}/\theta \tag{6.29}$$

Fundamental Driving-Driven Frequency (Neutron-Proton Mass) Ratio, Inverse

$$\frac{\omega_0}{\omega_p} = 1.0013784732..., \frac{\omega_p}{\omega_0} = 0.998623425... \tag{6.30}$$

Fundamental Driven-Reflected Wave Frequency (Proton-Electron Mass) Ratio, Inverse (Based on the CODATA Neutron-Electron Mass Ratio)

$$\frac{\omega_p}{\omega_e} = 1836.152075..., \frac{\omega_e}{\omega_p} = 0.000544617199... \tag{6.31}$$

Fundamental Driving-Reflected Wave Frequency (Neutron-Electron Mass) Ratio, Inverse (Based on the CODATA Proton-Electron Mass Ratio)

$$\frac{\omega_0}{\omega_e} = 1838.684256..., \frac{\omega_e}{\omega_0} = 0.0005438671685... \tag{6.32}$$

Differential Linear Inertial Density and Cosmic Expansion Rate

$$\partial_2 \boxminus \equiv \partial \lambda_0 \equiv X_e \equiv \boxminus \partial \kappa_0^2 \equiv c^{-2} \boxminus \partial \omega_0^2 = 2.35896879... \times 10^{-18} \, kg/m/s, \Delta m/s \tag{6.33}$$

Differential Expansion Force

$$\partial \boxminus_2 \equiv \partial \tau_0 = 0.212013542... N/s \tag{6.34}$$

Hubble Rate

$$H_0 = 72,791.17172... m/s/mps \tag{6.35}$$

Young's Modulus of the Spacetime Continuum

$$6\sqrt{3}\,_3\boxminus_2 \equiv Y = 8.041025... \times 10^{53} \, N/m^2 \tag{6.36}$$

Volume Potential Force Density

$$F_1 \equiv 6\sqrt{3}\,_4\boxminus_2 \equiv A_{Planck}^{-1} = 3.82871... \times 10^{69} \, N/m^3 \tag{6.37}$$

Appendix A - Direct Product and Inverse Square Law

At least two of the principle interactions between apparently discrete bodies are governed by this law which states that the strength or magnitude of the interaction, F_Q, between two bodies is directly proportional to the product of the quantitative measure of some quality of those bodies, Q, and inversely proportional to the square of the distance of their spatial separation, d. This finds mathematical expression as

$$F_Q = \left(\frac{Q_1 Q_2}{d^2}\right) k_Q, \qquad (7.1)$$

where k_Q is a constant of proportionality which may be a derivative with respect to some argument common to both side of the equation or simply a differential quantity which is integrated by the bracketed term.

If we assume that all bodies are composed of discrete portions or quanta, then the stated quality, Q_a, for each body, a, would be the number of such quanta, n_a, times the unit quality per quanta, q, or $Q_a = n_a q$. Similarly, the distance could be expressed in terms of some quantum unit r_q, so that $d = n_r r_q$. This gives

$$F_Q = \frac{n_1 n_2}{n_r^2}\left(\frac{q^2}{r_q^2} k_Q\right), \qquad (7.2)$$

and the term in brackets can be viewed as a quantum of the interaction magnitude or force, F_Q. According to Newton's second law, force is generally defined as the mass of a body times its acceleration, positive or negative as with

$$F = ma = mv'(t) = mr''(t) \qquad (7.3)$$

in which it is shown that acceleration is the change in velocity as a function of time, velocity being a change in position or state as a function of time. Mass is envisioned as a measure of the inertia or resistance to change in position of a body. In general, therefore, velocity can be viewed as the rate of change over time of any variable quality, acceleration as the rate of change in a variable rate of such changeable quality, and mass as an inverse measure of the susceptibility or a direct measure of the resistance to change of a second quality that is being changed. The third term in (7.3) indicates that force is a change in momentum over a period of time, where momentum, P_M, is such second quality, m, undergoing some change at a steady rate, v, which might possibly be zero

$$P_M = mr'(t) = mv \qquad (7.4)$$

Gravitation. The magnitude of gravitational attraction is directly proportional to the product of the quality of mass, M_a, of two interacting bodies and inversely proportional to the square of the distance, d, separating their centers of mass. This finds mathematical expression in Newton's law of universal gravitational attraction generally stated as

$$F_g = \frac{M_1 M_2}{d^2} G_N. \qquad (7.5)$$

where G_N is Newton's empirically determined gravitational constant.

In quantum terms this would be

$$F_g = \frac{n_1 n_2}{n_r^2}\left(\frac{m_q^2}{r_q^2}G_N\right). \tag{7.6}$$

Therefore, the number in brackets represents a quantum of gravity or

$$dG_q = \left(\frac{m_q^2}{r_q^2}G_N\right). \tag{7.7}$$

It is worth noting that if we use the CODATA value of the mass of the neutron for m_q, and the neutron Compton wavelength over 2π for r_q, dG_q resolves to that same value of r_q divided by $6\sqrt{3}$ within a factor of 1.000014648, which is within the margin of error for G_N. This value for dG_q is derived herein.

Electrostatics. The strength or magnitude of electrostatic interaction is directly proportional to the magnitude of the product of the quality of charge, q_a, of each of two interacting bodies and inversely proportional to the square of the distance separating their centers of charge. This finds mathematical expression in Coulomb's law of universal electrostatic interaction generally stated as

$$F_{es} = \frac{q_1 q_2}{d^2}k_\varepsilon = \frac{n_1 e_1 n_2 e_2}{d^2}\frac{1}{4\pi\varepsilon_0}. \tag{7.8}$$

The right most term uses quanta of charge, e, for the direct product, where e_1 and e_2 may be of like or opposite complex sense. Thus, if the sense of both is $-i$ or $+i$, the product will be negative, whereas if the senses are opposite, the product will be positive, the positive product representing an attractive and the negative, a repulsive, force. The constant of proportionality in this case, k_ε, is equal to the inverse of 4π times the dielectric constant or permittivity of the vacuum, ε_0. Applying the same logic to the inverse square term as above, we have

$$F_{es} = \frac{n_1 n_2}{n_r^2}\left(\frac{\pm e^2}{r_q^2}\frac{1}{4\pi\varepsilon_0}\right), \tag{7.9}$$

where once again the bracketed term can be seen as a quantum of force. Since force has the dimensions of $\frac{ml}{t^2}$ or mass times length over time squared, the numerator of that term might be imagined to have the dimensions of ml and the bottom of t^2. In fact, if ε_0 is given the dimensions of v^{-2}, or inverse velocity squared, this could be the case.

Electrodynamics. The strength of an electromotive force between two currents of charge is directly proportional to the currents or the quantity of moving charges per second of each and inversely proportional to their distance of separation. It is not strictly speaking an inverse square law. It finds mathematical expression in the following definition of the ampere or basic unit of current.

$$\mathbf{F}_{ba} = i_b \mathbf{L} \times \mathbf{B}_a, \text{ where } B_a = \frac{\mu_0 i_a}{2\pi d} \tag{7.10}$$

For two currents running parallel this becomes

$$\frac{F_{ba}}{L} = \frac{\mu_0 i_a i_b}{2\pi d} \qquad (7.11)$$

B_a is the magnetic field produced by current i_a at the location of current i_b, d is their distance of separation, μ_0 is the magnetic constant or the permeability of the vacuum and F_{ba} is the force exerted on length L of current i_b by the magnetic field B_a. That force is directed toward current i_a in the case of parallel currents, that is, those running in the same direction, and away from i_a in the case of anti-parallel currents, those running in opposite directions. It is noted that L and d are normal or perpendicular to each other. For an empirically determined value for F_{ba} of $+2 \times 10^{-7}$ Newton per meter length of current, L, at separation, d, of one meter, μ_0 is conventionally set at a magnitude of $4\pi \times 10^{-7}$ so that equal currents for both i_a and i_b, have a value of 1 Ampere. This is an arbitrary value selection for μ_0 to facilitate computation. As such, the product $\mu_0 \varepsilon_0$ is equal to the inverse of the speed of light squared or c^{-2}, so that if μ_0 is given another value, ε_0 and i will adjust accordingly. What is not arbitrary is the product of the number of charge quanta, $n_{e1} n_{e2}$, per length and separation of currents per period of time, required to produce an F_{ba} of $+2 \times 10^{-7}$ Newton.

Since L and d are both of the same magnitude, we can speculate that for any value of $L = d$, and $i_a = i_b = 1$, F_{ba} is invariant, including at the quantum level indicated in (7.1) above of r_q. With respect to (7.11) it is apparent that $\mu_0 i_a i_b$ has the dimensions of force. We have the option, however, of assigning the current the dimensions of the square root of force and making μ_0 a dimensionless number in keeping with the comments concerning (7.9) above or of making μ_0 an inverse quantum of force and letting the current represent a count of quanta of force, n_1 and n_2, as previously done above. Therefore we have

$$F_{ba} = \frac{\mu_0 i_a i_b}{2\pi} = 2\pi \alpha n_{e1} n_{e2}, \qquad (7.12)$$

where the current has been replaced by the number of elementary charges per volt and thereby, with unit resistance per ampere, ת, (tav), is an inertial constant of dimensions mass-distance, and α is the fine structure constant. We can then assign the dimension of inverse time to each charge count. In reality, ת is a second order force differential with respect to time or frequency which is integrated by the product of the count of the charges. The final two terms of (7.12) can then be rearranged to give the more familiar

$$\frac{i_a i_b}{n_{e1} n_{e2}} \frac{1}{4\pi \alpha \varepsilon_0} = \frac{e^2}{4\pi \alpha \varepsilon_0} = ת c^2 \qquad (7.13)$$

where the final term is equivalent to the quantum of action times the speed of light. It is apparent that the second term is equivalent to the bracketed term of (7.9) above with the substitution of α for the r_q^2. This gives us the option of assigning the square of some quantum unit-length dimension to α, which we will see can be done with some justification, or of making the current a count of flow per unit length. While current is generally conceived as the count of the number of charge carriers or charged "particles" operating at a given point over a unit of time, we can also think of it as the number per length of current. On a quantum level, with some rearrangement (7.13) then becomes, where α is dimensionless,

$$\frac{e^2}{r_q^2 4\pi\varepsilon_0} = \alpha\hbar c^2 \frac{1}{r_q^2} = \alpha\hbar\omega_q^2 = \alpha\hbar c^2 \kappa_q^2 \qquad (7.14)$$

and we find the equivalence of (7.9).

One final observation concerning dimensions involved in mathematical operations such as taking of the product or square root, is that while the product of two linear dimensions preserves the number of dimensions in the produced area, for example, the same is not necessarily true in the case of vector operations. Thus the cross product of two forces resulting in a torque retains the number of dimensions *(mlt⁻²)* as each of the two components, though the resulting product is not quite the same quality as its components. The decomposition of such product through the square root or other factoring thus correctly results in an apparent doubling of dimensions and not necessarily, for example, the square root of force for each component. Thus charge and current can be modeled as "forces" which operate tangentially to produce a radial "force" effect. The quotation marks are used to denote dimensional, though not necessarily technical, equivalence.

Appendix B – Wave Transmission at a Discontinuity

We might look to a function that concerns the reflection and transmission of waves at a discontinuity of the propagating medium, to better understand the boundary conditions governing charge generation. Such discontinuity might be a change in inertial density of the medium itself or a change in tension or perhaps shearing stress along its span. We can assume initially that the transverse displacement of the medium on both sides of the discontinuity is the same, so that the frequency of a wave crossing the boundary is unchanged. The following is taken from Elmore and Heald, <u>Physics of Waves</u>.

For a one dimensional model of wave transmission at a discontinuity, the boundary conditions require that in addition to the transmission of the initial wave with altered amplitude, a reflected wave must be generated. As a result of these conditions we have the following, in which A_1 is the amplitude of the incident wave, $\breve{A}_2$ is the complex amplitude of the transmitted wave, and κ_2, its wave number, $\breve{B}_2$ is the complex amplitude of the reflected wave, and κ_1 is the wave number of the initial and reflected waves.

$$A_1 + \breve{B}_1 = \breve{A}_2 \tag{7.15}$$

$$\kappa_1 A_1 + \kappa_1 \breve{B}_1 = \kappa_2 \breve{A}_2 \tag{7.16}$$

From this we can solve for the amplitude reflection, $\breve{R}_a$, and transmission, $\breve{T}_a$, coefficients or ratios, and for the power reflection, R_p, and transmission, T_p, coefficients, which depend only on the properties of the medium. We can see from (2.91) that if the frequency is constant, the impedance Z relationships will be identical to that of the wave numbers.

$$\breve{R}_a \equiv \frac{\breve{B}_1}{A_1} = \frac{\kappa_1 - \kappa_2}{\kappa_1 + \kappa_2} = \frac{Z_1 - Z_2}{Z_1 + Z_2} \tag{7.17}$$

$$\breve{T}_a \equiv \frac{\breve{A}_2}{A_1} = \frac{2\kappa_1}{\kappa_1 + \kappa_2} = \frac{2Z_1}{Z_1 + Z_2} \tag{7.18}$$

$$R_p \equiv \frac{\frac{1}{2}\lambda_1 c_1 \omega^2 B_1^2}{\frac{1}{2}\lambda_1 c_1 \omega^2 A_1^2} = \frac{B_1^2}{A_1^2} = \left(\frac{Z_1 - Z_2}{Z_1 + Z_2}\right)^2 \tag{7.19}$$

$$T_p \equiv \frac{\frac{1}{2}\lambda_2 c_2 \omega^2 A_2^2}{\frac{1}{2}\lambda_1 c_1 \omega^2 A_1^2} = \frac{\lambda_2 c_2 A_2^2}{\lambda_1 c_1 A_1^2} = \frac{4 Z_1 Z_2}{(Z_1 + Z_2)^2} \tag{7.20}$$

It should further be stated that the requirements of conservation of energy and power dictate that

$$R_P + T_P = 1 \tag{7.21}$$

Appendix C – Vector Orthogonality

The 4-vector equation can be generalized to n dimensions for an ortho-normal space as,

$$\sqrt{n-1}\, \underline{x}_{0i} = \sqrt{\underline{x}_n^2 - x_{0nth}^2} \tag{7.22}$$

Appendix D - Exponentiation

Calculus is the study of the rate of change in one variable quantity, conventionally denoted by a y, which is held to be a function, f, wholly or partially, of another variable, generally denoted by an x or sometimes a t. This underlying functional relationship between the variables is denoted by

$$y = f(x) \text{ or } y = f(t). \tag{7.23}$$

In the case of a partial function, a function of more than one variable, we write

$$y = f(x,t). \tag{7.24}$$

Thus, with (7.23) when $x = a$, $y = b$, and with (7.24) when $x = a$ and $t = b$, $y = c$. a, b and c are arbitrary symbols standing for unknown quantities of the stated variable x, t and y, and depending on the context and circumstance a, b and c may in fact be the same or of equal value.

The underlying functional relationship or function does not necessarily indicate that x causes y or that y is the operational function of x. While this may be so in the case of some physical and organizational conditions, in general terms the function simply indicates that when x has the value of a, y is always, within the context determined by f, uniquely observed to have the value of b.

Thus, given a right triangle of variable angle, α, but fixed, unit length hypotenuse, the cosine can be stated as a function of the length of the adjacent side, a, and the length of a can be stated as an inverse function of the cosine. In the language of mathematics, we would say that the cosine function maps the value of a onto the cosine and the inverse function maps the value of the cosine onto a. This concept of mapping reflects the fact that any function that we might consider can be visualized and charted against the backdrop of an orthogonal co-ordinate system.

Thus it may equally be true that

$$x = f(y) \text{ or } t = f(y). \tag{7.25}$$

Note that it is not generally stated that

$$x,t = f(y), \tag{7.26}$$

although that may in fact be the case. If x is the adjacent side of α and t is the opposite, they both vary with respect to some variation in the angle, $y = \alpha$.

While it may be of interest to know the value of y for any value of x or t, it is often of equal or greater interest to know the rate at which y is changing for any value of x or t. This rate of change or ratio of variability of one quantity with respect to another is known as the derivative function, f', of y with respect to x or t, or

$$\frac{dy}{dx} = f'(x) \text{ or } \frac{dy}{dt} = f'(t). \tag{7.27}$$

The quantity dy is the differential amount of change in y that occurs for every differential amount of change, dx, in x. While dy and dx are customarily envisioned as being infinitesimally small, they are generally not small by the same proportions, and are indeed expressed as the change in *y* in units of that quality for every change of <u>one unit</u> of *x*. Hence they are often used in partial derivative form as direction cosines, where by implication, ∂x would be the hypotenuse of a right triangle of unit value and ∂y is the adjacent side. Here *f'* constitutes another function of *x*, with the prime notation indicating that it is a derivative function of *f(x)*. The single prime is the rate of change in *f(x)* commensurate with a change in *x*. If it is a function with respect to time, *f(t)*, i.e. a rate of change over some unit of time, it is a speed, or if a direction is specified, a velocity.

If the rate of change in *f(x)* or *f(t)* is not steady or constant, then we have a second derivative of these functions, denoted by a double prime

$$\frac{d^2y}{dx^2} = f''(x) \text{ or } \frac{d^2y}{dt^2} = f''(t). \quad (7.28)$$

The change in velocity, acceleration, which is the second derivative with respect to time, $f''(t)$, is commonly encountered and understood. The equivalent with respect to *x*, also a type of acceleration, is a change in the intensity or magnitude of some derivative, $f'(x)$, with each change in *x*. Thus if $f'(x)$ represents the slope of a mountainside, the change in elevation per change in horizontal displacement, when the slope is a constant pitch, then $f''(x) = 0$, that is, it does not exist. If the slope gets steeper as the mountain is climbed, then the second derivative is positive. In physics this second derivative of *x* is called the Laplacian. A force embodying the inverse square law is an instance of the second derivative.

If the acceleration, the change in $f'(x)$ or $f'(t)$, is not constant, then we have a third derivative of these functions, denoted by the triple prime or

$$\frac{d^3y}{dx^3} = f'''(x) \text{ or } \frac{d^3y}{dt^3} = f'''(t). \quad (7.29)$$

With respect to $f''(t)$, this acceleration of acceleration is known as jerk. Any change from a position of rest involves an element of jerk, since the acceleration from zero to any finite velocity is not instantaneous or constant. With respect to a mountainside, if the slope increases exponentially with the climb, instead of at a steady rate of say 50 meters per kilometer of horizontal distance covered, then the third derivative is functioning.

In (7.28) and (7.29) it will be noticed that the differential with respect to *y* is preceded by the order of the derivative, while the differential with respect to *x* and *t* is followed by the order or exponent of the derivative. This is due to the fact that the latter variables are actual squares and cubes, that is powers of the differentials, while the order of the dependent differential of *y* indicates the change in *y* attributed to the independent variable of the same order. The dimensionality of *y* is always of what ever happens to be the inherent dimensionality of the quality *y* represents. If *y* is a force, d^ny will itself have units of force. Notice that $f^n(t)$, then would have units of force per time to the nth power.

A geometric example will perhaps make this clear. While the following may not be the customary context for the second and higher order derivatives, it is a legitimate instance of such. The equation for the volume of a cube, in which the volume, V, is a function of the length of one side, all sides being by definition equal, is

$$V = f(x) = x^3. \tag{7.30}$$

The inverse function is

$$x = f(V) = V^{\frac{1}{3}}. \tag{7.31}$$

A change in volume, dV, is still a three dimensional quality, and we might be tempted to say that it is equal to dx^3, which in a certain context it is. If we want to express that change as a derivative function, however, we must use the definition of a derivative, which gives the original function plus the differential change as

$$V + dV = (x + dx)^3 = x^3 + 3x^2 dx + 3x dx^2 + dx^3. \tag{7.32}$$

Subtracting (7.30), the terms in square brackets, gives the differential

$$V + dV - [V] = x^3 + 3x^2 dx + 3x dx^2 + dx^3 - [x^3] \tag{7.33}$$

$$dV = 3x^2 dx + 3x dx^2 + dx^3 \tag{7.34}$$

This last equation is a bit opaque, however, as it assumes that one of the corners of the cube is at the origin of a co-ordinate system and injects a corresponding bias into the derivation, which may or may not be warranted. If we locate the origin at the center of cube, while still assuming each side aligned with the co-ordinates, we must assign the differential change in x to each end of a length and we have instead of (7.32)

$$\begin{aligned}V + dV - [V] &= (x + 2dx)^3 - [x^3] \\ &= x^3 + 6x^2 dx + 12x dx^2 + 8dx^3 - [x^3]\end{aligned} \tag{7.35}$$

$$dV = 6x^2 dx + 12x dx^2 + 8dx^3 \tag{7.36}$$

From this last derivation, it is immediately clear that the differential volume is made up of the six sides of the cube, the twelve edges, and the eight vertices or corners, all times an infinitesimal length of varying orders. The first order derivative, corresponding to the six square sides of the cube of measure x^2 and the one usually deemed to have the greatest significance, is explicitly stated as

$$f'(x) = \frac{d^1 V}{dx^1} = 6x^2. \tag{7.37}$$

The second order derivative, corresponding with the twelve edges of the cube, consisting of twelve line segments of length x, is

$$f''(x) = \frac{d^2 V}{dx^2} = 12x^1. \tag{7.38}$$

The third order derivative, corresponding to the eight vertices, each of zero, or technically, vanishing dimension, is

$$f'''(x) = \frac{d^3 V}{dx^3} = 8x^0. \tag{7.39}$$

Thus for the total derivative or (increasing) change in V with respect to a change in x, we have

$$\sum_{n=1}^{3} \frac{d^n V}{dx^n} = \frac{d^1 V}{dx^1} + \frac{d^2 V}{dx^2} + \frac{d^3 V}{dx^3} = 6x^2 + 12x + 8. \qquad (7.40)$$

If V were decreasing, and dx were a decrement, the change would be

$$-\sum_{n=1}^{3} \frac{d^n V}{dx^n} = -\frac{d^1 V}{dx^1} + \frac{d^2 V}{dx^2} - \frac{d^3 V}{dx^3} = -6x^2 + 12x - 8. \qquad (7.41)$$

Note that the decrease from the initial condition is indicated by the negative sense of the summation, but that the magnitude of the sum in this case is of a positive 6 squares, minus 12 line segments, while adding back the 8 vertices. In more elucidating fashion the magnitude of the derivative becomes

$$\sum_{n=1}^{3} \frac{d^n V}{dx^n} = \frac{d^1 V}{dx^1} - \left(\frac{d^2 V}{dx^2} - \frac{d^3 V}{dx^3} \right) = 6x^2 - (12x - 8). \qquad (7.42)$$

While it is clear that the contribution of the 8 vertices is not effected by the value of x, it is obvious that as x increases, the contribution to the sum made by the edges increases linearly with x, while the contribution made by the surface squares increases exponentially, specifically by the power of 2.

If dx is not quite zero, but exceedingly small compared to x, then it is apparent that the order of the derivatives is a fair appraisal of each component's contribution to dV. The additional volume is predominantly surface differential. In fact, at the limit as dx approaches 0, each order is exponentially greater that the next order in succession. However, if the components are allowed to increase exponentially beyond the value of x, the situation inverts itself.

If we think of the original cube, still positioned about the origin, as having some very small unit edges of length x, where x is the smallest imaginable length, and make the change, dx, exceedingly great, in fact approaching infinity, then the first order of six squares constitutes the six sense-axes of a 3-D space, the second order, the twelve edges, constitutes the twelve quadrants of the x-y, y-z, and z-x planes, while the third order of the eight vertices become the 3-D octants of the co-ordinate system.

In the above cubic scenario, it is apparent that the numerical coefficients, which in a standard development of the calculus arise through the operation of the binomial expansion as with (7.32), are actually inherent aspects of the specific cubic geometry. The derivation consists of a division of each of the orders of differentiation by x, n times, where n indicates the order of each term. As such it represents a reduction in the power of each term by 1 for each time or order. Alternatively, we can view this as a multiplication for each instance of differentiation of $x^{-1}dx$. Thus, with the observation that a change in the volume of a cube must occur at its 3 boundary elements, i.e. faces, edges and vertices, (7.36) can be arrived at by

$$dV = \sum_{n=1}^{3} \frac{d^n V}{dx^n} dx^n = \sum_{n=1}^{3} \frac{{}^n V}{x^n} dx^n = \sum_{n=1}^{3} {}^n V \left(\frac{dx}{x}\right)^n$$

$$= 6x^3 \left(\frac{1}{x}dx\right)^1 + 12x^3 \left(\frac{1}{x}dx\right)^2 + 8x^3 \left(\frac{1}{x}dx\right)^3 \quad (7.43)$$

$$= 6x^2 dx + 12x dx^2 + 8 dx^3$$

Notice that this describes an arrangement of 27 cubes, 3 x 3 x 3, with the original cube of volume V at the center, and that the total number of elements in the surface is 26, corresponding to the 26 adjacent cubes. The process of differentiation reduces each of these cubes exponentially, according to its relationship to the center cube. The n in ${}^n V$ indicates both the magnitude and the geometry of the coefficients that arise through the polynomial expansion. As there are two boundaries to any interval x, and an interval of equal magnitude to x at each boundary, x_b, a binomial of power n is

$$(x + 2x_b)^n = 1x^n + (3^n - 1) x^{n-m} x_b^m. \quad (7.44)$$

As an example, a 4-D hypercube expansion is

$$(x + 2x_b)^4 = 1x^4 + 80 x^{n-m} x_b^m = 1x^4 + 8x^3 x_b^1 + 24 x^2 x_b^2 + 32 x^1 x_b^3 + 16 x^0 x_b^4. \quad (7.45)$$

Making the substitution, $x_b \underset{\lim \to 0}{=} dx$ for the case of a 4-D equivalent to (7.43) gives

$$(x + 2dx)^4 - x^4 = 80 x^{n-m} dx^m = 80 x^n \left(\frac{dx}{x}\right)^m$$

$$= 8x^4 \left(\frac{dx}{x}\right)^1 + 24 x^4 \left(\frac{dx}{x}\right)^2 + 32 x^4 \left(\frac{dx}{x}\right)^3 + 16 x^4 \left(\frac{dx}{x}\right)^4 \quad (7.46)$$

$$= 8x^3 dx^1 + 24 x^2 dx^2 + 32 x^1 dx^3 + 16 x^0 dx^4$$

As this is obviously a logarithmic operation, as indicated by the bracketed terms, where

$$x^n = y \therefore \log_x y = n \quad (7.47)$$

and as the derivative of the natural log is

$$d \ln x = d \log_e x = \frac{1}{x} dx, \quad (7.48)$$

we can recast (7.43) as

$$dV = \sum_{n=1}^{3} \frac{d^n V}{dx^n} dx^n = \sum_{n=1}^{3} {}^n V (d \ln x)^n$$

$$= 6x^3 (d \ln x)^1 + 12 x^3 (d \ln x)^2 + 8 x^3 (d \ln x)^3 \quad (7.49)$$

This last observation suggests a fundamental tie-in between the derivative of a polynomial function and that of the natural logarithm.

With this in mind, it is apparent that the matter of differentiation is closely related to the subject of exponentiation, and appears to consist of reduction by one power or order of exponentiation for each order of differentiation. In the case of the cube, it is clear that each order indicates an orthogonal reduction, from cube to square, to line segment to point.

We might wonder if this is the case for all functions of the form of (7.23). We might, for example, have a function that relates the perimeter of a square to its edge. The resulting equation and its derivative are

$$P = 4x$$
$$P + dP - [P] = 4(x + dx) - [4x]$$
$$dP = 4dx \qquad (7.50)$$
$$\frac{dP}{dx} = 4$$

What is different in this case, is that there is no apparent second derivative. In other words, for

$$\frac{dP}{dx} = f'(x) = 4 \qquad (7.51)$$

the function $f'(x)$ is not affected by the value of x. It is a constant and does not involve a rate of change of $f'(x)$ with respect to a change in x.

Any function that does not change has a first order derivative of 0. This does not mean, however, that it might not have a second order derivative. Metaphorically speaking, the water swirling around a drain might be described by some function that maps its motion around the plane of the water's surface. At the point at which it becomes vertical and disappears down the drain, effectively leaving the dimensional space of the tub, the derivative with respect to change in that space vanishes. Obviously there still must be some function describing its motion vertically and perhaps even horizontally once it has entered the plumbing system, albeit, in terms of that other dimension. The key is to realize that (7.51) actually should be written

$$\frac{dP}{dx} = f'(x) = 4x^0 = 4(1) = 4\ln_x x \qquad (7.52)$$

$$\therefore \frac{d^2P}{dxd\ln_x x} = 4 \qquad (7.53)$$

$$\frac{d^2P}{dx^2} = f''(x) = \frac{4}{x}$$

There is another condition, however, in which, although there is a changing rate of change, there is no <u>apparent</u> change in the rate of change, i.e. no <u>apparent</u> second derivative, and that is the case of the exponential function, specifically of the natural base e, inversely related to the natural logarithmic function. The exponential function is its own derivative, of whatever order we might envision, where

$$f(x) = y = e^x \text{ iff } \ln y = x \qquad (7.54)$$

so that

$$f'(x) = D_x[f(x)] = D_x e^x = e^x \qquad (7.55)$$

where D_x is a differential operator, that is, it operates on f to produce f'. Thus since

$$D_x''e^x = D_x\left[D_x e^x\right] = D_x\left[e^x\right] = e^x \tag{7.56}$$

the difference between any two orders of differentiation, in fact, between any order and the exponential function itself is

$$D_x''e^x - D_x e^x = D_x e^x - e^x = e^x - e^x = 0. \tag{7.57}$$

and

$$\frac{e^x}{D_x^n e^x} = 1 \tag{7.58}$$

We might wonder what significance this has, since subtracting one order derivative from another is rather like subtracting oranges from apples. They are two different types of entity, just as a point is a different type of entity than a line segment or length, which is itself different from an area, itself different from a volume. In fact, there are generally held to be an infinite number of points in a length, lengths in an area, and areas in a volume, i.e. of dimensions in the next higher order of dimension, so the subtraction of the lesser from the greater leaves the latter substantially unchanged.

This then is the point in (7.58). The dimensional identity of each order of differentiation is the same as that of the basic function, $f(x) = y$, since it is the exponent itself that is variable and not the base. If we compare this condition with that of the cube and its derivative orders, in which the relative contribution of each order to the overall change in V is dependent on the ratio of $dx^n:dx$, we see that for an exponential change, the relative contribution of each order at the limit is unaffected by the change in x, or as it is often the variable used in this context, t. The ratio in the exponential case is always 1, hence the apparent lack of change.

A doubling of the sum of the lengths of the edges and a quadrupling of each surface area results in an eightfold increase in the volume of the cube. Note that there is no change in the number of boundary elements and their angular configuration, which is the defining condition of the cube. Using a combinatorial or additive approach to creating a change in the cube, then, we see that it is more economical to augment the edges to move the vertices further apart, which defines the volume change, than to fill the cube with volume, since a unit of length is orders of magnitude less than a unit of volume. Yet there is no conformal or topological difference between a cube of unit volume and one of volume 8, something we intuitively understand. The matter of scale only attains significance within a combinatorial or economical approach.

Hence in a continuum analysis, where the elements of various dimensions are integrally related, i.e. non-combinatorially, if we had a cube experiencing a continuous exponential change, each of the elements in its boundary, the faces, edges and vertices, would increase proportionally to its order with the change in volume, each derivative order increasing in proportion as the exponent of x or t. In such event, using the value of $x = \sqrt[3]{V}$ as our standard, all orders show the same exponential change and the whole is relatively, or perhaps better stated, intrinsically unchanged. It is only within the context of some extrinsically determined property, such as some external standard of length or density, i.e. volume, surface or linear, that change is registered or observed.

This is intuitively understood, especially in the preparation of scaled engineering drawings and models and other graphic representations. It is also known rarely to occur in the physical world in which physical forms result from the combination of discrete units or building blocks of matter. Adult humans do not generally look like babies three to four times their original length. Equally true, most tree girth-to-height ratios increase with growth. On the other hand, most celestial bodies of any size assume a generally spherical shape, irrespective of their volume. Rephrasing (7.43)

$$dV = \sum_{n=1}^{3} \frac{d^n V}{dV^{\frac{n}{3}}} dV^{\frac{n}{3}} = \sum_{n=1}^{3} \frac{^n V}{V^{\frac{n}{3}}} dV^{\frac{n}{3}} = \sum_{n=1}^{3} {}^n V \left(\frac{dV^{\frac{1}{3}}}{V^{\frac{1}{3}}} \right)^n$$

$$= 6V \left(\frac{dV^{\frac{1}{3}}}{V^{\frac{1}{3}}} \right)^1 + 12V \left(\frac{dV^{\frac{1}{3}}}{V^{\frac{1}{3}}} \right)^2 + 8V \left(\frac{dV^{\frac{1}{3}}}{V^{\frac{1}{3}}} \right)^3 \qquad (7.59)$$

$$= 6 \left(V^{\frac{3}{3}-\frac{1}{3}} \right) dV^{\frac{1}{3}} + 12 \left(V^{\frac{3}{3}-\frac{2}{3}} \right) dV^{\frac{2}{3}} + 8 \left(V^{\frac{3}{3}-\frac{3}{3}} \right) dV^{\frac{3}{3}}$$

$$= 6 V^{\frac{2}{3}} dV^{\frac{1}{3}} + 12 V^{\frac{1}{3}} dV^{\frac{2}{3}} + 8 V^{\frac{0}{3}} dV^{\frac{3}{3}}$$

From another but still exponential perspective, in terms of our initially outlined derivative orders, this indicates that the magnitude of displacement, velocity, acceleration, and jerk might be equal or

$$f'''(x) - f''(x) = f''(x) - f'(x) = f'(x) - f(x) = 0. \qquad (7.60)$$

This is essentially the same equation as (7.57) and indicates that a relationship of this type is exponential in nature.

We do not have to look far for another familiar instance of such. While change is generally equated with motion and thereby with displacement or translational change of position, rotation presents an instance of motion without translational displacement, taking the position of the rotating body as a whole. It is in a sense change without change, and is elegantly presented using the Euler identity, which involves the exponential expansion of an imaginary logarithm or

$$e^{ix} = \cos x + i \sin x = y \qquad (7.61)$$

As with the exponential function of (7.54), the domain of x is the real number line, but in this case, instead of the range of $y > 0$, y oscillates over the range of $-1 \leq y \leq 1$ if we consider only the real component, for each change in x of 2π, our angles and the "natural" unit of x in this case being in radians. Otherwise y must be a complex number whose range is a circle centered on the x-y origin in the complex plane, in this case of implied radius r = 1. If we then map y to the real x-y plane, by multiplying y times its complex conjugate, where y then gives the radius r and x is the count of the rotations, the range of y will be a horizontal line crossing the y axis at y = r = 1. Note that a circle of fixed radius from the origin maps as a horizontal line segment of 2π length. Since the derivative of such a line is zero, the rate of change of the rotation is zero, at least in this mapping. The only variable in such case might be the velocity of the rotation which might be reflected in the scale, the density, of the real number line. Presumably a denser placement of the integers would represent a greater velocity.

The use of a radian as the unit measure of x makes the equation self normalizing, that is, it sets the x and y axes of a co-ordinate system against which we might plot the function to the same scale. Assuming a rotational amplitude or modulus, i.e. the radius, r, equal to the hypotenuse, selection of a unit value for the y axis for a cosine of 1 automatically dictates the unit length for the x axis, since a radius, r, and a rotational arc of one radian measured at a distance r from the center are of equal length. We can then envision x as the distance traveled by a point P on the circumference of a rotating disk or equator of a sphere of radius r, but it might simply be a point in space that is revolving about some center of oscillation which is also the polar origin.

Thus for any value of x,

$$\frac{x}{2\pi} = n + \frac{\varphi}{2\pi} \qquad (7.62)$$

where n is an integer number of rotational cycles, $-\infty < n < +\infty$, and where φ is a remainder angle or phase in which $0 < \varphi < 2\pi$. As we shall soon see, we might also state

$$\frac{x}{\frac{\pi}{2}} = n + \frac{2\varphi}{\pi} \qquad (7.63)$$

where n is the count of the number of times that $|y|$ equals one.

If we want to express x in some conventional unit such as meters, we simply multiply it by the number of meters per radian and x will be in units of meters. In such case, in order to convert x back to normalized units, (7.61) becomes

$$e^{i\kappa x} = \cos\theta + i\sin\theta = y \qquad (7.64)$$

where κ is the angular wave number or number of arc radians per unit of length and

$$\kappa = \frac{\theta}{x}. \qquad (7.65)$$

A similar approach for t gives us

$$e^{it} = \cos t + i\sin t = y \qquad (7.66)$$

where time is in natural units or radians, and for conversion from conventional units of time,

$$e^{i\omega t} = \cos\theta + i\sin\theta = y \qquad (7.67)$$

where ω is the angular frequency or number of arc radians per unit of time and

$$\omega = \frac{\theta}{t}. \tag{7.68}$$

Such a condition would apply to a standing wave of fixed angular frequency.

For a wave of fixed frequency traveling from a propagating source, we can combine the two to get

$$e^{i\theta} = e^{i(\kappa x + \omega t)} = \cos\theta + i\sin\theta = y \tag{7.69}$$

where it is understood that x and t are of ambivalent sense.

Finally by extending the exponent of e to complex numbers we have,

$$e^{r \pm i\theta} = e^r e^{\pm i\theta} = e^r e^{\pm i(\kappa x + \omega t)} = R(\cos\theta \pm i\sin\theta) = y \tag{7.70}$$

$$e^{-r \pm i\theta} = e^{-r} e^{\pm i\theta} = e^{-r} e^{\pm i(\kappa x + \omega t)} = \frac{1}{R}(\cos\theta \pm i\sin\theta) = y, \tag{7.71}$$

and we can see that (7.69) is simply a special case of these last two in which $R = 1$. Assuming that $R > 1$, using $y = r = \sqrt{a^2 - (ib)^2}$, (7.70) now maps to the x-y real plane as a horizontal line greater than $y = 1$, and (7.71) maps to a line between the x axis and the line $y = 1$. The argument, as the middle term of (7.64) is called, conceals the fact that the sense of the angle θ determines whether $i\sin\theta$ is positive, changing in a counterclockwise sense, or negative, in a clockwise sense. Thus we could apply a convention in which the negative sense of $i\sin\theta$ maps (7.70) and (7.71) to horizontal lines crossing the negative y axis. The rotation velocity and frequency would then switch sense.

Let us examine the function

$$y = f(W) = We^W. \tag{7.72}$$

Inverting the function so that

$$W = f(y) \tag{7.73}$$

gives

$$f(y)e^{f(y)} = y. \tag{7.74}$$

We can find that

$$f(y)e^{f(y)} = W(n)e^{W(n)} = y \tag{7.75}$$

where $W(n)$ is related to the Lambert W function and in fact is identical to that function for the principal branch, integer values $n > 0$, or Lambert $W(0, n>0)$. As will be shown, while the Lambert W function is complex for all values of $n < 0$, for all values of $-\infty \le n \le +\infty$ $W(n)$ is real and $W(n) = -W(-n)$. Further, we define

$$W(n) = n\ln x. \tag{7.76}$$

Therefore, (7.75) becomes

$$n\ln x e^{n\ln x} = n\ln x (x^n) = U_n n = y. \tag{7.77}$$

where U_n is the co-efficient of n needed to produce x for any value of y.

Inverting the function to its original

$$y = nx^n \ln x \tag{7.78}$$

and differentiating gives

$$\begin{aligned}dy &= (nx^n) d\ln x + (n\ln x) nx^{n-1} dx \\ &= (nx^n)\frac{dx}{x} + (n\ln x) nx^{n-1} dx\end{aligned} \tag{7.79}$$

with the derivative

$$\frac{dy}{dx} = (1 + n\ln x) nx^{n-1} \tag{7.80}$$

If we substitute for the natural log derivative, we have instead

$$\frac{dy}{d\ln x} = (1 + n\ln x) nx^n. \tag{7.81}$$

From (7.76) it follows that

$$\frac{dy}{dx} = (1 + W(n)) nx^{n-1} \tag{7.82}$$

and with a little foresight, we might imagine that this hides a complex function, as with

$$\begin{aligned}\frac{dy}{dx} &= (1 + iW(n)) nx^{n-1} \\ \frac{dy}{d\ln x} &= (1 + iW(n)) nx^n\end{aligned} \tag{7.83}$$

Returning to (7.76) for that case in which $U_n = 1$, $y = n$ and the normalized value of $W(n)$ is

$$W_0(n) = n \ln x_n, \tag{7.84}$$

where the n in the subscript of x relates that value of x as the unique normalizing value for $W_0(n)$ and we have

$$W_0(n)(x_n^n) = n \ln x_n (x_n^n) = n = \frac{y}{U_n}. \tag{7.85}$$

It follows that

$$\frac{W_0(n)}{n} = \ln x_n = \frac{1}{x_n^n} = \kappa_n^n = \frac{y}{nU_n x_n^n}. \tag{7.86}$$

As before, substituting t for x, the equivalent for (7.86) is

$$\frac{W_0(n)}{n} = \ln t_n = \frac{1}{t_n^n} = \omega_n^n = \frac{y}{nU_n t_n^n} \tag{7.87}$$

Finally, continuing in that vein

$$\omega_n = \frac{1}{t_n} = \left(\frac{W_0(n)}{n}\right)^{\frac{1}{n}} = \frac{1}{x_n} = \kappa_n \qquad (7.88)$$

$$= (\ln t_n)^{\frac{1}{n}} = (\ln x_n)^{\frac{1}{n}}$$

Since for that case in which $n = 0$, (7.86) becomes

$$\ln x_0 = \frac{W_0(0)}{0} = \frac{1}{x_0^0} = \kappa_0^{\,0} = 1, \qquad (7.89)$$

apparently

$$x_0 = e = e_0 \text{ and } \kappa_0 = e^{-1} = e_0^{-1} \qquad (7.90)$$

and the dividend in the first term of (7.86) must be a continuous function which approaches 0 as n approaches 0. Thus for any n, we have a fundamental base e_n in which it can be stated

$$\ln x_n = \ln_0 e_n = \frac{W_0(n)}{n} = e_n^{-n} = e_{-n}^{\,n} = e_{\kappa n}^{\,n}. \qquad (7.91)$$

and for the inverse of x_n

$$\ln x_n^{-1} = \ln_0 e_n^{-1} = \ln_0 e_{-n} = \frac{W_0(-n)}{n} = -e_n^{-n} = -e_{-n}^{\,n} = -e_{\kappa n}^{\,n}. \qquad (7.92)$$

where we define, for conceptual reasons,

$$e_n^{-n} \equiv e_{-n}^{\,n} \equiv e_{\kappa n}^{\,n}. \qquad (7.93)$$

It is noted that the natural log in the second term is specified to apply to the case of (7.89) so that for any value x, for the conventional natural log x, or $\ln x$,

$$\ln x = \ln_n x \bigg|_{n=0} = \ln_0 x. \qquad (7.94)$$

Thus for any e_n, we have

$$\ln_n e_n = e_n^{\,n} \ln_0 e_n = 1 \qquad (7.95)$$

and

$$\ln_n x = e_n^{\,n} \ln_0 x = y = U_n n. \qquad (7.96)$$

In continuation, we have

$$e_n^{\,y} = e_0^{\left(\frac{y}{e_n^{\,n}}\right)} = e_0^{\,y e_{-n}^{\,n}} = x, \qquad (7.97)$$

$$e_0^{\,y} = e_n^{\,y e_n^{\,n}} = e_{-n}^{\,-y e_n^{\,n}} = e_n^{\,y e_{-n}^{\,-n}} = e_{-n}^{\,-y e_{-n}^{\,-n}}, \qquad (7.98)$$

$$e_0^{\,-y} = e_{-n}^{\,y e_n^{\,n}} = e_{-n}^{\,y e_{-n}^{\,-n}} = e_n^{\,-y e_n^{\,n}} = e_n^{\,-y e_{-n}^{\,-n}} \qquad (7.99)$$

and

$$\ln_0 x = \frac{\ln_n x}{e_n^{\,n}} = e_{-n}^{\,n} \ln_n x = e_{-n}^{\,n} y = e_{-n}^{\,n} U_n n, \qquad (7.100)$$

It follows with respect to derivatives that

$$\frac{1}{x} = \frac{d\ln_0 x}{dx} = e_{-n}{}^n \frac{d\ln_n x}{dx}. \qquad (7.101)$$

It is further noted that

$$\ln_{-n} x = \left(\ln_n x\right)^{-1} \qquad (7.102)$$

For all integer values of $n > 0$, we redefine (7.85) and have

$$y = nU_n = n\ln_0 x\left(x^n\right) \qquad (7.103)$$

$$nx^n = \frac{nU_n}{\ln_0 x} = \frac{y}{\ln_0 x} \qquad (7.104)$$

$$x^n = \frac{U_n}{\ln_0 x} = \frac{y}{n\ln_0 x} \qquad (7.105)$$

and

$$\ln_0 x = \frac{U_n}{x^n} = \frac{y}{nx^n} \qquad (7.106)$$

Multiplying (7.104) by (7.101), gives

$$D_x\left(x^n\right) = nx^{n-1} = \left(\frac{1}{x}\right)nx^n = \frac{nU_n}{\ln_0 x}\left(\frac{e_{-n}{}^n d\ln_n x}{dx}\right) = \frac{y}{\ln_0 x}\left(\frac{e_{-n}{}^n d\ln_n x}{dx}\right) \qquad (7.107)$$

and it can be seen that the terms on the right are equal to the derivative of x^n.

With some rearrangement we have

$$x^n\left(n\frac{dx}{x}\right) = \frac{U_n}{\ln_0 x}\left(ne_{-n}{}^n d\ln_n x\right) = \frac{y}{n\ln_0 x^n}\left(ne_{-n}{}^n d\ln_n x\right). \qquad (7.108)$$

Since the terms in brackets in (7.108) are equal, it is apparent that the differential of any variable x of order n, of any function $f(x^n)$, is the product of that function and

$$n\frac{dx}{x} = nd\ln_0 x = ne_{-n}{}^n d\ln_n x = W_0(n)d\ln_n x \qquad (7.109)$$

where the term $ne_{-n}{}^n$ is equal to $W_0(n)$ and to the principal branch value of the Lambert W function for n. It follows that the derivative of x with respect to the nth natural log is

$$\frac{dx}{d\ln_n x} = \frac{xd\ln_0 x}{d\ln_n x} = e_{-n}{}^n x = \frac{W_0(n)}{n}x \qquad (7.110)$$

which normalized to the nth power would be

$$1_n = \frac{dx}{xd\ln_n x} = \frac{d\ln_0 x}{d\ln_n x} = e_{-n}{}^n = \frac{W_0(n)}{n} \qquad (7.111)$$

A factor for normalizing to the 0th power, therefore, would be, $e_n{}^n$, remembering (7.93), giving

$$1_0 = e_n{}^n \frac{dx}{xd\ln_n x} = e_n{}^n \frac{d\ln_0 x}{d\ln_n x} = e_n{}^n e_{-n}{}^n = e_n{}^n \frac{W_0(n)}{n} = \frac{W_0(n)}{n}\left(e_0{}^{W_0(n)}\right). \qquad (7.112)$$

where the last term is of the form of (7.75).

For a negative derivative

$$-1_n = \frac{-dx}{xd\ln_n x} = \frac{-d\ln_0 x}{d\ln_n x} = -e_{-n}{}^n = \frac{W_0(-n)}{n} \qquad (7.113)$$

the factor is $-e_n{}^n$, so that

$$1_0 = -e_n{}^n \frac{-dx}{xd\ln_n x} = -e_n{}^n \frac{-d\ln_0 x}{d\ln_n x} = -e_n{}^n\left(-e_{-n}{}^n\right)$$
$$= -e_n{}^n \frac{W_0(-n)}{n} = \frac{W_0(-n)}{n}\left(-e_0{}^{W_0(n)}\right) \qquad (7.114)$$

If we now introduce a rotational (imaginary) element into this condition, from (7.86) and the above development, we have

$$e_n{}^i = e_0{}^{ie_{-n}{}^n} = e_0{}^{\frac{W(in)}{n}} \therefore e_n{}^{in} = e_{in}{}^n = e_0{}^{ine_{-n}{}^n} = e_0{}^{W(in)} \qquad (7.115)$$

$$e_n{}^{-i} = e_0{}^{-ie_{-n}{}^n} = e_0{}^{\frac{W(-in)}{n}} \therefore e_n{}^{-in} = e_{-in}{}^n = e_0{}^{-ine_{-n}{}^n} = e_0{}^{W(-in)} \qquad (7.116)$$

in which (7.115) and (7.116) are complex conjugates, each representing a unit vector in the complex plane, so that

$$e_n{}^{in} e_n{}^{-in} = 1 + i0. \qquad (7.117)$$

It follows that

$$i1_n = \frac{idx}{xd\ln_n x} = \frac{id\ln_0 x}{d\ln_n x} = ie_{-n}{}^n = \frac{W_0(in)}{n}. \qquad (7.118)$$

implying

$$iW(n) = W(in). \qquad (7.119)$$

Thus with substitution from (7.96), using the normalizing factor $-ie_n{}^n$, (7.118) becomes

$$1_0 = -ie_n{}^n \frac{idx}{xd\ln_n x} = -ie_n{}^n \frac{id\ln_0 x}{d\ln_n x} = \left[-ie_n{}^n ie_{-n}{}^n\right]$$
$$= -ie_n{}^n \frac{W_0(in)}{n} = \frac{W_0(in)}{n}\left(-ie_0{}^{W_0(n)}\right) \qquad (7.120)$$

Similarly for a clockwise rotation,

$$-i1_n = \frac{-idx}{xd\ln_n x} = \frac{-id\ln_0 x}{d\ln_n x} = -ie_{-n}{}^n = \frac{W_0(-in)}{n} \qquad (7.121)$$

we have the normalizing factor, $ie_n{}^n$, and

$$1_0 = ie_n{}^n \frac{-idx}{xd\ln_n x} = ie_n{}^n \frac{-id\ln_0 x}{d\ln_n x} = \left[ie_n{}^n\left(-ie_{-n}{}^n\right)\right]$$
$$= ie_n{}^n \frac{W_0(-in)}{n} = \frac{W_0(-in)}{n}\left(ie_0{}^{W_0(n)}\right) \qquad (7.122)$$

In the above treatment of (7.118) through (7.122), we have used real normalizing factors with imaginary sense. It is further noted that the normalizations shown in the square brackets of (7.120) and (7.122), where the left $\mp e_n{}^n$ operates on the right initial condition, $\pm e_{-n}{}^n$, are instances of a type of complex inversion and are therefore conformal and are not instances of complex conjugation as shown in (7.117). In this latter case we interpret the subscript as the real exponent, n, of the nth exponential base and the superscript as the rotational or "imaginary" exponent. In this regards we will find that

$$e_n{}^{in} = e_{in}{}^n \tag{7.123}$$

so the determining indication for the rotational exponent is the presence of the i sense. Thus we have the following complex identities

$$e_n{}^{in} \equiv e_{-n}{}^{-in} \equiv e_{in}{}^n \equiv e_{-in}{}^{-n} = \overline{e_n{}^{-in}} \equiv \overline{e_{-n}{}^{in}} \equiv \overline{e_{in}{}^{-n}} \equiv \overline{e_{-in}{}^n} \tag{7.124}$$

where the terms on the right are the complex conjugates of those on the left, all of which represent unit vectors whose points lie on the unit circle. Thus it is not to be interpreted in the usual sense of a decomposed complex number, since

$$z_e = e^n + e^{in} \neq e_n{}^{in} = \left(e^n\right)^{in} \equiv \left(e^{in}\right)^n. \tag{7.125}$$

We might also surmise that the above normalizations are analytic. The normalizing factor inverts first with respect to sense of the nth degree of the exponential base, e_{-n}, on the unit circle, which amplifies the modulus or vector length of that base. This changes (7.117) to

$$e_n{}^{in} e_{-n}{}^{in} = e_0{}^{in} = 1 + i\sin\theta. \tag{7.126}$$

This is followed by inversion in the real axis to get

$$e_0{}^{-in} e_n{}^{in} e_{-n}{}^{in} = e_0{}^{-in} e_0{}^{in} = e_0{}^0 = 1 + i\sin\theta + (0 - i\sin\theta) = 1 + i0. \tag{7.127}$$

where it is clear that the 0th exponential base raised to the 0th power is a unit vector on the real, x axis. Thus complex normalization in this instance amounts to

$$e_{-n}{}^{in} e_n{}^{-in} = e_0{}^0 = 1 + i0 \tag{7.128}$$

and complex inversion to

$$\frac{1+i0}{e_{-n}{}^{in}} = \frac{e_0{}^0}{e_{-n}{}^{in}} = e_n{}^{-in} \tag{7.129}$$

which is an amplitwist as defined by Tristan Needham in <u>Visual Complex Analysis</u>. This is the case of the bracketed term of (7.120), which might be represented by multiplication of a point in the counterclockwise interior of the unit circle by its reflection in that circle followed by multiplication of the resulting vector and its complex conjugate. In the case of (7.122), we have such multiplication of a clockwise interior vector by a counterclockwise exterior vector, both cases resulting in a unit vector along the real axis.

With respect to (7.89), we see that what appears at first glance to be a singularity is in fact an identity of the 0th order. Remembering that the natural log function maps to the y axis and is therefore equivalent to the imaginary axis in the complex plane and $i\sin\theta$, using the normalization factor, $e_0{}^0 = 1$, and recalling that $x_n = e_n$

$$e_0^{+i0} \ln x_0 = e_0^{-i0} = \frac{W_0(0)}{0} e^{W(0)} = x_0^{+i0} x_0^{-i0} = 1 \pm i0. \tag{7.130}$$

The sense of the 0th powers can be seen as a vector potential or direction, similar to the assignment of charge sense in a static electrical or potential field.

Investigation will show that for any n or q, real or imaginary,

$$\left(\frac{dx}{xd\ln_0 x}\right) = \frac{W(n)}{n} e^{W(n)} = \frac{W(q)}{q} e^{W(q)} \tag{7.131}$$

Applying the above normalization factors to (7.83), we have

$$-ie_0^{W(n)} \frac{dy}{dx} = \left(-ie_0^{W(n)} + W(n)e_0^{W(n)}\right) nx^{n-1} = \left(-ie_n^{\ n} + n\right) nx^{n-1}$$
$$-ie_0^{W(n)} \frac{dy}{d\ln x} = \left(-ie_0^{W(n)} + W(n)e_0^{W(n)}\right) nx^n = \left(-ie_n^{\ n} + n\right) nx^n \tag{7.132}$$

The interpretation of this development is that while the qth exponential base to the qth power maps the real number line x to the positive real number line y, the qth exponential base to the iqth power or the iq^{th} exponential base to the qth power maps the real number line to the unit circle. Further, whereas there is an asymptote for the former in the direction of the negative x axis, the unit circle is both <u>asymptote and tangent</u> in either sense for the rotational mapping of e.

With respect to the integer orders of e, it is apparent that each represents a mapping to the real number line of an exponential change in n orthogonal spaces. Thus referring to (7.88) in the context of (7.112), we can state the following normalizations of an n dimensional t or x

$$1 = e_n \omega_n = \frac{e_n}{t_n} = \left(\frac{W_0(n)}{n}\right)^{\frac{1}{n}} \left(e_n^{\ n}\right)^{\frac{1}{n}} \quad \text{and} \tag{7.133}$$
$$= e_n \left(\ln t_n\right)^{\frac{1}{n}}$$

$$1 = e_n \kappa_n = \frac{e_n}{x_n} = \left(\frac{W_0(n)}{n}\right)^{\frac{1}{n}} \left(e_n^{\ n}\right)^{\frac{1}{n}} \tag{7.134}$$
$$= e_n \left(\ln x_n\right)^{\frac{1}{n}}$$

Fleshing this out for the first 4 orders of n with conjectured generalization at infinity, we have the following table, as generated by Maple, where it can be seen that a negative n is simply an inversion of $e_n^{\ n}$ to $e_n^{\ -n} = e_{-n}^{\ n}$.

$f(n)$ \ n	0	1	2	3	...	∞
e_n	2.718281828..	1.763222834..	1.531584394..	1.419024454..		1
$e_{-n} = e_n^{-1}$	0.367879441..	0.567143291..	0.652918640..	0.704709490..		1
e_n^n	1	1.763222834..	2.345750756..	2.857390779..		∞
$e_{-n}^n = e_n^{-n}$	1	0.567143291..	0.426302751..	0.349969632..		0
$\ln_0 e_n$	1	0.567143291..	0.426302751..	0.349969632..		0
$\ln_0 e_{-n}$	-1	-0.567143291..	-0.426302751..	-0.349969632..		0
$\ln_n e_n$	1	1	1	1		1
$\ln_n e_{-n}$	-1	-1	-1	-1		-1
$\ln_0 e_n^n = W(n)$	0	0.567143291..	0.852605502..	1.049908893..		∞
$\ln_0 e_{-n}^n = W(-n)$	0	-0.567143291..	-0.852605502..	-1.049908893..		$-\infty$
$\ln_n e_n^n$	0	1	2	3		∞
$\ln_n e_{-n}^n$	0	-1	-2	-3		$-\infty$

n	$f(n)$	$n \ln_0 e_n$	$in \ln_0 e_n$	$n \ln_0 ie_n$	$in \ln_0 ie_n$
1		0.5671...	i0.5671...	0.5671...$+i\pi/2 (=+1\, i\pi/2)$	$-\pi/2 + i$0.5671...
2		0.8526...	i0.8526...	0.8526...$+i\pi (=+2\, i\pi/2)$	$-\pi + i$0.8526...
3		1.0499...	i1.0499...	1.0499...$+i3\pi/2 (=+3\, i\pi/2)$	$-3\pi/2 + i$1.0499...
4		1.2021...	i1.2021...	1.2021...$+i\, 2\pi (=+4\, i\pi/2)$	$-2\pi + i$1.2021...
5		1.3067...	i1.3067...	1.3067...$+i5\pi/2 (=+5\, i\pi/2)$	$-5\pi/2 + i$1.3067...
6		1.4324...	i1.4324...	1.4324...$+i\, 3\pi (=+6\, i\pi/2)$	$-3\pi + i$1.4324...

In this final table, it is clear that the integers, n, are the count of the rotations of ½ π and of the powers and hence the number of orders of i, both indications of a degree of orthogonal structure.

The special case of e_2 is shown to be of fundamental significance to an understanding of the foundations of quantum mechanics. Thus

$$e_2 = \left(\ln e_2\right)^{-\frac{1}{2}}$$
$$= \frac{1}{\omega_2} = \partial t = \left(\ln \partial t\right)^{-\frac{1}{2}}$$
$$= \frac{1}{\kappa_2} = \partial x = \left(\ln \partial x\right)^{-\frac{1}{2}} \quad (7.135)$$
$$= 1.531584394...$$

With respect to a conservative 3-field, here shown for a stress, where the volume potential energy density is conserved, a logarithmic change of the tension fields in one dimension, leading to a change of opposite sense in the two shear fields,

$$\left(\ln \partial T_\xi\right)\left(\partial T_\eta\right)\left(\partial T_\zeta\right) = E_1 = 1 \tag{7.136}$$

is therefore

$$\partial T_{\eta,\zeta} = \left(\ln \partial T_\xi\right)^{-\frac{1}{2}} \tag{7.137}$$

which can be stated by using the coefficients as

$$e_2 \partial T_0 = \left(\ln e_2\right)^{-\frac{1}{2}} \partial T_0. \tag{7.138}$$

Assuming the change in the tension field is less than unity results in a negative logarithm and an orthogonal (imaginary) sense to the transverse fields, giving us

$$ie_2 \partial T_0 = \left(\ln e_2^{-1}\right)^{-\frac{1}{2}} \partial T_0 = i1.531584394...\partial T_0. \tag{7.139}$$

Bibliography

Astronomy and Astrophysics 338, 856-862 (1998), "Magnetically supported tori in active galactic nuclei", Lovelace, Romanova, and Biermann.

Exploring Black Holes, Taylor and Wheeler, Addison Wesley Longman, Inc, New York, 2000.

The Extravagant Universe, Kirshner, Princeton University Press, Princeton, NJ, 2002.

The Feynman Lectures on Physics :Commemorative Issue, Feynman, Leighton, Sands, Volume I, Addison-Wesley Publishing Company, Inc., Reading, Massachusetts 1963.

Fundamentals of Physics, Fifth Edition, Halliday, Resnick, Walker, John Wiley & Sons, Inc. New York, 1997.

Gravitation, Misner, Thorne, and Wheeler, W.H. Freeman and Company, New York, 1973.

Mathematical Methods for Physicists, Fifth Edition, Arfken and Weber, Harcourt Academic Press, New York, 2001.

Physics of Waves, Elmore and Heald, Dover Publications, Inc., New York, 1985. This was the primary source for wave, elasticity and tensor equations.

The Six Core Theories of Modern Physics, Stevens, The MIT Press, Cambridge, Massachusetts, 1995.

Three Roads to Quantum Gravity, Smolin, Basic Books, New York, 2001.

Visual Complex Analysis, Needham, Oxford University Press, Oxford, England, 1997

National Institute of Standards and Technology, These are the **2002 CODATA recommended values** of the fundamental physical constants, the latest CODATA values available. For additional information, including the bibliographic citation of the source article for the 1998 CODATA values, see P. J. Mohr and B. N. Taylor, "The 2002 CODATA Recommended Values of the Fundamental Physical Constants, Web Version 4.0," available at physics.nist.gov/constants. This database was developed by J. Baker, M. Douma, and S. Kotochigova. (National Institute of Standards and Technology, Gaithersburg, MD 20899, 9 December 2003).

Table of Nuclides, Nuclear Data Evaluation Lab., Korea Atomic Energy Research Institute (c) 2000-2002, http://yoyo.cc.monash.edu.au/~simcam/ton/index.html

R.R.Kinsey, et al.,*The NUDAT/PCNUDAT Program for Nuclear Data*, paper submitted to the 9 th International Symposium of Capture-Gamma ray Spectroscopy and Related

Topics, Budapest, Hungary, Octover 1996.Data extracted from NUDAT database (Jan. 14/1999)

Schwarzschild, Bertram, "Tiny Mirror Asymmetry in Electron Scattering Confirms the Inconstancy of the Weak Coupling Constant", Physics Today, September, 2005

Wapstra, A. H. and Bos, K., "The 1983 atomic-mass evaluation. I. Atomic mass table," Nucl. Phys. A 432, 1-54, 1985, quoted at http://hyperphysics.phy-astr.gsu.edu/hbase/nucene/nucbin2.html

Construction of the Natural Numbers from a Real Exponential Field

By Mart Gibson

July 19, 2007

Mart Gibson
P.O. Box 2358
Southern Pines, NC 28388
910-692-7462
hmg@uniservent.com

Copyright © Martin Gibson 2021 All Rights Reserved

Construction of the Natural Numbers from a Real Exponential Field

Consider the following geometric construction.

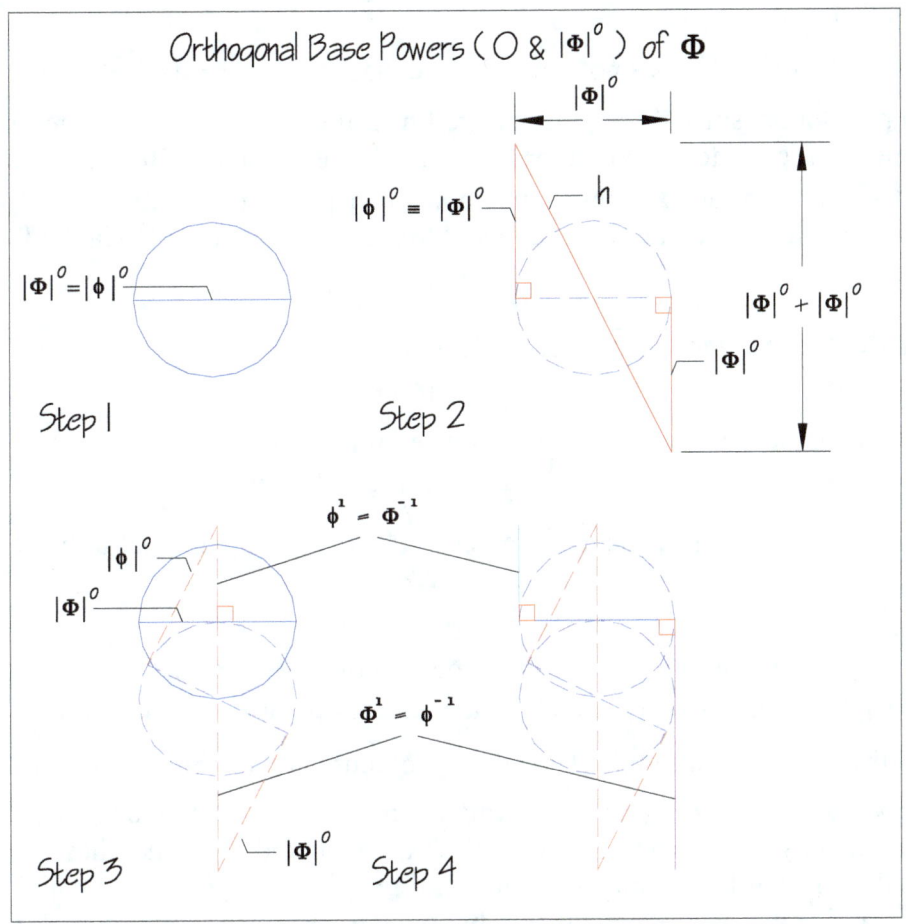

Step 1 – Construct a unit circle, that is a circle with unit diameter, d, and unit π circumference, c. Designate the unit base number, Φ, and the exponential unit, Φ^0. Designate the inverse unit base number, ϕ, and the inverse exponential unit, ϕ^0. Thus the magnitude of the diameter, which cannot be negative (and where the absolute brackets are added here for emphasis but will later be left off unless required for clarity), is

$$|d| = |\Phi|^0 = |\phi|^0 \tag{1.1}$$

It is noted that as exponential bases, these base numbers cannot be negative, though their exponents can be. We might also imagine that their exponential functions, $y_\Phi = f(x) = \Phi^x$ or $y_\phi = f(x) = \phi^x$, can be applied to a directional vector of any sense.

Step 2 – Extend two tangent line segments of unit length, orthogonal to and in opposite directions from opposing ends of the diameter. Construct a line segment joining the distal ends of the tangents, the hypotenuse, h, of a right triangle of sides $\Phi^0 + \Phi^0$ and Φ^0. The square of the hypotenuse is

$$h^2 = \left(\Phi^0 + \Phi^0\right)^{\Phi^0+\Phi^0} + \left(\Phi^0\right)^{\Phi^0+\Phi^0} \tag{1.2}$$

We will call $\Phi^0 = \phi^0$, the number 1 and $\Phi^0 + \Phi^0 = \phi^0 + \phi^0$, the number 2. Note, however, that

$$1 = \Phi^0 = \phi^0 \text{ does not necessarily equal } \left(\Phi^0\right)^{\Phi^0+\Phi^0} = \left(\phi^0\right)^{\phi^0+\phi^0} = 1^2 \tag{1.3}$$

in qualitative terms, since the first terms are linear intervals or spans of the real number line, while the second terms could represent a unit area, a span of the real number plane. Thus $\Phi^0 + \Phi^0 = 2$, as an exponent can denote an orthogonal condition, or a "square" number. It conveys a geometric component to the use of exponents via the Pythagorean theorem.

With respect to the extent of the unit circle above, then

$$c = \pi\left(\Phi^0\right) = \pi_c \tag{1.4}$$

while the extent of the sphere, s, with c as its extremal cross section is

$$s = \pi\left(\Phi^0\right)^2 = \pi_s \tag{1.5}$$

and it is obvious that, while the magnitudes of c and s are equal, qualitatively

$$\pi_c \neq \pi_s. \tag{1.6}$$

The above construction is equivalent to a rotation of Φ^0 about each end of the diameter, each through an angle of $+\frac{\pi}{2}$. Label the tangents by their magnitudes, the left tangent, $|\phi|^0$, and the right tangent, $|\Phi|^0$. Thus the hypotenuse can be thought of as a unit diameter stretched between and connecting the rotated ends of the tangents. Consider the following identity, in which the sense of the exponents indicates that these two terms are anti-parallel. This follows conventional use of imaginary exponents in which the sense of the exponent is rotational, where generally positive designates counterclockwise and negative, clockwise.

$$\phi^0 \equiv \Phi^{-0} \tag{1.7}$$

Here the imaginary notation in the exponent is masked, since a rotation of $+\frac{\pi}{2}$ is represented

$$\Phi^{+i} \tag{1.8}$$

and since the powers of exponents are additive, $\left(\Phi^{+i}\right)^2 = \Phi^{+i+i} = \Phi^{-}$

Instead as of rotation of diameters, we can also effect this construction with a rotation of two opposing, centrally directed radii, r, by moving their central points in opposite directions, orthogonally along the path of the resulting hypotenuse. We then have the condition of the dashed lines found in Step 3, which is the configuration of Step 2 rotated in the plane of the paper clockwise 0.4636476…radians. We find the radii have doubled their magnitudes in transforming to resulting tangents. The angle tangents of the distal vertices are correspondingly 1/2.

Step 3 – Centered at the intersection of the unit circle and the hypotenuse, adjacent to the tangent $|\phi|^0$, construct another unit circle with its diameter tangent to the first circle. The distance along the hypotenuse from this center to the vertex with $|\phi|^0$ will be

$$\phi^{\phi^0} \equiv \phi^1 = \Phi^{-\Phi^0} \equiv \Phi^{-1} \tag{1.9}$$

and the distance from that center along the hypotenuse to the other vertex, with the tangent $|\Phi|^0$ will be

$$\Phi^{\Phi^0} \equiv \Phi^1 = \phi^{-\phi^0} \equiv \phi^{-1}. \tag{1.10}$$

We now have the following identity

$$h^2 = \left(\Phi^0 + \Phi^0\right)^2 + \left(\Phi^0\right)^2 \equiv \left(\Phi^1 + \phi^1\right)^2 \tag{1.11}$$

Step 4 – Parallel to the hypotenuse and at the ends of the diameter of the second circle construct the tangents ϕ^1 and Φ^1. It is apparent from the above description that we have the following:

$$h = \sqrt{2^2 + 1^2} = \sqrt{5} \tag{1.12}$$

$$\phi^1 = \Phi^{-1} = \frac{\sqrt{5}}{2} - \frac{1}{2} = 0.618033989... \tag{1.13}$$

$$\Phi^1 = \phi^{-1} = \frac{\sqrt{5}}{2} + \frac{1}{2} = 1.618033989... \tag{1.14}$$

The above steps have created the base for an exponential, orthogonal expansion from a linear unitary condition. From a most fundamental rational operation of division, dividing a whole, 1, into halves, or of multiplication, doubling of a unit or 2, involving the most fundamental ratio, 1/2, we have arrived at what some have termed the most irrational of numbers. It is significant, however, that the primary triad of this particular rational group does not form a group under ordered addition, i.e.

$$a + b = c \tag{1.15}$$

where $a < b < c$, since

$$\tfrac{1}{2} + 1 \neq 2 \tag{1.16}$$

From another perspective, however, it is ½ and 2 which are the result of the rational operation of the bases Φ and ϕ, where

$$\phi = \frac{1}{\Phi} \tag{1.17}$$

This primary rational triad does form a group under ordered addition, since

$$\phi + 1 = \Phi \tag{1.18}$$

which can be generalized using the base Φ to the ordered orthogonal addition

$$\Phi^{-1} + \Phi^0 = \Phi^1 \tag{1.19}$$

and further, where q is any real number, to

$$\Phi^{q-1} + \Phi^q = \Phi^{q+1} \tag{1.20}$$

In fact (1.17) comprises the only set of three numbers, where
$$a = \frac{b}{c} \tag{1.21}$$
that satisfies the condition of (1.15). It is also of interest to our discussion that (1.17) as (1.21) is of the general form
$$\frac{1}{x^{\frac{1}{2}}} = \frac{x^{\frac{1}{2}}}{x}. \tag{1.22}$$
This implies that in some fashion, ϕ and Φ, respectively serve as surrogates for $\sqrt{1}$ and 1^2.

From (1.7) we can see that changing the rotational sense of the base results in mirror symmetry, while changing the ccw and cw conventions results in an inversion or rotation of π of the whole system.

This would be mildly interesting in itself, but when we add a third geometric dimension and generalize the linear powers of the base Φ, we make some compelling discoveries. By linear powers is meant a mapping of the exponential function of any base to the real number line. By contrast, a geometric mapping of n-integer exponents maps to an n-dimensional space, as indicated above.

In the above development, h is presented as a stretching or augmentation of the length of a unit diameter, Φ^0, for a unit π circle or 1-sphere manifold. It is also the diameter of a unit π 2-sphere manifold embedded in a 3-D Euclidean space, as mentioned. We could have drawn the steps shown on a unit square or 1-cube with an implied unit 2-cube in a 3-D space, in which case Step 2 would depict a 2-cube collapsed across one set of diagonals, as a box flattened under foot.

We can therefore think of $\Phi^{-1}\left(=\phi^1\right), \Phi^0\left(=\phi^{-0}\right)$, and $\Phi^1\left(=\phi^{-1}\right)$ as three bases of an orthogonal, exponential space. The unit 3-D space thus spanned has the same volume as a unit cube or 1^3. The magnitude of the vector sum of the three bases, however, instead of $\sqrt{3}$ as in the unit cube, in this case is $\sqrt{4} = 2$. For the purposes of this development, we will stipulate that any augmentation of the bases is exponential and not by addition or multiplication, that the middle base will remain a unit base, and that the volume spanned, that is the product of the three components, remains 1^3. That is
$$\Phi^{-q}\Phi^0\Phi^q = 1 \tag{1.23}$$
For that case in which $q = 0$, this remains the standard unit cube.

We will further limit ourselves to the condition where $q = n =$ an integer. It is immediately clear that a negative integer simply reverses the bases ϕ and Φ. We are interested in three orthogonal conditions, with two readings of each, as shown in the following diagram. Condition b consists of $b1$, in which ϕ and Φ are parallel, $b2$, in which ϕ and Φ are orthogonal, and $b3$, the condition established by Step 4 above, in which ϕ and Φ are anti-parallel. Condition a consists of the projection of $b1$, $b2$ and $b3$ into the plane normal to Φ^0, giving us $a1$, $a2$ and $a3$ respectively.

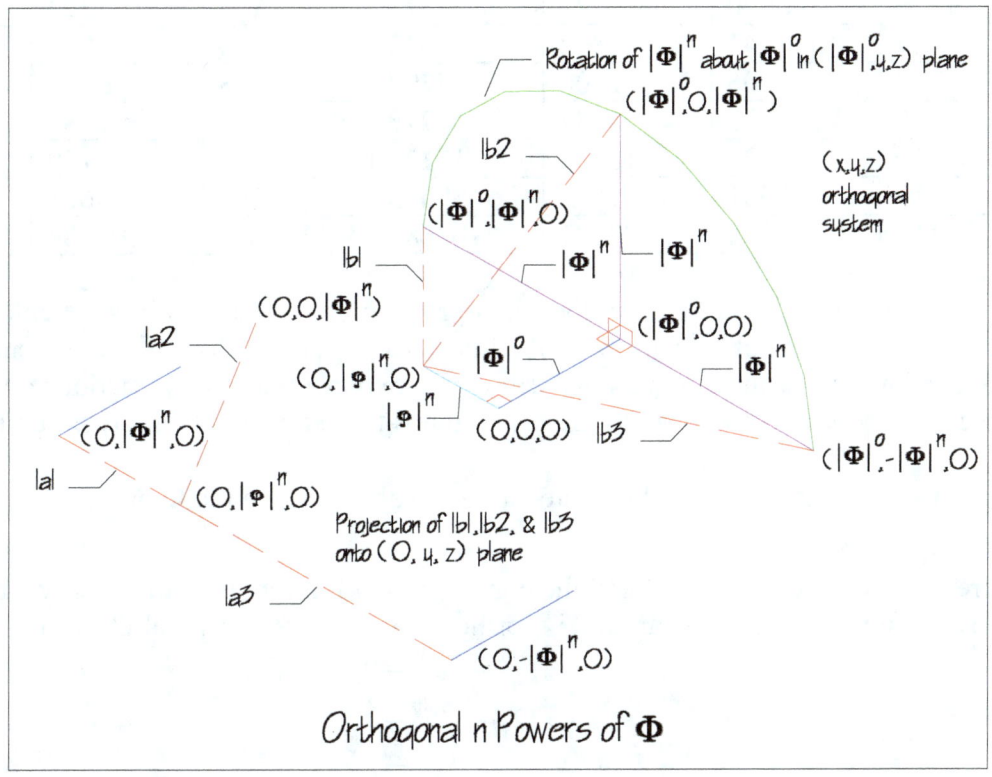

We are interested in the solutions for these 6 conditions for the values of n. The table below gives the first few of these, where the 6 conditions are identified as follows:

$$a1 \equiv \Phi^n - \phi^n \tag{1.24}$$

$$b1 \equiv \left[\left(\Phi^n - \phi^n\right)^2 + \left(\Phi^0\right)^2\right]^{\frac{1}{2}} \tag{1.25}$$

$$a2 \equiv \left[\left(\Phi^n\right)^2 + \left(\phi^n\right)^2\right]^{\frac{1}{2}} \tag{1.26}$$

$$b2 \equiv \left[\left(\Phi^n\right)^2 + \left(\Phi^0\right)^2 + \left(\phi^n\right)^2\right]^{\frac{1}{2}} \tag{1.27}$$

$$a3 \equiv \Phi^n + \phi^n \tag{1.28}$$

$$b3 \equiv \left[\left(\Phi^n + \phi^n\right)^2 + \left(\Phi^0\right)^2\right]^{\frac{1}{2}} \tag{1.29}$$

We have the following sequence, where the integers shown are the square of $a1$, $b1$, etc…for Φ to the n, thus should be read "the square root of (the integer) is $a1$, $b1$, etc…", for the first 13 integral values of n.

n		a1	b1	a2	b2	a3	b3
0		0	1	2	3	4	5
1		1	2	3	4	5	6
2		5	6	7	8	9	10

3	16	17	**18**	19	20	21
4	45	46	**47**	48	49	50
5	120	122	**123**	124	125	126
6	320	321	**322**	323	324	325
7	841	842	**843**	844	845	846
8	2205	2206	**2207**	2208	2209	2210
9	5776	5777	**5778**	5779	5780	5781
10	15125	15126	**15127**	15128	15129	15130
11	39601	39602	**39603**	39604	39605	39606
12	103680	103681	**103682**	103683	103684	103685

The bold figured values of $a2$ are (the square roots of) the even numbered elements of the Lucas series, starting at 0. This gives the linear values or lengths of the hypotenuse developed above through the various orthogonal exponential transformations outlined above. The square of these values therefore maps these integers to the real number <u>plane</u>.

From this relationship we might create a numerical system on the Φ base,

$$_a\Phi_b^n(n) \tag{1.30}$$

where n is the "decimal" place and the condition a or b might be indicated by a number or a sense sign, such as + = parallel, i = orthogonal, and − = anti-parallel, so that for $n = 0$

$$_+\Phi^0 = {_1\Phi^0} = 0 \tag{1.31}$$

$$\Phi^0_+ = \Phi^0_1 = 1 \tag{1.32}$$

$$_i\Phi^0 = {_2\Phi^0} = 2 \tag{1.33}$$

$$\Phi^0_i = \Phi^0_2 = 3 \tag{1.34}$$

$$_-\Phi^0 = {_3\Phi^0} = 4 \tag{1.35}$$

$$\Phi^0_- = \Phi^0_3 = 5 \tag{1.36}$$

Thus the number 1000 is variously represented, among other possibilities, as

$$_i\Phi^7 \, _i\Phi^5 \, _i\Phi^3 \, _+\Phi^3 = \Phi^7_- \, \Phi^5_- \, \Phi^3_- \, \Phi^1_- \, \Phi^0_- \tag{1.37}$$
$$843 + 123 + 18 + 16 = 846 + 126 + 21 + 6 + 1$$

The significance of this and the reason for its development here is conceptual and related to number theory. While the real number line is often represented as the "naturals" then by elaboration filled in with the "rationals", then finally made continuous by filling in the gaps, here we start by using a decidedly "irrational" base to generate a real plane continuum, by squaring the length of the hypotenuse of our continually transforming radial legs, from which the integers and thereby the rationals emerge as a function of the orthogonal positioning of the legs. This shows yet another geometric and orthogonal interpretation of exponentiation.

505

The Browser Economy

-:-

An Analysis for determining the

Optimization of Investment and Consumption Allocations

according to their

Valuation in a Market Economy

Mart Gibson
martin@uniservent.com
February 10, 2015

Copyright © Martin Gibson 2021 All Rights Reserved

The Browser Economy
Executive Summary
Martin Gibson

An effective state is fundamental to the operation of free markets, as essential as their ready and willing buyers and sellers. More than a necessary expense, the state is an integral institution of an economy's capital and like all capital must be adequately maintained for the efficient provision of final goods and services. The question for anyone with an interest in policy effectiveness in such provision is simple; along the spectrum of possible governmental scope of operations, from basic policing, to provision of infrastructure, to a comprehensive safety net, to public ownership of components of the productive apparatus, where should we place our policy target? We take it for granted that any implemented policy should be amended or discarded, if deemed ineffective. We feel some urgency in the determination of this target, given the recent global recession with its anemic recovery. We use the myth of a pre-market Browser Economy to elucidate the oft-unrecognized disparate nature of utility and value in current economic thinking.

It is the premise of this piece that the principal goal of policy in a modern economy must be to insure "life, liberty, and the pursuit of happiness" for its citizenry and welcome guests. It need not attempt to guarantee that happiness, in fact it cannot, but it should not allow that pursuit by one club to impede the pursuit by others. In a market economy in which virtually every good or service consumed in the satisfaction of this goal involves a monetary exchange and in which the pursuit of this goal by some private parties results over short time horizons in an upheaval of production, trade, and employment that for many is the only access to money for such exchange, maintenance of the necessary liquidity cannot be assured by private funding and is the ultimate responsibility of the public sector. In fact it is the responsibility of Congress alone according to the Constitution to "coin Money," and "regulate the Value thereof", though the general consensus has permitted the creation of money through the issuance of private banking debt since well before that document's creation.

We find in this study that for a given level of liquidity, as quantified by expenditures on final consumption and on capital goods and services, where capital expenditures include both public and private sectors, an optimum equilibrium ratio of 0.618... for

(1) final consumption to total expenditures

is equal to that of

(2) capital to final consumption expenditures.

Conditions favorable to overall economic growth, meaning a rise in the general standard of living, are indicated by ratios somewhat below the optimum for (1) and above that figure for (2).

Examination of World Bank data for the period 1970 to 2013 shows a ratio range for the world economy of a few percentage points below (1) and for the OECD nation average of a similar range, before rising above (1) in 2009. Some notable economies trending several points above the target for this duration are Greece, Mexico, and until 2004, Brazil and India. The U.S. trend rose above (1) in 1982 during the Reagan administration with the implementation of supply side policy and has risen gradually, with the exception of most of the Clinton tenure when it stabilized, to a current level of approximately 7 points above the mark.

With this optimization level in mind, we analyze the Z.1 Federal Reserve September 2014 data of U.S. sector and combined accounts, comparing the annual values for 1975, 2005, and 2013 as a percent of GDP and total asset value, with respect to structural changes based on those percentage differences between 1975 and 2005, 1975 and 2013, and 2005 and 2013. As compared with 1975, as of 2013 there was a 16.1% sector swing in percent of domestic net worth to households at 11.1% and ROW at 5.0%, from the business and public sectors, most noticeably in a reduction of Nonfinancial Noncorporate business of 3.9% and Federal government of 7.8%. We might expect this from supply side policies over this period, though the NN business figure is perhaps surprising, but what is more surprising, given the prognostications and promises of the Hayekian school, is that the return on total domestic wealth as measured by GDP as a percentage of Personal Sector asset value, decreased by 5.9% from to 21.0% in 1975 to 15.1% in 2005 before the financial crisis, before recovering slightly to a 5.4% differential in 2013 at 15.6% after modest market intervention. The differential figures using Personal Sector net worth are 0.2% less.

In final summaries of consolidated accounts, we add figures for land, infrastructure, and human capital to the national balance sheet to demonstrate the unfounded concern among some parties about the size of the national debt and to emphasize the lack of wisdom in failing to maintain public infrastructure and human capital. We also state why, in the context of free competition in the production of commodity final consumption products, labor is reduced to a commodity level, defined as compensation to abundant, fungible labor for no more than the cost of paycheck to paycheck costs of living. As a result, employers producing commodity goods for the global competitive markets cannot compensate such labor, regardless of skill set, for sunk education costs or long-term medical and retirement costs.

In light of these developments, United States policy recommendations are made for restructuring of business and personal taxes so as to target costs of common or public goods and services use and to get rid of arbitrary income taxes and thereby all loopholes with replacement through the use of electronic fiat currency as a Universal Basic Income equally for all citizens as a right of citizenship, to decouple long term social from immediate product costs in worker compensation, and to provide for a clear separation of basic public and premium private provision and insurance in health, education, property and finance, such as FDIC coverage. This latter matter indicates a separation of private and publicly backed banks and an end to "too big to fail" status. The intent of all envisioned policy is to free up entrepreneurial efforts to grow and succeed or fail on their own as private concerns and to provide for a rational, comprehensive public safety net, secured against ill-advised privatization.

The following is the link to the unpublished working paper:
https://uniservent.org/political-economy/
To the I.8.5+V Initiative:
https://uniservent.org/i-8-5-5-initiative/

The Browser Economy

Abstract

With respect to the current competitive global market, some axioms are stated concerning,

(1) the nature of market exchange for intermediate goods and services in which a good or service is said to trend toward commodity pricing given a surplus in supply and lack of entitlement of that good or service, and given a competitive market for the final good or service, resulting in
(2) a trend toward commoditization of labor of any surplus skill set, and the inability of the free and unregulated competitive market to provide for long term needs of any such employed labor, with a statement of
(3) the lack of productive value of investment in any financial asset that does not invoke productive human labor.

The implication of this last statement is that a fully automated economy operating without human labor cannot produce a good that has value in the market place for the simple reason that such economy cannot produce buyers with cash, with the result that the increased automation of production facilities trends back toward a browser economy, i.e. a market free economy as conceptualized herein.

An overview of the development of the production of consumption and capital goods and services from a natural pre-production and pre-economic valuation model is presented, including a conceptual development of pre-trade tokens of value into a system of monetary trade and of the division of labor into a system of entitlements based on that trade. A macro-econometric analysis suggests a natural optimization ratio between final consumption and combined final and capital goods and services, that is total production, below which investment is productive of overall growth and above which disinvestment leads to a stagnation for major sectors of the economy and real asset inflation for those with the financial assets to remain in the bidding for those real assets.

A look at data from the World Bank and the Federal Reserve System for the past four or so decades confirms this optimization level, where government disinvestment along with a running trade deficit from the advent of supply side policy implementations through 2005 over shadowed any increase in investment of the US private sectors, resulting in stagnation for over half the economy despite supply side forecasts. Failure to account for human capital in the national accounts is examined with its implications for misguided concerns about the public deficit and debt. Some policy implications addressing this misguided, anti-government bias are examined in the conclusion.

Foreword

This writing is an amalgamation of a couple of projects started over the past few years. The first of these was "The Browser Economy" which was intended to be a book length series of chapters alternating between an exposition of a mythical group of intelligent, spiritually aware, but what we might call technologically naïve humans in an initial environment of sufficiency and therefore pre-economic, and an exegesis of the mythology in light of our current economic condition. The intention was to weave the (hopefully) poetic with the prosaic and eschew any mathematical references, which are poetry to me, and in the process differentiate between the origins and nature of value and those of economic utility.

The second of these, which derived from the first, was an investigation of possible natural constraints to the allocation of production efforts between goods and services used for consumption, C, and investment, I, based on the fact that a common monetary valuation is applied to both categories, though C and I are fundamentally different in utilitarian genesis. Consumption is essentially grounded in human survival and the social <u>effort</u>, H_C, to draw sustenance as immediately as possible from the earth's natural bounty, while investment is essentially a combination of human ingenuity and effort, H_I, acting on those <u>natural resources</u>, R_N, to produce items for later consumption or to leverage current consumption in conditions of relative scarcity, natural or otherwise. The resulting investment goods <u>and services</u>, which includes the knowledge base or technologies, then become a hybrid, a manmade-natural resource, R_I, where total resource utilized, R, becomes a combination of R_N and R_I which we refer to as capital. Capital then is an effective multiplier of natural human effort and can in fact be less than one, as we shall see.

This second investigation is necessarily mathematical, and since consumption and investment are disparate activities, a primary school adage suggests itself, that you can only add things that are qualitatively alike and by logical implication only multiply things that are qualitatively and dimensionally disparate. Such investigation should attempt to see what happens if the combination of human effort and natural resources is modeled as a mathematical as well as economic product, that is as a cross-product, instead of conventionally in accounting terms as the summation of a commonly denominated monetary value. Still, since both C and I are so valued by a monetary resource of limited supply, those values must be subject to summation in the marketplace, so that in a market economy where trade is mediated by some token of value, v, the total production, P_G, should ideally equal the sum in an equilibrium condition, or

$$(_vH_C + {_vH_I}) \times {_vR} = {_vP_G} = {_vP_C} + {_vP_I} \,.$$

The problem is that only the consumption products, P_C, make it to final market to be exchanged for the value paid for making them, $_vH_C$. While in equilibrium, the value paid for investment efforts, $_vH_I$, should be equal to the capital that is newly produced, $_vP_I$, and to $_vR_I$ which is used up and must be replenished or maintained. However, in addition to the capital goods produced that show up in the intermediate market, there are necessary capital services that must be produced and maintained, that do not necessarily show up in the any market valuation, as part of $_vH_I = {_vR_I} = {_vP_I}$, as with government services and private financial transactions.

The result is the following, that has but one positive real solution, where I mean "real" in the mathematical sense in contrast with complex number notation,

$$\left({_vH_C} + {_vH_I} \right) \times {_vR} = {_vP_C} = {_vP_G} - {_vP_I}.$$

If $_vP_C$ is assumed to be equal to $_vH_C$ and set to 1, then the equilibrium value of $_vH_I = {_vR_I} = {_vP_I}$ is 0.618... which is also the ratio of consumption value, $_vP_C$, to total value, $_vP_G$, which we will call C_G, and serves as an attractor or optimization value toward which a system will gravitate unless disturbed by sectoral or political interests. In this regards, the GDP of an economy becomes a measure of $_vP_G$, the final consumption expenditure serves as a measure of $_vP_C$, and the difference, investment, government, and rest of the world accounts, serve as a measure of $_vP_I$.

Examination of international data from the World Bank, in which the world average level of final consumption has been just below C_G by an amount closely approximating the economic growth rate for the past 50 years, confirms the reasonableness of this hypothesis. A review of the Federal Reserve national account data for 1975, 2005 and 2013, supports this analysis and indicates the damaging effects to the US economy of disinvestment in both the public and private sectors, the latter by a shifts to imports and overseas production, resulting from a failure to adhere to this constraint.

In light of the current impasse in congress and the election results of 2014 and the likelihood that the incoming congress will want to continue on its misguided mission to dismantle the public sector, it is important to expedite publication of this information as I have not come across such analysis elsewhere. The nature of the comments on the disparate nature of consumption and investment in the analysis seemed to mesh with what I had written concerning the browser economy, so I decided to merge the two. The preface was then written to give some basis for what follows in the combined piece.

Preface

The browser economy presented here is important to study because it reduces all of modern economic activity to its fundamental components; (1) a naturally productive environment conducive to human survival, (2) human activity whose raison d'etre is the satisfaction of the physical needs for that survival, (3) human activity whose raison d'etre is the provision of emotional satisfaction, and which is largely directed to the acquisition and social exchange of objects or other facets of the environment in a system of value that conveys relational relevance and esteem within the community, (4) the application of innate organizational intelligence throughout the community whose reason d'etre is an understanding of and efficient and concerted operation within that environment for the satisfaction of (2) and (3), and finally the rationalized interplay of these first four components resulting in, (5) an increasingly effective manipulation of the facets of that environment so that the first component becomes a synthesis of both natural and human productivity.

The day in the life of the browser economy presented here represents that seminal moment when (4) comes to bear on (1) in such a manner as to mix the innate human capacities and propensities of (2) and (3), thereby initiating (5). It is my belief that most economic analysis fails to understand, or if understood, then understates or ignores that (3) is an innate motivating principle of the human condition essentially outside the realm of economics, separate and distinct from either (2) or (4). As a result, analysis from the left tends to deprecate (3) and see the answer to problems with (2) as being realized through a collective reorganization of (4) directed toward (2), from which the satisfactions of (3) naturally follow. Analysis from the right, on the other hand, sees problems with (2) as properly addressed by individual adjustments in (4) from which corrections in (2) and (3) will or will not follow depending on the individuals correct operation of (4) and tends to deprecate the effect of changes in (5) on (1, 2, 3, and 4).

Component 3 is the basis of currency and has an origin quite apart from any notion of its use in trade for necessary goods arising from component 2. The right of recreational activity along with that of possession and of sharing those possessions as one sees fit is a fundamental aspect of "liberty and the pursuit of happiness" embraced as a legitimate priority in the US and other democracies. Crystalized in the marketplace and trading floor as money, however, this emotional expression of validation adds a dimension to economic interaction that can obfuscate the rational production and consumption of goods and services. It is widely acknowledged that individuals will gladly give to others in perceived need out of their own larder for items they would eagerly hold for the highest bidder in the marketplace.

The notion of laissez faire has traction in public discourse for what it assumes about the inherent ethics of human beings, the actions of scoundrels notwithstanding, and not primarily for of what it says about the efficacy and effect on productivity of authoritative oversight of markets. Yet this latter rationale is the focus of most theorists of the right, who may or may not embrace the former, despite the fact that maintenance of a policing/military authority to deter common theft & robbery/pillage & plunder is axiomatic to their thinking.

The notion of redistribution of wealth, in particular of financial wealth through taxation, has traction because it acknowledges that a society in which the majority of individuals have no entitlement to employment to exchange for the money needed to purchase the necessities of life is antithetical to the first and foremost of the inalienable rights, that of "life". Such a society is rightly perceived to be anti-democratic and little more than the oligarchy it was designed to replace. Yet this thinking downplays the fact that such money has little value if the expertise and machinery to produce those necessities are disrupted or displaced by such redistribution.

The problems faced by the global market economy at this juncture are easy enough to state in light of the following axioms with respect to markets, in particular those for intermediate goods and services which involve most labor markets:

1. Over the long term, the market value of a good or service that is in short supply is determined by the market demand or willingness to pay for that good or service based on its perceived utility to the buyer for its end use or with an intention of incorporating it into a final product. To the degree that the market value exceeds the supplier's lowest willing sales price, it involves the payment of a market rent to the supplier, which he perceives as a market profit. The difference between the supplier's lowest willing sales price and the total cost, fixed plus variable, proportionally assigned to production of the good or service, is the operational profit of the supplier.

2. Over the long term, the market value of a good or service that is in abundant supply and fungible is determined by the costs the supplier must and is willing to spend in bringing the good or service to market, which for widely equivalent production techniques and costs is by its commodity pricing. To the degree that the market value is less than the buyer's highest willing purchase price for the good or service based on its perceived end use value or intermediate value in a final product, it involves the payment of a market rent-in-kind to the buyer, which he perceives eventually as a surplus value or market profit. Such commodity pricing will have no market profit for the supplier, though it may have an operational profit at the high end, but will cover at best only the cost of production on the low end.

3. Over the long term, in the absence of seller entitlements, defined as arising either naturally, i.e. due to geography, uniqueness, rarity of product, trade secrets, slight of hand, etc. or culturally, i.e. due to property, patent, copy rights, professional license, slight of hand, such goods and services become commoditized. If the final product of which such commodities are a component is not itself commoditized, the supply of such commodities will involve payment of a rent-in-kind to the buyer, otherwise the final product will be a commodity.

4. Long term equilibrium in a product market, therefore, occurs when commoditization of supply of intermediate or final products is achieved, and can include an operational, but no trade or market profit.

5. Labor, which is a service, *of any skill set* is subject to the above constraints, and over time will tend toward commoditization, in particular toward the low end of the compensation scale in which only the current costs make up its market value. Any historic costs of education and upbringing become sunk costs, while anticipated prospective medical and retirement costs cannot be commanded by such labor due to the fungibility of commodity labor and the fact that such costs are not currently incurred. Commodity labor, therefore, cannot provide for its own safety net, unless it is provided as an entitlement as part of its compensation or as a social program. As a result, free markets employing commodity labor cannot compensate for historic costs or prospective future costs for that labor, regardless of the desire or willingness of any employer to do so. Governmental structuring of labor compensation to account for such costs can be effective only to the degree that it is universally applied and to the degree that isolation against competition from products involving unstructured compensation is consistently assured.

6. Since labor in the broad sense of all skill sets, including holders of entitlement, is responsible for both intermediate and final production of and consumption demand for goods and services, at equilibrium cash income flow to such factors of final production should equal cash expenditure for final products. However, this is not necessarily the case for cash flows to and between owners of raw materials, intermediate or pre-existing goods or other instances of real capital, due to the retention of value in unconsumed real capital products and in cash as working capital or uninvested savings.

7. Financial savings do not constitute an economic investment unless they are used to employ labor, in the broad sense, in the production of new capital goods and services, which are then used in the production of final goods and services. If they are used to purchase existing capital goods or facilities, they constitute a transfer payment and do not create net new investment, unless the facilities were idle and are subsequently put to use with added labor in the production of final goods and services. Investment does not produce a return until such intermediate goods and services result in the production and sale of final goods and services. If it fails to result in such final production and is not utilized, it represents consumption or productive waste; as a transfer of title only, any gain in sales price over purchase price, net of maintenance and perceived as profit by the original owner, represents an inflation of asset price to the overall economy and does not constitute growth.

8. Thus financial arbitrage is a transfer of entitlement and does not of itself increase production, though it may increase the money supply if it involves the assumption of debt. Similarly, the sale of existing equities and liabilities and their appreciation are transfers of entitlement and do not in themselves increase production unless they initiate or increase funding of active productive operations.

9. In the long term, financial transactions, regardless of the parties involved, are productive only if they are integral to a long-term increase in production of real goods and services. In this regards, government or public expenditures are no different from private expenditures in their effect on production, and may represent an investment, productive or wasteful, consumption, wise or not, or a transfer payment, inflationary, stabilizing, or deflationary. As an investment it may or may not be integral to a long-term increase in the production of goods and services, depending on the effectiveness and wisdom of the investment; as a consumption expenditure, likewise. As a transfer payment, if a government transfer goes from a tax receipt, security sale, or fiat creation to a party that saves it as a financial asset, it is simply a transfer of entitlement and tends toward inflation of asset pricing. However, if it goes to a party that spends it on active investment or increased consumption, while a transfer of entitlement, it is one that leads to a net increase in production, or at the very least, the purchase of unsold inventory; i.e. to economic equilibrium or growth. In the case of transfer through taxation or security sales it is to the benefit of the transferee and of negative and positive benefit respectively to the other parties involved. In the case of fiat creation, judiciously applied, it is to the benefit of the transferee and to the producers of the good or service on which the transfer is spent. It is of potential negative benefit, and then relatively and not absolutely, only to those who profit primarily from their financial position in the marketplace.

10. As a result, government or public expenditure of funds raised through taxation, borrowing, or fiat are indeterminate with respect to any effect on private funding and expenditure, depending on the wisdom of the policies they implement. They are however, the only way that long-term costs of living, i.e. socialization or upbringing, education, medical, and retirement costs, can be financed for commodity labor.

With this axiomatic understanding, from recent history of the industrial, market economy:

A technology based decrease in transportation and communications costs leads to an extension of market development, domestic and global, made possible by a related increase in economies of scale. However, this requires more labor and attendant costs which, with resulting competition, forces businesses to reduce production costs. A related technology based innovation of plant and equipment results in productivity increases with employment of higher skilled workers who are initially in short supply and therefore command higher wages. Eventual rise in the supply of skilled workers, along with increasing automation of technology, leads to oversupply and a commoditization of labor in affected sectors, and eventually in long-term equilibrium in those competitive labor markets. Commodity pricing of labor covers only the costs of current production, i.e. week-to- week or at best month-to-month living costs, and so excludes payment of long term past investment costs, such as education, and future decommissioning costs, such as retirement and major medical. The current costs of commodity labor production, including current debt service, meanwhile adjust to whatever level the market will bear. Liquidity, beyond requirements for paycheck intervals, is squeezed from the holdings of such labor over time, which ceases to be a source of savings and capital formation, so that liquidity concentrates as business and higher skilled labor holdings. As the relative pool of commodity labor increases as a percentage of the overall market, the business holdings concentrate further, but these savings cannot be reinvested without the development of new markets and instead are used to bid up the price of existing and new non-productive assets, including financial assets which though ostensibly productive, can not be economically productive unless they engage new labor. As a result, the financial assets targeted must be of long horizon or derivative of an underlying asset that is of long horizon. When the climbing asking price for the non-productive asset can no longer be met, the resulting introspection reveals an asset bubble and the owners or managers of these holdings then turn to public assistance that they otherwise deprecate.

This is not disparaging of entrepreneurial efforts, rather of those interests that truly believe that money makes money or simply want others to believe it. Judicious use of money, employing individuals in an enterprise of shared value makes money, or better yet, makes a living. Arbitrage by itself does not. It simply transfers it from one pocket to another. What the arbitrageur does with the trade differential determines whether it represents investment or asset inflation.

This short piece will attempt to show that the dynamic just outlined is a natural consequence of real capital production and the attendant financial capital creation via debt and that it leads inevitably to the capacity for production of more goods and services with less human effort. It would be advantageous if that capacity were directed to the benefit of all, but the structure of valuations and pay in a fully rationalized market system will never allow it without public overview of the market place.

Such government involvement need not be of the type feared most by the right, confiscatory taxation or government usurpation of markets or property. Nor does the government need to borrow from its citizens and especially from foreign interests to finance its necessary expenditures. But it does need to insure that all citizens have sufficient liquidity to access to the necessities and basic satisfactions of life. In terms of the Declaration of Independence from oligarchic control; first, life with liberty for all, and thereby the pursuit of happiness.

We will try to address this quest, herein.

The Browser Economy

He awoke suddenly. The bower above remained obscure. He struggled to slip back into the dream, to pick up the curious conduit of the narrative before it evaporated in the coming light.

There had been this tantalizing fruit of a color he had never seen, hanging just out of reach. It was shiny, like the early sun when you could still gaze directly at its face. A long and rather large lizard at the edge of the brush distracted him momentarily. The lizard had been gazing at him for some time without interest, when it quickly retracted its legs and slithered off into the bushes . . . He now looked back with anticipation at the fruit and was annoyed to see the lizard had emerged in the tree above him and was devouring its strange produce. Initial alarm gave way to an abrupt rage, and he began jumping and grabbing for the legless lizard, when it intertwined itself several times before falling cool and hard and lifeless at his feet. Then he woke again.

Further attempts to re-enter the dream were pointless. He reran the images again and again through his mind, looking for connections to some more familiar territory, to no satisfaction. Finally, he put it aside. He would discuss it with the group at the evening lore fire. It would be a novel place of interest to conduct to their lore. He might even gain the satisfaction of garnering the lore tokens from the group for his contribution to their history.

Dawn was beginning to filter into the bower. No one else in his bower group stirred, but he did hear familiar rustling from the adjacent copse. Parting the bushes, he looked out across the field toward the ocean's edge in the direction of the growing light. A friend emerged from the bower to his left and raised both arms in the customary greeting. He returned the salute and made his way toward the stream at the far side of the field. Here he knelt down to quench his daybreak thirst and began to browse from the dark berry bushes on the far side of the stream.

The field and stream and surrounding groves gently unfurled from the western heights to the eastern sea. The group had no way of comprehending it in, but they where the sole human inhabitants of a large island near the earth's equator; sole across the extent of the island's geographic range and prehistoric age. The group's oral history held tales of other, distant places for browsing along the coast, and even suggested the insularity of the domain, but their biosphere being lush and generally unbroken, even the eldest members of the group had no direct knowledge of places more than a few dozen miles in either direction. Over this range there was no reorientation of the general shoreline with respect to the night sky.

Over time, the group slowly browsed their way back and forth across the slope and along the coast in response to changes in the immediate biosphere and cues from their lore. Being close to the equator, such time was normally measured in days and months, or rather cycles of the moon, since there was in daytime no discernable seasonal variation to mark out the passage of a year. There was a recognized variation in the arc of the sun as it crossed the sky, with a ranging of its zenith from what we recognize as north to south and back again. It was noted that this oscillation coincided with a variation in the direction and frequency of the afternoon rains, which came in over the heights, but as there was no winter, summer, spring or fall, this was not of mundane interest.

This solar oscillation in turn coincided with a cycle of the nighttime sky as judged by the regressive rising of familiar stars and their groupings over the course of approximately 13 lunar cycles, and this was of interest not to the group's day to day concerns but to its lore. As with the members of their group, the sun too was seen to undergo a birth and death, daily emerging always in full vigor, ever young. The moon was apparently much older. Since she was observed going through a gradual cycle of quiescence and brilliant activity every 28 days or so, progressing in her change day or night to emerge the next in anticipated increase or decrease from the one before, she appeared not to die. Her influence appeared to be continuous as evidenced by the monthly flow of the fertile females of the group. It was obvious by this connection that she was herself essentially female.

As for the rest of the night sky, the stars and their groups appeared to be immutable, other than in their time of arrival on the evening horizon. Their lore told them that the stars were the lore fires of their ancestors. Then there were the other strange inhabitants of the night sky. There were the ones that moved slowly back and forth among the stars, spreading tales among the groups, the swift, bright ones that streaked alone and sometimes in groups like birds among the tree tops, and finally the ones that existed primarily in lore, that streaked like the swift bright ones, but much slower, over many nights. These were important beings . . . but they were all important beings.

While the group had long ago harnessed fire for the lore sessions, they seldom cooked. There were no large animals on the island, and there was such an abundance of browse vegetation that the practice rarely suggested itself. Certain insects and grubs, bird and reptile eggs, even the occasional lizard, using care to avoid the salamanders, were generally eaten as they were found, raw. Nor was fire used for heat. The temperature ranged between 16 and 32 degrees Celsius along the coastal slopes, though it was known to be cooler in the heights. There was generally no need for clothing, though the group members did like to indulge in individual adornment, generally of an ephemeral nature.

According to the lore, fire had been recovered on several occasions in the aftermath of a lightning strike, before its need for wood and air was finally understood. Means of ignition from friction and sparking eventually developed, which combined with its obvious capacity to dispel the dark, made it a fixture of the lore sessions, a response to the nightly lighting of the ancestral stars.

He slowly worked his way down the stream, picking berries and the occasional fruit from the bushes and trees. As the sun arose higher from the sea before him, his mate and girl child and a few other members of the group joined him for the morning browse. A short while passed as the small group worked its way from the fruit to the nut trees on the near side of the stream, drank from the stream, and being satiated, walked down to the shore and began to look among the shells in the wash of the gently breaking waves for a novel find that could be added to their lore tokens.

Heads bent to the surf, they slowly made their way along the beach, nudging the whelk and tusk and cone and cockle and cowry shells with toe and hand, flipping them in a quick glance for any inner luster; and for a nicely worn hole at the hinge mound of the bivalves to accommodate a lanyard. The cowries could be a special treat, prized for their durability and sheen.

The group members had developed a token culture of trading these and other trinkets, so that those of a more durable or unusual nature came to be held with some admiration and esteem. They had no economic value, however, as we think of it, since the economic need, the sustenance of the group was wholly met by browsing. However, such tokens did make their head and hearts feel full, just as the fruit and berries and other food did for their bellies.

There was no collection, therefore no hoarding of foodstuffs, or preparation for hunting or warfare. Nor was there any need to protect against predators, as there were none. Most significantly from our modern perspective, there was no division in either the perception or the thinking of the group members between the production and consumption of economic goods or services. As a result, there was no division of activities in the group, beyond the natural ones that occurred due to childbearing and age. Browsing and in turn all activities of the group and group members were simply processes of "conduction". Thus browsing was a process of conducting food spirits through their systems, sleeping, of conducting their own spirits into the dream places, sexual coupling, of conducting the spirits of new members into the group to replace the ones who had been lead away into the dream places.

When they gathered at the end of the day to discuss anything unusual they had observed or experienced during the day or anything noteworthy from the dream places they had been to the previous night, they were conducting lore, or as we would describe it, making history, transforming current events into collective memory. And since memory showed up in the dream places, and vice versa, where they often met members of the group who had already conducted there at death, the group had long ago, always they said, come to identify memory and the dream places along with all other places of thought as a unified, pervasive, permanent though changeable, aspect of themselves, individually and as a group. So any new individual experiences of conducting and of browsing or of dream places were turned into new words and phrases to join all the others that already existed in the lore of group memory and in more protean form in group dream places. In fact, it was this process of conducting history that was their identity.

There was no absolute division of their world into physical and mental realms, waking and sleeping states of consciousness, social and psychological significance, economic and political endeavor, individual and group identity. It was one world of various places and times, some found in immediate experience and some found in memory, past, present, or future as memory and the process of thinking itself were indistinguishable.

It was in this context, after a day browsing and conducting searches for trinkets among the shells on the beach and chatting with his friends, that he would approach the group gathering that evening and tell them about his dream. He was eager to hear if anyone else had been to such a place, though he was sure he would have heard about it if they had.

The members of the group had no concept of dream interpretations such as we might have, no Freudian or Jungian or other conception of a dream object or event as a representation of an unrecognized process or occurrence in the awake state. They did have an intense interest in how one person's dream place might relate to or lead to someone else's dream experience or how a dream experience might lead to some experience or place in the daylight. It was the interwoven nature of memory and experience that drove their investigation.

As the group worked its way along the gentle surf, he noticed a bright reflection in the upper beach sand where a wave had just retreated. The glint did not dissipate with the wave, indicating it was not due to surface wetness alone, and he hurried toward it to examine. The object was shiny but roughly pitted and irregular, unlike most shells, and much of its extent appeared to be buried in the sand. He began to dig around it and after a very short moment was able to lift the shiny stone and look at it closely. It was heavier than any stone of comparable size that he had encountered and the shiny surfaces between the pits were the color of the sun in early morning after it had washed itself of its natal redness.

It was early afternoon and the group was beginning to get hungry once again, but he did not want to leave the spot, which might harbor more such stones. This find was his by lore agreement, but if he could uncover more he would be happy to share the tokens with the others. Suggesting that his mate and the others should go first to the grove, he continued to dig.

In a while the lunch group returned from the grove, and he noticed that his mate carried with her a mat made of palm fronds in which she had placed two mangoes and several handfuls of nuts and berries. He had relayed his waking dream to her earlier in the morning, including the peculiar way that the legless lizard had been able to weave itself together in almost a nest like fashion, and it had occurred to her to emulate the scene with the slender leaves. She laid the crude basket on the beach next to him and smiled as he began to eat. He returned her smile and after finishing the meal, placed the shiny new stone on the mat. They laughed.

The group continued to dig for quite a while, but no other shiny stones were to be found. His mate tried to give the one he had uncovered back to him, but he refused it. As the sun lowered in the west, they placed the pile of shells they had gathered earlier on the mat, folded it up and headed to the lore site.

That evening's gathering was more energetic than the norm. He told them first of his dream, of the strange lizard with no legs and of the miraculous fruit just beyond his reach and that the lizard had devoured before he could retrieve it. Then he told them about the shiny stone, about which of course they had already heard. He motioned to his mate, who held it out for everyone to see, then passed it around the fire. Each member of the group tested its heft in the palm of their hand, examined its luster in the light of the fire, and passed it on to the next member of the group, before it made its round back to her. She next passed the mat around, and found that among some of the members there was at least as much interest in its interwoven configuration as there was in the weight and brightness of the stone. They picked at it and flexed it, then placed their token shells in it, picked it up by opposing edges, stood up and swung it back and forth to their side with the delight of broad smiles and chuckles.

There was much discussion among the members that evening, and toward the end of the lore fire all agreed to a double award of the lore tokens. As they rose and left the fire to return to their bowers, the members circled past his seat and that of his mate, depositing at each seat collections of cowry shells as they so deemed was warranted, to him for his dream and the find of the shiny nugget and to her for the woven mat. When everyone had passed, they would have seen that her pile was slightly larger than his, had anyone noticed such things.

They returned to their bower and slept soundly. In the morning and before the morning browse, everyone in the group gathered early at the palmetto grove and began to make mats . . . and then a basket. Then they went for a browse.

Exegesis

The desire to satisfy basic human needs, both physical and psycho-social, inexorably drives economic activity back in the direction of its fundamental state in the browser economy. In that proverbial Eden there is no recognized intermediary between a desire and its satisfaction other than the time it takes to move to the location of natural produce or tokens of satiety, these latter being objects whose acquisition, and in some cases retention, result in emotional satisfaction. These latter items are socially recognized in general, although subject to individual appreciation.

An uncharacteristic dearth of immediately accessible produce in that Eden leads to the recognized development of technology in the form of productive tools and learned skills to mediate the hunt for and gathering of both remote natural produce and tokens of satiety. Chief among these technological innovations is a division of tasks according to the skills or status of the individual participants and the individual retention of at least some of the products of that skill, such as tools and habitat, along with any retained pure tokens of satiety, making barter and trade of goods possible. Second among these technologies, but first to be examined, is the use of broadly appreciated tokens for their exchange utility with natural produce, hunted and gathered, and with tools made by other individuals or groups.

The quantitative measure or value of exchange utility for such tokens is determined by negotiation of the parties to the exchange, which in turn make their determinations according to their perception of the relative abundance of the tokens of satiety and the items of consumable or productive utility to be exchanged. The persistence of such exchange utility results in a recognized mediation system, using persistent token exchange value, i.e. agreed utility or satisfaction of desire, for the acquisition and distribution of natural and human-made production.

The development and eventual rise of storable agricultural produce, which is secured in centrally managed granaries, gives rise to a second form of exchange, human-made tokens of dried clay or metal or other durable quality as a count of stored produce for use in redemption, as these tokens of account with right of redemption can be utilized for exchange in a similar manner as the tokens of satiety.

There is an inherent contradiction in the exchange utility of the tokens of satiety and the tokens of account, however, as the value persistence of the first is inherent in the token itself, while the value persistence of the second is dependent on the storage integrity of the granaries and the abundance of future harvests; that is satisfaction for the first is ultimately found in holding the token, while satisfaction of the second is ultimately found in exchanging the token for grain and is dependent on the value of production already achieved as well as of production yet to be realized.

As a result, value of the first is generally persistent as long as the token is appreciated or found to satisfy an acquisitive need, though it is incapable of being consumed, while value of the second can fluctuate depending on the storage integrity of the granaries and the success or failure of subsequent crops and is therefore more subject to speculation as to its consumption value at the time of redemption. This inherent contradiction is not removed by the coinage of precious metals, such metals until recent history being chiefly token of satisfaction and without further productive or consumption value.

Coinage, however, is subject to debasement through shaving and alloy, and requires authoritative validation at the marketplace in the exchange process, so the development of standards of weight and measure along with secure storage of specie in banks in a manner analogous to granary storage arises, with the use of scrip and banknotes as the corresponding tokens of account, which are themselves subject to speculation.

While stored grain as consumable produce has a fundamental value as stored human productive effort and survival need, stored specie has both a fundamental value as tokens of satiety and a secondary, derived value as tokens of account for the fundamental value of the grain or any other stored consumable products for which it was issued.

Therefore, if we assume that specie is initially coined and issued as tokens of account of some stored product by a governing authority and redeemed from circulation with the consumption of that product, the total sum of specie in circulation should be of the order of the amount of stored production, where circulation includes storage of specie in any banks other than those of the governing authority.

Any excess of specie in circulation over the store of production has inherent, though negotiable, value as token of satiety, but depending on the frequency of transactions no more than a reduced value for the redemption of stored production based on the ratio of the coinage in circulation to the amount of the stored production. In a similar manner, if the utility of the stored production is reduced by inherent deterioration, acts of nature or human activity, the value of the specie in circulation as token of account is similarly reduced.

In light of such excess, if the transactional frequency of redemption is for a minor portion of the overall store of production, such reduction in value will not be apparent to the transacting parties; however, if it is for a significant portion, it will become apparent and result in devaluation of a unit of specie as the diminished quantity of remaining stock of product will be less than the nominal value or quantity of the remaining specie.

On the other hand, if there are other resources, productive or consumable, natural or human-made, beyond authorized storage and with owners ready for their conveyance, or if the remaining holders of specie have no immediate need of redeeming their currency, the remaining specie may maintain its value or even increase in terms of its purchasing power. In fact, since in general there are always additional resources seeking a market, there may well be a desire for more specie than is provided by the governing authority, so that in the absence of additional issuance of specie by that authority such desire can only be met by borrowing from others holding specie, and generally this means through the lending of bank holdings.

As the production capacity of an economy exceeds the level of specie in circulation, it indicates a need for tokens of account in excess of tokens of satiety, and the validity of using scrip and banknotes as tokens of account becomes apparent, since they are without inherent consumption utility or rarity.

Since the issuance of scrip and banknotes and their use as tokens of account have the same effect in an exchange as tokens of satiety, they can be issued for lending with the same effect as issuance of specie, so that over time tokens of satiety become superfluous as reserve stores of value, unlike the stored production in granaries and elsewhere that maintains its fundamental consumption and productive utility and that the specie represents. As a result of this development, over time the total value of paper currency well exceeds that of specie or precious metal, bank reserve requirements can legally be met with reserve paper currency instead of specie, and the vast majority of the total of tokens of account in circulation exist as deposit accounting entries in the denomination of the paper currency.

In the final analysis in the modern economy, the paper currency is replaced by electronic accounting entries of suitable denomination and the monetary system becomes completely devoid of any relationship with a tangible token of satiety-in-its-own-right, though an account balance may retain a satiety of allusion by virtue of its universal appreciation and acceptance as a exchangeable token of account. Still, the final value of a modern monetary unit is devoid of inherent satisfaction and has only a negotiable value dependent on the perceived utility of current stored consumable and productive goods as well as the utility of consumable and productive goods and services yet to be created.

According to the Federal Reserve Statistical Release, dated September 18, 2014, for 2013, the reported value of financial assets in the household, nonfinancial corporate and nonfinancial noncorporate business sectors of the US economy is in the neighborhood of $85T with nonfinancial or tangible assets of around $57T. Assuming that the value in any financial asset is in its exchangeability for an existing or future tangible asset, consumable or productive, as quantified in its nominal value, at least one third of the stated financial asset value must necessarily be as claims against goods and services yet to be produced, and in fact the value of pension entitlements in the above three sectors is approximately $20T.

Financial assets are either tokens of unspecified and universal redemption, i.e. money or "legal tender for all debts, public and private", to quote the US paper currency, or certificate tokens of contractual obligation, such as performance, stocks, bonds, options, or other equity or liability shares, for specific redemption as goods, services, real estate or other financial assets; they are either universal or specific accounts of actual past or potential future transactions. The actual case is that these financial assets represent anticipated future as well as past production and as such are measures of both current real assets and projected future earnings and asset value.

The thing that bears emphasizing is that if we were to divide the financial assets by ten or multiply them by the same factor, the tangible physical or real asset base would remain the same, and its nominal valuation in terms of financial assets could be held to vary accordingly. However, holding the nominal value of real assets steady, along with employed factors of production, deflating the financial assets will have a tendency to reduce the production of new real assets and consumables, while inflating the financial assets will tend to increase their production.

The effect is simple; for the economy to run smoothly, the quantity of financial assets, particularly liquid assets, must be sufficient to first, engage the necessary factors of production, and second, provide them with capacity to purchase the resulting supply of production, both consumable and productive. In this regard in a market economy, any appreciation for money as a token of satiety, i.e. as something to be held, is irrelevant and can be counter productive, and it is principally the quantity of money as a token of account of transactional sufficiency that is relevant.

It follows that in a democratic society in which the chief function of government is preservation of the integrity of the social contract, which by inference indicates the maintenance of orderly markets and settlement of transactional disputes, one of the principal tools of government should be the oversight and provision of optimum liquidity in the marketplace.

With respect to access to means of consumption and production, as determined by ownership or entitlement to utility, goods are designated as either 1. excludable/2. non-excludable and A. rivalrous/B. non-rivalrous, so that we have 1A. private, 1B. club, 2A. common, and 2B. public goods.

Since money in an advanced market economy is an instrument of accounting and not a real asset in its own right, having utility only in the determination of relationships in exchange and contracting between individual parties both private and public, it cannot properly be considered a private good, which reserves the right of its full consumption to its individual owner. Money is therefore obviously not a private good as it is never consumed, and merely transfers its utility to another individual in the course of a transaction.

Money would not appear to be a club good since its use is non-excludable; one individual's use of the money in their possession may exclude others from the cotemporaneous use of that same batch of cash, but it does not exclude them from the use of money in general, or of that same batch of cash once it leaves its original holder, just as the air one individual breathes out may be inhaled by the person next to them, albeit laden with less oxygen and more carbon dioxide and who knows what else.

Money would not appear to be a common good, since its use is non-rivalrous as one individual's use on food at his private supper club does not prevent his neighbor's use of the same amount on drink at his private lounge. On the other hand, if two different goods are competing in the market place for the contents of the same purse, that would tend to indicate a type of rivalry.

We might be tempted to state that money is a public good, since there are cases where everyone has access to it and can use it without preventing its use by others; except for the fact that the use by some individuals of what is obviously not a private good can indeed result in a case where certain other individuals are denied access to its use. In a market where prices have been set by sellers for products that are not in short supply, the money used by one buyer is not rivalrous or excluding of others use, but if there are only a few products with open bidding, one well heeled individual can prevent the others from using their money. In this sense there are cases in which the accumulation of money in few hands effectively approximates it to a club good.

In fact, money does not completely fit any of these categories, but since its utility does not result in its consumption as does the utilization of a good or service and as it is not produced by any human factor of production, (people do not make money in the workplace, they provide their effort and make things in return for other people's money), we contend that at least a component of it should be treated as a public good where everyone has access to the basic amount needed to survive.

What money definitely is not, in its total supply, is a private good. Thus while money in a modern market economy is viewed by many if not most market participants as a private good with inherent value, in reality it is inherently non-private due to its exchange nature and without innate valuation.

We look now to the first of the previously mentioned new technologies arising from the browser economy, the breakdown of tasks into components for specialization and skill suitability and enhancement; the division of labor.

We can rightly believe that task specialization both results from and leads to the development of a status hierarchy within the browser economy. Ability to find the best food and the most comfortable sleeping bower would transfer to the benefit of those most familial to such apt individual and result in elevation of his or her esteem in their eyes. Inabilities in these regards would likely result in a contrary appraisal.

With the advent of hunting and gathering, tool making ability and hunting skills will have a similar effect, including choice of the best fire and hut site, whether given by the appreciative other members in the group or taken by the strong arm of individual outright. The development of such individual skills, if they are transferable, also accrues to the benefit of the whole group through the learning of those skills and eventually of related group skills, resulting in an increase in consumption and a rise in the asset base of both lead individuals and the group.

In times of plenty such skills may be of marginal benefit to the group as it reverts to browsing, but in times of want they can mean the difference between mild hunger and starvation, between life and death. In recurring times of plenty, they can mean reduced time spent providing for the necessities of life and additional time for sport and the pursuit of individual curiosity and interest with the potential for derivative benefit to the whole group. In general, times of plenty provide for individual pursuits and times of want require group retrenchment and reliance, though both want and plenty can coexist but in a modern economy with social stratification.

Evidently, the persistence of a hierarchy of skills and knowledge and of attendant social status leads in time to hereditarily established class and caste systems which can be susceptible to corruption and abuse. Entitlement, i.e. ownership and license rights, are a straightforward component of such systems, and also subject to abuse.

Still, individual ownership, an enlightened perspective might say stewardship, of natural and human modified resources is in general a more efficient and effective method for the <u>detailed</u> allocation of those resources to direct human effort toward productive ends than is social ownership. However, the <u>overall</u> allocation of those resources is a social ownership; individual ownership is a socially derived entitlement and must ultimately serve a socially agreed and acceptable end in terms of the social contract or it is no different from the presumptive divine right of monarchs that the democratic revolution, now well into its third century, set itself to upend.

Monarchy of the absolute type is simply private ownership of the entire economy, as the legendary King Arthur emboldened with the presumption "The king and the land are one", to good or ill effect. Free market economics presumes itself to prevent a resurgence of such monarchy, but without oversight, "free" markets, even in the absence of strong-arm tactics, work to the ascendance of those with competitive advantage, both through merit and through entitlement.

This is in part because a market by its nature is rarely a meeting of equals; generally the propensity to buy and the willingness to sell are not equal, even though an agreed price is reached. This is not generally recognized in theoretical constructs which idealize each transaction as an optimization for both parties. Either the buyer or the seller may have a greater urgency in concluding the transaction than the other party, though equivalence is possible and is usually suggested by both for a variety of reasons.

This may be true in a barter exchange as well, but in a monetary transaction the buyer is always a universal agent and can buy anything that his interest pursues and his purse can afford, while the seller, unless his product has universal appeal, can only approach a limited market that is interested in his product. In addition, a seller generally has expended money or effort in acquiring or producing his product and from the instant of bringing it to market and even before, desires to exchange it for the universality of cash, while the buyer will try to maintain his universality of cash until the urgency of consumption or acquisition is immediate. If the buyer has a specific and urgent need to fill, of course, the roles may be reversed.

This is also true in the employment of other people's labor, where the laborer, having already invested in food and shelter and clothing and transportation to make himself employable, is ready to exchange his efforts for cash, while the employer would rather delay that employment until he is confident in his ability to readily exchange the product of that employment into cash.

Depending on the nature of their resource allocation, over time such differentials in the market can lead to oligarchy or effective monarchy. Some enterprises are inherently monopolistic by the nature of their resource involvement, and as long they are do not operate contrary to the widely perceived public good can be allowed to remain unregulated. When they are antithetical to that good, in a modern democracy they should be regulated. Thus free market economics is a relative concept and not a prescription for anarchy. Governance is necessary to preserve the social contract, of which access to the necessities of living is an integral part.

As an organizing principle in the production of marketable goods and services, private enterprise is arguably the most efficient, since the owner(s) need not consult anyone else's opinion in operational matters, and also the most susceptible to abuse, for similar reasons. Public enterprise, both of government and publicly traded firms, is subject to the inefficiencies of bureaucracy and regulation, but is more open to public scrutiny in its operations, or at least should be in a democratic society.

Private concerns are therefore highly governed by both the skill and the ethics of the parties involved. They may be more apt or at least able to embrace technologies of proven efficiency, but in conditions of rapidly changing technology this can be disruptive of the general social good, aside from any questions of ethics, for a couple of reasons.

First, the introduction of such technology means that the same amount of production can be achieved with employment of less human effort. If such disemployment is broadly based throughout the economy, it can be remedied by a general reduction in the workweek and more leisure time at the same relative level of pay and consumption, and this has occurred over the past century or so in the western democracies. However, where there is unevenness in this process across different industries and market segments it is likely to result in areas of underemployment or unemployment, which in a market economy results in segments of the population being cut off from access to necessary resource consumption.

Second and less immediately obvious is that the benefits of increases in productivity, to the degree that the increases are embodied technologically in plant and equipment instead of in the specialized and relatively hard to reproduce skill of their operating workers, accrue to those who own the plant and equipment and not to the employees. No matter how technologically advanced the skillset, those who first master new technological skills and have an initial competitive advantage in the job market may become enterprise owners themselves; otherwise they gradually lose their ability to command a wage premium as such skills become commonplace or are superseded by newer technology.

In this dynamic, the monetary tokens of successful productive enterprise accumulate with those employers who own the technology, while slowing to a subsistence level for those they employ increasingly as commodity labor. So along with an overabundant supply of production and productive capacity, at least until it no longer makes sense to maintain all of it, the market economy paradoxically results in a condition in which there is insufficient pay or other funding of workers to allow them to purchase the goods and services that they produce and that their employers would like to sell back to them.

The excess cash accumulation of employers then results in a disinvestment in productive real capital, by exporting of operations to areas of lower labor cost, and to a rise in non-productive asset pricing. As savings has fewer productive investment opportunities, it pursues more expensive toys and emblems of status, both real and financial, eventually attracting foreign financial capital to the excitement in the process.

A review of the attached table of Household Final Consumption Expenditures for selected countries and the world shows the effects of this dynamic over the past thirty some years in the United States starting with the implementation of supply side thinking in fiscal policy. The ratio of consumption spending to total production, designated in the following as, C_G, has steadily climbed since the early '80s with a brief period of leveling off during the Clinton administration. It is important to remember that the final consumption expenditures are for all factors of production in the economy and not just what is traditionally thought of as labor, and include the personal expenditures of professionals, business owners and managers, and rentiers.

This is the period of increasing globalization, which is another way of saying, at least for the US, a period of disinvestment in domestic production, both directly by the export of production or indirectly by the import of finished goods. Both direct and indirect processes result in an export of monetary tokens, or at least their electronic equivalent.

These in turn have found there way back into the western developed economies, adding to the funds bidding up the asset prices and service costs of everything from real estate to corporate and financial equities to medical and related insurance costs to the funding of human capital in the form of college degrees. During this time, with the brief exception of a notable rise in income for all segments of the economy during the second term of the Clinton administration, as seen in the attached chart, the indexed income for the lowest 50% of households has been stagnant. Yet these households have continued to work and live in areas subject to the pressures of upward asset pricing of all types, many if not most with little access to non-market sources of consumption in the event of an economic downturn, i.e. as found in a rural environment.

The long-term trend is toward an increasingly productive, automated economy employing increasingly fewer skilled manufacturing workers and more semi-skilled retail and service workers at commodity priced wages. Once the differentials in wages between global segments have stabilized, incentives will increase for domestic reinvestment in manufacturing, but not necessarily with substantial wage increase.

We can expect that the income spread that now trends the 80th percentile closer to 90th than to the 50th will fall much closer to the 50th as the 90th falls further from the 95th, etc. These members of the upper middle class who make their livings as professionals and service consultants to the top fraction of a percent will become fewer in number and find their relative incomes drop. Small business will continue to fail or operate on tight budgets, offering little hiring to well paid positions.

This trend will continue as long as well-meaning voters on the right continue to believe that government spending is the source of their economic woes. Should the well-meaning voters on the left gain ascendency, believing that government taxation and redistribution of income or even wealth is the solution to their economic woes, the problems will be likely to continue, though in a different vein.

A final innovation away from the browser economy, taxation arose with the division of labor as a means, other than plunder and slavery, of supporting the leading strata of the society and their enforcement branches, noticeably as a tax-in-kind of agricultural produce, i.e. a share of the granary deposits, on the largely agricultural base. With the rise of industrial democracies, this process resulted in the tax on income as a means of assuring a reliable flow of cash for government operations, since most currencies at the time were specie or precious metal based and governments could not make such metals or metal based specie on demand.

Government expenses are either for investment and the purchase of goods and services, particularly in infrastructure and police/military sectors or for transfer payments to others, and in an economy expecting tokens of satiety for payment of goods, services or transfer payments, taxation-in-kind of this kind of token is necessary.

However, in an economy where money is denominated by a fiat currency and transactions consist of computerized accounting entries in an electronic medium, with government payments consisting of such electronic direct deposit entries, and where cash transactions represent a small fraction of overall exchange and coin transactions represent much, much less, such taxation is likely an anachronism. I say likely, since from a rational point of view it is an anachronism, but it is hard to gauge how every citizen would respond to such innovation.

The potential for abuse of the alternative direct "printing" of money, while real and to be avoided by the institution of proper policy and systematic oversight, is no more a danger than the abuse of value creation engaged in by the speculative financial industry prior to and, of course, still continuing since the crisis of 2008, or by electronic arbitrage in the stock, commodity and other markets' and with IPOs of enterprises of unproven value.

The complaint against such government monetary creation in the payment of expenses, as with similar complaints against quantitative easing, national debt and deficits, inflation, and of course, the Fed in general, comes from those with holdings of financial assets that would prefer not to see their value diluted by inflation. This is understandable. However, a key in successful management of any endeavor is in knowing how to interpret the effectiveness of ones policies and procedures and that involves understanding the feedback signs, so we turn to that next, with the aid of some mathematical analysis.

Analysis

Macroeconomic analysis attempts to shed light at a domestic or global level on the interplay of the production and consumption of goods and services, the financial structures that facilitate that interplay, and the monetary and fiscal systems that govern those structures. This overview is of another such attempt, yet with what is believed to be a novel twist. It uses as its point of departure the browser economy to which we return. In general experience, most developments of economic theory start with the concepts of supply and demand, which are expressions in transactional monetary terms of the corollary physical production and consumption of goods and services. The value of such goods and services is expressed as a quantity of some monetary unit that is exchanged from the demanding consumer to the supplying producer.

A monetary unit is taken as an axiom of that exchange without regard to a definitive understanding of money. It is generally assumed to be a medium of exchange, a unit of account, and a store of value. While we would accept the first two of these, we would stipulate regarding the last that it is only token of a store of goods or services that have intrinsic use value, having no intrinsic use value itself other than in anticipation and execution of an exchange. While this is generally understood, the logical implications of the distinction are insufficiently pursued. Such pursuit as just outlined indicates that lacking any intrinsic value, the value of an individually held store of money is relative to what the entire population of money holders do with their money and not just what the individual holder does, so that money cannot be defined correctly as a private good as most analysis does.

In contrast as we have seen, the browser economy has no monetary system. Its analysis begins instead with the necessary propensity to consume natural produce in order to maintain human life. In the browser economy, all production is that of nature which is immediately consumed by the human population, without a mediating production arena, market or currency. There is no demand in the usual economic sense, just a utilization of resources to satisfy a need or want. And while there is a supply of natural resources, there is no human productive effort embodied in that supply other than the physical movement of each individual to the location of the produce in order to consume it. This same supply of natural resources also provides tokens of emotional satisfaction or satiety for individuals that are socially recognized and appreciated, that is valued, and which form the basis for the eventual evolution of a system of money, but in the browser economy such system is separate from the consumption of basic human survival needs.

We start with the consumption of natural production rather than with demand and supply of human production, because we hold it to be the primary motivator of everything else in the economic universe; biological need and want, healthy and pathological, generates demand, and demand generates attempts to access, provide and control supply. While all analysis realizes this fact on some level, all too often economic modeling embraces blithe absurdities about monetarily expressed autogenous demand; as if factors of production such as labor can turn the "propensity" to consume off and on like a spigot to a reservoir of need during a financial crisis. As if a financial crisis, a scarcity of funding or more particularly for those who cause it, of returns on funding, is equivalent to an economic crisis, a scarcity of some necessary natural resource or factor of production, despite an obvious relationship.

We intend to show that the drive behind all modern economic activity, regardless of its self-perceived direction, is movement back in the direction of the browser economy; to the satisfying experience of living with a minimum of self-conscious effort in the company of a supportive family group or community. In dry economic terms we might say to the satisfaction of wants and needs via efficient production and consumption of goods and services in co-operation with our fellow factors of production. Such production is necessarily not "natural" production, since it is realized only through the acculturation of human effort in response to scarcity, but that effort is in turn increasingly less directly human through the growing productivity and inventiveness of human technology, culminating as of the present in the digitally mastered, productive environment in which any perceived scarcity is increasingly of human genesis.

For the browser economy, instead of the usual monetary income denominated production factors, from the microeconomic view, of labor, raw and intermediate materials, capital goods, rent and return on investment that go into the production function, we will start with a basic consumption or utility function, U, consisting only of the human, H, and natural resources, R, utilized in meeting the wants and needs of the economic population, all of which are human, i.e. they are households, not firms. This utility function can be written

$$U(H,R). \qquad (0.1)$$

In the browser economy, human consumption and productive effort are one and the same thing. Picking the berries off the bush, the fruit off the tree, and popping them in your mouth is both production and consumption in one motion. The human utility function, then, is measured or quantified by the human effort expended in consuming those resources, H_R, or by the quantity of natural resources that are thus consumed, R_N, or

$$U(H,R) = H_R = R_N = U. \qquad (0.2)$$

Note that since the same utility is expressed in each term, H_R and R_N, the human effort is equal to the resource consumed. In such consumption quantity we include any resource, human or natural, "wasted" in the utilization process. Note that we do not add the human and natural resources to arrive at the utility. This Utility is understood in the active voice of Humans picking and eating berries or in the passive voice of Berries being picked and eaten by humans. Note however that this equality of effort, resource, and utility does not indicate an identity of any of the components. According to customary mathematical logic you can only add quantities or equate sums of the same qualitative or "dimensional" units, i.e. apples and apples; unless you are counting up a more general qualitative unit, fruit, as in apples and pears and bananas. This forces us to make a couple of adjustments, which will prove to be helpful.

First, we can say that the value, v, expressed below as a pre-subscript, which is in essence a subjective, though quantifiable, property identified with a thing, the *value* of the three components is not only equal, it is an identity or

$$_vH_R \equiv {_vR_N} \equiv {_vU} . \tag{0.3}$$

In fact we can make a case that the economic utility of a good or service is its value. In the marketplace, something that is recognized as having no utility has no value. So by definition the effort required to utilize a resource is equal to its value or the effort would not be expended to utilize it; similarly so with any utility imputed to a natural resource, without which it would not be valued. Still, in spite of a shared quantitative value, the human effort, the natural resource, and the satisfaction of want or need inherent in its utility are three separate properties or dimensional qualities.

Now in addition to the effort involved in using the resource in its entirety, let us suppose that there is some effort necessary to effect its utilization that is, however, not completely used up. We are now moving from a browsing economy to capitalism, in this case of the hunting and gathering type. Instead of eating all the fruit and berries at their source, we construct some baskets to collect more than we can currently consume and take the saved surplus back to our camp and store in pottery we have made, to eat later. Instead of being satisfied with the occasional hand caught raw fish or rabbit, we build some nets and spears to catch enough to take home and cook. When we have consumed all the fruit and berries and fish and rabbits, we can now use these baskets and nets and spears once more to go back out and bring back natural produce to our camp. We won't have to expend as much effort the next time out, though we will have to maintain the fire and spend a little time mending the baskets and bowls and nets and sharpening the spears, and perhaps making a few more of each.

The value identity expressed in (0.3) is not so simple now. The total utility expressed by $_vU$, which includes the extra meals now possible back at the camp, is still equal to the total human effort expended, $_vH_R$, and the total of resources utilized, $_vR_N$. But the total utility is found in both the food consumed during the browsing and the food gathered and brought back for later consumption, which can be thought of as the *product* of the human effort *operating on* the natural resources.

This brings us to our second adjustment. This product operation can be expressed mathematically as the cross-product of $_vH_R$ on $_vR_N$, where this mathematical product is the economic product to be utilized, $_vP_U$, of the total food browsed, caught and gathered, or

$$_vH_R \times {_vR_N} = {_vP_U} = {_vU}. \tag{0.4}$$

In the case of a cross-product each of the three terms is dimensionally distinct, so that here, instead of an equation stating that the utility of an individual collecting apples equals the utility of apples collected equals the utility of a meal of apples, as with (0.2) we might have one stating that a human gatherer mashing up some picked apples produces apple sauce. But in our hunting and gathering party we have more than this, since in addition to the human effort expended, H_C, on hunting and gathering the food resources to be consumed on the expedition and later back at camp, we have the initial effort, H_I, invested in making the bowls and baskets and the masher and nets and spears from other natural resources. We have capital goods. And H_I includes the essential organizational understanding gained of the coordinated effort required to hunt and gather; beyond the social skills of browsing and the lore fire and in addition to learned individual skills.

Thus (0.4) can be elaborated as

$$\left({_vH_{C0}} + {_vH_{I0}} \right) \times {_vR_{N0}} = {_vP_{C0}} + {_vP_{I0}} = {_vU_0}, \tag{0.5}$$

where the efforts of this initial hunting and gathering party, E_0, (the subscript being a time sequence or cycle number), result in the production of consumable utility, $_vP_C$, in the form of food goods, and the production of tools and technological skills that can be reused on the next hunting and gathering effort, that is investment in productive or capital goods and services, $_vP_I$. That next party, E_1, has the advantage of reuse of the initial capital as a type of modified natural resource, R_I, where the new resource base for sequence 1 is

$$_vR_1 = {_vR_{N1}} + {_vR_{I1}}. \tag{0.6}$$

For highly non-durable production goods, the value to be used up in the next round of production, $n+1$, is roughly equal to the value of the tools produced, but not consumed, in the current round, n, or

$$_vR_{I(n+1)} \cong {_vP_{I(n)}}. \tag{0.7}$$

For highly durable production goods, that value is the value of the tools currently produced less the retained or unused value in the next round, $_vR_{+W(n+1)}$, or

$$_vR_{I(n+1)} = {_vP_{I(n)}} - {_vR_{+W(n+1)}}. \tag{0.8}$$

There must first be some tool mending and perhaps production of additional baskets, nets, spears and perhaps a club made, along with a discussion and refinement of hunting strategies so that the resulting production will be

$$(_vH_{C1} + {_vH_{I1}}) \times {_vR_1} = {_vP_{C1}} + {_vP_{I1}} = {_vU_1}, \tag{0.9}$$

After n sorties into the wild to hunt and gather, the group has bunches of baskets, and pottery to store what they catch and gather, and elaborate snares and atlatls, bows, slings and arrows in addition to the spears and nets and a nice tribal hut where they can review their organization, not to mention boast about their daring, hunting skills, and outrageous fortune. In other words, they have a total man-made resource base or wealth, $_vR_W$, invested or the total amount produced less the total consumed after n sorties of

$$_vR_W = {_vP_{I\Sigma(n)}} - {_vR_{I\Sigma(n)}}. \tag{0.10}$$

It then dawns on the elders of the group that the current utility they realize is not as expressed by (0.9). The capital goods investment of at least $_vP_{I(n)}$ is advisable in replacing that of $_vR_{I(n)}$ used in production period E_n, and while it is necessary for future production of $_vP_{C(n+1)}$ at the current level of $_vP_{C(n)}$, it or its equivalent is only utilized or consumed in that next period. Therefore the value of the current utility is to be found not in (0.9) but in the following, where the gross product for the period is $_vP_G$

$$(_vH_{C(n)} + {_vH_{I(n)}}) \times {_vR_{(n)}} = {_vP_{C(n)}} = {_vP_{G(n)}} - {_vP_{I(n)}} = {_vU_{(n)}}, \tag{0.11}$$

The group is spending more and more time keeping up their growing stock of capital goods, but they are also living better as they see it, i.e. consuming more in any current period. They have instituted a token system to keep track of everyone's contributions, so that everyone is paid according to his efforts and in turn can retrieve his share of the sortie production. The elders want to know the optimum proportion of pay between H_C and H_I, since it is time consuming to make and maintain tokens. Assuming it is equal to the current utility, $_vU$, and to the value of effort expended on that consumable production, $_vH_C$, they want to know for each token unit of consumable production, $_vH_C = {_vP_C} = 1\$$, what should be the <u>value</u> of effort (but not necessarily the time) in tokens, $\$$, invested in making tools and other production goods, $_vH_I$, assuming that it is just enough to replace, i.e. be equal to, the used-up production goods, $_vR_I$. In other words, solve

$$(1_{C(n)} + x_{I(n)}) \times x_{(n)} = 1_{C(n)} = (1+x)_{G(n)} - x_{I(n)} = 1_{(n)}, \tag{0.12}$$

All the terms to the right equal 1 so that we have, as a positive real solution

$$x^2 + x - 1 = 0 \tag{0.13}$$
$$x = 0.618033989...$$

meaning that for a given economic sequence or period at equilibrium, the ratio of capital goods and services to consumer goods production should conform to (0.13).

Plugging this figure into (0.12) gives the following

$$\left(1H_{C(n)} + 0.6...H_{I(n)}\right) \times 0.6...R_{(n)} = 1P_{C(n)} = 1.6...P_{G(n)} - 0.6...P_{I(n)} = 1U_{(n)}. \quad (0.14)$$

Therefore, in equilibrium, where capital is replenished at the rate of depletion, the ratio of consumption to total production, C_G, should be the same as capital to consumer production or

$$C_G = \frac{_vH_C}{_vH_C + _vH_I} = \frac{_vP_C}{_vP_G} = \frac{1}{1.618...} = 61.8...\% \quad (0.15)$$

with the ratio of investment in capital goods and services, i.e. real and human, to total production, I_G, of

$$I_G = \frac{_vH_I}{_vH_C + _vH_I} = \frac{_vP_I}{_vP_G} = \frac{0.618...}{1.618...} = 38.2...\% \quad (0.16)$$

Note that (0.16) is not a measure of relative energy or effort or time extended in producing P_I with respect to P_G; rather it is the relative trade *value*, $_vP_I$, that it must represent in order for the income stream paid for H_C to match the sales price of P_C. As stated in the attached Production Cycle table, if I_G is valued at less than (0.16), there will be a mismatch with H_C greater than P_C and several things can occur, which we will go into in a minute. In general there will be (a) a shortage of current consumer supply or (b) price inflation, depending on whether (i) the propensity to sell or (ii) the propensity to buy is greater. If I_G is valued at more than (0.16), there will be a contrary mismatch with H_C less than P_C, resulting in (a) a surplus of current supply, (though not necessarily availability) or (b) price reduction, this time depending on whether (i) the propensity to buy or (ii) the propensity to sell is greater.

Over the long haul, real economic growth can only occur through increases in size of the working population or increased productivity for a given population or a combination of both. In the short run, it can be achieved by a given population producing more (consumables) than it consumes. Therefore, for stable growth to occur, the increase in production due to productivity and/or human working population of H_C and H_I should be such that a ratio of consumer to total spending a few points below (0.15) is maintained. Thus the monetary value of the increase should be reflected in both H and P_C, so that instead of percentages with respect to P_C, (0.11) in monetary terms becomes

Production Cycle with respect to proportionality constraints of Consumption and Investment Valuation

	Hc	Hi	R	Pc	Pi	PG	Hc/(Hc+Hi)	Pc/PG		Excess	PI/PG	Cycle	Growth	Liquidity	Capacity	Public	General Environment
1	1	0.618	0.618	1.000	0.618	1.618	0.618	0.618		Eq	0.382	D	S	Eq	Opt	O,$,I,R,T	Stable
2	1	0.518	0.618	0.969	0.518	1.487	0.659	0.652	<	0.007 D	0.348	D	-	C-	O-	O,$+,I+,R+,T-	Below optimum slowdown
3	1	0.518	0.518	0.887	0.518	1.405	0.659	0.631	>	0.028 D	0.369	T	S	CS	OS	O,$+,I+,R+,T-	Trough or down plateau
4	1	0.618	0.518	0.915	0.618	1.533	0.618	0.597	<	0.021 D	0.403 +	I	+	C+	U+	O,$,I,R,T	Recovery growth
5	1	0.618	0.618	1.000	0.618	1.618	0.618	0.618		Eq	0.382	I	S	Eq	Opt	O,$,I,R,T	Stable
6	1	0.718	0.618	1.030 +	0.718	1.748	0.582	0.589	>	0.007 S	0.411 +	I	+	D+	U+	O,$-,I,R,T+	Salubrious growth
7	1	0.718	0.718	1.111 +	0.718	1.829	0.582	0.607	<	0.025 S	0.393 +	P	S	DS	US	O,$-,I-,R,T+	Peak, up plateau, overheating
8	1	0.618	0.718	1.078 +	0.618	1.696	0.618	0.636	>	0.018 S	0.364 -	D	-	D-	O-	O,$,I,R,T	Slow down
1	1	0.618	0.618	1.000	0.618	1.618	0.618	0.618		Eq	0.382	D	S	Eq	Opt	O,$,I,R,T	Stable

1 Hc	Current valuation for consumption goods and services sum of current production factors, here theoretically set to 1 for all evaluations	
2 Hi	Current valuation for capital goods and services current production factors	
3 R	Current imputed valuation of natural resources and previously produced / currently consumed capital	
4 Pc	Current valuation of consumption current final production - Sq Rt of ((Hc + Hi) x R)	
5 Pi	Current valuation of capital goods and services current current production - Hi	
6 PG	Current valuation of total goods and services current production - Pc + Pi	
7 Hc/(Hc+Hi)	Ratio of current consumption production to total production - production factor valuation	
8 Pc/PG	Ratio of current consumption production to total production - product valuation	
9 Excess	Difference in 7 and 8 as an excess of S = Supply or D = Demand, Eq = Equilibrium	
10 PI/PG	Ratio of current capital production to total production - product valuation	
11 Cycle	Position of row as a component of a production cycle, D = Decreasing, T = Trough or down plateau, I = Increasing, P = Peak or up plateau	
12 Growth	Trend in Growth as the directional change in Hi, S = Stable, - = decreasing, + = increasing	
13 Liquidity	Trend in market Liquidity with respect to C = Concentration or D = Diffusion of fund pools, Eq = Equilibrium, S = Stable, - = decreasing and + = increasing toward cycle Peak production	
14 Capacity	Trend in production Capacity with respect to O = Oversupply or U = Undersupply of Hc and/or Hi, Opt = Optimum, S = Stable, - = decreasing and + = increasing toward cycle Peak production	
15 Public	Advised Public involvement O = Oversight/general welfare, $ = Money supply, I = Infrastructure, R = Research/development, T = Use based taxes, for baseline, - = decreasing, and + = increasing involvement	
16 G. E.	Comments concerning basic economic environment	

$$\sqrt{(_sH_C + {_sH_I}) \times {_sR_I}} = {_sP_C} \qquad (0.17)$$

This is because the product, P_C, is a separate dimensional property than either H or R, in the same sense that a square foot is a different dimensional property than the two lateral linear feet that comprise it. The token valuations, however, are all of the same kind or property and the square root of the computed product valuation gives the product transactional value in token or monetary dimensions. Instead of the square roots, we might decompose the product into other roots to reflect a weighted value for the H and the R components, i.e. the income and the asset components. This is what is done by the market in non-equilibrium conditions both beneficial to growth and to its detriment, and not coincidentally to the labor portions both of H_C and H_I.

It should be noted and emphasized that while there is a rough and maybe not so rough equivalence between labor and H_C and between entitlement to returns on capital and resources and H_I, there is return on labor and on capital in both of H_C and H_I. In other words,

$$_vH_C = \left(_vH_{C_L} + {_vH_{C_C}}\right)$$
$$_vH_I = \left(_vH_{I_L} + {_vH_{I_C}}\right) \qquad (0.18)$$

Thus labor, L, and capital, C, are

$$_vL = \left(_vH_{C_L} + {_vH_{I_L}}\right)$$
$$_vC = \left(_vH_{C_C} + {_vH_{I_C}}\right) \qquad (0.19)$$

The chief distinction between labor and capital made here is not primarily in the usual sense of ownership or relation to the means of production. Rather it is that the value in labor, regardless of skill level or remuneration, is invested in the product at the point of production (and trade, distribution, etc.) and is therefore current and variable, while capital is human effort, newly and previously invested in productive property, that is invested at the start of a production process and throughout the process as needed, which maintains and oversees the necessary continuity of that process and is therefore of long-term utilization. Laborers, all humans obviously, are in turn maintained, produced day by day, by consumption and other utilization of the necessities of life. The engine that drives the economy then is not some abstract, microscopic appraisal of the marginal utility of multiple market choices throughout the day, but instead the biological drive of humans to fulfill their consumption needs through the process of producing those needs at the workplace.

From this, it is straightforward that the value in capital, $_vC$, is from the accumulation of unconsumed production over time in the wealth of the economy, $_vR_W$. This includes not just the remaining utility of productive plant and equipment and other resources, $_vP_{I\Sigma(n)}$, but also that of the furnished homes and automobiles and other durable products that are needed to "produce" the able bodied worker at the front door of the factory or office or store or fast food restaurant or educational or medical institution every morning, five or more days a week. The value of the labor, the human capital, $_vL$, is the value in the products from the factories and offices and stores and fast food restaurants and educational and medical institutions and invested in the experience of school and work and service to the community, embodied in the workers from cradle to grave, required to produce their human productive capacity that grows with that experience until the aging process begins to diminish its effectiveness. The current value utilized from this human capital is in the expenditures of food and drink and clothing and for the household and getting to and from work that is consumed or depleted over the course of each day, week or month.

There are two things we should address in regards to the production of human labor and human capital and its value in the market place. The first is that fungible labor skills that are in sufficient surplus become a commodity subject to commodity pricing. In the trend toward commodity pricing, market constraints indicate payment only for the cost of producing that good or service, which for labor is the day-to-day cost of food, shelter, clothing, and transportation to and from the workplace. It does not pay for the cost of schooling and upbringing, for medical care, especially major medical, or for retirement. Such human capital costs, like all capital costs, are sunk costs that cannot be recouped unless there is need for their replacement, i.e. in the manner in which $_vR_{In}$ is priced at its replacement cost of $_vP_{In}$. Since commodity labor can be replaced by another laborer of like kind, except in time of full employment, in the free market all that commodity labor can command is the ongoing current costs of producing it, of getting it back to the work place from week to week or month to month.

Similarly, if a productive enterprise is forced by competition to implement commodity pricing of its whole line of products, it may not be able to fund its capital replacement and will continue to run only as long as it can employ commodity labor. When there is a general market slowdown and when the utility of its physical capital is depleted or degraded, as when the roof is leaking everywhere and needs to be replaced, if there is no funding available internally or externally, it will have to cease operations. There may be land value or cash value in plant and equipment and unfinished materials, but unless someone else can buy the enterprise, repair the roof and make the operation profitable, these are sunk costs and are gone.

As a result of this fact, there is no way that a competitive free market can answer the problems of long term health care, education loan repayment, and retirement needs of commodity labor. Only industry sub-sectors that are not subject to commodity pressure can do so. It bears repeating, *there is no way*! In the United States, especially in the wake of World War II when it was relatively free of competition, such needs were provided for in varying degrees by both private and public sectors until globalization and competition with foreign goods forced down the price of domestic production and reduced the corresponding labor toward commodity pricing. As a result of globalization, commodity labor can now make less contribution to public sector protection, leaving the private sector employers, already cutting costs, to shoulder more of the responsibility. Tariffs could address this issue, but at the cost of potential foreign market share loss and trade rivalry.

This has nothing to do with the skill or education level of the labor. If the STEM programs for science, technology, engineering and mathematics succeed in providing an oversupply of graduates at a time when the human application of those skills is being widely superseded by foreign competition and advanced computerized equipment that can be operated by less skilled workers, except for a highly specialized subset for whom robotic replacement is infeasible, those graduates will find their compensation falling toward the commodity pricing of the less skilled workers. Their college fees will be a sunk and unrecoverable cost, and there will be no compensation for medical and retirement cost other than through public recourse.

There is nothing sinister about this, at least of intent, though there is a destructive effect. However, it does point to the fact that if you want everyone in your society to have at least a modicum of civic pride and sense of belonging, not to mention the basic necessities of life, you can't have open global trade and look for the "private" business sector or the "free" market to provide it for everyone. It just can't happen. Even if every employer is a decent, caring citizen, if he or she is competing against foreign products, which are often times not only cheaper, but better, he can't provide long term life care subsidization and still compete.

This application of the above relationship of consumption and investment valuation constraints is represented by the following Production Cycle table. For a unit level of valuation of production efforts for consumption final goods and services, H_C, and an assumed stable technological and population base, the table shows the effects of changes in capital goods investment resulting in the production cycle. As detailed there, the resulting excess in demand and supply, overall growth, liquidity needs and constraints, productive capacity and nature of required public involvement are shown. Each phase row represents a period of indeterminate duration and can last decades.

Phase 1, 5, and the final 1 represent periods of equilibrium as defined above through one cycle. While the 9 rows represent a hypothetical cycle of contraction, followed by return to equilibrium, then expansion and return, in actual practice contraction and return to equilibrium might be followed by a second contraction without any expansion and vice versa. So the Cycle designations of Decreasing or Increasing productive activity have relevance only in cases in which the equilibrium is surpassed in a movement of the cycle to the opposite side. The Excess column indicates that supply and demand for consumption goods is balanced or in Equilibrium, and the Liquidity column shows that the money supply facilitates this balance for these rows. Productive Capacity is Optimized, with a balance of capital goods and labor utilization. Finally, the Public expenditure column indicates a baseline involvement in Oversight and general welfare, maintenance of an optimum and adequate money supply, $\$$, investment in Infrastructure, Research and development, and Taxation with an emphasis on common goods and externality use-based taxation. A positive or negative sense after a designation indicates an increase or decrease in the level of involvement based on the perceived phase of the cycle.

Phase 2 is a reduction in investment, resulting in a decline in the value of consumable production, and leading eventually to a reduction in employment. In the aftermath of a financial bubble or crises, this will often occur due to momentum from phase 8. Imbalance in supply and demand is shown by excess or unmet demand as indicated by unemployment, negative growth, and a concentrating of liquidity, C-, while production capacity is oversupplied, O-, and therefore underutilized both in terms of labor and capital. Advised public involvement includes increased diffusion in the money supply and in expenditures for infrastructure, R&D, with a decrease in taxation.

Phase 3 and 7 represent a respective trough and a peak in a cycle, but can also represent contraction and expansion plateaus prior to repeated contraction and expansion. No growth, as defined by no change in the investment/consumption mix, is indicated as stable for these rows. This indicates that extended periods of underemployment and overemployment of labor and capital are feasible.

Phase 3, Trough production, Excess is a period maximum for unmet demand, not necessarily in the marketplace, since there are liquidity constraints to market access, but in terms of unmet needs and wants. As shown, liquidity is at maximum concentration absent public redistribution or offset expenditure or issuance, and productive capacity is oversupplied and therefore underutilized both in terms of labor and capital. Advised public involvement is a continuation of Phase 2.

Phase 4 represents a swing back in the positive direction of Growth as indicated by both the sense in that column and the caret indicating the direction of equilibrium under the P_C/P_G ratio and the continued strength of the unmet demand in the next column. The cycle is in an increasing production phase, with liquidity still concentrated but expanding, C+. Capacity is undersupplied based on demand, but is increasing, based on the level of P_I/P_G which is actually above the equilibrium level of Phase 5. Public economic involvement can move back toward the baseline.

In a dynamic recovery, Phase 5 is quickly overshot and Phase 4 moves into Phase 6 with a leading of supply over demand. This continued growth encourages assumption of debt and a resulting diffusion of liquidity, while capacity remains undersupplied, but increasing. At some point, advised public involvement may include contracting of the money supply, and a reduction in infrastructure, with an increasing of use based taxation.

With Phase 7, Peak production is marked by continued investment and risk of eventual oversupply of consumer products and/or inflation via diffuse and growing liquidity if demand rises to meet it, while capacity is strained with the eventual undersupply of labor and thereby a move toward more productive real capital. If money creation via debt is unrestrained, inflation can occur resulting in an eventual bubble in affected asset classes and sectors. Advised public involvement is a continuation of Phase 6, but with a rapid and steep transition through Phase 8 may need to be agile and reverse itself.

Phase 8 is perhaps the most crucial for public involvement as a result, in dealing with the private sector retrenchment from the peak. The tendency is for a well known shrinkage in liquidity and lending, along with an oversupply in capacity and inventory. The public goal should be a soft landing in Phase 1 and not an overshoot into Phase 2 recession or a Phase 3 depression.

It is the assertion of this development that the figures shown in (0.15) and (0.16) represent valid natural optimization constraints on C_G and I_G in an advance market economy and that the cyclic dynamics just outlined play out and intertwine for locales and firms and industries within greater domestic and global cycles.

If the elders of the hunting and gathering group wanted to verify their thinking and had access to the internet and looked at the accompanying figures from the World Bank website, they would see that the Household Final Consumption Expenditure, from the table so named, as a percent of GDP for the world ranged between 60.51% and 58.29% over a period from 59.13% in 1970 to 60.37% in 2012, indicating a complementary expenditure range on productive goods and services of 39.49% to 41.71%. This is a range of 1.3% to 3.5% below/above the equilibrium positions and indicates a corresponding range for net investment above equilibrium. This is in keeping with the world economic growth rate for that time period of 1% to 4%. In general, for an individual country, a figure below (0.15) indicates a financial surplus or investment and a figure above indicates a deficit or disinvestment.

Among various countries selected the range was from a high of 92.25% for consumption spending in Greece at the start of the Junta in 1968 to a low of 10.68% for oil rich Qatar in 2011. The three oil states shown, plus the autocracies of Singapore and China are all well below the equilibrium, reflecting the resource exploitation and exportation of the first three and presumably the state provision of or assistance with dwelling and other consumption products in all five. A review of this table shows a wide range across the selected nations within the context of a steady global condition of modest growth. It is instructive that Switzerland has maintained a level within a few percentage points above and generally below the 61.8% benchmark over the fifty-year period.

As seen, the United States in 2012 was at 68.64%, or 6.84% above equilibrium. Note in particular that the United States figures from 1965 to 1969, during the height of the Vietnam War, were slightly above (0.15), dipped below from 1970 to 1981, and then rose back above that benchmark with the advent of Reaganomics in 1982 where they have steadily risen to the present level. The level rose 1.72% between

1982 and the end of the Reagan presidency, 0.75% during the first Bush tenure, 0.53% the first Clinton year in office, remained essentially flat for 6 years, before rising sharply with the dotcom crash in 2000 for a total 1.63% during his two terms, rose 1.87% during Bush II, and 0.67% during the first Obama term.

There is much analytical fodder in this relationship. It clearly shows supply side fiscal theory for the wishful thinking and fallacy that it is, since the most stable period of investment in the US, which was actually a period of no increase in disinvestment, observed during the Clinton administration was also the only period of any significant general wage increase over the last 40 years as seen in the attached chart from the Economic Policy Institute. A rising tide may lift all boats, but it is quicker to swamp those of meager freeboard. Perhaps better said, it may lift all boats, except for the ones in dry dock. This last chart shows the income levels of the 95th percentile have increased by roughly 35% since 1973 (others sources show as much as 65%) compared to about 5% for the 50th percentile.

A look at the total net worth of households in the next data from the Federal Reserve System shows a net worth of a bit under $6 trillion in 1975 or 343% of GDP, while the 2013 net worth was almost $79 trillion or 470% of the GDP. An uncritical approach would interpret this period as one of real economic growth for the US, when in fact this analysis shows that it has been one of disinvestment for the population as a whole. The last period of real overall domestic investment as outlined for Phase 6 & 7 was in the aftermath of the Vietnam War in the 1970's, culminating as typified in high inflation. Instead of providing for a soft landing, however, the advised public policy was replaced by the mistaken notions of supply side, privatization thinking, resulting in over thirty years of Phase 2 and 3 stagnation, at least for over half the citizenry.

With respect to the above benchmark, this period of globalization has been one of overall domestic disinvestment in both public and private facilities. The movement of domestic capital overseas for the finished production of consumer goods, along with the importation of foreign produced goods and services, resulted in pressure on domestic wages in the relocating and domestic replacement industries toward commodity pricing. This in turn was made possible by the importation of basic consumer commodities, particularly textiles and household durables, followed eventually by food products, making such goods and the wages they attract stable over time. The only finished good that was not directly importable was housing, though it came to rely heavily on "illegal" imported, commodity labor, and fell prey to the same dynamic through low interest rates and under-supervised lending.

Over the same period of deregulation, services and production that were not easily exported, along with financial services that as a result of globalization were no longer inherently domestic, but retained a genetic disposition for the US, were able to grow and appreciate at a more rapid rate than the production assets of basic commodities which were tied to commodity labor. The result was a bid up in the pricing of medical and educational services, along with commercial and other

Household final consumption expenditure as % of GDP

Country Name	Code	1960	1961	1962	1963	1964	1965	1966	1967	1968	1969	1970	1971	1972	1973	1974	1975	1976	1977	1978	1979
World	WLD											59.13	59.00	58.87	58.29	58.48	59.50	59.16	59.19	58.90	58.97
United States	USA						62.75	62.10	61.94	62.28	62.36	60.21	60.03	60.00	59.58	60.17	61.15	61.25	61.20	60.52	60.39
United Kingdom	GBR						63.26	62.63	61.96	61.34	61.28	60.07	59.60	61.30	60.19	60.64	59.15	56.99	57.03	57.15	58.15
European Union	EUU											57.52	57.42	57.59	57.11	57.45	58.23	57.83	57.84	57.37	57.67
Germany	DEU											54.58	54.61	55.28	55.12	55.74	57.82	57.88	58.17	57.51	57.60
France	FRA											56.34	56.34	56.22	55.50	55.86	56.47	56.47	56.28	55.99	56.23
Canada	CAN						58.78	57.16	57.59	57.53	57.30	54.76	54.45	55.19	54.02	53.09	54.16	53.62	53.46	54.24	53.03
Switzerland	CHE						57.49	57.94	58.10	58.03	57.92	56.93	56.23	56.43	56.87	57.42	61.41	63.03	63.82	62.97	63.27
Israel	ISR	70.70	69.41	64.25	64.74	64.03	67.13	67.00	67.11	65.25	64.37	60.33	56.50	56.99	58.14	53.05	60.01	61.97	61.12	61.39	62.84
Ireland	IRL											68.81	67.90	64.97	58.14	68.03	63.72	61.97	64.17	63.82	65.08
Spain	ESP											63.24	63.59	63.11	62.71	63.01	63.18	64.52	64.03	63.05	63.57
Italy	ITA						57.82	59.10	59.44	57.73	57.60	58.40	58.58	58.84	58.89	58.29	59.84	58.83	58.54	57.31	57.77
Poland	POL																				
Portugal	PRT											65.37	67.96	63.64	63.47	70.57	77.57	74.19	69.93	66.45	66.28
Greece	GRC						47.77	87.66	71.92	95.25	80.84	64.63	63.02	60.17	56.82	61.63	62.13	60.48	63.57	63.57	62.99
Mexico	MEX	79.25	79.00	78.93	76.77	77.00	75.33	75.60	75.41	76.09	75.21	71.92	73.22	71.82	70.49	69.84	68.72	68.08	66.30	66.05	64.41
Brazil	BRA	66.28	64.06	70.04	67.93	68.98	66.92	68.12	71.54	70.56	66.77	68.55	69.58	69.66	67.54	70.74	66.49	68.76	69.16	68.56	69.55
Russian Federation	RUS																				
India	IND	80.83	79.49	78.00	76.90	77.63	76.22	77.08	78.12	77.82	76.20	75.37	73.41	74.06	75.13	73.51	72.54	71.03	72.05	70.34	70.03
China	CHN																				
Singapore	SGP	94.79	94.13	85.54	87.37	80.72	78.99	74.96	75.24	70.50	70.07	68.98	67.79	62.71	59.30	60.28	60.05	56.96	55.64	55.07	53.80
Saudi Arabia	SAU									33.14	35.47	34.86	28.15	24.92	21.18	11.01	16.56	16.95	26.28	31.80	31.49
United Arab Emirates	ARE																				
Qatar	QAT																				
High income	HIC											57.51	57.41	57.42	56.86	57.11	58.34	58.10	58.18	57.81	57.91
Upper middle income	UMC																				
Lower middle income	LMC	78.89	78.57	79.13	77.13	76.48	76.60	78.39	78.58	77.43	76.33	74.53	73.09	72.29	72.07	71.55	70.86	69.05	69.07	68.84	68.62
Low & middle income	LMY																				
Low income	LIC																				
Heavily indebted poor countries (HIPC)	HPC						74.96	74.32	74.19	72.86	71.92	71.04	72.77	72.03	71.39	71.35	74.06	73.31	72.44	72.71	73.03
High income: OECD	OEC											57.66	57.67	57.75	57.23	57.57	58.73	58.57	58.55	58.11	58.27
OECD members	OED											58.22	58.27	58.28	57.75	58.09	59.18	58.97	58.91	58.51	58.63
High income: nonOECD	NOC																				
Small states	SST																				
Sub-Saharan Africa (all income levels)	SSF	81.23	80.04	78.28	77.49	78.09	67.72	66.82	66.62	65.65	65.80	65.48	64.06	63.10	63.68	62.68	63.01	57.81	58.93	59.04	60.23
South Asia	SAS						76.64	77.27	78.56	77.64	76.73	76.18	74.10	74.59	76.65	76.16	75.66	62.73	59.54	60.41	59.09
																		74.04	73.99	73.11	73.19
Latin America & Caribbean (all income levels)	LCN	71.43	70.44	72.83	71.56	71.99	70.34	70.65	71.60	71.73	70.14	69.06	69.93	69.64	68.17	67.54	66.59	66.91	66.45	66.22	65.92
Middle East & North Africa (all income levels)	MEA									49.89	50.14	50.10	46.89	44.68	42.43	36.21	40.35	39.87	45.08	47.78	47.85

INDICATOR_CODE
NE.CON.PETC.ZS

INDICATOR_NAME
Household final consumption expenditure, etc. (% of GDP)

SOURCE_ORGANIZATION
World Bank national accounts data, and OECD National Accounts data files.

SOURCE_NOTE
Household final consumption expenditure (formerly private consumption) is the market value of all goods and services, including durable products (such as cars, washing machines, and home computers), purchased by households. It excludes purchases of dwellings but includes imputed rent for owner-occupied dwellings. It also includes payments and fees to governments to obtain permits and licenses. Here, household consumption expenditure includes the expenditures of nonprofit institutions serving households, even when reported separately by the country. This item also includes any statistical discrepancy in the use of resources relative to the supply of resources.

Household final consumption expenditure as % of GDP

Country Name	Code	1980	1981	1982	1983	1984	1985	1986	1987	1988	1989	1990	1991	1992	1993	1994	1995	1996	1997	1998	1999
World	WLD	59.44	59.28	59.56	60.06	59.53	59.69	59.74	59.63	59.47	59.10	59.07	59.31	59.50	59.73	59.66	59.63	59.91	59.67	60.04	60.24
United States	USA	61.29	60.34	62.00	62.85	61.82	62.64	63.14	63.49	63.72	63.50	63.98	64.14	64.47	65.00	64.87	65.03	65.04	64.60	64.95	65.35
United Kingdom	GBR	57.65	58.02	58.46	58.77	59.15	58.84	60.84	60.77	61.48	61.43	62.07	62.61	63.30	64.09	63.54	63.17	63.90	64.00	64.49	65.12
European Union	EUU	58.13	58.79	58.93	58.87	58.69	58.62	58.41	58.61	58.17	58.17	57.71	58.07	58.53	58.89	58.60	58.16	58.49	58.31	58.34	58.53
Germany	DEU	58.43	59.55	59.84	59.74	59.94	59.74	58.05	58.33	57.92	58.27	57.65	57.51	57.62	58.29	57.95	57.69	58.06	58.08	57.67	58.09
France	FRA	56.75	58.31	58.56	58.31	58.36	58.88	58.28	58.98	57.85	57.55	57.47	57.34	57.37	57.60	57.27	56.97	57.42	56.37	56.18	55.90
Canada	CAN	53.14	52.51	53.14	53.73	53.55	54.15	55.28	55.15	54.46	54.83	55.81	57.07	57.48	57.53	56.59	55.70	56.17	56.78	56.90	56.27
Switzerland	CHE	62.53	62.32	62.39	63.32	62.06	61.38	60.88	60.86	59.58	58.37	57.57	59.08	60.37	60.36	59.54	60.06	61.09	61.12	61.03	61.67
Israel	ISR	53.11	54.87	57.80	59.32	54.64	57.98	62.91	64.33	63.06	62.21	55.60	60.42	60.54	62.43	63.37	55.63	54.86	53.54	53.86	53.95
Ireland	IRL	65.17	65.29	59.54	59.56	59.07	59.80	59.95	59.58	59.87	59.74	57.33	57.73	57.62	56.00	55.87	53.02	51.90	49.49	49.62	47.57
Spain	ESP	64.57	64.97	64.65	63.99	62.60	62.33	61.60	61.59	60.85	61.11	60.55	60.46	61.09	60.97	60.75	60.08	59.83	59.66	59.51	59.59
Italy	ITA	59.03	59.08	59.25	58.51	58.60	58.57	58.58	58.33	57.85	58.36	57.20	57.48	58.43	57.95	58.46	58.32	57.86	58.45	59.13	59.83
Poland	POL											48.30	58.50	61.12	62.61	62.58	60.44	62.27	62.73	62.46	63.22
Portugal	PRT	65.33	66.74	66.79	67.97	70.45	68.57	65.91	64.85	64.38	63.80	64.38	65.22	66.08	67.84	66.93	65.24	65.25	64.21	63.28	63.51
Greece	GRC	65.14	66.74	66.09	67.53	64.86	63.83	64.58	69.27	69.35	70.49	71.84	71.56	73.34	73.69	73.48	73.29	73.93	72.48	72.08	71.14
Mexico	MEX	65.07	64.39	61.61	60.86	63.08	64.52	68.45	65.84	67.64	68.85	69.58	70.50	71.81	69.83	70.52	65.67	65.85	67.18	67.03	66.98
Brazil	BRA	69.71	67.98	69.57	71.24	70.35	65.78	67.77	62.26	59.49	54.13	59.30	61.57	61.52	60.08	59.64	62.46	64.66	64.88	64.33	64.73
Russian Federation	RUS										45.15	48.87	46.94	37.46	45.24	50.80	52.09	52.61	54.75	59.63	53.54
India	IND	74.86	70.63	69.38	71.38	69.00	67.39	66.19	66.21	65.36	65.05	64.61	65.87	65.02	67.13	66.14	64.03	68.22	65.11	65.60	62.26
China	CHN			49.55	49.97	49.80	51.35	49.40	48.99	50.01	49.98	46.98	45.61	45.90	41.91	41.13	42.68	43.51	43.35	43.96	45.13
Singapore	SGP	52.24	49.23	44.83	42.23	42.62	45.16	48.09	48.95	47.11	45.12	44.80	43.59	43.57	43.96	42.93	41.32	39.86	38.63	37.52	40.96
Saudi Arabia	SAU	25.08	26.62	36.43	45.20	47.87	53.22	54.96	53.33	53.26	51.29	46.68	44.24	44.43	48.04	47.80	46.91	43.92	42.31	45.99	41.79
United Arab Emirates	ARE																				
Qatar	QAT															29.91	32.05	27.28	22.70	25.31	21.11
High income	HIC	58.38	58.22	59.16	59.61	58.97	59.18	59.25	59.38	59.15	58.79	58.79	59.08	59.23	59.82	59.84	59.73	59.88	59.68	59.99	60.23
Upper middle income	UMC			58.67	59.27	59.39	59.54	59.73	58.18	58.35	58.19	57.74	57.58	58.18	55.92	55.50	55.99	56.33	56.39	56.80	57.20
Lower middle income	LMC	70.16	68.52	68.31	69.40	68.90	68.28	67.66	67.55	67.57	65.94	66.51	66.96	66.69	67.80	67.22	67.04	69.53	67.77	69.08	68.16
Low & middle income	LMY			61.43	62.15	62.13	62.07	62.04	60.81	60.92	60.54	60.33	60.34	60.71	59.32	58.85	59.19	60.03	59.66	60.24	60.32
Low income	LIC		82.64	83.68	83.15	83.47	82.08	80.60	80.95	79.84	80.96	80.66	81.09	80.82	81.67	80.44	80.99	80.85	81.12	80.06	79.89
Heavily indebted poor countries (HIPC)	HPC	76.29	77.37	77.90	77.68	76.63	76.38	74.88	76.09	78.35	79.21	76.92	78.38	79.91	79.60	78.16	78.33	77.98	78.53	78.45	78.99
High income: OECD	OEC	58.80	58.61	59.47	59.89	59.24	59.38	59.46	59.65	59.35	59.24	59.20	59.45	59.91	60.32	60.28	60.11	60.29	60.04	60.23	60.64
OECD members	OED	59.17	58.98	59.72	60.14	59.58	59.75	59.89	59.91	59.62	59.58	59.55	59.78	60.25	60.61	60.58	60.32	60.51	60.28	60.43	60.86
High income: nonOECD	NOC										50.34	51.52	52.46	47.18	50.87	52.14	53.01	52.71	53.29	55.86	53.00
Small states	SST	57.89	61.26	63.26	63.51	62.38	60.65	60.16	59.77	59.59	60.04	61.40	61.58	62.26	63.31	59.89	59.08	58.18	63.79	64.97	63.03
Sub-Saharan Africa (all income levels)	SSF	58.86	61.88	65.06	64.04	65.05	62.25	62.36	63.86	64.41	62.19	64.06	64.75	67.83	67.56	65.57	68.68	69.46	69.59	71.66	69.90
South Asia	SAS	77.02	73.22	72.33	73.70	71.81	70.55	69.23	68.83	68.11	67.58	67.38	67.70	67.07	68.91	67.97	66.39	69.84	67.40	67.31	65.07
Latin America & Caribbean (all income levels)	LCN	66.37	66.09	66.01	66.85	66.99	65.82	67.92	66.00	65.27	64.36	66.07	67.99	68.94	67.32	67.17	66.18	66.54	66.61	66.73	66.49
Middle East & North Africa (all income levels)	MEA	45.90	48.04	52.19	53.97	54.83	57.79	61.17	58.55	61.26	60.93	58.09	60.46	56.00	56.50	55.24	54.52	52.93	52.82	56.00	53.00

Household final consumption expenditure as % of GDP

Country Name	Code	2000	2001	2002	2003	2004	2005	2006	2007	2008	2009	2010	2011	2012	2013	Range High	Range Low	Since
World	WLD	59.92	60.42	60.51	60.28	59.77	59.58	59.09	58.97	59.36	60.26	59.94	60.29	60.37		60.51	58.29	1960
United States	USA	66.10	66.89	67.26	67.44	67.26	67.13	67.09	67.29	67.97	68.27	68.20	68.96	68.64		68.96	59.58	1965
United Kingdom	GBR	65.62	65.70	65.63	64.95	64.65	64.66	63.86	63.63	63.51	64.37	64.53	64.57	65.40	65.88	65.88	56.99	1965
European Union	EUU	58.66	58.62	58.38	58.40	58.12	58.24	57.73	57.03	57.32	58.46	58.42	58.48	58.62	58.66	58.93	57.03	1970
Germany	DEU	58.37	58.68	58.18	58.88	58.46	58.76	57.89	55.87	56.17	58.65	57.52	57.41	57.53	57.46	59.94	54.58	1970
France	FRA	56.20	56.51	56.43	56.84	56.62	56.91	56.73	56.49	56.88	58.10	58.09	57.72	57.68	57.46	58.98	55.50	1970
Canada	CAN	54.58	55.04	55.95	55.64	54.64	54.12	54.10	54.36	54.14	57.22	56.53	55.70	55.66	55.72	58.78	52.51	1965
Switzerland	CHE	60.58	60.60	60.57	60.79	60.40	59.88	58.21	56.81	56.47	58.47	57.94	57.33	57.41	57.28	63.82	56.23	1965
Israel	ISR	53.89	55.48	56.32	55.87	56.28	55.75	55.69	56.82	57.10	56.27	56.88	57.28	56.23		70.70	53.05	1960
Ireland	IRL	48.06	46.23	44.71	44.52	43.96	44.71	45.71	47.72	50.06	48.30	50.27	48.79	46.94		68.81	43.96	1970
Spain	ESP	59.70	59.14	58.34	57.65	57.93	57.77	57.45	57.41	57.21	56.62	57.87	58.57	59.33	59.25	64.97	56.62	1970
Italy	ITA	59.93	59.10	58.71	59.06	58.62	59.02	59.04	58.62	59.17	60.31	60.83	61.27	60.93	60.38	61.27	57.20	1965
Poland	POL	64.13	65.00	66.95	65.82	64.69	63.39	62.48	60.50	61.56	61.07	61.26	61.11	61.42	60.80	66.95	48.30	1990
Portugal	PRT	63.60	63.12	62.88	63.29	64.02	64.72	65.12	65.34	66.84	65.14	65.94	66.02	65.71	64.56	77.57	62.88	1970
Greece	GRC	69.70	69.95	70.34	67.97	67.66	69.79	69.70	69.59	72.34	72.35	73.40	74.60	73.64	72.41	95.25	47.77	1965
Mexico	MEX	66.82	68.93	70.00	68.24	68.43	68.48	67.31	67.65	66.99	66.61	67.52	67.42	66.27	67.18	79.25	60.86	1960
Brazil	BRA	64.35	63.47	61.72	61.93	59.78	60.27	60.30	59.90	58.93	61.11	59.64	60.33	62.62	62.62	71.54	54.13	1960
Russian Federation	RUS	46.19	48.94	51.20	49.85	49.87	49.36	48.71	49.92	47.43	52.85	50.58	48.43	49.09	51.99	59.63	37.46	1989
India	IND	64.22	62.93	64.10	63.12	58.37	57.59	56.96	55.69	58.61	57.18	56.40	58.57	60.28	61.77	80.83	55.69	1960
China	CHN	46.69	45.65	43.97	41.86	40.12	38.11	35.22	35.94	34.94	33.94	34.73	35.90	34.78	34.09	51.35	33.94	1982
Singapore	SGP	42.08	44.25	44.91	43.27	40.01	38.62	37.53	36.54	38.13	38.52	35.52	36.57	37.82	37.74	94.79	35.52	1960
Saudi Arabia	SAU	36.53	37.82	36.83	33.55	30.39	26.34	26.03	27.85	26.88	36.78	32.37	27.15	30.34	29.39	54.96	11.01	1968
United Arab Emirates	ARE			63.51	60.55	62.39	58.27	57.76	61.74	61.34	54.56	58.96	51.69	49.75		63.51	49.75	2001
Qatar	QAT	15.23	15.44	18.42	16.50	17.46	15.96	16.15	16.34	15.31	19.28	13.33	10.68	12.45		32.05	10.68	1994
High income	HIC	59.95	60.50	60.72	60.64	60.31	60.26	60.00	59.79	60.23	61.28	61.01	61.29	61.36		61.36	56.86	1970
Upper middle income	UMC	57.25	57.45	56.47	55.72	54.52	53.44	51.66	51.96	51.46	52.16	51.93	52.39	52.12	51.96	59.73	51.46	1982
Lower middle income	LMC	65.74	66.47	67.32	66.07	64.19	64.25	63.16	63.68	65.06	63.82	62.55	63.81	64.99	66.44	79.13	62.55	1960
Low & middle income	LMY	59.75	60.06	59.59	58.71	57.37	56.60	55.04	55.34	55.36	55.57	55.03	55.69	55.75	55.77	62.15	55.03	1982
Low income	LIC	78.79	78.73	80.33	79.47	78.78	79.51	79.88	79.44	80.62	79.33	77.98	79.17	78.03		83.68	77.98	1981
Heavily indebted poor countries (HIPC)	HPC	75.77	75.92	77.22	76.48	74.63	76.43	75.64	74.98	74.84	75.56	73.49	71.71	73.42		79.91	71.04	1965
High income: OECD	OEC	60.58	61.06	61.25	61.27	60.95	60.99	60.76	60.47	60.98	61.90	61.71	62.10	62.14		62.14	57.23	1970
OECD members	OED	60.83	61.30	61.52	61.53	61.23	61.27	61.00	60.75	61.20	62.10	61.93	62.34	62.33		62.34	57.75	1970
High income: nonOECD	NOC	48.87	51.54	52.01	50.39	49.92	48.30	47.67	48.68	48.22	51.16	49.47	47.76	48.18	50.63	55.86	47.18	1989
Small states	SST	60.09	61.13	60.86	59.02	60.39	57.21	56.77	60.18	64.49	69.46	68.70	68.86			69.46	56.77	1976
Sub-Saharan Africa (all income levels)	SSF	63.25	67.98	67.89	68.09	67.01	67.46	65.14	67.37	66.07	68.32	64.43	63.64	64.44	64.17	71.66	58.86	1965
South Asia	SAS	66.52	65.73	66.61	65.84	62.00	61.67	61.16	60.20	63.10	61.42	60.85	62.90	64.33	65.27	81.23	60.20	1960
Latin America & Caribbean (all income levels)	LCN	66.35	67.15	66.22	65.63	64.25	64.00	63.22	63.41	63.15	64.25	63.71	63.74	64.73	65.31	72.83	63.15	1960
Middle East & North Africa (all income levels)	MEA	48.30	51.47	51.45	49.81	49.20	46.26	45.03	45.96	45.00	49.07	47.62	44.55	45.32		61.26	36.21	1968

		95.25	Greece
		10.68	Qatar

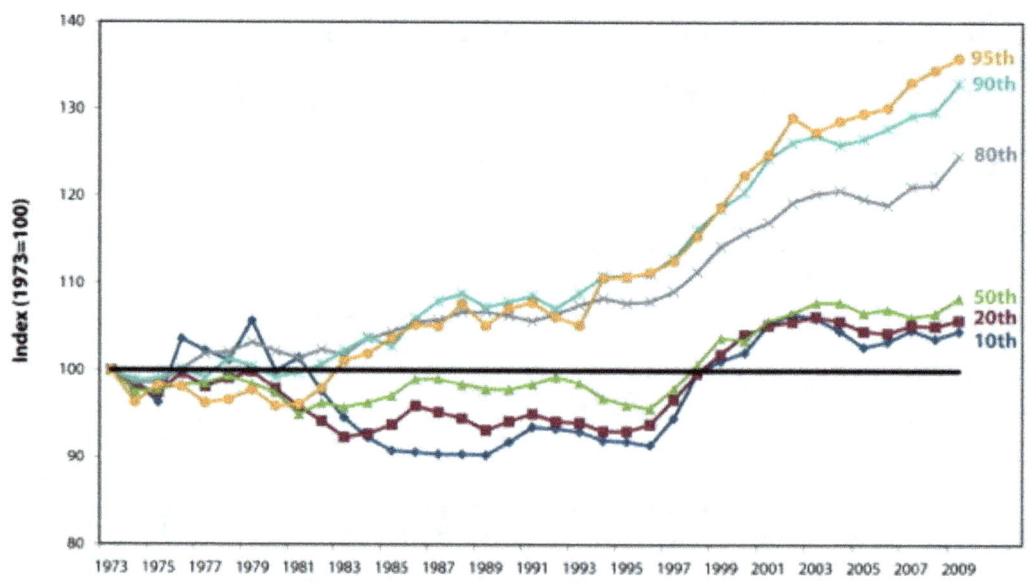

income producing and high end real estate and other assets, since in periods of disinvestment money is concentrated, and seeks narrower investment opportunities in real and financial asset purchase. In addition, the bidder base for much of these assets and services was global, compounding the appreciation.

The shift overseas for higher margins is inseparable from the flight from more expensive and restrictive domestic labor, with governmental regulation and taxation in an atmosphere of declining willingness for tariff protection, to an environment of newer and more productive facilities abroad with an often greater tolerance of externalities or social costs.

We turn now to the Federal Reserve Statistical Release of September 18, 2014, to see whether it bears this out. The attached spreadsheets taken from that release compare annual data for 1975 against 2005 and 2013 with respect to structural changes in the economy as measured by account percentages of the corresponding GDP and, as applicable, asset or net worth values. This information comes from the Z.1 Financial Accounts of the United States and incorporates the following:

Sheet 1	- Distribution of US Gross Domestic Product,
Sheet 2	- Distribution of US National Income,
Sheet 3	- US Savings and Investment by Sector

Balance Sheets of the seven sectors;

Sheet 4	- Household and Nonprofit Organizations,
Sheet 5	- Nonfinancial Corporate Business,
Sheet 6	- Nonfinancial Noncorporate Business,
Sheet 7-8	- Financial Business,
Sheet 9-10	- Federal Government,
Sheet 11	- State and Local Government,
Sheet 12-13	- Rest of the World
Sheet 14-15	- US Credit Market Debt and Total Liabilities and its Relation to Total Financial Assets
Sheet 16-21	- US Sector Account Breakdown Comparison (from Bal. Shts.)
Sheet 22	- US Assets and Liabilities of the Personal Sectors
Sheet 23	- US Assets of the Public and Private Sectors
Sheet 24-25	- US BS of the Federal Government with the addition of Human Capital at Market Value, Natural Resource and Infrastructure Value Allocations
Sheet 26	- US BS of State & Local Government with addition of Land and Infrastructure
Sheet 27	- US Assets of the Public and Private Sectors with the addition of Human Capital at Market Value, Natural Resource and Infrastructure Value Allocations to the Public Sectors
Sheet 28-29	- US BS of the Federal Government with the addition of Human Capital at Total Value, Natural Resource and Infrastructure Value Allocations

Sheet 30 - US Assets of the Public and Private Sectors with the addition of Human Capital at Total Value, Natural Resource and Infrastructure Value Allocations to the Public Sectors

1975 was the last year that the United States ran a trade surplus, approximately 1% of that year's GDP, and was at the end of the first round of inflation about three years before the major inflation of 1979 to 1981 and before the start of the Reagan years and the embrace of laissez-faire macroeconomic policy. 2005 was near the peak of the asset bubble after 25 years of the supply side reign and a couple of years before its anticipated and realized crash.

Sheet 1 is abstracted from table F.6 for the distribution of gross domestic product. The data is arranged by year with the first column showing the detail amounts in billions of US dollars. The annual GDP is at the top of the column and the column to its right shows the amount for each account line as a percentage of the GDP. 1975 is first, followed by a corresponding pair for 2005, which is then followed by a column stating the change in line percentage for 2005 from the 1975 proportion. This is followed by the amounts and percentages for 2013, with the percentage variation for 2013 from 2005 and finally from 1975. This arrangement will be followed for each of the sheets with the addition of columns showing each entry as a percentage of total asset, net worth or account value for the seven sector balance sheets, credit market debt sheets, and sector account comparisons respectively. These are followed by a sheet showing assets and liabilities of the personal sector, taken from table L.10 of the Fed release and one compiled from various tables of the assets and liabilities of the public and private sectors.

Finally, a similar sheet for the public and private sectors is shown with the inclusion of values in the public sector for land, infrastructure and human capital to complete the picture of the national economy. The infrastructure values are a broad brush stroke based on recent figures from the American Society of Civil Engineers, found at http://www.infrastructurereportcard.org/a/#p/grade-sheet/americas-infrastructure-investment-needs, for the estimate of repairs required to bring existing infrastructure by the year 2020 to a B grade level, defined as "good: adequate for now", from the current grade of D+, defined as "poor: at risk". These have been extrapolated to a value for the installed base, using the repair estimate of $3.6T as an assumed value of one half the cash value of the current infrastructure. This indicates a market value for the total infrastructure of $7.2T, of which one third has been allocated to federal and two thirds to state and local balance sheets. With a total real estate value for the private sector of $43.3T, this may be somewhat low, but is a necessary first guess. The ASCE had no figures for the installed base of national infrastructure.

Land was figured on 650M acres of public land at $3,000/acre, with an assumed state and local figure of one-tenth that amount, which is probably on the high side.

The human capital account, as shown in the final balance sheet of the federal government, is based on a report by Michael S. Christian, entitled "Human Capital Accounting in the United States, 1994-2006", published by the U.S. Department of Commerce, Bureau of Economic Analysis, which can be found at http://bea.gov/scb/pdf/2010/06%20June/0610_christian.pdf. The report derives a market value and a nonmarket value for the national human capital stock, and states the amounts over the years 1994-2006 in both real and nominal terms. The nominal values were used for 2005, for both market value at $200T and total value at $667T, and extrapolated to 1975 and 2013. These values are properly included in the public account as a statement of national productive capacity, whose market value is realized in future balance sheets of both public and private sectors. The figures derived by Christian are borne out by this analysis for the market value of human capital. The nonmarket value of human capital, which is for the productive capacity that is not reflected in earnings of the individuals and valued by non-market time, i.e. raising a family, social activity, charity work, is approximately 70% of the total, and indicates a national asset value for 2013 for the US of over a quadrillion dollars, of which the national debt was less than 1.4%. An amount for both documented and undocumented aliens has been added to these sheets for 2005 and 2013 as well.

With respect to GDP, the percentage for personal consumption expenditures in 1975 was 61.2%, very close to the theoretical optimum of (0.15) at 61.8% and just above the current world average. The Fed document is primarily concerned with financial flows, and there is only a basic breakdown for consumer spending, but we do see that non-durable goods, the things that are readily produced and consumed over the short term was 20.7%, with services at 32.1%, making up over half of the GDP, with a balance of 8.4% for durable goods. The remaining 38.8% of GDP was for private investment at 15.2%, net exports at 0.9% and government consumption, federal at 7.9% and state and local at 9.8%, and government investment, federal at 2.4% and state and local at 2.6%, for a total government expenditure of 22.7%, over half of which was state and local.

We will deal with government first. After thirty years of laissez-faire policy and in the midst of the Iraq war, the 2005 total government expenditures had fallen to 19% of GDP or by 3.6%. Of this total, 3.0% was federal and 0.6% was state and local. In 2005 consumption expenditures had risen to 67.2% or by 6.0% from 1975, though paradoxically nondurable goods, a mix of the basics and no doubt some of the finer things in life, had fallen by 5.8% to 14.9% of the total or by 38.9% from its original proportion. Services rose by 11.6% to 43.6% of GDP or by 36.1% from the original. Consumer durables, cars and refrigerators and the like, stayed about even, rising only 0.2%.

The difference in personal consumption expenditures as a percentage of GDP was offset by a major shift in the balance of trades of 6.5% to a deficit 5.5% of the total, a reduction in public expenditures of 3.6%, and an increase in gross private domestic investment of 4.1%. Contrary to supply side prognostications, however, over this 30 year period in which fixed investment rose 3.2%, most of which was residential investment by the household sector, that is a homeowner investment increase of 2.5% of GDP and nonprofit investment in households of 0.3%, nonfinancial corporate business investment rose by only 0.1% and nonresidential noncorporate business, of which mom and pop businesses constitute a significant part, actually fell by 0.4% of GDP. Fixed investment for financial institutions did rise by 0.6% and business inventories rose 0.8% indicating excess capacity as outlined previously. (An "x" to the left of a percentage change column in these tables merely indicates a point of interest. Percentage change totals may vary by 0.1% due to rounding discrepancies.) In the public sectors, reductions consisted of 2.6% in government consumption expenditures and 1.1% in government investments.

By 2013, in the aftermath of the financial crises of 2007 and 2008, consumer expenditures had risen another 1.3% to a new high of 68.5% of GDP. Durable goods fell this time by 1.2%, though nondurables rose by 0.6%. Both were still below their 1975 levels, nondurables by 5.2% of GDP. The bright spot remained services, medical, legal, financial, etc, up another 1.9% to 45.5% of GDP. Government managed to drop another 0.3%, all of it and more by state and local government. The federal government had a net increase of 0.1%. Private domestic investment adjusted by divesting down 3.5% of GDP, principally in the residential sector, 3.1%, along with nonprofits, 0.2%, and financial institutions, 0.2%, but corporate and noncorporate business did improve slightly, by 0.3%. And the balance of trades deficit fell by 2.5% of GDP. Private inventories rose in absolute terms, but remained level as a percent of GDP.

What does the related distribution of national income tell us? Between 1975 and 2005, wages and other labor income fell by 2.6% of GDP and another 1.2% by 2013. In contrast, combined proprietor's and corporation's business and rental income rose by 4.1% of GDP over the thirty years, but even more telling, in the aftermath of the crisis it rose by 3.5% between 2005 and 2013, in almost a quarter the time of the first period. After-tax profits for corporations went up 1.0% of GDP from 2005 to 2013. Tellingly, from 1975 to 2013, proprietorship profits rose by only 1.0%, while rental incomes rose by 2.2% and corporate profits rose 4.3% of GDP. Most significantly, domestic financial and rest of the world accounts were 2.0% and 1.5% of the corporate total. The only negative on the earnings side was in interest income, which has fallen by 2.8% of GDP over the 38-year period.

A look at the savings and investment picture reveals that there was only a slight increase in savings as a percentage of GDP for domestic business between 1975 and 2005, up just 0.5%, with a slightly greater adjustment in investment of 1.3% of GDP. The household picture was much different, however. While the household investment picture rose by 2.8% over this period, the savings fell by 7.6% of GDP and household borrowing switched from a net lending position of 7.2% to one of borrowing at 1.8% a shift of almost 9.0% of GDP. The popular portrayal of this fact is that the citizenry is made up of undisciplined spenders, but the correct analysis is that globalization has made it difficult to make ends meet for the majority of workers, while those with the wherewithal found it advisable to borrow to capture the ongoing rise in asset appreciation.

The period between 2005 and 2013 has resulted in disinvestment across the board, more heavily for households and institutions than for business, but the former sector has cut borrowing by 4.5% and as of 2013 saves more than nonfinancial corporations.

Let's look at the balance sheets now. In 1975 total household and nonprofit assets were 388.4% of the year's GDP. In 2005 the same sector assets were 565.5% of that current year GDP and for 2013 552.7%. For the nonfinancial corporate business sector, the sector assets as a percent of GDP for the same years were 172.1%, 199.1% and 208.4% and for the nonfinancial noncorporate business sector, i.e. proprietorships, were 88.7%. 96.5% and 89.6%, all reasonably stable. In contrast, financial business assets of 193.1% in 1975 more than doubled to 438.9% in 2005 and 492.9% in 2013.

We can look at this another way and think of GDP as a return on the total assets of the economy by inverting the percentages. In this case, for the household sector, 1975 was a better year at 25.7% than 2005 at 17.7% though 2013 was an improvement over 2005 at 18.1%. For the nonfinancial corporate sector things were similar, starting out at in 1975 at 58.1%, with 2005 at 50.2%, but continued dropping slightly to 48.0% in 2013. For proprietorships, however, the thing is more interesting, starting out in 1975 at 112.8%, dropping to 103.6% for 2005, but rising back almost to the initial rate at 111.6% in 2013. For financial businesses, the figures are similar to the nonfinancial corporate sector, starting out at 51.8% in 1975, but then drop significantly to 22.8% in 2005 and to 20.3% in 2013. In other words, by this broad-brush approach, all sectors where 8 to 10 points more productive during 1975 than during 2005 and only the corporate sector contributions, nonfinancial and financial, were less productive in 2013 than in 2005. The financial sector in particular appears less productive and bloated in 2005 and 2013 compared with 1975.

In his recent book, Capital in the Twenty-First Century, Thomas Piketty uses a similar approach in contrasting an economy's wealth to the GDP, though he uses the closely related and approximate figure of National Income instead of the GDP. Though I am but half way through a read of this commendable work as of this writing, he reports national wealth over a three hundred year span for several important domestic economies to show the level of wealth in relation to national income. The implication appears to be that income tends toward a natural level of return on capital and thereby wealth generation, except for conditions of governmental intervention, notably around the period of World War I and II. While this is no doubt true, the position taken in this analysis is somewhat reversed in that the return on national capital in unregulated markets is shown to be appreciably less that the return on that capital with proper public oversight and intervention. The distribution of that return is necessarily different, but such oversight is more productive and more egalitarian. Less oversight and involvement leads to commoditization of labor, disinvestment in human capital and infrastructure and inflation of private asset valuation.

As for the asset breakdown, for households and nonprofits over the 30-year period from 1975 to 2005, nonfinancial assets rose by 3.7% of total sector asset value, principally in real estate at 6.9%, though this fell for nonprofits, and consumer durable goods fell by 3.3%. Interestingly, this rise in nonfinancials for the sector and the corresponding offset by the same amount in a decline in financial assets, contrasts with the large contrary decrease in nonfinancial assets in the nonfinancial corporate and noncorporate sectors at 15.0% and 19.0% respectively. While there were noteworthy increases in the proportions of mutual fund shares, followed by pension entitlements, corporate equities, municipals, and a few others, these were more than offset, significantly by noncorporate business equity, time, savings and checking deposits. Liabilities were an additional 4.8% of asset value in 2005, primarily due to mortgages, though consumer credit was slightly lower, so that net worth was 4.8% less as well. Perhaps most interesting in the corporate balance sheet is the Miscellaneous assets line which rose from 7.3% in 1975 to 22.9% in 2005 and to 21.6% or over one third the corporate equities market value in 2013. Miscellaneous liabilities followed suit at roughly half the asset value.

For the financial business balance sheet, in which nonfinancial assets represent 2.7% of total asset value in 1975, followed by 2.3% in 2005 and 1.9% in 2013, not surprisingly net worth fell from 4.8% of total asset value in 1975 to a negative 2.0% in 2005 before the crash with a slight recovery to a negative 0.6% in 2013.

Yet by 2013 things are dramatically different for the household balance sheet, with an 8.7% swing in asset valuation from nonfinancial to financial assets, the first of which dropped from 38.35% to 29.9% of total asset value for households and nonprofits, the total asset value having appreciated $18.6T from $74.1T to $92.7T or by 25% of its 2005 value. According to the Fed's report, the nonfinancial assets depreciated by $863B but the financial assets grew by $19.5T, more or less equally in pension entitlements, mutual funds, corporate equities, and liquidity. What does this tell us about the valuation of the pension entitlements, mutual funds, and corporate equities down the road? The ratio of financial to nonfinancial assets has gone from 1.60 to 2.34 in 8 years. In the same period of time, the equity in the noncorporate business sector, which includes most of the small businesses, which do most of the hiring, has gone up only $585B.

Pension entitlements have to be turned into cash at some point, as do mutual funds and equities. What are the underlying assets? Mutual funds and equities are easy enough to understand if they represent a productive investment somewhere, something making consumables that can be sold for profit. The pension entitlements in 2013 were at $19.9T, presumably deriving their value from future corporate earnings. Corporate net worth is $19.1T, but it must continue paying dividends if it is to keep up its market value, rationally speaking. And during this time, 2005 to

2013, when the financial asset component of household equity has effectively risen by $19T, while real asset value diminished by $0.9T, the financial sector on which this wealth is ostensibly based, has managed to cut its negative net worth by only $0.7T. But then one person's liability is (or was) someone else's asset. Fortunately (or not, based on your perspective), one person's asset does not have to be another person's liability.

This is perhaps not as gloomy as it appears since financial assets consist primarily of two types, those which are valued for their immediate or short term use as a medium of exchange for goods and services and those that exist as a combination of an intended long term store of value and source of future cash flow for future exchange for goods and services. As such the value of the financial business and its negative net worth is primarily in its future appreciation and dispensing of cash, still in 8 years it has only appreciated 0.6%. We might be tempted to shake this off as a result of the financial crisis of 2008 were it not for the fact that during the 30 years of the heyday of supply side policy, it has lost 6.9% of its relative asset value and entered into negative net worth territory at the same time going from a positive 9.3% of GDP to a negative 9.0% of GDP. The supply siders would like to convince us and perhaps even themselves that the financial crisis and its "worsening" is a result of government debt and deficits and the public sector getting in the way of the private, but it is the lack of investing in the public sector that has brought this on. Any improvement since the crisis is the result of government intervention, which could and should have been much stronger.

Now look at Sheet 14 for US Credit Market Debt where the net worth of all three sectors, instead of assets, is displayed across the top. This net worth line is not in the corresponding Fed table, but has been added to elucidate. Look at the inverted figure of net worth as a percentage of GDP, in other words GDP as a return on net worth. For 1975 it is 19.3%, for 2005, 15.4% and for 2013, up a bit at 15.7%. If the stock market keeps going up, with more guys needing pensions, that number is going to go down even further. Now look at the corporate equities line near the bottom. 1975 equities were valued at roughly 50% of GDP, 2005 were at 157% even though the return on net worth for the economy as a whole near the peak of the greatest run in asset value in recent history at 15.4% was 4% less than it was in 1975 when the equity value was only 50% and constrained by the government strangle hold on the economy. Now look at 2013 with equity valued at 200%, yet return on aggregate net worth is only 0.3% higher than in 2005. Is any body listening?

Now for a structural analysis, look at Sheet 16 of the US Sectors Accounts table where the values for the seven sectors of the economy are arranged for assets, with nonfinancial and financial breakdown, followed by liabilities and new worth. Note that the second percentage column and the corresponding differentials state each sector as a percentage of each particular account total. For total assets, 2005 sector percentages before the crisis were all down, with the exception of financial business and the rest of the world account, which were up 9.7% and 4.3% respectively from

1975 proportions. We might be tempted to think this simply means that the economy doesn't need as much of the other things, but remember that financial instruments are ultimately only worth what ever goods and services they can be exchanged for somewhere down the line.

Looking at the nonfinancial assets tells a somewhat different story, with nonfinancial business assets down 9.3%, government property down 6.2% and financial real assets up only 0.9%. Rest of the world does not have any real assets. The household sector is up 14.6% due to the housing bubble and the general run up of real estate and major durable pricing. The household financial assets are down almost as much at 13.6%, due in no small part to the down turn in deposits seen in the financial assets breakdown, and indicating the liquidity crisis that soon followed. The financial sector financial assets on the other hand are up 7.1% along with the 5.9% increase in rest of the world holdings.

In fact, the financial sector along with the necessarily related rest of the world sector are the only sectors that grew across all accounts in their share of structural change, including that of liabilities. This trend continues to the 2013 figures. However with regards to net worth, as previously indicated, for 2005 the financial sector shows a decrease in share, even greater than the federal government. The exception is the rest of world account, which continues to grow, presumably a result of the emergence of China and equity holders in the advanced economies. But the pull back of the financial sector with respect to the structure in 1975 while the federal government position is three times as extended as it was in 2005 is where we should be concerned. These figures show that a financial bubble still exists and that there will be even more intensified, though more ill advised, attempts to blame it on government spending and the accumulation of public debt.

Equally significant is the change in net worth accounts as a percent of the total for the rest of the sectors. In contrast to 1975, only two sectors have increased their share of the total net worth of the nation. The first is the household sector by 11.1% and the second is the rest of the world by 5.0%. The household sector, of course, doesn't mean everyone in that sector has benefited as is well documented. We are stating nothing new here. What is perhaps more starkly presented is the 3.9% decrease in proprietorships and the 1.7% decrease in nonfinancial corporate percentage since 1975. What this shows is the concentration and privatization of both public <u>and</u> private wealth to interests both domestic and international.

To shine some sunlight on this situation, we go to Sheet 22 for the US Assets and Liabilities of the Personal Sector table taken from the Fed release, table L.10. That table consolidates household, nonprofit and nonfinancial noncorporate balance sheets, leaving out the corporate, nonfinancial and financial, other world, and of course the public sectors. We have added a column in the financial assets and liabilities accounts to indicate the presumed nature of the financial asset as to whether it represents liquid or long term assets. Basically the liquid assets are those included in the Deposits account on Sheet 18, page 5 of the Financial Assets

Breakdown table. Note the asset, liability and net worth totals for these two sectors. For 2013 the total asset value is $107.7T up from $86.7T in 2005 after 8 years of very little real growth. This does not include the $34.9T in nonfinancial corporate assets. The liability for 2013 is $19.8T with net worth of $87.9T, which with a corporate net worth from the earlier table of $19.1T puts us back up to $107.0 net worth for the three sectors.

We will use this information to consolidate our balance sheets for both the private and public sectors and then for the overall national balance sheet. We will need to state a few axioms, first one being that real assets have a real value that is exclusively based on the human effort required to produce or reproduce them and they have a financial value that is relative to what they can be traded for in money in the marketplace, and the second being that in the market place all liabilities are financial and not of in-kind goods and services, even though the value in long-term instruments is their exchangeability for consumer products. In the end this doesn't matter as collateral transfer due to default merely changes the party that has title to the property and does not change the value of the property itself. On the other hand it does matter in that the utility of the collateral or defaulted contractual agreement will likely be markedly different in effect on the debtor and on the creditor.

However, for our present analytical purpose, liabilities for one consolidated sector are exclusively financial assets to some other sector or sectors, so we will start by arriving at a net financial asset account for each sector in order to consolidate all sectors. We will thereby dispense with the liabilities accounts and add the corporate nonfinancial and net financial assets to the above personal sector table to arrive at a total for the domestic private sectors. This includes both nonfinancial and financial sectors, the latter of which has a deficit for the financial assets for 2005 and 2013.

We will then assume that the rest of the world accounts are private, at least with respect to the US economy and add the financial assets (there are no rest of world nonfinancial assets shown) to the domestic private sectors to arrive at a total value of assets for the private sectors. The rest of the world assets show up as negatives for 2005 and 2013 as they indicate foreign owned equity of domestic assets.

A similar treatment is performed for the two public sectors, federal and state and local governments, for nonfinancial and financial assets with totals and then for a consolidation of public and private. The results can be viewed in the table US Assets and Liabilities of the Public and Private Sectors and related tables of Sheets 23 to 30.

Note first that the total asset value and therefore net worth (liabilities have been deducted from financial assets remember) for the national private sector after including the rest of the world is $93.1T as opposed to the $107.0 for the three principal sectors mentioned above. The public sector has a net asset value of minus $3.5T, which some would interpret as a liability. Added to the private sector value this gives a total national net worth of $89.6T. Note also that the financial business sector contribution to this total is a negative half a trillion dollars. Of this total the nonfinancial asset contribution is $71.9T and the financial asset contribution is $17.7T of which apparently $6.3T consists of checkable deposits and currency and the balance is other sorts of deposits of which there was $19.4T less the checkable deposits and currency prior to consolidation. This indicates the total net financial assets, if all were held to be liquid or money, backed by the public sector, are under the value stated in the Fed figures by $1.7T.

Looking at the consolidation this way basically assumes that all forward appreciating financial assets, i.e obligations of the private sectors, are currently worthless, or to put it in more nuanced terms, of uncertain future value. However, the productive apparatus, homes, factories, cars, farms, oil fields, electrical grid, software, etc., of which there is $71.9T worth, remains in play going forward to provide income for the citizenry and goods and services for them to buy. A complaint could be raised that the vast majority of the $192.1T financial assets found on Sheet 16 represent obligations for future payment of liabilities, of which there are $156.6T, leaving a net financial asset value of $35.5T, and cannot just be "netted" out. This is almost precisely twice the consolidated asset value of $17.7T stated above. But if $0.9T worth of real assets can be replaced by $19T worth of financial assets held by a financial sector that has a negative net worth, one must ask if a stampede toward liquidity couldn't do just that, a liquidity backed by the public sector as insurer of last resort, in which the economic needs of the present are deemed to vastly outweigh those of an unknown future.

Putting ones faith in money is fine as long as everyone agrees everything else is running smoothly, but that hasn't been the case for a while. Let's complete the picture. The public sector accounts have nothing in them for the vast infrastructure in place and necessary for a functioning modern economy, infrastructure at both a federal and at a state and local level that could never be installed by and which so far has not been adequately maintained by private funding.

In addition to infrastructure there is approximately 650 million acres of publicly owned land and attendant natural resources in the US in addition to a smaller amount of state and local land. A large portion of this land was bought by the far sighted founding father Thomas Jefferson, not without some resistance on the grounds that it was "unconstitutional", to quote the Wikipedia entry concerning the matter, "... though opposition was ultimately not widespread. Jefferson agreed that the U.S. Constitution did not contain provisions for acquiring territory, but decided to go ahead with the purchase anyway ...". Those founding fathers, for all their faults, had some gall.

Finally, there is no accounting in the balance sheets for human capital, without which we cannot achieve next quarter's projected figures. We have added these three components to Sheets 24-25, in a revised balance sheet of the federal government as described earlier to arrive at a revised statement of US Assets and Liabilities of the Public and Private Sectors. This appears in the public sector instead of the private sector for the simple reason that unlike the regulations in place at the founding of the constitution and in fact codified in that document, a human being can no longer claim another human as an asset and therefore human capital cannot be included in a private balance sheet but certainly can in a consolidated statement of assets. This does not mean that human capital is envisioned as a government entitlement, to those who might so fear, but it is certainly not a private property entitlement, other than to one's own self. Registration of human capital stock in the economy belongs in the consolidated public and private sector balance sheet as an entitlement of citizenship or guest worker status.

The first of these revised sheets uses the Christian evaluation for the market value of human capital stock for 2005 and extrapolates forward and backward for the other two years. The interesting thing is that the total asset value, public and private, arrived at by this method for 1975, almost forty years ago or one series of working lives assuming a working life is forty years long, at $60.2T is reasonably close to the $89.6T figure for 2013 arrived at without using the human capital figures. And if the total human capital value is used as in Sheet 30 for 1975 resulting in a national asset value of $169.4T, the figure is to the other side of the 2013 figure and the Sector Account table net worth for all sectors of $107.4T, indicating that the human capital derivation used by Christian is reasonable. Based on the total asset value of public and private sectors without the human capital accounts at $89.6T, this indicates that Christian's market value figure as extrapolated to 1975 is actually a bit low, with implications for the 2005 and 2013 values. On the basis of this second total capital figure, the total asset value of the US national public and private sectors in 2013 was $1,153.0T or almost $1.2Q. This dwarfs the $16.4T federal, state and local debt for this period, which was between 5.1% to 1.5% of the total asset value, using the market and the total human capital stock values for the range of values.

Also of interest is the imputed change in productivity of that capital based on this analysis, using the GDP as a measure of the return on total capital base. In 1975 the return was 3.4% using the market value of human capital and 1.1% using the total human capital. In 2013 the figures were 5.2% and 1.6%. This is a range of around 67% improvement in productivity for the whole economy over the 38-year span. A human capital stock of $46.8T in 1975 with $5.9T worth of real productive assets, (assuming all real assets enhance human life and thereby productivity, which is not quite true,), $0.4T in fiat money, and another $0.9T in negotiable assets, in 38 years produced $71.9T in real productive assets, an increase of $66.0T, with an additional $6.0T in checkable deposits and currency and another $12.0T in near money, plus an additional $170.0+T of tokens of satisfaction of unknown future value, plus a human capital stock of market and nonmarket value worth an additional $1 quadrillion.

Like any capital, the human type requires proper maintenance and utilization to stay in working order. The idea that any amount of current financial assets, (the net amount being much less than the $192T shown on the national books), can of themselves create more wealth and provide an income stream for tens of millions of retires is of course ridiculous. It cannot guarantee that natural or more financial disasters or war or plague or social decay won't deplete the human capital to the point that it will barely support the working population, let alone the retirement and rentier population. Yet if it is properly educated, acculturated, and filled with enthusiasm for living, there is no reason that the next 40 years can't result in a transformation of the production of the necessary goods and services and reduction of the necessary work week that will usher in a new, technological browser economy, with plenty of opportunity left for those whose idea of play is entrepreneurial work. That is, provided that idea of work doesn't play out as most of us watching a handful of others playing poker with the national financial system.

US GDP and Wealth Comparisons 1975, 2005, 2013 - Sheet 1

Re: Distribution of US Gross Domestic Product
F.6 Comparison 1975/2005/2013

GDP Billions of dollars	1975	% GDP	2005	% GDP	+/- GDP from 1975	2013	% GDP	+/- % GDP from 2005	+/- % GDP from 1975
GDP	1688.9		13093.7			16768.0			
Personal consumption expenditures	1032.8	61.2%	8794.1	67.2%	6.0% x	11484.3	68.5%	1.3%	7.3% x
Durable goods	142.2	8.4%	1127.2	8.6%	0.2%	1249.3	7.5%	(1.2%)	(1.0%) x
Nondurable goods	349.2	20.7%	1953.0	14.9%	(5.8%) x	2601.9	15.5%	0.6%	(5.2%) x
Services	541.4	32.1%	5713.8	43.6%	11.6% x	7633.2	45.5%	1.9%	13.5% x
Gross private domestic investment	257.3	15.2%	2527.1	19.3%	4.1% x	2648.0	15.8%	(3.5%)	0.6%
Fixed Investment	263.5	15.6%	2467.5	18.8%	3.2% x	2573.9	15.4%	(3.5%)	(0.3%)
Nonresidential	196.8	11.7%	1611.5	12.3%	0.7%	2054.0	12.2%	(0.1%)	0.6%
Household sector (nonprofits)	11.0	0.7%	126.1	1.0%	0.3%	136.0	0.8%	(0.2%)	0.2%
Nonfinancial corporate business	143.8	8.5%	1132.2	8.6%	0.1%	1475.9	8.8%	0.2%	0.3%
Nonfinancial noncorporate business	29.0	1.7%	173.2	1.3%	(0.4%) x	240.7	1.4%	0.1%	(0.3%)
Financial institutions	12.9	0.8%	180.0	1.4%	0.6% x	201.5	1.2%	(0.2%)	0.4%
Residential	66.7	3.9%	856.1	6.5%	2.6%	519.9	3.1%	(3.4%)	(0.8%)
Household sector	53.1	3.1%	738.9	5.6%	2.5% x	433.4	2.6%	(3.1%)	(0.6%)
Nonfinancial corporate business	0.5	0.0%	6.0	0.0%	0.0%	(8.1)	(0.0%)	(0.1%)	(0.1%)
Nonfinancial noncorporate business	13.0	0.8%	108.7	0.8%	0.1%	79.9	0.5%	(0.4%)	(0.3%)
REITs	0.1	0.0%	2.5	0.0%	0.0%	14.7	0.1%	0.1%	0.1%
Change in private inventories	(6.3)	(0.4%)	59.6	0.5%	0.8%	74.1	0.4%	(0.0%)	0.8%
Nonfinancial corporate business	(8.3)	(0.5%)	52.5	0.4%	0.9%	56.7	0.3%	(0.1%)	0.8%
Nonfinancial noncorporate business	2.1	0.1%	7.1	0.1%	(0.1%) x	17.4	0.1%	0.0%	(0.0%)
Net US exports of goods and services	16.0	0.9%	(721.2)	(5.5%)	(6.5%) x	(508.2)	(3.0%)	2.5%	(4.0%) x
Exports	138.7	8.2%	1308.9	10.0%	1.8%	2262.2	13.5%	3.5%	5.3%
Imports	(122.7)	(7.3%)	(2030.1)	(15.5%)	(8.2%) x	(2770.4)	(16.5%)	(1.0%)	(9.3%)
Government consumption exp & gross inv	382.9	22.7%	2493.7	19.0%	(3.6%) x	3143.9	18.7%	(0.3%)	(3.9%) x
Consumption expenditures	298.5	17.7%	1980.0	15.1%	(2.6%) x	2547.6	15.2%	0.1%	(2.5%) x
Federal	132.9	7.9%	723.4	5.5%	(2.3%) x	963.0	5.7%	0.2%	(2.1%)
State and local	165.6	9.8%	1256.6	9.6%	(0.2%)	1584.5	9.4%	(0.1%)	(0.4%)
Gross investment	84.4	5.0%	513.6	3.9%	(1.1%) x	596.3	3.6%	(0.4%)	(1.4%) x
Federal	39.8	2.4%	222.9	1.7%	(0.7%)	268.5	1.6%	(0.1%)	(0.8%)
State and local	44.6	2.6%	290.8	2.2%	(0.4%)	327.8	2.0%	(0.3%)	(0.7%)
Net US income receipts from rest of world	13.0	0.8%	92.6	0.7%	(0.1%) x	224.3	1.3%	0.6%	0.6%
US income receipts	28.0	1.7%	578.8	4.4%	2.8% x	810.4	4.8%	0.4%	3.2%
US income payments	(15.0)	(0.9%)	(483.2)	(3.7%)	(2.8%) x	(586.1)	(3.5%)	0.2%	(2.6%)
GNP GDP + net US Income receipts	1701.9	100.8%	13186.3	100.7%	(0.1%)	16992.4	101.3%	0.6%	0.6%

Source: Z.1 Financial Accounts of the United States, Federal Reserve Statistical Release, September 18, 2014, F.6 Historical Annuals, 1975-1984 and 2005-2013

US GDP and Wealth Comparisons 1975, 2005, 2013 - Sheet 2

Re: Distribution of US National Income
F.7 Comparison 1975/2005/2013

GDP Billions of dollars	1975	% GDP	2005	% GDP	+/- GDP from 1975	2013	% GDP	+/- % GDP from 2005	+/- % GDP from 1975
	1688.9		13093.7			16768.0			
National Income	1451.2	85.9%	11239.8	85.8%	(0.1%)	14577.1	86.9%	1.1%	1.0%
Compensation of employees	950.2	56.3%	7086.8	54.1%	(2.1%)	8844.8	52.7%	(1.4%)	(3.5%)
Wages and other labor income	903.5	53.5%	6658.7	50.9%	(2.6%) x	8318.6	49.6%	(1.2%)	(3.9%) x
Employer social insurance contributions	46.7	2.8%	428.1	3.3%	0.5%	526.1	3.1%	(0.1%)	0.4%
Proprietor's Income*	118.2	7.0%	979.0	7.5%	0.5% x	1336.6	8.0%	0.5%	1.0% x
Rental Income of persons*	22.5	1.3%	238.4	1.8%	0.5% x	595.8	3.6%	1.7%	2.2% x
Corporate profits*	138.9	8.2%	1477.7	11.3%	3.1% x	2106.9	12.6%	1.3%	4.3% x
Corporate profits with inventory adjustments	144.2	8.5%	1621.2	12.4%	3.8%	2238.7	13.4%	1.0%	4.8%
Profits before tax	154.8	9.2%	1653.3	12.6%	3.5% x	2235.3	13.3%	0.7%	4.2%
Domestic nonfinancial	119.7	7.1%	1004.6	7.7%	0.6% x	1298.9	7.7%	0.1%	0.7%
Domestic financial	20.4	1.2%	409.7	3.1%	1.9% x	533.5	3.2%	0.1%	2.0%
Rest of the world	14.6	0.9%	239.1	1.8%	1.0%	403.1	2.4%	0.6%	1.5%
Less:									
Taxes on corporate income	51.6	3.1%	412.4	3.1%	0.1%	474.3	2.8%	(0.3%)	(0.2%)
Domestic nonfinancial	41.9	2.5%	271.9	2.1%	(0.4%)	329.3	2.0%	(0.1%)	(0.5%)
Domestic financial	9.7	0.6%	140.5	1.1%	0.5%	144.9	0.9%	(0.2%)	0.3%
Equals:									
Profits after tax	103.1	6.1%	1240.9	9.5%	3.4% x	1761.1	10.5%	1.0%	4.4%
Net dividends	22.0	1.3%	580.5	4.4%	3.1% x	959.6	5.7%	1.3%	4.4%
Domestic nonfinancial	24.6	1.5%	170.8	1.3%	(0.2%) x	536.9	3.2%	1.9%	1.7%
Domestic financial	1.8	0.1%	148.9	1.1%	1.0% x	289.8	1.7%	0.6%	1.6%
Rest of the world	6.6	0.4%	260.9	2.0%	1.6% x	132.9	0.8%	(1.2%)	0.4%
Undistributed profits	70.2	4.2%	660.4	5.0%	0.9%	801.5	4.8%	(0.3%)	0.6%
Domestic nonfinancial	53.2	3.1%	561.9	4.3%	1.1%	432.6	2.6%	(1.7%)	(0.6%)
Domestic financial	8.9	0.5%	120.3	0.9%	0.4%	98.7	0.6%	(0.3%)	0.1%
Rest of the world	8.1	0.5%	(21.8)	(0.2%)	(0.6%)	270.2	1.6%	1.8%	1.1%
Inventory valuation adjustment	(10.5)	(0.6%)	(32.1)	(0.2%)	0.4%	3.3	0.0%	0.3%	0.6%
Capital consumption adjustment	(5.4)	(0.3%)	(143.5)	(1.1%)	(0.8%)	(131.8)	(0.8%)	0.3%	(0.5%)
Domestic nonfinancial	(4.9)	(0.3%)	(101.5)	(0.8%)	(0.5%)	(88.3)	(0.5%)	0.2%	(0.2%)
Domestic financial	(0.5)	(0.0%)	(41.9)	(0.3%)	(0.3%)	(43.5)	(0.3%)	0.1%	(0.2%)
Net Interest and miscellaneous payments	85.9	5.1%	496.8	3.8%	(1.3%) x	499.8	3.0%	(0.8%)	(2.1%)
Taxes on production and imports	135.3	8.0%	934.5	7.1%	(0.9%)	1162.4	6.9%	(0.2%)	(1.1%)
Less Subsidies	4.5	0.3%	60.9	0.5%	0.2%	60.2	0.4%	(0.1%)	0.1%
Business current transfer payments (net)	9.0	0.5%	93.9	0.7%	0.2%	120.6	0.7%	0.0%	0.2%
Current surplus of government enterprises	(4.4)	(0.3%)	(6.4)	(0.0%)	0.2%	(29.6)	(0.2%)	(0.1%)	0.1%
Plus:									
Private consumption of fixed capital	178.8	10.6%	1653.3	12.6%	2.0% x	2120.8	12.6%	0.0%	2.1%
Government consumption of fixed capital	58.7	3.5%	346.7	2.6%	(0.8%) x	506.4	3.0%	0.4%	(0.5%)
Statistical discrepancy	13.2	0.8%	(35.5)	(0.3%)	(1.1%)	(211.9)	(1.3%)	(1.0%)	(2.0%)
Less:									
Net US income receipts from rest of the world	13.0	0.8%	92.6	0.7%	(0.1%)	224.3	1.3%	0.6%	0.6%

* with inventory & capital consumption adjustments

Source: Z.1 Financial Accounts of the United States, Federal Reserve Statistical Release, September 18, 2014, F.7 Historical Annuals, 1975-1984 and 2005-2013

US GDP and Wealth Comparisons 1975, 2005, 2013 - Sheet 3

Re: US Savings and Investment by Sector
F.8 Comparison 1975/2005/2013

GDP Billions of dollars	1975	% GDP	2005	% GDP	+/- from 1975	2013	% GDP	+/- % GDP from 2005	+/- % GDP from 1975
	1688.9		13093.7			16768.0			
Gross saving	348.3	20.6%	2333.4	17.8%	(2.8%)	3034.1	18.1%	0.3%	(2.5%)
Net saving	110.7	6.6%	354.1	2.7%	(3.9%)	406.8	2.4%	(0.3%)	(4.1%)
Net private saving	213.3	12.6%	722.8	5.5%	(7.1%) x	1281.1	7.6%	2.1%	(5.0%)
Domestic business	54.3	3.2%	484.8	3.7%	0.5%	673.0	4.0%	0.3%	0.8%
Nonfinancial corporate business	45.5	2.7%	411.7	3.1%	0.5%	571.3	3.4%	0.3%	0.7%
Financial business	8.7	0.5%	73.1	0.6%	0.0%	101.7	0.6%	0.0%	0.1%
Households and institutions	159.0	9.4%	237.9	1.8%	(7.6%) x	608.1	3.6%	1.8%	(5.8%)
Net government saving	(102.5)	(6.1%)	(371.3)	(2.8%)	3.2%	(874.3)	(5.2%)	(2.4%)	0.9%
Federal	(97.0)	(5.7%)	(304.7)	(2.3%)	3.4%	(649.1)	(3.9%)	(1.5%)	1.9%
State and local	(5.6)	(0.3%)	(66.6)	(0.5%)	(0.2%)	(225.1)	(1.3%)	(0.8%)	(1.0%)
Consumption of fixed capital	237.5	14.1%	1982.0	15.1%	1.1%	2627.2	15.7%	0.5%	1.6%
Private	178.8	10.6%	1635.3	12.5%	1.9%	2120.8	12.6%	0.2%	2.1%
Domestic business	147.9	8.8%	1273.2	9.7%	1.0%	1693.7	10.1%	0.4%	1.3%
Households and institutions	30.9	1.8%	362.1	2.8%	0.9%	427.1	2.5%	(0.2%)	0.7%
Government	58.7	3.5%	346.7	2.6%	(0.8%)	506.4	3.0%	0.4%	(0.5%)
Federal	37.8	2.2%	189.4	1.4%	(0.8%)	268.0	1.6%	0.2%	(0.6%)
State and local	21.0	1.2%	157.3	1.2%	(0.0%)	238.4	1.4%	0.2%	0.2%
Gross domestic investment =, Capital acct & net lend	361.5	21.4%	2297.9	17.5%	(3.9%) x	2822.1	16.8%	(0.7%)	(4.6%)
Gross domestic investment	341.7	20.2%	3040.7	23.2%	3.0% x	3244.3	19.3%	(3.9%)	(0.9%)
Gross private domestic investment	257.3	15.2%	2527.1	19.3%	4.1% x	2648.0	15.8%	(3.5%)	0.6%
Domestic business	193.1	11.4%	1662.1	12.7%	1.3%	2078.6	12.4%	(0.3%)	1.0%
Households and institutions	64.2	3.8%	865.0	6.6%	2.8%	569.4	3.4%	(3.2%)	(0.4%)
Gross government investment	84.4	5.0%	513.6	3.9%	(1.1%) x	596.3	3.6%	(0.4%)	(1.4%)
Federal	39.8	2.4%	222.9	1.7%	(0.7%)	268.5	1.6%	(0.1%)	(0.8%)
State and local	44.6	2.6%	290.8	2.2%	(0.4%)	327.8	2.0%	(0.3%)	(0.7%)
Capital account transactions	0.1	0.0%	(12.9)	(0.1%)	(0.1%)	0.8	0.0%	0.1%	0.0%
Net lending or borrowing	19.8	1.2%	(730.0)	(5.6%)	(6.7%) x	(423.0)	(2.5%)	3.1%	(3.7%)
Private	143.3	8.5%	(173.7)	(1.3%)	(9.8%) x	534.5	3.2%	4.5%	(5.3%)
Domestic business	22.5	1.3%	60.3	0.5%	(0.9%)	80.6	0.5%	0.0%	(0.9%)
Households and institutions	120.8	7.2%	(234.0)	(1.8%)	(8.9%) x	454.0	2.7%	4.5%	(4.4%)
Government	(123.6)	(7.3%)	(556.3)	(4.2%)	3.1% x	(957.5)	(5.7%)	(1.5%)	1.6%
Federal	(103.4)	(6.1%)	(396.6)	(3.0%)	3.1% x	(705.6)	(4.2%)	(1.2%)	1.9%
State and local	(20.1)	(1.2%)	(159.7)	(1.2%)	(0.0%)	(252.0)	(1.5%)	(0.3%)	(0.3%)
Statistical discrepancy	13.2	0.8%	(35.5)	(0.3%)	(1.1%) x	(211.9)	(1.3%)	(1.0%)	(2.0%)
Disaster losses	0.0	0.0%	110.4	0.8%	0.8%	0	0.0%	(0.8%)	0.0%

Source: Z.1 Financial Accounts of the United States, Federal Reserve Statistical Release, September 18, 2014, F.8 Historical Annuals, 1975-1984 and 2005-2013

US GDP and Wealth Comparisons 1975, 2005, 2013 - Sheet 4

Re: US Balance Sheet of Households and Nonprofit Organizations
B.100 Comparison 1975/2005/2013

GDP	Billions of dollars	1975	% GDP 25.7%	% Assets	2005	% GDP 17.7%	% Assets	+/- GDP from 1975	+/- Assets		2013	% GDP 18.1%	% Assets	+/- GDP from 2005	+/- Assets	+/- GDP from 1975	+/- Assets
		1688.9	388.4%		13093.7	565.5%		177.1%			16768.0	552.7%		(12.9%)	0.0%	164.2%	0.0%
Assets		6560.1	388.4%	100.0%	74048.8	565.5%	100.0%	177.1%			92669.2	552.7%	100.0%	(12.9%)	0.0%	164.2%	0.0%
	Nonfinancial assets	2286.5	135.4%	34.9%	28532.0	217.9%	38.5%	82.5%	3.7% x		27669.2	165.0%	29.9%	(52.9%)	(8.7%)	29.6%	(5.0%)
	Real estate	1684.2	99.7%	25.7%	24138.7	184.4%	32.6%	84.6%	6.9% x		22295.9	133.0%	24.1%	(51.4%)	(8.5%)	33.2%	(1.6%)
	Households	1413.7	83.7%	21.5%	22030.1	168.2%	29.8%	84.5%	8.2% x		19631.3	117.1%	21.2%	(51.2%)	(8.6%)	33.4%	(0.4%)
	Nonprofit organizations	270.5	16.0%	4.1%	2108.6	16.1%	2.8%	0.1%	(1.3%) x		2664.6	15.9%	2.9%	(0.2%)	0.0%	(0.1%)	(1.2%)
	Equipment (nonprofits)	19.2	1.1%	0.3%	206.1	1.6%	0.3%	0.4%	(0.0%)		306.9	1.8%	0.3%	0.3%	0.1%	0.7%	0.0%
	Intellectual property products (nonprofits)	5.1	0.3%	0.1%	79.5	0.6%	0.1%	0.3%	0.0%		124.1	0.7%	0.1%	0.1%	0.0%	0.4%	0.1%
	Consumer durable goods	578.0	34.2%	8.8%	4107.8	31.4%	5.5%	(2.9%)	(3.3%) x		4942.2	29.5%	5.3%	(1.9%)	(0.2%)	(4.7%)	(3.5%)
	Financial assets	4273.7	253.0%	65.1%	45516.7	347.6%	61.5%	94.6%	(3.7%) x		65000.1	387.6%	70.1%	40.0%	8.7%	134.6%	5.0%
	Deposits	924.5	54.7%	14.1%	6257.5	47.8%	8.5%	(6.9%)	(5.6%) x		9630.5	57.4%	10.4%	9.6%	1.9%	2.7%	(3.7%)
	Foreign deposits	0.0	0.0%	0.0%	63.8	0.5%	0.1%	0.5%	0.1%		52.5	0.3%	0.1%	(0.2%)	(0.0%)	0.3%	0.1%
	Checkable deposits and currency	158.7	9.4%	2.4%	285.9	2.2%	0.4%	(7.2%)	(2.0%) x		1035.0	6.2%	1.1%	4.0%	0.7%	(3.2%)	(1.3%)
	Time and savings deposits	762.2	45.1%	11.6%	4965.0	37.9%	6.7%	(7.2%)	(4.9%) x		7397.8	44.1%	8.0%	6.2%	1.3%	(1.0%)	(3.6%)
	Money market fund shares	3.7	0.2%	0.1%	942.7	7.2%	1.3%	7.0%	1.2% x		1145.2	6.8%	1.2%	(0.4%)	(0.0%)	6.6%	1.2%
	Credit market instruments	320.8	19.0%	4.9%	3452.8	26.4%	4.7%	7.4%	(0.2%) x		3848.8	23.0%	4.2%	(3.4%)	(0.5%)	4.0%	(0.7%)
	Open market paper	19.7	1.2%	0.3%	98.4	0.8%	0.1%	(0.4%)	(0.2%)		15.0	0.1%	0.0%	(0.7%)	(0.1%)	(1.1%)	(0.3%)
	Treasury securities	111.8	6.6%	1.7%	425.7	3.3%	0.6%	(3.4%)	(1.1%) x		949.0	5.7%	1.0%	2.4%	0.4%	(1.0%)	(0.7%)
	Agency and GSE backed securities	7.6	0.4%	0.1%	587.4	4.5%	0.8%	4.0%	0.7% x		97.7	0.6%	0.1%	(3.9%)	(0.7%)	0.1%	(0.0%)
	Municipal securities	66.8	4.0%	1.0%	1600.6	12.2%	2.2%	8.3%	1.1% x		1617.7	9.6%	1.7%	(2.6%)	(0.4%)	5.7%	0.7%
	Corporate and foreign bonds	64.3	3.8%	1.0%	591.4	4.5%	0.8%	0.7%	(0.2%)		1004.6	6.0%	1.1%	1.5%	0.3%	2.2%	0.1%
	Other loans and advances	0.0	0.0%	0.0%	8.7	0.1%	0.0%	0.1%	0.0%		25.9	0.2%	0.0%	0.1%	0.0%	0.2%	0.0%
	Mortgages	50.5	3.0%	0.8%	140.8	1.1%	0.2%	(1.9%)	(0.6%)		79.8	0.5%	0.1%	(0.6%)	(0.1%)	(2.5%)	(0.7%)
	Consumer credit (student loans)	0.0	0.0%	0.0%	0	0.0%	0.0%	0.0%	0.0%		59.1	0.4%	0.1%	0.4%	0.1%	0.4%	0.1%
	Corporate equities (market value)	584.6	34.6%	8.9%	8025.4	61.3%	10.8%	26.7%	1.9% x		12451.3	74.3%	13.4%	13.0%	2.6%	39.6%	4.5%
	Mutual fund shares	38.7	2.3%	0.6%	3527.1	26.9%	4.8%	24.6%	4.2% x		7152.4	42.7%	7.7%	15.7%	3.0%	40.4%	7.1%
	Security credit	4.7	0.3%	0.1%	623.4	4.8%	0.8%	4.5%	0.8% x		815.5	4.9%	0.9%	0.1%	0.0%	4.6%	0.8%
	Life insurance reserves	168.6	10.0%	2.6%	1082.6	8.3%	1.5%	(1.7%)	(1.1%)		1232.8	7.4%	1.3%	(0.9%)	(0.1%)	(2.6%)	(1.2%)
	Pension entitlements	1037.1	61.4%	15.8%	13466.3	102.8%	18.2%	41.4%	2.4% x		19886.1	118.6%	21.5%	15.7%	3.3%	57.2%	5.7%
	Equity in noncorporate business	1150.6	68.1%	17.5%	8473.0	64.7%	11.4%	(3.4%)	(6.1%)		9057.7	54.0%	9.8%	(10.7%)	(1.7%)	(14.1%)	(7.8%)
	Miscellaneous assets	44.0	2.6%	0.7%	608.7	4.6%	0.8%	2.0%	0.2%		924.9	5.5%	1.0%	0.9%	0.2%	2.9%	0.3%
Liabilities		763.8	45.2%	11.6%	12162.4	92.9%	16.4%	47.7%	4.8% x		13801.2	82.3%	14.9%	(10.6%)	(1.5%)	37.1%	3.2%
	Credit market instruments	736.9	43.6%	11.2%	11721.3	89.5%	15.8%	45.9%	4.6% x		13179.2	78.6%	14.2%	(10.9%)	(1.6%)	35.0%	3.0%
	Home mortgages	459.1	27.2%	7.0%	8912.7	68.1%	12.0%	40.9%	5.0% x		9415.9	56.2%	10.2%	(11.9%)	(1.9%)	29.0%	3.2%
	Consumer credit	207.0	12.3%	3.2%	2320.6	17.7%	3.1%	5.5%	(0.0%)		3097.9	18.5%	3.3%	0.8%	0.2%	6.2%	0.2%
	Municipal securities	2.7	0.2%	0.0%	212.7	1.6%	0.3%	1.5%	0.2%		227.8	1.4%	0.2%	(0.3%)	(0.0%)	1.2%	0.2%
	Depository institution loans	21.5	1.3%	0.3%	-16.5	(0.1%)	(0.0%)	(1.4%)	(0.4%)		92.7	0.6%	0.1%	0.7%	0.1%	(0.7%)	(0.2%)
	Other loans and advances	31.5	1.9%	0.5%	119.0	0.9%	0.2%	(1.0%)	(0.3%)		141.3	0.8%	0.2%	(0.1%)	(0.0%)	(1.0%)	(0.3%)
	Commercial mortgages	15.0	0.9%	0.2%	172.8	1.3%	0.2%	0.4%	0.0%		203.5	1.2%	0.2%	(0.1%)	(0.0%)	0.3%	(0.0%)
	Security credit	8.6	0.5%	0.1%	232.4	1.8%	0.3%	1.3%	0.2%		339.2	2.0%	0.4%	0.2%	0.1%	1.5%	0.2%
	Trade payables	10.6	0.6%	0.2%	186.3	1.4%	0.3%	0.8%	0.1%		255.0	1.5%	0.3%	0.1%	0.0%	0.9%	0.1%
	Deferred and unpaid life insurance premiums	7.7	0.5%	0.1%	22.4	0.2%	0.0%	(0.3%)	(0.1%)		27.9	0.2%	0.0%	(0.0%)	(0.0%)	(0.3%)	(0.1%)
Net worth		5796.4	343.2%	88.4%	61886.4	472.6%	83.6%	129.4%	(4.8%) x		78868.0	470.3%	85.1%	(2.3%)	1.5%	127.1%	(3.3%)

Source: Z.1 Financial Accounts of the United States, Federal Reserve Statistical Release, September 18, 2014, B.100 Historical Annuals, 1975-1984 and 2005-2013

US GDP and Wealth Comparisons 1975, 2005, 2013 - Sheet 5

Re: US Balance Sheet of Nonfinancial Corporate Business
B.102 Comparison 1975/2005/2013

	1975	% GDP	% Assets	2005	% GDP	% Assets	+/- GDP from 1975	+/- Assets	2013	% GDP	% Assets	+/- GDP from 2005	+/- Assets	+/- GDP from 1975	+/- Assets
GDP	1688.9	58.1%		13093.7	50.2%		27.0%		16768.0	48.0%		9.3%	0.0%	36.3%	0.0%
Assets	2906.1	172.1%	100.0%	26068.7	199.1%	100.0%	(18.0%)	(19.0%) x	34941.9	208.4%	100.0%	2.5%	(1.2%)	(15.5%)	(20.2%)
Nonfinancial assets	2131.8	126.2%	73.4%	14168.0	108.2%	54.3%	(8.8%)	(9.9%) x	18561.3	110.7%	53.1%	0.5%	(1.1%)	(8.3%)	(11.0%)
Real estate	1170.9	69.3%	40.3%	7924.3	60.5%	30.4%	(7.4%)	(6.3%) x	10236.1	61.0%	29.3%	0.2%	(0.5%)	(7.2%)	(6.7%)
Equipment	550.2	32.6%	18.9%	3297.8	25.2%	12.7%	(7.4%)		4261.7	25.4%	12.2%	1.3%	0.4%	5.3%	1.9%
Intellectual property products	103.6	6.1%	3.6%	1322.0	10.1%	5.1%	4.0%	1.5% x	1916.9	11.4%	5.5%	0.4%			
Inventories	307.2	18.2%	10.6%	1623.9	12.4%	6.2%	(5.8%)	(4.3%) x	2146.6	12.8%	6.1%	0.4%	(0.1%)	(5.4%)	(4.4%)
Financial assets	774.2	45.8%	26.6%	11900.6	90.9%	45.7%	45.0%	19.0% x	16380.6	97.7%	46.9%	6.8%	1.2%	51.8%	20.2%
Foreign deposits	2.6	0.2%	0.1%	63.6	0.5%	0.2%	0.3%	0.2%	87.8	0.5%	0.3%	0.0%	0.0%	0.4%	0.2%
Checkable deposits and currency	58.7	3.5%	2.0%	268.0	2.0%	1.0%	(1.4%)	(1.0%)	357.3	2.1%	1.0%	0.1%	(0.0%)	(1.3%)	(1.0%)
Time and savings deposits	24.1	1.4%	0.8%	450.4	3.4%	1.7%	2.0%	0.9%	649.5	3.9%	1.9%	0.4%	0.1%	2.4%	1.0%
Money market fund shares	0.0	0.0%	0.0%	352.2	2.7%	1.4%	2.7%	1.4%	521.1	3.1%	1.5%	0.4%	0.1%	3.1%	1.5%
Security repurchase agreements	1.2	0.1%	0.0%	14.6	0.1%	0.1%	0.0%	0.0%	8.8	0.1%	0.0%	(0.1%)	(0.0%)	(0.0%)	(0.0%)
Credit market instruments	61.7	3.7%	2.1%	340.4	2.6%	1.3%	(1.1%)	(0.8%)	169.0	1.0%	0.5%	(1.6%)	(0.8%)	(2.6%)	(1.6%)
Commercial paper	8.4	0.5%	0.3%	111.0	0.8%	0.4%	0.4%	0.1%	38.0	0.2%	0.1%	(0.6%)	(0.3%)	(0.3%)	(0.2%)
Treasury securities	14.3	0.8%	0.5%	52.1	0.4%	0.2%	(0.4%)	(0.3%)	40.2	0.2%	0.1%	(0.2%)	(0.1%)	(0.6%)	(0.4%)
Agency and GSE backed securities	3.3	0.2%	0.1%	17.4	0.1%	0.1%	(0.1%)	(0.0%)	9.4	0.1%	0.0%	(0.1%)	(0.0%)	(0.1%)	(0.1%)
Municipal securities	4.8	0.3%	0.2%	32.1	0.2%	0.1%	(0.0%)	(0.0%)	12.1	0.1%	0.0%	(0.2%)	(0.1%)	(0.2%)	(0.1%)
Mortgages	9.7	0.6%	0.3%	68.3	0.5%	0.3%	(0.1%)	(0.1%)	25.8	0.2%	0.1%	(0.4%)	(0.2%)	(0.4%)	(0.3%)
Consumer credit	21.2	1.3%	0.7%	59.6	0.5%	0.2%	(0.8%)	(0.5%)	43.5	0.3%	0.1%	(0.2%)	(0.1%)	(1.0%)	(0.6%)
Mutual fund shares	0.9	0.1%	0.0%	134.4	1.0%	0.5%	1.0%	0.5%	211.4	1.3%	0.6%	0.2%	0.1%	1.2%	0.6%
Trade receivables	271.4	16.1%	9.3%	2108.2	16.1%	8.1%	0.0%	(1.3%)	2469.4	14.7%	7.1%	(1.4%)	(1.0%)	(1.3%)	(2.3%)
US direct investment abroad	141.0	8.3%	4.9%	2205.7	16.8%	8.5%	8.5%	3.6% x	4370.0	26.1%	12.5%	9.2%	4.0%	17.7%	7.7%
Miscellaneous assets	212.6	12.6%	7.3%	5963.2	45.5%	22.9%	33.0%	15.6% x	7536.1	44.9%	21.6%	(0.6%)	(1.3%)	32.4%	14.3%
Liabilities	1092.4	64.7%	37.6%	11181.8	85.4%	42.9%	20.7%	5.3% x	15827.8	94.4%	45.3%	9.0%	2.4%	29.7%	7.7%
Credit market instruments	572.7	33.9%	19.7%	5260.7	40.2%	20.2%	6.3%	0.5%	7118.0	42.4%	20.4%	2.3%	0.2%	8.5%	0.7%
Commercial paper	9.6	0.6%	0.3%	90.1	0.7%	0.3%	0.1%	0.0%	144.5	0.9%	0.4%	0.2%	0.1%	0.3%	0.1%
Municipal securities	6.7	0.4%	0.2%	227.3	1.7%	0.9%	1.3%	0.6%	518.5	3.1%	1.5%	1.4%	0.6%	2.7%	1.3%
Corporate bonds	253.8	15.0%	8.7%	2662.6	20.3%	10.2%	5.3%	1.5%	4134.5	24.7%	11.8%	4.3%	1.6%	9.6%	3.1%
Depository institution loans	141.9	8.4%	4.9%	590.9	4.5%	2.3%	(3.9%)	(2.6%)	654.5	3.9%	1.9%	(0.6%)	(0.4%)	(4.5%)	(3.0%)
Other loans and advances	52.7	3.1%	1.8%	903.4	6.9%	3.5%	3.8%	1.7%	1063.5	6.3%	3.0%	(0.6%)	(0.4%)	3.2%	1.2%
Mortgages	107.9	6.4%	3.7%	786.4	6.0%	3.0%	(0.4%)	(0.7%)	602.5	3.6%	1.7%	(2.4%)	(1.3%)	(2.8%)	(2.0%)
Trade payables	176.9	10.5%	6.1%	1700.5	13.0%	6.5%	2.5%	0.4%	1968.2	11.7%	5.6%	(1.2%)	(0.9%)	1.3%	(0.5%)
Taxes payable	22.3	1.3%	0.8%	86.2	0.7%	0.3%	(0.7%)	(0.4%)	44.6	0.3%	0.1%	(0.4%)	(0.2%)	(1.1%)	(0.6%)
Foreign direct investment in US	21.5	1.3%	0.7%	1549.4	11.8%	5.9%	10.6%	5.2% x	2610.3	15.6%	7.5%	3.7%	1.5%	14.3%	6.7%
Miscellaneous liabilities	299.1	17.7%	10.3%	2584.9	19.7%	9.9%	2.0%	(0.4%)	4086.7	24.4%	11.7%	4.6%	1.8%	6.7%	1.4%
Net worth (market value)	1813.7	107.4%	62.4%	14886.9	113.7%	57.1%	6.3%	(5.3%) x	19114.1	114.0%	54.7%	0.3%	(2.4%)	6.6%	(7.7%)
Corporate profits	138.9	8.2%	4.8%	1477.7	11.3%	5.7%	3.1%	0.9%	2106.9	12.6%	6.0%	1.3%	0.4%	4.3%	1.3%
Return on equity	138.9		7.7%	1477.1		9.9%	0.0%	2.3%	2106.9		11.0%		1.1%		3.4%
Profits after tax	103.1	0.0%	3.5%	1240.9	9.5%	4.8%	9.5%	1.2%	1761.1	10.5%	5.0%	1.0%	0.3%	10.5%	1.5%
Return on equity after tax	103.1		5.7%	1240.9		8.3%	0.0%	2.7%	1761.1		9.2%		0.9%		3.5%

Source: Z.1 Financial Accounts of the United States, Federal Reserve Statistical Release, September 18, 2014, B.102 Historical Annuals, 1975-1984 and 2005-2013

US GDP and Wealth Comparisons 1975, 2005, 2013 - Sheet 6

Re: US Balance Sheet of Nonfinancial Noncorporate Business
B.103 Comparison 1975/2005/2013

	1975	% GDP	% Assets	2005	% GDP	% Assets	+/- GDP from 1975	+/- Assets	2013	% GDP	% Assets	+/- GDP from 2005	+/- Assets	+/- GDP from 1975	+/- Assets
GDP Billions of dollars	1688.9	112.8%		13093.7	103.6%		7.9%	0.0%	16768.0	111.6%		(7.0%)	0.0%	0.9%	0.0%
Assets															
Nonfinancial assets	1497.7	88.7%	100.0%	12641.7	96.5%	100.0%			15023.1	89.6%	100.0%				
Real estate	1414.7	83.8%	94.5%	10044.9	76.7%	79.5%	(7.0%)	(15.0%) x	11168.7	66.6%	74.3%	(10.1%)	(5.1%)	(17.2%)	(20.1%)
Residential	1197.0	70.9%	79.9%	9099.8	69.5%	72.0%	(1.4%)	(7.9%) x	9864.4	58.8%	65.7%	(10.7%)	(6.3%)	(12.0%)	(14.3%)
Nonresidential	578.7	34.3%	38.6%	5822.7	44.5%	46.1%	10.2%	7.4% x	5513.4	32.9%	36.7%	(11.6%)	(9.4%)	(1.4%)	(1.9%)
Equipment	618.3	36.6%	41.3%	3277.1	25.0%	25.9%	(11.6%)	(15.4%) x	4351.0	25.9%	29.0%	0.9%	3.0%	(10.7%)	(12.3%)
Residential	128.3	7.6%	8.6%	578.9	4.4%	4.6%	(3.2%)	(4.0%)	791.7	4.7%	5.3%	0.3%	0.7%	(2.9%)	(3.3%)
Nonresidential	8.2	0.5%	0.5%	41.7	0.3%	0.3%	(0.2%)	(0.2%)	43.3	0.3%	0.3%	(0.1%)	(0.0%)	(0.2%)	(0.3%)
Intellectual property products	120.1	7.1%	8.0%	537.2	4.1%	4.2%	(3.0%)	(3.8%)	748.3	4.5%	5.0%	0.4%	0.7%	(2.6%)	(3.0%)
Inventories	9.3	0.6%	0.6%	147.8	1.1%	1.2%	0.6%	0.5% x	215.3	1.3%	1.4%	0.2%	0.3%	0.7%	0.8%
	80.1	4.7%	5.3%	218.4	1.7%	1.7%	(3.1%)	(3.6%)	297.3	1.8%	2.0%	0.1%	0.3%	(3.0%)	(3.4%)
Financial assets	83.0	4.9%	5.5%	2596.8	19.8%	20.5%	14.9%	15.0% x	3854.4	23.0%	25.7%	3.2%	5.1%	18.1%	20.1%
Checkable deposits and currency	24.3	1.4%	1.6%	374.2	2.9%	3.0%	1.4%	1.3%	540.5	3.2%	3.6%	0.4%	0.6%	1.8%	2.0%
Time and savings deposits	8.1	0.5%	0.5%	324.1	2.5%	2.6%	2.0%	2.0%	372.7	2.2%	2.5%	(0.3%)	(0.1%)	1.7%	1.9%
Money market fund shares	0.0	0.0%	0.0%	69.0	0.5%	0.5%	0.5%	0.5%	80.4	0.5%	0.5%	(0.0%)	(0.0%)	0.5%	0.5%
Credit market instruments	7.2	0.4%	0.5%	96.8	0.7%	0.8%	0.3%	0.3%	100.2	0.6%	0.7%	(0.1%)	(0.1%)	0.2%	0.2%
Treasury securities	1.3	0.1%	0.1%	56.2	0.4%	0.4%	0.4%	0.4%	50.3	0.3%	0.3%	(0.1%)	(0.1%)	0.2%	0.2%
Municipal securities	0.0	0.0%	0.0%	4.4	0.0%	0.0%	0.0%	0.0%	5.9	0.0%	0.0%	0.0%	0.0%	0.0%	0.0%
Mortgages	3.3	0.2%	0.2%	36.2	0.3%	0.3%	0.1%	0.1%	44.0	0.3%	0.3%	0.0%	0.0%	0.1%	0.1%
Consumer credit	2.7	0.2%	0.2%	0.0	0.0%	0.0%	(0.2%)	(0.2%)	0.0	0.0%	0.0%	0.0%	0.0%	(0.2%)	(0.2%)
Trade receivables	14.6	0.9%	1.0%	430.9	3.3%	3.4%	2.4%	2.4%	558.0	3.3%	3.7%	0.0%	0.3%	2.5%	2.7%
Miscellaneous assets	28.8	1.7%	1.9%	1301.8	9.9%	10.3%	8.2%	8.4% x	2202.6	13.1%	14.7%	3.2%	4.4%	11.4%	12.7%
Insurance receivables	17.5	1.0%	1.2%	99.3	0.8%	0.8%	(0.3%)	(0.4%)	108.5	0.6%	0.7%	(0.1%)	(0.1%)	(0.4%)	(0.4%)
Equity investment in GSEs	1.6	0.1%	0.1%	4.1	0.0%	0.0%	(0.1%)	(0.1%)	7.8	0.0%	0.1%	0.0%	0.0%	(0.0%)	(0.1%)
Other	9.7	0.6%	0.6%	1198.4	9.2%	9.5%	8.6%	8.8% x	2086.3	12.4%	13.9%	3.3%	4.4%	11.9%	13.2%
Liabilities	349.3	20.7%	23.3%	4180.5	31.9%	33.1%	11.2%	9.7% x	5983.1	35.7%	39.8%	3.8%	6.8%	15.0%	16.5%
Credit market instruments	287.6	17.0%	19.2%	2898.0	22.1%	22.9%	5.1%	3.7% x	4180.0	24.9%	27.8%	2.8%	4.9%	7.9%	8.6%
Depository institution loans	56.7	3.4%	3.8%	670.5	5.1%	5.3%	1.8%	1.5%	1121.8	6.7%	7.5%	1.6%	2.2%	3.3%	3.7%
Other loans and advances	30.0	1.8%	2.0%	134.6	1.0%	1.1%	(0.7%)	(0.9%)	186.9	1.1%	1.2%	0.1%	0.2%	(0.7%)	(0.8%)
Mortgages	201.0	11.9%	13.4%	2092.9	16.0%	16.6%	4.1%	3.1% x	2871.4	17.1%	19.1%	1.1%	2.6%	5.2%	5.7%
Trade payables	28.4	1.7%	1.9%	334.5	2.6%	2.6%	0.9%	0.7%	441.4	2.6%	2.9%	0.1%	0.3%	1.0%	1.0%
Taxes payable	6.6	0.4%	0.4%	86.7	0.7%	0.7%	0.3%	0.2%	105.4	0.6%	0.7%	(0.0%)	0.0%	0.2%	0.3%
Foreign direct investment in US	0.6	0.0%	0.0%	3.3	0.0%	0.0%	(0.0%)	(0.0%)	7.3	0.0%	0.0%	0.0%	0.0%	0.0%	0.0%
Miscellaneous liabilities	26.1	1.5%	1.7%	857.9	6.6%	6.8%	5.0%	5.0% x	1249.0	7.4%	8.3%	0.9%	1.5%	5.9%	6.6%
Net worth	1148.5	68.0%	76.7%	8461.2	64.6%	66.9%	(3.4%)	(9.8%) x	9040.1	53.9%	60.2%	(10.7%)	(6.8%)	(14.1%)	(16.5%)
Proprietor's Income	118.2	7.0%	7.9%	979.0	7.5%	7.7%	0.5%	(0.1%) x	1336.6	8.0%	8.9%	0.5%	1.2%	1.0%	1.0%
Return on equity	118.2		10.3%	979.0		11.6%		1.3% x	1336.6		14.8%		3.2%		4.5%

Source: Z.1 Financial Accounts of the United States, Federal Reserve Statistical Release, September 18, 2014, B.103 Historical Annuals, 1975-1984 and 2005-2013

US GDP and Wealth Comparisons 1975, 2005, 2013 - Sheet 7

Re: US Balance Sheet of Financial Business
S.6.a Comparison 1975/2005/2013

Billions of dollars	1975	% GDP	% Assets	2005	% GDP	% Assets	+/- GDP from 1975	+/- Assets	2013	% GDP	% Assets	+/- GDP from 2005	+/- Assets	+/- GDP from 1975	+/- Assets
GDP	1688.9			13093.7					16768.0						
Assets	3261.1	193.1%	100.0%	57468.5	438.9%	100.0%	245.8%		82657.4	492.9%	100.0%	54.0%	0.0%	299.9%	0.0%
Nonfinancial assets	86.9	5.1%	2.7%	1306.8	10.0%	2.3%	4.8%	(0.4%)	1584.2	9.4%	1.9%	(0.5%)	(0.4%)	4.3%	(0.7%)
Real estate	55.7	3.3%	1.7%	751.8	5.7%	1.3%	2.4%	(0.4%)	899.2	5.4%	1.1%	(0.4%)	(0.2%)	2.1%	(0.6%)
Equipment	27.8	1.6%	0.9%	445.4	3.4%	0.8%	1.8%	(0.1%)	519.0	3.1%	0.6%	(0.3%)	(0.1%)	1.4%	(0.2%)
Intellectual property products	3.4	0.2%	0.1%	109.5	0.8%	0.2%	0.6%	0.1%	166.1	1.0%	0.2%	0.2%	0.0%	0.8%	0.1%
Financial assets	3174.2	187.9%	97.3%	56161.7	428.9%	97.7%	241.0%	0.4% x	81073.2	483.5%	98.1%	54.6%	0.4% x	295.6%	0.7%
Monetary gold	11.6	0.7%	0.4%	11.0	0.1%	0.0%	(0.6%)	(0.3%)	11.0	0.1%	0.0%	(0.0%)	(0.0%)	(0.6%)	(0.3%)
Currency and deposits	87.4	5.2%	2.7%	764.8	5.8%	1.3%	0.7%	(1.3%)	3513.1	21.0%	4.3%	15.1%	2.9%	15.8%	1.6%
Debt securities	777.4	46.0%	23.8%	14758.1	112.7%	25.7%	66.7%	1.8% x	22426.0	133.7%	27.1%	21.0%	1.5% x	87.7%	3.3%
SDR certificates	0.5	0.0%	0.0%	2.2	0.0%	0.0%	(0.0%)	(0.0%)	5.2	0.0%	0.0%	0.0%	0.0%	0.0%	(0.0%)
Open market paper	33.2	2.0%	1.0%	1124.7	8.6%	2.0%	6.6%	0.9%	720.0	4.3%	0.9%	(4.3%)	(1.1%)	2.3%	(0.1%)
Treasury securities	214.3	12.7%	6.6%	1647.4	12.6%	2.9%	(0.1%)	(3.7%)	4900.5	29.2%	5.9%	16.6%	3.1%	16.5%	(0.6%)
Agency and GSE backed securities	83.4	4.9%	2.6%	4140.2	31.6%	7.2%	26.7%	4.6% x	6310.9	37.6%	7.6%	6.0%	0.4%	32.7%	5.1%
Municipal securities	146.3	8.7%	4.5%	1346.1	10.3%	2.3%	1.6%	(2.1%) x	1946.9	11.6%	2.4%	1.3%	0.0% x	2.9%	(2.1%)
Corporate and foreign bonds	257.6	15.3%	7.9%	5614.6	42.9%	9.8%	27.6%	1.9% x	7191.6	42.9%	8.7%	0.0%	(1.1%) x	27.6%	0.8%
Nonmarketable government securities	42.1	2.5%	1.3%	882.9	6.7%	1.5%	4.3%	0.2% x	1352.3	8.1%	1.6%	1.3%	0.1% x	5.6%	0.3%
Loans	1330.3	78.8%	40.8%	19781.7	151.1%	34.4%	72.3%	(6.4%)	22325.8	133.1%	27.0%	(17.9%)	(7.4%)	54.4%	(13.8%)
Short term	640.1	37.9%	19.6%	8152.1	62.3%	14.2%	24.4%	(5.4%) x	9509.1	56.7%	11.5%	(5.5%)	(2.7%) x	18.8%	(8.1%)
Long term (mortgages)	690.2	40.9%	21.2%	11629.6	88.8%	20.2%	48.0%	(0.9%) x	12816.7	76.4%	15.5%	(12.4%)	(4.7%) x	35.6%	(5.7%)
Equity and investment fund shares	290.9	17.2%	8.9%	15535.6	118.6%	27.0%	101.4%	18.1%	24957.4	148.8%	30.2%	30.2%	3.2%	131.6%	21.3%
Corporate equities	221.4	13.1%	6.8%	10376.3	79.2%	18.1%	66.1%	11.3% x	15853.9	94.5%	19.2%	15.3%	1.1% x	81.4%	12.4%
Mutual fund shares	3.4	0.2%	0.1%	2180.7	16.7%	3.8%	16.5%	3.7% x	3555.2	21.2%	4.3%	4.5%	0.5% x	21.0%	4.2%
Money market fund shares	0.0	0.0%	0.0%	516.4	3.9%	0.9%	3.9%	0.9%	651.7	3.9%	0.8%	(0.1%)	(0.1%)	3.9%	0.8%
Equity in government sponsored enterprises	2.7	0.2%	0.1%	42.0	0.3%	0.1%	0.2%	(0.0%) x	33.4	0.2%	0.0%	(0.1%)	(0.0%) x	0.0%	(0.0%)
US direct investment abroad	8.5	0.5%	0.3%	446.0	3.4%	0.8%	2.9%	0.5%	914.0	5.5%	1.1%	2.0%	0.3%	4.9%	0.8%
Stock in Federal Reserve Banks	0.9	0.1%	0.0%	13.5	0.1%	0.0%	0.0%	(0.0%)	27.5	0.2%	0.0%	0.1%	0.0%	0.1%	0.0%
Investment in subsidiaries	54.0	3.2%	1.7%	1960.6	15.0%	3.4%	11.8%	1.8%	3921.7	23.4%	4.7%	8.4%	1.3%	20.2%	3.1%
Insurance, pension and std guarantee schemes	581.5	34.4%	17.8%	2304.7	17.6%	4.0%	(16.8%)	(13.8%) x	4102.5	24.5%	5.0%	6.9%	1.0% x	(10.0%)	(12.9%)
Other accounts receivable	95.1	5.6%	2.9%	3005.7	23.0%	5.2%	17.3%	2.3%	3737.4	22.3%	4.5%	(0.7%)	(0.7%)	16.7%	1.6%
Total liabilities and net worth	3261.1	193.1%	100.0%	57468.5	438.9%	100.0%	245.8%	0.0%	82657.4	492.9%	100.0%	54.0%	0.0%	299.9%	0.0%
Liabilities	3104.1	183.8%	95.2%	58641.8	447.9%	102.0%	264.1%	6.9%	83173.9	496.0%	100.6%	48.2%	(1.4%)	312.2%	5.4%
Currency and deposits	1241.7	73.5%	38.1%	8082.9	61.7%	14.1%	(11.8%)	(24.0%) x	15865.2	94.6%	19.2%	32.9%	5.1% x	21.1%	(18.9%)
Debt securities	219.8	13.0%	6.7%	11964.7	91.4%	20.8%	78.4%	14.1% x	12888.1	76.9%	15.6%	(14.5%)	(5.2%) x	63.8%	8.9%
Agency and GSE backed securities	107.3	6.4%	3.3%	6140.7	46.9%	10.7%	40.5%	7.4% x	7769.7	46.3%	9.4%	(0.6%)	(1.3%)	40.0%	6.1%
Corporate bonds	56.1	3.3%	1.7%	4653.8	35.5%	8.1%	32.2%	6.4%	4718.3	28.1%	5.7%	(7.4%)	(2.4%)	24.8%	4.0%
Commercial paper	56.4	3.3%	1.7%	1170.2	8.9%	2.0%	5.6%	0.3%	400.2	2.4%	0.5%	(6.6%)	(1.6%)	(1.0%)	(1.2%)
Loans	118.7	7.0%	3.6%	5029.8	38.4%	8.8%	31.4%	5.1%	4888.8	29.2%	5.9%	(9.3%)	(2.8%)	22.1%	2.3%
Short term	117.2	6.9%	3.6%	4884.3	37.3%	8.5%	30.4%	4.9% x	4686.6	27.9%	5.7%	(9.4%)	(2.8%) x	21.0%	2.1%
Long term (mortgages)	1.5	0.1%	0.0%	145.5	1.1%	0.3%	1.0%	0.2%	202.2	1.2%	0.2%	0.1%	(0.0%)	1.1%	0.2%

US GDP and Wealth Comparisons 1975, 2005, 2013 - Sheet 8

	1975	% GDP	% Assets	2005	% GDP	% Assets	+/- GDP from 1975	+/- Assets	2013	% GDP	% Assets	+/- GDP from 2005	+/- Assets	+/- GDP from 1975	+/- Assets
Equity and investment fund shares	184.8	10.9%	5.7%	15146.2	115.7%	26.4%	104.7%	20.7%	25253.9	150.6%	30.6%	34.9%	4.2%	139.7%	24.9%
Money market fund shares	3.7	0.2%	0.1%	1993.1	15.2%	3.5%	15.0%	3.4%	2678.3	16.0%	3.2%	0.8%	(0.2%)	15.8%	3.1%
Corporate equity issues	75.0	4.4%	2.3%	4631.9	35.4%	8.1%	30.9%	5.8% x	6465.6	38.6%	7.8%	3.2%	(0.2%)	34.1%	5.5%
Mutual fund shares	43.0	2.5%	1.3%	6045.6	46.2%	10.5%	43.6%	9.2% x	11544.8	68.9%	14.0%	22.7%	3.4% x	66.3%	12.6%
Equity in government sponsored enterprises	4.4	0.3%	0.1%	46.7	0.4%	0.1%	0.1%	(0.1%)	41.7	0.2%	0.1%	(0.1%)	(0.0%)	(0.0%)	(0.1%)
Foreign direct investment in the US	3.2	0.2%	0.1%	353.2	2.7%	0.6%	2.5%	0.5%	559.3	3.3%	0.7%	0.6%	0.1%	3.1%	0.6%
Equity in noncorporate business	2.1	0.1%	0.1%	11.7	0.1%	0.0%	(0.0%)	(0.0%)	17.6	0.1%	0.0%	0.0%	0.0%	(0.0%)	(0.0%)
Investment by parent	52.5	3.1%	1.6%	2050.4	15.7%	3.6%	12.6%	2.0%	3919.0	23.4%	4.7%	7.7%	1.2%	20.3%	3.1%
Stock in Federal Reserve Banks	0.9	0.1%	0.0%	13.5	0.1%	0.0%	0.0%	(0.0%)	27.5	0.2%	0.0%	0.1%	0.0%	0.1%	0.0%
Insurance, pension and std guarantee schemes	1277.5	75.6%	39.2%	15953.5	121.8%	27.8%	46.2%	(11.4%) x	23046.8	137.4%	27.9%	15.6%	0.1% x	61.8%	(11.3%)
Other accounts payable	61.6	3.6%	1.9%	2464.7	18.8%	4.3%	15.2%	2.4%	1231.0	7.3%	1.5%	(11.5%)	(2.8%)	3.7%	(0.4%)
Net worth (market value)	157.1	9.3%	4.8%	(1173.3)	(9.0%)	(2.0%)	(18.3%)	(6.9%)	(516.5)	(3.1%)	(0.6%)	5.9%	1.4%	(12.4%)	(5.4%)
Corporate profits		0.0%	0.0%		0.0%	0.0%	0.0%	0.0% x		0.0%	0.0%	0.0%	0.0% x	0.0%	0.0%

Source: Z.1 Financial Accounts of the United States, Federal Reserve Statistical Release, September 18, 2014, S.6.a Historical Annuals, 1975-1984 and 2005-2013

US GDP and Wealth Comparisons 1975, 2005, 2013 - Sheet 9

Re: US Balance Sheet of Federal Government
S.7.a Comparison 1975/2005/2013

	1975	% GDP	% Assets	2005	% GDP	% Assets	+/- GDP from 1975	+/- Assets	2013	% GDP	% Assets	+/- GDP from 2005	+/- Assets	+/- GDP from 1975	+/- Assets
GDP	1688.9	246.5%		13093.7	426.5%				16768.0	341.8%					
Billions of dollars															
Assets	685.1	40.6%	100.0%	3070.2	23.4%	100.0%	(17.1%)		4905.4	29.3%	100.0%	5.8%	0.0%	(11.3%)	0.0%
Nonfinancial assets	564.1	33.4%	82.3%	2426.5	18.5%	79.0%	(14.9%)	(3.3%) x	3191.7	19.0%	65.1%	0.5%	(14.0%) x	(14.4%)	(17.3%)
Real estate	288.1	17.1%	42.1%	1176.5	9.0%	38.3%	(8.1%)	(3.7%)	1472.1	8.8%	30.0%	(0.2%)	(8.3%)	(8.3%)	(12.0%)
Structures	288.1	17.1%	42.1%	1176.5	9.0%	38.3%	(8.1%)	(3.7%)	1472.1	8.8%	30.0%	(0.2%)	(8.3%)	(8.3%)	(12.0%)
Equipment	131.8	7.8%	19.2%	520.7	4.0%	17.0%	(3.8%)	(2.3%)	732.2	4.4%	14.9%	0.4%	(2.0%)	(3.4%)	(4.3%)
Intellectual property products	144.2	8.5%	21.0%	729.3	5.6%	23.8%	(3.0%)	2.7%	987.4	5.9%	20.1%	0.3%	(3.6%)	(2.6%)	(0.9%)
Financial assets	121.0	7.2%	17.7%	643.7	4.9%	21.0%	(2.2%)	3.3% x	1713.7	10.2%	34.9%	5.3%	14.0% x	3.1%	17.3%
Monetary gold and SDRs	2.3	0.1%	0.3%	8.2	0.1%	0.3%	(0.1%)	(0.1%)	55.2	0.3%	1.1%	0.3%	0.9%	0.2%	0.8%
Monetary gold	0.0	0.0%	0.0%	0.0	0.0%	0.0%	0.0%	0.0%	0.0	0.0%	0.0%	0.0%	0.0%	0.0%	0.0%
SDR holdings	2.3	0.1%	0.3%	8.2	0.1%	0.3%	(0.1%)	(0.1%)	55.2	0.3%	1.1%	0.3%	0.9%	0.2%	0.8%
Currency and deposits	17.5	1.0%	2.6%	68.0	0.5%	2.2%	(0.5%)	(0.3%) x	221.8	1.3%	4.5%	0.8%	2.3% x	0.3%	2.0%
Official foreign currencies	0.0	0.0%	0.0%	18.7	0.1%	0.6%	0.1%	0.6%	23.6	0.1%	0.5%	(0.0%)	(0.1%)	0.1%	0.5%
Reserve position in IMF (net)	2.2	0.1%	0.3%	8.1	0.1%	0.3%	(0.1%)	(0.1%)	31.0	0.2%	0.6%	0.1%	0.4%	0.1%	0.3%
Currency and transferable deposits	11.2	0.7%	1.6%	37.0	0.3%	1.2%	(0.4%)	(0.4%)	163.0	1.0%	3.3%	0.7%	2.1%	0.3%	1.7%
Time and savings deposits	0.6	0.0%	0.1%	1.4	0.0%	0.0%	(0.0%)	(0.0%)	1.5	0.0%	0.0%	(0.0%)	(0.0%)	(0.0%)	(0.1%)
Nonofficial foreign currencies	3.6	0.2%	0.5%	2.6	0.0%	0.1%	(0.2%)	(0.4%)	2.6	0.0%	0.1%	(0.0%)	(0.0%)	(0.2%)	(0.5%)
Debt securities	0.0	0.0%	0.0%	0.0	0.0%	0.0%	0.0%	0.0%	0.6	0.0%	0.0%	0.0%	0.0%	0.0%	0.0%
Agency and GSE backed securities	0.0	0.0%	0.0%	0.0	0.0%	0.0%	0.0%	0.0%	0.0	0.0%	0.0%	0.0%	0.0%	0.0%	0.0%
Corporate and foreign bonds	0.0	0.0%	0.0%	0.0	0.0%	0.0%	0.0%	0.0%	0.6	0.0%	0.0%	0.0%	0.0%	0.0%	0.0%
Loans	85.5	5.1%	12.5%	271.7	2.1%	8.8%	(3.0%)	(3.6%)	1039.2	6.2%	21.2%	4.1%	12.3%	1.1%	8.7%
Short term	66.4	3.9%	9.7%	195.2	1.5%	6.4%	(2.4%)	(3.3%)	923.7	5.5%	18.8%	4.0%	12.5%	1.6%	9.1%
Consumer credit	0	0.0%	0.0%	89.8	0.7%	2.9%	0.7%	2.9%	729.8	4.4%	14.9%	3.7%	12.0%	4.4%	14.9%
Other loans and advances	66.4	3.9%	9.7%	105.3	0.8%	3.4%	(3.1%)	(6.3%)	193.9	1.2%	4.0%	0.4%	0.5%	(2.8%)	(5.7%)
Long term (mortgages)	19.1	1.1%	2.8%	76.6	0.6%	2.5%	(0.5%)	(0.3%)	115.5	0.7%	2.4%	0.1%	(0.1%)	(0.4%)	(0.4%)
Equity and investment fund shares	4.1	0.2%	0.6%	43.2	0.3%	1.4%	0.1%	0.8%	98.9	0.6%	2.0%	0.3%	0.6%	0.3%	1.4%
Corporate equities	0.0	0.0%	0.0%	0.0	0.0%	0.0%	0.0%	0.0%	35.1	0.2%	0.7%	0.2%	0.7%	0.2%	0.7%
Equity in international organizations	4.1	0.2%	0.6%	43.2	0.3%	1.4%	0.1%	0.8%	59.5	0.4%	1.2%	0.0%	(0.2%)	0.1%	0.6%
Equity in government sponsored enterprises	0.0	0.0%	0.0%	0.0	0.0%	0.0%	0.0%	0.0%	0.0	0.0%	0.0%	0.0%	0.0%	0.0%	0.0%
Equity in investment under Public-Private IP	0.0	0.0%	0.0%	0.0	0.0%	0.0%	0.0%	0.0%	4.4	0.0%	0.1%	0.0%	0.1%	0.0%	0.1%
Other accounts receivable	11.5	0.7%	1.7%	252.5	1.9%	8.2%	1.2%	6.5%	298.1	1.8%	6.1%	(0.2%)	(2.1%)	1.1%	4.4%
Trade receivables	6.5	0.4%	0.9%	71.0	0.5%	2.3%	0.2%	1.4%	48.8	0.3%	1.0%	(0.3%)	(1.3%)	(0.1%)	0.0%
Taxes receivables	5.0	0.3%	0.7%	91.6	0.7%	3.0%	0.4%	2.3%	165.8	1.0%	3.4%	0.3%	0.4%	0.7%	2.7%
Other (miscellaneous assets)	0.0	0.0%	0.0%	89.9	0.7%	2.9%	0.7%	2.9% x	83.4	0.5%	1.7%	(0.2%)	(1.2%) x	0.5%	1.7%

US GDP and Wealth Comparisons 1975, 2005, 2013 - Sheet 10

	1975	% GDP	% Assets	2005.0	% GDP	% Assets	+/- GDP from 1975	+/- Assets	2013.0	% GDP	% Assets	+/- GDP from 2005	+/- Assets	+/- GDP from 1975	+/- Assets
Total liabilities and net worth	685.1	40.6%	100.0%	3070.2	23.4%	100.0%	(17.1%)	0.0%	4905.4	29.3%	100.0%	5.8%	0.0%	(11.3%)	0.0%
Liabilities	931.6	55.2%	136.0%	7370.3	56.3%	240.1%	1.1%	104.1%	16132.6	96.2%	328.9%	39.9%	88.8%	41.1%	192.9%
SDR allocations	2.7	0.2%	0.4%	7.0	0.1%	0.2%	(0.1%)	(0.2%)	54.4	0.3%	1.1%	0.3%	0.9%	0.2%	0.7%
Currency and deposits	8.2	0.5%	1.2%	27.5	0.2%	0.9%	(0.3%)	(0.3%) x	25.6	0.2%	0.5%	(0.1%)	(0.4%) x	(0.3%)	(0.7%)
Debt securities	485.4	28.7%	70.9%	5587.0	42.7%	182.0%	13.9%	111.1%	13710.3	81.8%	279.5%	39.1%	97.5%	53.0%	208.6%
SDR certificates	0.5	0.0%	0.1%	2.2	0.0%	0.1%	(0.0%)	(0.0%)	5.2	0.0%	0.1%	0.0%	0.0%	0.0%	0.0%
Treasury securities including savings bonds	434.9	25.8%	63.5%	4678.0	35.7%	152.4%	10.0%	88.9% x	12328.3	73.5%	251.3%	37.8%	99.0% x	47.8%	187.8%
Federal agency securities	7.9	0.5%	1.2%	23.8	0.2%	0.8%	(0.3%)	(0.4%)	24.5	0.1%	0.5%	(0.0%)	(0.3%)	(0.3%)	(0.7%)
Nonmarketable sec held by pension plans	42.1	2.5%	6.1%	882.9	6.7%	28.8%	4.3%	22.6%	1352.3	8.1%	27.6%	1.3%	(1.2%)	5.6%	21.4%
Loans (mortgages)	1.1	0.1%	0.2%	0.0	0.0%	0.0%	(0.1%)	(0.2%)	0.0	0.0%	0.0%	0.0%	0.0%	(0.1%)	(0.2%)
Insurance, pension and std guarantee schemes	421.4	25.0%	61.5%	1540.9	11.8%	50.2%	(13.2%)	(11.3%)	2089.3	12.5%	42.6%	0.7%	(7.6%)	(12.5%)	(18.9%)
Insurance reserves	10.2	0.6%	1.5%	42.7	0.3%	1.4%	(0.3%)	(0.1%)	50.3	0.3%	1.0%	(0.0%)	(0.4%)	(0.3%)	(0.5%)
Retiree Health Care Funds	0.0	0.0%	0.0%	75.4	0.6%	2.5%	0.6%	2.5%	246.9	1.5%	5.0%	0.9%	2.6%	1.5%	5.0%
Claims of pension fund on sponsor	411.2	24.3%	60.0%	1422.8	10.9%	46.3%	(13.5%)	(13.7%) x	1792.1	10.7%	36.5%	(0.2%)	(9.8%) x	(13.7%)	(23.5%)
Other accounts payable	12.9	0.8%	1.9%	208.0	1.6%	6.8%	0.8%	4.9%	253.0	1.5%	5.2%	(0.1%)	(1.6%)	0.7%	3.3%
Trade payables	11.8	0.7%	1.7%	202.6	1.5%	6.6%	0.8%	4.9%	250.7	1.5%	5.1%	(0.1%)	(1.5%)	0.8%	3.4%
Other (miscellaneous liabilities)	1.1	0.1%	0.2%	5.4	0.0%	0.2%	(0.0%)	0.0%	2.3	0.0%	0.0%	(0.0%)	(0.1%)	(0.1%)	(0.1%)
Net worth (market value)	(246.5)	(14.6%)	(36.0%)	(4300.1)	(32.8%)	(140.1%)	(18.2%)	(104.1%) x	(11227.2)	(67.0%)	(228.9%)	(34.1%)	(88.8%) x	(52.4%)	(192.9%)

Source: Z.1 Financial Accounts of the United States, Federal Reserve Statistical Release, September 18, 2014, S.7.a Historical Annuals, 1975-1984 and 2005-2013

US GDP and Wealth Comparisons 1975, 2005, 2013 - Sheet 11

Re: US Balance Sheet of State and Local Government
S.8.a Comparison 1975/2005/2013

		1975	% GDP	% Assets		2005	% GDP	% Assets	+/- GDP from 1975	+/- Assets	2013	% GDP	% Assets	+/- GDP from 2005	+/- Assets	+/- GDP from 1975	+/- Assets
GDP	Billions of dollars	1688.9	158.7%			13093.7	152.7%		2.5%	0.0%	16768.0	132.6%		9.9%	0.0%	12.4%	0.0%
Assets		1064.5	63.0%	100.0%		8576.2	65.5%	100.0%	2.5%	0.0%	12641.4	75.4%	100.0%	9.9%	0.0%	12.4%	0.0%
	Nonfinancial assets	918.2	54.4%	86.3%		6259.0	47.8%	73.0%	(6.6%)	(13.3%)	9703.2	57.9%	76.8%	10.1%	3.8%	3.5%	(9.5%)
	Real estate	887.4	52.5%	83.4%		5976.8	45.6%	69.7%	(6.9%)	(13.7%)	9335.6	55.7%	73.8%	10.0%	4.2%	3.1%	(9.5%)
	Structures	887.4	52.5%	83.4%		5976.8	45.6%	69.7%	(6.9%)	(13.7%)	9335.6	55.7%	73.8%	10.0%	4.2%	3.1%	(9.5%)
	Land	0.0	0.0%	0.0%		0.0	0.0%	0.0%	0.0%	0.0%	0.0	0.0%	0.0%	0.0%	0.0%	0.0%	0.0%
	Equipment	25.5	1.5%	2.4%		206.9	1.6%	2.4%	0.1%	0.0%	247.6	1.5%	2.0%	(0.1%)	(0.5%)	(0.0%)	(0.4%)
	Intellectual property products	5.3	0.3%	0.5%		75.3	0.6%	0.9%	0.3%	0.4%	120.0	0.7%	0.9%	0.1%	0.1%	0.4%	0.5%
	Financial assets	146.4	8.7%	13.8%		2317.2	17.7%	27.0%	9.0%	13.3%	2938.2	17.5%	23.2%	(0.2%)	(3.8%)	8.9%	9.5%
	Currency and deposits	60.3	3.6%	5.7%		234.8	1.8%	2.7%	(1.8%)	(2.9%)	425.8	2.5%	3.4%	0.7%	0.6%	(1.0%)	(2.3%)
	Currency and transferable deposits	13.4	0.8%	1.3%		66.0	0.5%	0.8%	(0.3%)	(0.5%)	123.9	0.7%	1.0%	0.2%	0.2%	(0.1%)	(0.3%)
	Time and savings deposits	46.9	2.8%	4.4%		168.8	1.3%	2.0%	(1.5%)	(2.4%)	301.9	1.8%	2.4%	0.5%	0.4%	(1.0%)	(2.0%)
	Debt securities	51.0	3.0%	4.8%		1217.4	9.3%	14.2%	6.3%	9.4%	1347.7	8.0%	10.7%	(1.3%)	(3.5%)	5.0%	5.9%
	Open market paper	0.0	0.0%	0.0%		153.3	1.2%	1.8%	1.2%	1.8%	76.9	0.5%	0.6%	(0.7%)	(1.2%)	0.5%	0.6%
	Treasury securities	27.8	1.6%	2.6%		512.3	3.9%	6.0%	2.3%	3.4%	593.4	3.5%	4.7%	(0.4%)	(1.3%)	1.9%	2.1%
	Agency and GSE backed securities	18.2	1.1%	1.7%		413.4	3.2%	4.8%	2.1%	3.1%	490.9	2.9%	3.9%	(0.2%)	(0.9%)	1.8%	2.2%
	Municipal securities	5.0	0.3%	0.5%		6.9	0.1%	0.1%	(0.2%)	(0.4%)	13.9	0.1%	0.1%	0.0%	0.0%	(0.2%)	(0.4%)
	Corporate and foreign bonds	0.0	0.0%	0.0%		131.5	1.0%	1.5%	1.0%	1.5%	172.5	1.0%	1.4%	0.0%	(0.2%)	1.0%	1.4%
	Loans	10.7	0.6%	1.0%		288.9	2.2%	3.4%	1.6%	2.4%	346.3	2.1%	2.7%	(0.1%)	(0.6%)	1.4%	1.7%
	Short term	-2.1	(0.1%)	(0.2%)		130.0	1.0%	1.5%	1.1%	1.7%	132.5	0.8%	1.0%	(0.2%)	(0.5%)	0.9%	1.2%
	Long term (mortgages)	12.8	0.8%	1.2%		158.9	1.2%	1.9%	0.5%	0.7%	213.9	1.3%	1.7%	0.1%	(0.2%)	0.5%	0.5%
	Equity and investment fund shares	0.0	0.0%	0.0%		246.6	1.9%	2.9%	1.9%	2.9%	415.4	2.5%	3.3%	0.6%	0.4%	2.5%	3.3%
	Money market fund shares	0.0	0.0%	0.0%		89.9	0.7%	1.0%	0.7%	1.0%	166.6	1.0%	1.3%	0.3%	0.3%	1.0%	1.3%
	Corporate equities	0.0	0.0%	0.0%		116.0	0.9%	1.4%	0.9%	1.4%	167.6	1.0%	1.3%	0.1%	(0.0%)	1.0%	1.3%
	Mutual fund shares	0.0	0.0%	0.0%		40.7	0.3%	0.5%	0.3%	0.5%	81.2	0.5%	0.6%	0.2%	0.2%	0.5%	0.6%
	Other accounts receivable	24.4	1.4%	2.3%		329.4	2.5%	3.8%	1.1%	1.5%	403.0	2.4%	3.2%	(0.1%)	(0.7%)	1.0%	0.9%
	Trade receivables	15.2	0.9%	1.4%		142.5	1.1%	1.7%	0.2%	0.2%	168.7	1.0%	1.3%	(0.1%)	(0.3%)	0.1%	(0.1%)
	Taxes receivables	9.2	0.5%	0.9%		102.3	0.8%	1.2%	0.2%	0.3%	123.8	0.7%	1.0%	(0.0%)	(0.2%)	0.2%	0.1%
	Other (miscellaneous assets)	0.1	0.0%	0.0%		84.6	0.6%	1.0%	0.6%	1.0%	110.5	0.7%	0.9%	0.0%	(0.1%)	0.7%	0.9%
Total liabilities and net worth		1064.5	63.0%	100.0%		8576.2	65.5%	100.0%	2.5%	0.0%	12641.4	75.4%	100.0%	9.9%	0.0%	12.4%	0.0%
Liabilities		349.5	20.7%	32.8%		3385.9	25.9%	39.5%	5.2%	6.6%	4931.0	29.4%	39.0%	3.5%	(0.5%)	8.7%	6.2%
	Debt securities (municipals)	213.6	12.6%	20.1%		2579.2	19.7%	30.1%	7.1%	10.0%	2924.9	17.4%	23.1%	(2.3%)	(6.9%)	4.8%	3.1%
	Short term	18.6	1.1%	1.7%		42.5	0.3%	0.5%	(0.8%)	(1.3%)	45.3	0.3%	0.4%	(0.1%)	(0.1%)	(0.8%)	(1.4%)
	Other	195.0	11.5%	18.3%		2536.7	19.4%	29.6%	7.8%	11.3%	1879.6	11.2%	14.9%	(8.2%)	(14.7%)	(0.3%)	(3.4%)
	Loans (short term)	5.8	0.3%	0.5%		10.6	0.1%	0.1%	(0.3%)	(0.4%)	16.2	0.1%	0.1%	0.0%	0.0%	(0.2%)	(0.4%)
	Insurance, pension and std guarantee schemes (Claims of pension fund on sponsor)	92.7	5.5%	8.7%		314.0	2.4%	3.7%	(3.1%)	(5.0%)	1204.1	7.2%	9.5%	4.8%	5.9%	1.7%	0.8%
	Other accounts payable (trade payables)	37.4	2.2%	3.5%		482.0	3.7%	5.6%	1.5%	2.1%	785.8	4.7%	6.2%	1.0%	0.6%	2.5%	2.7%
Net worth (market value)		715.0	42.3%	67.2%		5190.3	39.6%	60.5%	(2.7%)	(6.6%)	7710.4	46.0%	61.0%	6.3%	0.5%	3.6%	(6.2%)

Source: Z.1 Financial Accounts of the United States, Federal Reserve Statistical Release, September 18, 2014, S.8.a Historical Annuals, 1975-1984 and 2005-2013

US GDP and Wealth Comparisons 1975, 2005, 2013 - Sheet 12

Re: S.9.a US Balance Sheet of Rest of the World
Comparison 1975/2005/2013

Billions of dollars	1975	% GDP	% Assets	2005	% GDP	% Assets	+/- GDP from 1975	+/- Assets	2013	% GDP	% Assets	+/- GDP from 2005	+/- Assets	+/- GDP from 1975	+/- Assets
GDP	1688.9	826.3%		13093.7	122.6%				16768.0	79.5%					
Assets															
Nonfinancial assets	204.4	12.1%	100.0%	10678.1	81.6%	100.0%	69.4%	0.0%	21093.4	125.8%	100.0%	44.2%	0.0%	113.7%	0.0%
Real estate	0	0.0%	0.0%		0.0%	0.0%	0.0%	0.0%		0.0%	0.0%	0.0%	0.0%	0.0%	0.0%
Structures	0	0.0%	0.0%		0.0%	0.0%	0.0%	0.0%		0.0%	0.0%	0.0%	0.0%	0.0%	0.0%
Land		0.0%	0.0%		0.0%	0.0%	0.0%	0.0%		0.0%	0.0%	0.0%	0.0%	0.0%	0.0%
Equipment	0	0.0%	0.0%		0.0%	0.0%	0.0%	0.0%		0.0%	0.0%	0.0%	0.0%	0.0%	0.0%
Intellectual property products	0	0.0%	0.0%		0.0%	0.0%	0.0%	0.0%		0.0%	0.0%	0.0%	0.0%	0.0%	0.0%
Financial assets	204.4	12.1%	100.0%	10678.1	81.6%	100.0%	69.4%	0.0%	21093.4	125.8%	100.0%	44.2%	0.0%	113.7%	0.0%
SDR allocations	2.7	0.2%	1.3%	7.0	0.1%	0.1%	(0.1%)	(1.3%)	54.4	0.3%	0.3%	0.3%	0.2%	0.2%	(1.1%)
Currency and deposits	40.2	2.4%	19.7%	560.1	4.3%	5.2%	1.9%	(14.4%)	1429.8	8.5%	6.8%	4.2%	1.5%	6.1%	(12.9%)
Currency	9.9	0.6%	4.8%	280.4	2.1%	2.6%	1.6%	(2.2%)	491.9	2.9%	2.3%	0.8%	(0.3%)	2.3%	(2.5%)
Transferable deposits	13.7	0.8%	6.7%	19.7	0.2%	0.2%	(0.7%)	(6.5%)	98.4	0.6%	0.5%	0.4%	0.3%	(0.2%)	(6.2%)
Time deposits	22.6	1.3%	11.1%	223.2	1.7%	2.1%	0.4%	(9.0%)	442.8	2.6%	2.1%	0.9%	0.0%	1.3%	(9.0%)
Net interbank items due from US banks	-6.0	(0.4%)	(2.9%)	36.7	0.3%	0.3%	0.6%	3.3%	396.7	2.4%	1.9%	2.1%	1.5%	2.7%	4.8%
Debt securities	88.0	5.2%	43.1%	4980.2	38.0%	46.6%	32.8%	3.6%	9592.0	57.2%	45.5%	19.2%	(1.2%)	52.0%	2.4%
Open market paper	5.3	0.3%	2.6%	156.8	1.2%	1.5%	0.9%	(1.1%)	101.7	0.6%	0.5%	(0.6%)	(1.0%)	0.3%	(2.1%)
Treasury securities	65.5	3.9%	32.0%	1984.4	15.2%	18.6%	11.3%	(13.5%)	5794.9	34.6%	27.5%	19.4%	8.9%	30.7%	(4.6%)
Agency and GSE backed securities	2.7	0.2%	1.3%	1006.1	7.7%	9.4%	7.5%	8.1%	885.3	5.3%	4.2%	(2.4%)	(5.2%)	5.1%	2.9%
Municipal securities	0.0	0.0%	0.0%	29.0	0.2%	0.3%	0.2%	0.3%	76.1	0.5%	0.4%	0.2%	0.1%	0.5%	0.4%
Corporate bonds	14.5	0.9%	7.1%	1803.8	13.8%	16.9%	12.9%	9.8%	2734.0	16.3%	13.0%	2.5%	(3.9%)	15.4%	5.9%
Loans (short term)	0.8	0.0%	0.4%	867.6	6.6%	8.1%	6.6%	7.7%	874.4	5.2%	4.1%	(1.4%)	(4.0%)	5.2%	3.8%
Security repurchases	0.5	0.0%	0.2%	705.0	5.4%	6.6%	5.4%	6.4%	734.1	4.4%	3.5%	(1.0%)	(3.1%)	4.3%	3.2%
Loans to US corporate business	0.3	0.0%	0.1%	132.5	1.0%	1.2%	1.0%	1.1%	140.4	0.8%	0.7%	(0.2%)	(0.6%)	0.8%	0.5%
Equity and investment fund shares	58.6	3.5%	28.7%	4210.0	32.2%	39.4%	28.7%	10.8%	8998.4	53.7%	42.7%	21.5%	3.2%	50.2%	14.0%
Money market fund shares	0.0	0.0%	0.0%	23.0	0.2%	0.2%	0.2%	0.2%	113.3	0.7%	0.5%	0.5%	0.3%	0.7%	0.5%
Corporate equities	33.4	2.0%	16.3%	2118.4	16.2%	19.8%	14.2%	3.5%	5163.7	30.8%	24.5%	14.6%	4.6%	28.8%	8.1%
Mutual fund shares	0.0	0.0%	0.0%	162.7	1.2%	1.5%	1.2%	1.5%	544.5	3.2%	2.6%	2.0%	1.1%	3.2%	2.6%
Foreign direct investment in the US	25.2	1.5%	12.3%	1906.0	14.6%	17.8%	13.1%	5.5%	3176.9	18.9%	15.1%	4.4%	(2.8%)	17.5%	2.7%
Other accounts receivable	14.2	0.8%	6.9%	55.3	0.4%	0.5%	(0.4%)	(6.4%)	144.4	0.9%	0.7%	0.4%	0.2%	0.0%	(6.3%)

US GDP and Wealth Comparisons 1975, 2005, 2013 - Sheet 13

	1975	% GDP	% Assets	2005	% GDP	% Assets	+/- GDP from 1975	+/- Assets	2013	% GDP	% Assets	+/- GDP from 2005	+/- Assets	+/- GDP from 1975	+/- Assets
Total liabilities and net worth	204.4	12.1%	100.0%	10678.1	81.6%	100.0%	69.4%	0.0%	21093.4	125.8%	100.0%	44.2%	0.0%	113.7%	0.0%
Liabilities	288.5	17.1%	141.1%	9027.4	68.9%	84.5%	51.9%	(56.6%)	16709.8	99.7%	79.2%	30.7%	(5.3%)	82.6%	(61.9%)
SDR holdings	2.3	0.1%	1.1%	8.2	0.1%	0.1%	(0.1%)	(1.0%)	55.2	0.3%	0.3%	0.3%	0.2%	0.2%	(0.9%)
Currency and deposits	11.6	0.7%	5.7%	1213.8	9.3%	11.4%	8.6%	5.7%	1035.6	6.2%	4.9%	(3.1%)	(6.5%)	5.5%	(0.8%)
Official foreign currencies	0.1	0.0%	0.0%	37.6	0.3%	0.4%	0.3%	0.3%	47.4	0.3%	0.2%	(0.0%)	(0.1%)	0.3%	0.2%
Reserve position in IMF (net)	2.2	0.1%	1.1%	8.0	0.1%	0.1%	(0.1%)	(1.0%)	30.8	0.2%	0.1%	0.1%	0.1%	0.1%	(0.9%)
Us private deposits	5.8	0.3%	2.8%	1165.5	8.9%	10.9%	8.6%	8.1%	954.8	5.7%	4.5%	(3.2%)	(6.4%)	5.4%	1.7%
Nonofficial foreign currencies	3.6	0.2%	1.8%	2.6	0.0%	0.0%	(0.2%)	(1.7%)	2.6	0.0%	0.0%	(0.0%)	(0.0%)	(0.2%)	(1.7%)
Debt securities	27.1	1.6%	13.3%	1208.8	9.2%	11.3%	7.6%	(1.9%)	2657.3	15.8%	12.6%	6.6%	1.3%	14.2%	(0.7%)
Commercial paper	0.6	0.0%	0.3%	384.0	2.9%	3.6%	2.9%	3.3%	407.0	2.4%	1.9%	(0.5%)	(1.7%)	2.4%	1.6%
Bonds	26.5	1.6%	13.0%	824.8	6.3%	7.7%	4.7%	(5.2%)	2250.4	13.4%	10.7%	7.1%	2.9%	11.9%	(2.3%)
Loans (short term)	68.2	4.0%	33.4%	523.3	4.0%	4.9%	(0.0%)	(28.5%)	1030.8	6.1%	4.9%	2.2%	(0.0%)	2.1%	(28.5%)
Security repurchases	0.0	0.0%	0.0%	381.0	2.9%	3.6%	2.9%	3.6%	721.6	4.3%	3.4%	1.4%	(0.1%)	4.3%	3.4%
Other loans and advances	46.6	2.8%	22.8%	31.9	0.2%	0.3%	(2.5%)	(22.5%)	32.2	0.2%	0.2%	(0.1%)	(0.1%)	(2.6%)	(22.6%)
Depository institution loans	21.6	1.3%	10.6%	110.4	0.8%	1.0%	(0.4%)	(9.5%)	276.7	1.7%	1.3%	0.8%	0.3%	0.4%	(9.3%)
Nonofficial foreign currency (swap lines)	0.0	0.0%	0.0%	0.0	0.0%	0.0%	0.0%	0.0%	0.3	0.0%	0.0%	0.0%	0.0%	0.0%	0.0%
Equity and investment fund shares	166.7	9.9%	81.6%	6039.8	46.1%	56.6%	36.3%	(25.0%)	11885.6	70.9%	56.3%	24.8%	(0.2%)	61.0%	(25.2%)
Corporate equities	9.6	0.6%	4.7%	3317.7	25.3%	31.1%	24.8%	26.4%	6444.2	38.4%	30.6%	13.1%	(0.5%)	37.9%	25.9%
US government equity in IBRD, etc.	4.1	0.2%	2.0%	43.2	0.3%	0.4%	0.1%	(1.6%)	59.5	0.4%	0.3%	0.0%	(0.1%)	0.1%	(1.7%)
US direct investment abroad	149.5	8.9%	73.1%	2651.7	20.3%	24.8%	11.4%	(48.3%)	5284.0	31.5%	25.1%	11.3%	0.2%	22.7%	(48.1%)
Investment by holding companies	3.5	0.2%	1.7%	27.1	0.2%	0.3%	(0.0%)	(1.5%)	97.8	0.6%	0.5%	0.4%	0.2%	0.4%	(1.2%)
Other accounts payable	12.5	0.7%	6.1%	33.6	0.3%	0.3%	(0.5%)	(5.8%)	45.3	0.3%	0.2%	0.0%	(0.1%)	(0.5%)	(5.9%)
Net worth (market value)	(84.1)	(5.0%)	(41.1%)	1650.7	12.6%	15.5%	17.6%	56.6%	4383.6	26.1%	20.8%	13.5%	5.3%	31.1%	61.9%

Source: Z.1 Financial Accounts of the United States, Federal Reserve Statistical Release, September 18, 2014, S.9.a Historical Annuals, 1975-1984 and 2005-2013

US GDP and Wealth Comparisons 1975, 2005, 2013 - Sheet 14

Re: US Credit Market Debt and Total Liabilities and its Relation to Total Financial Assets
L.4&5 Comparison 1975/2005/2013

Billions of dollars

	1975	% GDP	% NW	2005	% GDP	% NW	+/- GDP from 1975	+/- NW	2013	% GDP	% NW	+/- GDP from 2005	+/- NW	+/- GDP from 1975	+/- NW
GDP	1688.9	19.3%		13093.7	15.4%		132.4%		16768.0	15.7%		(12.7%)		119.7%	
Net Worth Household & Nonprofits, Nonfinancial Corporate & Noncorporate Business *	8758.6	518.6%	100.0%	85234.5	651.0%	100.0%	161.9%	18.8% x	107022.2	638.3%	100.0%	21.3%	4.3% x	183.1%	23.1%
Credit Market Debt	2616.9	154.9%	29.9%	41480.9	316.8%	48.7%	161.9%	18.8% x	56685.7	338.1%	53.0%	21.3%	4.3% x	183.1%	23.1%
Open market paper	66.6	3.9%	0.8%	1644.2	12.6%	1.9%	8.6%	1.2%	951.6	5.7%	0.9%	(6.9%)	(1.0%)	1.7%	0.1%
Treasury securities	434.9	25.8%	5.0%	4678.0	35.7%	5.5%	10.0%	0.5% x	12328.3	73.5%	11.5%	37.8%	6.0% x	47.8%	6.6%
Agency and GSE backed securities	115.2	6.8%	1.3%	6164.5	47.1%	7.2%	40.3%	5.9% x	7794.1	46.5%	7.3%	(0.6%)	0.1% x	39.7%	6.0%
Municipal securities	223.0	13.2%	2.5%	3019.3	23.1%	3.5%	9.9%	1.0% x	3671.2	21.9%	3.4%	(1.2%)	(0.1%) x	8.7%	0.9%
Corporate and foreign bonds	336.4	19.9%	3.8%	8141.2	62.2%	9.6%	42.3%	5.7% x	11103.2	66.2%	10.4%	4.0%	0.8% x	46.3%	6.5%
Depository institution loans	262.3	15.5%	3.0%	1583.0	12.1%	1.9%	(3.4%)	(1.1%)	2508.8	15.0%	2.3%	2.9%	0.5%	(0.6%)	(0.7%)
Other loans and advances	186.0	11.0%	2.1%	1819.8	13.9%	2.1%	2.9%	0.0%	1935.0	11.5%	1.8%	(2.4%)	(0.3%)	0.5%	(0.3%)
Mortgages	785.6	46.5%	9.0%	12110.3	92.5%	14.2%	46.0%	5.2% x	13295.5	79.3%	12.4%	(13.2%)	(1.8%) x	32.8%	3.5%
Consumer credit	207.0	12.3%	2.4%	2320.6	17.7%	2.7%	5.5%	0.4%	3097.9	18.5%	2.9%	0.8%	0.2%	6.2%	0.5%
Total Liabilities and its Relation to Total Financial Assets															
Total credit market debt	2616.9	154.9%	29.9%	41480.9	316.8%	48.7%	161.9%	18.8% x	56685.7	338.1%	53.0%	21.3%	4.3% x	183.1%	23.1%
US official reserve assets	7.3	0.4%	0.1%	60.9	0.5%	0.1%	0.0%	(0.0%)	187.7	1.1%	0.2%	0.7%	0.1%	0.7%	0.1%
SDR certificates	0.5	0.0%	0.0%	2.2	0.0%	0.0%	(0.0%)	(0.0%)	5.2	0.0%	0.0%	0.0%	0.0%	0.0%	(0.0%)
Treasury currency	8.2	0.5%	0.1%	27.5	0.2%	0.0%	(0.3%)	(0.1%)	25.6	0.2%	0.0%	(0.1%)	(0.0%)	(0.3%)	(0.1%)
Foreign deposits	5.8	0.3%	0.1%	1165.5	8.9%	1.4%	8.6%	1.3% x	954.8	5.7%	0.9%	(3.2%)	(0.5%) x	5.4%	0.8%
Net interbank liabilities	23.5	1.4%	0.3%	46.0	0.4%	0.1%	(1.0%)	(0.2%)	2795.4	16.7%	2.6%	16.3%	2.6% x	15.3%	2.3%
Checkable deposits and currency	331.1	19.6%	3.8%	1527.2	11.7%	1.8%	(7.9%)	(2.0%) x	3186.5	19.0%	3.0%	7.3%	1.2% x	(0.6%)	(0.8%)
Small time and savings deposits	726.5	43.0%	8.3%	4598.1	35.1%	5.4%	(7.9%)	(2.9%) x	8110.8	48.4%	7.6%	13.3%	2.2% x	5.4%	(0.7%)
Large time deposits	158.3	9.4%	1.8%	1892.4	14.5%	2.2%	5.1%	0.4%	1762.0	10.5%	1.6%	(3.9%)	(0.6%)	1.1%	(0.2%)
Money market fund shares	3.7	0.2%	0.0%	1993.1	15.2%	2.3%	15.0%	2.3%	2678.3	16.0%	2.5%	0.8%	0.2%	15.8%	2.5%
Security repurchase agreements	60.9	3.6%	0.7%	3756.6	28.7%	4.4%	25.1%	3.7% x	3652.9	21.8%	3.4%	(6.9%)	(1.0%)	18.2%	2.7%
Mutual fund shares	43.0	2.5%	0.5%	6045.6	46.2%	7.1%	43.6%	6.6% x	11544.8	68.9%	10.8%	22.7%	3.7% x	66.3%	10.3%
Security credit	24.6	1.5%	0.3%	893.3	6.8%	1.0%	5.4%	0.8%	1236.2	7.4%	1.2%	0.6%	0.1%	5.9%	0.9%
Life insurance reserves	168.6	10.0%	1.9%	1082.6	8.3%	1.3%	(1.7%)	(0.7%)	1416.5	8.4%	1.3%	0.2%	0.1%	(1.5%)	(0.6%)
Pension entitlements	1037.1	61.4%	11.8%	13466.3	102.8%	15.8%	41.4%	4.0% x	19886.1	118.6%	18.6%	15.7%	2.8% x	57.2%	6.7%
Trade payables	278.6	16.5%	3.2%	2982.6	22.8%	3.5%	6.3%	0.3%	3760.9	22.4%	3.5%	(0.3%)	0.0%	5.9%	0.3%
Taxes payable	31.3	1.9%	0.4%	214.5	1.6%	0.3%	(0.2%)	(0.1%)	90.9	0.5%	0.1%	(1.1%)	(0.2%)	(1.3%)	(0.3%)
Miscellaneous	1091.9	64.7%	12.5%	12195.8	93.1%	14.3%	28.5%	1.8% x	17190.4	102.5%	16.1%	9.4%	1.8% x	37.9%	3.6%
Total Liabilities	6729.5	398.5%	76.8%	97988.8	748.4%	115.0%	349.9%	38.1% x	143631.9	856.6%	134.2%	108.2%	19.2% x	458.1%	57.4%
Financial assets not included in liabilities:															
** Gold	11.6	0.7%	0.1%	11.0	0.1%	0.0%	(0.6%)	(0.1%)	11.0	0.1%	0.0%	(0.0%)	(0.0%)	(0.6%)	(0.1%)
** Corporate equities by sector holding (MV)	839.4			20636.1					33671.6						
Corporate equities held - Households & nonprofits	584.6	34.6%	6.7%	8025.4	61.3%	9.4%	26.7%	2.7%	12451.3	74.3%	11.6%	13.0%	2.2%	39.6%	5.0%
Corporate equities held - Federal Government	0.0	0.0%	0.0%	0.0	0.0%	0.0%	0.0%	0.0%	35.1	0.2%	0.0%	0.2%	0.0%	0.2%	0.0%
Corporate equities held - Financial	221.4	13.1%	2.5%	10376.3	79.2%	12.2%	66.1%	9.6% x	15853.9	94.5%	14.8%	15.3%	2.6% x	81.4%	12.3%
Corporate equities held - State & local	0.0	0.0%	0.0%	116.0	0.9%	0.1%	0.9%	0.1%	167.6	1.0%	0.2%	0.1%	0.0%	1.0%	0.2%
Corporate equities held - Rest of the World	33.4	2.0%	0.4%	2118.4	16.2%	2.5%	14.2%	2.1%	5163.7	30.8%	4.8%	14.6%	2.3%	28.8%	4.4%
Corporate equities by sector issuing (MV)	839.4			20636.2					33671.5						
Corporate equities issued - Nonfinancial	754.8	44.7%	8.6%	12686.6	96.9%	14.9%	52.2%	6.3% x	20761.7	123.8%	19.4%	26.9%	4.5% x	79.1%	10.8%
Corporate equities issued - Financial	75.0	4.4%	0.9%	4631.9	35.4%	5.4%	30.9%	4.6% x	6465.6	38.6%	6.0%	3.2%	0.6% x	34.1%	5.2%
Corporate equities issued - Rest of the World	9.6	0.6%	0.1%	3317.7	25.3%	3.9%	24.8%	3.8% x	6444.2	38.4%	6.0%	13.1%	2.1% x	37.9%	5.9%
** Household equity in noncorporate business	1150.6	68.1%	13.1%	8473.0	64.7%	9.9%	(3.4%)	(3.2%) x	9057.7	54.0%	8.5%	(10.7%)	(1.5%) x	(14.1%)	(4.7%)

US GDP and Wealth Comparisons 1975, 2005, 2013 - Sheet 15

	1975	% GDP	% NW	2005	% GDP	% NW	+/- GDP from 1975	+/- NW from 1975	2013	% GDP	% NW	+/- GDP from 2005	+/- NW from 2005	+/- GDP from 1975	+/- NW from 1975
**** Liabilities not identified as assets:															
Treasury currency	-2.0	(0.1%)	(0.0%)	-9.1	(0.1%)	(0.0%)	0.0%	0.0%	-19.9	(0.1%)	(0.0%)	(0.0%)	(0.0%)	(0.0%)	0.0%
Foreign deposits	3.2	0.2%	0.0%	962.3	7.3%	1.1%	7.2%	1.1% x	780.8	4.7%	0.7%	(2.7%)	(0.4%) x	4.5%	0.7%
Net interbank transactions	-12.8	(0.8%)	(0.1%)	-60.7	(0.5%)	(0.1%)	0.3%	0.1%	76.9	0.5%	0.1%	0.9%	0.1%	1.2%	0.2%
Security repurchase agreements	6.1	0.4%	0.1%	322.2	2.5%	0.4%	2.1%	0.3% x	38.6	0.2%	0.0%	(2.2%)	(0.3%) x	(0.1%)	(0.0%)
Taxes payable	17.2	1.0%	0.2%	20.6	0.2%	0.0%	(0.9%)	(0.2%)	-198.7	(1.2%)	(0.2%)	(1.3%)	(0.2%)	(2.2%)	(0.4%)
Miscellaneous	27.4	1.6%	0.3%	-3945.0	(30.1%)	(4.6%)	(31.8%)	(4.9%) x	-6604.6	(39.4%)	(6.2%)	(9.3%)	(1.5%) x	(41.0%)	(6.5%)
Floats not included in assets:															
Checkable deposits: State and local government	5.1	0.3%	0.1%	0.0	0.0%	0.0%	(0.3%)	(0.1%)	0.0	0.0%	0.0%	0.0%	0.0%	(0.3%)	(0.1%)
Federal government	-0.3	(0.0%)	(0.0%)	1.6	0.0%	0.0%	0.0%	0.0%	1.2	0.0%	0.0%	(0.0%)	(0.0%)	0.0%	0.0%
Private domestic	24.2	1.4%	0.3%	7.4	0.1%	0.0%	(1.4%)	(0.3%)	3.7	0.0%	0.0%	(0.0%)	(0.0%)	(1.4%)	(0.3%)
Trade credit	-50.9	(3.0%)	(0.6%)	-5.2	(0.0%)	(0.0%)	3.0%	0.6%	240.6	1.4%	0.2%	1.5%	0.2%	4.4%	0.8%
Total identified to sectors as assets ‡‡	8776.9	519.7%	100.2%	129814.8	991.4%	152.3%	471.7%	52.1% x	192053.6	1145.4%	179.5%	153.9%	27.1% x	625.7%	79.2%

* Total of sector new worths taken from associated balance sheets
** "Total identified to sectors as assets" equals "Total Liabilities" plus accounts preceded by **

Source: Z.1 Financial Accounts of the United States, Federal Reserve Statistical Release, September 18, 2014, L.4 & L.5 Historical Annuals, 1975-1984 and 2005-2013

US GDP and Wealth Accounts
Re: US Sector Accounts
Various Comparison 1975/2005/2013

US GDP and Wealth Comparisons 1975, 2005, 2013 - Sheet 16

Billions of dollars	1975	% GDP	% Acct	2005	% GDP	% Acct	+/- GDP from 1975	+/- Acct	2013	% GDP	% Acct	+/- GDP from 2005	+/- Acct	+/- GDP from 1975	+/- Acct
Assets															
Households & nonprofits	6560.1	388.4%	40.5%	74048.8	565.5%	38.5%	177.1%	(2.1%)	92669.2	552.7%	35.1%	(12.9%)	(3.3%)	164.2%	(5.4%)
Nonfinancial Corporate Business	2906.1	172.1%	18.0%	26068.7	199.1%	13.5%	27.0%	(4.4%)	34941.9	208.4%	13.2%	9.3%	(0.3%)	36.3%	(4.7%)
Nonfinancial Noncorporate Business	1497.7	88.7%	9.3%	12641.7	96.5%	6.6%	7.9%	(2.7%)	15023.1	89.6%	5.7%	(7.0%)	(0.9%)	0.9%	(3.6%)
Financial Business	3261.1	193.1%	20.2%	57468.5	438.9%	29.8%	245.8%	9.7%	82657.4	492.9%	31.3%	54.0%	1.5%	299.9%	11.2%
Federal Government	685.1	40.6%	4.2%	3070.2	23.4%	1.6%	(17.1%)	(2.6%)	4905.4	29.3%	1.9%	5.8%	0.3%	(11.3%)	(2.4%)
State & Local Government	1064.5	63.0%	6.6%	8576.2	65.5%	4.5%	2.5%	(2.1%)	12641.4	75.4%	4.8%	9.9%	0.3%	12.4%	(1.8%)
Rest of the World	204.4	12.1%	1.3%	10678.1	81.6%	5.5%	69.4%	4.3%	21093.4	125.8%	8.0%	44.2%	2.4%	113.7%	6.7%
	16179.0		100.0%	192552.2		100.0%			263931.8		100.0%				
Nonfinancial assets															
Households & nonprofits	2286.5	135.4%	30.9%	28532.0	217.9%	45.5%	82.5%	14.6%	27669.2	165.0%	38.5%	(52.9%)	(7.0%)	29.6%	7.6%
Nonfinancial Corporate Business	2131.8	126.2%	28.8%	14168.0	108.2%	22.6%	(18.0%)	(6.2%)	18561.3	110.7%	25.8%	2.5%	3.2%	(15.5%)	(3.0%)
Nonfinancial Noncorporate Business	1414.7	83.8%	19.1%	10044.9	76.7%	16.0%	(7.0%)	(3.1%)	11168.7	66.6%	15.5%	(10.1%)	(0.5%)	(17.2%)	(3.6%)
Financial Business	86.9	5.1%	1.2%	1306.8	10.0%	2.1%	4.8%	0.9%	1584.2	9.4%	2.2%	(0.5%)	0.1%	4.3%	1.0%
Federal Government	564.1	33.4%	7.6%	2426.5	18.5%	3.9%	(14.9%)	(3.8%)	3191.7	19.0%	4.4%	0.5%	0.6%	(14.4%)	(3.2%)
State & Local Government	918.2	54.4%	12.4%	6259.0	47.8%	10.0%	(6.6%)	(2.4%)	9703.2	57.9%	13.5%	10.1%	3.5%	3.5%	1.1%
Rest of the World	0.0	0.0%	0.0%	0.0	0.0%	0.0%	0.0%	0.0%	0.0	0.0%	0.0%	0.0%	0.0%	0.0%	0.0%
	7402.2		100.0%	62737.2		100.0%			71878.3		100.0%				
Financial assets															
Households & nonprofits	4273.7	253.0%	48.7%	45516.7	347.6%	35.1%	94.6%	(13.6%)	65000.1	387.6%	33.8%	40.0%	(1.2%)	134.6%	(14.8%)
Nonfinancial Corporate Business	774.2	45.8%	8.8%	11900.6	90.9%	9.2%	45.0%	0.3%	16380.6	97.7%	8.5%	6.8%	(0.6%)	51.8%	(0.3%)
Nonfinancial Noncorporate Business	83.0	4.9%	0.9%	2596.8	19.8%	2.0%	14.9%	1.1%	3854.4	23.0%	2.0%	3.2%	0.0%	18.1%	1.1%
Financial Business	3174.2	187.9%	36.2%	56161.7	428.9%	43.3%	241.0%	7.1%	81073.2	483.5%	42.2%	54.6%	(1.0%)	295.6%	6.0%
Federal Government	121.0	7.2%	1.4%	643.7	4.9%	0.5%	(2.2%)	(0.9%)	1713.7	10.2%	0.9%	5.3%	0.4%	3.1%	(0.5%)
State & Local Government	146.4	8.7%	1.7%	2317.2	17.7%	1.8%	9.0%	0.1%	2938.2	17.5%	1.5%	(0.2%)	(0.3%)	8.9%	(0.1%)
Rest of the World	204.4	12.1%	2.3%	10678.1	81.6%	8.2%	69.4%	5.9%	21093.4	125.8%	11.0%	44.2%	2.8%	113.7%	8.7%
	8776.9		100.0%	129814.8		100.0%			192053.6		100.0%				
Liabilities															
Households & nonprofits	763.8	45.2%	11.1%	12162.4	92.9%	11.5%	47.7%	0.4%	13801.2	82.3%	8.8%	(10.6%)	(2.7%)	37.1%	(2.3%)
Nonfinancial Corporate Business	1092.4	64.7%	15.9%	11181.8	85.4%	10.6%	20.7%	(5.3%)	15827.8	94.4%	10.1%	9.0%	(0.4%)	29.7%	(5.8%)
Nonfinancial Noncorporate Business	349.3	20.7%	5.1%	4180.5	31.9%	3.9%	11.2%	(1.1%)	5983.1	35.7%	3.8%	3.8%	(0.1%)	15.0%	(1.3%)
Financial Business	3104.1	183.8%	45.1%	58641.8	447.9%	55.3%	264.1%	10.2%	83173.7	496.0%	53.1%	48.2%	(2.2%)	312.2%	8.0%
Federal Government	931.6	55.2%	13.5%	7370.3	56.3%	7.0%	1.1%	(6.6%)	16132.6	96.2%	10.3%	39.9%	3.3%	41.1%	(3.2%)
State & Local Government	349.5	20.7%	5.1%	3385.9	25.9%	3.2%	5.2%	(1.9%)	4931.0	29.4%	3.1%	3.5%	(0.0%)	8.7%	(1.9%)
Rest of the World	288.5	17.1%	4.2%	9027.4	68.9%	8.5%	51.9%	4.3%	16709.8	99.7%	10.7%	30.7%	2.2%	82.6%	6.5%
	6879.2		100.0%	105950.1		100.0%			156559.4		124.5%				
Net Worth															
Households & nonprofits	5796.4	343.2%	62.3%	61886.4	472.6%	71.5%	129.4%	9.1%	78868.0	470.3%	73.5%	(2.3%)	2.0%	127.1%	11.1%
Nonfinancial Corporate Business	1813.7	107.4%	19.5%	14886.9	113.7%	17.2%	6.3%	(2.3%)	19114.1	114.0%	17.8%	0.3%	0.6%	6.6%	(1.7%)
Nonfinancial Noncorporate Business	1148.5	68.0%	12.3%	8461.2	64.6%	9.8%	(3.4%)	(2.6%)	9040.1	53.9%	8.4%	(10.7%)	(1.4%)	(14.1%)	(3.9%)
Financial Business	157.1	9.3%	1.7%	(1173.3)	(9.0%)	(1.4%)	(18.3%)	(3.0%)	(516.5)	(3.1%)	(0.5%)	5.9%	0.9%	(12.4%)	(2.2%)
Federal Government	(246.5)	(14.6%)	(2.7%)	(4300.1)	(32.8%)	(5.0%)	(18.2%)	(2.3%)	(11227.2)	(67.0%)	(10.5%)	(34.1%)	(5.5%)	(52.4%)	(7.8%)
State & Local Government	715.0	42.3%	7.7%	5190.3	39.6%	6.0%	(2.7%)	(1.7%)	7710.4	46.0%	7.2%	6.3%	1.2%	3.6%	(0.5%)
Rest of the World	(84.1)	(5.0%)	(0.9%)	1650.7	12.6%	1.9%	17.6%	2.8%	4383.6	26.1%	4.1%	13.5%	2.2%	31.1%	5.0%
	9300.1		100.0%	86602.1		100.0%			107372.5		100.0%				

US GDP and Wealth Comparisons 1975, 2005, 2013 - Sheet 17

Nonfinancial Asset Breakdown

Billions of dollars

	1975	% GDP	% Acct	2005	% GDP	% Acct	+/- GDP from 1975	+/- Acct	2013	% GDP	% Acct	+/- GDP from 2005	+/- Acct	+/- GDP from 1975	+/- Acct
Real Estate															
Households & nonprofits	1684.2	99.7%	31.9%	24138.7	184.4%	49.2%	84.6%	17.3%	22295.9	133.0%	41.2%	(51.4%)	(8.0%)	33.2%	9.3%
Nonfinancial Corporate Business	1170.9	69.3%	22.2%	7924.3	60.5%	16.1%	(8.8%)	(6.0%)	10236.1	61.0%	18.9%	0.5%	2.8%	(8.3%)	(3.2%)
Nonfinancial Noncorporate Business	1197.0	70.9%	22.7%	9099.8	69.5%	18.5%	(1.4%)	(4.1%)	9864.4	58.8%	18.2%	(10.7%)	(0.3%)	(12.0%)	(4.4%)
Financial Business	55.7	3.3%	1.1%	751.8	5.7%	1.5%	2.4%	0.5%	899.2	5.4%	1.7%	(0.4%)	0.1%	2.1%	0.6%
Federal Government	288.1	17.1%	5.5%	1176.5	9.0%	2.4%	(8.1%)	(3.1%)	1472.1	8.8%	2.7%	(0.2%)	0.3%	(8.3%)	(2.7%)
State & Local Government	887.4	52.5%	16.8%	5976.8	45.6%	12.2%	(6.9%)	(4.6%)	9335.6	55.7%	17.3%	10.0%	5.1%	3.1%	0.5%
Rest of the World	0.0	0.0%	0.0%	0.0	0.0%	0.0%	0.0%	0.0%	0.0	0.0%	0.0%	0.0%	0.0%	0.0%	0.0%
	5283.3		100.0%	49067.9		100.0%			54103.3		100.0%				
Equipment, Durable Goods, & Inventories															
Households & nonprofits*	597.2	35.4%	32.3%	4313.9	32.9%	38.5%	(2.4%)	6.2%	5249.1	31.3%	36.8%	(1.6%)	(1.6%)	(4.1%)	4.5%
Nonfinancial Corporate Business*	857.4	50.8%	46.4%	4921.7	37.6%	43.9%	(13.2%)	(2.5%)	6408.3	38.2%	45.0%	0.6%	1.1%	(12.5%)	(1.4%)
Nonfinancial Noncorporate Business*	208.4	12.3%	11.3%	797.3	6.1%	7.1%	(6.3%)	(4.2%)	1089.0	6.5%	7.6%	0.4%	0.5%	(5.8%)	(3.6%)
Financial Business	27.8	1.6%	1.5%	445.4	3.4%	4.0%	1.8%	2.5%	519.0	3.1%	3.6%	(0.3%)	(0.3%)	1.4%	2.1%
Federal Government	131.8	7.8%	7.1%	520.7	4.0%	4.6%	(3.8%)	(2.5%)	732.2	4.4%	5.1%	0.4%	0.5%	(3.4%)	(2.0%)
State & Local Government	25.5	1.5%	1.4%	206.9	1.6%	1.8%	0.1%	0.5%	247.6	1.5%	1.7%	(0.1%)	(0.1%)	(0.0%)	0.4%
Rest of the World	0.0	0.0%	0.0%	0.0	0.0%	0.0%	0.0%	0.0%	0.0	0.0%	0.0%	0.0%	0.0%	0.0%	0.0%
	1848.1		100.0%	11205.9		100.0%			14245.2		100.0%				
Intellectual property products															
Households & nonprofits	5.1	0.3%	1.9%	79.5	0.6%	3.2%	0.3%	1.3%	124.1	0.7%	3.5%	0.1%	0.3%	0.4%	1.6%
Nonfinancial Corporate Business	103.6	6.1%	38.2%	1322.0	10.1%	53.7%	4.0%	15.4%	1916.9	11.4%	54.3%	1.3%	0.6%	5.3%	16.1%
Nonfinancial Noncorporate Business	9.3	0.6%	3.4%	147.8	1.1%	6.0%	0.6%	2.6%	215.3	1.3%	6.1%	0.2%	0.1%	0.7%	2.7%
Financial Business	3.4	0.2%	1.3%	109.5	0.8%	4.4%	0.6%	3.2%	166.1	1.0%	4.7%	0.2%	0.3%	0.8%	3.5%
Federal Government	144.2	8.5%	53.2%	729.3	5.6%	29.6%	(3.0%)	(23.6%)	987.4	5.9%	28.0%	0.3%	(1.6%)	(2.6%)	(25.3%)
State & Local Government	5.3	0.3%	2.0%	75.3	0.6%	3.1%	0.3%	1.1%	120.0	0.7%	3.4%	0.1%	0.3%	0.4%	1.4%
Rest of the World	0.0	0.0%	0.0%	0.0	0.0%	0.0%	0.0%	0.0%	0.0	0.0%	0.0%	0.0%	0.0%	0.0%	0.0%
	270.9		100.0%	2463.4		100.0%			3529.8		100.0%				

US GDP and Wealth Comparisons 1975, 2005, 2013 - Sheet 18

Financial Asset Breakdown

Billions of dollars	1975	% GDP	% Acct	2005	% GDP	% Acct	+/- GDP from 1975	+/- Acct	2013	% GDP	% Acct	+/- GDP from 2005	+/- Acct	+/- GDP from 1975	+/- Acct
Deposits, incl currency, time, savings & money market shares															
Households & nonprofits	924.5	54.7%	70.8%	6257.5	47.8%	60.2%	(6.9%)	(10.6%)	9630.5	57.4%	49.7%	9.6%	(10.6%)	2.7%	(21.1%)
Nonfinancial Corporate Business	86.6	5.1%	6.6%	1148.8	8.8%	11.1%	3.6%	4.4%	1624.5	9.7%	8.4%	0.9%	(2.7%)	4.6%	1.7%
Nonfinancial Noncorporate Business	32.4	1.9%	2.5%	767.3	5.9%	7.4%	3.9%	4.9%	993.6	5.9%	5.1%	0.1%	(2.3%)	4.0%	2.6%
Financial Business	99.0	5.9%	7.6%	775.8	5.9%	7.5%	0.1%	(0.1%)	3524.1	21.0%	18.2%	15.1%	10.7%	15.2%	10.6%
Federal Government	19.8	1.2%	1.5%	76.2	0.6%	0.7%	(0.6%)	(0.8%)	277.0	1.7%	1.4%	1.1%	0.7%	0.5%	(0.1%)
State & Local Government	60.3	3.6%	4.6%	234.8	1.8%	2.3%	(1.8%)	(2.4%)	425.8	2.5%	2.2%	0.7%	(0.1%)	(1.0%)	(2.4%)
Rest of the World	83.1	4.9%	6.4%	1127.1	8.6%	10.9%	3.7%	4.5%	2914.0	17.4%	15.0%	8.8%	4.2%	12.5%	8.7%
	1305.7		100.0%	10387.5		100.0%			19389.5		100.0%				
Checkable deposits and currency															
Households & nonprofits	158.7	9.4%	42.1%	285.9	2.2%	13.6%	(7.2%)	(28.4%)	1035.0	6.2%	16.4%	4.0%	2.7%	(3.2%)	(25.7%)
Nonfinancial Corporate Business	58.7	3.5%	15.6%	268.0	2.0%	12.8%	(1.4%)	(2.8%)	357.3	2.1%	5.7%	0.1%	(7.1%)	(1.3%)	(9.9%)
Nonfinancial Noncorporate Business	24.3	1.4%	6.4%	374.2	2.9%	17.9%	1.4%	11.4%	540.5	3.2%	8.5%	0.4%	(9.3%)	1.8%	2.1%
Financial Business	87.4	5.2%	23.2%	764.8	5.8%	36.5%	0.7%	13.3%	3513.1	21.0%	55.6%	15.1%	19.1%	15.8%	32.4%
Federal Government	11.2	0.7%	3.0%	37.0	0.3%	1.8%	(0.4%)	(1.2%)	163.0	1.0%	2.6%	0.7%	0.8%	0.3%	(0.4%)
State & Local Government	13.4	0.8%	3.6%	66.0	0.5%	3.1%	(0.3%)	(0.4%)	123.9	0.7%	2.0%	0.2%	(1.2%)	(0.1%)	(1.6%)
Rest of the World	23.6	1.4%	6.3%	300.1	2.3%	14.3%	0.9%	8.1%	590.3	3.5%	9.3%	1.2%	(5.0%)	2.1%	3.1%
	377.3		100.0%	2096.0		100.0%			6323.1		100.0%				
Credit market instruments* and Debt securities & Loans*															
Households & nonprofits*	320.8	19.0%	11.7%	25184.3	26.4%	36.0%	7.4%	24.3%	34558.5	23.0%	34.4%	(3.4%)	(1.7%)	4.0%	22.6%
Nonfinancial Corporate Business*	61.7	3.7%	2.3%	340.4	2.6%	0.5%	(1.1%)	(1.8%)	169.0	1.0%	0.2%	(1.6%)	(0.3%)	(2.6%)	(2.1%)
Nonfinancial Noncorporate Business*	7.2	0.4%	0.3%	96.8	0.7%	0.1%	0.3%	(0.1%)	100.2	0.6%	0.1%	(0.1%)	(0.0%)	0.2%	(0.2%)
Financial Business**	2107.7	31.0%	77.1%	32140.7	245.5%	46.0%	214.5%	(31.1%)	43030.8	256.6%	42.8%	11.2%	(3.2%)	225.6%	(34.3%)
Federal Government**	85.5	5.3%	3.1%	314.9	2.4%	0.5%	(2.9%)	(2.7%)	1138.7	6.8%	1.1%	4.4%	0.7%	1.5%	(2.0%)
State & Local Government**	61.7	3.7%	2.3%	1752.9	13.4%	2.5%	9.7%	0.3%	2109.4	12.6%	2.1%	(0.8%)	(0.4%)	8.9%	(0.2%)
Rest of the World**	88.8	8.7%	3.2%	10057.8	76.8%	14.4%	68.1%	11.1%	19464.8	116.1%	19.4%	39.3%	5.0%	107.4%	16.1%
	2733.4		100.0%	69887.8		100.0%			100571.4		100.0%				
Corporate equities, Life Insurance & Noncorporate equity* and Equity and Investment fund shares*															
Households & nonprofits*	1778.6	115.3%	83.4%	20648.9	166.0%	50.6%	50.7%	(32.8%)	29476.9	183.1%	45.9%	17.2%	(4.6%)	67.9%	(37.4%)
Nonfinancial Corporate Business*	0.9	0.1%	0.0%	134.4	1.0%	0.3%	1.0%	0.3%	211.4	1.3%	0.3%	0.2%	0.0%	1.2%	0.3%
Nonfinancial Noncorporate Business*															
Financial Business**	290.9	17.2%	13.6%	15535.6	118.6%	38.1%	101.4%	24.4%	24957.4	148.8%	38.9%	30.2%	0.8%	131.6%	25.3%
Federal Government**	4.1	0.2%	0.2%	43.2	0.3%	0.1%	0.1%	(0.1%)	98.9	0.6%	0.2%	0.3%	0.0%	0.3%	(0.0%)
State & Local Government**	0.0	0.0%	0.0%	246.6	1.9%	0.6%	1.9%	0.6%	415.4	2.5%	0.6%	0.6%	0.0%	2.5%	0.6%
Rest of the World**	58.6	3.5%	2.7%	4210.0	32.2%	10.3%	28.7%	7.6%	8998.4	53.7%	14.0%	21.5%	3.7%	50.2%	11.3%
	2133.1		100.0%	40818.7		100.0%			64158.4		100.0%				

US GDP and Wealth Comparisons 1975, 2005, 2013 - Sheet 19

Financial Asset Breakdown (continued)

Billions of dollars	1975	% GDP	% Acct	2005	% GDP	% Acct	+/- GDP from 1975	+/- Acct	2013	% GDP	% Acct	+/- GDP from 2005	+/- Acct	+/- GDP from 1975	+/- Acct
Pensions & Life Insurance reserves															
Households & nonprofits	1205.7	71.4%	67.5%	14548.9	111.1%	86.3%	39.7%	18.9%	21118.9	125.9%	83.7%	14.8%	(2.6%)	54.6%	16.3%
Nonfinancial Corporate Business															
Nonfinancial Noncorporate Business															
Financial Business	581.5	34.4%	32.5%	2304.7	17.6%	13.7%	(16.8%)	(18.9%)	4102.5	24.5%	16.3%	6.9%	2.6%	(10.0%)	(16.3%)
Federal Government															
State & Local Government															
Rest of the World															
	1787.2		100.0%	16853.6		100.0%			25221.4		100.0%				
Receivables															
Households & nonprofits															
Nonfinancial Corporate Business	271.4	16.1%	62.9%	2108.2	16.1%	34.1%	0.0%	(28.8%)	2469.4	14.7%	32.4%	(1.4%)	(1.7%)	(1.3%)	(30.5%)
Nonfinancial Noncorporate Business	14.6	0.9%	3.4%	430.9	3.3%	7.0%	2.4%	3.6%	558.0	3.3%	7.3%	0.0%	0.4%	2.5%	3.9%
Financial Business	95.1	5.6%	22.1%	3005.7	23.0%	48.6%	17.3%	26.6%	3737.4	22.3%	49.1%	(0.7%)	0.5%	16.7%	27.1%
Federal Government	11.5	0.7%	2.7%	252.5	1.9%	4.1%	1.2%	1.4%	298.1	1.8%	3.9%	(0.2%)	(0.2%)	1.1%	1.3%
State & Local Government	24.4	1.4%	5.7%	329.4	2.5%	5.3%	1.1%	(0.3%)	403.0	2.4%	5.3%	(0.1%)	(0.0%)	1.0%	(0.4%)
Rest of the World	14.2	0.8%	3.3%	55.3	0.4%	0.9%	(0.4%)	(2.4%)	144.4	0.9%	1.9%	0.4%	1.0%	0.0%	(1.4%)
	431.2			6182.0					7610.3						
Miscellaneous assets															
Households & nonprofits	44.0	2.6%	15.4%	608.7	4.6%	7.7%	2.0%	(7.7%)	924.9	5.5%	8.7%	0.9%	0.9%	2.9%	(6.7%)
Nonfinancial Corporate Business	212.6	12.6%	74.5%	5963.2	45.5%	75.7%	33.0%	1.2%	7536.1	44.9%	70.7%	(0.6%)	(5.1%)	32.4%	(3.8%)
Nonfinancial Noncorporate Business	28.8	1.7%	10.1%	1301.8	9.9%	16.5%	8.2%	6.4%	2202.6	13.1%	20.7%	3.2%	4.1%	11.4%	10.6%
Financial Business															
Federal Government															
State & Local Government															
Rest of the World															
	285.4		100.0%	7873.7		100.0%			10663.6		100.0%				

US GDP and Wealth Comparisons 1975, 2005, 2013 - Sheet 20

Liabilities Breakdown

Billions of dollars	1975	% GDP	% Acct	2005	% GDP	% Acct	+/- GDP from 1975	+/- Acct from 1975	2013	% GDP	% Acct	+/- GDP from 2005	+/- Acct from 2005	+/- GDP from 1975	+/- Acct from 1975
Credit market instruments* and Debt securities & Loans Equity and Investment fund shares**															
Households & nonprofits*	753.2	43.6%	17.2%	11976.1	89.5%	15.4%	45.9%	(1.8%)	13546.3	78.6%	11.6%	(10.9%)	(3.9%)	35.0%	(5.7%)
Nonfinancial Corporate Business*	572.7	33.9%	13.1%	5260.7	40.2%	6.8%	6.3%	(6.3%)	7118.0	42.4%	6.1%	2.3%	(0.7%)	8.5%	(7.0%)
Nonfinancial Noncorporate Business*	287.6	17.0%	6.6%	2898.0	22.1%	3.7%	5.1%	(2.8%)	4180.0	24.9%	3.6%	2.8%	(0.2%)	7.9%	(3.0%)
Financial Business** (includes currency & equity)	1765.0	93.6%	40.4%	40223.6	191.5%	51.9%	98.0%	11.5%	58896.0	200.6%	50.3%	9.1%	(1.6%)	107.1%	9.9%
Federal Government**	497.4	29.5%	11.4%	5621.5	42.9%	7.2%	13.5%	(4.1%)	13790.3	82.2%	11.8%	39.3%	4.5%	52.8%	0.4%
State & Local Government**	219.4	13.0%	5.0%	2589.8	19.8%	3.3%	6.8%	(1.7%)	2941.1	17.5%	2.5%	(2.2%)	(0.8%)	4.5%	(2.5%)
Rest of the World**	275.9	6.5%	6.3%	8993.9	22.6%	11.6%	16.1%	5.3%	16664.5	28.5%	14.2%	5.9%	2.6%	22.0%	7.9%
	4371.2		100.0%	77563.6		100.0%			117136.2		100.0%				
Mortgages (included in above)															
Households & nonprofits	474.1	28.1%	60.3%	9085.5	69.4%	75.0%	41.3%	14.7%	9619.4	57.4%	72.4%	(12.0%)	(2.7%)	29.3%	12.0%
Home mortgages	459.1	27.2%	58.4%	8912.7	68.1%	73.6%	40.9%	15.2%	9415.9	56.2%	70.8%	(11.9%)	(2.8%)	29.0%	12.4%
Commercial mortgages	15.0	0.9%	1.9%	172.8	1.3%	1.4%	0.4%	(0.5%)	203.5	1.2%	1.5%	(0.1%)	0.1%	0.3%	(0.4%)
Nonfinancial Corporate Business	107.9	6.4%	13.7%	786.4	6.0%	6.5%	(0.4%)	(7.2%)	602.5	3.6%	4.5%	(2.4%)	(2.0%)	(2.8%)	(9.2%)
Nonfinancial Noncorporate Business	201.0	11.9%	25.6%	2092.9	16.0%	17.3%	4.1%	(8.3%)	2871.4	17.1%	21.6%	1.1%	4.3%	5.2%	(4.0%)
Financial Business	1.5	0.1%	0.2%	145.5	1.1%	1.2%	1.0%	1.0%	202.2	1.2%	1.5%	0.1%	0.3%	1.1%	1.3%
Federal Government	1.1	0.1%	0.1%	0.0	0.0%	0.0%	(0.1%)	(0.1%)	0.0	0.0%	0.0%	0.0%	0.0%	(0.1%)	(0.1%)
State & Local Government															
Rest of the World	785.6		100.0%	12110.3		100.0%			13295.5		100.0%				
Pensions & Life Insurance reserves															
Households & nonprofits*															
Nonfinancial Corporate Business*															
Nonfinancial Noncorporate Business*															
Financial Business*	1277.5	75.6%	71.3%	15953.5	121.8%	89.6%	46.2%	18.3%	23046.8	137.4%	87.5%	15.6%	(2.1%)	61.8%	16.2%
Federal Government**	421.4	25.0%	23.5%	1540.9	11.8%	8.7%	(13.2%)	(14.9%)	2089.3	12.5%	7.9%	0.7%	(0.7%)	(12.5%)	(15.6%)
State & Local Government**	92.7	5.5%	5.2%	314.0	2.4%	1.8%	(3.1%)	(3.4%)	1204.1	7.2%	4.6%	4.8%	2.8%	1.7%	(0.6%)
Rest of the World**															
	1791.6		100.0%	17808.4		100.0%			26340.2		100.0%				
Accounts payable															
Households & nonprofits	10.6	0.6%	2.9%	186.3	1.4%	3.3%	0.8%	0.5%	255.0	1.5%	5.0%	0.1%	1.6%	0.9%	2.1%
Nonfinancial Corporate Business	199.2	11.8%	54.0%	1786.7	13.6%	32.0%	1.9%	(21.9%)	2012.8	12.0%	39.2%	(1.6%)	7.2%	0.2%	(14.7%)
Nonfinancial Noncorporate Business	35.0	2.1%	9.5%	421.2	3.2%	7.5%	1.1%	(1.9%)	546.8	3.3%	10.7%	0.0%	3.1%	1.2%	1.2%
Financial Business	61.6	3.6%	16.7%	2464.7	18.8%	44.2%	15.2%	27.5%	1231.0	7.3%	24.0%	(11.5%)	(20.2%)	3.7%	7.3%
Federal Government	12.9	0.8%	3.5%	208.0	1.6%	3.7%	0.8%	0.2%	253.0	1.5%	4.9%	(0.1%)	1.2%	0.7%	1.4%
State & Local Government	37.4	2.2%	10.1%	482.0	3.7%	8.6%	1.5%	(1.5%)	785.8	4.7%	15.3%	1.0%	6.7%	2.5%	5.2%
Rest of the World	12.5	0.7%	3.4%	33.6	0.3%	0.6%	(0.5%)	(2.8%)	45.3	0.3%	0.9%	0.0%	0.3%	(0.5%)	(2.5%)
	369.2		100.0%	5582.5		100.0%			5129.7		100.0%				

US GDP and Wealth Comparisons 1975, 2005, 2013 - Sheet 21

Liabilities Breakdown (continued)

Billions of dollars	1975	% GDP	% Acct	2005	% GDP	% Acct	+/- GDP from 1975	+/- Acct	2013	% GDP	% Acct	+/- GDP from 2005	+/- Acct	+/- GDP from 1975	+/- Acct
Miscellaneous liabilities															
Households & nonprofits															
Nonfinancial Corporate Business	320.6	19.0%	92.3%	4134.3	31.6%	82.8%	12.6%	(9.6%)	6697	39.9%	84.2%	8.4%	1.4%	21.0%	(8.1%)
Nonfinancial Noncorporate Business	26.7	1.6%	7.7%	861.2	6.6%	17.2%	5.0%	9.6%	1256.3	7.5%	15.8%	0.9%	(1.4%)	5.9%	8.1%
Financial Business															
Federal Government															
State & Local Government															
Rest of the World	347.3		100.0%	4995.5		100.0%			7953.3		100.0%				

US GDP and Wealth Comparisons 1975, 2005, 2013 - Sheet 22

Re: US Assets and Liabilities of the Personal Sector
L.10 Comparison 1975/2005/2013

		1975	% GDP	% Assets	2005	% GDP	% Assets	+/- GDP from 1975	+/- Assets	2013	% GDP	% Assets	+/- GDP from 2005	+/- Assets	+/- GDP from 1975	+/- Assets
GDP	Billions of dollars	1688.9	21.0%		13093.7	15.1%		185.0%		16768.0	15.6%		(19.8%)		165.1%	0.0%
Assets		8057.8	477.1%	100.0%	86690.5	662.1%	100.0%	75.5%		107692.3	642.2%	100.0%	(19.8%)	0.0%		
	Nonfinancial assets (BS-HH & BS-NN)	3701.2	219.1%	45.9%	38576.9	294.6%	44.5%	75.5%	(1.4%)	38837.9	231.6%	36.1%	(63.0%)	(8.4%)	12.5%	(9.9%)
	Real estate	2881.2	170.6%	35.8%	33238.5	253.9%	38.3%	83.3%	2.6%	32160.3	191.8%	29.9%	(62.1%)	(8.5%)	21.2%	(5.9%)
	Households	1992.4	118.0%	24.7%	27852.8	212.7%	32.1%	94.7%	7.4%	25144.7	150.0%	23.3%	(62.8%)	(8.8%)	32.0%	(1.4%)
	Nonprofit organizations & Nonresidential	888.8	52.6%	11.0%	5385.7	41.1%	6.2%	(11.5%)	(4.8%)	7015.6	41.8%	6.5%	0.7%	0.3%	(10.8%)	(4.5%)
	Consumer durable goods	578.0	34.2%	7.2%	4107.8	31.4%	4.7%	(2.9%)	(2.4%)	4942.2	29.5%	4.6%	(1.9%)	(0.1%)	(4.7%)	(2.6%)
	Equipment (nonprofits & noncorporate bus)	147.5	8.7%	1.8%	785	6.0%	0.9%	(2.7%)	(0.9%)	1098.6	6.6%	1.0%	0.6%	0.1%	(2.2%)	(0.8%)
	Inventories (noncorporate business)	80.1	4.7%	1.0%	218.4	1.7%	0.3%	(3.1%)	(0.7%)	297.3	1.8%	0.3%	0.1%	0.0%	(3.0%)	(0.7%)
	Intellectual property products (nonprofits)	14.4	0.9%	0.2%	227.3	1.7%	0.3%	0.9%	0.1%	339.4	2.0%	0.3%	0.3%	0.1%	1.2%	0.1%
	Financial assets (BS-HH & BS-NN #s equal Fed)	3208.2	190.0%	39.8%	39652.3	302.8%	45.7%	112.9%	5.9%	59814.3	356.7%	55.5%	53.9%	9.8%	166.8%	15.7%
L	Deposits	956.9	56.7%	11.9%	7024.8	53.7%	8.1%	(3.0%)	(3.8%)	10624.1	63.4%	9.9%	9.7%	1.8%	6.7%	(2.0%)
L	Foreign deposits	0.0	0.0%	0.0%	63.8	0.5%	0.1%	0.5%	0.1%	52.5	0.3%	0.0%	(0.2%)	(0.0%)	0.3%	0.0%
L	Checkable deposits and currency	183	10.8%	2.3%	660.1	5.0%	0.8%	(5.8%)	(1.5%)	1575.5	9.4%	1.5%	4.4%	0.7%	(1.4%)	(0.8%)
L	Time and savings deposits	770.3	45.6%	9.6%	5289.1	40.4%	6.1%	(5.2%)	(3.5%)	7770.5	46.3%	7.2%	5.9%	1.1%	0.7%	(2.3%)
L	Money market fund shares	3.7	0.2%	0.0%	1011.7	7.7%	1.2%	7.5%	1.1%	1225.6	7.3%	1.1%	(0.4%)	(0.0%)	7.1%	1.1%
F	Securities	894.8	53.0%	11.1%	14916.4	113.9%	17.2%	60.9%	6.1%	23343.9	139.2%	21.7%	25.3%	4.5%	86.2%	10.6%
F	Open market paper	19.7	1.2%	0.2%	98.4	0.8%	0.1%	(0.4%)	(0.1%)	15.0	0.1%	0.0%	(0.7%)	(0.1%)	(1.1%)	(0.2%)
F	Treasury securities	113.1	6.7%	1.4%	481.9	3.7%	0.6%	(3.0%)	(0.8%)	999.3	6.0%	0.9%	2.3%	0.4%	(0.7%)	(0.5%)
F	Agency and GSE backed securities	7.6	0.4%	0.1%	587.4	4.5%	0.7%	4.0%	0.6%	97.7	0.6%	0.1%	(3.9%)	(0.6%)	0.1%	0.0%
F	Municipal securities	66.8	4.0%	0.8%	1605	12.3%	1.9%	8.3%	1.0%	1623.6	9.7%	1.5%	(2.6%)	(0.3%)	5.7%	0.7%
F	Corporate and foreign bonds	64.3	3.8%	0.8%	591.4	4.5%	0.7%	0.7%	(0.1%)	1004.6	6.0%	0.9%	1.5%	0.3%	2.2%	0.1%
F	Corporate equities (market value)	584.6	34.6%	7.3%	8025.4	61.3%	9.3%	26.7%	2.0%	12451.3	74.3%	11.6%	13.0%	2.3%	39.6%	4.3%
F	Mutual fund shares	38.7	2.3%	0.5%	3527.1	26.9%	4.1%	24.6%	3.6%	7152.4	42.7%	6.6%	15.7%	2.6%	40.4%	6.2%
M	Security credit	4.7	0.3%	0.1%	4.7	0.0%	0.0%	(0.2%)	(0.1%)	4.7	0.0%	0.0%	(0.0%)	(0.0%)	(0.3%)	(0.1%)
F	Life insurance reserves	168.6	10.0%	2.1%	1082.6	8.3%	1.2%	(1.7%)	(0.8%)	1232.8	7.4%	1.1%	(0.9%)	(0.1%)	(2.6%)	(0.9%)
F	Pension entitlements	1037.1	61.4%	12.9%	13466.3	102.8%	15.5%	41.4%	2.7%	19886.1	118.6%	18.5%	15.7%	2.9%	57.2%	5.6%
F	Life Insurance Companies	72.3	4.3%	0.9%	2004.9	15.3%	2.3%	11.0%	1.4%	2816.0	16.8%	2.6%	1.5%	0.3%	12.5%	1.7%
F	Private pension funds	314.5	18.6%	3.9%	5435.9	41.5%	6.3%	22.9%	2.4%	8141.8	48.6%	7.6%	7.0%	1.3%	29.9%	3.7%
F	Governments	650.2	38.5%	8.1%	6025.5	46.0%	7.0%	7.5%	(1.1%)	8927.4	53.2%	8.3%	7.2%	1.3%	14.7%	0.2%
?	Miscellaneous and other assets	150.7	8.9%	1.9%	3162.3	24.2%	3.6%	15.2%	1.8%	4727.4	28.2%	4.4%	4.0%	0.7%	19.3%	2.5%
Liabilities		1113.0	65.9%	13.8%	16342.9	124.8%	18.9%	58.9%	5.0%	19784.3	118.0%	18.4%	(6.8%)	(0.5%)	52.1%	4.6%
F	Home mortgages*	459.1	27.2%	5.7%	8912.7	68.1%	10.3%	40.9%	4.6%	9415.9	56.2%	8.7%	(11.9%)	(1.5%)	29.0%	3.0%
F	Other mortgages*	216.0	12.8%	2.7%	2265.7	17.3%	2.6%	4.5%	(0.1%)	3074.9	18.3%	2.9%	1.0%	0.2%	5.5%	0.2%
M	Consumer credit	207.0	12.3%	2.6%	2320.6	17.7%	2.7%	5.5%	0.1%	3097.9	18.5%	2.9%	0.8%	0.2%	6.2%	0.3%
M	Policy loans*	31.5	1.9%	0.4%	119.0	0.9%	0.1%	(1.0%)	(0.3%)	141.3	0.8%	0.1%	(0.1%)	(0.0%)	(1.0%)	(0.3%)
M	Security credit	8.6	0.5%	0.1%	232.4	1.8%	0.3%	1.3%	0.2%	339.2	2.0%	0.3%	0.2%	0.0%	1.5%	0.2%
?	Other liabilities*	190.9	11.3%	2.4%	2492.4	19.0%	2.9%	7.7%	0.5%	3715.2	22.2%	3.4%	3.1%	0.6%	10.9%	1.1%
Net worth		6944.8	411.2%	86.2%	70347.6	537.3%	81.1%	126.1%	(5.0%)	87908.0	524.3%	81.6%	(13.0%)	0.5%	113.1%	(4.6%)

* Fed figures in table L.10 differ slightly for Home and Other mortgages, but total is the same. Similar with Policy loans and other liabilities. These flow from BS-HH & BS-NN.

L = Liquidity held for near term transactions, F = Securities held for longer term cash flow, i.e. provision of liquidity, M = Mixed L & F, ? = unknown

Source: Z.1 Financial Accounts of the United States, Federal Reserve Statistical Release, September 18, 2014, L.10 Historical Annuals, 1975-1984 and 2005-2013

US GDP and Wealth Comparisons 1975, 2005, 2013 - Sheet 23

Re: US Assets of the Public & Private Sectors
L.10+ Comparison 1975/2005/2013

		1975	% GDP	% Assets	2005	% GDP	% Assets	+/- GDP from 1975	+/- Assets	2013	% GDP	% Assets	+/- GDP from 2005	+/- Assets	+/- GDP from 1975	+/- Assets
GDP	Billions of dollars	1688.9	21.5%		13093.7	17.7%		79.0%		16768.0	18.0%		(37.4%)		41.5%	
Total Assets (public & private sectors)		8319.7	492.6%	100.0%	74839.3	571.6%	100.0%	40.9%	(5.1%)	89565.3	534.1%	100.0%	(50.5%)	(3.6%)	(9.6%)	(8.7%)
Nonfinancial assets public & private sectors		7402.2	438.3%	89.0%	62737.2	479.1%	83.8%	38.1%	5.1%	71878.3	428.7%	80.3%	13.1%	3.6%	51.2%	8.7%
Net financial assets public & private sectors		917.5	54.3%	11.0%	12102.1	92.4%	16.2%			17686.8	105.5%	19.7%				
Assets (public sectors)		468.6	27.7%	100.0%	890.2	6.8%	100.0%	(20.9%)		(3516.8)	(21.0%)	100.0%	(27.8%)		(48.7%)	
Nonfinancial assets public sectors		1482.3	87.8%	316.3%	8685.5	66.3%	975.7%	(21.4%)	659.4%	12894.9	76.9%	(366.7%)	10.6%	1342.3%	(10.9%)	(683.0%)
Nonfinancial assets Federal government		564.1	33.4%	120.4%	2426.5	18.5%	272.6%	(14.9%)	152.2%	3191.7	19.0%	(90.8%)	0.5%	363.3%	(14.4%)	(211.1%)
Nonfinancial assets State & local government		918.2	54.4%	195.9%	6259.0	47.8%	703.1%	(6.6%)	507.2%	9703.2	57.9%	(275.9%)	10.1%	979.0%	3.5%	(471.9%)
Net financial assets public sectors		(1013.7)	(60.0%)	(216.3%)	(7795.3)	(59.5%)	(875.7%)	0.5%	(659.4%)	(16411.7)	(97.9%)	466.7%	(38.3%)	1342.3%	(37.9%)	683.0%
Net financial assets Federal government		(810.6)	(48.0%)	(173.0%)	(6726.6)	(51.4%)	(755.6%)	(3.4%)	(582.6%)	(14418.9)	(86.0%)	410.0%	(34.6%)	1165.6%	(38.0%)	583.0%
Net financial assets state & local government		(203.1)	(12.0%)	(43.3%)	(1068.7)	(8.2%)	(120.1%)	3.9%	(76.7%)	(1992.8)	(11.9%)	56.7%	(3.7%)	176.7%	0.1%	100.0%
Assets (private sectors)		7851.1	464.9%	100.0%	73949.1	564.8%	100.0%	99.9%		93081.9	555.1%	100.0%	(9.7%)		90.3%	
Nonfinancial assets (Domestic private sectors)		5919.9	350.5%	75.4%	54051.7	412.8%	73.1%	62.3%	(2.3%)	58983.4	351.8%	63.4%	(61.0%)	(9.7%)	1.2%	(12.0%)
Households & nonprofits nonfinancial assets		3701.2	219.1%	47.1%	38576.9	294.6%	52.2%	75.5%	5.0%	38837.9	231.6%	41.7%	(63.0%)	(10.4%)	12.5%	(5.4%)
Nonfinancial corporate business nonfin assets		2131.8	126.2%	27.2%	14168.0	108.2%	19.2%	(18.0%)	(8.0%)	18561.3	110.7%	19.9%	2.5%	0.8%	(15.5%)	(7.2%)
Financial business nonfinancial assets		86.9	5.1%	1.1%	1306.8	10.0%	1.8%	4.8%	0.7%	1584.2	9.4%	1.7%	(0.5%)	(0.1%)	4.3%	0.6%
Nonfinancial assets (breakdown)																
Real estate		4107.8	243.2%	52.3%	41914.6	320.1%	56.7%	76.9%	4.4%	43295.6	258.2%	46.5%	(61.9%)	(10.2%)	15.0%	(5.8%)
Households		1992.4	118.0%	25.4%	27852.8	212.7%	37.7%	94.7%	12.3%	25144.7	150.0%	27.0%	(62.8%)	(10.7%)	32.0%	1.6%
Nonprofit organizations & nonresidential		888.8	52.6%	11.3%	5385.7	41.1%	7.3%	(11.5%)	(4.0%)	7015.6	41.8%	7.5%	0.7%	0.3%	(10.8%)	(3.8%)
Nonfinancial corporate business		1170.9	69.3%	14.9%	7924.3	60.5%	10.7%	(8.8%)	(4.2%)	10236.1	61.0%	11.0%	0.5%	0.3%	(8.3%)	(3.9%)
Financial business		55.7	3.3%	0.7%	751.8	5.7%	1.0%	2.4%	0.3%	899.2	5.4%	1.0%	(0.4%)	(0.1%)	2.1%	0.3%
Durable goods and equipment		1303.5	77.2%	16.6%	8636.0	66.0%	11.7%	(11.2%)	(4.9%)	10821.5	64.5%	11.6%	(1.4%)	(0.1%)	(12.6%)	(5.0%)
Consumer durable goods		578.0	34.2%	7.4%	4107.8	31.4%	5.6%	(2.9%)	(1.8%)	4942.2	29.5%	5.3%	(1.9%)	(0.2%)	(4.7%)	(2.1%)
Nonprofits & nonfinancial noncorporate bus		147.5	8.7%	1.9%	785	6.0%	1.1%	(2.7%)	(0.8%)	1098.6	6.6%	1.2%	0.6%	0.1%	(2.2%)	(0.7%)
Nonfinancial corporate business		550.2	32.6%	7.0%	3297.8	25.2%	4.5%	(7.4%)	(2.5%)	4261.7	25.4%	4.6%	0.2%	0.1%	(7.2%)	(2.4%)
Financial business		27.8	1.6%	0.4%	445.4	3.4%	0.6%	1.8%	0.2%	519.0	3.1%	0.6%	(0.3%)	(0.0%)	1.4%	0.2%
Inventories		387.3	22.9%	4.9%	1842.3	14.1%	2.5%	(8.9%)	(2.4%)	2443.9	14.6%	2.6%	0.5%	0.1%	(8.4%)	(2.3%)
Nonfinancial noncorporate business		80.1	4.7%	1.0%	218.4	1.7%	0.3%	(3.1%)	(0.7%)	297.3	1.8%	0.3%	0.1%	0.0%	(3.0%)	(0.7%)
Nonfinancial corporate business		307.2	18.2%	3.9%	1623.9	12.4%	2.2%	(5.8%)	(1.7%)	2146.6	12.8%	2.3%	0.4%	0.1%	(5.4%)	(1.6%)
Intellectual property products		121.4	7.2%	1.5%	1658.8	12.7%	2.2%	5.5%	0.7%	2422.4	14.4%	2.6%	1.8%	0.4%	7.3%	1.1%
Nonprofits & nonfinancial noncorporate bus		14.4	0.9%	0.2%	227.3	1.7%	0.3%	0.9%	0.1%	339.4	2.0%	0.4%	0.3%	0.1%	1.2%	0.2%
Nonfinancial corporate business		103.6	6.1%	1.3%	1322.0	10.1%	1.8%	4.0%	0.5%	1916.9	11.4%	2.1%	1.3%	0.3%	5.3%	0.7%
Financial business		3.4	0.2%	0.0%	109.5	0.8%	0.1%	0.6%	0.1%	166.1	1.0%	0.2%	0.2%	0.0%	0.8%	0.1%
Net financial assets domestic private sectors		1847.1	109.4%	23.5%	21548.1	164.6%	29.1%	55.2%	5.6%	38482.1	229.5%	41.3%	64.9%	12.2%	120.1%	17.8%
Net financial assets HH & NN business		2095.2	124.1%	26.7%	23309.4	178.0%	31.5%	54.0%	4.8%	40030.0	238.7%	43.0%	60.7%	11.5%	114.7%	16.3%
Net financial assets NC business		(318.2)	(18.8%)	(4.1%)	718.8	5.5%	1.0%	24.3%	5.0%	552.8	3.3%	0.6%	(2.2%)	(0.4%)	22.1%	4.6%
Net financial assets financial business		70.1	4.2%	0.9%	(2480.1)	(18.9%)	(3.4%)	(23.1%)	(4.2%)	(2100.7)	(12.5%)	(2.3%)	6.4%	1.1%	(16.7%)	(3.1%)
Net financial assets private sectors		1931.2	114.3%	24.6%	19897.4	152.0%	26.9%	37.6%	2.3%	34098.5	203.4%	36.6%	51.4%	9.7%	89.0%	12.0%
Net financial assets rest of the world		84.1	5.0%	1.1%	(1650.7)	(12.6%)	(2.2%)	(17.6%)	(3.3%)	(4383.6)	(26.1%)	(4.7%)	(13.5%)	(2.5%)	(31.1%)	(5.8%)

Source: Z.1 Financial Accounts of the United States, Federal Reserve Statistical Release, September 18, 2014, Various tables Historical Annuals, 1975-1984 and 2005-2013.

US GDP and Wealth Comparisons 1975, 2005, 2013 - Sheet 24

Re: MV US Balance Sheet of Federal Government with the addition of Human Capital at Market Value, Natural Resource and Infrastructure Value Allocations
S.7.a Comparison 1975/2005/2013

		1975	% GDP	% Assets	2005	% GDP	% Assets	+/- GDP from 1975	+/- Assets	2013	% GDP	% Assets	+/- GDP from 2005	+/- Assets	+/- GDP from 1975	+/- Assets
GDP	Billions of dollars	1688.9	3.4%		13093.7	5.9%				16768.0	5.1%					
Assets		49520.1	2932.1%	100.0%	223520.2	1707.1%	100.0%	(1225.0%)		325680.4	1942.3%	100.0%	235.2%	0.0%	(989.8%)	0.0%
	Nonfinancial assets	49399.1	2924.9%	99.8%	222876.5	1702.2%	99.7%	(1222.8%)	(0.0%)	323966.7	1932.1%	99.5%	229.9%	(0.2%)	(992.9%)	(0.3%)
	Real estate	2323.1	137.6%	4.7%	5426.5	41.4%	2.4%	(96.1%)	(2.3%)	6197.1	37.0%	1.9%	(4.5%)	(0.5%)	(100.6%)	(2.8%)
	Structures	288.1	17.1%	0.6%	1176.5	9.0%	0.5%	(8.1%)	(0.1%)	1472.1	8.8%	0.5%	(0.2%)	(0.1%)	(8.3%)	(0.1%)
*	Land	535.0	31.7%	1.1%	1950.0	14.9%	0.9%	(16.8%)	(0.2%)	2325.0	13.9%	0.7%	(1.0%)	(0.2%)	(17.8%)	(0.4%)
**	Infrastructure	1500.0	88.8%	3.0%	2300.0	17.6%	1.0%	(71.2%)	(2.0%)	2400.0	14.3%	0.7%	(3.3%)	(0.3%)	(74.5%)	(2.3%)
	Equipment	131.8	7.8%	0.3%	520.7	4.0%	0.2%	(3.8%)	(0.0%)	732.2	4.4%	0.2%	0.4%	(0.0%)	(3.4%)	(0.0%)
	Intellectual property products	144.2	8.5%	0.3%	729.3	5.6%	0.3%	(3.0%)	0.0%	987.4	5.9%	0.3%	0.3%	(0.0%)	(2.6%)	0.0%
***	Human capital	46800.0	2771.0%	94.5%	216200.0	1651.2%	96.7%	(1119.9%)	2.2%	316050.0	1884.8%	97.0%	233.7%	0.3%	(886.2%)	2.5%
***	Human capital of citizenry	46800.0	2771.0%	94.5%	200000.0	1527.5%	89.5%	(1243.6%)	(5.0%)	294000.0	1753.3%	90.3%	225.9%	0.8%	(1017.7%)	(4.2%)
***	Human capital of aliens		0.0%	0.0%	8100.0	61.9%	3.6%	61.9%	3.6%	11025.0	65.8%	3.4%	3.9%	(0.2%)	65.8%	3.4%
***	Human capital of undocumented aliens		0.0%	0.0%	8100.0	61.9%	3.6%	61.9%	3.6%	11025.0	65.8%	3.4%	3.9%	(0.2%)	65.8%	3.4%
	Financial assets	121.0	7.2%	0.2%	643.7	4.9%	0.3%	(2.2%)	0.0%	1713.7	10.2%	0.5%	5.3%	0.2%	3.1%	0.3%
	Monetary gold and SDRs	2.3	0.1%	0.0%	8.2	0.1%	0.0%	(0.1%)	(0.0%)	55.2	0.3%	0.0%	0.3%	0.0%	0.2%	0.0%
	Monetary gold	0.0	0.0%	0.0%	0	0.0%	0.0%	0.0%	0.0%	0	0.0%	0.0%	0.0%	0.0%	0.0%	0.0%
	SDR holdings	2.3	0.1%	0.0%	8.2	0.1%	0.0%	(0.1%)	(0.0%)	55.2	0.3%	0.0%	0.3%	0.0%	0.2%	0.0%
	Currency and deposits	17.5	1.0%	0.0%	68	0.5%	0.0%	(0.5%)	(0.0%)	221.8	1.3%	0.1%	0.8%	(0.0%)	0.3%	0.0%
	Official foreign currencies	0.0	0.0%	0.0%	18.7	0.1%	0.0%	0.1%	(0.0%)	23.6	0.1%	0.0%	(0.0%)	(0.0%)	0.1%	0.0%
	Reserve position in IMF (net)	2.2	0.1%	0.0%	8.1	0.1%	0.0%	(0.1%)	(0.0%)	31	0.2%	0.0%	0.1%	(0.0%)	0.1%	0.0%
	Currency and transferable deposits	11.2	0.7%	0.0%	37	0.3%	0.0%	(0.4%)	(0.0%)	163	1.0%	0.1%	0.7%	(0.0%)	0.3%	0.0%
	Time and savings deposits	0.6	0.0%	0.0%	1.4	0.0%	0.0%	(0.0%)	(0.0%)	1.5	0.0%	0.0%	(0.0%)	(0.0%)	(0.0%)	(0.0%)
	Nonofficial foreign currencies	3.6	0.2%	0.0%	2.6	0.0%	0.0%	(0.2%)	(0.0%)	2.6	0.0%	0.0%	(0.0%)	(0.0%)	(0.2%)	(0.0%)
	Debt securities	0.0	0.0%	0.0%	0	0.0%	0.0%	0.0%	0.0%	0.6	0.0%	0.0%	0.0%	0.0%	0.0%	0.0%
	Agency and GSE backed securities	0.0	0.0%	0.0%	0	0.0%	0.0%	0.0%	0.0%	0	0.0%	0.0%	0.0%	0.0%	0.0%	0.0%
	Corporate and foreign bonds	0.0	0.0%	0.0%	0	0.0%	0.0%	0.0%	0.0%	0.6	0.0%	0.0%	0.0%	0.0%	0.0%	0.0%
	Loans	85.5	5.1%	0.2%	271.7	2.1%	0.1%	(3.0%)	(0.1%)	1039.2	6.2%	0.3%	4.1%	0.2%	1.1%	0.1%
	Short term	66.4	3.9%	0.1%	195.2	1.5%	0.1%	(2.4%)	(0.0%)	923.7	5.5%	0.3%	4.0%	0.2%	1.6%	0.1%
	Consumer credit	0	0.0%	0.0%	89.8	0.7%	0.0%	0.7%	0.0%	729.8	4.4%	0.2%	3.7%	0.2%	4.4%	0.2%
	Other loans and advances	66.4	3.9%	0.1%	105.3	0.8%	0.0%	(3.1%)	(0.1%)	193.9	1.2%	0.1%	0.4%	0.0%	(2.8%)	(0.1%)
	Long term (mortgages)	19.1	1.1%	0.0%	76.6	0.6%	0.0%	(0.5%)	(0.0%)	115.5	0.7%	0.0%	0.1%	0.0%	(0.4%)	(0.0%)
	Equity and investment fund shares	4.1	0.2%	0.0%	43.2	0.3%	0.0%	0.1%	0.0%	98.9	0.6%	0.0%	0.3%	0.0%	0.3%	0.0%
	Corporate equities	0.0	0.0%	0.0%	0	0.0%	0.0%	0.0%	0.0%	35.1	0.2%	0.0%	0.2%	0.0%	0.2%	0.0%
	Equity in international organizations	4.1	0.2%	0.0%	43.2	0.3%	0.0%	0.1%	0.0%	59.5	0.4%	0.0%	0.0%	0.0%	0.1%	0.0%
	Equity in government sponsored enterprises	0.0	0.0%	0.0%	0	0.0%	0.0%	0.0%	0.0%	0	0.0%	0.0%	0.0%	0.0%	0.0%	0.0%
	Equity in investment under Public-Private IP	0.0	0.0%	0.0%	0	0.0%	0.0%	0.0%	0.0%	4.4	0.0%	0.0%	0.0%	0.0%	0.0%	0.0%
	Other accounts receivable	11.5	0.7%	0.0%	252.5	1.9%	0.1%	1.2%	0.1%	298.1	1.8%	0.1%	(0.2%)	(0.0%)	1.1%	0.1%
	Trade receivables	6.5	0.4%	0.0%	71	0.5%	0.0%	0.2%	0.0%	48.8	0.3%	0.0%	(0.3%)	(0.0%)	(0.1%)	0.0%
	Taxes receivables	5.0	0.3%	0.0%	91.6	0.7%	0.0%	0.4%	0.0%	165.8	1.0%	0.1%	0.3%	0.0%	0.7%	0.0%
	Other (miscellaneous assets)	0.0	0.0%	0.0%	89.9	0.7%	0.0%	0.7%	0.0%	83.4	0.5%	0.0%	(0.2%)	(0.0%)	0.5%	0.0%

*Resource value of land was figured for 650M acres of federal land at $3,000 acre in 2005 and extrapolated for 1975 and 2013.

**American Society of Civil Engineers reports funding needs to raise US infrastructure to Grade B status of $3.6T by 2020. Assuming depreciation of 33% gives current cash value of $7.2T and it could be greater. One third of this amount has been assigned to the Federal and two thirds to the State & local accounts for 2013 and extrapolated backward for 2005 and 1975.

***Figures in Nonfinancial assets Federal government include human capital based on figures from "Human Capital Accounting in the United States, 1994-2006" by Michael S. Christian, Survey of Current Business, June 2010, U.S. Department of Commerce, Bureau of Economic Analysis. Taken from Table 1. Human Capital Stock for 2005, Nominal, Total column and extrapolated for 1975 and 2013 and for alien population of 12M each category 2005 and 2013.

Source: Z.1 Financial Accounts of the United States, Federal Reserve Statistical Release, September 18, 2014, S.7.a Historical Annuals, 1975-1984 and 2005-2013

US GDP and Wealth Comparisons 1975, 2005, 2013 - Sheet 25

	1975	% GDP	% Assets	2005	% GDP	% Assets	+/- GDP from 1975	+/- Assets	2013	% GDP	% Assets	+/- GDP from 2005	+/- Assets	+/- GDP from 1975	+/- Assets
Total liabilities and net worth	49520.1	2932.1%	100.0%	223520.2	1707.1%	100.0%	(1225.0%)	0.0%	325680.4	1942.3%	100.0%	235.2%	0.0%	(989.8%)	0.0%
Liabilities	931.6	55.2%	1.9%	7370.3	56.3%	3.3%	1.1%	1.4%	16132.6	96.2%	5.0%	39.9%	1.7%	41.1%	3.1%
SDR allocations	2.7	0.2%	0.0%	7	0.1%	0.0%	(0.1%)	(0.0%)	54.4	0.3%	0.0%	0.3%	0.0%	0.2%	0.0%
Currency and deposits	8.2	0.5%	0.0%	27.5	0.2%	0.0%	(0.3%)	(0.0%)	25.6	0.2%	0.0%	(0.1%)	(0.0%)	(0.3%)	(0.0%)
Debt securities	485.4	28.7%	1.0%	5587	42.7%	2.5%	13.9%	1.5%	13710.3	81.8%	4.2%	39.1%	1.7%	53.0%	3.2%
SDR certificates	0.5	0.0%	0.0%	2.2	0.0%	0.0%	(0.0%)	(0.0%)	5.2	0.0%	0.0%	0.0%	0.0%	0.0%	0.0%
Treasury securities including savings bonds	434.9	25.8%	0.9%	4678	35.7%	2.1%	10.0%	1.2%	12328.3	73.5%	3.8%	37.8%	1.7%	47.8%	2.9%
Federal agency securities	7.9	0.5%	0.0%	23.8	0.2%	0.0%	(0.3%)	(0.0%)	24.5	0.1%	0.0%	(0.0%)	(0.0%)	(0.3%)	(0.0%)
Nonmarketable securities held by pension pla	42.1	2.5%	0.1%	882.9	6.7%	0.4%	4.3%	0.3%	1352.3	8.1%	0.4%	1.3%	0.0%	5.6%	0.3%
Loans (mortgages)	1.1	0.1%	0.0%	0	0.0%	0.0%	(0.1%)	(0.0%)	0	0.0%	0.0%	0.0%	0.0%	(0.1%)	(0.0%)
Insurance, pension and std guarantee schemes	421.4	25.0%	0.9%	1540.9	11.8%	0.7%	(13.2%)	(0.2%)	2089.3	12.5%	0.6%	0.7%	(0.0%)	(12.5%)	(0.2%)
Insurance reserves	10.2	0.6%	0.0%	42.7	0.3%	0.0%	(0.3%)	(0.0%)	50.3	0.3%	0.0%	(0.0%)	(0.0%)	(0.3%)	(0.0%)
Retiree Health Care Funds	0.0	0.0%	0.0%	75.4	0.6%	0.0%	0.6%	0.0%	246.9	1.5%	0.1%	0.9%	0.0%	1.5%	0.1%
Claims of pension fund on sponsor	411.2	24.3%	0.8%	1422.8	10.9%	0.6%	(13.5%)	(0.2%)	1792.1	10.7%	0.6%	(0.2%)	(0.1%)	(13.7%)	(0.3%)
Other accounts payable	12.9	0.8%	0.0%	208	1.6%	0.1%	0.8%	0.1%	253	1.5%	0.1%	(0.1%)	(0.0%)	0.7%	0.1%
Trade payables	11.8	0.7%	0.0%	202.6	1.5%	0.1%	0.8%	0.1%	250.7	1.5%	0.1%	(0.1%)	(0.0%)	0.8%	0.1%
Other (miscellaneous liabilities)	1.1	0.1%	0.0%	5.4	0.0%	0.0%	(0.0%)	0.0%	2.3	0.0%	0.0%	(0.0%)	(0.0%)	(0.1%)	(0.0%)
Net worth (market value)	48588.5	2876.9%	98.1%	216149.9	1650.8%	96.7%	(1226.1%)	(1.4%)	309547.8	1846.1%	95.0%	195.3%	(1.7%)	(1030.9%)	(3.1%)

Source: Z.1 Financial Accounts of the United States, Federal Reserve Statistical Release, September 18, 2014, S.7.a Historical Annuals, 1975-1984 and 2005-2013

US GDP and Wealth Comparisons 1975, 2005, 2013 - Sheet 26

Re: LI US Balance Sheet of State and Local Government with addition of Land and Infrastructure
S.8.a Comparison 1975/2005/2013

	1975	% GDP	% Assets	2005	% GDP	% Assets	+/- GDP from 1975	+/- Assets	2013	% GDP	% Assets	+/- GDP from 2005	+/- Assets	+/- GDP from 1975	+/- Assets
GDP	1688.9			13093.7	97.9%				16768.0	93.8%					
Billions of dollars															
Assets	4118.1	243.8%	100.0%	13371.2	102.1%	100.0%	(141.7)	0.0%	17873.9	106.6%	100.0%	4.5%	0.0%	(137.2)	0.0%
Nonfinancial assets	3971.7	235.2%	96.4%	11054.0	84.4%	82.7%	(150.7)	(13.8%)	14935.7	89.1%	83.6%	4.7%	0.9%	(146.1)	(12.9%)
Real estate	3940.9	233.3%	95.7%	10771.8	82.3%	80.6%	(151.1)	(15.1%)	14568.1	86.9%	81.5%	4.6%	0.9%	(146.5)	(14.2%)
Structures	887.4	52.5%	21.5%	5976.8	45.6%	44.7%	(6.9%)	23.2%	9335.6	55.7%	52.2%	10.0%	7.5%	3.1%	30.7%
* Land & Infrastructure	3053.5	180.8%	74.1%	4795.0	36.6%	35.9%	(144.2)	(38.3%)	5232.5	31.2%	29.3%	(5.4%)	(6.6%)	(149.6)	(44.9%)
Equipment	25.5	1.5%	0.6%	206.9	1.6%	1.5%	0.1%	0.9%	247.6	1.5%	1.4%	(0.1%)	(0.2%)	(0.0%)	0.8%
Intellectual property products	5.3	0.3%	0.1%	75.3	0.6%	0.6%	0.3%	0.4%	120.0	0.7%	0.7%	0.1%	0.1%	0.4%	0.5%
Financial assets	146.4	8.7%	3.6%	2317.2	17.7%	17.3%	9.0%	13.8%	2938.2	17.5%	16.4%	(0.2%)	(0.9%)	8.9%	12.9%
Currency and deposits	60.3	3.6%	1.5%	234.8	1.8%	1.8%	(1.8%)	0.3%	425.8	2.5%	2.4%	0.7%	0.6%	(1.0%)	0.9%
Currency and transferable deposits	13.4	0.8%	0.3%	66.0	0.5%	0.5%	(0.3%)	0.2%	123.9	0.7%	0.7%	0.2%	0.2%	(0.1%)	0.4%
Time and savings deposits	46.9	2.8%	1.1%	168.8	1.3%	1.3%	(1.5%)	0.1%	301.9	1.8%	1.7%	0.5%	0.4%	(1.0%)	0.6%
Debt securities	51.0	3.0%	1.2%	1217.4	9.3%	9.1%	6.3%	7.9%	1347.7	8.0%	7.5%	(1.3%)	(1.6%)	5.0%	6.3%
Open market paper	0.0	0.0%	0.0%	153.3	1.2%	1.1%	1.2%	1.1%	76.9	0.5%	0.4%	(0.7%)	(0.7%)	0.5%	0.4%
Treasury securities	27.8	1.6%	0.7%	512.3	3.9%	3.8%	2.3%	3.2%	593.4	3.5%	3.3%	(0.4%)	(0.5%)	1.9%	2.6%
Agency and GSE backed securities	18.2	1.1%	0.4%	413.4	3.2%	3.1%	2.1%	2.6%	490.9	2.9%	2.7%	(0.2%)	(0.3%)	1.8%	2.3%
Municipal securities	5.0	0.3%	0.1%	6.9	0.1%	0.1%	(0.2%)	(0.1%)	13.9	0.1%	0.1%	0.0%	0.0%	(0.2%)	(0.0%)
Corporate and foreign bonds	0.0	0.0%	0.0%	131.5	1.0%	1.0%	1.0%	1.0%	172.5	1.0%	1.0%	0.0%	0.0%	1.0%	1.0%
Loans	10.7	0.6%	0.3%	288.9	2.2%	2.2%	1.6%	1.9%	346.3	2.1%	1.9%	(0.1%)	(0.2%)	1.4%	1.7%
Short term	-2.1	(0.1%)	(0.1%)	130.0	1.0%	1.0%	1.1%	1.0%	132.5	0.8%	0.7%	(0.2%)	(0.2%)	0.9%	0.8%
Long term (mortgages)	12.8	0.8%	0.3%	158.9	1.2%	1.2%	0.5%	0.9%	213.9	1.3%	1.2%	0.1%	0.0%	0.5%	0.9%
Equity and investment fund shares	0.0	0.0%	0.0%	246.6	1.9%	1.8%	1.9%	1.8%	415.4	2.5%	2.3%	0.6%	0.5%	2.5%	2.3%
Money market fund shares	0.0	0.0%	0.0%	89.9	0.7%	0.7%	0.7%	0.7%	166.6	1.0%	0.9%	0.3%	0.3%	1.0%	0.9%
Corporate equities	0.0	0.0%	0.0%	116.0	0.9%	0.9%	0.9%	0.9%	167.6	1.0%	0.9%	0.1%	0.1%	1.0%	0.9%
Mutual fund shares	0.0	0.0%	0.0%	40.7	0.3%	0.3%	0.3%	0.3%	81.2	0.5%	0.5%	0.2%	0.1%	0.5%	0.5%
Other accounts receivable	24.4	1.4%	0.6%	329.4	2.5%	2.5%	1.1%	1.9%	403.0	2.4%	2.3%	(0.1%)	(0.2%)	1.0%	1.7%
Trade receivables	15.2	0.9%	0.4%	142.5	1.1%	1.1%	0.2%	0.7%	168.7	1.0%	0.9%	(0.1%)	(0.1%)	0.1%	0.6%
Taxes receivables	9.2	0.5%	0.2%	102.3	0.8%	0.8%	0.2%	0.5%	123.8	0.7%	0.7%	(0.0%)	(0.1%)	0.2%	0.5%
Other (miscellaneous assets)	0.1	0.0%	0.0%	84.6	0.6%	0.6%	0.6%	0.6%	110.5	0.7%	0.6%	0.0%	0.0%	0.7%	0.6%
Total liabilities and net worth	4118.1	243.8%	100.0%	13371.2	102.1%	100.0%	(141.7)	0.0%	17873.9	106.6%	100.0%	4.5%	0.0%	(137.2)	0.0%
Liabilities	349.5	20.7%	8.5%	3385.9	25.9%	25.3%	5.2%	16.8%	4931.0	29.4%	27.6%	3.5%	2.3%	8.7%	19.1%
Debt securities (municipals)	213.6	12.6%	5.2%	2579.2	19.7%	19.3%	7.1%	14.1%	2924.9	17.4%	16.4%	(2.3%)	(2.9%)	4.8%	11.2%
Short term	18.6	1.1%	0.5%	42.5	0.3%	0.3%	(0.8%)	(0.1%)	45.3	0.3%	0.3%	(0.1%)	(0.1%)	(0.8%)	(0.2%)
Other	195.0	11.5%	4.7%	2536.7	19.4%	19.0%	7.8%	14.2%	1879.6	11.2%	10.5%	(8.2%)	(8.5%)	(0.3%)	5.8%
Loans (short term)	5.8	0.3%	0.1%	10.6	0.1%	0.1%	(0.3%)	(0.1%)	16.2	0.1%	0.1%	0.0%	0.0%	(0.2%)	(0.1%)
Insurance, pension and std guarantee schemes (Claims of pension fund on sponsor)	92.7	5.5%	2.3%	314.0	2.4%	2.3%	(3.1%)	0.1%	1204.1	7.2%	6.7%	4.8%	4.4%	1.7%	4.5%
Other accounts payable (trade payables)	37.4	2.2%	0.9%	482.0	3.7%	3.6%	1.5%	2.7%	785.8	4.7%	4.4%	1.0%	0.8%	2.5%	3.5%
Net worth (market value)	3768.6	223.1%	91.5%	9985.3	76.3%	74.7%	(146.9%)	(16.8%)	12942.9	77.2%	72.4%	0.9%	(2.3%)	(146.0%)	(19.1%)

*Balance of infrastructure per note on Federal balance sheet plus 10% of land estimate for that sheet.

Source: Z.1 Financial Accounts of the United States, Federal Reserve Statistical Release, September 18, 2014, S.8.a Historical Annuals, 1975-1984 and 2005-2013

US GDP and Wealth Comparisons 1975, 2005, 2013 - Sheet 27

Re: MV US Assets of the Public & Private Sectors with the addition of Human Capital at Market Value, Natural Resource and Infrastructure Value Allocations to the Public Sectors
L.10+ Comparison 1975/2005/2013

	1975	% GDP	% Assets	2005	% GDP	% Assets	+/- GDP from 1975	+/- Assets	2013	% GDP	% Assets	+/- GDP from 2005	+/- Assets	+/- GDP from 1975	+/- Assets
GDP Billions of dollars	1688.9	3.4%		13093.7	5.9%				16768.0	5.2%					
Total Assets (public & private sectors)	60208.2	3564.9%	100.0%	300084.3	2291.8%	100.0%	(1273.1%)		415572.6	2478.4%	100.0%	186.5%		(1086.6%)	(2.7%)
Nonfinancial assets public & private sectors	59290.7	3510.6%	98.5%	287982.2	2199.4%	96.0%	(1311.2%)	(2.5%)	397885.8	2372.9%	95.7%	173.5%	(0.2%)	(1137.7%)	(2.7%)
Net financial assets public & private sectors	917.5	54.3%	1.5%	12102.1	92.4%	4.0%	38.1%	2.5%	17686.8	105.5%	4.3%	13.1%	0.2%	51.2%	2.7%
Assets (public sectors)	52357.1	3100.1%	100.0%	226135.2	1727.1%	100.0%	(1373.0%)		322490.7	1923.3%	100.0%	196.2%		(1176.8%)	
Nonfinancial assets public sectors	53370.8	3160.1%	101.9%	233930.5	1786.6%	103.4%	(1373.5%)	1.5%	338902.4	2021.1%	105.1%	234.5%	1.6%	(1139.0%)	3.2%
Nonfinancial assets Federal government	49399.1	2924.9%	94.4%	222876.5	1702.2%	98.6%	(1222.8%)	4.2%	323966.7	1932.1%	100.5%	229.9%	1.9%	(992.9%)	6.1%
Nonfinancial assets State & local government	3971.7	235.2%	7.6%	11054.0	84.4%	4.9%	(150.7%)	(2.7%)	14935.7	89.1%	4.6%	4.7%	(0.3%)	(146.1%)	(3.0%)
Net financial assets public sectors	(1013.7)	(60.0%)	(1.9%)	(7795.3)	(59.5%)	(3.4%)	0.5%	(1.5%)	(16411.7)	(97.9%)	(5.1%)	(38.3%)	(1.6%)	(37.9%)	(3.2%)
Net financial assets Federal government	(810.6)	(48.0%)	(1.5%)	(6726.6)	(51.4%)	(3.0%)	(3.4%)	(1.4%)	(14418.9)	(86.0%)	(4.5%)	(34.6%)	(1.5%)	(38.0%)	(2.9%)
Net financial assets state & local government	(203.1)	(12.0%)	(0.4%)	(1068.7)	(8.2%)	(0.5%)	3.9%	(0.1%)	(1992.8)	(11.9%)	(0.6%)	(3.7%)	(0.1%)	0.1%	(0.2%)
Assets (private sectors)	7851.1	464.9%	100.0%	73949.1	564.8%	100.0%	99.9%		93081.9	555.1%	100.0%	(9.7%)		90.3%	
Nonfinancial assets (Domestic private sectors)	5919.9	350.5%	75.4%	54051.7	412.8%	73.1%	62.3%	(2.3%)	58983.4	351.8%	63.4%	(61.0%)	(9.7%)	1.2%	(12.0%)
Households & nonprofits nonfinancial assets	3701.2	219.1%	47.1%	38576.9	294.6%	52.2%	75.5%	5.0%	38837.9	231.6%	41.7%	(63.0%)	(10.4%)	12.5%	(5.4%)
Nonfinancial corporate business nonfin assets	2131.8	126.2%	27.2%	14168.0	108.2%	19.2%	(18.0%)	(8.0%)	18561.3	110.7%	19.9%	2.5%	0.8%	(15.5%)	(7.2%)
Financial business nonfinancial assets	86.9	5.1%	1.1%	1306.8	10.0%	1.8%	4.8%	0.7%	1584.2	9.4%	1.7%	(0.5%)	(0.1%)	4.3%	0.6%
Nonfinancial assets (breakdown)															
Real estate	4107.8	243.2%	52.3%	41914.6	320.1%	56.7%	76.9%	4.4%	43295.6	258.2%	46.5%	(61.9%)	(10.2%)	15.0%	(5.8%)
Households	1992.4	118.0%	25.4%	27852.8	212.7%	37.7%	94.7%	12.3%	25144.7	150.0%	27.0%	(62.8%)	(10.7%)	32.0%	1.6%
Nonprofit organizations & nonresidential	888.8	52.6%	11.3%	5385.7	41.1%	7.3%	(11.5%)	(4.0%)	7015.6	41.8%	7.5%	0.7%	0.3%	(10.8%)	(3.8%)
Nonfinancial corporate business	1170.9	69.3%	14.9%	7924.3	60.5%	10.7%	(8.8%)	(4.2%)	10236.1	61.0%	11.0%	0.5%	0.3%	(8.3%)	(3.9%)
Financial business	55.7	3.3%	0.7%	751.8	5.7%	1.0%	2.4%	0.3%	899.2	5.4%	1.0%	(0.4%)	(0.1%)	2.1%	0.3%
Durable goods and equipment	1303.5	77.2%	16.6%	8636.0	66.0%	11.7%	(11.2%)	(4.9%)	10821.5	64.5%	11.6%	(1.4%)	(0.1%)	(12.6%)	(5.0%)
Consumer durable goods	578.0	34.2%	7.4%	4107.8	31.4%	5.6%	(2.9%)	(1.8%)	4942.2	29.5%	5.3%	(1.9%)	(0.2%)	(4.7%)	(2.1%)
Nonprofits & nonfinancial noncorporate bus	147.5	8.7%	1.9%	785	6.0%	1.1%	(2.7%)	(0.8%)	1098.6	6.6%	1.2%	0.6%	0.1%	(2.2%)	(0.7%)
Nonfinancial corporate business	550.2	32.6%	7.0%	3297.8	25.2%	4.5%	(7.4%)	(2.5%)	4261.7	25.4%	4.6%	0.2%	0.1%	(7.2%)	(2.4%)
Financial business	27.8	1.6%	0.4%	445.4	3.4%	0.6%	1.8%	0.2%	519.0	3.1%	0.6%	(0.3%)	(0.0%)	1.4%	0.2%
Inventories	387.3	22.9%	4.9%	1842.3	14.1%	2.5%	(8.9%)	(2.4%)	2443.9	14.6%	2.6%	0.5%	0.1%	(8.4%)	(2.3%)
Nonfinancial noncorporate business	80.1	4.7%	1.0%	218.4	1.7%	0.3%	(3.1%)	(0.7%)	297.3	1.8%	0.3%	0.1%	0.0%	(3.0%)	(0.7%)
Nonfinancial corporate business	307.2	18.2%	3.9%	1623.9	12.4%	2.2%	(5.8%)	(1.7%)	2146.6	12.8%	2.3%	0.4%	0.1%	(5.4%)	(1.6%)
Intellectual property products	121.4	7.2%	1.5%	1658.8	12.7%	2.2%	5.5%	0.7%	2422.4	14.4%	2.6%	1.8%	0.4%	7.3%	1.1%
Nonprofits & nonfinancial noncorporate bus	14.4	0.9%	0.2%	227.3	1.7%	0.3%	0.9%	0.1%	339.4	2.0%	0.4%	0.3%	0.1%	1.2%	0.2%
Nonfinancial corporate business	103.6	6.1%	1.3%	1322.0	10.1%	1.8%	4.0%	0.5%	1916.9	11.4%	2.1%	1.3%	0.3%	5.3%	0.7%
Financial business	3.4	0.2%	0.0%	109.5	0.8%	0.1%	0.6%	0.1%	166.1	1.0%	0.2%	0.2%	0.0%	0.8%	0.1%
Net financial assets domestic private sectors	1847.1	109.4%	23.5%	21548.1	164.6%	29.1%	55.2%	5.6%	38482.1	229.5%	41.3%	64.9%	12.2%	120.1%	17.8%
Net financial assets HH & NN business	2095.2	124.1%	26.7%	23309.4	178.0%	31.5%	54.0%	4.8%	40030.0	238.7%	43.0%	60.7%	11.5%	114.7%	16.3%
Net financial assets NC business	(318.2)	(18.8%)	(4.1%)	718.8	5.5%	1.0%	24.3%	5.0%	552.8	3.3%	0.6%	(2.2%)	(0.4%)	22.1%	4.6%
Net financial assets financial business	70.1	4.2%	0.9%	(2480.1)	(18.9%)	(3.4%)	(23.1%)	(4.2%)	(2100.7)	(12.5%)	(2.3%)	6.4%	1.1%	(16.7%)	(3.1%)
Net financial assets private sectors	1931.2	114.3%	24.6%	19897.4	152.0%	26.9%	37.6%	2.3%	34098.5	203.4%	36.6%	51.4%	9.7%	89.0%	12.0%
Net financial assets rest of the world	84.1	5.0%	1.1%	(1650.7)	(12.6%)	(2.2%)	(17.6%)	(3.3%)	(4383.6)	(26.1%)	(4.7%)	(13.5%)	(2.5%)	(31.1%)	(5.8%)

Figures in Nonfinancial assets Federal government include human capital based on figures from "Human Capital Accounting in the United States, 1994-2006" by Michael S. Christian, Survey of Current Business, June 2010, U.S. Department of Commerce, Bureau of Economic Analysis

Source: Z.1 Financial Accounts of the United States, Federal Reserve Statistical Release, September 18, 2014, Various tables Historical Annuals, 1975-1984 and 2005-2013.

US GDP and Wealth Comparisons 1975, 2005, 2013 - Sheet 28

Re: TV US Balance Sheet of Federal Government with the addition of Human Capital, Natural Resource and Infrastructure Value Allocations
S.7.a Comparison 1975/2005/2013

	Billions of dollars	1975	% GDP	% Assets	2005	% GDP	% Assets	+/- GDP from 1975	+/- Assets	2013	% GDP	% Assets	+/- GDP from 2005	+/- Assets	+/- GDP from 1975	+/- Assets
GDP		1688.9	1.1%		13093.7	1.8%				16768.0	1.6%					
Assets		158720.1	9397.8%	100.0%	728320.2	5562.4%	100.0%	(3835.5%)		1063080.4	6339.9%	100.0%	777.6%	0.0%	(3057.9%)	0.0%
	Nonfinancial assets	158599.1	9390.7%	99.9%	727676.5	5557.5%	99.9%	(3833.2%)	(0.0%)	1061366.7	6329.7%	99.8%	772.3%	(0.1%)	(3061.0%)	(0.1%)
	Real estate	2323.1	137.6%	1.5%	5426.5	41.4%	0.7%	(96.1%)	(0.7%)	6197.1	37.0%	0.6%	(4.5%)	(0.2%)	(100.6%)	(0.9%)
	Structures	288.1	17.1%	0.2%	1176.5	9.0%	0.2%	(8.1%)	(0.0%)	1472.1	8.8%	0.1%	(0.2%)	(0.0%)	(8.3%)	(0.0%)
*	Land	535.0	31.7%	0.3%	1950.0	14.9%	0.3%	(16.8%)	(0.1%)	2325.0	13.9%	0.2%	(1.0%)	(0.0%)	(17.8%)	(0.1%)
**	Infrastructure	1500.0	88.8%	0.9%	2300.0	17.6%	0.3%	(71.2%)	(0.6%)	2400.0	14.3%	0.2%	(3.3%)	(0.1%)	(74.5%)	(0.7%)
	Equipment	131.8	7.8%	0.1%	520.7	4.0%	0.1%	(3.8%)	(0.0%)	732.2	4.4%	0.1%	0.4%	(0.0%)	(3.4%)	(0.0%)
	Intellectual property products	144.2	8.5%	0.1%	729.3	5.6%	0.1%	(3.0%)	0.0%	987.4	5.9%	0.1%	0.3%	(0.0%)	(2.6%)	0.0%
***	Human capital	156000.0	9236.8%	98.3%	721000.0	5506.5%	99.0%	(3730.3%)	0.7%	1053450.0	6282.5%	99.1%	776.0%	0.1%	(2954.3%)	0.8%
***	Human capital of citizenry	156000.0	9236.8%	98.3%	667000.0	5094.1%	91.6%	(4142.7%)	(6.7%)	980000.0	5844.5%	92.2%	750.4%	0.6%	(3392.3%)	(6.1%)
***	Human capital of aliens		0.0%	0.0%	27000.0	206.2%	3.7%	206.2%	3.7%	36750.0	219.2%	3.5%	13.0%	(0.3%)	219.2%	3.5%
***	Human capital of undocumented aliens		0.0%	0.0%	27000.0	206.2%	3.7%	206.2%	3.7%	36700.0	218.9%	3.5%	12.7%	(0.3%)	218.9%	3.5%
	Financial assets	121.0	7.2%	0.1%	643.7	4.9%	0.1%	(2.2%)	0.0%	1713.7	10.2%	0.2%	5.3%	0.1%	3.1%	0.1%
	Monetary gold and SDRs	2.3	0.1%	0.0%	8.2	0.1%	0.0%	(0.1%)	(0.0%)	55.2	0.3%	0.0%	0.3%	0.0%	0.2%	0.0%
	Monetary gold	0.0	0.0%	0.0%	0	0.0%	0.0%	0.0%	0.0%	0	0.0%	0.0%	0.0%	0.0%	0.0%	0.0%
	SDR holdings	2.3	0.1%	0.0%	8.2	0.1%	0.0%	(0.1%)	(0.0%)	55.2	0.3%	0.0%	0.3%	0.0%	0.2%	0.0%
	Currency and deposits	17.5	1.0%	0.0%	68	0.5%	0.0%	(0.5%)	(0.0%)	221.8	1.3%	0.0%	0.8%	0.0%	0.3%	0.0%
	Official foreign currencies	0.0	0.0%	0.0%	18.7	0.1%	0.0%	0.1%	0.0%	23.6	0.1%	0.0%	(0.0%)	(0.0%)	0.1%	0.0%
	Reserve position in IMF (net)	2.2	0.1%	0.0%	8.1	0.1%	0.0%	(0.1%)	(0.0%)	31	0.2%	0.0%	0.1%	0.0%	0.1%	0.0%
	Currency and transferable deposits	11.2	0.7%	0.0%	37	0.3%	0.0%	(0.4%)	(0.0%)	163	1.0%	0.0%	0.7%	0.0%	0.3%	0.0%
	Time and savings deposits	0.6	0.0%	0.0%	1.4	0.0%	0.0%	(0.0%)	(0.0%)	1.5	0.0%	0.0%	(0.0%)	(0.0%)	(0.0%)	0.0%
	Nonofficial foreign currencies	3.6	0.2%	0.0%	2.6	0.0%	0.0%	(0.2%)	(0.0%)	2.6	0.0%	0.0%	(0.0%)	(0.0%)	(0.2%)	(0.0%)
	Debt securities	0.0	0.0%	0.0%	0	0.0%	0.0%	0.0%	0.0%	0.6	0.0%	0.0%	0.0%	0.0%	0.0%	0.0%
	Agency and GSE backed securities	0.0	0.0%	0.0%	0	0.0%	0.0%	0.0%	0.0%	0	0.0%	0.0%	0.0%	0.0%	0.0%	0.0%
	Corporate and foreign bonds	0.0	0.0%	0.0%	0	0.0%	0.0%	0.0%	0.0%	0.6	0.0%	0.0%	0.0%	0.0%	0.0%	0.0%
	Loans	85.5	5.1%	0.1%	271.7	2.1%	0.0%	(3.0%)	(0.0%)	1039.2	6.2%	0.1%	4.1%	0.1%	1.1%	0.0%
	Short term	66.4	3.9%	0.0%	195.2	1.5%	0.0%	(2.4%)	(0.0%)	923.7	5.5%	0.1%	4.0%	0.1%	1.6%	0.0%
	Consumer credit	0	0.0%	0.0%	89.8	0.7%	0.0%	0.7%	0.0%	729.8	4.4%	0.1%	3.7%	0.1%	4.4%	0.1%
	Other loans and advances	66.4	3.9%	0.0%	105.3	0.8%	0.0%	(3.1%)	(0.0%)	193.9	1.2%	0.0%	0.4%	0.0%	(2.8%)	(0.0%)
	Long term (mortgages)	19.1	1.1%	0.0%	76.6	0.6%	0.0%	(0.5%)	(0.0%)	115.5	0.7%	0.0%	0.1%	0.0%	(0.4%)	(0.0%)
	Equity and investment fund shares	4.1	0.2%	0.0%	43.2	0.3%	0.0%	0.1%	0.0%	98.9	0.6%	0.0%	0.3%	0.0%	0.3%	0.0%
	Corporate equities	0.0	0.0%	0.0%	0	0.0%	0.0%	0.0%	0.0%	35.1	0.2%	0.0%	0.2%	0.0%	0.2%	0.0%
	Equity in international organizations	4.1	0.2%	0.0%	43.2	0.3%	0.0%	0.1%	0.0%	59.5	0.4%	0.0%	0.0%	(0.0%)	0.1%	0.0%
	Equity in government sponsored enterprises	0.0	0.0%	0.0%	0	0.0%	0.0%	0.0%	0.0%	0	0.0%	0.0%	0.0%	0.0%	0.0%	0.0%
	Equity in investment under Public-Private IP	0.0	0.0%	0.0%	0	0.0%	0.0%	0.0%	0.0%	4.4	0.0%	0.0%	0.0%	0.0%	0.0%	0.0%
	Other accounts receivable	11.5	0.7%	0.0%	252.5	1.9%	0.0%	1.2%	0.0%	298.1	1.8%	0.0%	(0.2%)	(0.0%)	1.1%	0.0%
	Trade receivables	6.5	0.4%	0.0%	71	0.5%	0.0%	0.2%	0.0%	48.8	0.3%	0.0%	(0.3%)	(0.0%)	(0.1%)	0.0%
	Taxes receivables	5.0	0.3%	0.0%	91.6	0.7%	0.0%	0.4%	0.0%	165.8	1.0%	0.0%	0.3%	0.0%	0.7%	0.0%
	Other (miscellaneous assets)	0.0	0.0%	0.0%	89.9	0.7%	0.0%	0.7%	0.0%	83.4	0.5%	0.0%	(0.2%)	(0.0%)	0.5%	0.0%

*Resource value of land was figured for 650M acres of federal land at $3,000 acre in 2005 and extrapolated for 1975 and 2013.

**American Society of Civil Engineers reports funding needs to raise US infrastructure to Grade B status of $3.6T by 2020. Assuming depreciation of 33% gives current cash value of $7.2T and it could be much greater, extrapolated back.

***Figures in Nonfinancial assets Federal government include human capital based on figures from "Human Capital Accounting in the United States, 1994-2006" by Michael S. Christian, Survey of Current Business, June 2010, U.S. Department of Commerce, Bureau of Economic Analysis. Taken from Table 1. Human Capital Stock for 2005, Nominal, Total column and extrapolated for 1975 and 2013 and for alien population of 12M each category 2005 and 2013.

Source: Z.1 Financial Accounts of the United States, Federal Reserve Statistical Release, September 18, 2014, S.7.a Historical Annuals, 1975-1984 and 2005-2013

US GDP and Wealth Comparisons 1975, 2005, 2013 - Sheet 29

	1975	% GDP	% Assets	2005	% GDP	% Assets	+/- GDP from 1975	+/- Assets	2013	% GDP	% Assets	+/- GDP from 2005	+/- Assets	+/- GDP from 1975	+/- Assets
Total liabilities and net worth	158720.1	9397.8%	100.0%	728320.2	5562.4%	100.0%	(3835.5%)	0.0%	1063080.4	6339.9%	100.0%	777.6%	0.0%	(3057.9%)	0.0%
Liabilities	931.6	55.2%	0.6%	7370.3	56.3%	1.0%	1.1%	0.4%	16132.6	96.2%	1.5%	39.9%	0.5%	41.1%	0.9%
SDR allocations	2.7	0.2%	0.0%	7	0.1%	0.0%	(0.1%)	(0.0%)	54.4	0.3%	0.0%	0.3%	0.0%	0.2%	0.0%
Currency and deposits	8.2	0.5%	0.0%	27.5	0.2%	0.0%	(0.3%)	(0.0%)	25.6	0.2%	0.0%	(0.1%)	(0.0%)	(0.3%)	(0.0%)
Debt securities	485.4	28.7%	0.3%	5587	42.7%	0.8%	13.9%	0.5%	13710.3	81.8%	1.3%	39.1%	0.5%	53.0%	1.0%
SDR certificates	0.5	0.0%	0.0%	2.2	0.0%	0.0%	(0.0%)	(0.0%)	5.2	0.0%	0.0%	0.0%	0.0%	0.0%	0.0%
Treasury securities including savings bonds	434.9	25.8%	0.3%	4678	35.7%	0.6%	10.0%	0.4%	12328.3	73.5%	1.2%	37.8%	0.5%	47.8%	0.9%
Federal agency securities	7.9	0.5%	0.0%	23.8	0.2%	0.0%	(0.3%)	(0.0%)	24.5	0.1%	0.0%	(0.0%)	(0.0%)	(0.3%)	(0.0%)
Nonmarketable securities held by pension pla	42.1	2.5%	0.0%	882.9	6.7%	0.1%	4.3%	0.1%	1352.3	8.1%	0.1%	1.3%	0.0%	5.6%	0.1%
Loans (mortgages)	1.1	0.1%	0.0%	0	0.0%	0.0%	(0.1%)	(0.0%)	0	0.0%	0.0%	0.0%	0.0%	(0.1%)	(0.0%)
Insurance, pension and std guarantee schemes	421.4	25.0%	0.3%	1540.9	11.8%	0.2%	(13.2%)	(0.1%)	2089.3	12.5%	0.2%	0.7%	(0.0%)	(12.5%)	(0.1%)
Insurance reserves	10.2	0.6%	0.0%	42.7	0.3%	0.0%	(0.3%)	(0.0%)	50.3	0.3%	0.0%	(0.0%)	(0.0%)	(0.3%)	(0.0%)
Retiree Health Care Funds	0.0	0.0%	0.0%	75.4	0.6%	0.0%	0.6%	0.0%	246.9	1.5%	0.0%	0.9%	0.0%	1.5%	0.0%
Claims of pension fund on sponsor	411.2	24.3%	0.3%	1422.8	10.9%	0.2%	(13.5%)	(0.1%)	1792.1	10.7%	0.2%	(0.2%)	(0.0%)	(13.7%)	(0.1%)
Other accounts payable	12.9	0.8%	0.0%	208	1.6%	0.0%	0.8%	0.0%	253	1.5%	0.0%	(0.1%)	(0.0%)	0.7%	0.0%
Trade payables	11.8	0.7%	0.0%	202.6	1.5%	0.0%	0.8%	0.0%	250.7	1.5%	0.0%	(0.1%)	(0.0%)	0.8%	0.0%
Other (miscellaneous liabilities)	1.1	0.1%	0.0%	5.4	0.0%	0.0%	(0.0%)	0.0%	2.3	0.0%	0.0%	(0.0%)	(0.0%)	(0.1%)	(0.0%)
Net worth (market value)	157788.5	9342.7%	99.4%	720949.9	5506.1%	99.0%	(3836.6%)	(0.4%)	1046947.8	6243.7%	98.5%	737.6%	(0.5%)	(3099.0%)	(0.9%)

Source: Z.1 Financial Accounts of the United States, Federal Reserve Statistical Release, September 18, 2014, S.7.a Historical Annuals, 1975-1984 and 2005-2013

US GDP and Wealth Comparisons 1975, 2005, 2013 - Sheet 30

Re: TV US Assets of the Public & Private Sectors with the addition of Human Capital, Natural Resource and Infrastructure Value Allocations to the Public Sectors
L.10+ Comparison 1975/2005/2013

	1975	% GDP	% Assets	2005	% GDP	% Assets	+/- GDP from 1975	+/- Assets	2013	% GDP	% Assets	+/- GDP from 2005	+/- Assets	+/- GDP from 1975	+/- Assets
GDP Billions of dollars	1688.9	1.1%		13093.7	1.8%				16768.0	1.6%					
Total Assets (public & private sectors)	169408.2	10030.7%	100.0%	804884.3	6147.1%	100.0%	(3883.6%)		1152972.6	6876.0%	100.0%	728.9%		(3154.7%)	
Nonfinancial assets public & private sectors	168490.7	9976.4%	99.5%	792782.2	6054.7%	98.5%	(3921.7%)	(1.0%)	1135285.8	6770.5%	98.5%	715.9%	(0.0%)	(3205.8%)	(1.0%)
Net financial assets public & private sectors	917.5	54.3%	0.5%	12102.1	92.4%	1.5%	38.1%	1.0%	17686.8	105.5%	1.5%	13.1%	0.0%	51.2%	1.0%
Assets (public sectors)	161557.1	9565.8%	100.0%	730935.2	5582.3%	100.0%	(3983.5%)		1059890.7	6320.9%	100.0%	738.6%		(3244.9%)	
Nonfinancial assets public sectors	162570.8	9625.8%	100.6%	738730.5	5641.9%	101.1%	(3984.0%)	0.4%	1076302.4	6418.8%	101.5%	776.9%	0.5%	(3207.1%)	0.9%
Nonfinancial assets Federal government	158599.1	9390.7%	98.2%	727676.5	5557.5%	99.6%	(3833.2%)	1.4%	1061366.7	6329.7%	100.1%	772.3%	0.6%	(3061.0%)	2.0%
Nonfinancial assets State & local government	3971.7	235.2%	2.5%	11054.0	84.4%	1.5%	(150.7%)	(0.9%)	14935.7	89.1%	1.4%	4.7%	(0.1%)	(146.1%)	(1.0%)
Net financial assets public sectors	(1013.7)	(60.0%)	(0.6%)	(7795.3)	(59.5%)	(1.1%)	0.5%	(0.4%)	(16411.7)	(97.9%)	(1.5%)	(38.3%)	(0.5%)	(37.9%)	(0.9%)
Net financial assets Federal government	(810.6)	(48.0%)	(0.5%)	(6726.6)	(51.4%)	(0.9%)	(3.4%)	(0.4%)	(14418.9)	(86.0%)	(1.4%)	(34.6%)	(0.4%)	(38.0%)	(0.9%)
Net financial assets state & local government	(203.1)	(12.0%)	(0.1%)	(1068.7)	(8.2%)	(0.1%)	3.9%	(0.0%)	(1992.8)	(11.9%)	(0.2%)	(3.7%)	(0.0%)	0.1%	(0.1%)
Assets (private sectors)	7851.1	464.9%	100.0%	73949.1	564.8%	100.0%	99.9%		93081.9	555.1%	100.0%	(9.7%)		90.3%	
Nonfinancial assets (Domestic private sectors)	5919.9	350.5%	75.4%	54051.7	412.8%	73.1%	62.3%	(2.3%)	58983.4	351.8%	63.4%	(61.0%)	(9.7%)	1.2%	(12.0%)
Households & nonprofits nonfinancial assets	3701.2	219.1%	47.1%	38576.9	294.6%	52.2%	75.5%	5.0%	38837.9	231.6%	41.7%	(63.0%)	(10.4%)	12.5%	(5.4%)
Nonfinancial corporate business nonfin assets	2131.8	126.2%	27.2%	14168.0	108.2%	19.2%	(18.0%)	(8.0%)	18561.3	110.7%	19.9%	2.5%	0.8%	(15.5%)	(7.2%)
Financial business nonfinancial assets	86.9	5.1%	1.1%	1306.8	10.0%	1.8%	4.8%	0.7%	1584.2	9.4%	1.7%	(0.5%)	(0.1%)	4.3%	0.6%
Nonfinancial assets (breakdown)															
Real estate	4107.8	243.2%	52.3%	41914.6	320.1%	56.7%	76.9%	4.4%	43295.6	258.2%	46.5%	(61.9%)	(10.2%)	15.0%	(5.8%)
Households	1992.4	118.0%	25.4%	27852.8	212.7%	37.7%	94.7%	12.3%	25144.7	150.0%	27.0%	(62.8%)	(10.7%)	32.0%	1.6%
Nonprofit organizations & nonresidential	888.8	52.6%	11.3%	5385.7	41.1%	7.3%	(11.5%)	(4.0%)	7015.6	41.8%	7.5%	0.7%	0.3%	(10.8%)	(3.8%)
Nonfinancial corporate business	1170.9	69.3%	14.9%	7924.3	60.5%	10.7%	(8.8%)	(4.2%)	10236.1	61.0%	11.0%	0.5%	0.3%	(8.3%)	(3.9%)
Financial business	55.7	3.3%	0.7%	751.8	5.7%	1.0%	2.4%	0.3%	899.2	5.4%	1.0%	(0.4%)	(0.1%)	2.1%	0.3%
Durable goods and equipment	1303.5	77.2%	16.6%	8636.0	66.0%	11.7%	(11.2%)	(4.9%)	10821.5	64.5%	11.6%	(1.4%)	(0.1%)	(12.6%)	(5.0%)
Consumer durable goods	578.0	34.2%	7.4%	4107.8	31.4%	5.6%	(2.9%)	(1.8%)	4942.2	29.5%	5.3%	(1.9%)	(0.2%)	(4.7%)	(2.1%)
Nonprofits & nonfinancial noncorporate bus	147.5	8.7%	1.9%	785	6.0%	1.1%	(2.7%)	(0.8%)	1098.6	6.6%	1.2%	0.6%	0.1%	(2.2%)	(0.7%)
Nonfinancial corporate business	550.2	32.6%	7.0%	3297.8	25.2%	4.5%	(7.4%)	(2.5%)	4261.7	25.4%	4.6%	0.2%	0.1%	(7.2%)	(2.4%)
Financial business	27.8	1.6%	0.4%	445.4	3.4%	0.6%	1.8%	0.2%	519.0	3.1%	0.6%	(0.3%)	(0.0%)	1.4%	0.2%
Inventories	387.3	22.9%	4.9%	1842.3	14.1%	2.5%	(8.9%)	(2.4%)	2443.9	14.6%	2.6%	0.5%	0.1%	(8.4%)	(2.3%)
Nonfinancial nnoncorporate business	80.1	4.7%	1.0%	218.4	1.7%	0.3%	(3.1%)	(0.7%)	297.3	1.8%	0.3%	0.1%	0.0%	(3.0%)	(0.7%)
Nonfinancial corporate business	307.2	18.2%	3.9%	1623.9	12.4%	2.2%	(5.8%)	(1.7%)	2146.6	12.8%	2.3%	0.4%	0.1%	(5.4%)	(1.6%)
Intellectual property products	121.4	7.2%	1.5%	1658.8	12.7%	2.2%	5.5%	0.7%	2422.4	14.4%	2.6%	1.8%	0.4%	7.3%	1.1%
Nonprofits & nonfinancial noncorporate bus	14.4	0.9%	0.2%	227.3	1.7%	0.3%	0.9%	0.1%	339.4	2.0%	0.4%	0.3%	0.1%	1.2%	0.2%
Nonfinancial corporate business	103.6	6.1%	1.3%	1322.0	10.1%	1.8%	4.0%	0.5%	1916.9	11.4%	2.1%	1.3%	0.3%	5.3%	0.7%
Financial business	3.4	0.2%	0.0%	109.5	0.8%	0.1%	0.6%	0.1%	166.1	1.0%	0.2%	0.2%	0.0%	0.8%	0.1%
Net financial assets domestic private sectors	1847.1	109.4%	23.5%	21548.1	164.6%	29.1%	55.2%	5.6%	38482.1	229.5%	41.3%	64.9%	12.2%	120.1%	17.8%
Net financial assets HH & NN business	2095.2	124.1%	26.7%	23309.4	178.0%	31.5%	54.0%	4.8%	40030.0	238.7%	43.0%	60.7%	11.5%	114.7%	16.3%
Net financial assets NC business	(318.2)	(18.8%)	(4.1%)	718.8	5.5%	1.0%	24.3%	5.0%	552.8	3.3%	0.6%	(2.2%)	(0.4%)	22.1%	4.6%
Net financial assets financial business	70.1	4.2%	0.9%	(2480.1)	(18.9%)	(3.4%)	(23.1%)	(4.2%)	(2100.7)	(12.5%)	(2.3%)	6.4%	1.1%	(16.7%)	(3.1%)
Net financial assets private sectors	1931.2	114.3%	24.6%	19897.4	152.0%	26.9%	37.6%	2.3%	34098.5	203.4%	36.6%	51.4%	9.7%	89.0%	12.0%
Net financial assets rest of the world	84.1	5.0%	1.1%	(1650.7)	(12.6%)	(2.2%)	(17.6%)	(3.3%)	(4383.6)	(26.1%)	(4.7%)	(13.5%)	(2.5%)	(31.1%)	(5.8%)

Figures in Nonfinancial assets Federal government include human capital based on figures from "Human Capital Accounting in the United States, 1994-2006" by Michael S. Christian, Survey of Current Business, June 2010, U.S. Department of Commerce, Bureau of Economic Analysis

Source: Z.1 Financial Accounts of the United States, Federal Reserve Statistical Release, September 18, 2014, Various tables Historical Annuals, 1975-1984 and 2005-2013.

Conclusion

A review of these national accounts gives support to the description given above of the production phases and their relationship to the optimal consumption - capital expenditure mix. The mistaken notion that government participation in the economy is non-productive is clearly shown to be false by a thorough analysis of the data of the past 40 years. This does not mean that some of the government regulation is not onerous, but misguided regulations should be replaced by effective ones, not by a lack thereof. Despite the fact that government participation in the GDP and net worth percentage of the national economy has decreased respectively by 3.9% and by 8.3% from 1975 to 2013, business percentage, both financial and nonfinancial, corporate and noncorporate, of net worth of the economy has decreased by almost as much at 7.8%. The offsetting winners have been foreign interests at 5.0% and high net worth private individuals at 11.1% gain, hardly the lift-all-boats that was boasted by supply side theorists. Equally unimpressive is the 6.6% decrease in ownership of nonfinancial, productive goods assets of both nonfinancial business sectors. Coupled with the 3.2% reduction in federal government nonfinancial assets and a 7.6% increase in high-income household nonfinancial assets, principally in real estate and durable goods, this indicates a disinvestment in the economy between 1975 and 2013 in line with the increase of 7.3% in the personal consumption expenditures from 61.2% to 68.5%. Personal consumption has been going up for the fortunate few, but not for the 50% plus of the population that has had 40 years of stagnate wages.

This trend must be reversed and the consumption spending index brought back to the levels of 1975 if we are to arrest the deteriorating condition in education, medical care, retirement and national infrastructure. Acts of terrorism understandably get the press attention, but an individual is more likely to die in an infrastructure failure than at the hands of a religious fanatic. The cost of the misguided Iraq War would have gone a long way toward correcting the infrastructure deficiencies with no loss of national security and probably made the same contractors happy.

The objection to doing anything about this arises from the inability or unwillingness to address the question of where the money comes from, yet the only answer is deceptively simple. It starts with an understanding of what money is. The fact that people are attached to it emotionally, clinging to it as tokens of emotional satisfaction, obscures the fact that in the modern world it is simply a tool of accounting. It is not a thing of inherent value, a store of energy that can be used run the economic engine. For a mechanical analogy, it is more like the motor oil in the lubrication system of a car or the air in the pneumatic system used to inflate its tires. Too much or too little of either in the fluid reservoirs makes for inefficient operation.

This analysis has attempted to show that the reservoir of value for the US economy is not the financial assets that have mushroomed over the past 40 years, which can disappear as quickly as they came, but rather the vast pool of human capital that is presently being underutilized and wasted. Such underutilization and waste is not the fault of business owners and managers per se, responding as they must to global competition, but it does not help that many sit on the sidelines while economic claptrap designed to elevate personal greed to a prerogative of the constitution is put into policy. The responsibility lies with the policy makers themselves, but they, in turn are elected by polarized and vocal groups that put special interest over the common good.

The left wants to tax or borrow. The right doesn't want to be taxed or to borrow, since they fear the latter may affect their own ability to borrow. They pontificate about the federal government's lack of fiscal restraint with the old saw that the government should be as fiscally responsible as its citizens; yet how many of them own a home with no mortgage that wasn't inherited. If they were to run their own households the way they demand the government run its house, they would have to charge their children for dinner and room rent, TV and computer time, not to mention the cell phones, require them don only fashion-free clothing unless they can pony up the premium, and make them go without seeing the doctors because they couldn't pay for the visits themselves. You get the picture.

The current polarization has substance. The problems of the blue states are generally those of urban developed areas while those of the red states are likely to originate in a more recent rural history. The fallback possibilities during economic hard times are prone to depend more on individual initiative and effort in a rural environment than in a dense urban one, making its population suspect of a policy from a remote federal bureaucracy run by people with an urban pedigree. It has not always been this way. That bureaucracy was a lifeline to these rural areas in the Great Depression, as it was to the Eastern banking establishments.

As stated, the answer is simple, but demands discipline. Fundamentally overhaul or phase out the income tax, retaining only those taxes that cover actual public costs, including compensation minimums and supporting tariffs, and phase in payment for necessary public programs, such as infrastructure, research and development, and the social safety net with electronic fiat money, recognizing that the money supply, if considered an economic good at all, is more public than private, more a common than a club good. From this recognition, provide the proper oversight of its operation and optimization. Allow private interests to provide a premium marketplace for whatever education, insurance, housing, retirement or other club-good programs they desire, but with zero public recourse in the event of their failure, no matter what size the failure. In financial matters someone's loss is someone else's gain, but this is not so in the resulting economic loss. Limit the size of financial establishments, but not their ability to work transparently together. Reign in the ridiculous compensation of management.

Publicly develop the human capital base through decentralized, community based education, employment training and basic medical care programs. Revamp mortgage rules so that once a banking establishment, which not so incidentally is insured by the public through the FDIC, has qualified a buyer and closed the deal, the property cannot so easily be foreclosed, while also removing any regulatory compulsion to lend when the bank deems a buyer unqualified. In this regards, in the recent fiasco it would have been a lot cheaper for the public to provide mortgage insurance for everyone than to allow the development of a private, opaque credit default swap market. Implement a guest worker visa program. Decriminalize vice, but properly penalize and tax its damaging excesses. Develop quick response reprisals for foreign belligerents that are aimed at proportional response, not necessarily aimed at eradication, which avoid prolonged engagement, and offer negotiation. Etcetera, etcetera, and so forth.

The position taken here is not that markets should be arbitrarily controlled and regulated. The position is that in a competitive free market system under resource constraint and therefore faced with excess labor, not all goods and services will be produced and distributed by that system as needed due to the fact that labor is subject to the same commodity pressures as any other good or service. What is required, therefore, is not a redistribution of wealth from the haves to the have-nots that is disruptive of established and adequately functioning productive organizations, but a provision of liquidity distribution, albeit with necessary overall programs for the engagement of equivalent productive, meaningful employment. Such guarantees indicate that in the absence of real productive employment opportunity, as in the recent crisis, willing citizens should not be denied access to the liquidity necessary to maintain a decent life.

It is my belief that the liquidity structure outlined here offers a real liquidity optimization constraint in C_G and I_G and a target for policy implementation. The public *as public* should invest 4% of 2013 GDP in infrastructure and R&D for each of the next 6 years out of fiat liquidity, which is in line with ASCE suggestions, and private business should invest 2.7% of GDP in domestic investment toward the target reduction in net consumption spending to 61.8%. If private business won't invest the 2.7%, then the public should do the rest, and going forward provide adequately for the basics in education, medical, and appropriate end of life care for those for whom the free market fails to provide.

If the human capital of the United States, its citizenry and guest workers, is properly engaged, enhanced, and enriched, through enlightened public–private initiatives, there is no cause for pessimism.

PGFWABF,
PHACHB,
PHAYHH,
PFSAHG.

Citations:

Charts and Tables:

"Household final consumption expenditures as % of GDP", Source: World Bank national accounts data and OECD national accounts data files, http://data.worldbank.org/indicator/NE.CON.PETC.ZS

"Indexed Stagnation of Wages by Percentile 1973 – 2009", Source: EPI analysis of U.S. Census Bureau, Current Population Survey, Outgoing rotations group. Indexed Stagnation.jpg, www.washingtonpost.com

"US GDP and Wealth Comparisons 1975, 2005, 2013", Source: Z.1 Financial Accounts of the United States, Federal Reserve Statistical Release, September 18, 2014, Historical Annuals, 1975-1984 and 2005-2013, http://www.federalreserve.gov/releases/z1/Current/data.htm

Infrastructure and human capital information:

Infrastructure: American Society of Civil Engineers, http://www.infrastructurereportcard.org/a/#p/grade-sheet/americas-infrastructure-investment-needs

Human Capital: "Human Capital Accounting in the United States, 1994-2006", a report by Michael S. Christian, published by the U.S. Department of Commerce, Bureau of Economic Analysis, http://bea.gov/scb/pdf/2010/06%20June/0610_christian.pdf

Text:

Thomas Piketty, Capital in the Twenty-First Century, Harvard: Belknap Press, 2014.

A Critique of Neoliberal Economics

Part I – Quantitative Analysis & Assumptions

Capital as Position in Ergodic Economic Modeling

-:-

Hierarchical Position and Focused Rationality
in an Analysis of Stocks & Flows

US Federal Reserve 4Q 1989 & 4Q 2019
Household Income and Net Worth Data

Martin Gibson
martin@uniservent.org
December 10, 2020

Capital as Position in Ergodic Economic Modeling

Hierarchical Position and Focused Rationality in an Analysis of Stocks & Flows

US Federal Reserve 4Q 1989 & 4Q 2019 Household Income and Net Worth Data

Martin Gibson, UniServEnt.org

Abstract

Current macroeconomic structural analysis is heavily constrained quantitatively and qualitatively by nominal valuation of sector and subsector accounts in terms of the history of market transactions and a forecast of capital changes based on the sector trends. A focus on private sector rates of return, nominal and real, on GDP and on capital held as financial and real equities and interpreted as a prime indicator of the health of a national economy, ignores key factors as evidenced by the market collapse of 2008.

Apparently missing from this analysis are the following;

1. The focus on maximization of returns as real over nominal rates encourages the reduction in taxes (for public outlays of indeterminant private benefit) and labor costs (deemed to be inflationary), with little consideration of the macroeconomic benefit derived. This is a straw man based on the true nature of modern monetary policy in which public outlays as issuance of a national fiat currency involve no direct borrowing or income tax transfer from the private sector. Instead of a currency flow from available private to deficient public sectors as generally portrayed, the use of borrowing and taxation is ostensibly meant to retire money thought to be circulating or looking for investment in the private sectors to offset the fiat issuance of public funding, to thereby avoid inflationary pressure from too much money chasing too few goods and services in either sector, though the obverse appears to be the case based on the ongoing lack of a balanced budget and almost zero inflationary pressure, at least for basic goods and services.

2. The true structural picture of the economy is only seen by an analysis of sector and subsector changes over time as percentages of total national production with respect to capital and consumption flows and to stocks rather than for sector changes measured against same sector baselines.[i] This can be addressed by a variant of analytical modeling which ergodically constrains the average flows and stocks of individual economic microstates exhibiting focused rationality as decision making groups over time to the average of those microstate flows and stocks of the ensembled macrostate of that economy at a given point in time.

---

[i] https://uniservent.org/political-economy/ Link to working paper, The Browser Economy, Martin Gibson

3. There is no allowance in the national accounting for valuation of human capital on either a market-valued (MHC) or non-market-valued (NHC) basis. A 2010 study determines a range of 25-30% MHC to 75-70% NHC.[ii]

Comparison of unbiased ergodicity in non-market and of weighted ergodicity in market modeling of human capital with 4Q 1989 and 4Q 2019 Household Income and Net Worth US Fed Data gives credence to this approach as indicated in quantitative analysis of the first section and summed up in the qualitative political economic analysis of the second.

[ii] Human Capital: "Human Capital Accounting in the United States, 1994-2006", a report by Michael S. Christian, published by the U.S. Department of Commerce, Bureau of Economic Analysis.

Some Axioms concerning Economic Activity and Capital

The word, capital, is from the Latin for 'head', *caput*; related to a common Latin stem for 'roomy, the ability to hold', *capax*, related to 'capacity', the ability to receive, contain, perform, or function in a position or role, both physical and mental. The word, power, is also from the Latin, *posse* or *potens*, by way of the Anglo-French, *poer*, as 'to be able, to have power'. It is therefore somewhat redundant, if not tautological, to state that social power and the cultured ability to direct the energy coming from any natural potential as economic activity, comes to be represented by way of cranial capacity as the experienced mental ability of collectively engaged human beings.

In an economic context, **capital** is the rationally and emotionally effective productive capacity and potential, the **power,** of human beings working with the natural resources at hand, by their individual and collective **labor**, to maintain their place of **living** in the world. An economic interchange among human beings involves an expenditure of energy and incorporation of material in the supply, production, warehousing, and distribution of the final consumable **goods and services (G&S)** required for living; as an integral corollary this includes the percentage of that energy and material that is required for the creation, maintenance, and replenishment of any intermediate goods and services necessary for production and distribution of the final consumables. All these intermediate goods and services are forms of real capital and necessarily include the creation, maintenance, and replenishment by way of the sustenance required of and for the human beings involved in the living experience.

A portion of these interchanges between human beings are performed within the context of a market mechanism as a trade of G&S for ostensibly equivalent **tokens of satisfaction ($)** as instruments of financial capital in the form of money, promissory notes, certificates of extant stock or future flows of assets, or other tabulation of accounting **value**, so that such portions of both intermediate and final G&S have a representational, stock-valued human entitlement as **market-valued human capital (MHC)**. The remaining portion of the interchanges that are productive of intermediate and final G&S are assignable as **non-market-valued human capital (NHC)** which can only be estimated based on broad equivalences between market and non-market components. These NHC interactions are the earthly product of Nature's bounty, God's grace, however you may choose to designate God as the endless, supreme potential source for change, and community nurture that must be sustained from cradle to grave by means of both MHC and NHC; this NHC as might be indicated with "all men (and by implication, women) are created equal, … endowed by their Creator with certain unalienable Rights, that among these are Life, Liberty, and the Pursuit of happiness" in the US Declaration of Independence and which we might deem applicable internationally.

Certain accounting in the US economy of recent years puts the ratio of MHC:NHC average in a range of 0.43:1 to 0.33:1, where 1 is an implied statistical average of NMC.[1] From this thinking as we will detail, MHC of 0.43 to 0.33, or 0.38$^{+/-}$ 0.05, can be theoretically distributed along a spectrum with each individual's percentage of the total market valued capital of an economy, all greater than 0.0 (since a negative as debt would indicate involuntary servitude or Shylock's pound of flesh), while the NHC is simply the per capita average of the total NHC accounted as 1.0.

Capital then, all of which is an intermediate good or service, can be viewed and valued as: **human**—consisting of the physical and metaphysical skills and capabilities of people both MHC and NHC; **real**—consisting of the physical resources, both natural and technologically developed, available for human utilization through productive human effort; and **financial**—consisting of various methods devised for directing and accounting for the past and future production, distribution and consumption of goods and services required or otherwise desired by society.

The referenced power can be viewed qualitatively as: **social**—consisting of the skilled capacity, expertise, or position to organize and direct a desired productive enterprise; **energy source**—consisting of the available, technologically developed and thermodynamically defined capacity to fuel useful work, both mechanically and through the ecological and nutritional needs of human living; and individual **labor power**—consisting of the mental and physical ability to provide and sustain useful and necessary, thereby meaningful, work.

Labor can be viewed along a spectrum as: **self-initiated**—consisting of the employment and direction of the work of oneself and/or others; **employed**—consisting of the voluntary provision of one's own work effort at the direction of others; and **involuntarily servile**—consisting of wage, penal, chattel, or ensnared bondage under conditions of coerced toil.

Living can be viewed along a spectrum from a **risk of enslavement** to material wretchedness due to personal behavior or the depravity and disregard of others up to an **opportunity for liberation** of spiritual perception and intuitional creativity.

Capital and labor, both as an expression of power, is at work under any condition or state: be it tribalist, feudalist, capitalist, socialist, or communist; monarchist, oligopolist, liberalist, or anarchist; browser, hunter-gatherer, agricultural, industrial, or cybernetic. This dynamic of capital production and use is not confined to 'capitalism'; nor absent under 'socialism'. The political organization of that state is determined in proportion and valued degree by those who are positioned to exercise the power inherent in the instruments of both capital and labor over a range of **hierarchical structure** and **decision making**. That decision making is characterized herein as a choice defined by the motivation, expertise, capability, and position of the individual or affiliated group as a function of **focused rationality**, even if that reason is confined to the daily decision of how or whether to continue with a meaningless job, with no apparent alternatives in sight, in order to survive.

Quantitative Analysis—Weighted Ergodic Modeling of Economic Activity

An important aspect of the expertise involved in the social structuring of decision making is the ability of individuals to formulate a mental understanding consciously or semi-consciously for their beneficial participation in the group ecology. When this formulation is consciously directed toward the end of providing a group understanding of some natural dynamic which may or not include the group dynamic, with a measure of quantifiable specificity and predictability in terms of some stated fundamental constraints, we often call the formulation a **model**; sometimes a toy or mental model to distinguish the formulation from the real system, the 'reality' which it is intended to represent. It is important to emphasize that the model may interact with that reality, but it can never fully define that reality which necessarily includes any number of unknown causes and effects.

Such modeling generally posits some conserved potential for change in a system of elements or **microstates** within a field that is continuous across the expanse of the field which may or may not comprise its own boundaries and which may be observed to constitutes a **macrostate** of the system, or a portion of the system, at a given point in time. Such microstates may be independently constituted or qualitatively and quantitatively emergent phenomena of the field in response to endogenous or exogenous change in the system. Change may be deemed random for observable portions of the system, but boundary and conservation assumptions dictate that the system requires certain invariants and as a whole is necessarily deterministic. **It follows that if the average of the microstates of the system are invariant over time, though exhibiting periodic characteristic variability, while the average of the macrostate of the system exhibits the same invariance at a point in time, the model is said to be ergodic.**

As modeled here, the sum of the ongoing instances or microstates of focused rationality produce an ensemble or macrostate of the political economic system which is ergodic, where ergodicity is defined as the deterministic probability for the average of the individual **microstates over time** to be constrained or otherwise indicative of the average of the **macrostate at a given point in time**. We might surmise this not to be the case for macroeconomic modeling, but will find that if we factor in a weighting of the focused rationality of the microstate's decision making to an otherwise strictly ergodic distribution, the evolution of a hierarchical structure applicable to national economic data emerges and offers a reasonable level of confirmation of this line of thinking as graphically supported.

Each factored weighting of a microstate, m_M, in macrostate, M_n, is represented here as the extant result after n iterations of a differential binary branching from an immediately prior condition of evolved $n-1$ microstate value. For a weighting of zero, that is, with each flip of an unbiased coin, starting with a value of 1, after n flips we arrive at a probability as a microstate value of 2^{-n} for each of the 2^n iterated microstates, where the macrostate value remains an invariant 1.0. As an alternative expression of such unbiased distribution, we can use an invariant microstate value of 1.0 for each subsequent iteration, so that for 2^n iterated microstates, the macrostate value is 2^n and the probability for each microstate remains 2^{-n} as a percentage of the total.

The selection of n as the number of iterations indicates the hierarchy of the probability distribution and the granularity of the probability density of the microstates. For our purpose in this discussion we have used $n = 5$, where macrostate M_5 consists of 32 microstates enumerated in binary form from the **caput** or head end of a distribution in the Appendix, Charts 1 - 11 positioned starting at the top of column I and running down in descending order, as $m_5 11111$, $m_5 11110$, ... $m_5 00001$, $m_5 00000$, to end at the **cauda** or tail end of the distribution. This M_5 distribution is sufficient for our purposes as a proof of concept of this modeling approach.

Depending on the granularity in application of the model, each microstate can represent an individual decision maker or an aggregate of related decision makers. In the application used here, each of the 32 microstates of M_5 represents 3.125% of the decision makers of all ages from among the total US population of the reported data. With respect to the distribution structure, each microstate, $m_5 i^5$ as stated above, can be weighted by a factor, wf, from the range of -1 to +1, where each $m_5 i^5$ equals $m_4 i^4 (1+wf)$ for each caput and $m_4 i^4 (1-wf)$ for each cauda. A range less than zero simply inverts the weighting to the cauda end. The proportional scale of hierarchical divergence within the macrostate is directly related to the magnitude of the factor, wf, as expressed with each microstate as a percentage of the whole.

Therefore, each microstate transaction is conservative, in that the amount added to $m_4 iiii$ to produce $m_5 iiii 1$ is deducted from $m_4 iiii$ to produce $m_5 iiii 0$, where $2(m_4 iiii) = m_5 iiii 1 + m_5 iiii 0$. In this regard, though M_5 does represent a logically evolved macrostate, the n iterations prior to M_5 do not necessarily represent a historical sequence. In the context of focused rationality in a market, each pair ($m_5 iiii 1$, $m_5 iiii 0$) represents a market valued economic exchange in which the focus of the caput, $m_5 iiii 1$, of an underlying economic activity, $m_4 iiii$, represents the market position of a price maker of the transaction by virtue of a superior position, and the focus of the cauda, $m_5 iiii 0$, represents that of a price taker, each coming from a different perspective on satisfaction or with a different rationale for the exchange.

In transactions which we might view as *idiotic* **unfocused rationality** or as *malignant* **focused irrationality**, we might expect a departure from ergodicity as an expression of 'noise' in the system. In the case of unfocused rationality, a lack of experience-based skill or expertise results in the inability of an otherwise sound logic to focus on what is pertinent and what is extraneous in a given economic exchange, generally to the detriment of the cauda of a trade. In the case of focused irrationality, an accurate assessment of the parameters and conditions of a trade based on experience is coupled with an illogical course of action by a party to the transaction, which if it becomes habitual will work to the detriment of that party and perhaps the other party as well. We would anticipate such exchange more likely to be initiated to the intended, but not the resultant, benefit of the caput. Then there is the case of **unfocused irrationality**, chaos applied to chaos resulting in more chaos, as when a novice at the racetrack puts his life savings on a hunch and losses it all.

The applicability of weighted ergodicity to the topic of capital as power should become evident in the hierarchical structure of capital as the emergent function of the weighting factor is examined. The power of capital, the ability to use capital, is essentially a matter of various positions, some of which are: entitlement position–by birth, license, expertise;

market position–by geography, ecology, population; financial position–by network, leverage, liquidity.

Primary to all these positions is that of attention in the head of an individual, in the caput as **focused Logic** which is needed to understand and make the most of one's circumstance in the world; this, instead of focusing on the emotions in the gut. Without a focus on that Logic, desired progress in the use of real and financial capital will likely remain difficult. As that understanding is sufficiently developed, the emotions can be used to motivate the individual, through a willingness to take on circumstantial risk in life and thereby develop the specific expertise required to make a living. Note that I have not said 'to make money'. Those motivated first and primarily by the desire to 'make money' without first understanding the social implications of their productive enterprise run the risk of succumbing to greed and ultimately failure.

Second is the position in which the individual finds or places himself as a member of a primary group to which they existentially identify, be it family, friends, community, church, trade, labor union, profession, ethnic, or other social group. This primary group might be represented in this modeling as a microstate in a weighted ergodic macrostate. If that microstate is itself governed by a corresponding focused Logic or rationality, the individual will resonate with the group and capitalize on that membership in terms of human capital, market and especially non-market valued. If the microstate of the group is either unfocused in the rational head or focused on the irrational gut, or God forbid, unfocused in the irrational gut, the focused Logic of the individual should look elsewhere for the exercise of capital.

Third is the position of the individual in the hierarchical structure of the macrostate, generally in terms of a local, state, regional, national, or international political economy. Generally, the more caputs there are in an individual's microstate, and therefore the more elevated the microstate tier, the better will be the individual's position in gaining access to and control of capital and thereby of economic and financial power, for good or ill. This is so regardless of the degree of weighting on a spectrum of 0 to 100% for an individual microstate, which necessarily varies between microstate transactions, but which we can generalize for modeling purposes for the whole of a macrostate. Unfortunately, that weighting can have adverse effects as the macrostate factor moves toward the 100% end of the spectrum, particularly if the whole macrostate trends toward unfocused irrationality or what I will call ergodidiocy.

In the appendix,[2] each chart represents a 5th iteration macrostate of a binary branching from an initial base unit, where each branching is equivalent to the flip potential of an unbiased coin, with the two sides designated by a caput or head as 1 and by a cauda or tail as 0. The resulting probability distribution results in 32 microstates enumerated in descending order in column *J* where the resulting value for each microstate of the iteration is sequenced from left to right, next to the corresponding probability value in column *I* truncated to 2 decimals. This Fifth Iteration distribution is represented graphically to the right under columns *O-Y*.

Chart 1 indicates an Ergodic Binary Distribution: Unweighted Probability where the probability of each microstate is 0.03125 of the total distribution of 1.0.

Chart 2 indicates the same information in Ergodic Binary Distribution: Weighted 0% as effectively unweighted, but with each microstate value as an invariant 1.0 of the total distribution of 32.0 for the macrostate, which reiterates the same uniform probability density and percentage microstate of 3.125% of the total. The invariant 1.0 of the microstate evolution over iteration sequencing and therefore over time being equal to the average of the macrostate at the nth iteration and therefore at a point in time, makes the structure strictly and, based on a weighting of 0%, trivially ergodic.

Chart 3, Ergodic Binary Distribution: Weighted 10%, shows the emergence of a hierarchical structure weighted toward the caput end of each branched microstate and away from the cauda end as a result of various forms of market or command based direction of individual or group economic activity. In this presentation, each microstate represents a summation of economic activity from planning & development, acquisition of materials and agricultural products, application of productive technology, employment of a spectrum of labor skills, through the production and distribution of intermediate and final goods and services, all according to a valued scope determined by the location and scale of the microstate operations and in particular by the expertise of the public and private parties in the focus and reason they bring to the task of providing a livelihood for the microstate.

The remainder of Charts 4 through 11 give a qualitative sense of the effects of a quantitative increase in the empirically determined weighting factors.

—

The fractal nature of the distribution is evident. From the spread sheets not shown, starting with the 2nd iteration, the top valued caput, 11^2, and bottom valued cauda, 00^0, bracket the two intermediate microstates which are intermediately and equally valued with the second tier, at 10^1 and 01^1. Here the apparent exponents of each microstate are used as indices to indicate the hierarchy tier level of a microstate by number as the strength of the caput decision makers in the microstate. A similar bracketing by branched intermediate, composite microstates is apparent, where the hierarchical fractal nature can be read left to right as well as top to bottom, in the following iterations of

M_3 as

$[111^3, 110^2]$,
$[101^2, 100^1]$,
$[011^2, 010^1]$,
$[001^1, 000^0]$,

with the 8 microstates sorted by tier valuations as
1 @ $[111]^3$,
3 @ $[110, 101, 011]^2$,
3 @ $[100, 010, 001]^1$,
1 @ $[000]^0$

M_4 as

$[1111^4, 1110^3, 1101^3, 1100^2]$,
$[1011^3, 1010^2, 1001^2, 1000^1]$,
$[0111^3, 0110^2, 0101^2, 0100^1]$,
$[0011^2, 0010^1, 0001^1, 0000^0]$,

with the 16 microstates sorted by tier valuations as
1 @ $[1111]^4$,
4 @ $[1110, 1101, 1011, 0111]^3$,
6 @ $[1100, 1010, 1001, 0110, 0101, 0011]^2$,
4 @ $[1000, 0100, 0010, 0001]^1$,
1 @ $[0000]^0$,

M_5 as

$[11111^5, 11110^4, 11101^4, 11100^3, 11011^4, 11010^3, 11001^3, 11000^2]$,
$[10111^4, 10110^3, 10101^3, 10100^2, 10011^3, 10010^2, 10001^2, 10000^1]$,
$[01111^4, 01110^3, 01101^3, 01100^2, 01011^3, 01010^2, 01001^2, 01000^1]$,
$[00111^3, 00110^2, 00101^2, 00100^1, 00011^2, 00010^1, 00001^1, 00000^0]$,

with the 32 microstates sorted by tier valuations as
1 @ $[11111]^5$,
5 @ $[11110, 11101, 11011, 10111, 01111]^4$,
10 @ $[11100, 11010, 11001, 10110, 10101, 10011, 01110, 01101, 01011, 00111]^3$,
10 @ $[11000, 10100, 10010, 10001, 01100, 01010, 01001, 00110, 00101, 00011]^2$,
5 @ $[10000, 01000, 00100, 00010, 00001]^1$,
1 @ $[00000]^0$,

From this schema we can state that while the value of the NHC for a given microstate as an individual or group is the average of macrostate M_n valued in total as 2^n and equal to 1.0, the MHC of an individual microstate of tier $M_n i^m$ will be as valued by the strength of the number of caput in each microstate designation as weighted by wf for the macrostate.

Once again, such microstate exchange in a weighted ergodic structure can be modeled to represent millions of decision makers, for example, over a period such as a financial quarter of national accounting. The strength of the human capital and human controlled real and financial capital, and thereby the power of a given microstate is represented by the magnitude of the tier indices, as augmented by the wf stated after the colon along a spectrum of 0% to 100%, so that the weighting for each of the following Ergodic Binary Distributions in the Charts can be expressed as:

$M_{5:00}$, $M_{5:10}$, $M_{5:38.19}$, $M_{5:50}$, $M_{5:60}$, $M_{5:72}$, $M_{5:86.4}$, $M_{5:92.8}$, $M_{5:92.8R}$, & $M_{5:100.0}$.

The strength of the hierarchical tiers with a microstate base value of 1.0 can be designated as follows, where the upper indices indicates the strength or intensity of the caput and the lower indices indicates the intensity of the cauda, here shown with a weighting factor of 10%.

$M_{n:wf} i^m{}_{(n-m)} =$
$M_n i^{m:wf} = M_n i_{(n-m):wf} = m_5 iiiii$
$M_5 i^{5:10} = M_5 i_{0:10} = 1.61$
$M_5 i^{4:10} = M_5 i_{1:10} = 1.32$
$M_5 i^{3:10} = M_5 i_{2:10} = 1.08$
$M_5 i^{2:10} = M_5 i_{3:10} = 0.88$
$M_5 i^{1:10} = M_5 i_{4:10} = 0.72$
$M_5 i^{0:10} = M_5 i_{5:10} = 0.59$

These weighted distributions are sorted and compared in the following tables in which the hierarchical tiers are indicated by the rows in the left hand column for the column weightings as indicated in the top row. Values in the non-weighted $M_{5:00}$ column indicate the microstate values of Non-market-valued Human Capital of macrostate average or 1.0. All other values represent Market-valued Human Capital for the microstates based on a theoretically modeled or empirically derived data set. Blue lines angled up from left to right above each cell indicate an increasing value with respect to the lower weighted cell to the left and the red diagonals indicate a decreasing value for both tier microstate and the corresponding sums. Cells located beneath the bold red bars indicate a MHC for a microstate weighting or tier summation approaching zero.

—

Weighting Tier	$M_{5:00}$	$M_{5:10}$	$M_{5:3819}$	$M_{5:50}$	$M_{5:60}$	$M_{5:72}$	$M_{5:864}$	$M_{5:928}$	$M_{5:1.0}$	$M_{5:928R}$
$1@\ i^5_0$	1.0	1.61	5.04	7.59	10.49	15.05	22.50	26.61	32.00	0.00
$5@\ i^4_1$	1.0	1.32	2.25	2.53	2.62	2.45	1.64	1.00	0.00	0.00
$10@\ i^3_2$	1.0	1.08	1.01	0.84	0.66	0.40	0.12	0.04	0.00	0.00
$10@\ i^2_3$	1.0	0.88	0.45	0.28	0.16	0.06	0.01	0.00	0.00	0.04
$5@\ i^1_4$	1.0	0.72	0.20	0.09	0.04	0.01	0.00	0.00	0.00	1.00
$1@\ i^0_5$	1.0	0.59	0.09	0.03	0.01	0.00	0.00	0.00	0.00	26.61

Table 1 – Weighted Distributions per Tier of M_5, microstate NHC average 1.0

Weighting Tier	$M_{5:00}$	$M_{5:10}$	$M_{5:3819}$	$M_{5:50}$	$M_{5:60}$	$M_{5:72}$	$M_{5:864}$	$M_{5:928}$	$M_{5:1.0}$	$M_{5:928R}$
$\Sigma\ i^5_0$	1.0	1.61	5.04	7.59	10.49	15.05	22.50	26.61	32.00	0.00
$\Sigma\ i^4_1$	5.0	6.60	11.25	12.65	13.10	12.25	8.20	5.00	0.00	0.00
$\Sigma\ i^3_2$	10.0	10.80	10.10	8.40	6.60	4.00	1.20	0.20	0.00	0.00
$\Sigma\ i^2_3$	10.0	8.80	4.50	2.80	1.60	0.60	0.10	0.00	0.00	0.04
$\Sigma\ i^1_4$	5.0	3.60	1.00	0.45	0.20	0.05	0.00	0.00	0.00	1.00
$\Sigma\ i^0_5$	1.0	0.59	0.09	0.03	0.01	0.00	0.00	0.00	0.00	26.61
	32.0	32.0	32.0	32.0	32.0	32.0	32.0	32.0	32.0	

Table 2 – Weighted Distributions per Tier Sum of M_5, microstate NHC average 1.0

Weighting Tier	$M_{5:00}$	$M_{5:10}$	$M_{5:3819}$	$M_{5:50}$	$M_{5:60}$	$M_{5:72}$	$M_{5:864}$	$M_{5:928}$	$M_{5:1.0}$	$M_{5:928R}$
$\Sigma\ i^5_0$	3.1%	5.0%	15.8%	23.7%	32.8%	47.1%	70.3%	83.2%	100.0%	0.0%
$\Sigma\ i^4_1$	15.6%	20.7%	35.2%	39.5%	40.9%	38.3%	25.6%	15.6%	0.0%	0.0%
$\Sigma\ i^3_2$	31.3%	33.8%	31.5%	26.3%	20.6%	12.5%	3.8%	1.2%	0.0%	0.0%
$\Sigma\ i^2_3$	31.3%	27.5%	14.1%	8.9%	5.0%	1.9%	0.3%	0.0%	0.0%	1.2%
$\Sigma\ i^1_4$	15.6%	11.2%	3.1%	1.5%	0.6%	0.2%	0.0%	0.0%	0.0%	15.6%
$\Sigma\ i^0_5$	3.1%	1.8%	0.3%	0.1%	0.0%	0.0%	0.0%	0.0%	0.0%	83.2%
	100%	100%	100%	100%	100%	100%	100%	100%	100%	100%
US Fed Household Data conforms with the following distributions.										
			4Q 1989 Income	4Q 1989 N W & 4Q 2019 Income	4Q 2019 N W & 20?? Income	20?? Net Worth				

Table 3 – Weighted Distributions per Tier Sum of M_5 microstate percentages

The results of this comparison are reported as in Table 4 and as follows in the detail description.

Weight Tier	$M_{5:00}$	$M_{5:10}$	$M_{5:50}$	4Q 1989 Income adjusted	4Q 1989 Net Worth adjusted	$M_{5:60}$	4Q 2019 Income adjusted	4Q 2019 Net Worth adjusted	$M_{5:70}$
$\Sigma\, i^5{}_0$	3.1%	5.0%	23.7%	24.9%	36.5%	32.8%	33.9%	46.0%	44.4%
$\Sigma\, i^4{}_1$	15.6%	20.7%	39.5%	34.8%	37.0%	41.0%	37.5%	34.7%	39.2%
$\Sigma\, i^3{}_2$	31.3%	33.8%	26.4%	24.7%	22.6%	20.5%	20.1%	17.9%	13.8%
$\Sigma\, i^2{}_3$	31.3%	27.5%	8.8%	13.0%	3.4%	5.1%	6.9%	1.2%	2.4%
$\Sigma\, i^1{}_4$	15.6%	11.2%	1.5%	2.5%	0.5%	0.6%	1.5%	0.2%	0.2%
$\Sigma\, i^0{}_5$	3.1%	1.8%	0.1%	0.1%	0.0%	0.0%	0.1%	0.0%	0.0%
	100%	100%	100%	100%	100%	100%	100%	100%	100%
				US Fed adjusted data	US Fed adjusted data		US Fed adjusted data	US Fed adjusted data	

Comparison of modeled distributions and US Fed Household Income and Net Worth data for 4Q 1989 and 4Q 2019

Table 4 – Comparison of modeled distributions and US Fed Household Data

Comparison of Model and US Federal Reserve Data for 4Q 1989 and 2019

For a check of the validity of this modeling of weighted ergodicity based on focused rationality in the decision making of the microstates as applied to macroeconomic systems, we have compared the distribution of three of these weightings with Federal Reserve data for US Household Income flows and Net Worth stocks. [3]

To do this we first adjusted the Fed sorting categories as tiers, I^n, for Household Income and Net Worth from the fourth quarter data for 1989 and 2019 as follows.

The source data for income is grouped in skewed fashion in six groupings as:

	I^5	I^4	I^3	I^2	I^1	I^0
Income*	Top 1%	80-99%	60-80%	40-60%	20-40%	0-20%
4Q 1989	17.3%	43.6%	16.5%	12.3%	7.4%	3.0%
4Q 2019	25.9%	46.8%	14.6%	6.9%	3.9%	1.8%
*"Distributional National Accounts". federalreserve.gov. June 19, 2020. Retrieved August 20, 2020.						

Table 5a – Household Income as reported by the Fed

The source data for net worth is grouped similarly in four groupings as:

	I^5	I^4	I^3	I^0
Net Worth*	Top 1%	90-99%	50-90%	Bottom 50%
4Q 1989	23.6%	37.3%	35.2%	3.9%
4Q 2019	32.7%	37.2%	28.7%	1.4%
*"Distributional National Accounts". federalreserve.gov. June 19, 2020. Retrieved August 20, 2020.				

Table 6a – Household Net Worth as reported by the Fed

In order to compare the data with the model expected values, we have made adjustments by extrapolating the data as reported by the Fed to interpolate the percentages of the six tiers of the model as shown in Diagrams 1 & 2 of the Appendix. It bears noting that the skewing is biased toward the caput end of the model, which in this case is represented at the left end of the row structure. Using the trapezoidal rule for approximating a definite integral, the data for the Fed tiers is treated as an integral of the area under the stepped graph in blue for each tier, I^n, as the product of the percentile length of the base, L^n, and the height, computed as the mean of the tier range, which in the case of the stepped graph can be indicated by the mean, $J^n{}_m$. The Fed tier area is then converted to a right trapezoidal configuration to reflect a differential increase toward the caput end, starting from the cauda end of I^0 at zero to twice the average at the caput end of that same tier.

That value, J^0, then is used as the cauda end of the next tier, I^n, that is, I^1 for the income Diagram 1 and I^3 for the net worth Diagram 2. The resulting slope of the line which is shown in red intersecting the mid points of the blue stepped graph, the height of the areas defining the trapezoid for each of the tiers, i^n, of this model is computed, allowing for the first at the cauda end as a triangle starting at zero. Areas of the original six and four-tiered configuration are thereby extrapolated from I^n and interpolated into i^n of the model accordingly, for comparison with the values anticipated by the weighted model. The values of the tiered microstates of the model are then expressed as functions of the Fed tier variables as products of I^n and the coefficients arrived at by reducing the tier transformation constraints as shown in Diagrams 1 and 2, with the adjusted values shown for the 4Q 1989 and 4Q 2019 Household Income and Net Worth in Tables 5b and 6b.

The adjusted values as determined for i^5, i^4, and i^3 of these four data sets are then used and stated as tier totals of the ergodic modeling to derive the various weightings as indicated in bold, along with the values for the corresponding tiers and the divergence of these model values from the adjusted data sets. There is only one corresponding value in each model for the top caput tier, i^5, along the theoretically possible continuum from 0% to 100%. There are two possible matches for the i^4 tier valuation as shown, one on the lower increasing arc of the continuum and one on the higher decreasing arc. The weighting point of inflection along this arc has been determined to be 61.80% as shown, valued at 40.9%. The inflection point for the i^3 tier shows a weighting of 20% as the maximum valuation of 34.6%, which is considerably above the matched tier valuations of any of the data sets. In other words, the valuations of all four data sets for the i^3 tier are on the downward arc of that tier as shown. All of the three lower tier valuations, i^2, i^1, and i^0, are on a descending arc of the valuations in these data sets, as they are across the full spectrum of the model weighting from 0% to 100%, indicating that they are not determinative of a caput position in such market valuation.

It bears emphasizing that this comparison of data is sorted by tier valuation, and as a result obscures some of the dynamics and direct understanding of this fractal nature of the modeling which is more accessible in the Charts 1 through 11 below. As seen there, it is clear that the lower 16 microstates populated from the cauda and which result from the first iteration or flip of a coin cannot be equated with the lower three tiers of the sorted distribution.

Of the five microstates included in the i^4 tier, one arises in the lower cauda by virtue of its position at $m_5[01111^4]$.

Of the ten microstates included in the i^3 tier, four arise in the lower cauda by virtue of their positions at $m_5[01110^3, 01101^3, 01011^3, 00111^3]$.

Of the ten microstates included in the i^2 tier, six arise in the lower cauda by virtue of their positions at $m_5[01100^2, 01010^2, 01001^2, 00110^2, 00101^2, 00011^2]$.

Of the five microstates included in the i^1 tier, four arises in the lower cauda by virtue of their position at $m_5[01000^1, 00100^1, 00010^1, 00001^1]$.

And of course, the unique compositions of $m_5[00000^0]$ and $m_5[11111^5]$ remain unique to their initial caput and cauda positions from the initial iteration.

The macroeconomic significance of the unsorted weighted ergodic distribution is obscured by the usual sorting of national and international data into percentage groupings based on market valuation. This would become abundantly clear if the macrostate structure of M_5 were to be expanded to around $M_{28.3}$ to account for one microstate for each member of the US population or to around $M_{32.87}$ for the global population, so that each individual would assume a position in the model related to their focused/unfocused proximity to other rational/irrational decision makers. As it is, with an M_5 model sorted by market valuations of income and net worth, we can still gain a reasonable understanding of macroeconomic dynamics along with its policy implications.

With the summaries of Table 4 and in the detail of Tables 5b & 6 b – Analysis of conformation of Fed Income and Net Worth data with weighted ergodic modeling, we can begin to draw some inferences from the weighted ergodic distributions. First, as we have stated, position of the caput end of the distribution, for reasons we have discussed, has a marked advantage in the decision making of the market space with respect to material acquisition, production and distribution of intermediate and final goods and services over the cauda end. What the related Charts 3 through 11 indicate, in addition to the magnitude of the tier weighting factors as a percentage of the total macrostate value, is that the distribution of that value among the tiers is related to the proximate position of each microstate within the macrostate hierarchy.

As might be expected, the valuation of stocks as net worth in Table 6b is more heavily weighted than the income valuation of the same time periods, indicating a flow of value as income from the cauda to the caput, both as a source and a sink as a net worth stock for that flow. It is noted that the weighting percentage for the net worth figures is approximately 20% higher than the weighting of the corresponding year income data. We might surmise that this is in line with the markup in overhead expenses, rents and fees and profits from operations and trade that accrue to businesses, corporate and noncorporate alike, associated with the maintenance of ongoing operations. Households on the cauda end of the spectrum do not incur these costs, which are primarily consumption related and devoid of extensive capital costs. In a free market system in which employment of labor relies heavily on the production of commodities for final distribution and sale, income flow to that labor will tend to be at commodity labor rates which, absent statutory entitlement, is insufficient to cover extensive capital costs, which necessarily means public or private safety net costs.

What is equally noteworthy is that the weighting of the data for 4Q 2019 is approximately 20% greater (10% nominally) than that of the 4Q 1989, for both the income and net worth data. This indicates the further commoditization of such labor over time throughout the lower cauda of the distribution and intensifies the ergodic weighting.

The structural changes associated with this weighting change are of interest. As a base line, let us first look at Table 4 at the model distribution of column $M_{5:10}$, where the weighting to the caput side of any decision making is 10%. This is in line with the biblical referenced, cultural concept and wisdom of a tithe to what is essentially the governing component of the public sector, over the strictly egalitarian distribution of $M_{5:00}$. As seen in Chart 3, this represents a 61% surplus, but not necessarily a surfeit, over the average microstate value of i^5 or $m_5[11111^5]$, which might be used for a combination of private goods needed by that microstate and publicly needed by the macrostate. The next tier at i^4 has a 32% surplus over the represented 15.6% of the total population, tier i^3 has an 8% surplus for 31.3% of the total population, mirrored with another 31.3% by a roughly equal deficit for i^2 at 12%, a tier i^1 population of 15.6% with a 28% deficiency and a bottom tier i^0 cauda of 3.1% of the population with a 41% deficiency. All-in-all this represents a reasonably egalitarian, middle class life for a macrostate.

The 1989 and 2019 distributions have left this bucolic notion far behind. The i^5 tier for these adjusted data sets and model weightings, representing 3.1% of the population, show a roughly 10% increase of net worth over income and of both of these categories over the 30-year growth of the US economy. The lowest 3 tiers together, or half the population, show a decrease of roughly 75% between income and net worth valuations and approximately 50% decrease over the 30-year span, where the i^0 tiers are merely subsistent. This is clearly evident in the i^2 tier bottom half of the middle class.

In contrast, the top tier of the 62.6% population of the middle class, i^3, indicates the ability to hold on to most of their net worth, but the overall fall in proportional income and net worth over the 30 years is evident. The upper middle class, i^4, show the capacity to add to net worth out of income according to this reading of the 1989 data, and to an increase in income between 1989 and 2019. However, the net worth as a percentage of the total declined slightly over that time, and the net worth over income percentage had started to decline as of 2019. If this trend were to continue, only the top tier, i^5, as 3.1% referenced by the percentage of the total population, would be able to maintain their relative standard of living.

If we look at the detail provided by Tables 5b for income and 6b for net worth, after analyzing the model weighting in sequence for each of the matched i^5, i^4, and i^3 tier adjusted values, and then at a selected general best fit to the M_5 model, we find that this is accomplished primarily by an adjustment to the i^4 tier, 15.6% of the macrostate, and we might say at their expense to the tune of 3.5% to 4.7% as measured against a uniform model weighting of the respective data. Note that in all four data sets in which a match has been made to conform to the i^4 adjusted tier data, the divergence of the model is significantly higher in the tiers below i^4, while significantly lower in i^5, for the 'weighted to i^4, lower' rows. It remains higher for i^4 when considered as a deflection point for the 'weighted maximum i^4', and reverses to become lower in the tiers below i^4, while significantly higher in i^5, for the 'weighted to i^4, lower' rows of the tables.

Income* M_5	i^5	i^4	i^3	i^2	i^1	i^0
Adjustments to I^n tiers	3.125%	15.625%	31.25%	31.25%	15.625%	3.125%
$i^5 - 3.125\%$	$= I^5 + (0.21117*I^4 - 0.18873*(I^3 - I^2 + I^1 - I^0))$					
$i^4 - 15.625\%$	$= I^4 - (0.21117 + 0.00432)*I^4 - (0.18873 - 0.11677)*(I^3 - I^2 + I^1 - I^0)$					
$i^3 - 31.25\%$	$= (.75*I^2 - .5*I^1 + .5*I^0) + I^3 + (0.00432*I^4 + 0.11677*(I^3 - I^2 + I^1 - I^0))$					
$i^2 - 31.25\%$	$= 0.12109*I^0 + I^1 + (.25*I^2 + .5*I^1 - .5*I^0)$					
$i^1 - 15.625\%$	$= 0.85449*I^0$					
$i^0 - 3.125\%$	$= 0.24414*I^0$					
4Q 1989 Tier Adjusted	**24.9%**	**34.8%**	**24.7%**	**13.0%**	**2.5%**	**0.1%**
51.45 weighted to i^5	24.9	39.9	25.6	8.2	1.3	0.1
		− 5.1	− 0.9	+ 4.8	+ 1.2	
37.72 weighted to i^4, lower	15.3	**34.8**	31.8	14.5	3.3	0.3
	+ 9.6		− 7.1	− 1.5	− 0.8	-0.2
61.80 weighted maximum i^4	34.7	**40.9**	19.3	4.6	0.5	0.0
(stated for reference)	− 9.8	− 6.1	+ 5.4	+ 8.4	+ 2.0	+ 0.1
77.63 weighted to i^4, higher	55.3	**34.8**	8.8	1.1	0.0	0.0
	− 30.4	V	+ 15.9	+ 11.9	+ 2.5	+ 0.1
53.00 weighted to i^3	26.2	40.2	**24.7**	7.6	1.2	0.1
	− 1.3	− 5.4		+ 5.4	+ 1.3	
20.00 weighted maximum i^3	7.8	25.9	**34.6**	23.0	7.7	1.0
(stated for reference)	+ 17.1	+ 8.9	− 9.9	− 10.0	− 5.2	− 0.9
50.00 weighted	23.7	39.5	26.4	8.8	1.5	0.1
	+ 1.2	− 4.7	− 1.7	+ 4.2	+ 1.0	
	24.9	34.8	24.7	13.0	2.5	0.1
4Q 2019 Tier Adjusted	**33.9%**	**37.5%**	**20.1%**	**6.9%**	**1.5%**	**0.1%**
61.09 weighted to i^5	33.9	40.9	19.8	4.8	0.6	0.0
		− 3.4	+ 0.3	+ 2.1	+ 0.9	+ 0.1
43.66 weighted to i^4, lower	19.1	**37.5**	29.4	11.5	2.3	0.2
	+ 14.8		− 9.3	− 4.6	− 0.8	− 0.1
61.80 weighted maximum i^4	34.7	**40.9**	19.3	4.6	0.5	0.0
(stated for reference)	− 0.8	− 3.4	+ 0.8	+ 2.3	+ 1.0	+ 0.1
73.54 weighted to i^4, higher	49.2	**37.5**	11.4	1.7	0.2	0.0
	− 15.3	V	+ 8.7	+ 5.2	+ 1.3	+ 0.1
60.6 weighted to i^3	33.4	41.0	**20.1**	4.9	0.6	0.0
	+ 0.5	− 3.5		+ 2.0	+ 0.9	+ 0.1
20.00 weighted maximum i^3	7.8	25.9	**34.6**	23.0	7.7	1.0
(stated for reference)	+ 26.1	+ 11.6	− 14.5	− 16.1	− 6.2	− 0.9
60.00 weighted	32.8	41.0	20.5	5.1	0.6	0.0
	+ 1.1	− 3.5	− 0.4	+ 1.8	+ 0.9	+ 0.1
	33.9	37.5	20.1	6.9	1.5	0.1

(Adjustments in parenthesis account for skewing from adjacent tiers)
*As adjusted for comparison with the model

Table 5b – Analysis of conformation of Fed Income data with weighted ergodic modeling

Martin Gibson, UniServEnt.org – 12/10/2020

Net Worth* M_5	i^5	i^4	i^3	i^2	i^1	i^0
Adjustments to I^a tiers	3.125%	15.625%	31.25%	31.25%	15.625%	3.125%
i^5 – 3.125%	= $I^5 + 0.41648*I^4 – 0.08116*I^3 + 0.06493*I^0$					
i^4 – 15.625%	= $0.58352*I^4 + (0.38965 + 0.08116)*I^3 – (0.27346 + 0.06493)*I^0$					
i^3 – 31.25%	= $0.61031*I^3 + 0.27346*I^0$					
i^2 – 31.25%	= $0.85938*I^0$					
i^1 – 15.625%	= $0.13672*I^0$					
i^0 – 3.125%	= $0.00390*I^0$					
4Q 1989 Tier Adjusted	**36.5%**	**37.0%**	**22.6%**	**3.4%**	**0.5%**	**0.0%**
63.50 weighted to i^5	**36.5**	40.7	18.2	4.1	0.5	0.0
		– 3.7	+ 4.4	– 0.7		
42.40 weighted to i^4	18.3	**37.0**	29.9	12.1	2.5	0.2
	+18.2		– 7.3	– 8.7	– 2.0	-0.2
42.40 weighted to i^4, lower	18.3	**37.0**	29.9	12.1	2.5	0.2
	+18.2		– 7.3	– 8.7	– 2.0	-0.2
61.80 weighted maximum i^4	34.7	**40.9**	19.3	4.6	0.5	0.0
(stated for reference)	+ 1.8	– 3.9	+ 3.3	+ 1.2		
74.41 weighted to i^4, higher	50.4	**37.0**	10.9	1.6	0.1	0.0
	– 13.9	V	+ 11.7	+ 1.8	+ 0.4	
56.61 weighted to i^3	29.4	40.8	**22.6**	6.3	0.9	0.0
	+ 7.1	– 3.8		– 2.9	– 0.4	
20.00 weighted maximum i^3	7.8	25.9	**34.6**	23.0	7.7	1.0
(stated for reference)	+ 28.7	+ 11.1	– 12.0	– 19.6	– 7.2	– 1.0
60.00 weighted	32.8	41.0	20.5	5.1	0.6	0.0
	+ 3.7	– 4.0	+ 2.1	– 1.7	– 0.1	
	36.5%	37.0%	22.6%	3.4%	0.5%	0.0%
4Q 2019 Tier Adjusted	**46.0%**	**34.7%**	**17.9%**	**1.2%**	**0.2%**	**0.0%**
71.23 weighted to i^5	**46.0**	38.6	13.0	2.2	0.2	0.0%
		– 3.9	+ 4.9	– 1.0		
37.05 weighted to i^4, lower	15.1	**34.7**	31.9	14.6	3.4	0.3
	+ 30.9		– 14.0	– 13.4	– 3.2	– 0.3
61.80 weighted maximum i^4	34.7	40.9	19.3	4.6	0.5	0.0
(stated for reference)	+ 11.3	– 6.2	– 1.4	– 3.4	– 0.3	
77.76 weighted to i^4, higher	55.4	**34.7**	8.7	1.1	0.1	0.0
	– 9.4	V	+ 9.2	+ 0.1	+ 0.1	
63.95 weighted to i^3	37.0	40.7	**17.9**	4.0	0.4	0.0
	+ 9.0	– 6.0		– 2.8	– 0.2	
20.00 weighted maximum i^3	7.8	25.9	**34.6**	23.0	7.7	1.0
(stated for reference)	+ 38.2	+ 8.8	– 16.7	– 21.8	– 7.5	– 1.0
70.00 weighted	44.4	39.2	13.8	2.4	0.2	0.0
	1.6	– 4.5	+ 4.1	– 1.2		
	46.0	34.7	17.9	1.2	0.2	0.0

*As adjusted for comparison with the model

Table 6b – Analysis of conformation of Fed Net Worth data with weighted ergodic modeling

From the consideration of the divergence of the data set values from the model analysis we have selected the weighted values as shown of 50% and 60% for 1989 and 2019 Income and of 60% and 70% for 1989 and 2019 Net Worth comparison with the data which is repeated in summary form here.

Income	Top 3.1%	15.6%	31.3%	31.3%	15.6%	3.1%
4Q 1989*	24.9	34.8	24.7	13.0	2.5	0.1
50%	23.7	39.5	26.4	8.8	1.5	0.1
Divergence	+ 1.2	− 4.7	− 1.7	+ 4.2	+ 1.0	
4Q 2019*	33.9	37.5	20.1	6.9	1.5	0.1
60%	32.8	41.0	20.5	5.1	0.6	0.0
Divergence	+ 1.1	− 3.5	− 0.4	+ 1.8	+ 0.9	+ 0.1
*As adjusted for comparison with the model						

Table 5c – Conformation of Fed Income data with weighted ergodic modeling

Net Worth	Top 3.1%	15.6%	31.3%	31.3%	15.6%	3.1%
4Q 1989*	36.5	37.0	22.6	3.4	0.5	0.0
60%	32.8	41.0	20.5	5.1	0.6	0.0
Divergence	+ 3.7	− 4.0	+ 2.1	− 1.7	− 0.1	
4Q 2019*	46.0	34.7	17.9	1.2	0.2	0.0
70%	44.4	39.2	13.8	2.4	0.2	0.0
Divergence	1.6	− 4.5	+ 4.1	− 1.2		
*As adjusted for comparison with the model						

Table 6c – Conformation of Fed Net Worth data with weighted ergodic modeling

The divergence of the Fed adjusted tier data from the ergodic model weighted conformation in Tables 5c and 6c demonstrates a related consistency in the tiered structure of the US economy. The positive divergence figures indicate a tier summation of the US data over the expected model values, while the negative figures indicate an under-representation of those expectations. A consistency is seen in the general quantity and the arithmetic sense of the income and net worth figures from the 1989 to 2019 data and in a tier contrast between the income and net worth data.

Fine tuning of the specific general macrostate weightings selected for each of the four data sets would alter the tier divergence amounts, but not affect the implicit significance of the tier-to-tier relationships resulting from the hierarchical focus and rationale of the economics behind the underlying microstate decision making. The result of this decision making is a structure that is theoretically amenable via increasing granularity to Fourier analysis and synthesis; this, in terms of ergodic modeling through an understanding of individual microstate weighting based on focused rationality.

The implication of the selected fits is that while the top i5 tier is the primary beneficiary of growth of the weighting, due to their Caput position in the control of capital, the microstates of the i4 tier are the primary supporters of the system by being the principle funders of flows to the upper and lower tiers. This of course reflects the dynamics of the internal revenue system from the perspective of a weighted ergodic economic analysis. From this perspective, the top 3.1% in i5 with the weighting benefit of tax accountancy, pay no net taxes, while the next two tiers in the upper or caput half, and primarily the i4 tier, pay most of the rest.

The political imperatives of i5 tier with annual incomes over $330k and of the i4 tier, those with current annual incomes in the $150k to $330k range, are, or should be, self-evident. In an atmosphere of laissez faire political economic assumptions, that is, to reduce individual income tax outlays as much as possible.

From the perspective of a transformation from Chart 3 and 4 of an M5:10 and M5:38.19 macrostates, however, the i4 tier is the true beneficiary of the income flowing from the real and human capital stock of the lower tier i0, i1, i2, and once the M5:38.19 weighting is surpassed, i3 microstates. All of this is necessary to sustain the M5:50, M5:60, and M5:70, macrostates of the data sets, and bolsters the primary caput position of i5.

This does not mean that the value defined in the hierarchical distribution of a weighted macrostate is not reasonable, but it does speak to the inability, given the context of increasing ergodic weighting, of the lower cauda microstates to provide the necessary income to maintain their human capital livelihood positions within a market based environment, that is through reliance on a private safety net for satisfaction of long term health, education, and retirement needs. As depicted in this modeling of the data sets presented, in the absence of adequate income flows to the lower cauda for immediate and long term life requirements due to positions of birth and familial circumstance or reduction due to business dislocations from any source, there is only one recourse to life needs in an advanced market economy without access to economic gleaning. Such recourse requires the provision of ongoing liquidity to these tiers.

Per currently policy, provision of that liquidity to the top two tiers provides no access to that liquidity for immediate needs of the lower 80% of the macrostate. Historical methods of taxation and borrowing are logically invalid as insufficient and grounded in a lack of understanding of liquidity primarily as a means of accounting for distribution of life consumption and not as an incentive to produce, particularly when the production capacity is controlled not by the final consumer in need of goods and services, but by the upper caputs that control the capital.

Conclusion

Given the confirmation of this approach of weighted ergodic modeling for an understanding of the domestic macroeconomic structure as evidenced by the US Fed data presented here, there are only two ways to address the trending hierarchical imbalance that has precipitated and been exacerbated by the current covid-19 crisis; through the control by the consumer/producer citizens of the means of production in order to provide their life needs or by entitlement to the necessary liquidity for purchase of the immediate and long term needs produced by both private and public capital based on sustainability of the recourse to production. When the necessary recourse is profitable and not exploitative, private initiative would be a favored approach. When the equally necessary recourse is not privately profitable, public initiative is favored, including where necessary, redirection of entitlement to public use via eminent domain.

Discussion is invited and warranted.

— — —

Charts of Various Weighted Ergodic Models

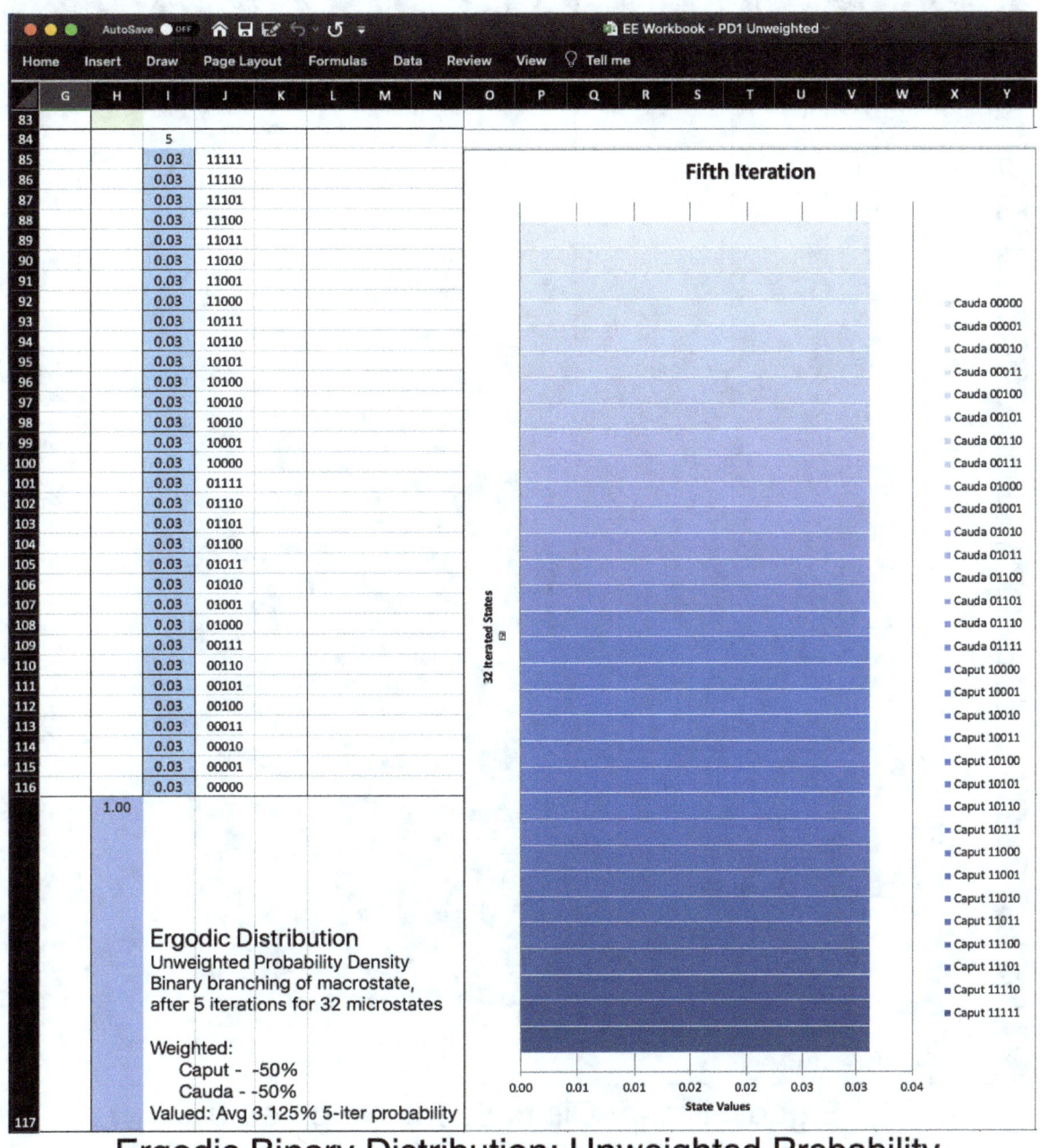

Chart 1

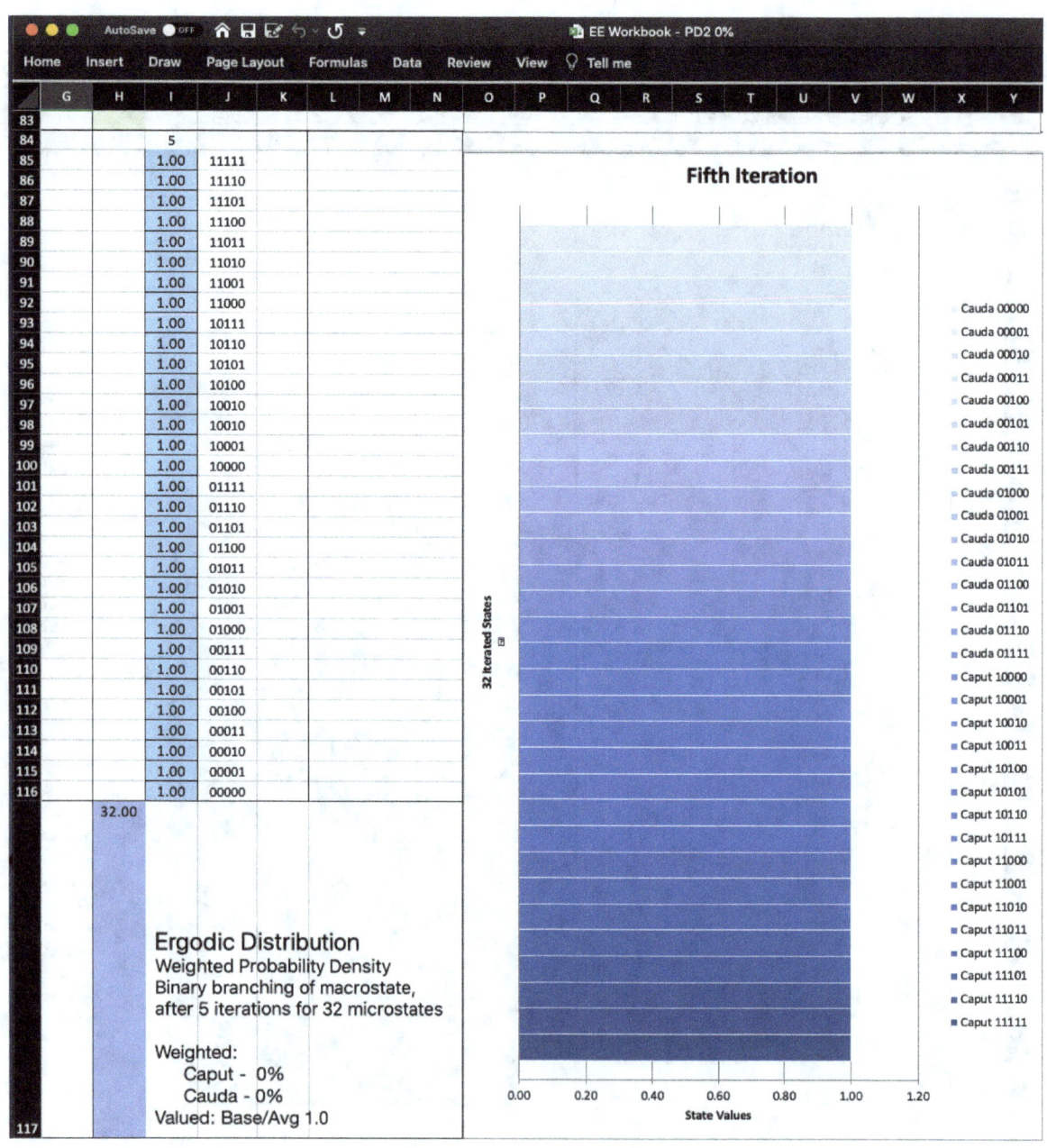

Ergodic Binary Distribution: Weighted 0%
Strictly egalitarian economy

Chart 2

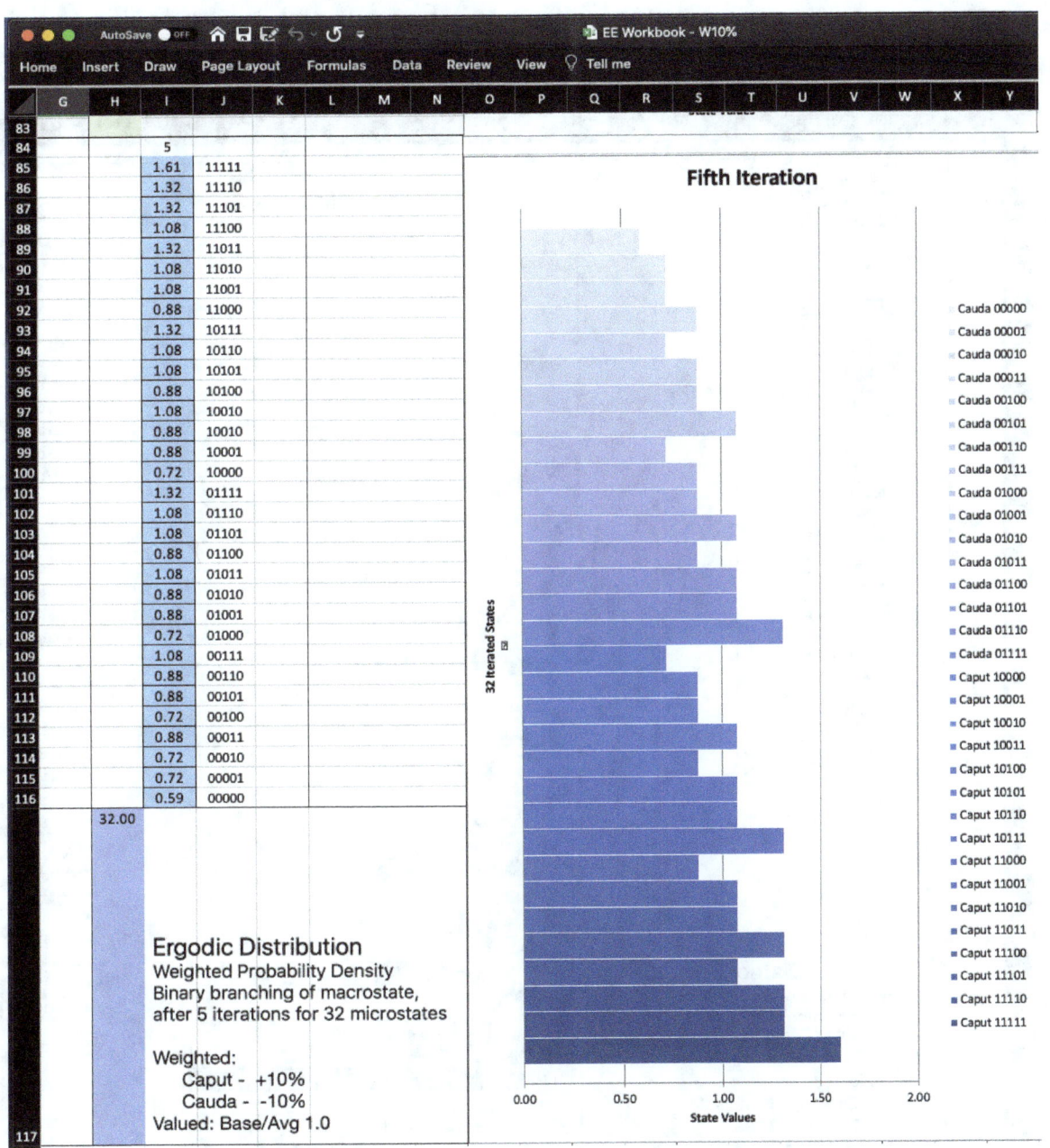

Ergodic Binary Distribution: Weighted 10%
Egalitarian, middle class economy in which 93.6% of microstates are within approximately 0.3 of the average 1.0

Chart 3

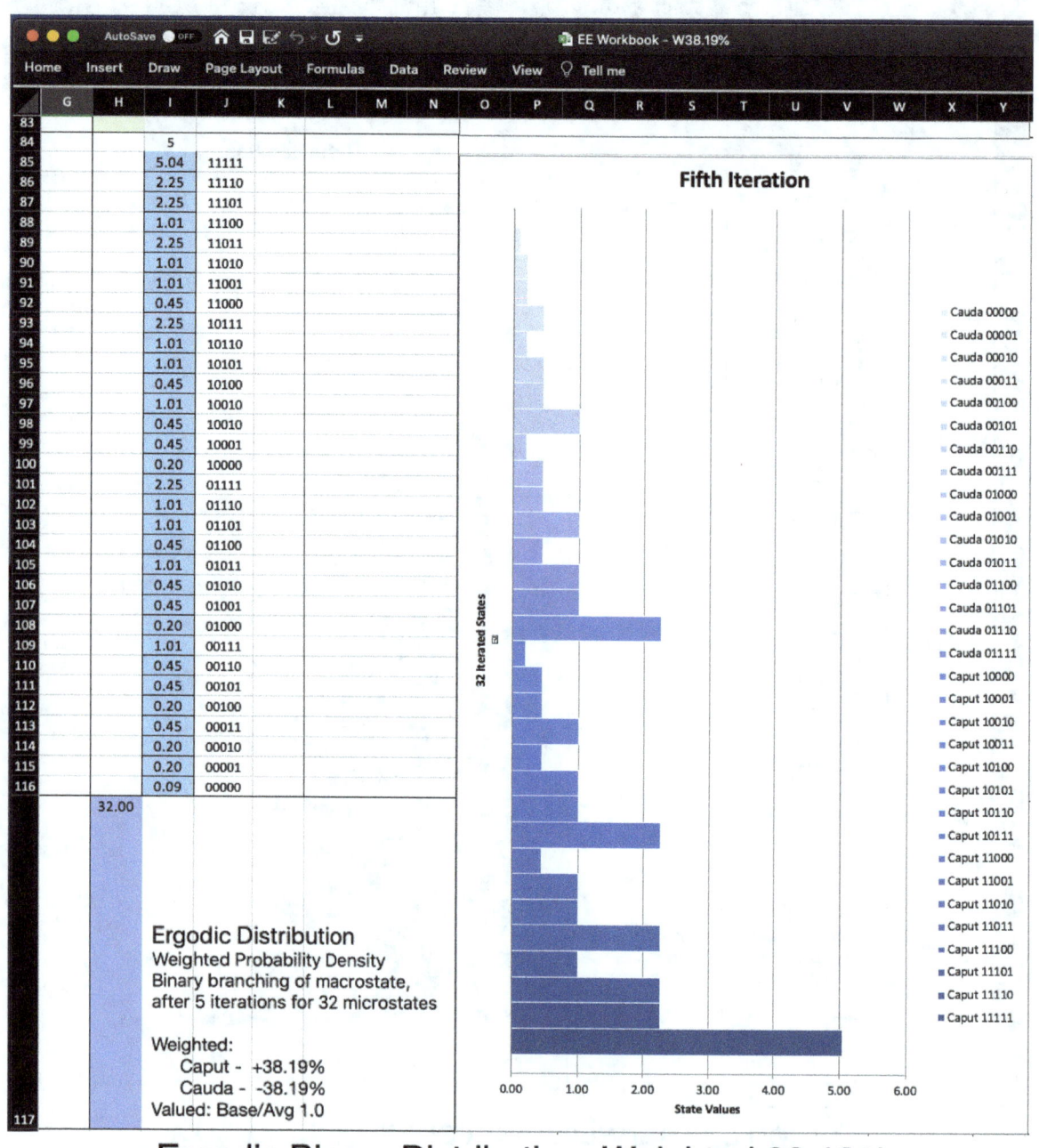

Ergodic Binary Distribution: Weighted 38.19%
Inflection Point for Upper Mid Level falling below average = 1.0, as Lower Mid < 0.5

Chart 4

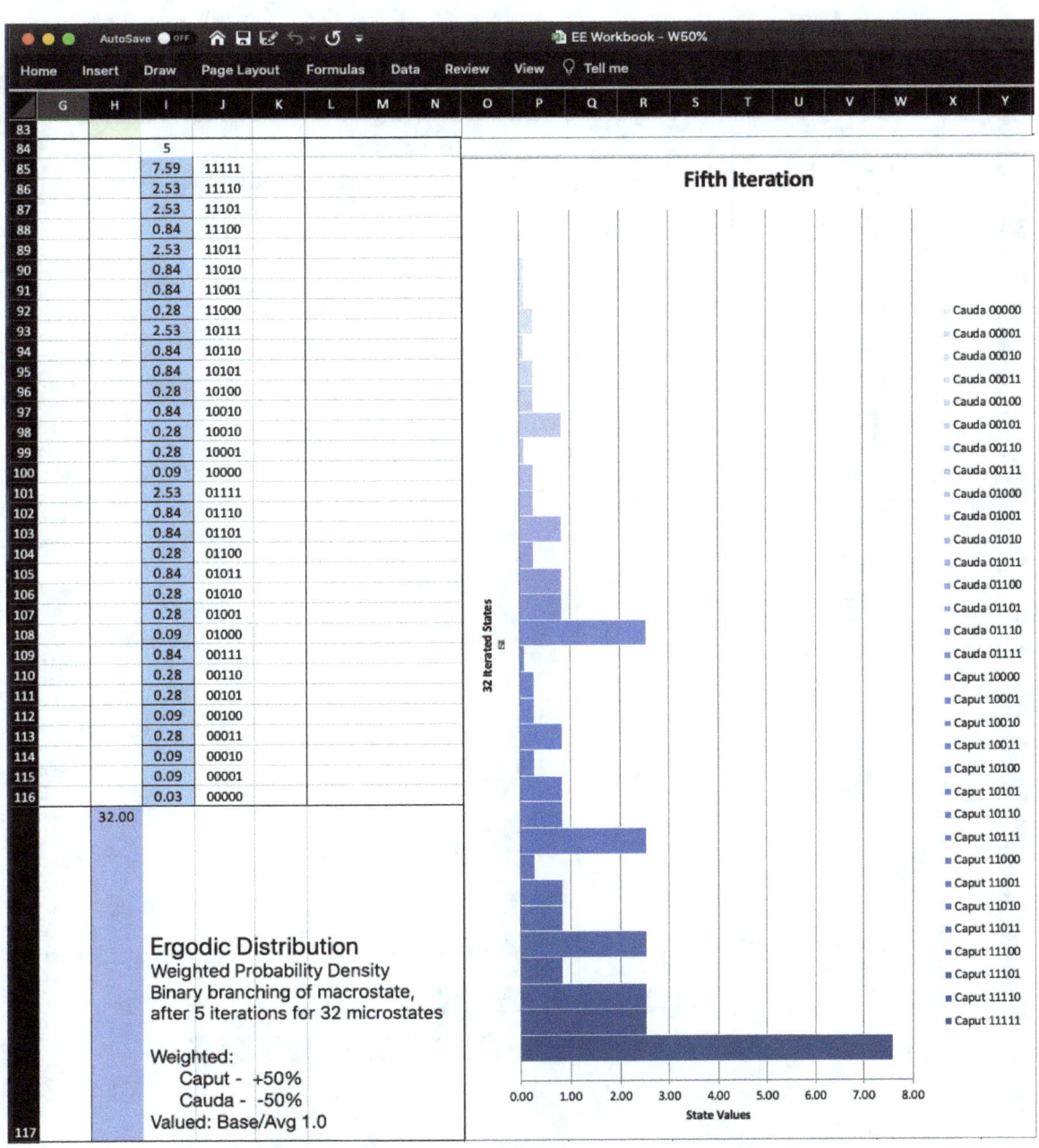

Ergodic Binary Distribution: Weighted 50%
Conforms to US Fed Household Data 4Q 1989 Income

Chart 5

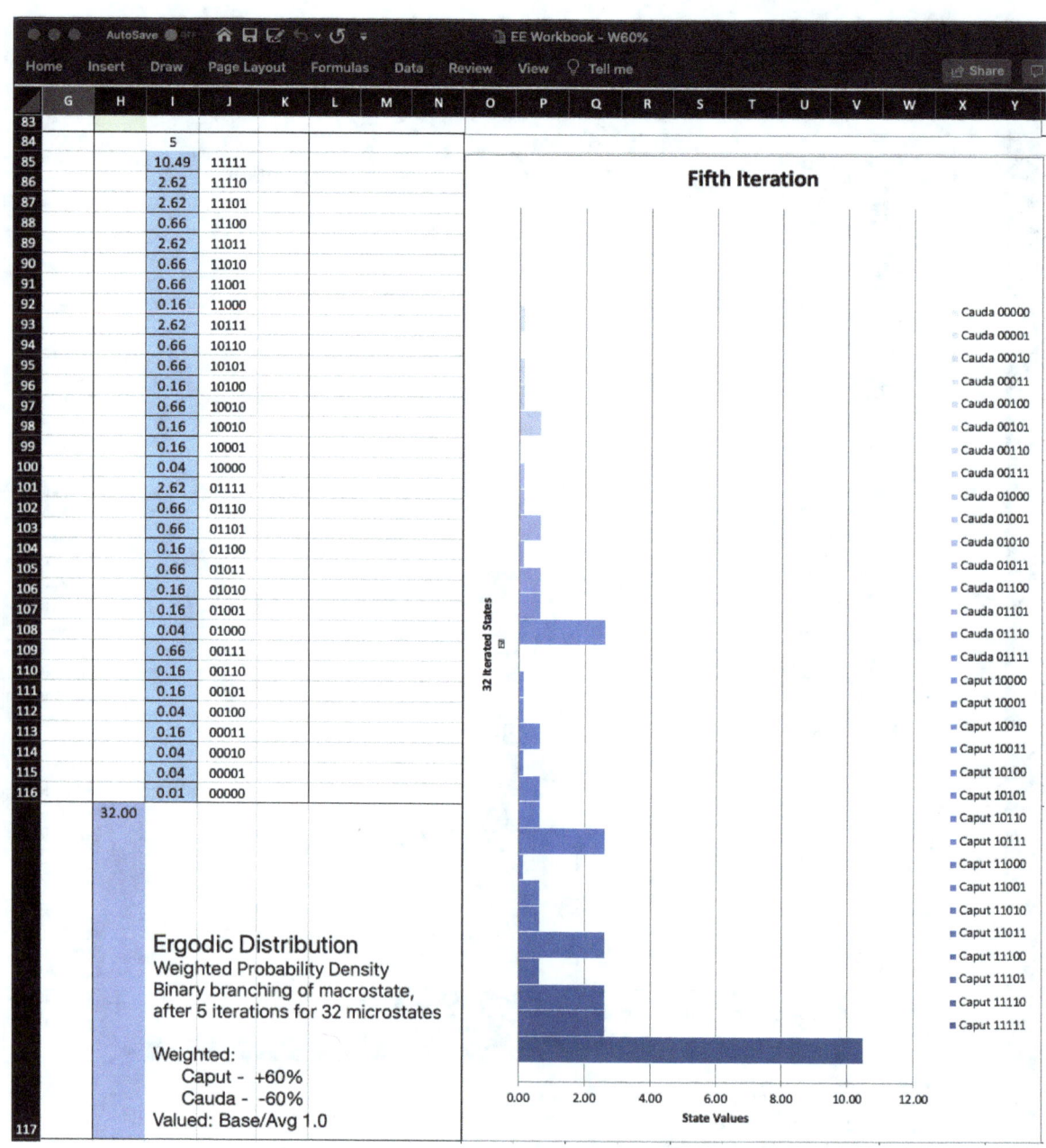

Ergodic Binary Distribution: Weighted 60%
Conforms to US Fed Household Data 4Q 1989 Net Worth & 4Q 2019 Income

Chart 6

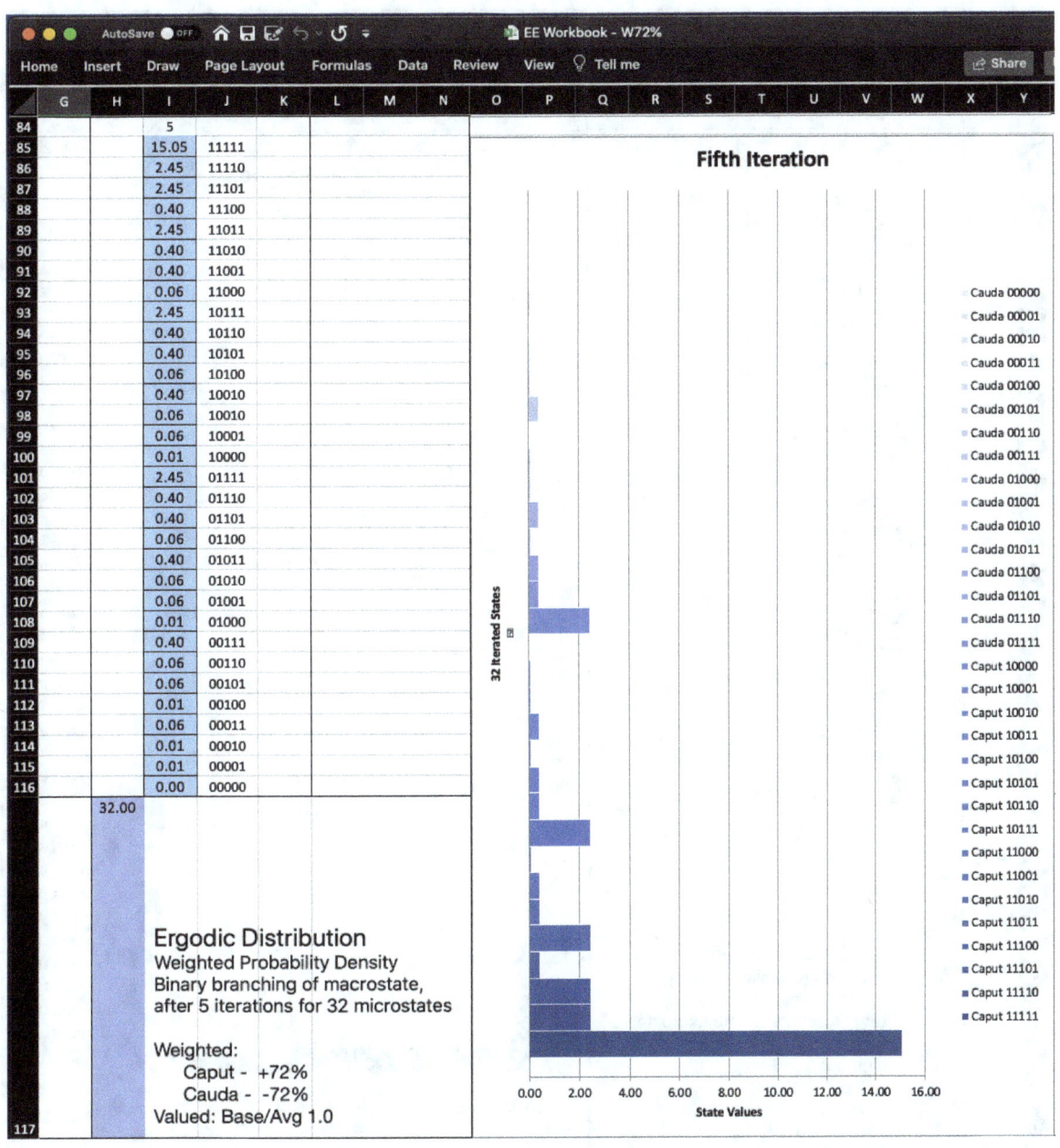

Ergodic Binary Distribution: Weighted 72%
Conforms to US Fed Household Data 4Q 2019 Net Worth & Projected Future Income

Chart 7

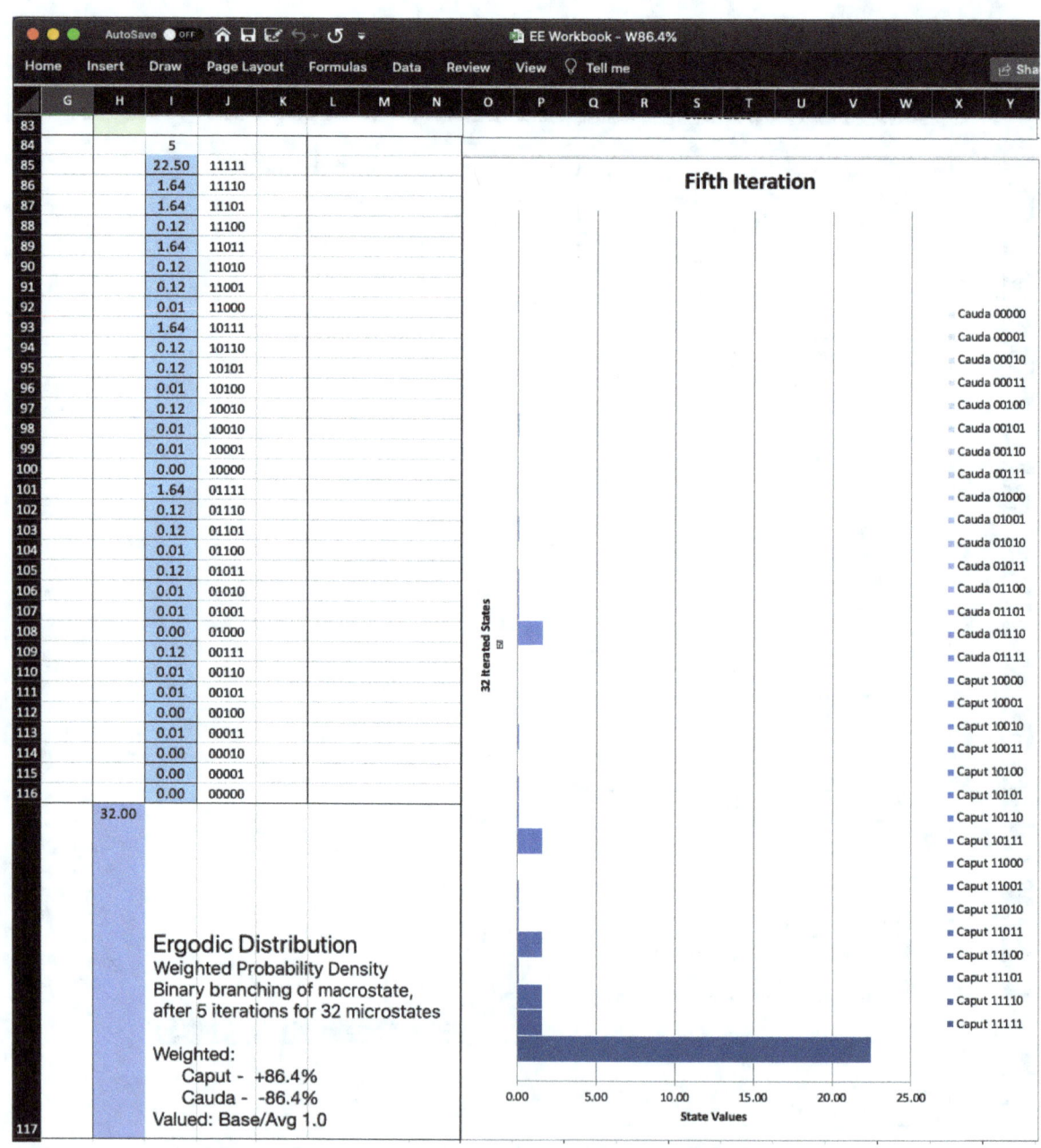

Ergodic Binary Distribution: Weighted 86.4%
US Fed Household Projected Net Worth based on Projected Future Income

Chart 8

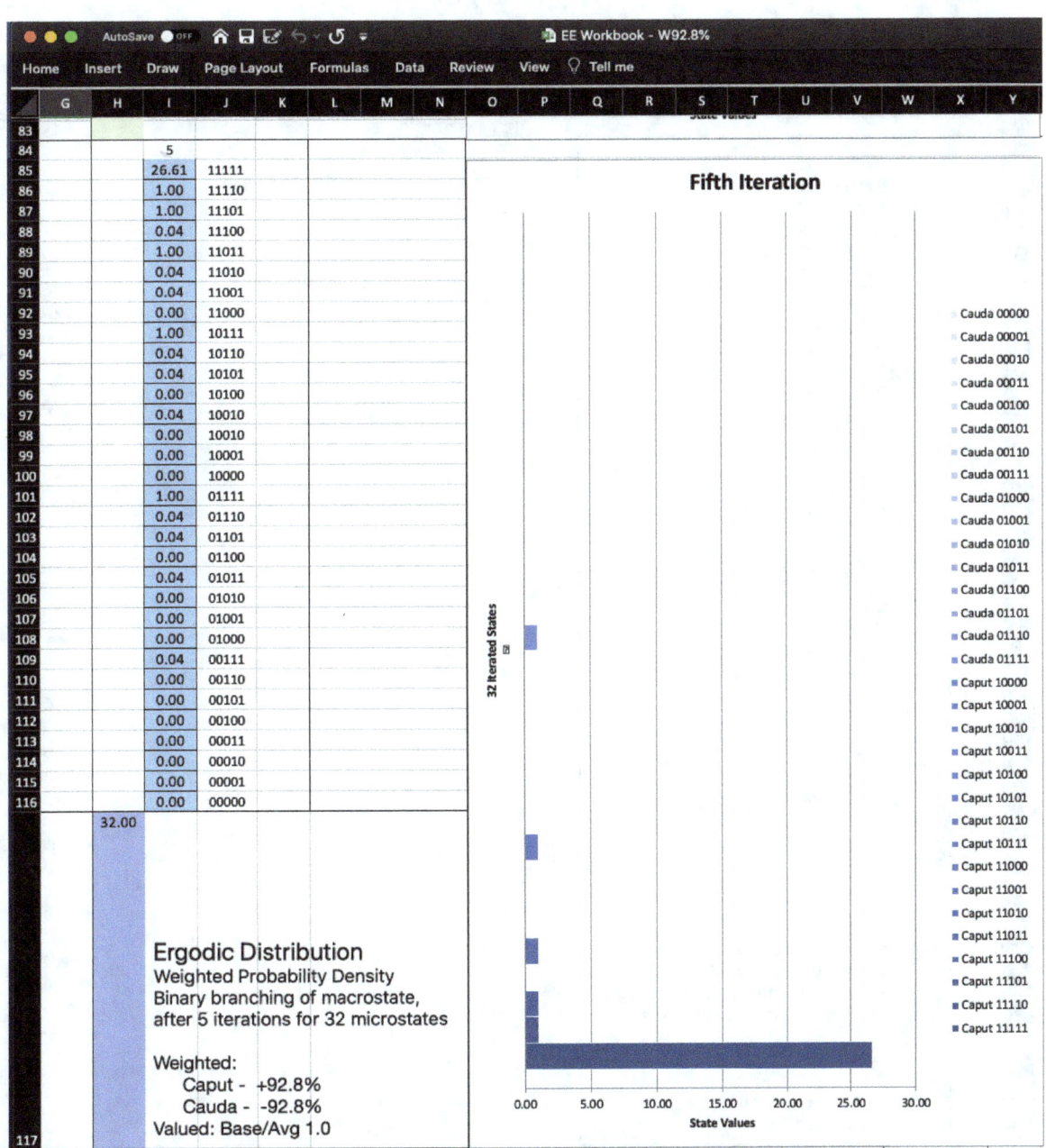

Ergodic Binary Distribution: Weighted 92.8%
Inflection Point for Mid Upper 15.6% falling below average = 1.0,
Top 3.2% has 26.61 of 32 total, Low Upper 31.2% avg < .04, Bottom Half vanishes

Chart 9

Martin Gibson, UniServEnt.org – 12/10/2020

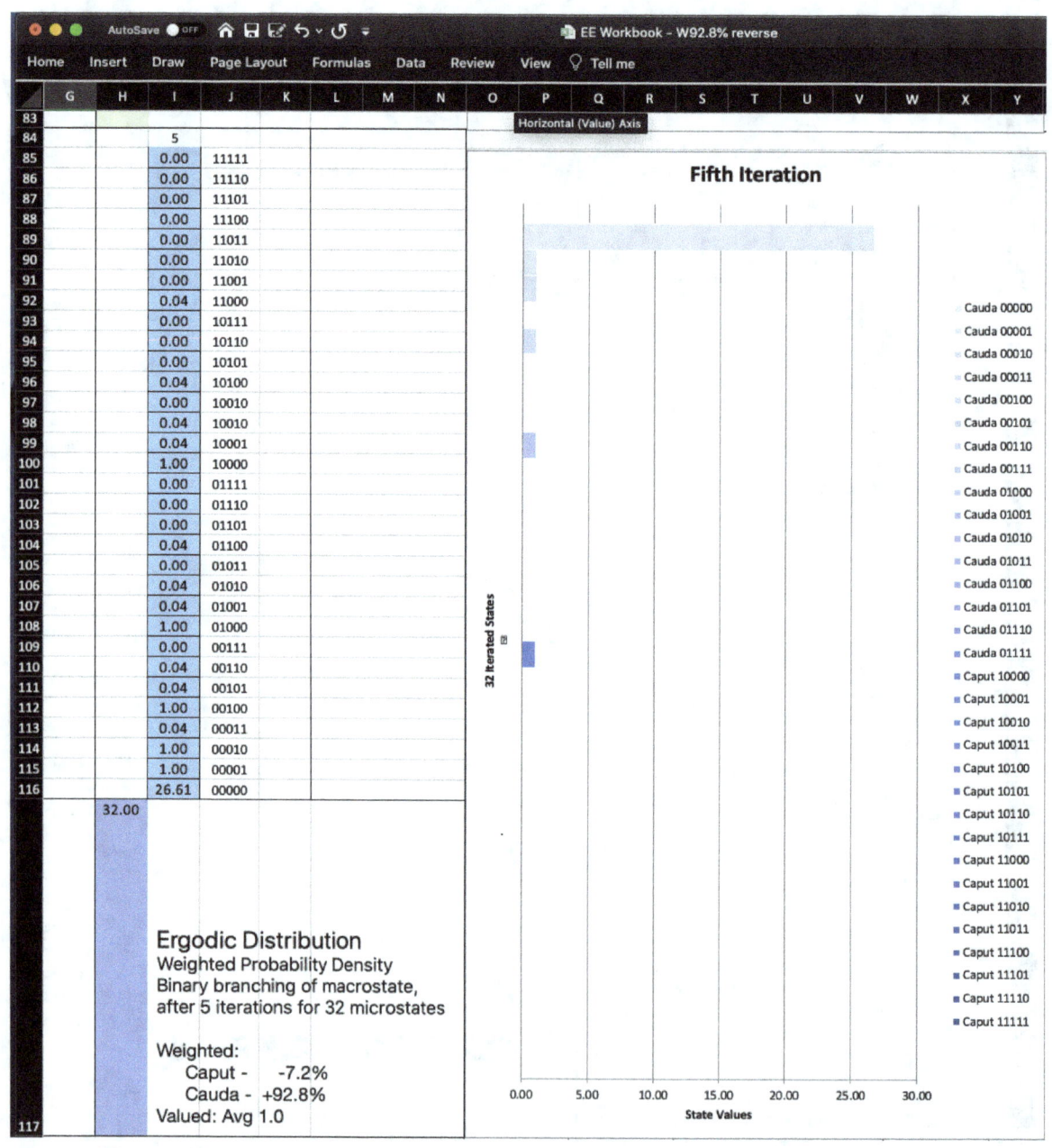

Ergodic Binary Distribution: Weighted 92.8% Reverse
Inflection Point as before simply showing equivalence of
'wrong' capital-consumption mix perhaps of a Stalinist state

Chart 10

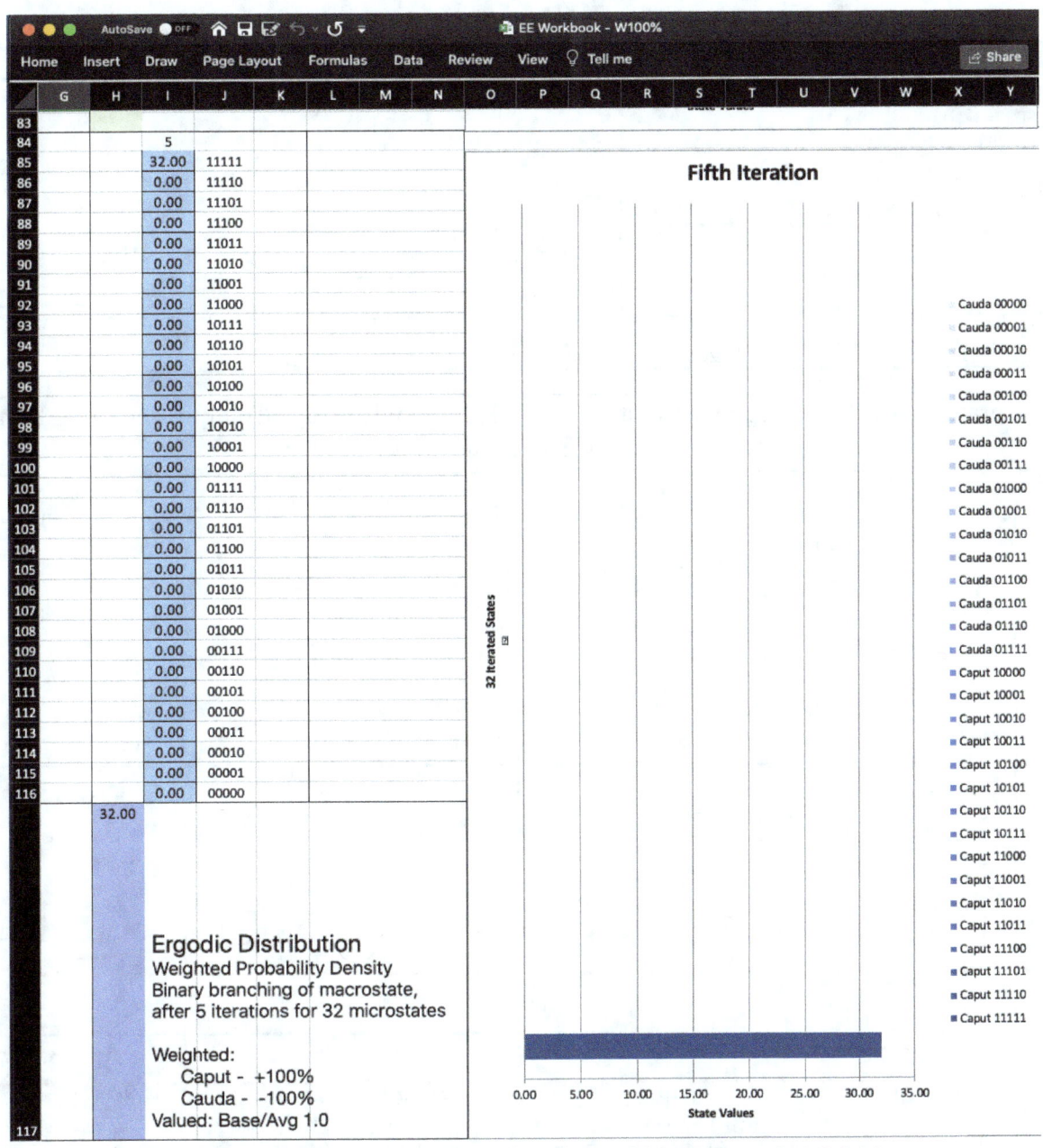

Ergodic Binary Distribution: Weighted 100%
Top 3.2% have 100% of Net Worth and market valued Income

Chart 1

Tables for converting US Fed Household Income and Net Worth data tiers to model configuration

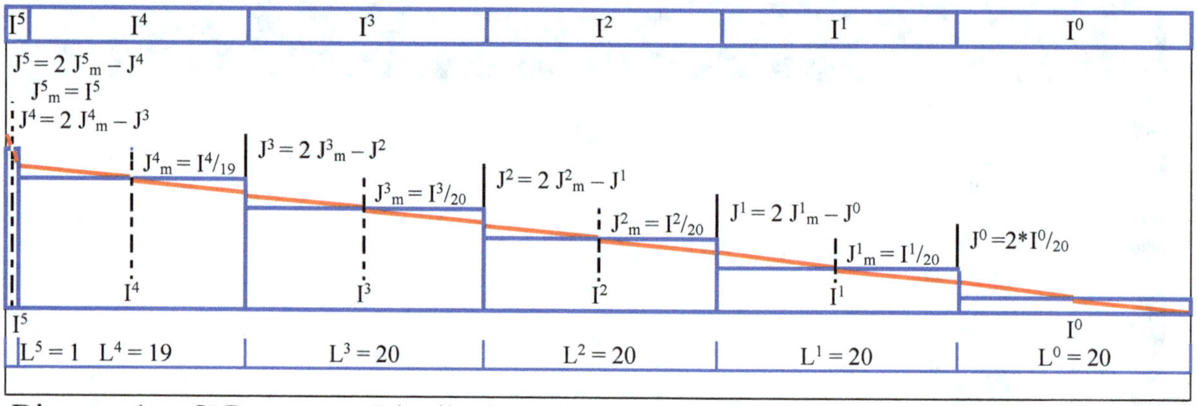

I^5	I^4	I^3	I^2	I^1	I^0

$J^5 = 2 J^5_m - J^4$
$J^5_m = I^5$
$J^4 = 2 J^4_m - J^3$
$J^4_m = I^4/19$
$J^3 = 2 J^3_m - J^2$
$J^3_m = I^3/20$
$J^2 = 2 J^2_m - J^1$
$J^2_m = I^2/20$
$J^1 = 2 J^1_m - J^0$
$J^1_m = I^1/20$
$J^0 = 2*I^0/20$

I^5	I^4	I^3	I^2	I^1	I^0
$L^5 = 1$	$L^4 = 19$	$L^3 = 20$	$L^2 = 20$	$L^1 = 20$	$L^0 = 20$

Diagram 1a – I^n Percentage Distribution of US Household Income Total for reporting from US Fed Data

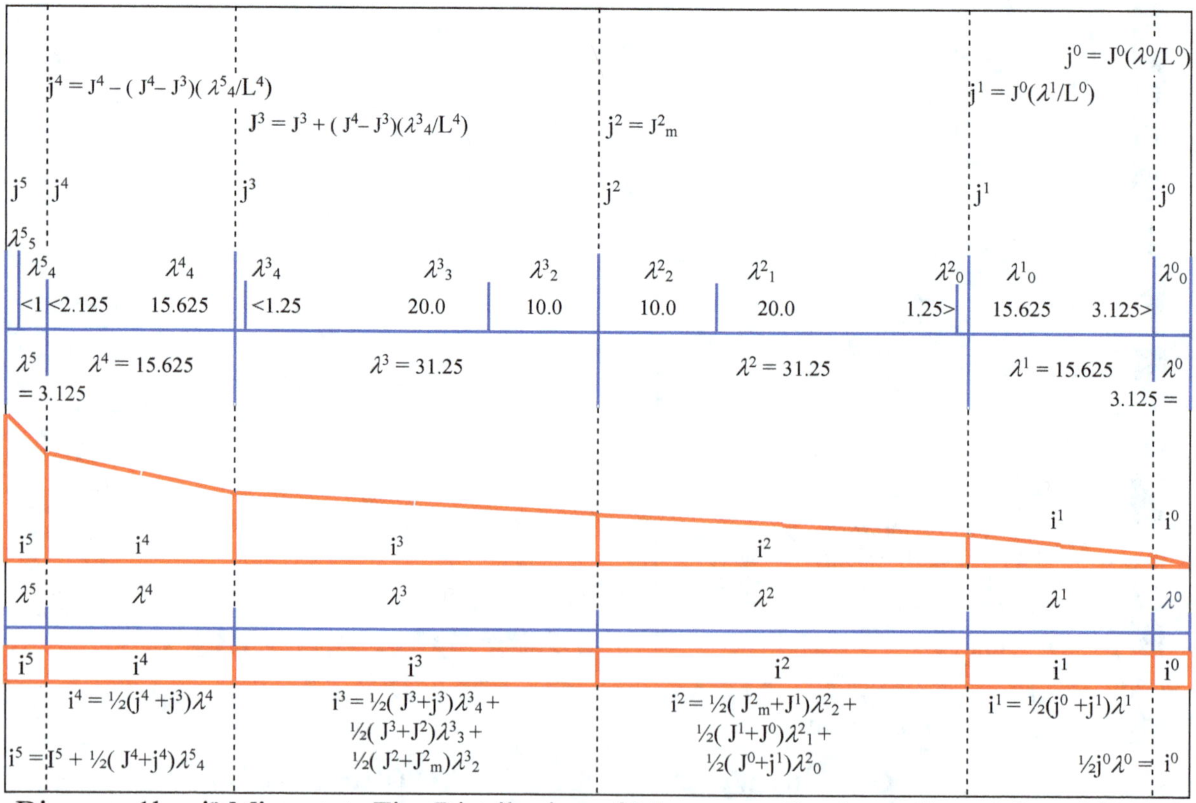

$j^4 = J^4 - (J^4 - J^3)(\lambda^5{}_4/L^4)$
$j^3 = J^3 + (J^4 - J^3)(\lambda^3{}_4/L^4)$
$j^2 = J^2_m$
$j^1 = J^0(\lambda^1/L^0)$
$j^0 = J^0(\lambda^0/L^0)$

j^5	j^4	j^3	j^2	j^1	j^0
$\lambda^5{}_5$					
$\lambda^5{}_4$	$\lambda^4{}_4$	$\lambda^3{}_4$ $\lambda^3{}_3$ $\lambda^3{}_2$	$\lambda^2{}_2$ $\lambda^2{}_1$	$\lambda^2{}_0$ $\lambda^1{}_0$	$\lambda^0{}_0$
<1	<2.125 15.625	<1.25 20.0 10.0	10.0 20.0	1.25> 15.625	3.125>
$\lambda^5 = 3.125$	$\lambda^4 = 15.625$	$\lambda^3 = 31.25$	$\lambda^2 = 31.25$	$\lambda^1 = 15.625$	$\lambda^0 = 3.125$

i^5	i^4	i^3	i^2	i^1	i^0
λ^5	λ^4	λ^3	λ^2	λ^1	λ^0

i^5	i^4	i^3	i^2	i^1	i^0

$i^5 = J^5 + ½(J^4+j^4)\lambda^5{}_4$
$i^4 = ½(j^4+j^3)\lambda^4$
$i^3 = ½(J^3+j^3)\lambda^3{}_4 + ½(J^3+J^2)\lambda^3{}_3 + ½(J^2+J^2_m)\lambda^3{}_2$
$i^2 = ½(J^2_m+J^1)\lambda^2{}_2 + ½(J^1+J^0)\lambda^2{}_1 + ½(J^0+j^1)\lambda^2{}_0$
$i^1 = ½(j^0+j^1)\lambda^1$
$i^0 = ½j^0\lambda^0$

Diagram 1b – i^n Microstate Tier Distribution of Macrostate Total as adjusted from US Fed Percentages

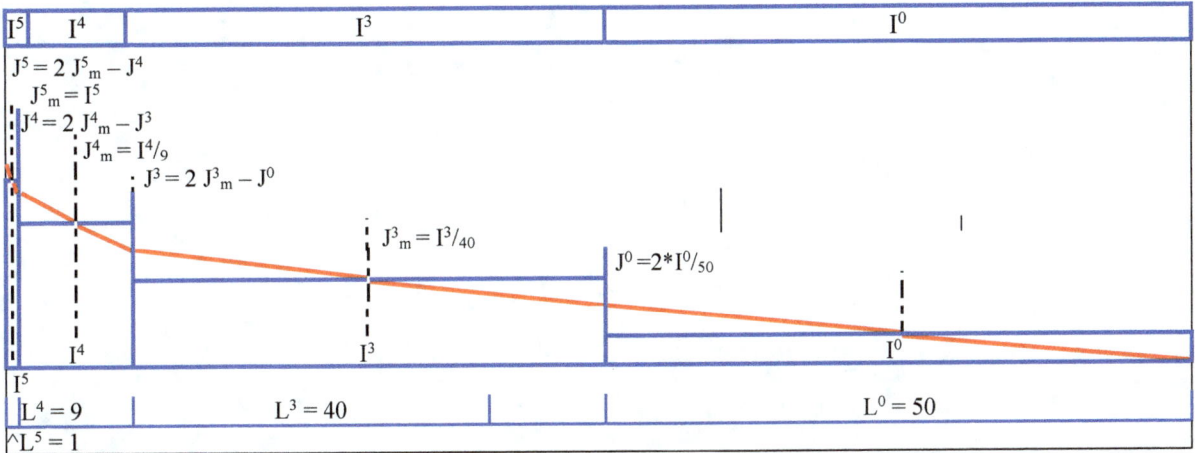

Diagram 2a – I^n Percentage Distribution of US Household Net Worth Total for reporting from US Fed Data

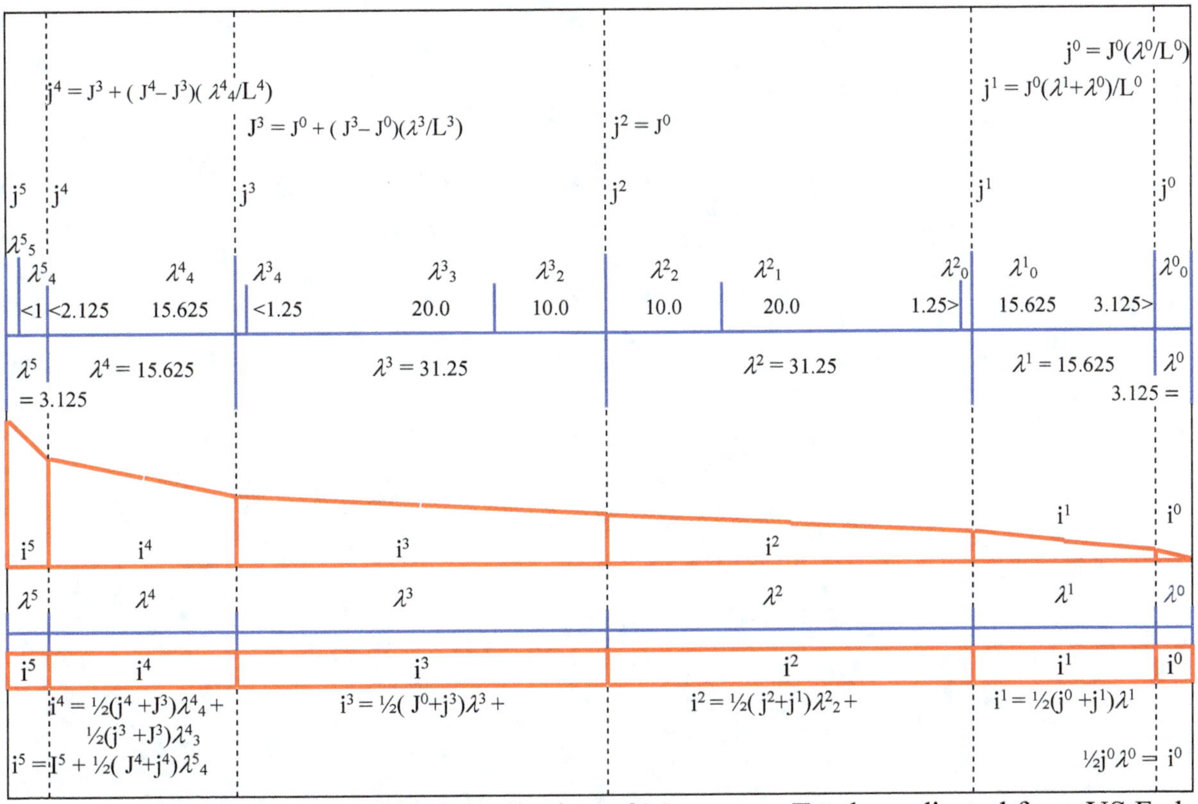

Diagram 2b – i^n Microstate Tier Distribution of Macrostate Total as adjusted from US Fed Percentages

Martin Gibson, UniServEnt.org – 12/10/2020

Citations

[1] Human Capital: "Human Capital Accounting in the United States, 1994-2006", a report by Michael S. Christian, published by the U.S. Department of Commerce, Bureau of Economic Analysis. https://www.bea.gov/research/papers/2010/human-capital-accounting-united-states-1994-2006

[2] https://uniservent.org/ergodidiocy/e-economics/ This link to the UniServEnt.org webpage entitled Ergodid Economics has access to various links, including a video presentation of the subject material and a download of the Excel spreadsheet from which the charts were derived.

[3] "Distributional National Accounts". federalreserve.gov. June 19, 2020. Retrieved August 20, 2020. https://www.federalreserve.gov/releases/z1/dataviz/dfa/distribute/table/

A Critique of Neoliberal Economics

Part II – Qualitative Analysis & Assumptions

Capital as Money, Focused Rationality, and Hierarchical Position

-:-

The Ideals of Omnipotent Money, Omniscient Decision Making, & Omnipresent Freedom bound by Material Conditions

The Dialectics of Neutral Monism
in the Historical Philosophy of Marx, Plato, and Christ

Martin Gibson
martin@uniservent.org
April 30, 2021

Capital as Money, Focused Rationality, and Hierarchical Position

The Ideals of Omnipotent Money, Omniscient Decision Making, & Omnipresent Freedom bound by Material Conditions

The Dialectics of Neutral Monism in the Historical Philosophy of Marx, Plato, and Christ

Martin Gibson, UniServEnt.org

Abstract

Current macroeconomic structural analysis is heavily constrained quantitatively and qualitatively by nominal valuation of sector and subsector accounts in terms of the history of market transactions and a forecast of capital changes based on the sector trends. A focus on private sector rates of return, nominal and real, on GDP and on capital held as financial and real equities and interpreted as a prime indicator of the health of a national economy, ignores key factors as evidenced by the market collapse of 2008.

Apparently missing from this analysis are the following:
1. The focus on maximization of returns as real over nominal rates encourages the reduction in taxes (for public outlays of indeterminant private benefit) and labor costs (deemed to be inflationary), with little consideration of the macroeconomic benefit derived. This is a straw man based on the true nature of modern monetary policy in which public outlays as issuance of a national fiat currency involve no direct borrowing or income tax transfer from the private sector. Instead of a currency flow from available private to deficient public sectors as generally portrayed, the use of borrowing and taxation is ostensibly meant to retire money thought to be circulating or looking for investment in the private sectors to offset the fiat issuance of public funding, to thereby avoid inflationary pressure from too much money chasing too few goods and services in either sector, though the obverse appears to be the case based on the ongoing lack of a balanced budget and almost zero inflationary pressure, at least for basic goods and services.
2. The true structural picture of the economy is only seen by an analysis of sector and subsector changes over time as percentages of total national production with respect to capital and consumption flows and to stocks rather than for sector changes measured against same sector baselines.[1] This can be addressed by a variant of analytical modeling which ergodically constrains the average flows and stocks of individual economic microstates exhibiting focused rationality as decision making groups over time to the average of those microstate flows and stocks of the ensembled macrostate of that economy at a given point in time.
3. There is no allowance in the national accounting for valuation of human capital on either a market-valued (MHC) or non-market-valued (NHC) basis. A 2010 study determines a range of 25-30% MHC to 75-70% NHC.[2]

---

[1] https://uniservent.org/political-economy/ Link to working paper, The Browser Economy, Martin Gibson
[2] Human Capital: "Human Capital Accounting in the United States, 1994-2006", a report by Michael S. Christian, published by the U.S. Department of Commerce, Bureau of Economic Analysis.

Comparison of unbiased ergodicity in non-market and of weighted ergodicity in market modeling of human capital with 4Q 1989 and 4Q 2019 Household Income and Net Worth US Fed Data gives credence to this approach as indicated in quantitative analysis of the first section and summed up in the qualitative political economic analysis of the second.

Some Axioms concerning Economic Activity and Capital

The word, capital, is from the Latin for 'head', *caput*; related to a common Latin stem for 'roomy, the ability to hold', *capax*, related to 'capacity', the ability to receive, contain, perform, or function in a position or role, both physical and mental. The word, power, is also from the Latin, *posse* or *potens*, by way of the Anglo-French, *poer*, as 'to be able, to have power'. It is therefore somewhat redundant, if not tautological, to state that social power and the cultured ability to direct the energy coming from any natural potential as economic activity, comes to be represented by way of cranial capacity as the experienced mental ability of collectively engaged human beings.

In an economic context, **capital** is the rationally and emotionally effective productive capacity and potential, the **power,** of human beings working with the natural resources at hand, by their individual and collective **labor**, to maintain their place of **living** in the world. An economic interchange among human beings involves an expenditure of energy and incorporation of material in the supply, production, warehousing, and distribution of the final consumable **goods and services (G&S)** required for living; as an integral corollary this includes the percentage of that energy and material that is required for the creation, maintenance, and replenishment of any intermediate goods and services necessary for production and distribution of the final consumables. All these intermediate goods and services are forms of real capital and necessarily include the creation, maintenance, and replenishment by way of the sustenance required of and for the human beings involved in the living experience.

A portion of these interchanges between human beings are performed within the context of a market mechanism as a trade of G&S for ostensibly equivalent **tokens of satisfaction ($)** as instruments of financial capital in the form of money, promissory notes, certificates of extant stock or future flows of assets, or other tabulation of accounting **value**, so that such portions of both intermediate and final G&S have a representational, stock-valued human entitlement as **market-valued human capital (MHC)**. The remaining portion of the interchanges that are productive of intermediate and final G&S are assignable as **non-market-valued human capital (NHC)** which can only be estimated based on broad equivalences between market and non-market components. These NHC interactions are the earthly product of Nature's bounty, God's grace, however you may choose to designate God as the endless, supreme potential source for change, and community nurture that must be sustained from cradle to grave by means of both MHC and NHC; this NHC as might be indicated with "all men (and by implication, women) are created equal, ... endowed by their Creator with certain unalienable Rights, that among these are Life, Liberty, and the Pursuit of happiness" in the US Declaration of Independence and which we might deem applicable internationally.

Certain accounting in the US economy of recent years puts the ratio of MHC:NHC average in a range of 0.43:1 to 0.33:1, where 1 is an implied statistical average of NMC.[i] From this thinking as we will detail, MHC of 0.43 to 0.33, or 0.38+/- 0.05, can be theoretically distributed along a spectrum with each individual's percentage of the total market valued capital of an economy, all greater than 0.0 (since a negative as debt would indicate involuntary servitude or Shylock's pound of flesh), while the NHC is simply the per capita average of the total NHC accounted as 1.0.

Capital then, all of which is an intermediate good or service, can be viewed and valued as: **human**—consisting of the physical and metaphysical skills and capabilities of people both MHC and NHC; **real**—consisting of the physical resources, both natural and technologically developed, available for human utilization through productive human effort; and **financial**—consisting of various methods devised for directing and accounting for the past and future production, distribution and consumption of goods and services required or otherwise desired by society.

The referenced power can be viewed qualitatively as: **social**—consisting of the skilled capacity, expertise, or position to organize and direct a desired productive enterprise; **energy source**—consisting of the available, technologically developed and thermodynamically defined capacity to fuel useful work, both mechanically and through the ecological and nutritional needs of human living; and individual **labor power**—consisting of the mental and physical ability to provide and sustain useful and necessary, thereby meaningful, work.

Labor can be viewed along a spectrum as: **self-initiated**—consisting of the employment and direction of the work of oneself and/or others; **employed**—consisting of the voluntary provision of one's own work effort at the direction of others; and **involuntarily servile**—consisting of wage, penal, chattel, or ensnared bondage under conditions of coerced toil.

Living can be viewed along a spectrum from a **risk of enslavement** to material wretchedness due to personal behavior or the depravity and disregard of others up to an **opportunity for liberation** of spiritual perception and intuitional creativity.

Capital and labor, both as an expression of power, is at work under any condition or state be it; tribalist, feudalist, capitalist, socialist, or communist; monarchist, oligopolist, liberalist, or anarchist; browser, hunter-gatherer, agricultural, industrial, or cybernetic. This dynamic of capital production and use is not confined to 'capitalism'; nor absent under 'socialism'. The political organization of that state is determined in proportion and valued degree by those who are positioned to exercise the power inherent in the instruments of both capital and labor over a range of **hierarchical structure** and **decision making**. That decision making is characterized herein as a choice defined by the motivation, expertise, capability, and position of the individual or affiliated group as a function of **focused rationality**, even if that reason is confined to the daily decision of how or whether to continue with a meaningless job, with no apparent alternatives in sight, in order to survive.

Qualitative Analysis and its Implications

The motivation, organization, resource utilization and participation of individuals and affiliated groups in economic activity—in the interactive production, storage, exchange, and consumption of goods and services—does not happen without much thought. Since virtually all human living involves the use of economic production in its ongoing activity, it is easy to believe that key components of that activity provide the principle definition of that life as homo economicus. Yet the rest of the biosphere manages to live and some to thrive with little thought beyond the next instinctive move. Clearly, there is an innate motivation, understanding of organization, wisdom of resource utilization, and intuitive participation of one and all in economic activity that arises from an essential source—naturally evolved, divinely inspired, or both. Economic thinking carries this essential activity beyond its source; it is an adaptive social process of focusing and deciding who and how best to produce, store, exchange, and consume—immediately or intermediately as capitalized—goods and services. Political thinking is the social process of persuading or commanding other individuals to identify and cooperate with any of the various parties that disagree on the who and the how.

Political economic thinking is a historical process that has focused on certain ongoing observations concerning societal attempts to understand—first, the essential nature and genesis of human life, and second, the experiential nature of knowledge motivated by emotional concern and trepidation due to observed changes and differentiation in the size, health, technical abilities, access to resources, risk–opportunity of involvement–exploitation, and conflict of endogenous and exogenous populations. This thinking has been formalized in various cosmological forms, starting in antiquity, to explain and control the forces of nature and the location of human activity within it. With increasing understanding and direction of those thermodynamic forces to mechanical advantage and profit in the rise of the industrial age, and the theoretical application of natural laws to practical capital-intensive manufacturing and innovative research & development in the physical and biological sciences, the growing rationalization by the scientific method has fostered progressively sophisticated attempts at quantitative analysis and modeling of human social organization in the field of economics.

The parallel and synergistic development of physical and fiscal technology has produced ground-breaking, if not earth-shaking, advancement in the service of humanity, but this has not been without peril.

Some Analytical Assumptions of Current Economic Thinking

Much of current neoclassical macroeconomic modeling along with congruent national public policy is based on several generally recognized but mistakenly conflated assumptions: **the notion of money as omnipotence**—as the ultimate source of power in driving economic activity in its ability to maximize utility and the generation of growth and profits; **the notion of decision making as omniscience**—as a function of an exhaustive awareness and wisdom in directing that money flow through optimum valuation to maximize economic activity; and **the notion of freedom as omnipresence**—as the inherent capability to pursue that maximized economic activity to the optimum location and environment and at the most appropriate time.

Under these assumptions, money in the possession of people is generally considered to be the essential ingredient to any productive project, without which nothing worth doing is possible, rather than the people themselves in the possession of productive knowledge and skill. Money in the possession of people with productive projects in mind is generally considered to be sufficient to plan and oversee the implementation of microeconomic projects for the public good, rather than the necessity of having people with an understanding of economic feasibility in positions of directing macroeconomic policy. Money used in the operation of such projects under the control and to the benefit of such decision makers is generally considered to be the necessary and sufficient condition to guarantee such benefits to the greatest number of people, rather than to the benefit of all the people that are involved and affected by that decision making. By implication, the assumption is that anything worth doing will attract the necessary capital so that a lack of capital is prima fascia indication that it is not worth doing, any unfunded innovative insight and entrepreneurial capability to the contrary.

The Notion of Money as Omnipotence

The first and foremost mistaken attitude is that 'money' is a store of value, analogous to a store of energy, that given sufficient quantity has an inherent potential to effect economic change apart from any real productive resource, where 'real' means any physical component or energy resource, including human, incorporated in an intermediate capital or final good or service. In contrast, money is not like a source of energy that can be used to fuel some economic apparatus, where one simply plugs into the grid and some good or service pops out at the end of the production line, ready for consumption; as if we only had enough money, we could produce anything we wanted. Money is essentially a universal call option, but one without a universally dedicated put; there may be specific puts, but none that are universal. If no one wants to sell what you want to buy, your call is powerless, but if no one wants to buy what you want to sell, your product is valueless.

The first notion of money is understood to have arisen in Mesopotamia as a record of credit for grain harvested by individual and family farmers and deposited under the auspices of a theocracy in a communal granary in return for a small portable clay tablet as a record of credit of the transaction. The grain, the real productive resource involved in the transaction, once placed in the granary represented a debt of the theocracy, where it could be retrieved in an appropriate quantity by surrender of the tablet to the granary, where debt in such case is an accounting of an asset with the granary as bailee.

Over time an unredeemed tablet could also be exchanged as a token of satisfaction for other goods and services with other members of the community, without drawing down grain from the granary reserves. This meant that the store of value in the granary as a basket of grain assumed a derivative, representational value in tablet form that was detached from the underlying real, utilitarian value of the harvested and stored grain. The tablet/token thereby became associated by custom as a real store of value that could be exchanged for real goods and services.

A token that is used to retrieve grain at some point in the future is of course not redeemable for the same kernels of grain that the original farmer deposited or for which they received credit in the form of the tablet. I learned this at an early age when I went to retrieve some shiny silver dollars that I had accumulated as gifts from family friends and deposited to open up a savings bank account, only to be shocked in disappointment when the withdrawal was later made in uninteresting reserve notes. That grain may have been consumed long since, but assuming that the grain is fungible—indistinguishable among any of the grain in the granary—the same token will be tradable for the same amount of grain as deposited before or since the original grain was deposited.

A problem occurs if the grain in the granary is damaged or cannot be fully replenished due to harvest failure from some natural or man-made disaster such as flooding, fire, or warfare. Then the nominal value of the token, based on the credit/debt relationship between the accounting on the clay and the real productive grain in the granary, is compromised. Transactions between individuals involving the exchange of an existing granary token of previous grain deposit for other types of goods or services that do not anticipate an immediate need for grain on the part of the token recipient may continue, but at some point the general knowledge of the crop or granary damage will affect the willingness of an individual to accept a token in a transaction, at least as the token was initially valued.

Money, then, originated as a credit accounting instrument to represent a real productive resource as food, stored as a community good for retrieval on demand in a publicly sanctioned institution, where the actual good to be retrieved as valued for consumption was deemed to be currently extant or understood in good faith to be part of a process for producing and storing that good in a customary manner on an ongoing basis. Such money as a token of satisfaction, therefore originally represented a stored, real good of utilitarian value that was already produced. It was generally not considered to be a call option for production of a good or service that had yet to be produced.

The value of a credit token or accounting instrument is susceptible to decrease or increase as a result of damage to granary contents or to a crop in the field. In that case, when the accounting is strictly representative of the grain as a share of a real good, the nominal tablet value as the number of baskets of grain exceeds the supply expressed as the number of real baskets available. In this case, when the accounting is subject to the authority of a theocracy in control of the grain dispensary, the credit tokens may be devalued by a rationed exchange of the nominally same token for a reduced amount of grain in the basket. If the exchange is price controlled by the authority, the first farmers to the granary get the grain and those that are late to retrieve the goods may be out of luck, in possession of only worthless tokens and no grain. In the absence of such authority, there may be a run on the granary 'bank', in which case, if they are available, more tokens may be offered for exchange than is warranted for each full basket of grain as a form of price inflation.

Note that this price inflation can only occur if there are farmers in possession of a number of tokens nominally valued in excess of their normal immediate needs for grain, which thereby allows them to retrieve a normal basket of grain while in short supply at an elevated price. If the supply of grain exceeds the demand at any point in time, there will be no impetus for grain price inflation and the excess tokens will be free to circulate for exchange of other goods and services, some of which may yet to be produced.

For the same general amount of a farmer's labor, a grain yield will vary from harvest to harvest, as a function of the seasons, pests, and weather variability. In times of plenty, the granaries are full, and there is an increase in the number, or nominal value, of grain tokens along with the corresponding surfeit of grain. In these conditions the nominal and real price for grain remains matched and low, and farmers may even find time to pursue other pastimes such as building ziggurats, in which case the excess number of tokens will find a varied secondary value in various secondary transactions, each related to the degree of satisfaction imagined as being realized by purchase of the secondary goods and services produced by others. The inflation of pricing of any secondary goods and services not tied to the supply of grain or other related commodities required for human life such as water, housing, shoes, or clothing, is of secondary concern to the value of grain tokens for the general population of farmers. The cost of the draperies in the theocrat's newest addition to his ziggurat will have little effect on the farmer's livelihood—until there is a plague or famine or invading horde that upsets the underlying value at the granary.

For the theocrats, tokens of clay tablets for grain were not nearly as useful as bullion and coin tokens of precious and utilitarian metal for trade in buying silk for drapes and robes or for exotic food and spice or for paying farmers in the off season for the time and material to craft any number of tools, artifacts, and weapons for building ziggurats or to fend off the invading armies that wanted to take over and occupy those ziggurats. Unlike clay tablets which are cumbersome to carry and can break, tokens of precious or utilitarian metal had a durability and value density that was more portable and retained some utility as a store of value as a real productive resource, if we are willing to call gold jewelry productive. Money as tokens of precious metal is different from money as grain credit tokens, in that the store of value is embodied in the precious metal token itself and is more durable. However, it is subject to loss or theft, while the store of value embodied in the grain token, is remotely securable, though subject to a risk of spoilage and depreciation.

This difference in durability and remote storing between tokens of grain credit and tokens of precious metals gave rise to a hybrid form of money, that of reserve, and eventually fractional reserve, banking. Tokens of credit for life sustaining grain deposited in a granary, inherently consumable and depreciable, yet reproducible in partial measure as seed stock, evolved into bank reserve notes and checks in circulation in a commercial banking system, backed by secured reserves of coin and bullion, in which the real resource as capital is not the incorruptible precious metal with only marginal productive utility, but rather the future productive capacity of the human beings who through their indebtedness fund the unsecured portion of the fractional reserve system.

The paradoxical nature of this money system is that the precious metal resource as a store of value while durable, cannot be consumed, and is inert and lifeless, and in that sense, unproductive. Precious metal money is of utility satisfaction only as a form of socially accepted adornment and status, whereas grain credit as money, though subject to deterioration like human life, is a resource that must be consumed, and in turn replanted and worked for replenishment in order to survive and is therefore precious to life. Grain incorporates the power of the sun as a real productive store of value that sustains life. Gold reflects the same solar power, but only as a glamorous and unproductive illusion.

The story of this money paradox does not end with reserve banking backed by an unproductive precious metal, however. The grain, harvested by the farmers of Mesopotamia by their own efforts and deposited as reserves credited to their own account in granaries under the custodial eye of their local theocrat, was the principle means of sustaining the livelihood of the community. A public monetary system of tokens of credit backed by grain reserves, essentially a system of exchange accounting backed by the human capital of the community in the form of the farming livelihood, made this possible. These tokens of credit, a form of financial capital, are logical tools used to direct the interaction of human beings in the development and use of real productive resources, the other necessary components of capital, toward the real end-product, human life.

The evolution of precious metal reserve banking from its beginnings in grain credit depositories has a certain sparse logic. Backed by the social security of full granaries, times of agricultural plenty allowed the governing theocratic elites of a community to employ farmers freed from agricultural subsistence labor to work on erecting theocratic edifices for the pursuit of theological statecraft and trade, with the resulting conquest and defense deployments that such theocracy and commerce suggest to the theocratic elites. Payment to these non-agricultural workers for the needs of the state facilitated a coinage in a durable, portable currency in the place of cumbersome tablets for grain, and in time to bailed deposits of precious metals in banks which paralleled the deposits of barley in granaries under the same theocratic control, both of which depend on the reliability of a credit mechanism.

With the realization in some societies of the redundancy of a parallel system of tokens, the equal weight of a metal coin and a basket of grain came to be denominated by a common term, the shekel, and the clay tablet become an anachronism. The conflated notion of a credit token in the form of coin for demand against the grain in the granaries, led to acceptance of demand drafts recognized for payment against precious metal debt accounts in banks. At this point in the evolution of the monetary system, the credit accounts still consisted of money representing a real product already produced by human effort, be it in the form of grain or a useable metal and deposited as a fungible bailment in a secure location.

At some point the notion of fractional reserve banking of those products occurred to some bright group of individuals, based on the realization that most of the gold reserves sitting in the bank, similar to grain sitting in a granary, was not used in circulation for coins or withdrawn for payment of accounts, and much of such payment to third parties by coin in circulation was superseded by a written draft transfer in the settling of accounts. It became apparent that it was feasible to extend credit on account in a bank, generally in return for an assignable interest in some secondary property, so that the credit instrument, as a token withdrawn for circulation, represented *a real asset that had yet to be produced.*

Such credit essentially represented a promise to pay, a debt, by creating a yet to be produced good or service in the community. In the event of no assigned interest or collateral and often despite the existence of such assignment, this development opened the door to increased risk, but it also opened the door to opportunity through the incentive of employment of idle farmers or other workers, the true source of all capital—the same capital that planted and harvested and stored the grain in Mesopotamia, which is to say human capital, motivated to work in order to sustain life.

Initially in the evolution of money, most of the goods produced and consumed to sustain that life did not require an exchange of money. Most of the food, the clothing, the house and farm tools, the buildings, along with the irrigation ditches were created by the farming family and community directly from the natural resources at hand. The few tokens in the farmer's household possession were credit tokens acquired in return for grain deposits made in the community granary from the farmer's own production. While in the possession of the community members, these tokens might circulate in the transfer of goods within the community, but when the grain was retrieved from the granary and the credit was satisfied, the clay token might be destroyed, or like a metal coin or bullion, retired from circulation and secured under the granary lock, to be recast or reissued upon deposit of grain with a new harvest.

A surplus of shekels sitting in the granary vault would be an attractive nuisance to the treasury in charge and on occasion would end up being lent to a reliable borrower in return for future repayment of the principle amount, in some cases plus an interest fee for the use of the money. In the absence of a real security or transfer of commensurate property to the lender as part of the lending agreement, the principle lent represented a debt instrument in circulation owed to the lender, instead of a credit instrument as a granary token for bailment owed by the lender. It represented property as capital goods yet to be created.

This is so regardless of whether the debt token is repaid to the granary, remains in circulation in good standing, or is in default. All of the money reserves that originated in the granary as metal tokens, if denominated as a shekel on the basis of an anticipated weight equivalence to a harvested basket of grain, is only valid as valued upon deposit of that basket by the farmer in the granary and issuance of the credit token in trade. If used in circulation instead of being redeemed for grain, the credit token becomes indistinguishable from the debt token lent by the same granary while circulating.

If the borrower commissions a carpenter to build a chest in return for the debt token to the satisfaction of the borrower, and the carpenter commissions a tailor to weave and sew a coat in return for the debt token to the satisfaction of the carpenter, and the tailor commissions a wheelwright to repair his wagon in return for the debt token to the satisfaction of the tailor, and the granary in all due rights calls the debt token, the wheelwright has a problem; and perhaps so do all the rest, depending on the civil code in place.

If the granary calls and retrieves the debt token, the wheelwright will lose the value of the work he performed for the tailor, and either he or the original borrower may be required to pay interest on the debt. The wheelwright may have the option of performing an equivalent amount of work for the granary in order to retain the debt token, but then he will have performed the work twice. The wheelwright instead may be able to keep the debt token, if the borrower is required to give up his chest to the granary, in which case the tokened debt is satisfied. The wheelwright's token thereby loses its debt or credit bias and it becomes simply a token of satisfaction subject to the free acceptance of the next tradesperson or buyer.

From a theoretical economic perspective, in the absence of the initial issuance of debt, the chest, the coat, and the wagon repair may not have been produced. On the other hand, they or equivalent products may have been produced outside of any monetary exchange. The same work and same products are capable of being produced under either motivation, as a monetary exchange or as a gift or favor. Facilitated by that issuance, however, the chest, the coat, and the wagon repair <u>are</u> produced, though the ownership of the chest and the financial condition of the wheelwright (and perhaps the granary) are undetermined.

An alternative to a call and retrieval of the debt token would be to leave it to continue in circulation indefinitely, with the resulting creation of more and better chests, coats, and wagon repairs, and with the lending of additional debt tokens, even more products than would have been made otherwise. The continued circulation with its added production can remain viable as long as the employment of the community's market-valued human capital (MHC) for this production does not deflect employment from the necessary subsistence farming of non-market-valued human capital (NHC) which supports the community or does not otherwise precipitate a shortage and competition for grain that inflates the price of that grain and related essential goods. If the community possesses widespread skills and non-market access to resources to produce the needed chests, coats, and wagon repairs without an intermediating market signal, as priced into borrowed tokens, to trigger demand for goods and services, it will continue to make these goods and perform these services as a function of the community's NHC through coordinated group effort, barter, and the interpersonal promise of private recompense.

From this basic evolution we can delineate two distinct but intertwined systems of distributive valuation as a monetary system including intermingled token money in circulation; that of non-market-valued human capital development founded in a communal system with <u>credit token equivalence of existing grain deposits</u> held in the granary requiring ongoing replenishment by non-market-valued present and future community efforts for future community consumption, and that of market-valued human capital development founded in a market system with <u>debt token equivalence of future promise to pay</u> represented initially by precious metal tokens held as security in commercial banks against future production and distribution of intermediate capital and final consumption goods and services of any type.

From this observation it bears emphasizing that for the monetary system as a whole, any market valuation of the <u>debt token equivalence as future promise to pay</u> is ultimately secured as an insurer of last resort by the ability of the community to maintain the viability of the non-market valuation of any system of <u>credit token equivalence of existing grain deposits or the modern equivalence of any basket of goods and services required to justly maintain human life.</u>

The expansion in scale of operations that comes with agricultural plenty and technologically facilitated population growth and the resulting distribution of classifiable, specialized skills across the community, tends over time to intensify the needs and demand for market human capital and to de-emphasize and even disregard the importance of non-market human capital. The burgeoning needs, for wage money that comes with industrial rationalization, for the ability to pool that money for consistent cash flows, for the funding of productive real capital investment in plant and technology, when successfully met can create an illusion of omnipotence, associated with the glamourous control of amassed money as a source of power. Instead it is the massed coordinated effort and initiative of both NHC and MHC of all skill levels that is the true source of power in any such productive endeavor, which only incidentally requires the mesmerizing pooling and flow of cash to direct and maintain the effort toward the intended outcome. It is the population of human capital with sufficient skills and needs in a sufficiently rich resource environment that determines the wisdom and potency of such endeavor.

The Notion of Decision Making as Omniscience

In the absence of experienced wisdom in the governance of economic policy, the effect of this glamour of money as the source of power is to decrease valuation of human capital in productive service which has no operational market price and an increasing valuation *and pricing* of human capital in such service in proportion to its operational market price, which sets up a self-fulfilling dynamic of expectation and realization.

In a competitive end-use environment with market-valued service in excess supply, that service valuation is reduced to a commodity level, where commodity pricing is determined by the lowest cost for the flow of replacement goods to and for that commodity, which may or may not be sustainable, and may or may not provide for long term care of that service. Non-market-valued human capital is essentially native and community based, dependent on habitual learning over extended periods of time for the transmission of customary skills benefitting the community. Community systems of decision making, including the control of land and community real property in granaries represented by credit tokens in circulation, initially are not governed and valued to a great extent by market mechanisms; with the exception, perhaps only, of human slavery.

Historically within the context of an agrarian community or tribe, we might presume that significant deference in decision making was extended to the elders of the group, in particular to the acknowledged leaders in work, in innovative and constructive problem solving, in warfare, in administration of group cohesion, and as seers of religious vision. That natural deference leads to special rivalrous and eventually excludable access to resources for leading elders in what comes to be defined as private property, starting perhaps as theocratic property, where the principle leadership positions in and of the community grant entitlement rights derived from their public positions. Based on a presumption of skilled leadership entitlement, we have the emergence of a hierarchical position from an egalitarian beginning as both cause and effect of non-market-valued human capital with control of real capital as property, financial capital as money wealth, and human capital as wage or indentured servants, serfs, or slaves.

The effectiveness of decision making by leaders of agrarian theocracies, using their understanding of non-market human capital in the allocation of resources within the community, would not appear to be significantly altered if that understanding came from a technologically advanced economy in which the human capital is valued by market transactions. Qualitative understanding of human capabilities is itself a qualitative human capability; *quantitative understanding* of human capabilities is also a qualitative human capability, distinct from any capacity to count or work with numbers in evaluating economic development and change.

And yet the capability of judging the efficacy of economic policy decision making in either private or public sector funding is subject to the allure and unrecognized corrupting effect of self-deception when faced with a perceived access to control of large sums of money for any purpose. Such allure does not preclude a pursuit of that access, but it does require a perspicacity concerning one's intent and transparency of those intentions in that pursuit. Absent that understanding, what is essentially a proliferation of banking debt money intended and used for private investment in future production and collateralized by the public good of the community, is seen myopically in the obverse, as monetized credit of private interests held against the indebtedness of the public sector, which it is not.

The notion that accumulations of wealth from whatever sources as a measure of market-valued human capital should be considered of equal value as a measure of non-market-valued human capital in the community is the consequence of this self-deception and hubris. Thus, quantitative decision making that leads to private wealth is easily accepted as the result of wisdom in the marketplace without adequate understanding of one's own position or the position of others. When that lack of understanding of oneself is applied to the market valuation of other human beings in any service capacity, the implication is that the non-market-valued human capital of others as a share in the public value of the community should be measured according to the market valuation of those others. This assumption of status as a private entitlement absolves such decision makers from the effects of exploitation of their own position in the market, to the point of reducing those without value, without income or net worth in the community, to involuntary servitude or worse, as unvalued human beings.

The Notion of Freedom as Omnipresence

The communities that took root in Mesopotamia are presumed to have started out as generally egalitarian tribal societies, governed by a group of family and tribal elders that were and still are closely responsive to the needs of that community, needs heavily constrained in their production by the primitive technology and access to resources of the region. Such tribes may have a bicameral leadership with a strong executive and a visionary seer to guide the operation of the community. On occasion, these two functions can be found in one leader that gives rise to the development of a theocratic social structure. That governing theocratic hierarchy may retain a measure of the egalitarian principles of the community, but in time, facilitated by customs of heredity, the governing interest concentrates in the hands of the hierarchy elite. The quality of hierarchical interest and control exercised by the elite might range from that of benefit to that of dereliction of the community.

Judging from history, this appears to be the case with hereditary leadership positions in particular, such as monarchy of the absolute type. Under this structure, the uncertainty of the quality of the community's life becomes a reflection of the intelligence and whim of the lineage of a dynasty. This same qualitative range in decision making is observed across the entire historical record of quantitatively described hierarchical development, from the most centralized monarchy to the most widely distributed anarchy, since the qualitatively wise operation of that hierarchy is dependent on the quantity of intelligently aware members in key positions of hierarchical power.

This is true even if the monarchy consists of a population of one, as a hermit living remotely in the wild. If he has performed the sufficient preparations in acquiring the necessary tools, skills, and especially the attitude before secluding himself in the natural environment, he will probably survive quite well, but if he finds himself suddenly alone, like a castaway on a deserted island, his solitary kingdom will be in for a test. The same thinking might apply to a migrating community beyond the frontier of an established state. The readiness of the individuals for the experience of such a diaspora relies heavily on an accurate understanding of the new terrain and of the interrelationships of the members of the group, of their abilities and of their capacities to adapt and perform as anticipated.

The success of a sojourn in such circumstance, as with the progeny of Aeneas of Troy, in turn is dependent on encountering a hospitable environment and a degree of openness in communication between the individuals and family groups which tends toward a democratic sharing of power and governance by the people. An inhospitable environment and fear of external and internal vulnerability on the part of the members causes a natural canvass of the community for the best or most appropriate leaders to address any perceived threats, where the leaders best able to address the perceived proximal threat may not be the ones best suited for grappling with more distal concerns. Unfortunately, the account of the democratic (on the part of the patricians only) elevation of Lucius Quinctius Cincinnatus to the position of Dictator in the early Roman Republic, in the face of external and internal threats on two occasions followed quickly by his quitclaim and return to private life, is an abjuration of authority too seldom found in the historical record.

In antiquity, the logic touching on the notion of freedom found expression in Plato's Republic,[3] comprised of 10 books, which depicts an ideal form of life cycle of an ideal city state Republic through the interaction of the ideal souls that inhabit it, where the integrity of those souls contribute to and are sustained by the strength, texture, and color of the city's social tapestry. In terms of the thinking of Plato, an ideal form, referred to eponymously as a Platonic form or Form, is a Principle—a Good or a Truth—that has Reality beyond any material representation of that Principle at a historical, current, or future point in time or space, but which can be depicted or referenced by such material representation as a historical, current, or imagined future instance of that Principle; for example, a Planetary Orb as distinguished by the material earth on which we find ourselves in this discussion.

Central to the ongoing vitality of the city is the judicious education of the citizenry, who are deemed in the Platonic view to be immortal tripartite souls, possessing reason, the passions or emotions, and physical appetites, and subject to metempsychosis or reincarnation. Both male and female candidates selected for training as guardians of the city are educated in certain ideals, the virtues, of Wisdom, Courage, Justice, and Temperance, in gymnastics and martial arts, in what we would call STEM subjects, in philosophy and logic, and in dialectics. Natural, cultural, and military crises, along with failure to maintain the required education for the guardians of the city and other elements of social dislocation can lead to a deterioration of health and happiness in the life of the city.

In the ideal State—perhaps as an initial idyllic state—those enlightened Principles, learned and promulgated by a wise and just, non-hereditary aristocracy with a philosopher king chosen from among the guardians, maintain the health and happiness of the polity. In Book VIII, Four Forms of Government, over time and in response to crises of external or internal urgency, a timocracy of honorable theocratic, political, and military leaders comes to the fore to manage the affairs of the Republic, which if successful in addressing the crises through trade and conquest, acquires and accumulates wealth in the form of natural, produced, and human resources, the latter as freemen and/or slaves.

Increased access to productive resources and the profitable growth of the general community leads to an increasingly self-involved oligarchy, whose growing wealth develops or intensifies the desire for hereditary propertied privilege. Again over time,

---

[3] This writer's treatment of the substance of Plato's work is not intended as academic endorsement or critique of the Republic or Plato's body of work; rather it is intended to be a constructive ideal part of the dialectical synthesis of this writing. In this regard and in keeping with some academic protocols, references to ideal or archetypal forms, usually absent the 'ideal' adjective, will be capitalized in this section and material, existential instances will be in lower case. The Forms are deemed to comprise a metaphysical, Essential, logical Reality or Logic that persists beyond any conscious recognition, though accessible to perception by the Soul, embodied or not embodied.

propertied control and acquisitiveness, if un-tempered by theocratic ideals, tends to kleptocracy that results by a revolt of the exploited and undervalued people of the community in the establishment of democracy and desire for freedom by those so dispossessed. In the absence of the education on which the happiness and well-being of the city is based, as given to the original guardians of the city, a final decline in the civic virtues occurs with defilement of democracy and freedom of the people and the elevation of an autocrat who by vice of his own moral debasement and that of the society becomes a tyrant. In time, either by birth of an heir to the autocracy or by conquest of a rival state, the city is graced by the arrival of another philosopher king, perhaps an avatar or messiah motivated by similar Platonic principles, to rejuvenate the process.

Central to Plato's notion is that the greatest freedom and the greatest happiness is achieved for the city and for each soul when each citizen is doing what is in his nature to do, and that is achieved when each soul is wise enough to be led by certain divinely endowed or inspired principles of Love and Reason in pursuit of his civic duty, and not by the passions or physical appetites. The notion of the philosopher king and the tripartite soul is echoed a century or so later in the Hindu scripture of the Bhagavad Gita, where the part of the philosopher king is reflected by the Lord Krishna as an avatar of Vishnu, and in which the reincarnating soul, with a similar set of three elements of a conflicting nature, is represented by Arjuna, the warrior prince, and freedom and happiness is understood to follow from adherence to the dharma—the correct way to live. The notion of the philosopher king finds resonance another few centuries later with the gospel of Jesus Christ as the prophetic Son of Man, head of the spiritual kingdom of God; and again, with Islam and the death of the prophet Muhammad, in the establishment of the Rashidun Caliphate in 632.

In contrast to the Platonic form of an ideal dialectic between the freedom of the tripartite souls of the citizenry and the happiness of the city as it plays out in the social history of the state, we can also look at the Marxist historical political dialectic between individuals, affiliated according to their access to life supporting material resources and conditions, in rivalrous ownership or control over their means and methods of economic production. In this comparison there is a generally straightforward correspondence between Plato's timocracy, oligarchy, and democracy and Marx's feudalism, capitalism, and socialism, though some interesting interweaving of the two narratives suggests itself. There is perhaps less direct correspondence between the Platonic and the Marxist vision with respect to the ideal initial state of the Just philosopher king or the degenerate state of Tyranny. Still, a pre-feudal agrarian community and the utopian post-capitalist classless vision of Marx can find resonance in the Platonic ideal of the Republic, and Plato's Tyranny can be understood as a degeneration resulting from the internecine struggle of the feudal aristocratic, commercial capitalist, or proletarian laboring classes, this last case being of particular interest in the manipulation of racial and servile fears.

What is significantly missing from Marxist analysis in this contrast perhaps is a mature, nuanced psychological understanding of the meaningful existence, and we might even say essential importance, of the psychic component of the city state played by the Tripartite Soul in the Platonic form. This is the metaphysical ideal component of the analysis generally missing from any materialist, modern scientific analysis, which by the very nature of the scientific method attempts to close off and exclude, from a scientific model, any and all axiomatic components which might be considered to be generated by autonomous sentient entities.

This is, of course, a logical impossibility, since it is the consciousness of human beings that is responsible for the creation of any and all analytical modeling or of any dialectical, material development as found in the historical record. All such modeling is at best an approximation of what is deemed to be going on in other people's heads based on the modelers' understanding of their observations of human experience. As a function of political economic class affiliation, whether a soul is thought of in a Platonic sense as a conscious being that exists in some ethereal realm prior to birth, with memory of prior experience that is removed or obscured from recollection with that birth until return to that ethereal realm with death, or whether a soul is thought of in a Western sense, based on faith or agnostic science as a function of either a divine or of an unconscious evolutionary process, as a being that is imbued by an unknown source at some point at or after conception with an emergent sense of a distinct separate self that is structured by life experience over time, individual human life unfolds in a social context within a natural environment that is structured to sustain that life when augmented by the ongoing exercise of individual and collective human effort.

The interaction of that effort with nature requires conscious thought and an initiating conscious will. That thought involves the process of a dialectic, a type of conversation between an individual and their social group and nature, like a child learning to walk. Crawling, or perhaps simply sitting in the case of the truly gifted, is the initial condition of a dialectic, which we refer to as a 'thesis'. Children have identified with and seen other humans walking, so they have a concept of a final condition of walking, which we refer to as a 'synthesis'. To get to walking from crawling, a child has to go through an intermediate process of learning to walk which we refer to as an 'antithesis', which is in some sense an opposite to the thesis. The thesis, sitting or crawling on the floor, is safe and comfortable and, as long as the child is fed and diapered, free of risk. The antithesis is perceived to be adventurous or precarious.

I don't have a personal recollection of the point in my childhood development at which I understood the difference between a thought as a mental picture of myself causing something to happen and the self-initiated happening itself unmediated by such a picture, but obviously it did and does occur to all healthily functioning human beings. Here I am referring to an instance of picturing doing something first, such as diving off a high-dive board before ever performing the first dive, since it should be obvious that with practice one quits picturing or thinking about the activity beforehand and just exercises the will to dive.

At this point in the learning process a child—whether represented by a clear mental picture or not—has a notion as a dialectical ideal, as an internal dialog of interacting steps involved in learning to walk, but it is just a dialectical ideal. It is only through the process of practicing the steps that the process becomes synthesized as tangibly real, where we can call it a material dialectical process.

The antithesis, which involves learning to stand and take the first steps, includes the very self-conscious element of risk, the concern of falling back down on the floor. That fall is usually not hurtful or harmful, but the apprehensive sitter/crawler doesn't know that to be true during the thesis stage and so may avoid learning to stand—until the desire to grab something up high and out of reach overwhelms the fear of falling back down, and the crawler holds on to something, perhaps an adult's leg, and stands up… before realizing his material antithetical upright position and quickly sitting back down. The process is then repeated until the child learns to maintain the upright position and learns to take a few steps while holding on to the parents leg or hand, falls a few more times, and tries again until the antithetical process of learning to walk is completed and absorbed in the ability of the child to walk in the final material synthesis, reified as the observed process of a child walking.

Not all processes are as simple and easy to describe or understand, or perhaps as universal as the dialectic of learning to walk. In a process of any degree of complexity, however, it helps to start by breaking down any observed real, aka material, process to as fine a degree as needed for a mental, aka ideal, understanding of the process to emerge. The dialectic is an ideal process, a process of mental change, in the Platonic sense of an ideal form, which is general and not specific to any particular instance of the form in space and time; the ideal dialectic can be specified as that of any child that ever was or will be learning to walk, not just you or me as children as individual instances of material dialectics. So that mental change happens by necessity within a material environment by way of observing and by various means of acting in and on that environment, and such change grants a Specificity to the General Ideal Dialectic while also giving it a historical representation as a material dialectical synthesis.

The Dialectic may be completely internal as a mental dialog in which the only material aspect (apart from the obvious neurological activity of the brain) is a pictorial representation of features of the physical environment. If someone is thinking about whether a mathematical problem has a solution, first they form an *ideal thesis* which frames the problem as thoroughly as necessary, then they think through a series of steps that prove not to be solutions, as an *ideal antithesis,* until a solution is found or proved to be insoluble as an *ideal synthesis* of the process. (Technically, any ideal thesis at least initially tends to have a specific material representation of an ideal antithesis in order to define a synthesis, since an antithesis is necessarily a qualification of a more generalized thesis.)

In a non-interactive analysis of a physical process, one starts with an observed condition as a *material thesis*, studies it with an *ideal antithesis* as a model of what might be going on, to arrive at an *ideal synthesis* which is a better understanding of the observed condition. In an interactive analysis of the same process, one starts as before with an observed condition of the *material thesis*, investigates it with the same *ideal antithesis* as a model of how it might be improved, and materially interacts to produce a *material synthesis* incorporating a better understanding of the original and the revised condition.

In an experimental test of a model, one starts with a model of what is anticipated as an *ideal thesis*, examines it under controlled physical conditions with a *material antithesis* designed to test its failure parameters, to verify or revise the model's validity as an *ideal synthesis* of understanding. In a constructive process, one starts with an engineered design or construction model of what is going to be built as an *ideal thesis*, constrains it with a *material antithesis* in an assembly of disparate components to controlled in situ conditions, to produce over time a *material synthesis* as an engineered structure.

Finally, we might examine two agrarian communities in ancient Mesopotamia, one on the banks of the Tigris and one on the Euphrates in which the theocratic elites are both determined to build really nice ziggurats as part of the *material thesis* of each of their cultures. The objective conditions for each community are the same, with the same general number of farmers, masons, and other laborers along with baskets of grain in their granaries, but one of the theocracies worships the sun god and the other of them enjoys the fellowship of the moon goddess.

The sun workers have been told to labor every day but Sunday, and the moon community, which for whatever reason really likes to stay busy, has learned to take Monday off. This is fine until one day the sun priest hears that the moon commune is working on Sunday in their community, (without letting the sun workers know that the moon community takes Monday off), so he tells the sun laborers they need to start toiling on Sunday as well, thereby setting up a *material antithesis* for the sun theocracy.

Meanwhile, the moon priestess hears about the sun community work increase and discusses it with the moon community, so they can decide whether they need to keep up with the sun folks. They reach a consensus to meet the challenge without adding another day by adjusting their work schedule according to the phases of the moon. The moon commune thereby creates an analogous *material antithesis* in the ziggurat building schedule, but the twist in the schedule makes the workload feel different. In fact, the *material antithesis* seems to have an *ideal antithetical component* to it that makes the moon commune happier.

When the sun community finds out how happy the moon commune still appears to be, the sun workers revolt and take both Sunday and Monday off to establish a five-day workweek and select new sun priests as a new *material synthesis* for the sun community. The moon commune continues along happily, having converted their *material+ideal antithesis* into a new *ideal-material synthesis*, (where a different *ideal-material synthesis* is technically the case for the sun community as well).

The moral of the story is, of course, even dialectical materialism can have an ideal happy ending, if all the members of the community are entitled to a voice in the control of the means of production and a say in construction of the ziggurat, <u>and</u> if a woman is in charge. A *dialectic is essentially ideal,* necessarily with material representation and observation, understanding, and identifiable implications, and is not a mindless, ineluctable process of class struggle, except perhaps for the class theocratic theoretician. As defined by Plato rather than Marx, the synthesis of any political economic process in the mind of the souls of a community is an expression of the individual's and the community's Will-to-Good, focused through a qualitative filter of descending interest from the sustained exercise of wise Justice, through Honor, Prosperity, and Freedom, to the irrational caprice of raw political Power.

According to Plato, if the souls of the community value Justice, they will focus their efforts to establish and find justice in the rules and laws of that community by educating themselves to understand the nature and purpose of reason, the passions, and the physical appetites, i.e. to understand themselves, and therefore to select the best experts from among their fellow citizens to maintain Justice as guardians of the Republic.

If the Republic is threatened by existential crisis, internal or external, the citizens come to value Honor as the primary Good and focus their efforts on administration of military protection and policing of the community and the republic becomes a Timocracy, ruled by generals in war and public safety; albeit with a loss of some commitment by the authorities to the Principles of Justice in the process.

Over time as the crisis abates, and the timocracy fosters commercial activity internally and conquest or trade with other communities externally, the citizens come to value the Good of Prosperity and focus on pursuing commercial opportunity over the risk of crisis, and the republic becomes an Oligarchy, ruled by Family interests in the judicious operations of economic production and commerce; albeit perhaps with a loss of some commitment or educated understanding by the families to the Principles of Justice and Honor in the process.

Over time as sound commerce and/or kleptocracy concentrates wealth in the families of oligarchs to the exclusion of working individual citizens who have been subjected to loss, indebtedness, and servitude due—in some cases—to circumstances beyond their control, those citizens come to value Individual Freedom to focus on transactional control of those circumstances over the vicissitudes of prosperity authored by others, and the republic becomes a Democracy, historically through a process of liberalizations to include women and other disenfranchised and enslaved individuals; albeit with a risk of further deterioration in the commitment and educated understanding of some individuals to the Principles of Justice, of Honor and of the basis of Prosperity in the process. In this democratic state therefore, the Individual soul tends to be governed by personal transactional reason, pursuit of personal passions, and satisfaction of personal physical appetites in both the private and public realm, even to the detriment of concerns for family, civil authority, or the learned opinion of experts.

Over time, if these Principles are not maintained by the republic, including an understanding of the nature of and part played by the tripartite souls in the ongoing maintenance of these Principles, these Principles can be replaced by existential fear in the event or prospect of further internal and external crises—real, imagined, or manufactured.

Then the final Good in the public mind required for the preservation of Plato's Republic is understood to be Power, which becomes the primary focus of the soul's reason as the citizens look for any one individual or group, human or divine, that can lead the republic back to the Just State of the Philosopher King. If there are enough of those who have retained a true understanding of these Principles in a Democracy, such souls will be able to align their interests with the intent of the republic and reassert the primary Good of Justice. If the weighted interest of those uneducated and unenlightened souls is focused on the Good of Power in hopes of warding off a calamity to themselves or to the republic, tyros as amateurs and juveniles in statecraft run the risk of failing by choosing a charismatic, but still tyronic leader, from among the group of similar tyronic individuals due to a combination of their own tyronic personal reasoning, tyronic pursuit of personal passions, or attempt to satisfy their own tyronic personal appetites; in other words, due to their own youth or inexperience. The result as expected is Tyranny of the one or few who may be educated enough to gain power—in truth no one ever gains Power, which is the rule of God alone, though it can be administered if Just by the enlightened guardians—but not enough to understand or administer Justice.

In ideal theory or in material practice, the liberating effect of human consciousness is not guaranteed by an adherence to Platonic formal ideas in the implementation of a material program or by an insistence on interpretation according to Marxist dogma when faced with material conditions running consistently counter to theoretical expectations. Identification with class can heavily influence how one develops and mobilizes those class interests for prosperity, freedom, and political power, in the manner of a Marx, but individual perspective transcending class identity can do even more to transform or synthesize those concerns across class interests in the manner of a Plato by the experience of pursuing, doing, and achieving what is Good. Thus, liberation is achieved by doing what is necessary and just and thereby Necessarily Just.

In Plato's Republic, the souls of the city are driven by their tripartite nature, that of reason, of motivating passions, and of physical appetite, to seek what is good for the city. That Good is found in doing what is just and what is Just is for each citizen to do what they do best. Knowing what is Just requires expertise in understanding and wisdom that can only come from years of experience and study on the part of the guardians of the city, both male and female, whose positions are de jure non-hereditary and non-propertied, and whose Soul Satisfaction is the Love of Reason and wisdom in the exercise of expertise and a marked restraint of the passions and the physical appetites. It does, however, take a minimum level of satisfaction of the emotional and physical appetites for all citizens and welcome guests for them to survive and begin to thrive.

The following Table 7 gives an overview of one possible hybrid interlinking of Platonic and Marxian forms as an aid in the analysis of this discussion of omnipresent freedom, ergodic economic modeling, and the pursuit of social justice through public policy. A contrasting Marxist view might start by positing individuals with a similar tripartite nature of human intellect, emotions, and physical survival needs satisfied by the common good of a classless agrarian community, in which the good of justice is found by the concept of 'from each according to his ability, to each according to his need', where the

Social Good Nominal Goal (Just/Unjust)	Rules as Nominal (Generalized)	Rule by	Soul Satisfaction Focus of Rationality, Passion, & Appetites	State Dialectic
Justice (Innate Wisdom)	**Philosopher King &** **Aristocracy** (Theocracy)	**Experts:** Native & Formal education Non-propertied (de jure) Non-hereditary (de jure)	Wisdom & Love of Reason	Plato–Ideal Marx–Ideal not Materialist as Synthesis of class struggle
Ideal Thetical as Agrarian Communalism, Philocracy or Theocracy; **Ideal-Material Synthetical** to all four devolved hierarchies as a Social Democracy, Techno-Meritocratic Anarchy, or Urban–State Communism.			
Honor (Administered Justice)	**Timocracy** (Technocracy)	**Military & police:** Experience & formal education w/Reliance on Experts Propertied (de facto) Hereditary (de facto)	Administrative Reason Motivational Passion	Plato–Ideal Marx–Material objective conditions
Material Antithetical to Agrarian Communalism under crisis and long term stress as Feudalism; **Material Antithetical-Synthetical** to Oligarchy, Democracy & Tyranny and hybrids & **Material Synthetical** to Feudal involving commercial growth, **Ideal-Material Antithetical** under stress as Fascism-Totalitarian Socialism.			
Prosperity (Commercial Justice)	**Oligarchy** (Meritocracy)	**Families:** All education regimes Propertied (de jure) Hereditary (de jure)	Commercial Reason Acquisitive Passion Physical Appetites	Plato–Ideal Marx–Material objective conditions
Material Antithetical to Feudalism through technological growth and meritocracy as Commercialism; **Material Synthetical** to Timocracy as maturing Commercialism, Resource Exploitation, & Nation building.			
Freedom (Injustice of externalities)	**Democracy** (Anarchy)	**Individuals:** All education regimes Propertied (de jure & de facto) Hereditary (de jure & de facto)	Transactional Reason Personal Passion Personal Appetites	Plato–Ideal Marx–Material objective conditions
Ideal-Material Antithetical to Timocracy and Oligarchy through accelerating technological growth and diffuse meritocracy as international Commercialism and resource Exploitation and liberalized Nation building; **Material Synthetical** through international Consumerism in international financial, real, and human capital concentration and disintermediation of national polities as global Capitalism; **Ideal-Material Auto-antithetical** and thereby internally contradictory conflict between international Market-valued Human Capital (MHC) and national Non-market-valued Human Capital (NHC) as public safety net amelioration or cynically orchestrated Populism.			
Power (Injustice)	**Tyranny** (Autocracy) (Kleptocracy)	**Tyrant Dictator:** Native education/non-expert Propertied (kleptocratic) Non-hereditary (kleptocratic)	Tyronic Reason - Existential, may be (Unfocused & Irrational) Tyronic Passion Tyronic Appetites	Plato–Ideal Marx–Ideal not Material as identified with intra-class degenerate conditions
Ideal-Material Antithetical to all other State forms, in response to external and internal stress and crises including the tyronic or degenerate condition of other States', as a Dictatorship, which may or may not be Kleptocratic, and which differs from what might be considered the ideal State of the Philosopher King by being ruled by the passions and appetites of a tyro lacking wisdom, which in the Platonic form means one unschooled or uninitiated in the nature of reasoned justice; **Material Not Synthetical** as a degenerate final State of social good or justice for citizen soul & autocrat.			

Table 7 – Hybrid of Platonic City State and Marxist Nation State Forms

wisdom of the community is well understood by the common experience of the individual members. The beginnings of any hierarchical structure would be found in the elevation of individuals with unique talents in a merit-based theocracy to the formal status of a hereditary class. As a complementary utopian end-vision of a historical dialectical materialist process, this might be much in harmony with Plato's ideal Republic.

In Plato, the well-being of the city declines in the degree that the citizen souls, who maintain what is Good and Just by learning to work at what they do best, become displaced and replaced in that work by any number of crises that shift the reliance on the expertise of the guardians in the pursuit of what is morally just to the administrative justice of the military–police in the pursuit of the principle measure of Good of the city, which is Honor, often cast as patriotism. Expertise is gained by military experience and formal training and reliance on the advice of schooled experts from the ruling elites, who presumably assume the de facto proprietary and hereditary entitlements of their status. The Soul Satisfactions of these timocratic elites would be a type of administrative reason motivated by sense of Duty, but with a marked restraint of the unruly passions and the physical appetites.

Contrasting with Marx, in response to external and internal crises as a material antithesis affecting the ongoing production and satisfaction of survival needs, a similar creation and elevation in the status of a military–policing capacity with a hereditary component at the top as a synthesis, would indicate the emergence of a feudal ruling class structure. The emergence and presence of class structure is central to Marxist thinking, with a de-emphasis and sometimes dismissal of the importance of any individual members conscious participation in the rising nation state, but it still understands what is recognized as Good within any ruling elites and the ruled, which in the case of timocracy from original feudalism to vestigial gangs, is Honor. With regards to Honor, in a timocracy one can still generally expect a sense of noblesse oblige toward the peasantry and other under classes, though that begins to disappear, for example, with the first of the Enclosure Acts that provided labor for the looms of England and the rise of an oligarchy.

In Plato, with the generational passing of successful asset accumulation arising from commerce or pillage of warfare, significantly to the inclusion of slave labor, in a timocracy the Good of the community becomes that of Prosperity, particularly of the ruling elites, whose sense of Justice is degraded to a type of commercial justice in which the measure of success is valued in monetary terms. The ruling state of an oligarchy is that of Family meritocracy, where the ruling elites are the commercially successful families with members of all types of educational experience and increasingly de jure property and hereditary positions. The Soul Satisfactions of the oligarchic elites would be those of commercial reason, acquisitive passions, and a restrained indulgence of epicurean physical appetites.

With Marx, the Platonic form applies almost word for word, to which we would emphasize again the presence of class structure in determining the Good of the state as Prosperity measured in wealth, with the assumption that the success of the bourgeois elites falls naturally—in the minds of the bourgeoisie at least—to the benefit of the working classes by unconscious largesse of those elites, by virtue of their concept of commercial justice, in which all souls rise to their just position in life. This gives rise to a discussion of Freedom as a Good and de jure Democracy as an Ideal, among members of the ruling and the subject, working classes.

In Plato, the excesses of oligarchic self-indulgence and self-interest become socialized across the various economic strata as personal transactional reason, pursuit of personal passions, and the satisfaction of personal appetites, so that the citizen soul that is not yet enfranchised glimpses first the possibility and then the necessity of the Good of personal, Individual Freedom and demands the rule of law in the establishment of a Universal Democracy to that end. The Individual, who can come from any educational experience or background, with relationships of property and heredity through a combination of de jure and de facto entitlements, becomes the *end* Good of the republic itself from this perspective, to the point that in the minds of some, the notion of Omnipresent Freedom, if practiced, would lead to Anarchy—liberating to some, enslaving to others. In the absence of the unifying justice of the guardians, what starts out as an intramural scrum for the republic risks descent into a winner-take-all free-for-all.

With Marx, the problem with this notion of freedom is that the individual is bound to the earth, to the material needs for oxygen and agricultural sustenance, to the warmth of hearth and home, to the need for reliance on social organization in the ongoing provision of those needs; where the just freedom for some found in independence from or within the oligarchic republic, unjustly enforces through the omnipresence of the marketplace the external public costs of contracts between private individuals in a laissez-faire Democracy. Given the familial hereditary property rights of an Oligarchy, a transition to a Democracy that relies heavily on the wage market for production and distribution of goods and services while retaining those property rights only intensifies the skew of existing property distribution.

The rationality of the market place Freedom intensifies the existing bifurcation of that market place into a top tier of families and individuals that control the pricing of financial, real, and market-valued human capital and the hereditary ownership of financial and real property, with an ancillary second tier that caters and aspires to and, in some cases, achieves the top, and the bottom tiers which are increasingly dispossessed and reduced to commodity wages, if they can get them, priced for the purchase of commodity consumption goods by the commodity labor market. Those hapless, increasingly homeless individuals, freed from the savage security of wage or chattel servitude, absent a response to a call for help by the elites, are left with only one hope—revolution.

That revolution can come in the form of Tyranny as an unjust despotic regime, using violence or a threat of violence to subjugate a community of tyros or it can come as a revival of the Guardians of the Republic through a return to Justice for all, or perhaps in the manner of the individual messiahs, human or spiritual, or a team of experts as a political movement or party, including those with technological knowledge and the wisdom to implement that knowledge for the amelioration of intra and extra community conflict.

The term 'tyrannical' would appear to come from the same Greek root word as 'tyro', though the readily available etymological dictionaries ascribe the second term as originating in the Latin 'tiro' for a 'young soldier, recruit'.[4] With respect to Plato in his Republic, this indicates the rule over a community of souls inexperienced in an understanding of Justice, who are swayed by a leader of similar inexperience, of the same understanding. From Wikipedia, 'Tyrant' we have the following quotation, "John Locke as part of his argument against the "Divine Right of Kings" in his book *Two Treatises of Government* defines it this way: "Tyranny is the exercise of power beyond right, which nobody can have a right to; and this is making use of the power any one has in his hands, not for the good of those who are under it, but for his own private, separate advantage.""; all of which sounds currently familiar.

A young soul gains *knowledge* of what is meant by a 'good' or a good 'thing', or for that matter a 'bad' or a bad 'thing', by learning about that 'thing' within the identified context of some material experience in time and space, often in common with others identified as a group or a class, but it is only when an individual soul's experience results in that soul's *understanding* that such a 'good' represents a qualitative aspect of reality that is beyond its or any historical context, that such material thing is recognized as representing an *ideal* 'Good'; in which the material experience of receiving justice is perceived as Just, in the highest Platonic social sense. This realization of an ideal Good is best and perhaps only understood by those souls who experience that Good as perception of an Ideal, and not simply a contextual experience of some material instance of fortune, good or bad.

As the highest Good of the ideal Republic is Justice, those souls who have experienced an emotionally profound justice or injustice in their life will have less trouble understanding Justice, even in its lack, as an ideal, a real Good that for the republic has significant material benefits that must be protected and preserved. A soul with such understanding shows Justice to others, even when threatened with intimidation and retribution by the unjust.

Those soldiers and workers who have observed or experienced a reverential honor operating in a chain of command of public or private authority will have little trouble understanding Honor as an operational ideal that as a military or political leader has material significance to the preservation of the republic. A soul with such understanding shows Honor to others, even if wronged and surrounded by others calling for a dereliction of Duty.

Those souls who have had the opportunity for invigorating and rewarding work and trade with family members and friends in a common enterprise that brings a shared prosperity with an ability to minimize the downside risk will easily understand the benefit of

---

[4] definition from https://www.merriam-webster.com/dictionary/tyro, "a beginner in learning; novice" as nuanced in the synonym as "implies inexperience often combined with audacity with resulting crudeness or blundering"

Prosperity as an ideal of wellbeing, even wealth, whether it be from social position, the pursuit of industry, or the grace of luck. A perceived potential for upside risk, a perception of lack of opportunity for growth, a knowledge of exposure or subjection to exploitation or fraud, all without alternative opportunity or from being in a disadvantaged position, however, is not so conducive to an understanding of the ideal of Prosperity as a result of Industry.

Given that the family wealth of those in the community is experienced proportionally to the wealth of the entire Republic, as Opportunity increases in proportion to the accumulation and concentration of a family's capital as the ideal of Family Prosperity, the supplementary Risk of disadvantage, loss, stagnation, lack, poverty, all forms of material restriction increases in accelerating fashion to the detriment of the overall community ideal. A soul with understanding of the vagaries of success that leads to Prosperity has a choice—to deal ethically with others in the community to uphold the ideals of Justice and Honor to work toward the common good or to deal with personal self-interest to maximize material position and status. Working toward the common Prosperity does not preclude a justly compensated meritocracy, but it does indicate a wise policy mitigation of the bifurcating market effects of Monetary-Omnipotence applied with Rational-Omniscience to achieve Freedom-Omnipresence in the marketplace, that is a mitigation of impersonalized, rationalized Greed.

The individual souls that lack the security of material prosperity and chafe under the restrictions that such lack imposes, learn from that experience, perhaps more readily than any other, the connection between the material, visceral good of its anticipation in the desire for freedom from want or harsh working, living, or political conditions, and the ideal Good of Freedom, as a realized state of the soul or, if yet unrealized, as motivation for that ultimate realization, and as a steadfast commitment to maintaining the realization at any cost. Those souls that are new to freedom or are in jeopardy of loosing it are most passionate about its protection. Those who are accommodated to its lack from long suffering may be less so until the possibility through circumstance becomes acute. The Good of Freedom perceived as an ideal to be realized is not the same as an understanding of the nature of the material conditions under which that ideal realization is possible.

The intention to understand, to create and maintain an awareness of, and to work to implement any community Good—Justice, Honor, Prosperity, Freedom, Power, along with the many other beneficial ideals—indicates a level of participation in the life of the Republic beyond a simple observance of community Goodwill as an ideal. We can refer to this intentional pursuit as the operation of a Will-to-Good on the part of the souls and their community. These acts of Will—of Will-to-Justice, of Will-to-Honor, of Will-to-Prosperity, of Will-to-Freedom, and of Will-to-Power—are more than a physical or emotional passion; they provide the raison d'etre and passion for Life of the community, the Republic.

The Will-to-Prosperity *in the community* demands an adherence to Justice *in* the interaction of all citizens and to Honor *in* the administration of duty of the same, if prosperity *of the community* is to be achieved and maintained. Prosperity of the community rests on a proper understanding of human value—including in an economic sense of human capital as a public Good, particularly with respect to the understanding of the nature of Non-market-valued (NHC) versus Market-valued Human Capital (MHC). Since human capital as a private Good, apart from one's own person, is chattel, indentured, contract, and in some cases wage servitude, if not under the ideals of Plato, without doubt it is inadmissible under the ideal of Freedom, in any modern-day republic.

Human capital must show up in the national accounts of a Republic if it is to have any quantifiably understandable effect on a nations policy, and the only logical place is as a public Good, in which each and every citizen has a per capita valuation of NHC for uncompensated production from birth through death valued in national summation on a current market equivalence and a similar per capita valuation of MHC for production compensated and summed as a national accounting of all market transactions. In terms of the developed theoretical modeling of unweighted ergodicity economics, since the NHC portion lacks, as it ideally should, an individual citizen accounting of the NHC valuation, it should remain and be treated as a per capita average of the total. The MHC valuation can then be seen to conform to the weighted ergodicity modeling distributions of the previous section, in accordance with the market transacted data for national income and net worth in the earlier Quantitative Analysis part of this essay.

The Will-to-Prosperity of the community on this basis, implies a Will-to-Freedom and Will-to-Power of the individual that would prevent the Will-to-Power of a cabal or tyrant, and thereby preclude a tyrannical Will-to-Honor of "rightist" military state Fascism or "leftist" police state Socialism, either of which excludes the personal Will-to-Freedom and Will-to-Power of a functioning democracy. Note that the economy of such a viable democracy would incorporate no program for the expropriation or control of any private means of production, unsubsidized by public funding, that does not infringe on the freedom and well-being of the community by the imposition of external social costs and denial of basic entitlements.

As a generality, the principle publicly owned productive means required by such a democracy incorporating the ideals of a Republic, in this case the "United States, in order to form a more perfect Union, establish Justice, insure domestic Tranquility, provide for the common defence, promote the general Welfare, and secure the Blessings of Liberty" are those which are natural monopolies or are part of the common natural resource wealth and those required for the maintenance of these ideals. They are also those which are not being or cannot be sustained without the long term licensed incentive of private profit, including transparently administered 1) personal and commercial transportation, communication, and transactional technology necessary in the production and distribution of goods and services for the ongoing formation of a more perfect Union, 2) provision of civil, criminal, and mental health courts and ancillary facilities and operations required to establish and maintain Justice throughout the Union, 3) provision of civil, criminal, and mental health facilities and their operations required to insure domestic Tranquility throughout the Union, 4) provision of the standing army, navy, air, marines, special force, intelligence, and cybersecurity services required for defense of the Union, 5) provision of education, health–wellbeing, and life care in and of the communities for the promotion of the general Welfare of the Union, and 6) provision of mechanisms, facilities, and technology for secure election and legal proceedings as required to secure the Blessings of Liberty for all citizens and welcome guests of the Union.

Included in this would be the transparent, publicly owned monetary system; that public monetary system does not own or claim title to any pools of debt token currency in any private hands, which necessarily is free to circulate through that system, but it does claim the sole entitlement to the issuance of credit token currency in the form of a universal non-merit-based income and for public funding on a national, state, and local level as required to secure the ideals of the Republic, in this case of the United States. It would allow the funding of common and public aspects of that social safety net and other entrepreneurial initiatives in addition to private initiatives, provided full transparency is provided for any co-mingling of funding.

Prosperity of the community, as stated here, means neither an abrogation of hereditary property rights nor a relinquishment of eminent domain entitlement and envisions a nullification of the national income tax and an end to national borrowing as currently practiced. For a comparison of the effects of taxation, borrowing, or the issuance of fiat currency on private and public sector economic activity, please view the link to download the UniServEnt app at https://uniservent.org/basic-income-study-slide-0/.

There is one further Good that is Essential to this discussion, perhaps beyond the discussion of Plato or Marx, but not beyond that of theology, and that is Truth and the Will-to-Truth. The Will-to-Truth, like all terms used here as the expression of an intention to achieve a desired Good, is an active commitment to gain knowledge in the material conditions of the truth of a matter within the context of the affairs of the community, and thereby gain understanding of the operation of the ideal of Truth in that matter. As with all Goods, what is achieved for the Republic is done so only through the realization of such Truth in the minds and hearts of the souls as guardians of the Republic in sufficient numbers. The implementation of any Good first requires an understanding and application of the ideal of Truth by the citizens as it concerns the material or existential truth of some matter in its effect on that Good in the life of the Republic.

The distinction between the material and the ideal components of a dialectic as discussed can be reflected in a distinction between an existential and an essential truth, between existential and essential reality. Material or existential truth is a matter of experience borne knowledge, of facts on the ground, so to speak. Ideal or Essential Truth is a matter of understanding borne of and confirmed by that experiential knowledge and brings Wisdom and Power in and from the heavens, also so to speak.

The term 'Will-to-Truth' will immediately bring to the minds of some of those familiar with the work of Friedrich Nietzsche, his concept of the Will-to-Power defined as "(in the philosophy of Nietzsche) the self-assertive creative drive in all individuals, regarded as the supreme quality of the superman."[5]—except that it is not really Supreme. This Will-to-Power of the superman might be as understood by the philosopher king, if constrained

---

[5] Dictionary.com, LLC, Version 8.9.2.1(11.2.0)

in its expression in the individual citizen soul to the ideals of Good, as exemplary in its implementation as Justice in the life of the Republic; in upholding Honor among the citizenry; in capitalizing on that Power in maintaining the Prosperity of the public; and in sustaining the Will-to-Freedom for all. Said Supremacy is in the Will-to-Truth.

When that Will-to-Power is applied by a tyro, however, it produces no Ubermensch, no Superman; regardless of high-minded intent and spirited rhetoric, the lack of true understanding of the novice—in both leader and follower alike—preys inevitably and unjustly on the best and the weakest of the citizenry, to produce a Republic of Banality, led by a banal tyrant in a banal tyrannical regime. The Tyranny of Das Dritte Reich was founded on the stiff artistic sketches and celluloid motions and notions of a cabal of self-mesmerizing, self-conscious cupidity, revolting to the light naughtiness of bourgeois cabaret with vicious sturm und drang.

All such adolescent tyro tyranny starts with lying to oneself—the ultimate expression of Omnipresent Freedom, of creating an alternative version of the world centered about oneself—needily underestimating one's capacity in essence while overcompensating for the perceived lack in exhibition. It then proceeds with an attempt to convince or convict the world at large of the lie. It responds—it thinks out of Freedom—but rather out of slavery to the appearance of adversity in the world, perceived as the work of an adversary—the Adversary—in reality assuming that position for itself, assuming the Will-to-Power over any inconvenient fact, whether real materially–conditionally or real as a manifestation of transcendent, ideal Truth. The tyro's Will-to-Power can have an effect on material reality, but it has none on essential reality, on Truth which it cannot assail, and therefore stands as antithesis to the Will-to-Truth and to the Power already realized by the philosopher king and guardians of the Republic.

Lying—we are not talking about 'white lies' here, those which are devoid of personal invective—bearing false witness against one's neighbor as the Bible states, as in a court of law or in a public arena, is not good and is not a Good; it is not just or honorable, does not enhance the prosperity and freedom of the community and is only done to evade the power of the court or public opinion or to enhance the personal power and illusive freedom of the one who is supposed to be a witness to the Truth. In the words attributed to Jesus Christ, in the Gospel of Mark, 3:29, the one unforgiveable sin is that of blaspheming the Holy Spirit, which is the Spirit of Truth. The Spirit of Truth—the Essence of Truth, the Essential Truth—is more than a material, factual truth and more than an ideal Good; from the perspective of Christian theology through the trinitarian topological lens of a logician, the Spirit of Truth—of Reality—is omnipresent in cosmic space and time as the essential *extended field of Living*—the Matirx, the womb—the unrecognized female aspect of the Divine Trinity, in which all phenomenal material experience occurs, where the Father is the *source of all Living* as the essential cause of all experience of that Life and the androgynous Son-Daughter of Man—and of God—is the reified *sink of all Living* toward which the entirety of that experience is drawn in submission for ultimate redemptive understanding in the fullness of time.

To blaspheme the Truth is to express an unwillingness or inability to submit to such understanding of the reality of Truth. Citizens bear false witness against their neighbor with harmful intent, but more often in error due to a lack of understanding of the Truth that operates behind the material context of adversity and motivates a denial of that Truth. The problem for the tyro, leader and follower alike, is confronting the driving force behind adversity and the lie it is propagated to address, which is fear. Adversity met with courage and expertise is an opportunity. That same adversity met with fear and a lack of knowledge of how to proceed is an existential risk that carries beyond the grave; and that requires a willingness of the tyro to reason the situation for themself, to discern the truthfulness of any circumstance and any leader in order to determine the correct way out of the labyrinth of that risk. Motivation to be truthful is difficult enough in the absence of urgency but becomes nearly impossible when fearful concern becomes paranoia. "Where there is no vision, the people perish: but he that keepeth the law, happy is he."[6]

Implementation of the Will-to-Truth on the other hand—rather than of the Will-to-Power vicariously as a tyro or inherently as a tyrant—motivates a realization on the part of individual souls toward the reification as the Philosopher King and Guardians, which in turn works to the Good of the ideal Republic. To quote again, this time from Moses in Deuteronomy 8:3, "that he might make thee know that man doth not live by bread only, but by every word that proceedeth out of the mouth of the LORD[7] doth man live," which is to say by the Spirit of Truth, the Essential Truth, the Annunciated Breath of Truth, the Holy Spirit who only speaks Truth and can only be understood as Truth by those who muster the Will to understand the Truth, who submit to the Will-to-Truth; that man is more than a material, biological being, and is in truth a Soul—a Child of Man and God.

The Platonic Reality of the Soul is idealized in the notion of the Omnipresent Freedom of the individual as a maximizer of economic activity, of the Good of Prosperity, but its glamorization is contradicted by the material reality of the commoditization of labor and subsistence living resulting from the rationalized externalities of the marketplace in a modern neoclassical/laissez faire economy. Such contradiction is nothing new to the body of work of Karl Marx in his empirically driven attempts to extract the ideal of Justice from the capitalist drive for Prosperity—and too often greed—and whom we might assume was familiar with Plato's thinking in the Republic. Marx is often maligned as an atheist, based on the statement, much-quoted out of context, that religion "is the opium of the people", which is used as prima facie evidence that nothing else coming from that source has value. No one but Marx can know if the notion of his atheism was true. He appears to have thought of himself as an empiricist but listed to the rationalist side. While a certain degree of agnosticism appears to have been present in his writing, what is equally apparent is that the heft of his materialism could not expunge its idealist core.

Dialectical materialism, the state ideology of the Soviet Union, dictated atheism as a direct conclusion of this historical narrative concerning the Christian church, coming

---

[6] Proverbs 29:18, KJV

[7] Generally given as 'Lord' in the KJV, the Hebrew uses the unspoken term or Tetragrammaton as 'That which Is, Was, and Will Be', the Self Existent One, Essence, Supreme Being, as the term for God.

through Marx, Feuerbach, Hegel and other philosophers before him, though there are other branches of Marxist thought that are neither atheistic nor materialist. Here, materialism is defined as a material organization in nature of sufficient complexity as required to give rise to conscious awareness and the development of thought, and only thereby and thereafter to any ideal conception of Platonic Forms such as Justice. However, such materialism does not preclude the reality of such ideal Forms as features of a metaphysical ontology, either monistic or dualistic. It simply states the assumption that any ideational capacity is not accessible until a threshold level of complexity is reached, which is obviously to be found in human beings, as by analogy, is the physical reality that light of the sun cannot be registered with the human eye until that eye has been created by Nature and/or Nature's God. With respect to a teleology, any materialist philosophy is agnostic at worst and will always be incapable of making a definitive affirmative statement on atheism—in short, the only way to know and show that God, as Supreme Conscious Being, does not exist is to become that God—which is not possible, being an axiomatic self-contradiction, especially if the Divine Spark was there all along.

A more extensive quote of Marx in fuller context, as found in the introduction to an 1843 work, concerning the then current state of philosophy in Germany, not published until after his death in <u>A Contribution to the Critique of Hegel's Philosophy of Right</u>, is worth stating here, as it makes a clear statement antithetical to the common use of the above quotation. It is in fact a statement of an essential, spiritual aspiration of humanity to transcend the state and the society 'which is an inverted consciousness of the world', through 'the struggle against that (inverted) world whose spiritual aroma is religion,' as 'the criticism of that vale of tears of which religion is the halo.' The emphasis has been added by this writer in the second paragraph and by Wikipedia in the fourth. It bears emphasizing that 'religion' as human aspiration and 'religion' as the operation of the state are fundamentally distinct and different, and that the 'opium of the people', as stated elsewhere, should be understood in terms of the time of his writing as anodynic rather than a form of anesthetic.

"For Germany, the *criticism of religion* has been essentially completed, and the criticism of religion is the prerequisite of all criticism.

"The *profane* existence of error is compromised as soon as its *heavenly oratio pro aris et focis* ["speech for the altars and hearths," i.e., for God and country] has been refuted. Man, who has found only the *reflection* of himself in the fantastic reality of heaven, where he sought a superman, will no longer feel disposed to find the *mere appearance* of himself, the non-man [*Unmensch*], where he seeks and must seek his true reality.

"The foundation of irreligious criticism is: Man makes religion, religion does not make man. Religion is, indeed, the self-consciousness and self-esteem of man who has either not yet won through to himself or has already lost himself again. But man is no abstract being squatting outside the world. Man is the world of man – state, society. This state and this society produce religion, which is an inverted consciousness of the world, because they are an inverted world. *Religion is the general theory of this world, its encyclopaedic compendium, its logic in popular form, its spiritual point d'honneur, its enthusiasm, its moral sanction, its solemn complement, and its universal basis of consolation and justification. It is the fantastic realization of the human essence since the human essence has not acquired any true reality.* The struggle against religion is, therefore, indirectly the struggle against that world whose spiritual aroma is religion.

"Religious suffering is, at one and the same time, the expression of real suffering and a protest against real suffering. *Religion is the sigh of the oppressed creature, the heart of a heartless world, and the soul of soulless conditions. It is the opium of the people.*

"The abolition of religion as the illusory happiness of the people is the demand for their real happiness. To call on them to give up their illusions about their condition is to call on them to give up a condition that requires illusions. The criticism of religion is, therefore, in embryo, the criticism of that vale of tears of which religion is the halo." [8]

In other words, the critique of religion and its faith is due to doubt that arises from the wretchedness of the socially determined human condition, the fall of Man to this earth from the halo of the heavens, in which is to be found '*the fantastic realization of the human essence since the human essence has not acquired any true reality*' prior to his, prior to humankind's resurrection.

It follows that the reality of the tripartite soul as previously defined according to Plato is not contradicted by the mental, emotional, and physical, much less the essential, spiritual nature of human beings in the concepts of Marx, though perhaps the notion of metempsychosis would be surprising to find in his work; as it would be if found in the canon of Christianity after the Second Council of Constantinople in AD 553, making anathema of the concept of pre-existence of souls which had been stated by the early church father, Origen, some three hundred years before in Alexandria. There is nothing to suggest that there is no *spiritual motivation* to the cause of realizing the ideal Freedom in the work of Marx. What is at first blush unique and sets his approach to philosophy apart from Plato's guardianship of the philosopher king is his statement concerning the relationship between the philosopher and the proletariat from the end of the Introduction, concerning the social and political conditions in Germany in 1843, which he has been contrasting with the history of franchize improvements in France.

"The only liberation of Germany which is *practically* possible is liberation from the point of view of *that* theory which declares man to be the supreme being for man. Germany can emancipate itself from the Middle Ages only if it emancipates itself at the same time from the *partial* victories over the *Middle Ages*. In Germany, *no* form of bondage can be broken without breaking *all* forms of bondage. Germany, which is renowned for its *thoroughness*, cannot make a revolution unless it is a *thorough* one. The *emancipation of the German* is the *emancipation of man*. The *head* of this emancipation is *philosophy*, its *heart* the *proletariat*. Philosophy cannot realize itself without the transcendence of the proletariat, and the proletariat cannot transcend itself without the realization of philosophy.

---

[8] Marx, K. 1976. *Introduction to A Contribution to the Critique of Hegel's Philosophy of Right*. Collected Works, v. 3. New York.

"When all the inner conditions are met, the *day of the German resurrection* will be heralded by the *crowing of the cock of Gaul*."[9]

For all that might be said about the historical materialism of Karl Marx, which is much, he is a philosopher of idealism, of the Will-to-Truth, to the end. 'Philosophy cannot realize itself without the transcendence of the proletariat, and the proletariat cannot transcend itself without the realization of philosophy,' sounds straight out of Plato's Republic, updated to the industrial age, and when we recall that the primary purpose of the Guardians of the Republic is the ongoing education of the citizenry in the realization of philosophy as the Spirit of Truth to become Just by doing what each Soul does best, both sound true.

We might thus rephrase this statement as, 'The Love of Truth cannot realize itself without the resurrection of the community of the faithful, and that community of the faithful cannot resurrect itself without the Love of Truth.' This sounds like the early Christian church which from the day of Pentecost where filled with the Holy Spirit and "did testify and exhort, saying 'Save yourselves from this untoward generation'", "and all that believed were together, and had all things common; and sold their possessions and goods, and parted them to all men, *as every man had need*."[10]

It is no stretch to add the teachings of Christ to this correlation of the ideal dialectic of Plato and the historical materialism of Marx. As the events of history testify to the material struggles of the community of faithful in this 'inverted world', they testify to the struggles of the Christian Philosopher King in the material 'vale of tears' according to the early church guardians; as the ideal Goods of Truth and Justice were deemed by Plato to reside eternally with the Gods—the Elohiym, collectively Gods–Goddesses, of Genesis 1:1 and not the translated Monarchic deity of King James—to be realized from time to time through the education and authority of the Guardians, the ideal of the divine Son of Man is deemed by the community of the faithful to ever be revealed and to realize itself in this 'inverted world' through the anointing of its heart of the community—Marx might say 'its heart, the proletariat'—by the Spirit of Truth.

Was Marx therefore a prophet? Perhaps—pluribus, unum—materially human and ideally a guardian of humanity, like many others trying to discover and pass on the Truth; as of course was Jesus Christ and the Buddha, and all the many, many other Philosopher Kings; all of whom lead by example more than precept to forswear all untruth; and to submit all questionable matters to the Spirit of Truth, all ideals of the Good to Justice for all.

The illusive glamor of the Will-to-Power to attain ownership of omnipotent money, omniscient rationality, and omnipresent freedom in the marketplace works against the Will-to-Truth and the Good of Justice in the community. It corrodes the body politic and enslaves first the tyronic and ultimately the tyrannic Souls of society that identify with

---

[9] Ibidem
[10] Acts 2:40, 2:44-45, KJV

the substance of any lies and illusions perpetrated individually and as a class. It does not matter if these lies are made out of existential fear or covetous aggrandizement, if they are reciprocated. They hamper communication and cooperation and end up in destruction.

We have gone into some detail in this analysis of the work of Plato and of Marx and finally of the life of Jesus Christ—as reported by the early Christian church fathers who themselves were influenced in some regards by Plato and his student Aristotle—because of the significant effect of these three historical figures on the notion of Freedom in politics and the economy across the globe, an effect that continues to intensify to the present day. The ideals of the Constitution of the United States go back through Europe and the Enlightenment to Aristotle and Plato and Christianity. To that same heritage starting in the middle of the 19th century, the ideals of Marx and others cloaked as they were in various strands of materialism, were added in Eurasia and areas of most of the rest of world, but by and large excluded from the United States. It is not insignificant that the relative absence of Marxist thinking in the US was in no small degree complemented by the presence of independent protestant evangelical organization. The bulk of the remaining global ideological affiliations of Islam and Hinduism-Buddhism were wrapped under European colonial idealism for various periods of time over the past 200 years.

The fact that dialectical thinking has long been a part of my approach to logical dynamics is responsible for the past dozen or so pages of this analysis. Neither Plato nor Marx were in the sights of this section on Freedom as Omnipresence when it started, though no doubt just below the horizon, but it emerged quickly in light of the strong dialectics of both, naturally contrasted by the focus on the ideal forms of Plato and on the importance of material circumstance of Marx. To this we add the moral authority and moderating example of Jesus Christ. Whether the reader views Christ as the Son of God, one of the three persons of the Holy Trinity, as a liberated mendicant in the manner of a Buddha, as a historical healer and Jewish Rabbi, as a representational account of the Platonic ideal of the Enlightened Philosopher King, or perhaps as some combination of all the above, no single figure has been a greater motivating factor or had more adherents to the cause of peace—or been more misunderstood and misused in the cause of strife.

In the discussion we will think of this third figure in terms of the Philosopher King, since from a Platonic perspective, any material entity is recognized in terms of the ideal Form that the entity essentially represents. Coming on a vacant campsite, a four-legged stool and a 16" boulder can both be recognized ideally, with equal utility, as a Seat, one of which is also recognized teleologically as an embodiment of Human Ingenuity, commonly constructed and understood by others. The historical existence of Jesus is recognized ideally, by anyone sufficiently familiar with the historical record, as an Anointed Soul, as a Son of God or Son of Man depending on the reference, also recognized teleologically as an Embodiment of the Omnipotent Source of Life, understood by a common Humanity of Souls. Some people recognize the potential for that Soul's anointing in all of Humanity, as did Plato and others believing in metempsychosis. As do I.

A discussion of metempsychosis—of the existence, not to mention pre-existence, or further re-existence of the Soul—may constitute a strange and unexpected place for such apparition, here in an exposition of political economics. But then who could anticipate the current divisions in the US polity resulting from the variety of understanding of this subject? So, let's dig in a bit deeper, starting with an overview.

The official position of the Soviet Union at the time of its dissolution was atheist, as currently with the People's Republic of China. A similar, but unstated position might be made for much of the world's scientific community, depending on the discipline, where the principle ideology of physics and therefore astrophysics is the big bang and, at the other end of the spectrum, where the principle ideology of biology and therefore microbiology is Darwinian evolution of species.

With respect to astrophysics, no professionally published author claims to have an axiomatic understanding of the genesis of quantum phenomena or the relationship between said quanta and the cosmic venue within which it appears to the technologically enhanced senses. With respect to microbiology, to my knowledge no published scientific authority has an understanding of the necessary organizational principle or principal that accounts for the chicken-or-egg-first phenomenology on a molecular level. That is the teleological fact that DNA requires a lipid cell membrane as an environmental constraint and gateway by which to assemble the necessary biochemicals to self-replicate and produce the proteins for RNA, which requires RNA to create and maintain the membrane required, etc.

It apparently never occurs to the uninitiated a-teleologist that the organizing principal in any process of intelligent choice, as focused observation and intentional activity, is a quantum image of God that he sees in the mirror when he brushes his teeth, seamlessly involved in the process of animation.

The other extreme understanding of the nature of the Soul comes from the politically motivated apostacy of Justinian I at the Second Council of Constantinople with the institution of what has become traducianism or creationism; with traducianism the Soul is deemed to originate as biblically indicated by God with the creation of Adam and Eve and passed down and individuated in tandem with the body at each generation, whereas with creationism the Soul is deemed to be created by God at the point of conception. Hence, we have the right-to-life–right-to-choose political turmoil in the United States, due to the creationist view of the Soul among many of the Christian congregations, which tends to conform, but by no means exclusively, with the young earth creationist view of cosmology. Traducianism is a bit more complicated and largely not widely taught in Christian theology, though still productive of some insight with thoughtful consideration.

In the middle perhaps, can be found the various forms of understanding of the pre-existence of the soul, formally in the work of Plato and other Greek schools, in the Hindu-Buddhist tradition, and in Islam; and in many of the informal beliefs of enlightened individuals who are open to both scientific and spiritual discussion and perception, including some Christian. In this group we might even find a few political economists. It will also include many of those who come to their understanding as a result of reason, of intuition—of the Buddhist type—and of revelation of the initiatory type such is described in Christian terms as the anointing, the christening, of the Spirit of Truth; hence the origin of the name for the Christian religion, as Jesus was so christened.

That the notion of the existence of such revelation and of what is indeed revealed will appear and sound inscrutable to those who have not had such experience is not surprising; it is further confounding as such revelation varies in import and scope of meaning among those to whom such is revealed. Such revelation may be as simple as the aha moment when a key understanding of some subject of study comes to the student after sufficient meditation, formal or informal, for example as the solution to a math problem; or it may come in the form of lucid and enlightening dreams and visions as in the anecdotal case of the chemist August Kekule's vision of the ouroboros, the ancient symbol of a snake seizing its tale, from which Kekule was able to deduce the ring form of benzene; it may come in the enlightenment of Gautama seated beneath the bohdi tree; it may come in the form of the Holy Spirit symbolized by the dove that descended on Jesus of Nazareth as he emerged from the water with the baptism by John the Baptist at the start of his earthly mission. There are few experiences, in fact perhaps none, that can be reduced in understanding to an interaction of inert material things; even the physical study of billiards requires a human being at the dumb end of the pool stick to initiate a change in the state of the tabletop.

A mundane case of this inscrutability might be found in the attitude of a pre-pubescent juvenile. The sassy laughter brought about by a first exposure to the idea of sexual coupling is only erased by the momentous advent of hormonal arousal and its eventual consummation; or if not then, perhaps to be replaced by the self-conscious comedy of knowing. In the growing mind, some things must be experienced to be believed and in time understood.

If the Spirit of Truth as found in the 'peace that transcends Understanding'[11], as the confidence, serenity, and equanimity that comes from realizing the monistic indivisibility, yet ordered submission of Nature to Natures' God, of material observational to ideal motivational Reality, of existential Principles to essential Principals—if that Spirit is found motivating the Will-to-Truth residing in the individual Soul, in an analogous manner to which the endocrinal spirits motivate the individual to the ecstasy of coupling, but on a higher rung of the perceptual ladder, that individual will realize themself to be Essential—Spiritual—not through the ecstasy of a material body, but rather as the vivid material expression in time and space of an ideal Soul which is transcendent in its communal ecstasy—as a child of God, and as a philosopher king and guardian of the Republic of Plato, and even as a Marxist philosopher-activist leader of the heart of the proletariat. And how do such Souls lead but by the example of reified ideals.

For the Christian, this asks quite simply that one live the example of Christ; by living by the Spirit of Truth, which means with humility, by understanding there is always in that

---

[11] As referenced in Philippians 4:7

Spirit, That which is greater than oneself, "for my Father is greater than I"[12]; by living the Truth, which means by maintaining and exhibiting integrity, "the words that I speak unto you I speak not of myself: but the Father that dwelleth in me, he doeth the works."[13]; by Loving the Truth in all of Life, rather than fearing and hating those that fear and hate you and others, by Living that Love of Truth to the benefit of everyone that is able and willing to follow the example. Read the Gospel of John, Chapter 14; read the Gospel of John, Chapter 17; read the whole of the Gospel of John.

For the follower in the footsteps of any Philosopher King—any 'Lover of Wisdom' and 'Truthful Authority' who so leads by their ideal example—the Will-to-Truth requires the development of discernment with regard to any phenomena, in learning to separate the ideal, essential kernel of understanding of that experience from the material, existential chaff in which the packaged phenomena is presented to the senses. It is the Soul, as an integral aspect and interactive recipient of such presentations, that must carefully unfold the package to retrieve the contents of Life's experience.

That phenomenal packaging is the physical, material forms of Nature—seen through the quantifying eyes of humanity in the flat-earth plains of Mesopotamia or before in Africa; to evolve over time to include the celestial bodies of the Ptolemaic, Copernican–Galilean, and Hubble universes on the grand scale; to the single cell organisms and subsidiary structures of Hooke, van Leeuwenhoek, again Galileo, and others on the microscopic level.

The qualitative contents of that packaging are the higher mental, ideal Forms indicated by Plato, Aristotle, and many other philosophers and psychologists—understood through their evolving dialectic and branching once again, to Galileo, and to Bacon, Newton, Maxwell, Planck, Einstein, Bohr and others in the sciences; to Descartes, Locke, Hume, Kant, Bentham, Smith, Mill, Ricardo, Hegel, Marx, Keynes, Hayek, and many others in philosophy, political economy, and sociology; to Freud, Jung, and others in psychology; to Gautama Buddha, Jesus Christ, Muhammed ibn 'Abdullah and others in revelatory faith.

The material forms of Nature and the ideal Forms of the abstract Mind are things that are recognizable as tangible, or more accurately detectable, in their separate, unique ways. Material, physical forms are accessible to the eye and hand directly or representationally as malleable images of the mind. Ideal Forms can be similarly *represented* as images to the mind or as concrete objects in material form, but only as *representations* in time and space for realities in Essence and transcendental in their Abstraction. At first glance these ideal Forms, as the Goods of Plato's Republic, might appear to be 'unreal'—they are certainly immaterial in the usual sense of the word—but if they are recognized and understood by the individual soul, they are in truth intimately tangible, as they are felt *within* the Soul itself, and that tangibility is the recognition of a Reality from a source that may initiate, but is more than, a biological emotion as a response; just as the recognized *representational* images of the mind are registered, constrained, and manipulated in the

---

[12] From John 14:28
[13] From John 14:10

Soul which in turn serves as the mental source, as software so to speak, that initiates the neural network and thereby the biological activity of, and is more than, the human body.

The Soul on the other hand is formless, as is Essence—Whole, Holy, and Truthful Spirit—which is the sentient source of all souls, of all Living, and in fact is the source of all phenomenal appearance of material forms, mental–representational forms, and ideal Forms, known and unknown, as well. This essential Reality is both One Thing and No-Thing; it is a monism. It is the Essence of Essences, the Supreme Being of noetic Godhead. But while the Spirit is Formless and Boundless—a topologist might call it an 'n-dimensional manifold without boundary'—the formless human Soul is self-bounded in the extents of its own experience—perhaps a 'fiber of an n-dimensional fiber bundle'—while being completely embedded in the boundless, limitless, continuous extents of Spiritual Sentience and Power. (We elsewhere show, with the necessary quantitative analytical support, that the foundation of all physical phenomena as rest mass and messenger quanta on which all material aggregations and structures are based, is a mathematically defined function of the phenomenology of isotropic expansion of this continuum at the Hubble rate.)

To say that the sentient Soul and Spirit are *formless* is not to say that they are void; they may be void of inherently permanent form, but not of what was at one time in the physical sciences called 'substance'. The innate formlessness of the Soul retains the nature of Sentient Potential to register, effect, and assume various protean forms—material, representational, and ideal—of ambient change. It is the realization of this state of Being as a Soul that is the goal of all human effort and endeavor, consciously recognized or not, which motivates all of us as the notion of Omnipresent Freedom, valid when properly recognized, but which blocks many of us through the invalid wisdom of any notion of a pursuit of Omnipotent Money as inordinate acquisitiveness.

It is the spiritual nature of Omnipresent Freedom that motivates a quantitative analysis of Capital as Power (and weighted position) in Ergodic Economic Modeling to mushroom into this unintended, long winded qualitative exposition on the ideal notions of tyronic Power within the context of Platonic, Marxian, and Christian philosophy. It is covetousness that has given rise to the exploitation of some human beings by others with the means, opportunity, and motivation for that exploitation, and it is such covetousness that must be replaced by enlightened initiative in the entitlement to and production of the necessary goods and services of living for the benefit of all, if the contradictions within the system that has lead to social conflict of a Christian, Marxian, or Platonic nature are to be shunted to constructive public and private Good.